ADVANCED ENGINEERING MATHEMATICS
with MATLAB®

SECOND EDITION

ADVANCED ENGINEERING MATHEMATICS
with MATLAB®

SECOND EDITION

Dean G. Duffy

CHAPMAN & HALL/CRC

A CRC Press Company
Boca Raton   London   New York   Washington, D.C.

Library of Congress Cataloging-in-Publication Data

Catalog record is available from the Library of Congress

Visit the CRC Press Web site at www.crcpress.com

© 2003 by Chapman & Hall/CRC

No claim to original U.S. Government works
International Standard Book Number 1-58488-349-9
Printed in the United States of America   2 3 4 5 6 7 8 9 0
Printed on acid-free paper

Contents

Dedicated to the Brigade of Midshipmen
and the Corps of Cadets

Acknowledgments

I would like to thank the many midshipmen and cadets who have taken engineering mathematics from me. They have been willing or unwilling guinea pigs in testing out many of the ideas and problems in this book.

Special thanks go to Dr. Mike Marcozzi for his many useful and often humorous suggestions for improving this book. Many of the plots and calculations were done using MATLAB.

MATLAB is a registered trademark of
The MathWorks Inc.
24 Prime Park Way
Natick, MA 01760–1500
Phone: (508) 647–7000
Email: info@mathworks.com
www.mathworks.com

Finally, I would like to express my appreciation to all those authors and publishers who allowed me the use of their material from the scientific and engineering literature.

Introduction

This book is an updated and expanded version of my *Advanced Engineering Mathematics*. I have taken this opportunity to correct misprints, rewrite some of the text, and include new examples, problems, and projects. Of equal importance, however, is the addition of three new chapters so that the book can now be used in a wide variety of differential equations and engineering mathematics courses. These courses normally occur after classes on the calculus of single, multivariable, and vector-valued functions.

The book begins with complex variables. All students need to know how to do simple arithmetic operations involving complex numbers; this is presented in the first two sections of Chapter 1. The remaining portions of this chapter focus on contour integration. This material should be taught if the course is devoted to transform methods.

After this introduction, subsequent chapters or sections follow from the goals of the course. In its broadest form, there are two general tracks:

Differential Equations Course: Most courses on differential equations cover three general topics: fundamental techniques and concepts, Laplace transforms, and separation of variable solutions to partial differential equations.

The course begins with first- and higher-order ordinary differential equations, Chapters 2 and 3, respectively. After some introductory remarks, Chapter 2 devotes itself to presenting general methods for solving first-order ordinary differential equations. These methods include separation of variables, employing the properties of homogeneous, linear, and exact differential equations, and finding and using integrating factors.

The reason most students study ordinary differential equations is for their use in elementary physics, chemistry, and engineering courses. Because these differential equations contain constant coefficients, we focus on how to solve them in Chapter 3, along with a detailed analysis of the simple, damped, and forced harmonic oscillator. Furthermore, we include the commonly employed techniques of undetermined coefficients and variation of parameters for finding particular solutions. Finally, the special equation of Euler and Cauchy is included because of its use in solving partial differential equations in spherical coordinates.

After these introductory chapters, the course would next turn to Laplace transforms. Laplace transforms are useful in solving nonhomogeneous differential equations where the initial conditions have been specified and the forcing function "turns on and off." The general properties are explored in §§6.1 to 6.7; the actual solution technique is presented in §§6.8 and 6.9.

Most differential equations courses conclude with a taste of partial differential equations via the method of separation of variables. This topic usually begins with a quick introduction to Fourier series, §§4.1 to 4.4, followed by separation of variables as it applies to the heat (§§11.1–11.3), wave (§§10.1–10.3), or Laplace's equation (§§12.1–12.3). The exact equation that is studied depends upon the future needs of the students.

Engineering Mathematics Course: This book can be used in a wide variety of engineering mathematics classes. In all cases the student should have seen most of the material in Chapters 2 and 3. There are at least three possible combinations:

• *Option A*: The course is a continuation of a calculus reform sequence where elementary differential equations have been taught. This course begins with Laplace transforms and separation of variables techniques for the heat, wave, and/or Laplace equations, as outlined above. The course then concludes with either vector calculus or linear algebra. Vector calculus is presented in Chapter 13 and focuses on the gradient operator as it applies to line integrals, surface integrals, the divergence theorem, and Stokes' theorem. Chapter 14 presents linear algebra as a method for solving systems of linear equations and includes such topics as matrices, determinants, Cramer's rule, and the solution of systems of ordinary differential equations via the classic eigenvalue problem.

• *Option B*: This is the traditional situation where the student has already studied differential equations in another course before he takes engineering mathematics. Here separation of variables is retaught from the general viewpoint of eigenfunction expansions. Sections 9.1–9.3 explain how any piecewise continuous function can be reexpressed in an eigenfunction expansion using eigenfunctions from the classic Sturm-Liouville problem. Furthermore, we include two sections which focus on Bessel functions (§9.5) and Legendre polynomials (§9.4). These eigenfunctions appear in the solution of partial differential equations in cylindrical and spherical coordinates, respectively.

The course then covers linear algebra and vector calculus as given in Option A.

• *Option C*: I originally wrote this book for an engineering mathematics course given to sophomore and junior communication, systems, and electrical engineering majors at the U.S. Naval Academy. In this case, you would teach all of Chapter 1 with the possible exception of §1.10 on Cauchy principal-value integrals. This material was added to prepare the student for Hilbert transforms.

Because most students come to this course with a good knowledge of differential equations, we begin with Fourier series, Chapter 4, and proceed through Chapter 8. Chapter 5 generalizes the Fourier series to aperiodic functions and introduces the Fourier transform in Chapter 5. This leads naturally to Laplace transforms, Chapter 6. Throughout these chapters, I make use of complex variables in the treatment and inversion of the transforms.

With the rise of digital technology and its associated difference equations, a version of the Laplace transform, the z-transform, was developed. Chapter 7 introduces the z-transform by first giving its definition and then developing some of its general properties. We also illustrate how to compute the inverse by long division, partial fractions, and contour integration. Finally, we use z-transforms to solve difference equations, especially with respect to the stability of the system.

Finally, I added a new chapter on the Hilbert transform. With the explosion of interest in communications, today's engineer must have a command of this transform. The Hilbert transform is introduced in §8.1 and its properties are explored in §8.2. Two important applications of Hilbert transforms are introduced in §§8.3 and 8.4, namely the concept of analytic signals and the Kramers-Kronig relationship.

In addition to the revisions of the text and topics covered in this new addition, I now incorporate the mathematical software package MATLAB to reinforce the concepts that are taught. The power of MATLAB is its ability to quickly and easily present results in a graphical format. I have exploited this aspect and now included code (scripts) so that the student can explore the solution for a wide variety of parameters and different prospectives.

Of course this book still continues my principle of including a wealth of examples from the scientific and engineering literature. The answers to the odd problems are given in the back of the book while worked solutions to all of the problems are available from the publisher. Most of the MATLAB scripts may be found at www.crcpress.com under Electronic Products Downloads & Updates.

Chapter 1
Complex Variables

The theory of complex variables was originally developed by mathematicians as an aid in understanding functions. Functions of a complex variable enjoy many powerful properties that their real counterparts do not. That is *not* why we will study them. For us they provide the keys for the complete mastery of transform methods and differential equations.

In this chapter all of our work points to one objective: integration on the complex plane by the method of residues. For this reason we minimize discussions of limits and continuity which play such an important role in conventional complex variables in favor of the computational aspects. We begin by introducing some simple facts about complex variables. Then we progress to differential and integral calculus on the complex plane.

1.1 COMPLEX NUMBERS

A *complex number* is any number of the form $a + bi$, where a and b are real and $i = \sqrt{-1}$. We denote any member of a *set* of complex numbers by the *complex variable* $z = x + iy$. The real part of z, usually denoted by $\text{Re}(z)$, is x while the imaginary part of z, $\text{Im}(z)$, is y. The *complex conjugate*, $\bar{z}$ or z^*, of the complex number $a + bi$ is $a - bi$.

Complex numbers obey the fundamental rules of algebra. Thus, two complex numbers $a + bi$ and $c + di$ are equal if and only if $a = c$ and $b = d$.

1

Just as real numbers have the fundamental operations of addition, subtraction, multiplication, and division, so too do complex numbers. These operations are defined:

Addition

$$(a + bi) + (c + di) = (a + c) + (b + d)i \qquad (1.1.1)$$

Subtraction

$$(a + bi) - (c + di) = (a - c) + (b - d)i \qquad (1.1.2)$$

Multiplication

$$(a + bi)(c + di) = ac + bci + adi + i^2bd = (ac - bd) + (ad + bc)i \qquad (1.1.3)$$

Division

$$\frac{a + bi}{c + di} = \frac{a + bi}{c + di}\frac{c - di}{c - di} = \frac{ac - adi + bci - bdi^2}{c^2 + d^2} = \frac{ac + bd + (bc - ad)i}{c^2 + d^2}. \qquad (1.1.4)$$

The *absolute value* or *modulus* of a complex number $a + bi$, written $|a + bi|$, equals $\sqrt{a^2 + b^2}$. Additional properties include:

$$|z_1 z_2 z_3 \cdots z_n| = |z_1||z_2||z_3| \cdots |z_n| \qquad (1.1.5)$$

$$|z_1/z_2| = |z_1|/|z_2| \quad \text{if} \quad z_2 \neq 0 \qquad (1.1.6)$$

$$|z_1 + z_2 + z_3 + \cdots + z_n| \leq |z_1| + |z_2| + |z_3| + \cdots + |z_n| \qquad (1.1.7)$$

and

$$|z_1 + z_2| \geq |z_1| - |z_2|. \qquad (1.1.8)$$

The use of inequalities with complex variables has meaning only when they involve absolute values.

It is often useful to plot the complex number $x + iy$ as a point (x, y) in the xy-plane, now called the *complex plane*. Figure 1.1.1 illustrates this representation.

This geometrical interpretation of a complex number suggests an alternative method of expressing a complex number: the polar form. From the polar representation of x and y,

$$x = r\cos(\theta) \quad \text{and} \quad y = r\sin(\theta), \qquad (1.1.9)$$

where $r = \sqrt{x^2 + y^2}$ is the *modulus*, *amplitude*, or *absolute value* of z and θ is the *argument* or *phase*, we have that

$$z = x + iy = r[\cos(\theta) + i\sin(\theta)]. \qquad (1.1.10)$$

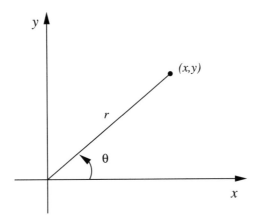

Figure 1.1.1: The complex plane.

However, from the Taylor expansion of the exponential in the real case,

$$e^{i\theta} = \sum_{k=0}^{\infty} \frac{(\theta i)^k}{k!}. \tag{1.1.11}$$

Expanding (1.1.11),

$$e^{i\theta} = 1 - \frac{\theta^2}{2!} + \frac{\theta^4}{4!} - \frac{\theta^6}{6!} + \cdots + i\left(\theta - \frac{\theta^3}{3!} + \frac{\theta^5}{5!} - \frac{\theta^7}{7!} + \cdots\right) \tag{1.1.12}$$

$$= \cos(\theta) + i\sin(\theta). \tag{1.1.13}$$

Equation (1.1.13) is *Euler's formula*. Consequently, we may express (1.1.10) as

$$z = re^{i\theta}, \tag{1.1.14}$$

which is the *polar form* of a complex number. Furthermore, because

$$z^n = r^n e^{in\theta} \tag{1.1.15}$$

by the law of exponents,

$$z^n = r^n[\cos(n\theta) + i\sin(n\theta)]. \tag{1.1.16}$$

Equation (1.1.16) is *De Moivre's theorem*.

- **Example 1.1.1**

  Let us simplify the following complex number:

  $$\frac{3-2i}{-1+i} = \frac{3-2i}{-1+i} \times \frac{-1-i}{-1-i} = \frac{-3-3i+2i+2i^2}{1+1} = \frac{-5-i}{2} = -\frac{5}{2} - \frac{i}{2}. \tag{1.1.17}$$

• **Example 1.1.2**

Let us reexpress the complex number $-\sqrt{6} - i\sqrt{2}$ in polar form. From (1.1.9) $r = \sqrt{6+2}$ and $\theta = \tan^{-1}(b/a) = \tan^{-1}(1/\sqrt{3}) = \pi/6$ or $7\pi/6$. Because $-\sqrt{6} - i\sqrt{2}$ lies in the third quadrant of the complex plane, $\theta = 7\pi/6$ and

$$-\sqrt{6} - i\sqrt{2} = 2\sqrt{2}e^{7\pi i/6}. \qquad (1.1.18)$$

Note that (1.1.18) is not a unique representation because $\pm 2n\pi$ may be added to $7\pi/6$ and we still have the same complex number since

$$e^{i(\theta \pm 2n\pi)} = \cos(\theta \pm 2n\pi) + i\sin(\theta \pm 2n\pi) = \cos(\theta) + i\sin(\theta) = e^{i\theta}. \quad (1.1.19)$$

For uniqueness we often choose $n = 0$ and define this choice as the *principal branch*. Other branches correspond to different values of n.

• **Example 1.1.3**

Find the curve described by the equation $|z - z_0| = a$.
From the definition of the absolute value,

$$\sqrt{(x - x_0)^2 + (y - y_0)^2} = a \qquad (1.1.20)$$

or

$$(x - x_0)^2 + (y - y_0)^2 = a^2. \qquad (1.1.21)$$

Equation (1.1.21), and hence $|z - z_0| = a$, describes a circle of radius a with its center located at (x_0, y_0). Later on, we shall use equations such as this to describe curves in the complex plane.

• **Example 1.1.4**

As an example in manipulating complex numbers, let us show that

$$\left| \frac{a + bi}{b + ai} \right| = 1. \qquad (1.1.22)$$

We begin by simplifying

$$\frac{a + bi}{b + ai} = \frac{a + bi}{b + ai} \times \frac{b - ai}{b - ai} = \frac{2ab}{a^2 + b^2} + \frac{b^2 - a^2}{a^2 + b^2}i. \qquad (1.1.23)$$

Therefore,

$$\left| \frac{a + bi}{b + ai} \right| = \sqrt{\frac{4a^2b^2}{(a^2 + b^2)^2} + \frac{b^4 - 2a^2b^2 + a^4}{(a^2 + b^2)^2}} = \sqrt{\frac{a^4 + 2a^2b^2 + b^4}{(a^2 + b^2)^2}} = 1.$$

$$(1.1.24)$$

MATLAB can also be used to solve this problem. Typing the commands

```
>> syms a b real
>> abs((a+b*i)/(b+a*i))
```

yields

```
ans =
1
```

Note that you must declare a and b real in order to get the final result.

Problems

Simplify the following complex numbers. Represent the solution in the Cartesian form $a + bi$. Check your answers using MATLAB.

1. $\dfrac{5i}{2+i}$

2. $\dfrac{5+5i}{3-4i} + \dfrac{20}{4+3i}$

3. $\dfrac{1+2i}{3-4i} + \dfrac{2-i}{5i}$

4. $(1-i)^4$

5. $i(1-i\sqrt{3})(\sqrt{3}+i)$

Represent the following complex numbers in polar form:

6. $-i$

7. -4

8. $2+2\sqrt{3}\,i$

9. $-5+5i$

10. $2-2i$

11. $-1+\sqrt{3}\,i$

12. By the law of exponents, $e^{i(\alpha+\beta)} = e^{i\alpha}e^{i\beta}$. Use Euler's formula to obtain expressions for $\cos(\alpha + \beta)$ and $\sin(\alpha + \beta)$ in terms of sines and cosines of α and β.

13. Using the property that $\sum_{n=0}^{N} q^n = (1-q^{N+1})/(1-q)$ and the geometric series $\sum_{n=0}^{N} e^{int}$, obtain the following sums of trigonometric functions:

$$\sum_{n=0}^{N} \cos(nt) = \cos\left(\frac{Nt}{2}\right) \frac{\sin[(N+1)t/2]}{\sin(t/2)}$$

and

$$\sum_{n=1}^{N} \sin(nt) = \sin\left(\frac{Nt}{2}\right) \frac{\sin[(N+1)t/2]}{\sin(t/2)}.$$

These results are often called *Lagrange's trigonometric identities*.

14. (a) Using the property that $\sum_{n=0}^{\infty} q^n = 1/(1-q)$, if $|q| < 1$, and the geometric series $\sum_{n=0}^{\infty} \epsilon^n e^{int}$, $|\epsilon| < 1$, show that

$$\sum_{n=0}^{\infty} \epsilon^n \cos(nt) = \frac{1 - \epsilon \cos(t)}{1 + \epsilon^2 - 2\epsilon \cos(t)}$$

and

$$\sum_{n=1}^{\infty} \epsilon^n \sin(nt) = \frac{\epsilon \sin(t)}{1 + \epsilon^2 - 2\epsilon \cos(t)}.$$

(b) Let $\epsilon = e^{-a}$, where $a > 0$. Show that

$$2 \sum_{n=1}^{\infty} e^{-na} \sin(nt) = \frac{\sin(t)}{\cosh(a) - \cos(t)}.$$

1.2 FINDING ROOTS

The concept of finding roots of a number, which is rather straightforward in the case of real numbers, becomes more difficult in the case of complex numbers. By finding the *roots* of a complex number, we wish to find all of the solutions w of the equation $w^n = z$, where n is a positive integer for a given z.

We begin by writing z in the polar form:

$$z = re^{i\varphi}, \tag{1.2.1}$$

while we write

$$w = Re^{i\Phi} \tag{1.2.2}$$

for the unknown. Consequently,

$$w^n = R^n e^{in\Phi} = re^{i\varphi} = z. \tag{1.2.3}$$

We satisfy (1.2.3) if

$$R^n = r \quad \text{and} \quad n\Phi = \varphi + 2k\pi, \quad k = 0, \pm 1, \pm 2, \ldots, \tag{1.2.4}$$

because the addition of any multiple of 2π to the argument is also a solution. Thus, $R = r^{1/n}$, where R is the uniquely determined real positive root, and

$$\Phi_k = \frac{\varphi}{n} + \frac{2\pi k}{n}, \quad k = 0, \pm 1, \pm 2, \ldots. \tag{1.2.5}$$

Because $w_k = w_{k\pm n}$, it is sufficient to take $k = 0, 1, 2, \ldots, n-1$. Therefore, there are exactly n solutions:

$$w_k = Re^{\Phi_k i} = r^{1/n} \exp\left[i\left(\frac{\varphi}{n} + \frac{2\pi k}{n}\right)\right] \tag{1.2.6}$$

with $k = 0, 1, 2, \ldots, n - 1$. They are the n roots of z. Geometrically we can locate these solutions w_k on a circle, centered at the point $(0,0)$, with radius R and separated from each other by $2\pi/n$ radians. These roots also form the vertices of a regular polygon of n sides inscribed inside of a circle of radius R. (See Example 1.2.1.)

In summary, the method for finding the n roots of a complex number z_0 is as follows. First, write z_0 in its polar form: $z_0 = re^{i\varphi}$. Then multiply the polar form by $e^{2i\pi k}$. Using the law of exponents, take the $1/n$ power of both sides of the equation. Finally, using Euler's formula, evaluate the roots for $k = 0, 1, \ldots, n - 1$.

- **Example 1.2.1**

Let us find all of the values of z for which $z^5 = -32$ and locate these values on the complex plane.

Because

$$-32 = 32e^{\pi i} = 2^5 e^{\pi i}, \tag{1.2.7}$$

$$z_k = 2\exp\left(\frac{\pi i}{5} + \frac{2\pi i k}{5}\right), \qquad k = 0, 1, 2, 3, 4, \tag{1.2.8}$$

or

$$z_0 = 2\exp\left(\frac{\pi i}{5}\right) = 2\left[\cos\left(\frac{\pi}{5}\right) + i\sin\left(\frac{\pi}{5}\right)\right], \tag{1.2.9}$$

$$z_1 = 2\exp\left(\frac{3\pi i}{5}\right) = 2\left[\cos\left(\frac{3\pi}{5}\right) + i\sin\left(\frac{3\pi}{5}\right)\right], \tag{1.2.10}$$

$$z_2 = 2e^{\pi i} = -2, \tag{1.2.11}$$

$$z_3 = 2\exp\left(\frac{7\pi i}{5}\right) = 2\left[\cos\left(\frac{7\pi}{5}\right) + i\sin\left(\frac{7\pi}{5}\right)\right] \tag{1.2.12}$$

and

$$z_4 = 2\exp\left(\frac{9\pi i}{5}\right) = 2\left[\cos\left(\frac{9\pi}{5}\right) + i\sin\left(\frac{9\pi}{5}\right)\right]. \tag{1.2.13}$$

Figure 1.2.1 shows the location of these roots in the complex plane.

- **Example 1.2.2**

Let us find the cube roots of $-1 + i$ and locate them graphically.
Because $-1 + i = \sqrt{2}\exp(3\pi i/4)$,

$$z_k = 2^{1/6}\exp\left(\frac{\pi i}{4} + \frac{2i\pi k}{3}\right), \qquad k = 0, 1, 2, \tag{1.2.14}$$

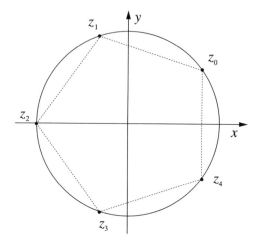

Figure 1.2.1: The zeros of $z^5 = -32$.

or

$$z_0 = 2^{1/6} \exp\left(\frac{\pi i}{4}\right) = 2^{1/6}\left[\cos\left(\frac{\pi}{4}\right) + i\sin\left(\frac{\pi}{4}\right)\right], \qquad (1.2.15)$$

$$z_1 = 2^{1/6} \exp\left(\frac{11\pi i}{12}\right) = 2^{1/6}\left[\cos\left(\frac{11\pi}{12}\right) + i\sin\left(\frac{11\pi}{12}\right),\right] \qquad (1.2.16)$$

and

$$z_2 = 2^{1/6} \exp\left(\frac{19\pi i}{12}\right) = 2^{1/6}\left[\cos\left(\frac{19\pi}{12}\right) + i\sin\left(\frac{19\pi}{12}\right)\right]. \qquad (1.2.17)$$

Figure 1.2.2 gives the location of these zeros on the complex plane.

• **Example 1.2.3**

    The routine `solve` in MATLAB can also be used to compute the roots of complex numbers. For example, let us find all of the roots of $z^4 = -a^4$.
    The MATLAB commands are as follows:

```
>> syms a z
>> solve(z^4+a^4)
```
This yields the solution

```
ans=
[  (1/2*2^(1/2)+1/2*i*2^(1/2))*a]
[ (-1/2*2^(1/2)+1/2*i*2^(1/2))*a]
[  (1/2*2^(1/2)-1/2*i*2^(1/2))*a]
```

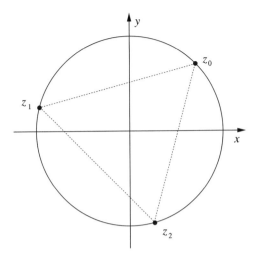

Figure 1.2.2: The zeros of $z^3 = -1 + i$.

[(-1/2*2^(1/2)-1/2*i*2^(1/2))*a]

Problems

Extract all of the possible roots of the following complex numbers. Verify your answer using MATLAB.

1. $8^{1/6}$ 2. $(-1)^{1/3}$ 3. $(-i)^{1/3}$ 4. $(-27i)^{1/6}$

5. Find algebraic expressions for the square roots of $a - bi$, where $a > 0$ and $b > 0$.

6. Find all of the roots for the algebraic equation $z^4 - 3iz^2 - 2 = 0$. Then check your answer using `solve` in MATLAB.

7. Find all of the roots for the algebraic equation $z^4 + 6iz^2 + 16 = 0$. Then check your answer using `solve` in MATLAB.

1.3 THE DERIVATIVE IN THE COMPLEX PLANE: THE CAUCHY-RIEMANN EQUATIONS

In the previous two sections, we introduced complex arithmetic. We are now ready for the concept of function as it applies to complex variables.

We already defined the complex variable $z = x + iy$, where x and y are variable. We now introduce another complex variable $w = u + iv$ so that for each value of z there corresponds a value of $w = f(z)$. From all of the possible complex functions that we might invent, we focus on those functions where

z-plane   w-plane

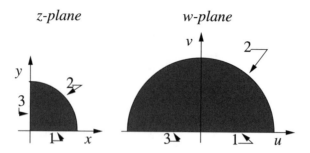

Figure 1.3.1: The complex function $w = z^2$.

for each z there is one, and only one, value of w. These functions are *single-valued*. They differ from functions such as the square root, logarithm, and inverse sine and cosine, where there are multiple answers for each z. These *multivalued functions* do arise in various problems. However, they are beyond the scope of this book and we shall always assume that we are dealing with single-valued functions.

A popular method for representing a complex function involves drawing some closed domain in the z-plane and then showing the corresponding domain in the w-plane. This procedure is called *mapping* and the z-plane illustrates the *domain* of the function while the w-plane illustrates its *image* or *range*. Figure 1.3.1 shows the z-plane and w-plane for $w = z^2$; a pie-shaped wedge in the z-plane maps into a semicircle on the w-plane.

• **Example 1.3.1**

Given the complex function $w = e^{-z^2}$, let us find the corresponding $u(x, y)$ and $v(x, y)$.

From Euler's formula,

$$w = e^{-z^2} = e^{-(x+iy)^2} = e^{y^2 - x^2} e^{-2ixy} = e^{y^2 - x^2}[\cos(2xy) - i\sin(2xy)]. \tag{1.3.1}$$

Therefore, by inspection,

$$u(x, y) = e^{y^2 - x^2} \cos(2xy), \quad \text{and} \quad v(x, y) = -e^{y^2 - x^2} \sin(2xy). \tag{1.3.2}$$

Note that there is no i in the expression for $v(x, y)$. The function $w = f(z)$ is single-valued because for each distinct value of z, there is an unique value of $u(x, y)$ and $v(x, y)$.

• **Example 1.3.2**

As counterpoint, let us show that $w = \sqrt{z}$ is a multivalued function.

We begin by writing $z = re^{i\theta + 2\pi ik}$, where $r = \sqrt{x^2 + y^2}$ and $\theta = \tan^{-1}(y/x)$. Then,

$$w_k = \sqrt{r} e^{i\theta/2 + \pi ik}, \qquad k = 0, 1, \tag{1.3.3}$$

or

$$w_0 = \sqrt{r}\left[\cos(\theta/2) + i\sin(\theta/2)\right] \qquad \text{and} \qquad w_1 = -w_0. \tag{1.3.4}$$

Therefore,

$$u_0(x, y) = \sqrt{r}\cos(\theta/2), \qquad v_0(x, y) = \sqrt{r}\sin(\theta/2), \tag{1.3.5}$$

and

$$u_1(x, y) = -\sqrt{r}\cos(\theta/2), \qquad v_1(x, y) = -\sqrt{r}\sin(\theta/2). \tag{1.3.6}$$

Each solution w_0 or w_1 is a *branch* of the multivalued function $\sqrt{z}$. We can make $\sqrt{z}$ single-valued by restricting ourselves to a single branch, say w_0. In that case, the $\text{Re}(w) > 0$ if we restrict $-\pi < \theta < \pi$. Although this is not the only choice that we could have made, it is a popular one. For example, most digital computers use this definition in their complex square root function. The point here is our ability to make a multivalued function single-valued by defining a particular branch.

Although the requirement that a complex function be single-valued is important, it is still too general and would cover all functions of two real variables. To have a useful theory, we must introduce additional constraints. Because an important property associated with most functions is the ability to take their derivative, let us examine the derivative in the complex plane.

Following the definition of a derivative for a single real variable, the derivative of a complex function $w = f(z)$ is defined as

$$\frac{dw}{dz} = \lim_{\Delta z \to 0} \frac{\Delta w}{\Delta z} = \lim_{\Delta z \to 0} \frac{f(z + \Delta z) - f(z)}{\Delta z}. \tag{1.3.7}$$

A function of a complex variable that has a derivative at every point within a region of the complex plane is said to be *analytic* (or *regular* or *holomorphic*) over that region. If the function is analytic everywhere in the complex plane, it is *entire*.

Because the derivative is defined as a limit and limits are well behaved with respect to elementary algebraic operations, the following operations carry over from elementary calculus:

$$\frac{d}{dz}\left[cf(z)\right] = cf'(z), \qquad c \text{ a constant} \tag{1.3.8}$$

$$\frac{d}{dz}\left[f(z) \pm g(z)\right] = f'(z) \pm g'(z) \tag{1.3.9}$$

$$\frac{d}{dz}\left[f(z)g(z)\right] = f'(z)g(z) + f(z)g'(z) \tag{1.3.10}$$

$$\frac{d}{dz}\left[\frac{f(z)}{g(z)}\right] = \frac{g(z)f'(z) - g'(z)f(z)}{g^2(z)} \tag{1.3.11}$$

$$\frac{d}{dz}\left\{f[g(z)]\right\} = f'[g(z)]g'(z), \qquad \text{the chain rule.} \tag{1.3.12}$$

Another important property that carries over from real variables is l'Hôspital rule: Let $f(z)$ and $g(z)$ be analytic at z_0, where $f(z)$ has a zero[1] of order m and $g(z)$ has a zero of order n. Then, if $m > n$,

$$\lim_{z \to z_0} \frac{f(z)}{g(z)} = 0; \tag{1.3.13}$$

if $m = n$,

$$\lim_{z \to z_0} \frac{f(z)}{g(z)} = \frac{f^{(m)}(z_0)}{g^{(m)}(z_0)}; \tag{1.3.14}$$

and if $m < n$,

$$\lim_{z \to z_0} \frac{f(z)}{g(z)} = \infty. \tag{1.3.15}$$

● **Example 1.3.3**

Let us evaluate $\lim_{z \to i}(z^{10} + 1)/(z^6 + 1)$. From l'Hôspital rule,

$$\lim_{z \to i} \frac{z^{10} + 1}{z^6 + 1} = \lim_{z \to i} \frac{10z^9}{6z^5} = \frac{5}{3} \lim_{z \to i} z^4 = \frac{5}{3}. \tag{1.3.16}$$

So far, we introduced the derivative and some of its properties. But how do we actually know whether a function is analytic or how do we compute its derivative? At this point we must develop some relationships involving the known quantities $u(x, y)$ and $v(x, y)$.

We begin by returning to the definition of the derivative. Because $\Delta z = \Delta x + i\Delta y$, there is an infinite number of different ways of approaching the limit $\Delta z \to 0$. Uniqueness of that limit requires that (1.3.7) must be independent of the manner in which Δz approaches zero. A simple example is to take Δz in the x-direction so that $\Delta z = \Delta x$; another is to take Δz in the y-direction so that $\Delta z = i\Delta y$. These examples yield

$$\frac{dw}{dz} = \lim_{\Delta z \to 0} \frac{\Delta w}{\Delta z} = \lim_{\Delta x \to 0} \frac{\Delta u + i\Delta v}{\Delta x} = \frac{\partial u}{\partial x} + i\frac{\partial v}{\partial x} \tag{1.3.17}$$

---

[1] An analytic function $f(z)$ has a zero of order m at z_0 if and only if $f(z_0) = f'(z_0) = \cdots = f^{(m-1)}(z_0) = 0$ and $f^{(m)}(z_0) \neq 0$.

Figure 1.3.2: Although educated as an engineer, Augustin-Louis Cauchy (1789–1857) would become a mathematician's mathematician, publishing 789 papers and 7 books in the fields of pure and applied mathematics. His greatest writings established the discipline of mathematical analysis as he refined the notions of limit, continuity, function, and convergence. It was this work on analysis that led him to develop complex function theory via the concept of residues. (Portrait courtesy of the Archives de l'Académie des sciences, Paris.)

and

$$\frac{dw}{dz} = \lim_{\Delta z \to 0} \frac{\Delta w}{\Delta z} = \lim_{\Delta y \to 0} \frac{\Delta u + i\Delta v}{i\Delta y} = \frac{\partial v}{\partial y} - i\frac{\partial u}{\partial y}. \qquad (1.3.18)$$

In both cases we are approaching zero from the positive side. For the limit to be unique and independent of path, (1.3.17) must equal (1.3.18), or

$$\frac{\partial u}{\partial x} = \frac{\partial v}{\partial y} \qquad \text{and} \qquad \frac{\partial u}{\partial y} = -\frac{\partial v}{\partial x}. \qquad (1.3.19)$$

These equations which u and v must both satisfy are the *Cauchy-Riemann* equations. They are necessary but not sufficient to ensure that a function is differentiable. The following example illustrates this.

- **Example 1.3.4**

Consider the complex function

$$w = \begin{cases} z^5/|z|^4, & z \neq 0 \\ 0, & z = 0. \end{cases} \qquad (1.3.20)$$

Figure 1.3.3: Despite his short life, (Georg Friedrich) Bernhard Riemann's (1826–1866) mathematical work contained many imaginative and profound concepts. It was in his doctoral thesis on complex function theory (1851) that he introduced the Cauchy-Riemann differential equations. Riemann's later work dealt with the definition of the integral and the foundations of geometry and non-Euclidean (elliptic) geometry. (Portrait courtesy of Photo AKG, London.)

The derivative at $z = 0$ is given by

$$\frac{dw}{dz} = \lim_{\Delta z \to 0} \frac{(\Delta z)^5 / |\Delta z|^4 - 0}{\Delta z} = \lim_{\Delta z \to 0} \frac{(\Delta z)^4}{|\Delta z|^4}, \tag{1.3.21}$$

provided that this limit exists. However, this limit does not exist because, in general, the numerator depends upon the path used to approach zero. For example, if $\Delta z = re^{\pi i/4}$ with $r \to 0$, $dw/dz = -1$. On the other hand, if $\Delta z = re^{\pi i/2}$ with $r \to 0$, $dw/dz = 1$.

Are the Cauchy-Riemann equations satisfied in this case? To check this, we first compute

$$u_x(0,0) = \lim_{\Delta x \to 0} \left(\frac{\Delta x}{|\Delta x|} \right)^4 = 1, \tag{1.3.22}$$

$$v_y(0,0) = \lim_{\Delta y \to 0} \left(\frac{i\Delta y}{|\Delta y|} \right)^4 = 1, \tag{1.3.23}$$

$$u_y(0,0) = \lim_{\Delta y \to 0} \operatorname{Re} \left[\frac{(i\Delta y)^5}{\Delta y |\Delta y|^4} \right] = 0, \tag{1.3.24}$$

and

$$v_x(0,0) = \lim_{\Delta x \to 0} \operatorname{Im} \left[\left(\frac{\Delta x}{|\Delta x|} \right)^4 \right] = 0. \tag{1.3.25}$$

Hence, the Cauchy-Riemann equations are satisfied at the origin. Thus, even though the derivative is not uniquely defined, (1.3.21) happens to have the same value for paths taken along the coordinate axes so that the Cauchy-Riemann equations are satisfied.

In summary, if a function is differentiable at a point, the Cauchy-Riemann equations hold. Similarly, if the Cauchy-Riemann equations are not satisfied at a point, then the function is not differentiable at that point. This is one of the important uses of the Cauchy-Riemann equations: the location of non-analytic points. Isolated nonanalytic points of an otherwise analytic function are called *isolated singularities*. Functions that contain isolated singularities are called *meromorphic*.

The Cauchy-Riemann condition can be modified so that it is sufficient for the derivative to exist. Let us require that u_x, u_y, v_x, and v_y be continuous in some region surrounding a point z_0 and satisfy the Cauchy-Riemann equations there. Then

$$f(z) - f(z_0) = [u(z) - u(z_0)] + i[v(z) - v(z_0)] \tag{1.3.26}$$

$$
\begin{aligned}
= [u_x(z_0)(x - x_0) + u_y(z_0)(y - y_0) + \epsilon_1(x - x_0) + \epsilon_2(y - y_0)] \\
+ i[v_x(z_0)(x - x_0) + v_y(z_0)(y - y_0) + \epsilon_3(x - x_0) + \epsilon_4(y - y_0)]
\end{aligned}
\tag{1.3.27}
$$

$$
\begin{aligned}
= [u_x(z_0) + iv_x(z_0)](z - z_0) + (\epsilon_1 + i\epsilon_3)(x - x_0) \\
+ (\epsilon_2 + i\epsilon_4)(y - y_0),
\end{aligned}
\tag{1.3.28}
$$

where we used the Cauchy-Riemann equations and $\epsilon_1, \epsilon_2, \epsilon_3, \epsilon_4 \to 0$ as $\Delta x, \Delta y \to 0$. Hence,

$$f'(z_0) = \lim_{\Delta z \to 0} \frac{f(z) - f(z_0)}{\Delta z} = u_x(z_0) + iv_x(z_0), \tag{1.3.29}$$

because $|\Delta x| \leq |\Delta z|$ and $|\Delta y| \leq |\Delta z|$. Using (1.3.29) and the Cauchy-Riemann equations, we can obtain the derivative from any of the following formulas:

$$\frac{dw}{dz} = \frac{\partial u}{\partial x} + i\frac{\partial v}{\partial x} = \frac{\partial v}{\partial y} - i\frac{\partial u}{\partial y}, \tag{1.3.30}$$

and

$$\frac{dw}{dz} = \frac{\partial v}{\partial y} + i\frac{\partial v}{\partial x} = \frac{\partial u}{\partial x} - i\frac{\partial u}{\partial y}. \tag{1.3.31}$$

Furthermore, $f'(z_0)$ is continuous because the partial derivatives are.

• **Example 1.3.5**

Let us show that $\sin(z)$ is an entire function.

$$w = \sin(z) \tag{1.3.32}$$
$$u + iv = \sin(x + iy) = \sin(x)\cos(iy) + \cos(x)\sin(iy) \tag{1.3.33}$$
$$= \sin(x)\cosh(y) + i\cos(x)\sinh(y), \tag{1.3.34}$$

because

$$\cos(iy) = \tfrac{1}{2}\left[e^{i(iy)} + e^{-i(iy)}\right] = \tfrac{1}{2}\left[e^y + e^{-y}\right] = \cosh(y), \tag{1.3.35}$$

and

$$\sin(iy) = \tfrac{1}{2i}\left[e^{i(iy)} - e^{-i(iy)}\right] = -\tfrac{1}{2i}\left[e^y - e^{-y}\right] = i\sinh(y), \tag{1.3.36}$$

so that

$$u(x, y) = \sin(x)\cosh(y), \quad \text{and} \quad v(x, y) = \cos(x)\sinh(y). \tag{1.3.37}$$

Differentiating both $u(x, y)$ and $v(x, y)$ with respect to x and y, we have that

$$\frac{\partial u}{\partial x} = \cos(x)\cosh(y), \qquad \frac{\partial u}{\partial y} = \sin(x)\sinh(y), \tag{1.3.38}$$

$$\frac{\partial v}{\partial x} = -\sin(x)\sinh(y), \qquad \frac{\partial v}{\partial y} = \cos(x)\cosh(y), \tag{1.3.39}$$

and $u(x, y)$ and $v(x, y)$ satisfy the Cauchy-Riemann equations for all values of x and y. Furthermore, u_x, u_y, v_x, and v_y are continuous for all x and y. Therefore, the function $w = \sin(z)$ is an entire function.

• **Example 1.3.6**

Consider the function $w = 1/z$. Then

$$w = u + iv = \frac{1}{x + iy} = \frac{x}{x^2 + y^2} - \frac{iy}{x^2 + y^2}. \tag{1.3.40}$$

Therefore,

$$u(x, y) = \frac{x}{x^2 + y^2}, \quad \text{and} \quad v(x, y) = -\frac{y}{x^2 + y^2}. \tag{1.3.41}$$

Now

$$\frac{\partial u}{\partial x} = \frac{(x^2 + y^2) - 2x^2}{(x^2 + y^2)^2} = \frac{y^2 - x^2}{(x^2 + y^2)^2}, \tag{1.3.42}$$

$$\frac{\partial v}{\partial y} = -\frac{(x^2 + y^2) - 2y^2}{(x^2 + y^2)^2} = \frac{y^2 - x^2}{(x^2 + y^2)^2} = \frac{\partial u}{\partial x}, \tag{1.3.43}$$

$$\frac{\partial v}{\partial x} = -\frac{0 - 2xy}{(x^2 + y^2)^2} = \frac{2xy}{(x^2 + y^2)^2}, \tag{1.3.44}$$

and

$$\frac{\partial u}{\partial y} = \frac{0 - 2xy}{(x^2 + y^2)^2} = -\frac{2xy}{(x^2 + y^2)^2} = -\frac{\partial v}{\partial x}. \tag{1.3.45}$$

The function is analytic at all points except the origin because the function itself ceases to exist when both x and y are zero and the modulus of w becomes infinite.

- **Example 1.3.7**

Let us find the derivative of $\sin(z)$.
Using (1.3.30) and (1.3.34),

$$\frac{d}{dz}\left[\sin(z)\right] = \frac{\partial u}{\partial x} + i\frac{\partial v}{\partial x} \tag{1.3.46}$$

$$= \cos(x)\cosh(y) - i\sin(x)\sinh(y) \tag{1.3.47}$$

$$= \cos(x + iy) = \cos(z). \tag{1.3.48}$$

Similarly,

$$\frac{d}{dz}\left(\frac{1}{z}\right) = \frac{y^2 - x^2}{(x^2 + y^2)^2} + \frac{2ixy}{(x^2 + y^2)^2} \tag{1.3.49}$$

$$= -\frac{1}{(x + iy)^2} = -\frac{1}{z^2}. \tag{1.3.50}$$

The results in the above examples are identical to those for z real. As we showed earlier, the fundamental rules of elementary calculus apply to complex differentiation. Consequently, it is usually simpler to apply those rules to find the derivative rather than breaking $f(z)$ down into its real and imaginary parts, applying either (1.3.30) or (1.3.31), and then putting everything back together.

An additional property of analytic functions follows by cross differentiating the Cauchy-Riemann equations or

$$\frac{\partial^2 u}{\partial x^2} = \frac{\partial^2 v}{\partial x \partial y} = -\frac{\partial^2 u}{\partial y^2}, \quad \text{or} \quad \frac{\partial^2 u}{\partial x^2} + \frac{\partial^2 u}{\partial y^2} = 0, \tag{1.3.51}$$

and

$$\frac{\partial^2 v}{\partial x^2} = -\frac{\partial^2 u}{\partial x \partial y} = -\frac{\partial^2 v}{\partial y^2}, \quad \text{or} \quad \frac{\partial^2 v}{\partial x^2} + \frac{\partial^2 v}{\partial y^2} = 0. \tag{1.3.52}$$

Any function that has continuous partial derivatives of second order and satisfies Laplace's equation (1.3.51) or (1.3.52) is called a *harmonic function.* Because both $u(x,y)$ and $v(x,y)$ satisfy Laplace's equation if $f(z) = u + iv$ is analytic, $u(x,y)$ and $v(x,y)$ are called *conjugate harmonic functions.*

● **Example 1.3.8**

Given that $u(x,y) = e^{-x}[x\sin(y) - y\cos(y)]$, let us show that u is harmonic and find a conjugate harmonic function $v(x,y)$ such that $f(z) = u + iv$ is analytic.

Because

$$\frac{\partial^2 u}{\partial x^2} = -2e^{-x}\sin(y) + xe^{-x}\sin(y) - ye^{-x}\cos(y), \tag{1.3.53}$$

and

$$\frac{\partial^2 u}{\partial y^2} = -xe^{-x}\sin(y) + 2e^{-x}\sin(y) + ye^{-x}\cos(y), \tag{1.3.54}$$

it follows that $u_{xx} + u_{yy} = 0$. Therefore, $u(x,y)$ is harmonic. From the Cauchy-Riemann equations,

$$\frac{\partial v}{\partial y} = \frac{\partial u}{\partial x} = e^{-x}\sin(y) - xe^{-x}\sin(y) + ye^{-x}\cos(y), \tag{1.3.55}$$

and

$$\frac{\partial v}{\partial x} = -\frac{\partial u}{\partial y} = e^{-x}\cos(y) - xe^{-x}\cos(y) - ye^{-x}\sin(y). \tag{1.3.56}$$

Integrating (1.3.55) with respect to y,

$$v(x,y) = ye^{-x}\sin(y) + xe^{-x}\cos(y) + g(x). \tag{1.3.57}$$

Using (1.3.56),

$$v_x = -ye^{-x}\sin(y) - xe^{-x}\cos(y) + e^{-x}\cos(y) + g'(x)$$
$$= e^{-x}\cos(y) - xe^{-x}\cos(y) - ye^{-x}\sin(x). \tag{1.3.58}$$

Therefore, $g'(x) = 0$ or $g(x) = $ constant. Consequently,

$$v(x,y) = e^{-x}[y\sin(y) + x\cos(y)] + \text{constant}. \tag{1.3.59}$$

Hence, for our real harmonic function $u(x,y)$, there are infinitely many harmonic conjugates $v(x,y)$ which differ from each other by an additive constant.

Problems

Show that the following functions are entire:

1. $f(z) = iz + 2$ 2. $f(z) = e^{-z}$

3. $f(z) = z^3$ 4. $f(z) = \cosh(z)$

Find the derivative of the following functions:

5. $f(z) = (1 + z^2)^{3/2}$ 6. $f(z) = (z + 2z^{1/2})^{1/3}$

7. $f(z) = (1 + 4i)z^2 - 3z - 2$ 8. $f(z) = (2z - i)/(z + 2i)$

9. $f(z) = (iz - 1)^{-3}$

Evaluate the following limits:

10. $\displaystyle\lim_{z \to i} \frac{z^2 - 2iz - 1}{z^4 + 2z^2 + 1}$ 11. $\displaystyle\lim_{z \to 0} \frac{z - \sin(z)}{z^3}$

12. Show that the function $f(z) = z^*$ is nowhere differentiable.

For each of the following $u(x, y)$, show that it is harmonic and then find a corresponding $v(x, y)$ such that $f(z) = u + iv$ is analytic.

13.
$$u(x, y) = x^2 - y^2$$

14.
$$u(x, y) = x^4 - 6x^2y^2 + y^4 + x$$

15.
$$u(x, y) = x \cos(x)e^{-y} - y \sin(x)e^{-y}$$

16.
$$u(x, y) = (x^2 - y^2)\cos(y)e^x - 2xy \sin(y)e^x$$

1.4 LINE INTEGRALS

So far, we discussed complex numbers, complex functions, and complex differentiation. We are now ready for integration.

Just as we have integrals involving real variables, we can define an integral that involves complex variables. Because the z-plane is two-dimensional there is clearly greater freedom in what we mean by a complex integral. For example, we might ask whether the integral of some function between points A and B depends upon the curve along which we integrate. (In general it does.) Consequently, an important ingredient in any complex integration is the *contour* that we follow during the integration.

The result of a line integral is a complex number or expression. Unlike its counterpart in real variables, there is no physical interpretation for this quantity, such as area under a curve. Generally, integration in the complex plane is an intermediate process with a physically realizable quantity occurring only after we take its real or imaginary part. For example, in potential fluid flow, the lift and drag are found by taking the real and imaginary part of a complex integral, respectively.

How do we compute $\int_C f(z)\, dz$? Let us deal with the definition; we illustrate the actual method by examples.

A popular method for evaluating complex line integrals consists of breaking everything up into real and imaginary parts. This reduces the integral to line integrals of real-valued functions which we know how to handle. Thus, we write $f(z) = u(x, y) + iv(x, y)$ as usual, and because $z = x + iy$, formally $dz = dx + i\, dy$. Therefore,

$$\int_C f(z)\, dz = \int_C [u(x, y) + iv(x, y)][dx + i\, dy] \tag{1.4.1}$$

$$= \int_C u(x, y)\, dx - v(x, y)\, dy + i \int_C v(x, y)\, dx + u(x, y)\, dy. \tag{1.4.2}$$

The exact method used to evaluate (1.4.2) depends upon the exact path specified.

From the definition of the line integral, we have the following self-evident properties:

$$\int_C f(z)\, dz = -\int_{C'} f(z)\, dz, \tag{1.4.3}$$

where C' is the contour C taken in the opposite direction of C and

$$\int_{C_1+C_2} f(z)\, dz = \int_{C_1} f(z)\, dz + \int_{C_2} f(z)\, dz. \tag{1.4.4}$$

• Example 1.4.1

Let us evaluate $\int_C z^*dz$ from $z = 0$ to $z = 4 + 2i$ along two different contours. The first consists of the parametric equation $z = t^2 + it$. The second consists of two "dog legs": the first leg runs along the imaginary axis from $z = 0$ to $z = 2i$ and then along a line parallel to the x-axis from $z = 2i$ to $z = 4 + 2i$. See Figure 1.4.1.

For the first case, the points $z = 0$ and $z = 4 + 2i$ on C_1 correspond to $t = 0$ and $t = 2$, respectively. Then the line integral equals

$$\int_{C_1} z^*dz = \int_0^2 (t^2 + it)^* \, d(t^2 + it) = \int_0^2 (2t^3 - it^2 + t)\, dt = 10 - \tfrac{8i}{3}. \tag{1.4.5}$$

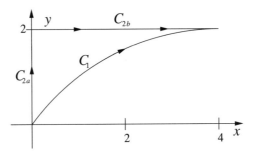

Figure 1.4.1: Contour used in Example 1.4.1.

The line integral for the second contour C_2 equals

$$\int_{C_2} z^* \, dz = \int_{C_{2a}} z^* \, dz + \int_{C_{2b}} z^* \, dz, \qquad (1.4.6)$$

where C_{2a} denotes the integration from $z = 0$ to $z = 2i$ while C_{2b} denotes the integration from $z = 2i$ to $z = 4 + 2i$. For the first integral,

$$\int_{C_{2a}} z^* \, dz = \int_{C_{2a}} (x - iy)(dx + i\,dy) = \int_0^2 y \, dy = 2, \qquad (1.4.7)$$

because $x = 0$ and $dx = 0$ along C_{2a}. On the other hand, along C_{2b}, $y = 2$ and $dy = 0$ so that

$$\int_{C_{2b}} z^* \, dz = \int_{C_{2b}} (x - iy)(dx + i\,dy) = \int_0^4 x \, dx + i \int_0^4 -2 \, dx = 8 - 8i. \quad (1.4.8)$$

Thus the value of the entire C_2 contour integral equals the sum of the two parts or $10 - 8i$.

The point here is that integration along two different paths has given us different results even though we integrated from $z = 0$ to $z = 4 + 2i$ both times. This result foreshadows a general result that is extremely important. Because the integrand contains nonanalytic points along and inside the region enclosed by our two curves, as shown by the Cauchy-Riemann equations, the results depend upon the path taken. Since complex integrations often involve integrands that have nonanalytic points, many line integrations depend upon the contour taken.

• **Example 1.4.2**

Let us integrate the *entire* function $f(z) = z^2$ along the two paths from $z = 0$ to $z = 2 + i$ shown in Figure 1.4.2. For the first integration, $x = 2y$

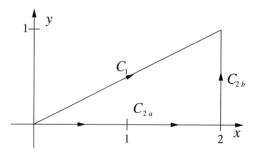

Figure 1.4.2: Contour used in Example 1.4.2.

while along the second path we have two straight paths: $z = 0$ to $z = 2$ and $z = 2$ to $z = 2 + i$.

For the first contour integration,

$$\int_{C_1} z^2 dz = \int_0^1 (2y + iy)^2 (2\,dy + i\,dy) \tag{1.4.9}$$

$$= \int_0^1 (3y^2 + 4y^2 i)(2\,dy + i\,dy) \tag{1.4.10}$$

$$= \int_0^1 6y^2\,dy + 8y^2 i\,dy + 3y^2 i\,dy - 4y^2\,dy \tag{1.4.11}$$

$$= \int_0^1 2y^2\,dy + 11y^2 i\,dy \tag{1.4.12}$$

$$= \tfrac{2}{3}y^3|_0^1 + \tfrac{11}{3}iy^3|_0^1 = \tfrac{2}{3} + \tfrac{11i}{3}. \tag{1.4.13}$$

For our second integration,

$$\int_{C_2} z^2\,dz = \int_{C_{2a}} z^2\,dz + \int_{C_{2b}} z^2\,dz. \tag{1.4.14}$$

Along C_{2a} we find that $y = dy = 0$ so that

$$\int_{C_{2a}} z^2\,dz = \int_0^2 x^2\,dx = \tfrac{1}{3}x^3|_0^2 = \tfrac{8}{3}, \tag{1.4.15}$$

and

$$\int_{C_{2b}} z^2\,dz = \int_0^1 (2 + iy)^2 i\,dy = i\left(4y + 2iy^2 - \frac{y^3}{3}\right)\bigg|_0^1 = 4i - 2 - \tfrac{i}{3}, \tag{1.4.16}$$

because $x = 2$ and $dx = 0$. Consequently,

$$\int_{C_2} z^2\,dz = \frac{2}{3} + \frac{11i}{3}. \tag{1.4.17}$$

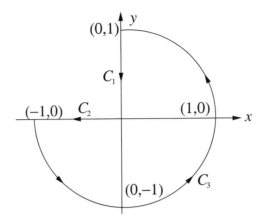

Figure 1.4.3: Contour used in Example 1.4.3.

In this problem we obtained the same results from two different contours of integration. Exploring other contours, we would find that the results are always the same; the integration is path-independent. But what makes these results path-independent while the integration in Example 1.4.1 was not? Perhaps it is the fact that the integrand is analytic everywhere on the complex plane and there are no nonanalytic points. We will explore this later.

Finally, an important class of line integrals involves *closed contours*. We denote this special subclass of line integrals by placing a circle on the integral sign: $\oint$. Consider now the following examples:

• **Example 1.4.3**

Let us integrate $f(z) = z$ around the closed contour shown in Figure 1.4.3.

From Figure 1.4.3,

$$\oint_C z\,dz = \int_{C_1} z\,dz + \int_{C_2} z\,dz + \int_{C_3} z\,dz. \qquad (1.4.18)$$

Now

$$\int_{C_1} z\,dz = \int_1^0 iy\,(i\,dy) = -\int_1^0 y\,dy = -\left.\frac{y^2}{2}\right|_1^0 = \frac{1}{2}, \qquad (1.4.19)$$

$$\int_{C_2} z\,dz = \int_0^{-1} x\,dx = \left.\frac{x^2}{2}\right|_0^{-1} = \frac{1}{2}, \qquad (1.4.20)$$

and

$$\int_{C_3} z\,dz = \int_{-\pi}^{\pi/2} e^{\theta i} i e^{\theta i}\,d\theta = \left.\frac{e^{2\theta i}}{2}\right|_{-\pi}^{\pi/2} = -1, \qquad (1.4.21)$$

where we used $z = e^{\theta i}$ around the portion of the unit circle. Therefore, the closed line integral equals zero.

• Example 1.4.4

Let us integrate $f(z) = 1/(z - a)$ around any circle centered on $z = a$. The Cauchy-Riemann equations show that $f(z)$ is a meromorphic function. It is analytic everywhere except at the isolated singularity $z = a$.

If we introduce polar coordinates by letting $z - a = re^{\theta i}$ and $dz = ire^{\theta i}d\theta$,

$$\oint_C \frac{dz}{z - a} = \int_0^{2\pi} \frac{ire^{\theta i}}{re^{\theta i}}\, d\theta = i \int_0^{2\pi} d\theta = 2\pi i. \qquad (1.4.22)$$

Note that the integrand becomes undefined at $z = a$. Furthermore, the answer is independent of the size of the circle. Our example suggests that when we have a closed contour integration it is the behavior of the function within the contour rather than the exact shape of the closed contour that is of importance. We will return to this point in later sections.

Problems

1. Evaluate $\oint_C (z^*)^2\, dz$ around the circle $|z| = 1$ taken in the counterclockwise direction.

2. Evaluate $\oint_C |z|^2\, dz$ around the square with vertices at $(0,0)$, $(1,0)$, $(1,1)$, and $(0,1)$ taken in the counterclockwise direction.

3. Evaluate $\int_C |z|\, dz$ along the right half of the circle $|z| = 1$ from $z = -i$ to $z = i$.

4. Evaluate $\int_C e^z\, dz$ along the line $y = x$ from $(-1, -1)$ to $(1, 1)$.

5. Evaluate $\int_C (z^*)^2\, dz$ along the line $y = x^2$ from $(0,0)$ to $(1,1)$.

6. Evaluate $\int_C z^{-1/2}\, dz$, where C is (a) the upper semicircle $|z| = 1$ and (b) the lower semicircle $|z| = 1$. If $z = re^{\theta i}$, restrict $-\pi < \theta < \pi$. Take both contours in the counterclockwise direction.

1.5 THE CAUCHY-GOURSAT THEOREM

In the previous section we showed how to evaluate line integrations by brute-force reduction to real-valued integrals. In general, this direct approach is quite difficult and we would like to apply some of the deeper properties of complex analysis to work smarter. In the remaining portions of this chapter we introduce several theorems that will do just that.

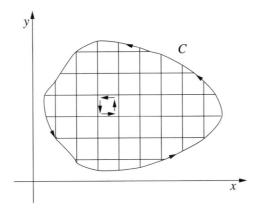

Figure 1.5.1: Diagram used in proving the Cauchy-Goursat theorem.

If we scan over the examples worked in the previous section, we see considerable differences when the function was analytic inside and on the contour and when it was not. We may formalize this anecdotal evidence into the following theorem:

Cauchy-Goursat theorem[2]: *Let $f(z)$ be analytic in a domain D and let C be a simple Jordan curve[3] inside D so that $f(z)$ is analytic on and inside of C. Then $\oint_C f(z)\,dz = 0$.*

Proof: Let C denote the contour around which we will integrate $w = f(z)$. We divide the region within C into a series of infinitesimal rectangles. See Figure 1.5.1. The integration around each rectangle equals the product of the average value of w on each side and its length,

$$\left[w + \frac{\partial w}{\partial x}\frac{dx}{2}\right]dx + \left[w + \frac{\partial w}{\partial x}dx + \frac{\partial w}{\partial(iy)}\frac{d(iy)}{2}\right]d(iy)$$

$$+ \left[w + \frac{\partial w}{\partial x}\frac{dx}{2} + \frac{\partial w}{\partial(iy)}d(iy)\right](-dx) + \left[w + \frac{\partial w}{\partial(iy)}\frac{d(iy)}{2}\right]d(-iy)$$

$$= \left(\frac{\partial w}{\partial x} - \frac{\partial w}{i\partial y}\right)(i\,dx\,dy). \tag{1.5.1}$$

Substituting $w = u + iv$ into (1.5.1),

$$\frac{\partial w}{\partial x} - \frac{\partial w}{i\,\partial y} = \left(\frac{\partial u}{\partial x} - \frac{\partial v}{\partial y}\right) + i\left(\frac{\partial v}{\partial x} + \frac{\partial u}{\partial y}\right). \tag{1.5.2}$$

---

[2] Goursat, E., 1900: Sur la définition générale des fonctions analytiques, d'après Cauchy. *Trans. Am. Math. Soc.*, **1**, 14–16.

[3] A Jordan curve is a simply closed curve. It looks like a closed loop that does not cross itself. See Figure 1.5.2.

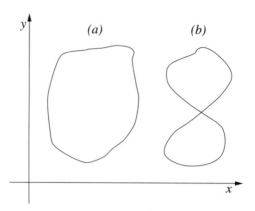

Figure 1.5.2: Examples of a (a) simply closed curve and (b) not simply closed curve.

Because the function is analytic, the right side of (1.5.1) and (1.5.2) equals zero. Thus, the integration around each of these rectangles also equals zero.

We note next that in integrating around adjoining rectangles we transverse each side in opposite directions, the net result being equivalent to integrating around the outer curve C. We therefore arrive at the result $\oint_C f(z)\, dz = 0$, where $f(z)$ is analytic within and on the closed contour. $\square$

The Cauchy-Goursat theorem has several useful implications. Suppose that we have a domain where $f(z)$ is analytic. Within this domain let us evaluate a line integral from point A to B along two different contours C_1 and C_2. Then, the integral around the closed contour formed by integrating along C_1 and then back along C_2, only in the opposite direction, is

$$\oint_C f(z)\, dz = \int_{C_1} f(z)\, dz - \int_{C_2} f(z)\, dz = 0 \qquad (1.5.3)$$

or

$$\int_{C_1} f(z)\, dz = \int_{C_2} f(z)\, dz. \qquad (1.5.4)$$

Because C_1 and C_2 are completely arbitrary, we have the result that if, in a domain, $f(z)$ is analytic, the integral between any two points within the domain is *path independent*.

One obvious advantage of path independence is the ability to choose the contour so that the computations are made easier. This obvious choice immediately leads to

The principle of deformation of contours: *The value of a line integral of an analytic function around any simple closed contour remains unchanged if we deform the contour in such a manner that we do not pass over a non-analytic point.*

• **Example 1.5.1**

Let us integrate $f(z) = z^{-1}$ around the closed contour C in the counterclockwise direction. This contour consists of a square, centered on the origin, with vertices at $(1,1)$, $(1,-1)$, $(-1,1)$, and $(-1,-1)$.

The direct integration of $\oint_C z^{-1} dz$ around the original contour is very cumbersome. However, because the integrand is analytic everywhere except at the origin, we may deform the origin contour into a circle of radius r, centered on the origin. Then, $z = re^{\theta i}$ and $dz = rie^{\theta i} d\theta$ so that

$$\oint_C \frac{dz}{z} = \int_0^{2\pi} \frac{rie^{\theta i}}{re^{\theta i}} d\theta = i \int_0^{2\pi} d\theta = 2\pi i. \qquad (1.5.5)$$

The point here is that no matter how bizarre the contour is, as long as it encircles the origin and is a simply closed contour, we can deform it into a circle and we get the same answer for the contour integral. This suggests that it is not the shape of the closed contour that makes the difference but whether we enclose any singularities [points where $f(z)$ becomes undefined] that matters. We shall return to this idea many times in the next few sections.

Finally, suppose that we have a function $f(z)$ such that $f(z)$ is analytic in some domain. Furthermore, let us introduce the analytic function $F(z)$ such that $f(z) = F'(z)$. We would like to evaluate $\int_a^b f(z)\, dz$ in terms of $F(z)$.

We begin by noting that we can represent F, f as $F(z) = U + iV$ and $f(z) = u + iv$. From (1.3.30) we have that $u = U_x$ and $v = V_x$. Therefore,

$$\int_a^b f(z)\, dz = \int_a^b (u + iv)(dx + i\, dy) \qquad (1.5.6)$$

$$= \int_a^b U_x\, dx - V_x\, dy + i \int_a^b V_x\, dx + U_x\, dy \qquad (1.5.7)$$

$$= \int_a^b U_x\, dx + U_y\, dy + i \int_a^b V_x\, dx + V_y\, dy \qquad (1.5.8)$$

$$= \int_a^b dU + i \int_a^b dV = F(b) - F(a) \qquad (1.5.9)$$

or

$$\int_a^b f(z)\, dz = F(b) - F(a). \qquad (1.5.10)$$

Equation (1.5.10) is the complex variable form of the fundamental theorem of calculus. Thus, if we can find the antiderivative of a function $f(z)$ that is analytic within a specific region, we can evaluate the integral by evaluating the antiderivative at the endpoints for any curves within that region.

• **Example 1.5.2**

Let us evaluate $\int_0^{\pi i} z \sin(z^2)\,dz$.

The integrand $f(z) = z \sin(z^2)$ is an entire function and has the antiderivative $-\frac{1}{2}\cos(z^2)$. Therefore,

$$\int_0^{\pi i} z \sin(z^2)\,dz = -\tfrac{1}{2}\cos(z^2)\big|_0^{\pi i} = \tfrac{1}{2}[\cos(0) - \cos(-\pi^2)] = \tfrac{1}{2}[1 - \cos(\pi^2)].$$

$$(1.5.11)$$

Problems

For the following integrals, show that they are path independent and determine the value of the integral:

1. $\displaystyle\int_{1-\pi i}^{2+3\pi i} e^{-2z}\,dz$

2. $\displaystyle\int_0^{2\pi} [e^z - \cos(z)]\,dz$

3. $\displaystyle\int_0^{\pi} \sin^2(z)\,dz$

4. $\displaystyle\int_{-i}^{2i} (z+1)\,dz$

1.6 CAUCHY'S INTEGRAL FORMULA

In the previous section, our examples suggested that the presence of a singularity within a contour really determines the value of a closed contour integral. Continuing with this idea, let us consider a class of closed contour integrals that explicitly contain a single singularity within the contour, namely $\oint_C g(z)\,dz$, where $g(z) = f(z)/(z - z_0)$, and $f(z)$ is analytic within and on the contour C. We closed the contour in the *positive sense* where the enclosed area lies to your left as you move along the contour.

We begin by examining a closed contour integral where the closed contour consists of the C_1, C_2, C_3, and C_4 as shown in Figure 1.6.1. The gap or cut between C_2 and C_4 is very small. Because $g(z)$ is analytic within and on the closed integral, we have that

$$\int_{C_1} \frac{f(z)}{z - z_0}\,dz + \int_{C_2} \frac{f(z)}{z - z_0}\,dz + \int_{C_3} \frac{f(z)}{z - z_0}\,dz + \int_{C_4} \frac{f(z)}{z - z_0}\,dz = 0. \quad (1.6.1)$$

It can be shown that the contribution to the integral from the path C_2 going into the singularity cancels the contribution from the path C_4 going away from the singularity as the gap between them vanishes. Because $f(z)$ is analytic at z_0, we can approximate its value on C_3 by $f(z) = f(z_0) + \delta(z)$, where δ is a small quantity. Substituting into (1.6.1),

$$\oint_{C_1} \frac{f(z)}{z - z_0}\,dz = -f(z_0) \int_{C_3} \frac{1}{z - z_0}\,dz - \int_{C_3} \frac{\delta(z)}{z - z_0}\,dz. \quad (1.6.2)$$

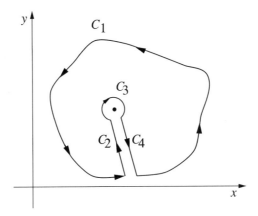

Figure 1.6.1: Diagram used to prove Cauchy's integral formula.

Consequently, as the gap between C_2 and C_4 vanishes, the contour C_1 becomes the closed contour C so that (1.6.2) may be written

$$\oint_C \frac{f(z)}{z - z_0}\, dz = 2\pi i f(z_0) + i \int_0^{2\pi} \delta\, d\theta, \qquad (1.6.3)$$

where we set $z - z_0 = \epsilon e^{\theta i}$ and $dz = i\epsilon e^{\theta i} d\theta$.

Let M denote the value of the integral on the right side of (1.6.3) and Δ equal the greatest value of the modulus of δ along the circle. Then

$$|M| < \int_0^{2\pi} |\delta|\, d\theta \leq \int_0^{2\pi} \Delta\, d\theta = 2\pi\Delta. \qquad (1.6.4)$$

As the radius of the circle diminishes to zero, Δ also diminishes to zero. Therefore, $|M|$, which is positive, becomes less than any finite quantity, however small, and M itself equals zero. Thus, we have that

$$f(z_0) = \frac{1}{2\pi i} \oint_C \frac{f(z)}{z - z_0}\, dz. \qquad (1.6.5)$$

This equation is *Cauchy's integral formula*. By taking n derivatives of (1.6.5), we can extend Cauchy's integral formula[4] to

$$f^{(n)}(z_0) = \frac{n!}{2\pi i} \oint_C \frac{f(z)}{(z - z_0)^{n+1}}\, dz \qquad (1.6.6)$$

---

[4] See Carrier, G. F., M. Krook, and C. E. Pearson, 1966: *Functions of a Complex Variable: Theory and Technique.* McGraw-Hill, pp. 39–40 for the proof.

for $n = 1, 2, 3, \ldots$. For computing integrals, it is convenient to rewrite (1.6.6) as

$$\oint_C \frac{f(z)}{(z - z_0)^{n+1}} \, dz = \frac{2\pi i}{n!} f^{(n)}(z_0). \tag{1.6.7}$$

• **Example 1.6.1**

Let us find the value of the integral

$$\oint_C \frac{\cos(\pi z)}{(z - 1)(z - 2)} \, dz, \tag{1.6.8}$$

where C is the circle $|z| = 5$. Using partial fractions,

$$\frac{1}{(z - 1)(z - 2)} = \frac{1}{z - 2} - \frac{1}{z - 1}, \tag{1.6.9}$$

and

$$\oint_C \frac{\cos(\pi z)}{(z - 1)(z - 2)} \, dz = \oint_C \frac{\cos(\pi z)}{z - 2} \, dz - \oint_C \frac{\cos(\pi z)}{z - 1} \, dz. \tag{1.6.10}$$

By Cauchy's integral formula with $z_0 = 2$ and $z_0 = 1$,

$$\oint_C \frac{\cos(\pi z)}{z - 2} \, dz = 2\pi i \, \cos(2\pi) = 2\pi i, \tag{1.6.11}$$

and

$$\oint_C \frac{\cos(\pi z)}{z - 1} \, dz = 2\pi i \cos(\pi) = -2\pi i, \tag{1.6.12}$$

because $z_0 = 1$ and $z_0 = 2$ lie inside C and $\cos(\pi z)$ is analytic there. Thus the required integral has the value

$$\oint_C \frac{\cos(\pi z)}{(z - 1)(z - 2)} \, dz = 4\pi i. \tag{1.6.13}$$

• **Example 1.6.2**

Let us use Cauchy's integral formula to evaluate

$$I = \oint_{|z|=2} \frac{e^z}{(z - 1)^2(z - 3)} \, dz. \tag{1.6.14}$$

We need to convert (1.6.14) into the form (1.6.7). To do this, we rewrite (1.6.14) as

$$\oint_{|z|=2} \frac{e^z}{(z - 1)^2(z - 3)} \, dz = \oint_{|z|=2} \frac{e^z/(z - 3)}{(z - 1)^2} \, dz. \tag{1.6.15}$$

Therefore, $f(z) = e^z/(z-3)$, $n = 1$ and $z_0 = 1$. The function $f(z)$ is analytic within the closed contour because the point $z = 3$ lies outside of the contour. Applying Cauchy's integral formula,

$$\oint_{|z|=2} \frac{e^z}{(z-1)^2(z-3)}\, dz = \frac{2\pi i}{1!}\frac{d}{dz}\left(\frac{e^z}{z-3}\right)\Bigg|_{z=1} \tag{1.6.16}$$

$$= 2\pi i\left[\frac{e^z}{z-3} - \frac{e^z}{(z-3)^2}\right]\Bigg|_{z=1} \tag{1.6.17}$$

$$= -\frac{3\pi i e}{2}. \tag{1.6.18}$$

Problems

Use Cauchy's integral formula to evaluate the following integrals. Assume all of the contours are in the positive sense.

1. $\oint_{|z|=1} \dfrac{\sin^6(z)}{z-\pi/6}\, dz$

2. $\oint_{|z|=1} \dfrac{\sin^6(z)}{(z-\pi/6)^3}\, dz$

3. $\oint_{|z|=1} \dfrac{1}{z(z^2+4)}\, dz$

4. $\oint_{|z|=1} \dfrac{\tan(z)}{z}\, dz$

5. $\oint_{|z-1|=1/2} \dfrac{1}{(z-1)(z-2)}\, dz$

6. $\oint_{|z|=5} \dfrac{\exp(z^2)}{z^3}\, dz$

7. $\oint_{|z-1|=1} \dfrac{z^2+1}{z^2-1}\, dz$

8. $\oint_{|z|=2} \dfrac{z^2}{(z-1)^4}\, dz$

9. $\oint_{|z|=2} \dfrac{z^3}{(z+i)^3}\, dz$

10. $\oint_{|z|=1} \dfrac{\cos(z)}{z^{2n+1}}\, dz$

Project: Computing Derivatives of Any Order of a Complex or Real Function

The most common technique for computing a derivative is finite differencing. Recently Mahajerin and Burgess[5] showed how Cauchy's integral formula can be used to compute the derivatives of any order of a complex or real function via numerical quadrature. In this project you will derive the algorithm, write code implementing it, and finally test it.

Step 1: Consider the complex function $f(z) = u + iv$ which is analytic inside the closed circular contour C of radius R centered at z_0. Using Cauchy's integral formula, show that

$$f^{(n)}(z_0) = \frac{n!}{2\pi R^n}\int_0^{2\pi} [u(x,y) + iv(x,y)][\cos(n\theta) - i\sin(n\theta)]\, d\theta,$$

---

[5] Reprinted from *Computers & Struct.*, **49**, E. Mahajerin and G. Burgess, An algorithm for computing derivatives of any order of a complex or real function, 385–387, ©1993, with permission from Elsevier Science.

where $x = x_0 + R\cos(\theta)$, and $y = y_0 + R\sin(\theta)$.

Step 2: Using five-point Gaussian quadrature, write code to implement the results from Step 1.

Step 3: Test out this scheme by finding the first, sixth, and eleventh derivative of $f(x) = 8x/(x^2+4)$ for $x = 2$. The exact answers are 0, 2.8125, and 1218.164, respectively. What is the maximum value of R? How does the accuracy vary with the number of subdivisions used in the numerical integration? Is the algorithm sensitive to the value of R and the number of subdivisions? For a fixed number of subdivisions, is there an optimal R?

1.7 TAYLOR AND LAURENT EXPANSIONS AND SINGULARITIES

In the previous section we showed what a crucial role singularities play in complex integration. Before we can find the most general way of computing a closed complex integral, our understanding of singularities must deepen. For this, we employ power series.

One reason why power series are so important is their ability to provide locally a general representation of a function even when its arguments are complex. For example, when we were introduced to trigonometric functions in high school, it was in the context of a right triangle and a real angle. However, when the argument becomes complex this geometrical description disappears and power series provide a formalism for defining the trigonometric functions, regardless of the nature of the argument.

Let us begin our analysis by considering the complex function $f(z)$ which is analytic everywhere on the boundary and the interior of a circle whose center is at $z = z_0$. Then, if z denotes any point within the circle, we have from Cauchy's integral formula that

$$f(z) = \frac{1}{2\pi i} \oint_C \frac{f(\zeta)}{\zeta - z}\, d\zeta = \frac{1}{2\pi i} \oint_C \frac{f(\zeta)}{\zeta - z_0} \left[\frac{1}{1 - (z - z_0)/(\zeta - z_0)} \right] d\zeta, \tag{1.7.1}$$

where C denotes the closed contour. Expanding the bracketed term as a geometric series, we find that

$$f(z) = \frac{1}{2\pi i} \left[\oint_C \frac{f(\zeta)}{\zeta - z_0}\, d\zeta + (z - z_0) \oint_C \frac{f(\zeta)}{(\zeta - z_0)^2}\, d\zeta + \cdots \right.$$
$$\left. + (z - z_0)^n \oint_C \frac{f(\zeta)}{(\zeta - z_0)^{n+1}}\, d\zeta + \cdots \right]. \tag{1.7.2}$$

Applying Cauchy's integral formula to each integral in (1.7.2), we finally obtain

$$f(z) = f(z_0) + \frac{(z - z_0)}{1!} f'(z_0) + \cdots + \frac{(z - z_0)^n}{n!} f^{(n)}(z_0) + \cdots \tag{1.7.3}$$

or the familiar formula for a Taylor expansion. Consequently, *we can expand any analytic function into a Taylor series.* Interestingly, the radius of convergence[6] of this series may be shown to be the distance between z_0 and the nearest nonanalytic point of $f(z)$.

- ## Example 1.7.1

Let us find the expansion of $f(z) = \sin(z)$ about the point $z_0 = 0$.

Because $f(z)$ is an entire function, we can construct a Taylor expansion anywhere on the complex plane. For $z_0 = 0$,

$$f(z) = f(0) + \frac{1}{1!}f'(0)z + \frac{1}{2!}f''(0)z^2 + \frac{1}{3!}f'''(0)z^3 + \cdots. \qquad (1.7.4)$$

Because $f(0) = 0$, $f'(0) = 1$, $f''(0) = 0$, $f'''(0) = -1$ and so forth,

$$f(z) = z - \frac{z^3}{3!} + \frac{z^5}{5!} - \frac{z^7}{7!} + \cdots. \qquad (1.7.5)$$

Because $\sin(z)$ is an entire function, the radius of convergence is $|z - 0| < \infty$, i.e., all z.

- ## Example 1.7.2

Let us find the expansion of $f(z) = 1/(1 - z)$ about the point $z_0 = 0$.

From the formula for a Taylor expansion,

$$f(z) = f(0) + \frac{1}{1!}f'(0)z + \frac{1}{2!}f''(0)z^2 + \frac{1}{3!}f'''(0)z^3 + \cdots. \qquad (1.7.6)$$

Because $f^{(n)}(0) = n!$, we find that

$$f(z) = 1 + z + z^2 + z^3 + z^4 + \cdots = \frac{1}{1-z}. \qquad (1.7.7)$$

Equation (1.7.7) is the familiar result for a geometric series. Because the only nonanalytic point is at $z = 1$, the radius of convergence is $|z - 0| < 1$, the unit circle centered at $z = 0$.

Consider now the situation where we draw two concentric circles about some arbitrary point z_0; we denote the outer circle by C while we denote the inner circle by C_1. See Figure 1.7.1. Let us assume that $f(z)$ is analytic inside the annulus between the two circles. Outside of this area, the function may or may not be analytic. Within the annulus we pick a point z and construct a small circle around it, denoting the circle by C_2. As the gap or *cut* in

---

[6] A positive number h such that the series diverges for $|z - z_0| > h$ but converges absolutely for $|z - z_0| < h$.

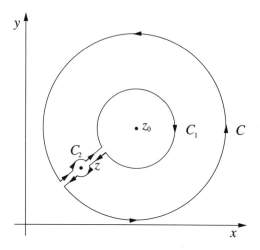

Figure 1.7.1: Contour used in deriving the Laurent expansion.

the annulus becomes infinitesimally small, the line integrals that connect the circle C_2 to C_1 and C sum to zero, leaving

$$\oint_C \frac{f(\zeta)}{\zeta - z}\, d\zeta = \oint_{C_1} \frac{f(\zeta)}{\zeta - z}\, d\zeta + \oint_{C_2} \frac{f(\zeta)}{\zeta - z}\, d\zeta. \qquad (1.7.8)$$

Because $f(\zeta)$ is analytic everywhere within C_2,

$$2\pi i f(z) = \oint_{C_2} \frac{f(\zeta)}{\zeta - z}\, d\zeta. \qquad (1.7.9)$$

Using the relationship:

$$\oint_{C_1} \frac{f(\zeta)}{\zeta - z}\, d\zeta = -\oint_{C_1} \frac{f(\zeta)}{z - \zeta}\, d\zeta, \qquad (1.7.10)$$

(1.7.8) becomes

$$f(z) = \frac{1}{2\pi i} \oint_C \frac{f(\zeta)}{\zeta - z}\, d\zeta + \frac{1}{2\pi i} \oint_{C_1} \frac{f(\zeta)}{z - \zeta}\, d\zeta. \qquad (1.7.11)$$

Now,

$$\frac{1}{\zeta - z} = \frac{1}{\zeta - z_0 - z + z_0} = \frac{1}{\zeta - z_0} \frac{1}{1 - (z - z_0)/(\zeta - z_0)} \qquad (1.7.12)$$

$$= \frac{1}{\zeta - z_0}\left[1 + \left(\frac{z - z_0}{\zeta - z_0}\right) + \left(\frac{z - z_0}{\zeta - z_0}\right)^2 + \cdots + \left(\frac{z - z_0}{\zeta - z_0}\right)^n + \cdots \right],$$

$$(1.7.13)$$

where $|z - z_0|/|\zeta - z_0| < 1$ and

$$\frac{1}{z - \zeta} = \frac{1}{z - z_0 - \zeta + z_0} = \frac{1}{z - z_0} \frac{1}{1 - (\zeta - z_0)/(z - z_0)} \tag{1.7.14}$$

$$= \frac{1}{z - z_0} \left[1 + \left(\frac{\zeta - z_0}{z - z_0} \right) + \left(\frac{\zeta - z_0}{z - z_0} \right)^2 + \cdots + \left(\frac{\zeta - z_0}{z - z_0} \right)^n + \cdots \right],$$

$$\tag{1.7.15}$$

where $|\zeta - z_0|/|z - z_0| < 1$. Upon substituting these expressions into (1.7.11),

$$f(z) = \left[\frac{1}{2\pi i} \oint_C \frac{f(\zeta)}{\zeta - z_0} \, d\zeta + \frac{z - z_0}{2\pi i} \oint_C \frac{f(\zeta)}{(\zeta - z_0)^2} \, d\zeta + \cdots \right.$$

$$\left. + \frac{(z - z_0)^n}{2\pi i} \oint_C \frac{f(\zeta)}{(\zeta - z_0)^{n+1}} \, d\zeta + \cdots \right]$$

$$+ \left[\frac{1}{z - z_0} \frac{1}{2\pi i} \oint_{C_1} f(\zeta) \, d\zeta + \frac{1}{(z - z_0)^2} \frac{1}{2\pi i} \oint_{C_1} f(\zeta)(\zeta - z_0) \, d\zeta + \cdots \right.$$

$$\left. + \frac{1}{(z - z_0)^n} \frac{1}{2\pi i} \oint_{C_1} f(\zeta)(\zeta - z_0)^{n-1} \, d\zeta + \cdots \right] \tag{1.7.16}$$

or

$$f(z) = \frac{a_1}{z - z_0} + \frac{a_2}{(z - z_0)^2} + \cdots + \frac{a_n}{(z - z_0)^n} + \cdots$$

$$+ b_0 + b_1(z - z_0) + \cdots + b_n(z - z_0)^n + \cdots. \tag{1.7.17}$$

Equation (1.7.17) is a *Laurent expansion.*[7] If $f(z)$ is analytic at z_0, then $a_1 = a_2 = \cdots = a_n = \cdots = 0$ and the Laurent expansion reduces to a Taylor expansion. If z_0 is a singularity of $f(z)$, then the Laurent expansion includes both positive and *negative* powers. The coefficient of the $(z - z_0)^{-1}$ term, a_1, is the *residue*, for reasons that will appear in the next section.

Unlike the Taylor series, there is no straightforward method for obtaining a Laurent series. For the remaining portions of this section we illustrate their construction. These techniques include replacing a function by its appropriate power series, the use of geometric series to expand the denominator, and the use of algebraic tricks to assist in applying the first two methods.

• Example 1.7.3

Laurent expansions provide a formalism for the classification of singularities of a function. *Isolated singularities* fall into three types; they are

---

[7] Laurent, M., 1843: Extension du théorème de M. Cauchy relatif à la convergence du développement d'une fonction suivant les puissances ascendantes de la variable *x*. *C. R. l'Acad. Sci.*, **17**, 938–942.

• *Essential Singularity:* Consider the function $f(z) = \cos(1/z)$. Using the expansion for cosine,

$$\cos\left(\frac{1}{z}\right) = 1 - \frac{1}{2!z^2} + \frac{1}{4!z^4} - \frac{1}{6!z^6} + \cdots \tag{1.7.18}$$

for $0 < |z| < \infty$. Note that this series never truncates in the inverse powers of z. Essential singularities have Laurent expansions which have an infinite number of inverse powers of $z - z_0$. The value of the residue for this essential singularity at $z = 0$ is zero.

• *Removable Singularity:* Consider the function $f(z) = \sin(z)/z$. This function has a singularity at $z = 0$. Upon applying the expansion for sine,

$$\frac{\sin(z)}{z} = \frac{1}{z}\left(z - \frac{z^3}{3!} + \frac{z^5}{5!} - \frac{z^7}{7!} + \frac{z^9}{9!} - \cdots\right) \tag{1.7.19}$$

$$= 1 - \frac{z^2}{3!} + \frac{z^4}{5!} - \frac{z^6}{7!} + \frac{z^8}{9!} - \cdots \tag{1.7.20}$$

for all z, if the division is permissible. We made $f(z)$ analytic by defining it by (1.7.20) and, in the process, removed the singularity. The residue for a removable singularity always equals zero.

• *Pole of order n:* Consider the function

$$f(z) = \frac{1}{(z-1)^3(z+1)}. \tag{1.7.21}$$

This function has two singularities: one at $z = 1$ and the other at $z = -1$. We shall only consider the case $z = 1$. After a little algebra,

$$f(z) = \frac{1}{(z-1)^3}\frac{1}{2+(z-1)} \tag{1.7.22}$$

$$= \frac{1}{2}\frac{1}{(z-1)^3}\frac{1}{1+(z-1)/2} \tag{1.7.23}$$

$$= \frac{1}{2}\frac{1}{(z-1)^3}\left[1 - \frac{z-1}{2} + \frac{(z-1)^2}{4} - \frac{(z-1)^3}{8} + \cdots\right] \tag{1.7.24}$$

$$= \frac{1}{2(z-1)^3} - \frac{1}{4(z-1)^2} + \frac{1}{8(z-1)} - \frac{1}{16} + \cdots \tag{1.7.25}$$

for $0 < |z - 1| < 2$. Because the largest inverse (negative) power is three, the singularity at $z = 1$ is a third-order pole; the value of the residue is $1/8$. Generally, we refer to a first-order pole as a *simple* pole.

• **Example 1.7.4**

Let us find the Laurent expansion for

$$f(z) = \frac{z}{(z-1)(z-3)} \qquad (1.7.26)$$

about the point $z = 1$.

We begin by rewriting $f(z)$ as

$$f(z) = \frac{1 + (z-1)}{(z-1)[-2 + (z-1)]} \qquad (1.7.27)$$

$$= -\frac{1}{2} \frac{1 + (z-1)}{(z-1)[1 - \frac{1}{2}(z-1)]} \qquad (1.7.28)$$

$$= -\frac{1}{2} \frac{1 + (z-1)}{(z-1)}[1 + \tfrac{1}{2}(z-1) + \tfrac{1}{4}(z-1)^2 + \cdots] \qquad (1.7.29)$$

$$= -\frac{1}{2} \frac{1}{z-1} - \frac{3}{4} - \frac{3}{8}(z-1) - \frac{3}{16}(z-1)^2 - \cdots \qquad (1.7.30)$$

provided $0 < |z-1| < 2$. Therefore we have a simple pole at $z = 1$ and the value of the residue is $-1/2$. A similar procedure would yield the Laurent expansion about $z = 3$.

• **Example 1.7.5**

Let us find the Laurent expansion for

$$f(z) = \frac{z^n + z^{-n}}{z^2 - 2z\cosh(\alpha) + 1}, \qquad \alpha > 0, \quad n \geq 0, \qquad (1.7.31)$$

about the point $z = 0$.

We begin by rewriting $f(z)$ as

$$f(z) = \frac{z^n + z^{-n}}{(z - e^\alpha)(z - e^{-\alpha})} = \frac{1}{2\sinh(\alpha)}\left(\frac{z^n + z^{-n}}{z - e^\alpha} - \frac{z^n + z^{-n}}{z - e^{-\alpha}}\right). \qquad (1.7.32)$$

Because

$$\frac{1}{z - e^\alpha} = -\frac{e^{-\alpha}}{1 - ze^{-\alpha}} = -e^{-\alpha}\left(1 + ze^{-\alpha} + z^2 e^{-2\alpha} + \cdots\right) \qquad (1.7.33)$$

if $|z| < e^\alpha$ and

$$\frac{1}{z - e^{-\alpha}} = -\frac{e^\alpha}{1 - ze^\alpha} = -e^\alpha\left(1 + ze^\alpha + z^2 e^{2\alpha} + \cdots\right) \qquad (1.7.34)$$

if $|z| < e^{-\alpha}$,

$$
\begin{aligned}
f(z) = \frac{e^{\alpha}}{2\sinh(\alpha)} \big(& z^n + z^{n+1}e^{\alpha} + z^{n+2}e^{2\alpha} + \cdots \\
& + z^{-n} + z^{1-n}e^{\alpha} + z^{2-n}e^{2\alpha} + \cdots \big) \\
- \frac{e^{-\alpha}}{2\sinh(\alpha)} \big(& z^n + z^{n+1}e^{-\alpha} + z^{n+2}e^{-2\alpha} + \cdots \\
& + z^{-n} + z^{1-n}e^{-\alpha} + z^{2-n}e^{-2\alpha} + \cdots \big), \quad (1.7.35)
\end{aligned}
$$

if $|z| < e^{-\alpha}$. Clearly we have a nth-order pole at $z = 0$. The residue, the coefficient of all of the z^{-1} terms in (1.7.35), is found directly and equals

$$
\mathrm{Res}[f(z); 0] = \frac{\sinh(n\alpha)}{\sinh(\alpha)}. \qquad (1.7.36)
$$

For complicated complex functions, it is very difficult to determine the nature of the singularities by finding the complete Laurent expansion and we must try another method. We shall call it "a poor man's Laurent expansion." The idea behind this method is the fact that we generally need only the first few terms of the Laurent expansion to discover its nature. Consequently, we compute these terms through the application of power series where we retain only the leading terms. Consider the following example.

● **Example 1.7.6**

Let us discover the nature of the singularity at $z = 0$ of the function

$$
f(z) = \frac{e^{tz}}{z\sinh(az)}, \qquad (1.7.37)
$$

where a and t are real.

We begin by replacing the exponential and hyperbolic sine by their Taylor expansion about $z = 0$. Then

$$
f(z) = \frac{1 + tz + t^2 z^2/2 + \cdots}{z(az + a^3 z^3/6 + \cdots)}. \qquad (1.7.38)
$$

Factoring out az in the denominator,

$$
f(z) = \frac{1 + tz + t^2 z^2/2 + \cdots}{az^2(1 + a^2 z^2/6 + \cdots)}. \qquad (1.7.39)
$$

Within the parentheses all of the terms except the leading one are small. Therefore, by long division, we formally have that

$$f(z) = \frac{1}{az^2}(1 + tz + t^2 z^2/2 + \cdots)(1 - a^2 z^2/6 + \cdots) \qquad (\textbf{1.7.40})$$

$$= \frac{1}{az^2}(1 + tz + t^2 z^2/2 - a^2 z^2/6 + \cdots) \qquad (\textbf{1.7.41})$$

$$= \frac{1}{az^2} + \frac{t}{az} + \frac{3t^2 - a^2}{6a} + \cdots. \qquad (\textbf{1.7.42})$$

Thus, we have a second-order pole at $z = 0$ and the residue equals t/a.

Problems

1. Find the Taylor expansion of $f(z) = (1 - z)^{-2}$ about the point $z = 0$.

2. Find the Taylor expansion of $f(z) = (z - 1)e^z$ about the point $z = 1$. [Hint: Don't find the expansion by taking derivatives.]

By constructing a Laurent expansion, describe the type of singularity and give the residue at z_0 for each of the following functions:

3. $f(z) = z^{10} e^{-1/z}; \quad z_0 = 0$ 4. $f(z) = z^{-3} \sin^2(z); \quad z_0 = 0$

5. $f(z) = \dfrac{\cosh(z) - 1}{z^2}; \quad z_0 = 0$ 6. $f(z) = \dfrac{z}{(z+2)^2}; \quad z_0 = -2$

7. $f(z) = \dfrac{e^z + 1}{e^{-z} - 1}; \quad z_0 = 0$ 8. $f(z) = \dfrac{e^{iz}}{z^2 + b^2}; \quad z_0 = bi$

9. $f(z) = \dfrac{1}{z(z - 2)}; \quad z_0 = 2$ 10. $f(z) = \dfrac{\exp(z^2)}{z^4}; \quad z_0 = 0$

1.8 THEORY OF RESIDUES

Having shown that around any singularity we may construct a Laurent expansion, we now use this result in the integration of closed complex integrals. Consider a closed contour in which the function $f(z)$ has a number of isolated singularities. As we did in the case of Cauchy's integral formula, we introduce a new contour C' which excludes all of the singularities because they are isolated. See Figure 1.8.1. Therefore,

$$\oint_C f(z)\,dz - \oint_{C_1} f(z)\,dz - \cdots - \oint_{C_n} f(z)\,dz = \oint_{C'} f(z)\,dz = 0. \qquad (\textbf{1.8.1})$$

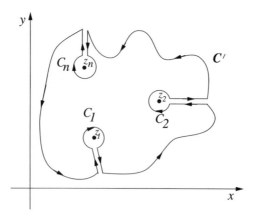

Figure 1.8.1: Contour used in deriving the residue theorem.

Consider now the mth integral, where $1 \leq m \leq n$. Constructing a Laurent expansion for the function $f(z)$ at the isolated singularity $z = z_m$, this integral equals

$$\oint_{C_m} f(z)\,dz = \sum_{k=1}^{\infty} a_k \oint_{C_m} \frac{1}{(z - z_m)^k}\,dz + \sum_{k=0}^{\infty} b_k \oint_{C_m} (z - z_m)^k\,dz. \quad (\mathbf{1.8.2})$$

Because $(z - z_m)^k$ is an entire function if $k \geq 0$, the integrals equal zero for each term in the second summation. We use Cauchy's integral formula to evaluate the remaining terms. The analytic function in the numerator is 1. Because $d^{k-1}(1)/dz^{k-1} = 0$ if $k > 1$, all of the terms vanish except for $k = 1$. In that case, the integral equals $2\pi i a_1$, where a_1 is the value of the residue for that particular singularity. Applying this approach to each of the singularities, we obtain

Cauchy's residue theorem[8]: *If $f(z)$ is analytic inside and on a closed contour C (taken in the positive sense) except at points $z_1, z_2, \ldots, z_n$ where $f(z)$ has singularities, then*

$$\oint_C f(z)\,dz = 2\pi i \sum_{j=1}^{n} \mathrm{Res}[f(z); z_j], \quad (\mathbf{1.8.3})$$

where $\mathrm{Res}[f(z); z_j]$ denotes the residue of the jth isolated singularity of $f(z)$ located at $z = z_j$.

---

[8] See Mitrinović, D. S., and J. D. Kečkić, 1984: *The Cauchy Method of Residues: Theory and Applications*. D. Reidel Publishing, 361 pp. Section 10.3 gives the historical development of the residue theorem.

• Example 1.8.1

Let us compute $\oint_{|z|=2} z^2/(z+1)\,dz$ by the residue theorem, assuming that we take the contour in the positive sense.

Because the contour is a circle of radius 2, centered on the origin, the singularity at $z = -1$ lies within the contour. If the singularity were not inside the contour, then the integrand would have been analytic inside and on the contour C. In this case, the answer would then be zero by the Cauchy-Goursat theorem.

Returning to the original problem, we construct the Laurent expansion for the integrand around the point $z = 1$ by noting that

$$\frac{z^2}{z+1} = \frac{[(z+1)-1]^2}{z+1} = \frac{1}{z+1} - 2 + (z+1). \tag{1.8.4}$$

The singularity at $z = -1$ is a simple pole and by inspection the value of the residue equals 1. Therefore,

$$\oint_{|z|=2} \frac{z^2}{z+1}\,dz = 2\pi i. \tag{1.8.5}$$

As it presently stands, it would appear that we must always construct a Laurent expansion for each singularity if we wish to use the residue theorem. This becomes increasingly difficult as the structure of the integrand becomes more complicated. In the following paragraphs we show several techniques that avoid this problem in practice.

We begin by noting that many functions that we will encounter consist of the ratio of two *polynomials*, i.e., rational functions: $f(z) = g(z)/h(z)$. Generally, we can write $h(z)$ as $(z - z_1)^{m_1}(z - z_2)^{m_2} \cdots$. Here we assumed that we divided out any common factors between $g(z)$ and $h(z)$ so that $g(z)$ does not vanish at $z_1, z_2, \ldots$. Clearly $z_1, z_2, \ldots$, are singularities of $f(z)$. Further analysis shows that the nature of the singularities are a pole of order m_1 at $z = z_1$, a pole of order m_2 at $z = z_2$, and so forth.

Having found the nature and location of the singularity, we compute the residue as follows. Suppose that we have a pole of order n. Then we know that its Laurent expansion is

$$f(z) = \frac{a_n}{(z-z_0)^n} + \frac{a_{n-1}}{(z-z_0)^{n-1}} + \cdots + b_0 + b_1(z-z_0) + \cdots. \tag{1.8.6}$$

Multiplying both sides of (1.8.6) by $(z - z_0)^n$,

$$\begin{aligned} F(z) &= (z-z_0)^n f(z) \\ &= a_n + a_{n-1}(z-z_0) + \cdots + b_0(z-z_0)^n + b_1(z-z_0)^{n+1} + \cdots. \end{aligned}$$
$$\tag{1.8.7}$$

Because $F(z)$ is analytic at $z = z_0$, it has the Taylor expansion

$$F(z) = F(z_0) + F'(z_0)(z - z_0) + \cdots + \frac{F^{(n-1)}(z_0)}{(n-1)!}(z - z_0)^{n-1} + \cdots. \quad (1.8.8)$$

Matching powers of $z - z_0$ in (1.8.7) and (1.8.8), the residue equals

$$\text{Res}[f(z); z_0] = a_1 = \frac{F^{(n-1)}(z_0)}{(n-1)!}. \quad (1.8.9)$$

Substituting in $F(z) = (z - z_0)^n f(z)$, we can compute the residue of a pole of order n by

$$\text{Res}[f(z); z_j] = \frac{1}{(n-1)!} \lim_{z \to z_j} \frac{d^{n-1}}{dz^{n-1}} \left[(z - z_j)^n f(z) \right]. \quad (1.8.10)$$

For a simple pole (1.8.10) simplifies to

$$\text{Res}[f(z); z_j] = \lim_{z \to z_j} (z - z_j) f(z). \quad (1.8.11)$$

Quite often, $f(z) = p(z)/q(z)$. From l'Hôspital's rule, it follows that (1.8.11) becomes

$$\text{Res}[f(z); z_j] = \frac{p(z_j)}{q'(z_j)}. \quad (1.8.12)$$

Remember that these formulas work only for finite-order poles. For an essential singularity we must compute the residue from its Laurent expansion; however, essential singularities are very rare in applications.

• **Example 1.8.2**

Let us evaluate

$$\oint_C \frac{e^{iz}}{z^2 + a^2}\, dz, \quad (1.8.13)$$

where C is any contour that includes both $z = \pm ai$ and is in the positive sense.

From Cauchy's residue theorem,

$$\oint_C \frac{e^{iz}}{z^2 + a^2}\, dz = 2\pi i \left[\text{Res}\left(\frac{e^{iz}}{z^2 + a^2}; ai\right) + \text{Res}\left(\frac{e^{iz}}{z^2 + a^2}; -ai\right) \right]. \quad (1.8.14)$$

The singularities at $z = \pm ai$ are simple poles. The corresponding residues are

$$\text{Res}\left(\frac{e^{iz}}{z^2 + a^2}; ai\right) = \lim_{z \to ai} (z - ai)\frac{e^{iz}}{(z - ai)(z + ai)} = \frac{e^{-a}}{2ia} \quad (1.8.15)$$

and

$$\text{Res}\left(\frac{e^{iz}}{z^2 + a^2}; -ai\right) = \lim_{z \to -ai} (z + ai)\frac{e^{iz}}{(z - ai)(z + ai)} = -\frac{e^{a}}{2ia}. \quad (1.8.16)$$

Consequently,

$$\oint_C \frac{e^{iz}}{z^2 + a^2}\, dz = -\frac{2\pi}{2a} \left(e^a - e^{-a}\right) = -\frac{2\pi}{a} \sinh(a). \quad (1.8.17)$$

- **Example 1.8.3**

Let us evaluate

$$\frac{1}{2\pi i} \oint_C \frac{e^{tz}}{z^2(z^2 + 2z + 2)}\, dz, \quad (1.8.18)$$

where C includes all of the singularities and is in the positive sense.

The integrand has a second-order pole at $z = 0$ and two simple poles at $z = -1 \pm i$ which are the roots of $z^2 + 2z + 2 = 0$. Therefore, the residue at $z = 0$ is

$$\text{Res}\left[\frac{e^{tz}}{z^2(z^2 + 2z + 2)}; 0\right] = \lim_{z \to 0} \frac{1}{1!}\frac{d}{dz}\left\{(z - 0)^2 \left[\frac{e^{tz}}{z^2(z^2 + 2z + 2)}\right]\right\} \quad (1.8.19)$$

$$= \lim_{z \to 0} \left[\frac{te^{tz}}{z^2 + 2z + 2} - \frac{(2z + 2)e^{tz}}{(z^2 + 2z + 2)^2}\right] = \frac{t - 1}{2}. \quad (1.8.20)$$

The residue at $z = -1 + i$ is

$$\text{Res}\left[\frac{e^{tz}}{z^2(z^2 + 2z + 2)}; -1 + i\right] = \lim_{z \to -1+i} [z - (-1 + i)]\frac{e^{tz}}{z^2(z^2 + 2z + 2)} \quad (1.8.21)$$

$$= \left(\lim_{z \to -1+i} \frac{e^{tz}}{z^2}\right)\left(\lim_{z \to -1+i} \frac{z + 1 - i}{z^2 + 2z + 2}\right) \quad (1.8.22)$$

$$= \frac{\exp[(-1 + i)t]}{2i(-1 + i)^2} = \frac{\exp[(-1 + i)t]}{4}. \quad (1.8.23)$$

Similarly, the residue at $z = -1 - i$ is

$$\text{Res}\left[\frac{e^{tz}}{z^2(z^2 + 2z + 2)}; -1 - i\right] = \lim_{z \to -1-i}[z - (-1 - i)]\frac{e^{tz}}{z^2(z^2 + 2z + 2)}$$

(1.8.24)

$$= \left(\lim_{z \to -1-i}\frac{e^{tz}}{z^2}\right)\left(\lim_{z \to -1-i}\frac{z + 1 + i}{z^2 + 2z + 2}\right)$$

(1.8.25)

$$= \frac{\exp[(-1 - i)t]}{(-2i)(-1 - i)^2} = \frac{\exp[(-1 - i)t]}{4}.$$

(1.8.26)

Then by the residue theorem,

$$\frac{1}{2\pi i}\oint_C \frac{e^{tz}}{z^2(z^2 + 2z + 2)}\,dz = \text{Res}\left[\frac{e^{tz}}{z^2(z^2 + 2z + 2)}; 0\right]$$

$$+ \text{Res}\left[\frac{e^{tz}}{z^2(z^2 + 2z + 2)}; -1 + i\right]$$

$$+ \text{Res}\left[\frac{e^{tz}}{z^2(z^2 + 2z + 2)}; -1 - i\right]$$

(1.8.27)

$$= \frac{t - 1}{2} + \frac{\exp[(-1 + i)t]}{4} + \frac{\exp[(-1 - i)t]}{4}$$

(1.8.28)

$$= \tfrac{1}{2}\left[t - 1 + e^{-t}\cos(t)\right].$$

(1.8.29)

Problems

Assuming that all of the following closed contours are in the positive sense, use the residue theorem to evaluate the following integrals:

1. $\displaystyle\oint_{|z|=1} \frac{z + 1}{z^4 - 2z^3}\,dz$

2. $\displaystyle\oint_{|z|=1} \frac{(z + 4)^3}{z^4 + 5z^3 + 6z^2}\,dz$

3. $\displaystyle\oint_{|z|=1} \frac{1}{1 - e^z}\,dz$

4. $\displaystyle\oint_{|z|=2} \frac{z^2 - 4}{(z - 1)^4}\,dz$

5. $\displaystyle\oint_{|z|=2} \frac{z^3}{z^4 - 1}\,dz$

6. $\displaystyle\oint_{|z|=1} z^n e^{2/z}\,dz, \quad n > 0$

7. $\displaystyle\oint_{|z|=1} e^{1/z}\cos(1/z)\,dz$

8. $\displaystyle\oint_{|z|=2} \frac{2 + 4\cos(\pi z)}{z(z - 1)^2}\,dz$

1.9 EVALUATION OF REAL DEFINITE INTEGRALS

One of the important applications of the theory of residues consists in the evaluation of certain types of real definite integrals. Similar techniques apply when the integrand contains a sine or cosine. See §3.4.

• Example 1.9.1

Let us evaluate the integral

$$\int_0^\infty \frac{dx}{x^2 + 1} = \frac{1}{2} \int_{-\infty}^\infty \frac{dx}{x^2 + 1}. \tag{1.9.1}$$

This integration occurs along the real axis. In terms of complex variables we can rewrite (1.9.1) as

$$\int_0^\infty \frac{dx}{x^2 + 1} = \frac{1}{2} \int_{C_1} \frac{dz}{z^2 + 1}, \tag{1.9.2}$$

where the contour C_1 is the line $\text{Im}(z) = 0$. However, the use of the residue theorem requires an integration along a closed contour. Let us choose the one pictured in Figure 1.9.1. Then

$$\oint_C \frac{dz}{z^2 + 1} = \int_{C_1} \frac{dz}{z^2 + 1} + \int_{C_2} \frac{dz}{z^2 + 1}, \tag{1.9.3}$$

where C denotes the complete closed contour and C_2 denotes the integration path along a semicircle at infinity. Clearly we want the second integral on the right side of (1.9.3) to vanish; otherwise, our choice of the contour C_2 is poor. Because $z = Re^{\theta i}$ and $dz = iRe^{\theta i} \, d\theta$,

$$\left| \int_{C_2} \frac{dz}{z^2 + 1} \right| = \left| \int_0^\pi \frac{iR\exp(\theta i)}{1 + R^2 \exp(2\theta i)} \, d\theta \right| \leq \int_0^\pi \frac{R}{R^2 - 1} \, d\theta, \tag{1.9.4}$$

which tends to zero as $R \to \infty$. On the other hand, the residue theorem gives

$$\oint_C \frac{dz}{z^2 + 1} = 2\pi i \, \text{Res}\left(\frac{1}{z^2 + 1}; i \right) = 2\pi i \lim_{z \to i} \frac{z - i}{z^2 + 1} = 2\pi i \times \frac{1}{2i} = \pi. \tag{1.9.5}$$

Therefore,

$$\int_0^\infty \frac{dx}{x^2 + 1} = \frac{\pi}{2}. \tag{1.9.6}$$

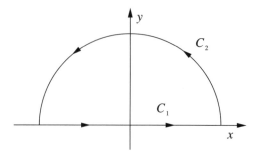

Figure 1.9.1: Contour used in evaluating the integral (1.9.1).

Note that we only evaluated the residue in the upper half-plane because it is the only one inside the contour.

This example illustrates the basic concepts of evaluating definite integrals by the residue theorem. We introduce a closed contour that includes the real axis and an additional contour. We must then evaluate the integral along this additional contour as well as the closed contour integral. If we properly choose our closed contour, this additional integral vanishes. For certain classes of general integrals, we shall now show that this additional contour is a circular arc at infinity.

Theorem: *If, on a circular arc C_R with a radius R and center at the origin, $zf(z) \to 0$ uniformly with $|z| \in C_R$ and as $R \to \infty$, then*

$$\lim_{R \to \infty} \int_{C_R} f(z)\, dz = 0. \tag{1.9.7}$$

The proof is as follows: If $|zf(z)| \le M_R$, then $|f(z)| \le M_R/R$. Because the length of C_R is αR, where α is the subtended angle,

$$\left| \int_{C_R} f(z)\, dz \right| \le \frac{M_R}{R}\, \alpha R = \alpha M_R \to 0, \tag{1.9.8}$$

because $M_R \to 0$ as $R \to \infty$. □

• **Example 1.9.2**

A simple illustration of this theorem is the integral

$$\int_{-\infty}^{\infty} \frac{dx}{x^2 + x + 1} = \int_{C_1} \frac{dz}{z^2 + z + 1}. \tag{1.9.9}$$

A quick check shows that $z/(z^2 + z + 1)$ tends to zero uniformly as $R \to \infty$. Therefore, if we use the contour pictured in Figure 1.9.1,

$$\int_{-\infty}^{\infty} \frac{dx}{x^2 + x + 1} = \oint_C \frac{dz}{z^2 + z + 1} = 2\pi i \operatorname{Res}\left(\frac{1}{z^2 + z + 1}; -\tfrac{1}{2} + \tfrac{\sqrt{3}}{2}i \right) \tag{1.9.10}$$

$$= 2\pi i \lim_{z \to -\frac{1}{2} + \frac{\sqrt{3}}{2}i} \left(\frac{1}{2z + 1} \right) = \frac{2\pi}{\sqrt{3}}. \tag{1.9.11}$$

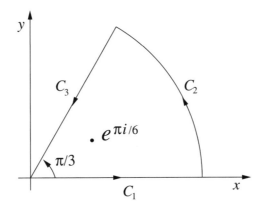

Figure 1.9.2: Contour used in evaluating the integral (1.9.13).

• **Example 1.9.3**

Let us evaluate

$$\int_0^\infty \frac{dx}{x^6 + 1}. \tag{1.9.12}$$

In place of an infinite semicircle in the upper half-plane, consider the following integral

$$\oint_C \frac{dz}{z^6 + 1}, \tag{1.9.13}$$

where we show the closed contour in Figure 1.9.2. We chose this contour for two reasons. First, we only have to evaluate one residue rather than the three enclosed in a traditional upper half-plane contour. Second, the contour integral along C_3 simplifies to a particularly simple and useful form.

Because the only enclosed singularity lies at $z = e^{\pi i/6}$,

$$\oint_C \frac{dz}{z^6 + 1} = 2\pi i \operatorname{Res}\left(\frac{1}{z^6 + 1}; e^{\pi i/6}\right) = 2\pi i \lim_{z \to e^{\pi i/6}} \frac{z - e^{\pi i/6}}{z^6 + 1} \tag{1.9.14}$$

$$= 2\pi i \lim_{z \to e^{\pi i/6}} \frac{1}{6z^5} = -\frac{\pi i}{3} e^{\pi i/6}. \tag{1.9.15}$$

Let us now evaluate (1.9.12) along each of the legs of the contour:

$$\int_{C_1} \frac{dz}{z^6 + 1} = \int_0^\infty \frac{dx}{x^6 + 1}, \tag{1.9.16}$$

$$\int_{C_2} \frac{dz}{z^6 + 1} = 0, \tag{1.9.17}$$

because of (1.9.7) and

$$\int_{C_3} \frac{dz}{z^6 + 1} = \int_\infty^0 \frac{e^{\pi i/3}\, dr}{r^6 + 1} = -e^{\pi i/3} \int_0^\infty \frac{dx}{x^6 + 1}, \tag{1.9.18}$$

since $z = re^{\pi i/3}$.

Substituting into (1.9.15),

$$\left(1 - e^{\pi i/3}\right) \int_0^\infty \frac{dx}{x^6 + 1} = -\frac{\pi i}{3} e^{\pi i/6} \tag{1.9.19}$$

or

$$\int_0^\infty \frac{dx}{x^6 + 1} = \frac{\pi i}{6} \frac{2i e^{\pi i/6}}{e^{\pi i/6} \left(e^{\pi i/6} - e^{-\pi i/6}\right)} = \frac{\pi}{6 \sin(\pi/6)} = \frac{\pi}{3}. \tag{1.9.20}$$

• **Example 1.9.4**

Rectangular closed contours are best for the evaluation of integrals that involve hyperbolic sines and cosines. To illustrate[9] this, let us evaluate the integral

$$2 \int_0^\infty \frac{\sin(ax) \sinh(x)}{[b + \cosh(x)]^2} \, dx = \int_{-\infty}^\infty \frac{\sin(ax) \sinh(x)}{[b + \cosh(x)]^2} \, dx \tag{1.9.21}$$

$$= \mathrm{Im}\left[\int_{-\infty}^\infty \frac{\sinh(x) e^{iax}}{[b + \cosh(x)]^2} \, dx\right], \tag{1.9.22}$$

where $a > 0$ and $b > 1$.

We begin by determining the value of

$$\oint_C \frac{\sinh(z) e^{iaz}}{[b + \cosh(z)]^2} \, dz$$

about the closed contour shown in Figure 1.9.3. Writing this contour integral in terms of the four line segments that constitute the closed contour, we have

$$\oint_C \frac{\sinh(z) e^{iaz}}{[b + \cosh(z)]^2} \, dz = \int_{C_1} \frac{\sinh(z) e^{iaz}}{[b + \cosh(z)]^2} \, dz + \int_{C_2} \frac{\sinh(z) e^{iaz}}{[b + \cosh(z)]^2} \, dz$$

$$+ \int_{C_3} \frac{\sinh(z) e^{iaz}}{[b + \cosh(z)]^2} \, dz + \int_{C_4} \frac{\sinh(z) e^{iaz}}{[b + \cosh(z)]^2} \, dz. \tag{1.9.23}$$

Because the integrand behaves as e^{-R} as $R \to \infty$, the integrals along C_2 and C_4 vanish. On the other hand,

$$\int_{C_1} \frac{\sinh(z) e^{iaz}}{[b + \cosh(z)]^2} \, dz = \int_{-\infty}^\infty \frac{\sinh(x) e^{iax}}{[b + \cosh(x)]^2} \, dx, \tag{1.9.24}$$

---

[9] This is a slight variation on a problem solved by Spyrou, K. J., B. Cotton, and B. Gurd, 2002: Analytical expressions of capsize boundary for a ship with roll bias in beam waves. *J. Ship Res.*, **46**, 167–174. Reprinted with the permission of the Society of Naval Architects and Marine Engineers (SNAME).

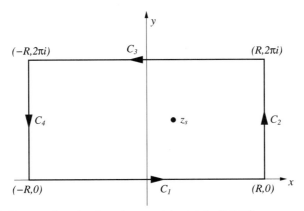

Figure 1.9.3: Rectangular closed contour used to obtain (1.9.32).

and

$$\int_{C_3} \frac{\sinh(z)e^{iaz}}{[b+\cosh(z)]^2}\,dz = -e^{-2\pi a}\int_{-\infty}^{\infty}\frac{\sinh(x)e^{iax}}{[b+\cosh(x)]^2}\,dx, \qquad (1.9.25)$$

because $\cosh(x+2\pi i)=\cosh(x)$ and $\sinh(x+2\pi i)=\sinh(x)$.

Within the closed contour C, we have a single singularity where $b+\cosh(z_s)=0$ or $e^{z_s}=-b-\sqrt{b^2-1}$ or $z_s=\ln(b+\sqrt{b^2-1})+\pi i$. To discover the nature of this singularity, we expand $b+\cosh(z)$ in a Taylor expansion and find that

$$b+\cosh(z) = \sinh(z_s)(z-z_s)+\tfrac{1}{2}\cosh(z_s)(z-z_s)^2+\cdots. \qquad (1.9.26)$$

Therefore, we have a second-order pole at $z=z_s$. Therefore, the value of the residue there is

$$\operatorname{Res}\left[\frac{\sinh(z)e^{iaz}}{[b+\cosh(z)]^2};z_s\right] = \lim_{z\to z_s}\frac{d}{dz}\left[\frac{\sinh(z)e^{iaz}}{\sinh^2(z_s)+\sinh(z_s)\cosh(z_s)(z-z_s)+\cdots}\right]$$
$$(1.9.27)$$

$$= \frac{ia\,e^{-\pi a}}{\sinh(z_s)}\exp[ia\cosh^{-1}(b)]. \qquad (1.9.28)$$

Therefore,

$$\int_{-\infty}^{\infty}\frac{\sinh(x)e^{iax}}{[b+\cosh(x)]^2}\,dx = -\frac{2\pi a\exp[-\pi a+ai\cosh^{-1}(b)]}{(1-e^{-2\pi a})\sinh(z_s)} \qquad (1.9.29)$$

$$= \frac{\pi a\exp[ai\cosh^{-1}(b)]}{\sqrt{b^2-1}\,\sinh(\pi a)}, \qquad (1.9.30)$$

because

$$\sinh(z_s) = \frac{1}{2}\left[-b-\sqrt{b^2-1}+\frac{1}{b+\sqrt{b^2-1}}\right] = -\sqrt{b^2-1}. \qquad (1.9.31)$$

Substituting (1.9.30) into (1.9.22) yields

$$\int_0^\infty \frac{\sin(ax)\,\sinh(x)}{[b+\cosh(x)]^2}\,dx = \frac{\pi a \sin[a\cosh^{-1}(b)]}{2\sqrt{b^2-1}\,\sinh(\pi a)}. \tag{1.9.32}$$

• **Example 1.9.5**

The method of residues is also useful in the evaluation of definite integrals of the form $\int_0^{2\pi} F[\sin(\theta),\cos(\theta)]\,d\theta$, where F is a quotient of polynomials in $\sin(\theta)$ and $\cos(\theta)$. For example, let us evaluate the integral[10]

$$I = \int_0^{2\pi} \frac{\cos^3(\theta)}{\cos^2(\theta) - a^2}\,d\theta, \qquad a > 1. \tag{1.9.33}$$

We begin by introducing the complex variable $z = e^{i\theta}$. This substitution yields the closed contour integral

$$I = \frac{1}{2i} \oint_C \frac{(z^2+1)^3}{(z^2+1)^2 - 4a^2 z^2}\frac{dz}{z^2}, \tag{1.9.34}$$

where C is a circle of radius 1 taken in the positive sense. The integrand of (1.9.34) has five singularities: a second-order pole at $z_5 = 0$ and simple poles located at

$$z_1 = -a - \sqrt{a^2-1}, \qquad z_2 = -a + \sqrt{a^2-1}, \tag{1.9.35}$$

$$z_3 = a - \sqrt{a^2-1}, \quad \text{and} \quad z_4 = a + \sqrt{a^2-1}. \tag{1.9.36}$$

Only the singularities z_2, z_3, and z_5 lie within C. Consequently, the value of I equals $2\pi i$ times the sum of the residues at these three singularities. The residues equal

$$\text{Res}\left\{\frac{(z^2+1)^3}{z^2[(z^2+1)^2 - 4a^2 z^2]}; -a + \sqrt{a^2-1}\right\}$$

$$= \lim_{z \to -a+\sqrt{a^2-1}} \frac{(z^2+1)^3}{z^2} \lim_{z \to -a+\sqrt{a^2-1}} \frac{z + a - \sqrt{a^2-1}}{(z^2+1)^2 - 4a^2 z^2} \tag{1.9.37}$$

$$= \lim_{z \to -a+\sqrt{a^2-1}} \frac{(z^2+1)^3}{4z^3(z^2+1-2a^2)} \tag{1.9.38}$$

$$= -\frac{a^2(a - \sqrt{a^2-1})^3}{(2a^2-1-2a\sqrt{a^2-1})(a^2-1-a\sqrt{a^2-1})}, \tag{1.9.39}$$

---

[10] Simplified version of an integral presented by Jiang, Q. F., and R. B. Smith, 2000: V-waves, bow shocks, and wakes in supercritical hydrostatic flow. *J. Fluid Mech.*, **406**, 27–53. Reprinted with the permission of Cambridge University Press.

$$\text{Res}\left\{\frac{(z^2+1)^3}{z^2[(z^2+1)^2-4a^2z^2]}; a-\sqrt{a^2-1}\right\}$$

$$=\lim_{z\to a-\sqrt{a^2-1}}\frac{(z^2+1)^3}{z^2}\lim_{z\to a-\sqrt{a^2-1}}\frac{z-a+\sqrt{a^2-1}}{(z^2+1)^2-4a^2z^2}\qquad(\mathbf{1.9.40})$$

$$=\lim_{z\to a-\sqrt{a^2-1}}\frac{(z^2+1)^3}{4z^3(z^2+1-2a^2)}\qquad(\mathbf{1.9.41})$$

$$=\frac{a^2(a-\sqrt{a^2-1})^3}{(2a^2-1-2a\sqrt{a^2-1})(a^2-1-a\sqrt{a^2-1})},\qquad(\mathbf{1.9.42})$$

and

$$\text{Res}\left\{\frac{(z^2+1)^3}{z^2[(z^2+1)^2-4a^2z^2]}; 0\right\}$$

$$=\lim_{z\to0}\frac{d}{dz}\left[\frac{(z^2+1)^3}{(z^2+1)^2-4a^2z^2}\right]\qquad(\mathbf{1.9.43})$$

$$=\lim_{z\to0}\frac{6z[(z^2+1)^4-4a^2z^2(z^2+1)^2]-4z(z^2+1)^3(z^2+1-2a^2)}{[(z^2+1)^2-4a^2z^2]^2}$$

$$\qquad(\mathbf{1.9.44})$$

$$=0.\qquad(\mathbf{1.9.45})$$

Summing the residues, we obtain 0. Therefore,

$$\int_0^{2\pi}\frac{\cos^3(\theta)}{\cos^2(\theta)-a^2}\,d\theta=0,\qquad a>1.\qquad(\mathbf{1.9.46})$$

Problems

Use the residue theorem to verify the following integrals:

1. $\displaystyle\int_0^\infty\frac{dx}{x^4+1}=\frac{\pi\sqrt{2}}{4}$

2. $\displaystyle\int_{-\infty}^\infty\frac{dx}{(x^2+4x+5)^2}=\frac{\pi}{2}$

3. $\displaystyle\int_{-\infty}^\infty\frac{x\,dx}{(x^2+1)(x^2+2x+2)}=-\frac{\pi}{5}$

4. $\displaystyle\int_0^\infty\frac{x^2}{x^6+1}\,dx=\frac{\pi}{6}$

5. $\displaystyle\int_0^\infty\frac{dx}{(x^2+1)^2}=\frac{\pi}{4}$

6. $\displaystyle\int_0^\infty\frac{dx}{(x^2+1)(x^2+4)^2}=\frac{5\pi}{288}$

7.

$$\int_{-\infty}^\infty\frac{x^2\,dx}{(x^2+a^2)(x^2+b^2)^2}=\frac{\pi}{2b(a+b)^2},\qquad a,b>0$$

8.

$$\int_0^\infty \frac{t^2}{(t^2+1)[t^2(a/h+1)+(a/h-1)]}\,dt = \frac{\pi}{4}\left[1 - \sqrt{\frac{a-h}{a+h}}\right], \quad a > h$$

9.

$$\int_0^{\pi/2} \frac{d\theta}{a + \sin^2(\theta)} = \frac{\pi}{2\sqrt{a+a^2}}, \quad a > 0$$

10.

$$\int_0^{\pi/2} \frac{d\theta}{a^2\cos^2(\theta) + b^2\sin^2(\theta)} = \frac{\pi}{2ab}, \quad b \geq a > 0$$

11.

$$\int_0^\pi \frac{\sin^2(\theta)}{a + b\cos(\theta)}\,d\theta = \frac{\pi}{b^2}\left(a - \sqrt{a^2-b^2}\right), \quad a > b > 0$$

12.

$$\int_0^{2\pi} \frac{e^{in\theta}}{1 + 2r\cos(\theta) + r^2}\,d\theta = 2\pi\frac{(-r)^n}{1-r^2}, \quad 1 > |r|, \quad n = 0, 1, 2, \ldots$$

13.

$$\int_0^{2\pi} \sin^{2n}(\theta)\,d\theta = \frac{2\pi(2n)!}{(2^n n!)^2}$$

14.

$$\int_{-\pi}^\pi \frac{\cos(n\theta)}{\cos(\theta) + \alpha}\,d\theta = 2\pi\frac{(-\alpha + \sqrt{\alpha^2-1})^n}{\sqrt{\alpha^2-1}}, \quad \alpha > 1, \quad n \geq 0$$

Hint:

$$\frac{2\sqrt{\alpha^2-1}}{z^2 + 2\alpha z + 1} = \frac{1}{z + \alpha - \sqrt{\alpha^2-1}} - \frac{1}{z + \alpha + \sqrt{\alpha^2-1}}$$

15.

$$\int_0^\pi \frac{\cos(n\theta)}{\cosh(\alpha) - \cos(\theta)}\,d\theta = \frac{\pi}{\sinh(\alpha)}e^{-n\alpha}, \quad \alpha \neq 0, \quad n \geq 0$$

Hint: See Example 1.7.5.

16. Show that

$$\int_0^\infty \frac{x^2}{(1-x^2)^2 + a^2 x^2}\,dx = \frac{\pi}{2|a|},$$

where a is real and not equal to zero. Hint: Show that the poles of

$$f(z) = \frac{z^2}{(1-z^2)^2 + a^2 z^2}$$

are simple and equal

$$z_n = \begin{cases} \pm\frac{1}{2}\left(\pm\sqrt{4-a^2}+|a|i\right), & \text{if} \quad 0 < |a| < 2, \\ \pm\frac{i}{2}\left(|a| \pm \sqrt{a^2-4}\right), & \text{if} \quad 2 < |a|. \end{cases}$$

If $|a| = 2$, we have second-order poles at $z_n = \pm i$.

17. Show that

$$\int_0^\infty \frac{\cos(ax)}{\cosh^2(bx)}\,dx = \frac{\pi a}{2b^2\,\sinh[a\pi/(2b)]}, \qquad a, b > 0.$$

Hint: Evaluate the closed contour integral

$$\oint_C \frac{e^{iaz}}{\cosh^2(bz)}\,dz,$$

where C is a *rectangular* contour with vertices at $(\infty, 0)$, $(-\infty, 0)$, $(\infty, \pi/b)$, and $(-\infty, \pi/b)$.

18. Show[11] that

$$\int_0^\infty \frac{dx}{\cosh(x)\cosh(x+a)} = \begin{cases} 2a/\sinh(a), & \text{if} \quad a \neq 0, \\ 2, & \text{if} \quad a = 2. \end{cases}$$

Hint: Evaluate the closed contour integral

$$\oint_C \frac{z}{\cosh(z)\cosh(z+a)}\,dz,$$

where C is a *rectangular* contour with vertices at $(\infty, 0)$, $(-\infty, 0)$, (∞, π), and $(-\infty, \pi)$.

19. During an electromagnetic calculation, Strutt[12] needed to prove that

$$\pi\frac{\sinh(\sigma x)}{\cosh(\sigma\pi)} = 2\sigma\sum_{n=0}^\infty \frac{\cos\left[\left(n+\frac{1}{2}\right)(x-\pi)\right]}{\sigma^2+\left(n+\frac{1}{2}\right)^2}, \qquad |x| \le \pi.$$

[11] Reprinted with permission from Yan, J. R., X. H. Yan, J. Q. You, and J. X. Zhong, 1993: On the interaction between two nonpropagating hydrodynamic solitons. *Phys. Fluids A*, **5**, 1651–1656. ©1993, American Institute of Physics.

[12] Strutt, M. J. O., 1934: Berechnung des hochfrequenten Feldes einer Kreiszylinderspule in einer konzentrischen leitenden Schirmhülle mit ebenen Deckeln. *Hochfrequenztechn. Elecktroak.*, **43**, 121–123.

Verify his proof by doing the following:

Step 1: Using the residue theorem, show that

$$\frac{1}{2\pi i} \oint_{C_N} \pi \frac{\sinh(xz)}{\cosh(\pi z)} \frac{dz}{z - \sigma} = \pi \frac{\sinh(\sigma x)}{\cosh(\sigma \pi)} - \sum_{n=-N-1}^{N} \frac{(-1)^n \sin\left[\left(n + \frac{1}{2}\right) x\right]}{\sigma - i\left(n + \frac{1}{2}\right)},$$

where C_N is a circular contour that includes the poles $z = \sigma$ and $z_n = \pm i\left(n + \frac{1}{2}\right)$, $n = 0, 1, 2, \ldots, N$.

Step 2: Show that in the limit of $N \to \infty$, the contour integral vanishes. [Hint: Examine the behavior of $z \sinh(xz)/[(z - \sigma)\cosh(\pi z)]$ as $|z| \to \infty$. Use (1.9.7) where C_R is the circular contour.]

Step 3: Break the infinite series in Step 1 into two parts and simplify.

    In the chapter on Fourier series, we shall show how we can obtain the same series by direct integration.

1.10 CAUCHY'S PRINCIPAL VALUE INTEGRAL

    The conventional definition of the integral of a function $f(x)$ of the real variable x over a finite interval $a \le x \le b$ assumes that $f(x)$ has a definite finite value at each point within the interval. We shall now extend this definition to cover cases when $f(x)$ is infinite at a finite number of points within the interval.

    Consider the case when there is only one point c at which $f(x)$ becomes infinite. If c is not an endpoint of the interval, we take two small positive numbers ϵ and η and examine the expression

$$\int_a^{c-\epsilon} f(x)\, dx + \int_{c+\eta}^b f(x)\, dx. \tag{1.10.1}$$

If (1.10.1) exists and tends to a unique limit as ϵ and η tend to zero independently, we say that the improper integral of $f(x)$ over the interval exists, its value being defined by

$$\int_a^b f(x)\, dx = \lim_{\epsilon \to 0} \int_a^{c-\epsilon} f(x)\, dx + \lim_{\eta \to 0} \int_{c+\eta}^b f(x)\, dx. \tag{1.10.2}$$

If, however, the expression does not tend to a limit as ϵ and η tend to zero independently, it may still happen that

$$\lim_{\epsilon \to 0} \left\{ \int_a^{c-\epsilon} f(x)\, dx + \int_{c+\epsilon}^b f(x)\, dx \right\} \tag{1.10.3}$$

exists. When this is the case, we call this limit the *Cauchy principal value* of the improper integral and denote it by

$$PV \int_a^b f(x)\,dx. \tag{1.10.4}$$

Finally, if $f(x)$ becomes infinite at an endpoint, say a, of the range of integration, we say that $f(x)$ is integrable over $a \le x \le b$ if

$$\lim_{\epsilon \to 0+} \int_{a+\epsilon}^b f(x)\,dx \tag{1.10.5}$$

exists.

• **Example 1.10.1**

Consider the integral $\int_{-1}^2 dx/x$. This integral does not exist in the ordinary sense because of the strong singularity at the origin. However, the integral would exist if

$$\lim_{\epsilon \to 0} \int_{-1}^{\epsilon} \frac{dx}{x} + \lim_{\delta \to 0} \int_{\delta}^2 \frac{dx}{x} \tag{1.10.6}$$

existed and had a unique value as ϵ and δ independently approach zero. Because this limit equals

$$\lim_{\epsilon,\delta \to 0} \left[\ln(\epsilon) + \ln(2) - \ln(\delta) \right] = \lim_{\epsilon,\delta \to 0} \left[\ln(2) - \ln(\delta/\epsilon) \right], \tag{1.10.7}$$

our integral would have the value of $\ln(2)$ if $\delta = \epsilon$. This particular limit is the Cauchy principal value of the improper integral which we express as

$$PV \int_{-1}^2 \frac{dx}{x} = \ln(2). \tag{1.10.8}$$

We can extend these ideas to complex integrals used to determine the value or principal value of an improper integral by Cauchy's residue theorem when the integrand has a singularity on the contour of integration. We avoid this difficulty by deleting from the area within the contour that portion which also lies within a small circle $|z - c| = \epsilon$ and then integrate around the boundary of the remaining region. This process is called *indenting* the contour.

The integral around the indented contour is calculated by the theorem of residues and then the radius of each indentation is made to tend to zero.

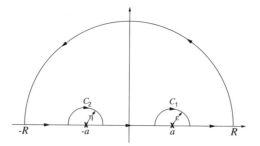

Figure 1.10.1: Contour C used in Example 1.10.2.

This process give the Cauchy principal value of the improper integral. The details of this method are shown in the following examples.

• **Example 1.10.2**

Let us show that

$$PV \int_{-\infty}^{\infty} \frac{\cos(x)}{a^2 - x^2}\, dx = \frac{\pi \sin(a)}{a}, \qquad a > 0. \qquad (1.10.9)$$

Consider the integral

$$\oint_C \frac{e^{iz}}{a^2 - z^2}\, dz, \qquad (1.10.10)$$

where the closed contour C consists of the real axis from $-R$ to R and a semicircle in the upper half of the z-plane where this segment is its diameter. See Figure 1.10.1. Because the integrand has poles at $z = \pm a$, which lie on this contour, we modify C by making an indentation of radius ϵ at a and another of radius η at $-a$. The integrand is now analytic within and on C and (1.10.10) equals zero by the Cauchy-Goursat theorem.

Evaluating each part of the integral (1.10.10), we have that

$$\int_0^\pi \frac{e^{iR\cos(\theta) - R\sin(\theta)}}{a^2 - R^2 e^{2\theta i}} iRe^{\theta i}\, d\theta + \int_{C_1} \frac{e^{iz}}{a^2 - z^2}\, dz + \int_{C_2} \frac{e^{iz}}{a^2 - z^2}\, dz$$

$$+ \int_{-R}^{-a-\eta} \frac{e^{ix}}{a^2 - x^2}\, dx + \int_{-a+\eta}^{a-\epsilon} \frac{e^{ix}}{a^2 - x^2}\, dx + \int_{a-\epsilon}^{R} \frac{e^{ix}}{a^2 - x^2}\, dx = 0,$$

$$(1.10.11)$$

where C_1 and C_2 denote the integrals around the indentations at a and $-a$, respectively. The modulus of the first term on the left side of (1.10.11) is less than $\pi R / (R^2 - a^2)$ so that this term tends to zero as $R \to \infty$. To evaluate C_1, we observe that $z = a + \epsilon e^{\theta i}$ along C_1, where θ decreases from π to 0.

Hence,

$$\int_{C_1} \frac{e^{iz}}{a^2 - z^2}\, dz = \lim_{\epsilon \to 0} \int_{\pi}^{0} \exp\left(ia + i\epsilon e^{\theta i}\right) \frac{\epsilon i e^{\theta i}}{-2a\epsilon e^{\theta i} - \epsilon^2 e^{2\theta i}}\, d\theta \qquad (1.10.12)$$

$$= \lim_{\epsilon \to 0} \int_{0}^{\pi} \exp\left(ia + i\epsilon e^{\theta i}\right) \frac{i}{2a + \epsilon e^{\theta i}}\, d\theta \qquad (1.10.13)$$

$$= \frac{\pi i e^{ia}}{2a}. \qquad (1.10.14)$$

Similarly,

$$\int_{C_2} \frac{e^{iz}}{a^2 - z^2}\, dz = -\frac{\pi i e^{-ia}}{2a}, \qquad (1.10.15)$$

as η tends to zero.

Upon letting $R \to \infty$, $\epsilon \to 0$, and $\eta \to 0$, we find that

$$PV \int_{-\infty}^{\infty} \frac{e^{ix}}{a^2 - x^2}\, dx = -\frac{\pi i}{2a}\left(e^{ia} - e^{-ia}\right) = \frac{\pi \sin(a)}{a}. \qquad (1.10.16)$$

Finally, equating the real and imaginary parts, we obtain

$$PV \int_{-\infty}^{\infty} \frac{\cos(x)}{a^2 - x^2}\, dx = \frac{\pi \sin(a)}{a}, \qquad PV \int_{-\infty}^{\infty} \frac{\sin(x)}{a^2 - x^2}\, dx = 0. \qquad (1.10.17)$$

• **Example 1.10.3**

Let us show that

$$\int_{-\infty}^{\infty} \frac{\sin(x)}{x}\, dx = \pi. \qquad (1.10.18)$$

Consider the integral

$$\oint_C \frac{e^{iz}}{z}\, dz, \qquad (1.10.19)$$

where the closed contour C consists of the real axis from $-R$ to R and a semicircle in the upper half of the z-plane where this segment is its diameter. Because the integrand has a pole at $z = 0$, which lies on the contour, we modify C by making an indentation of radius ϵ at $z = 0$. See Figure 1.10.2. Because e^{iz}/z is analytic along C,

$$\int_{0}^{\pi} e^{iR\cos(\theta) - R\sin(\theta)}\, i\, d\theta + \int_{-R}^{-\epsilon} \frac{e^{ix}}{x}\, dx + \int_{C_1} \frac{e^{iz}}{z}\, dz + \int_{\epsilon}^{R} \frac{e^{ix}}{x}\, dx = 0. \qquad (1.10.20)$$

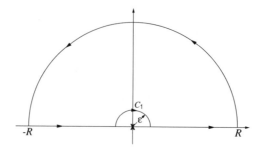

Figure 1.10.2: Contour C used in Example 1.10.3.

Since $e^{-R\sin(\theta)} < e^{-R\theta}$ for $0 < \theta < \pi$,

$$\left| \int_0^\pi e^{iR\cos(\theta) - R\sin(\theta)} i\, d\theta \right| \leq \int_0^\pi e^{-R\theta}\, d\theta = \frac{1 - e^{-\pi R}}{R} \qquad (1.10.21)$$

which tends to zero as $R \to \infty$. Therefore,

$$\int_{-\infty}^{-\epsilon} \frac{e^{ix}}{x}\, dx + \int_\epsilon^\infty \frac{e^{ix}}{x}\, dx = -\int_{C_1} \frac{e^{iz}}{z}\, dz. \qquad (1.10.22)$$

Now,

$$\int_{C_1} \frac{e^{iz}}{z}\, dz = \int_{C_1} \frac{dz}{z} + i\int_{C_1} dz - \int_{C_1} \frac{z}{2}\, dz + \cdots = -\pi i \qquad (1.10.23)$$

in the limit $\epsilon \to 0$ because $z = \epsilon e^{\theta i}$. Consequently, in the limit of $\epsilon \to 0$,

$$PV \int_{-\infty}^\infty \frac{e^{ix}}{x}\, dx = \pi. \qquad (1.10.24)$$

Upon separating the real and imaginary parts, we obtain

$$PV \int_{-\infty}^\infty \frac{\cos(x)}{x}\, dx = 0, \qquad \int_{-\infty}^\infty \frac{\sin(x)}{x}\, dx = \pi. \qquad (1.10.25)$$

Problems

1. Noting that

$$\int_0^{\theta - \epsilon} \frac{d\varphi}{\cos(\varphi) - \cos(\theta)} = \frac{1}{\sin(\theta)} \ln\left| \frac{\sin\left[\frac{1}{2}(\theta + \varphi)\right]}{\sin\left[\frac{1}{2}(\theta - \varphi)\right]} \right| \Bigg|_0^{\theta - \epsilon},$$

and

$$\int_{\theta+\epsilon}^{\pi} \frac{d\varphi}{\cos(\varphi) - \cos(\theta)} = \frac{1}{\sin(\theta)} \ln \left| \frac{\sin\left[\frac{1}{2}(\theta + \varphi)\right]}{\sin\left[\frac{1}{2}(\theta - \varphi)\right]} \right|_{\theta+\epsilon}^{\pi},$$

show that

$$PV \int_{0}^{\pi} \frac{d\varphi}{\cos(\varphi) - \cos(\theta)} = 0, \qquad 0 < \theta < \pi.$$

2. Using $f(z) = e^{i\pi z/2}/(z^2 - 1)$, show that

$$\int_{-\infty}^{\infty} \frac{\cos(\pi x/2)}{x^2 - 1} \, dx = -\pi.$$

3. Show that

$$\int_{-\infty}^{\infty} \frac{e^{ax} - e^{bx}}{1 - e^x} \, dx = \pi[\cot(a\pi) - \cot(b\pi)], \qquad 0 < a, b < 1.$$

Use a rectangular contour with vertices at $(-R, 0)$, $(R, 0)$, $(-R, \pi)$, and (R, π) with a semicircle indentation at the origin.

4. Show[13] that

$$\int_{-\infty}^{\infty} \frac{1 - \cos[2a(x + \zeta)]}{(x + \zeta)^2(x^2 + \alpha^2)} \, dx = \frac{\pi}{\alpha(\zeta^2 + \alpha^2)^2} \{2a\alpha(\zeta^2 + \alpha^2) + (\zeta^2 - \alpha^2)$$
$$- e^{-2a\alpha} \left[(\zeta^2 - \alpha^2)\cos(2a\zeta) + 2\alpha\zeta\sin(2a\zeta)\right]\},$$

where a, α, and ζ are real. Use a semicircular contour of infinite radius with the real axis as its diameter.

5. Using the complex function $e^{imz}/(z - a)$ and a closed contour similar to that shown in Figure 1.10.2, show that

$$PV \int_{-\infty}^{\infty} \frac{\cos(mx)}{x - a} \, dx = -\pi \sin(ma), \quad \text{and} \quad PV \int_{-\infty}^{\infty} \frac{\sin(mx)}{x - a} \, dx = \pi \cos(ma),$$

where $m > 0$ and a is real.

6. Using a closed contour similar to that shown in Figure 1.10.2 except that we now have two small semicircles around the singularities on the real axis, show that

$$PV \int_{-\infty}^{\infty} \frac{xe^{xi}}{x^2 - \pi^2} \, dx = -\pi i,$$

---

[13] Reprinted with permission from Ko, S. H., and A. H. Nuttall, 1991: Analytical evaluation of flush-mounted hydrophone array response to the Corcos turbulent wall pressure spectrum. *J. Acoust. Soc. Am.*, **90**, 579–588. ©1991, Acoustical Society of America.

and

$$PV \int_{-\infty}^{\infty} \frac{e^{imx}}{(x-1)(x-3)}\, dx = \frac{\pi i}{2}\left(e^{3mi} - e^{mi}\right), \qquad m > 0.$$

7. Redo Example 1.10.3 except the contour is now a rectangle with vertices at $\pm R$ and $\pm R + Ri$ indented at the origin.

8. Let the function $f(z)$ possess a simple pole with a residue $\mathrm{Res}[f(z);c]$ on a simply closed contour C. If C is indented at c, show that the integral of $f(z)$ around the indentation tends to $-\mathrm{Res}[f(z);c]\alpha i$ as the radius of the indentation tends to zero, α being the internal angle between the two parts of C meeting at c.

Chapter 2
First-Order Ordinary
Differential Equations

A *differential equation* is any equation that contains the derivatives or differentials of one or more dependent variables with respect to one or more independent variables. Because many of the known physical laws are expressed as differential equations, a sound knowledge of how to solve them is essential. In the next two chapters we present the fundamental methods for solving *ordinary differential equations*—a differential equation which contains only ordinary derivatives of one or more dependent variables. Later, in §5.6 and §6.8 we show how transform methods can be used to solve ordinary differential equations while systems of linear ordinary differential equations are treated in §14.6. Solutions for *partial differential equations*—a differential equation involving partial derivatives of one or more dependent variables of two or more *independent* variables—are given in Chapters 10, 11, and 12.

2.1 CLASSIFICATION OF DIFFERENTIAL EQUATIONS

Differential equations are classified three ways: by *type*, *order*, and *linearity*. There are two *types*: *ordinary* and *partial differential equations* which have already been defined. Examples of ordinary differential equations include

$$\frac{dy}{dx} - 2y = x, \tag{2.1.1}$$

$$(x - y)\, dx + 4y\, dy = 0, \tag{2.1.2}$$

$$\frac{du}{dx} + \frac{dv}{dx} = 1 + 5x, \tag{2.1.3}$$

and

$$\frac{d^2 y}{dx^2} + 2\frac{dy}{dx} + y = \sin(x). \tag{2.1.4}$$

On the other hand, examples of partial differential equations include

$$\frac{\partial u}{\partial x} + \frac{\partial u}{\partial y} = 0, \tag{2.1.5}$$

$$y\frac{\partial u}{\partial x} + x\frac{\partial u}{\partial y} = 2u, \tag{2.1.6}$$

and

$$\frac{\partial^2 u}{\partial t^2} + 2\frac{\partial u}{\partial t} = \frac{\partial^2 u}{\partial x^2}. \tag{2.1.7}$$

In the examples that we have just given, we have explicitly written out the differentiation operation. However, from calculus we know that dy/dx can also be written y'. Similarly the partial differentiation operator $\partial^4 u/\partial x^2 \partial y^2$ is sometimes written u_{xxyy}. We will also use this notation from time to time.

The *order* of a differential equation is given by the highest-order derivative. For example,

$$\frac{d^3 y}{dx^3} + 3\frac{d^2 y}{dx^2} + \left(\frac{dy}{dx}\right)^2 - y = \sin(x) \tag{2.1.8}$$

is a third-order ordinary differential equation. Because we can rewrite

$$(x + y)\, dy - x\, dx = 0 \tag{2.1.9}$$

as

$$(x + y)\frac{dy}{dx} = x \tag{2.1.10}$$

by dividing (2.1.9) by dx, we have a first-order ordinary differential equation here. Finally

$$\frac{\partial^4 u}{\partial x^2 \partial y^2} = \frac{\partial^2 u}{\partial t^2} \tag{2.1.11}$$

is an example of a fourth-order partial differential equation. In general, we can write a nth-order, ordinary differential equation as

$$f\left(x, y, \frac{dy}{dx}, \cdots, \frac{d^n y}{dx^n}\right) = 0. \tag{2.1.12}$$

The final classification is according to whether the differential equation is *linear* or *nonlinear*. A differential equation is *linear* if it can be written in the form:

$$a_n(x)\frac{d^n y}{dx^n} + a_{n-1}(x)\frac{d^{n-1}y}{dx^{n-1}} + \cdots + a_1(x)\frac{dy}{dx} + a_0(x)y = f(x). \quad \textbf{(2.1.13)}$$

Note that the linear differential equation (2.1.13) has two properties: (1) The dependent variable y and *all* of its derivatives are of first degree (the power of each term involving y is 1). (2) Each coefficient depends only on the independent variable x. Examples of linear first-, second-, and third-order ordinary differential equations are

$$(x+1)\,dy - y\,dx = 0, \quad \textbf{(2.1.14)}$$

$$y'' + 3y' + 2y = e^x, \quad \textbf{(2.1.15)}$$

and

$$x\frac{d^3 y}{dx^3} - (x^2+1)\frac{dy}{dx} + y = \sin(x), \quad \textbf{(2.1.16)}$$

respectively. If the differential equation is not linear, then it is *nonlinear*. Examples of nonlinear first-, second-, and third-order, ordinary differential equations are

$$\frac{dy}{dx} + xy + y^2 = x, \quad \textbf{(2.1.17)}$$

$$\frac{d^2 y}{dx^2} - \left(\frac{dy}{dx}\right)^5 + 2xy = \sin(x), \quad \textbf{(2.1.18)}$$

and

$$yy''' + 2y = e^x, \quad \textbf{(2.1.19)}$$

respectively.

At this point it is useful to highlight certain properties that all differential equations have in common regardless of their type, order and whether they are linear or not. First, it is not obvious that just because we can write down a differential equation that a solution exists. The *existence* of a solution to a class of differential equations constitutes an important aspect of the theory of differential equations. Because we are interested in differential equations that arise from applications, their solution should exist. In §2.2 we address this question further.

Quite often a differential equation has the solution $y = 0$, a *trivial* solution. For example, if $f(x) = 0$ in (2.1.13), a quick check shows that $y = 0$ is a solution. Trivial solutions are generally of little value.

Another important question is how many solutions does a differential equation have? In physical applications *uniqueness* is not important because, if we are lucky enough to actually find a solution, then its ties to a physical problem usually suggest uniqueness. Nevertheless, the question of uniqueness

is of considerable importance in the theory of differential equations. Uniqueness should not be confused with the fact that many solutions to ordinary
differential equations contain arbitrary constants much as indefinite integrals
in integral calculus. A solution to a differential equation that has no arbitrary
constants is called a *particular solution.*

• **Example 2.1.1**

Consider the differential equation

$$\frac{dy}{dx} = x + 1, \qquad y(1) = 2. \qquad (2.1.20)$$

This condition $y(1) = 2$ is called an *initial condition* and the differential equation plus the initial condition constitute an *initial-value problem.* Straightforward integration yields

$$y(x) = \int (x+1)\,dx + C = \tfrac{1}{2}x^2 + x + C. \qquad (2.1.21)$$

Equation (2.1.21) is the *general solution* to the differential equation (2.1.20)
because (2.1.21) is a solution to the differential equation for *every* choice of
C. However, if we now satisfy the initial condition $y(1) = 2$, we obtain a
particular solution. This is done by substituting the corresponding values of
x and y into (2.1.21) or

$$2 = \tfrac{1}{2}(1)^2 + 1 + C = \tfrac{3}{2} + C, \quad \text{or} \quad C = \tfrac{1}{2}. \qquad (2.1.22)$$

Therefore, the solution to the initial-value problem (2.1.20) is the particular
solution

$$y(x) = (x+1)^2/2. \qquad (2.1.23)$$

Finally, it must be admitted that most differential equations encountered
in the "real" world cannot be written down either explicitly or implicitly. For
example, the simple differential equation $y' = f(x)$ does not have an analytic solution unless you can integrate $f(x)$. This begs the question of why
it is useful to learn analytic techniques for solving differential equations that
often fail us. The answer lies in the fact that differential equations that we
can solve share many of the same properties and characteristics of differential equations which we can only solve numerically. Therefore, by working
with and examining the differential equations that we can solve exactly, we
develop our intuition and understanding about those that we can only solve
numerically.

Problems

Find the order and state whether the following ordinary differential equations are linear or nonlinear:

1. $y'/y = x^2 + x$

2. $y^2 y' = x + 3$

3. $\sin(y') = 5y$

4. $y''' = y$

5. $y'' = 3x^2$

6. $(y^3)' = 1 - 3y$

7. $y''' = y^3$

8. $y'' - 4y' + 5y = \sin(x)$

9. $y'' + xy = \cos(y'')$

10. $(2x + y)\,dx + (x - 3y)\,dy = 0$

11. $(1 + x^2)y' = (1 + y)^2$

12. $yy'' = x(y^2 + 1)$

13. $y' + y + y^2 = x + e^x$

14. $y''' + \cos(x)y' + y = 0$

15. $x^2 y'' + x^{1/2}(y')^3 + y = e^x$

16. $y''' + xy'' + e^y = x^2$

2.2 SEPARATION OF VARIABLES

The simplest method of solving a first-order ordinary differential equation, if it works, is *separation of variables*. It has the advantage of handling both linear and nonlinear problems, especially *autonomous equations*.[1] From integral calculus, we already met this technique when we solved the first-order differential equation

$$\frac{dy}{dx} = f(x). \qquad (2.2.1)$$

By multiplying both sides of (2.2.1) by dx, we obtain

$$dy = f(x)\,dx. \qquad (2.2.2)$$

At this point we note that the left side of (2.2.2) contains only y while the right side is purely a function of x. Hence, we can integrate directly and find that

$$y = \int f(x)\,dx + C. \qquad (2.2.3)$$

For this technique to work, we must be able to rewrite the differential equation so that all of the y dependence appears on one side of the equation while the x dependence is on the other. Finally we must be able to carry out the integration on both sides of the equation.

---

[1] An autonomous equation is a differential equation where the independent variable does not explicitly appear in the equation, such as $y' = f(y)$.

One of the interesting aspects of our analysis is the appearance of the arbitrary constant C in (2.2.3). To evaluate this constant we need more information. The most common method is to require that the dependent variable give a particular value for a particular value of x. Because the independent variable x often denotes time, this condition is usually called an *initial condition*, even in cases when the independent variable is not time.

● **Example 2.2.1**

Let us solve the ordinary differential equation

$$\frac{dy}{dx} = \frac{e^y}{xy}. \tag{2.2.4}$$

Because we can separate variables by rewriting (2.2.4) as

$$ye^{-y}\,dy = \frac{dx}{x}, \tag{2.2.5}$$

its solution is simply

$$-ye^{-y} - e^{-y} = \ln|x| + C \tag{2.2.6}$$

by direct integration.

● **Example 2.2.2**

Let us solve

$$\frac{dy}{dx} + y = xe^x y, \tag{2.2.7}$$

subject to the initial condition $y(0) = 1$.

Multiplying (2.2.7) by dx, we find that

$$dy + y\,dx = xe^x y\,dx, \tag{2.2.8}$$

or

$$\frac{dy}{y} = (xe^x - 1)\,dx. \tag{2.2.9}$$

A quick check shows that the left side of (2.2.9) contains only the dependent variable y while the right side depends solely on x and we have separated the variables onto one side or the other. Finally, integrating both sides of (2.2.9), we have

$$\ln(y) = xe^x - e^x - x + C. \tag{2.2.10}$$

Since $y(0) = 1$, $C = 1$ and

$$y(x) = \exp[(x - 1)e^x + 1 - x]. \tag{2.2.11}$$

In addition to the tried-and-true method of solving ordinary differential equation by hand, scientific computational packages such as MATLAB provide symbolic toolboxes that are designed to do the work for you. In the present case, typing

```
dsolve('Dy+y=x*exp(x)*y','y(0)=1','x')
```

yields

```
ans =
1/exp(-1)*exp(-x+x*exp(x)-exp(x))
```

which is equivalent to (2.2.11).

Our success here should not be overly generalized. Sometimes these toolboxes give the answer in a rather obscure form or they fail completely. For example, in the previous example, MATLAB gives the answer

```
ans =
-lambertw((log(x)+C1)*exp(-1))-1
```

The MATLAB function `lambertw` is Lambert's W function, where `w = lambertw(x)` is the solution to $we^w = x$. Using this definition, we can construct the solution as expressed in (2.2.6).

● **Example 2.2.3**

Consider the nonlinear differential equation

$$x^2 y' + y^2 = 0. \tag{2.2.12}$$

Separating variables, we find that

$$-\frac{dy}{y^2} = \frac{dx}{x^2}, \quad \text{or} \quad \frac{1}{y} = -\frac{1}{x} + C, \quad \text{or} \quad y = \frac{x}{Cx - 1}. \tag{2.2.13}$$

Equation (2.2.13) shows the wide variety of solutions possible for an ordinary differential equation. For example, if we require that $y(0) = 0$, then there are infinitely many different solutions satisfying this initial condition because C can take on any value. On the other hand, if we require that $y(0) = 1$, there is no solution because we cannot choose *any* constant C such that $y(0) = 1$. Finally, if we have the initial condition that $y(1) = 2$, then there is only one possible solution corresponding to $C = \frac{3}{2}$.

Consider now the trial solution $y = 0$. Does it satisfy (2.2.12)? Yes, it does. On the other hand, there is no choice of C which yields this solution. The solution $y = 0$ is called a *singular solution* to (2.2.12). *Singular solutions* are solutions to a differential equation which cannot be obtained from a solution with arbitrary constants.

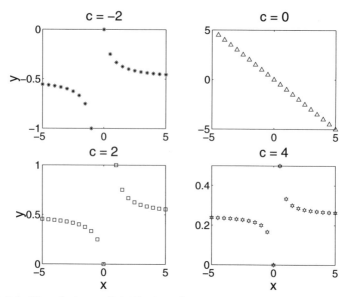

Figure 2.2.1: The solution to (2.2.13) when $C = -2, 0, 2, 4$.

Finally, we illustrate (2.2.13) using MATLAB. This is one of MATLAB's strengths — the ability to convert an abstract equation into a concrete picture. Here the MATLAB script

```
clear
hold on
x = -5:0.5:5;
for c = -2:2:4
y = x ./ (c*x-1);
if (c== -2) subplot(2,2,1), plot(x,y,'*')
    axis tight; title('c = -2'); ylabel('y','Fontsize',20); end
if (c== 0) subplot(2,2,2), plot(x,y,'^')
    axis tight; title('c = 0'); end
if (c== 2) subplot(2,2,3), plot(x,y,'s')
    axis tight; title('c = 2'); xlabel('x','Fontsize',20);
    ylabel('y','Fontsize',20); end
if (c== 4) subplot(2,2,4), plot(x,y,'h')
    axis tight; title('c = 4'); xlabel('x','Fontsize',20); end
end
```

yields Figure 2.2.1 which illustrates (2.2.13) when $C = -2, 0, 2$, and 4.

The previous example showed that first-order ordinary differential equations may have a unique solution, no solution, or many solutions. From a complete study[2] of these equations, we have the following theorem:

---

[2] The proof of the existence and uniqueness of first-order ordinary differential equations

Theorem: Existence and Uniqueness

Suppose some real-valued function $f(x, y)$ is continuous on some rectangle in the xy-plane containing the point (a, b) in its interior. Then the initial-value problem

$$\frac{dy}{dx} = f(x, y), \qquad y(a) = b, \tag{2.2.14}$$

has at least one solution on the same open interval I containing the point $x = a$. Furthermore, if the partial derivative $\partial f / \partial y$ is continuous on that rectangle, then the solution is unique on some (perhaps smaller) open interval I_0 containing the point $x = a$. □

• **Example 2.2.4**

Consider the initial-value problem $y' = 3y^{1/3}/2$ with $y(0) = 1$. Here $f(x, y) = 3y^{1/3}/2$ and $f_y = y^{-2/3}/2$. Because f_y is continuous over a small rectangle containing the point $(0, 1)$, there is a unique solution around $x = 0$, namely $y = (x + 1)^{3/2}$, which satisfies the differential equation and the initial condition. On the other hand, if the initial condition reads $y(0) = 0$, then f_y is *not* continuous on *any* rectangle containing the point $(0, 0)$ and there is no unique solution. For example, two solutions to this initial-value problem, valid on any open interval that includes $x = 0$, are $y_1(x) = x^{3/2}$ and

$$y_2(x) = \begin{cases} (x - 1)^{3/2}, & x \geq 1, \\ 0, & x < 1. \end{cases} \tag{2.2.15}$$

• **Example 2.2.5: Hydrostatic equation**

Consider an atmosphere where its density varies only in the vertical direction. The pressure at the surface equals the weight per unit horizontal area of all of the air from sea level to outer space. As you move upward, the amount of air remaining above decreases and so does the pressure. This is why we experience pressure sensations in our ears when ascending or descending in an elevator or airplane. If we rise the small distance dz, there must be a corresponding small decrease in the pressure, dp. This pressure drop must equal the loss of weight in the column per unit area, $-\rho g\, dz$. Therefore, the pressure is governed by the differential equation

$$dp = -\rho g\, dz, \tag{2.2.16}$$

commonly called the *hydrostatic equation*.

---

is beyond the scope of this book. See Ince, E. L., 1956: *Ordinary Differential Equations.* Dover Publications, Inc., Chapter 3.

To solve (2.2.16), we must express ρ in terms of pressure. For example, in an isothermal atmosphere at constant temperature T_s, the ideal gas law gives $p = \rho R T_s$, where R is the gas constant. Substituting this into (2.2.16) and separating variables yields

$$\frac{dp}{p} = -\frac{g}{RT_s}\,dz. \tag{2.2.17}$$

Integrating (2.2.17) gives

$$p(z) = p(0)\exp\left(-\frac{gz}{RT_s}\right). \tag{2.2.18}$$

Thus, the pressure (and density) of an isothermal atmosphere decreases exponentially with height. In particular, it decreases by e^{-1} over the distance RT_s/g, the so-called "scale height."

• Example 2.2.6: Terminal velocity

As an object moves through a fluid, its viscosity resists the motion. Let us find the motion of a mass m as it falls toward the earth under the force of gravity when the drag varies as the square of the velocity.

From Newton's second law, the equation of motion is

$$m\frac{dv}{dt} = mg - C_D v^2, \tag{2.2.19}$$

where v denotes the velocity, g is the gravitational acceleration, and C_D is the drag coefficient. We choose the coordinate system so that a downward velocity is positive.

Equation (2.2.19) can be solved using the technique of separation of variables if we change from time t as the independent variable to the distance traveled x from the point of release. This modification yields the differential equation

$$mv\frac{dv}{dx} = mg - C_D v^2, \tag{2.2.20}$$

since $v = dx/dt$. Separating the variables leads to

$$\frac{v\,dv}{1 - kv^2/g} = g\,dx, \tag{2.2.21}$$

or

$$\ln\left(1 - \frac{kv^2}{g}\right) = -2kx, \tag{2.2.22}$$

where $k = C_D/m$ and $v = 0$ for $x = 0$. Taking the inverse of the natural logarithm, we finally obtain

$$v^2(x) = \frac{g}{k}\left(1 - e^{-2kx}\right). \tag{2.2.23}$$

Thus, as the distance that the object falls increases, so does the velocity and it eventually approaches a constant value $\sqrt{g/k}$, commonly known as the *terminal velocity*.

Because the drag coefficient C_D varies with the superficial area of the object while the mass depends on the volume, k increases as an object becomes smaller, resulting in a smaller terminal velocity. Consequently, although a human being of normal size will acquire a terminal velocity of approximately 120 mph, a mouse, on the other hand, can fall any distance without injury.

• Example 2.2.7: Interest rate

Consider a bank account that has been set up to pay out a constant rate of P dollars per year for the purchase of a car. This account has the special feature that it pays an annual interest rate of r on the current balance. We would like to know the balance in the account at any time t.

Although financial transactions occur at regularly spaced intervals, an excellent approximation can be obtained by treating the amount in the account $x(t)$ as a continuous function of time governed by the equation

$$x(t + \Delta t) \approx x(t) + rx(t)\Delta t - P\Delta t, \qquad (2.2.24)$$

where we have assumed that both the payment and interest are paid in time increments of Δt. As the time between payments tends to zero, we obtain the first-order ordinary differential equation

$$\frac{dx}{dt} = rx - P. \qquad (2.2.25)$$

If we denote the initial deposit into this account by $x(0)$, then at any subsequent time

$$x(t) = x(0)e^{rt} - P\left(e^{rt} - 1\right)/r. \qquad (2.2.26)$$

Although we could compute $x(t)$ as a function of P, r, and $x(0)$, there are only three separate cases that merit our close attention. If $P/r > x(0)$, then the account will eventually equal zero at $rt = \ln\{P/\left[P - rx(0)\right]\}$. On the other hand, if $P/r < x(0)$, the amount of money in the account will grow without bound. Finally, the case $x(0) = P/r$ is the equilibrium case where the amount of money paid out balances the growth of money due to interest so that the account always has the balance of P/r.

• Example 2.2.8: Steady-state flow of heat

When the inner and outer walls of a body, for example the inner and outer walls of a house, are maintained at *different constant* temperatures, heat will flow from the warmer wall to the colder one. When each surface parallel to a wall has attained a constant temperature, the flow of heat has reached a steady state. In a steady state flow of heat, each surface parallel to a wall, because its temperature is now constant, is referred to as an isothermal surface. Isothermal surfaces at different distances from an interior wall will have different temperatures. In many cases the temperature of an isothermal surface is only a function of its distance x from the interior wall, and the rate of flow of heat Q in a unit time across such a surface is proportional both to the area A of the surface and to dT/dx, where T is the temperature of the isothermal surface. Hence,

$$Q = -\kappa A \frac{dT}{dx}, \tag{2.2.27}$$

where κ is called the thermal conductivity of the material between the walls.

In place of a flat wall, let us consider a hollow cylinder whose inner and outer surfaces are located at $r = r_1$ and $r = r_2$, respectively. At steady state, (2.2.27) becomes

$$Q_r = -\kappa A \frac{dT}{dr} = -\kappa(2\pi r L)\frac{dT}{dr}, \tag{2.2.28}$$

assuming no heat generation within the cylindrical wall.

We can find the temperature distribution inside the cylinder by solving (2.2.28) along with the appropriate conditions on $T(r)$ at $r = r_1$ and $r = r_2$ (the boundary conditions). To illustrate the wide choice of possible boundary conditions, let us require that inner surface is maintained at the temperature T_1. Along the outer surface we assume that heat is lost by convection to the environment which has the temperature T_∞. This heat loss is usually modeled by the equation

$$\kappa \left.\frac{dT}{dr}\right|_{r=r_2} = -h(T - T_\infty), \tag{2.2.29}$$

where $h > 0$ is the convective heat transfer coefficient. Upon integrating (2.2.28),

$$T(r) = -\frac{Q_r}{2\pi\kappa L}\ln(r) + C, \tag{2.2.30}$$

where Q_r is also an unknown. Substituting (2.2.30) into the boundary conditions, we obtain

$$T(r) = T_1 + \frac{Q_r}{2\pi\kappa L}\ln(r_1/r), \tag{2.2.31}$$

with

$$Q_r = \frac{2\pi\kappa L(T_1 - T_\infty)}{\kappa/r_2 + h\ln(r_2/r_1)}. \tag{2.2.32}$$

As r_2 increases, the first term in the denominator of (2.2.32) decreases while the second term increases. Therefore, Q_r has its largest magnitude when the denominator is smallest, assuming a fixed numerator. This occurs at the critical radius $r_{cr} = \kappa/h$, where

$$Q_r^{max} = \frac{2\pi\kappa L(T_1 - T_\infty)}{1 + \ln(r_{cr}/r_1)}. \tag{2.2.33}$$

• Example 2.2.9: Logistic equation

The study of population dynamics yields an important class of first-order, nonlinear, ordinary differential equations: the logistic equation. This equation arose in Pierre François Verhulst's (1804–1849) study of animal populations.[3] If $x(t)$ denotes the number of species in the population and k is the (constant) environment capacity (the number of species that can simultaneously live in the geographical region), then the logistic or Verhulst's equation is

$$x' = ax(k - x)/k, \tag{2.2.34}$$

where a is the population growth rate for a small number of species.

To solve (2.2.34), we rewrite it as

$$\frac{dx}{(1 - x/k)x} = \frac{dx}{x} + \frac{x/k}{1 - x/k}\, dx = r\, dt. \tag{2.2.35}$$

Integration yields

$$\ln|x| - \ln|1 - x/k| = rt + \ln(C), \tag{2.2.36}$$

or

$$\frac{x}{1 - x/k} = Ce^{rt}. \tag{2.2.37}$$

If $x(0) = x_0$,

$$x(t) = \frac{kx_0}{x_0 + (k - x_0)e^{-rt}}. \tag{2.2.38}$$

As $t \to \infty$, $x(t) \to k$, the asymptotically stable solution.

• Example 2.2.10: Chemical reactions

Chemical reactions are often governed by first-order ordinary differential equations. For example, first-order reactions, which describe reactions of the form $A \xrightarrow{k} B$, yield the differential equation

$$-\frac{1}{a}\frac{d[A]}{dt} = k[A], \tag{2.2.39}$$

where k is the rate at which the reaction is taking place. Because for every molecule of A that disappears one molecule of B is produced, $a = 1$ and (2.2.39) becomes

$$-\frac{d[A]}{dt} = k[A]. \tag{2.2.40}$$

---

[3] Verhulst, P. F., 1838: Notice sur la loi que la population suit dans son accroissement. *Correspond. Math. Phys.*, **10**, 113–121.

Integration of (2.2.40) leads to

$$-\int \frac{d[\mathrm{A}]}{[\mathrm{A}]} = k \int dt. \tag{2.2.41}$$

If we denote the initial value of [A] by $[\mathrm{A}]_0$, then integration yields

$$-\ln[\mathrm{A}] = kt - \ln[\mathrm{A}]_0, \tag{2.2.42}$$

or

$$[\mathrm{A}] = [\mathrm{A}]_0 e^{-kt}. \tag{2.2.43}$$

The exponential form of the solution suggests that there is a *time constant* τ which is called the *decay time* of the reaction. This quantity gives the time required for the concentration of decrease by $1/e$ of its initial value $[\mathrm{A}]_0$. It is given by $\tau = 1/k$.

Turning to second-order reactions, there are two cases. The first is a reaction between two identical species: $\mathrm{A} + \mathrm{A} \overset{k}{\to}$ products. The rate expression here is

$$-\frac{1}{2}\frac{d[\mathrm{A}]}{dt} = k[\mathrm{A}]^2. \tag{2.2.44}$$

The second case is an overall second-order reaction between two unlike species, given by $\mathrm{A} + \mathrm{B} \overset{k}{\to} \mathrm{X}$. In this case, the reaction is first order in each of the reactants A and B and the rate expression is

$$-\frac{d[\mathrm{A}]}{dt} = k[\mathrm{A}][\mathrm{B}]. \tag{2.2.45}$$

Turning to (2.2.44) first, we have by separation of variables

$$-\int_{[\mathrm{A}]_0}^{[\mathrm{A}]} \frac{d[\mathrm{A}]}{[\mathrm{A}]^2} = 2k \int_0^t d\tau, \tag{2.2.46}$$

or

$$\frac{1}{[\mathrm{A}]} = \frac{1}{[\mathrm{A}]_0} + 2kt. \tag{2.2.47}$$

Therefore, a plot of the inverse of A versus time will yield a straight line with slope equal to $2k$ and intercept $1/[\mathrm{A}]_0$.

With regard to (2.2.45), because an increase in X must be at the expense of A and B, it is useful to express the rate equation in terms of the concentration of X, $[\mathrm{X}] = [\mathrm{A}]_0 - [\mathrm{A}] = [\mathrm{B}]_0 - [\mathrm{B}]$, where $[\mathrm{A}]_0$ and $[\mathrm{B}]_0$ are the initial concentrations. Then, (2.2.45) becomes

$$\frac{d[\mathrm{X}]}{dt} = k\left([\mathrm{A}]_0 - [\mathrm{X}]\right)\left([\mathrm{B}]_0 - [\mathrm{X}]\right). \tag{2.2.48}$$

Separation of variables leads to

$$\int_{[X]_0}^{[X]} \frac{d\xi}{([A]_0 - \xi)([B]_0 - \xi)} = k \int_0^t d\tau. \qquad (2.2.49)$$

To integrate the left side, we rewrite the integral

$$\int \frac{d\xi}{([A]_0 - \xi)([B]_0 - \xi)} = \int \frac{d\xi}{([A]_0 - [B]_0)([B]_0 - \xi)}$$
$$- \int \frac{d\xi}{([A]_0 - [B]_0)([A]_0 - \xi)}. \qquad (2.2.50)$$

Carrying out the integration,

$$\frac{1}{[A]_0 - [B]_0} \ln\left(\frac{[B]_0[A]}{[A]_0[B]}\right) = kt. \qquad (2.2.51)$$

Again the reaction rate constant k can be found by plotting the data in the form of the left side of (2.2.51) against t.

Problems

For Problems 1–10, solve the following ordinary differential equations by separation of variables. Then use MATLAB to plot your solution. Try and find the symbolic solution using MATLAB's `dsolve`.

1. $\dfrac{dy}{dx} = xe^y$

2. $(1 + y^2)\, dx - (1 + x^2)\, dy = 0$

3. $\ln(x)\dfrac{dx}{dy} = xy$

4. $\dfrac{y^2}{x}\dfrac{dy}{dx} = 1 + x^2$

5. $\dfrac{dy}{dx} = \dfrac{2x + xy^2}{y + x^2 y}$

6. $\dfrac{dy}{dx} = (xy)^{1/3}$

7. $\dfrac{dy}{dx} = e^{x+y}$

8. $\dfrac{dy}{dx} = (x^3 + 5)(y^2 + 1)$

9. Solve the initial-value problem

$$\frac{dy}{dt} = -ay + \frac{b}{y^2}, \qquad y(0) = y_0,$$

where a and b are constants.

10. Setting $u = y - x$, solve the first-order ordinary differential equation

$$\frac{dy}{dx} = \frac{y - x}{x^2} + 1.$$

11. Using the hydrostatic equation, show that the pressure within an atmosphere where the temperature decreases uniformly with height, $T(z) = T_0 - \Gamma z$, varies as

$$p(z) = p_0 \left(\frac{T_0 - \Gamma z}{T_0} \right)^{g/(R\Gamma)},$$

where p_0 is the pressure at $z = 0$.

12. Using the hydrostatic equation, show that the pressure within an atmosphere with the temperature distribution

$$T(z) = \begin{cases} T_0 - \Gamma z, & 0 \le z \le H, \\ T_0 - \Gamma H, & H \le z, \end{cases}$$

is

$$p(z) = p_0 \begin{cases} \left(\dfrac{T_0 - \Gamma z}{T_0} \right)^{g/(R\Gamma)}, & 0 \le z \le H, \\ \left(\dfrac{T_0 - \Gamma H}{T_0} \right)^{g/(R\Gamma)} \exp\left[-\dfrac{g(z - H)}{R(T_0 - \Gamma H)} \right], & H \le z, \end{cases}$$

where p_0 is the pressure at $z = 0$.

13. The voltage V as a function of time t within an electrical circuit[4] consisting of a capacitor with capacitance C and a diode in series is governed by the first-order ordinary differential equation

$$C\frac{dV}{dt} + \frac{V}{R} + \frac{V^2}{S} = 0,$$

where R and S are positive constants. If the circuit initially has a voltage V_0 at $t = 0$, find the voltage at subsequent times.

14. A glow plug is an electrical element inside a reaction chamber which either ignites the nearby fuel or warms the air in the chamber so that the ignition will occur more quickly. An accurate prediction of the wire's temperature is important in the design of the chamber.

Assuming that heat convection and conduction are not important,[5] the temperature T of the wire is governed by

$$A\frac{dT}{dt} + B(T^4 - T_a^4) = P,$$

---

[4] See Aiken, C. B., 1938: Theory of the diode voltmeter. *Proc. IRE*, **26**, 859–876.

[5] Taken from Clark, S. K., 1956: Heat-up time of wire glow plugs. *Jet Propulsion*, **26**, 278–279.

where A equals the specific heat of the wire times its mass, B equals the product of the emissivity of the surrounding fluid times the wire's surface area times the Stefan-Boltzmann constant, T_a is the temperature of the surrounding fluid, and P is the power input. The temperature increases due to electrical resistance and is reduced by radiation to the surrounding fluid.

Show that the temperature is given by

$$\frac{4B\gamma^3 t}{A} = 2\left[\tan^{-1}\left(\frac{T}{\gamma}\right) - \tan^{-1}\left(\frac{T_0}{\gamma}\right)\right] - \ln\left[\frac{(T-\gamma)(T_0+\gamma)}{(T+\gamma)(T_0-\gamma)}\right],$$

where $\gamma^4 = P/B + T_a^4$ and T_0 is the initial temperature of the wire.

15. Let us denote the number of tumor cells by $N(t)$. Then a widely used deterministic tumor growth law[6] is

$$\frac{dN}{dt} = bN\ln(K/N),$$

where K is the largest tumor size and $1/b$ is the length of time required for the specific growth to decrease by $1/e$. If the initial value of $N(t)$ is $N(0)$, find $N(t)$ at any subsequent time t.

16. The drop in laser intensity in the direction of propagation x due to one and two-photon absorption in photosensitive glass is governed[7] by

$$\frac{dI}{dx} = -\alpha I - \beta I^2,$$

where I is the laser intensity, α and β are the single-photon and two-photon coefficients, respectively. Show that the laser intensity distribution is

$$I(x) = \frac{\alpha I(0)e^{-\alpha x}}{\alpha + \beta I(0)\left(1 - e^{-\alpha x}\right)},$$

where $I(0)$ is the laser intensity at the entry point of the media, $x = 0$.

17. The third-order reaction $A + B + C \xrightarrow{k} X$ is governed by the kinetics equation

$$\frac{d[X]}{dt} = k\left([A]_0 - [X]\right)\left([B]_0 - [X]\right)\left([C]_0 - [X]\right),$$

---

[6] Reprinted from *Math. Biosci.*, **61**, F. B. Hanson and C. Tier, A stochastic model of tumor growth, 73–100, ©1982, with permission from Elsevier Science.

[7] Reprinted with permission from Weitzman, P. S., and U. Österberg, 1996: Two-photon absorption and photoconductivity in photosensitive glasses. *J. Appl. Phys.*, **79**, 8648–8655. ©1996, American Institute of Physics.

where $[A]_0$, $[B]_0$, and $[C]_0$ denote the initial concentration of A, B, and C, respectively. Find how $[X]$ varies with time t.

18. The reversible reaction $A \underset{k_2}{\overset{k_1}{\rightleftharpoons}} B$ is described by the kinetics equation[8]

$$\frac{d[X]}{dt} = k_1 \left([A]_0 - [X]\right) - k_2 \left([B]_0 + [X]\right),$$

where $[X]$ denotes the increase in the concentration of B while $[A]_0$ and $[B]_0$ are the initial concentrations of A and B, respectively. Find $[X]$ as a function of time t. Hint: Show that this differential equation can be written

$$\frac{d[X]}{dt} = (k_1 - k_2)\left(\alpha + [X]\right), \qquad \alpha = \frac{k_1[A]_0 - k_2[B]_0}{k_1 + k_2}.$$

2.3 HOMOGENEOUS EQUATIONS

A *homogeneous ordinary differential equation* is a differential equation of the form

$$M(x,y)\,dx + N(x,y)\,dy = 0, \qquad (2.3.1)$$

where both $M(x,y)$ and $N(x,y)$ are homogeneous functions of the same degree n. That means: $M(tx, ty) = t^n M(x,y)$ and $N(tx, ty) = t^n N(x,y)$. For example, the ordinary differential equation

$$\left(x^2 + y^2\right) dx + \left(x^2 - xy\right) dy = 0 \qquad (2.3.2)$$

is a homogeneous equation because both coefficients are homogeneous functions of degree 2:

$$M(tx, ty) = t^2 x^2 + t^2 y^2 = t^2 \left(x^2 + y^2\right) = t^2 M(x,y), \qquad (2.3.3)$$

and

$$N(tx, ty) = t^2 x^2 - t^2 xy = t^2 \left(x^2 - xy\right) = t^2 N(x,y). \qquad (2.3.4)$$

Why is it useful to recognize homogeneous ordinary differential equations? Let us set $y = ux$ so that (2.3.2) becomes

$$\left(x^2 + u^2 x^2\right) dx + \left(x^2 - ux^2\right)(u\,dx + x\,du) = 0. \qquad (2.3.5)$$

Then,

$$x^2(1 + u)\,dx + x^3(1 - u)\,du = 0, \qquad (2.3.6)$$

---

[8] See Küster, F. W., 1895: Ueber den Verlauf einer umkehrbaren Reaktion erster Ordnung in homogenem System. *Zeit. Physik. Chem.*, **18**, 171–179.

$$\frac{1-u}{1+u}\,du + \frac{dx}{x} = 0, \qquad (2.3.7)$$

or

$$\left(-1 + \frac{2}{1+u}\right) du + \frac{dx}{x} = 0. \qquad (2.3.8)$$

Integrating (2.3.8),

$$-u + 2\ln|1+u| + \ln|x| = \ln|c|, \qquad (2.3.9)$$

$$-\frac{y}{x} + 2\ln\left|1 + \frac{y}{x}\right| + \ln|x| = \ln|c|, \qquad (2.3.10)$$

$$\ln\left[\frac{(x+y)^2}{cx}\right] = \frac{y}{x}, \qquad (2.3.11)$$

or

$$(x+y)^2 = cxe^{y/x}. \qquad (2.3.12)$$

Problems

First show that the following differential equations are homogeneous and then find their solution. Then use MATLAB to plot your solution. Try and find the symbolic solution using MATLAB's `dsolve`.

1. $(x+y)\dfrac{dy}{dx} = y$

2. $(x+y)\dfrac{dy}{dx} = x - y$

3. $2xy\dfrac{dy}{dx} = -(x^2 + y^2)$

4. $x(x+y)\dfrac{dy}{dx} = y(x-y)$

5. $xy' = y + 2\sqrt{xy}$

6. $xy' = y - \sqrt{x^2 + y^2}$

7. $y' = \sec(y/x) + y/x$

8. $y' = e^{y/x} + y/x.$

2.4 EXACT EQUATIONS

Consider the multivariable function $z = f(x,y)$. Then the total derivative is

$$dz = \frac{\partial f}{\partial x}\,dx + \frac{\partial f}{\partial y}\,dy = M(x,y)\,dx + N(x,y)\,dy. \qquad (2.4.1)$$

If the solution to a first-order ordinary differential equation can be written as $f(x,y) = c$, then the corresponding differential equation is

$$M(x,y)\,dx + N(x,y)\,dy = 0. \qquad (2.4.2)$$

How do we know if we have an *exact equation* (2.4.2)? From the definition of $M(x, y)$ and $N(x, y)$,

$$\frac{\partial M}{\partial y} = \frac{\partial^2 f}{\partial y \partial x} = \frac{\partial^2 f}{\partial x \partial y} = \frac{\partial N}{\partial x}, \tag{2.4.3}$$

if $M(x, y)$ and $N(x, y)$ and their first-order partial derivatives are continuous. Consequently, if we can show that our ordinary differential equation is exact, we can integrate

$$\frac{\partial f}{\partial x} = M(x, y) \qquad \text{and} \qquad \frac{\partial f}{\partial y} = N(x, y) \tag{2.4.4}$$

to find the solution $f(x, y) = c$.

- **Example 2.4.1**

Let us check and see if

$$[y^2 \cos(x) - 3x^2 y - 2x] \, dx + [2y \sin(x) - x^3 + \ln(y)] \, dy = 0 \tag{2.4.5}$$

is exact.

Since $M(x, y) = y^2 \cos(x) - 3x^2 y - 2x$, and $N(x, y) = 2y \sin(x) - x^3 + \ln(y)$, we find that

$$\frac{\partial M}{\partial y} = 2y \cos(x) - 3x^2, \tag{2.4.6}$$

and

$$\frac{\partial N}{\partial x} = 2y \cos(x) - 3x^2. \tag{2.4.7}$$

Because $N_x = M_y$, (2.4.5) is an exact equation.

- **Example 2.4.2**

Because (2.4.5) is an exact equation, let us find its solution. Starting with

$$\frac{\partial f}{\partial x} = M(x, y) = y^2 \cos(x) - 3x^2 y - 2x, \tag{2.4.8}$$

direct integration gives

$$f(x, y) = y^2 \sin(x) - x^3 y - x^2 + g(y). \tag{2.4.9}$$

Substituting (2.4.9) into the equation $f_y = N$, we obtain

$$\frac{\partial f}{\partial y} = 2y \sin(x) - x^3 + g'(y) = 2y \sin(x) - x^3 + \ln(y). \tag{2.4.10}$$

Thus, $g'(y) = \ln(y)$, or $g(y) = y\ln(y) - y + C$. Therefore, the solution to the ordinary differential equation (2.4.5) is

$$y^2 \sin(x) - x^3 y - x^2 + y\ln(y) - y = c. \tag{2.4.11}$$

• **Example 2.4.3**

Consider the differential equation

$$(x + y)\,dx + x\ln(x)\,dy = 0 \tag{2.4.12}$$

on the interval $(0, \infty)$. A quick check shows that (2.4.12) is not exact since

$$\frac{\partial M}{\partial y} = 1, \qquad \text{and} \qquad \frac{\partial N}{\partial x} = 1 + \ln(x). \tag{2.4.13}$$

However, if we multiply (2.4.12) by $1/x$ so that it becomes

$$\left(1 + \frac{y}{x}\right) dx + \ln(x)\,dy = 0, \tag{2.4.14}$$

then this modified differential equation is exact because

$$\frac{\partial M}{\partial y} = \frac{1}{x}, \qquad \text{and} \qquad \frac{\partial N}{\partial x} = \frac{1}{x}. \tag{2.4.15}$$

Therefore, the solution to (2.4.12) is

$$x + y\ln(x) = C. \tag{2.4.16}$$

This mysterious function that converts an inexact differential equation into an exact one is called an *integrating factor*. Unfortunately there is no general rule for finding one unless the equation is linear.

Problems

Show that the following equations are exact. Then solve them, using MATLAB to plot them. Finally try and find the symbolic solution using MATLAB's dsolve.

1. $2xyy' = x^2 - y^2$

2. $(x + y)y' + y = x$

3. $(y^2 - 1)\,dx + [2xy - \sin(y)]\,dy = 0$

4. $[\sin(y) - 2xy + x^2]\,dx + [x\cos(y) - x^2]\,dy = 0$

5. $-y\,dx/x^2 + (1/x + 1/y)\,dy = 0$ 6. $(3x^2 - 6xy)\,dx - (3x^2 + 2y)\,dy = 0$

7. $y\sin(xy)\,dx + x\sin(xy)\,dy = 0$ 8. $(2xy^2 + 3x^2)\,dx + 2x^2y\,dy = 0$

9. $(2xy^3 + 5x^4y)\,dx + (3x^2y^2 + x^5 + 1)\,dy = 0$

10. $(x^3 + y/x)\,dx + [y^2 + \ln(x)]\,dy = 0$

11. $[x + e^{-y} + x\ln(y)]\,dy + [y\ln(y) + e^x]\,dx = 0$

12. $\cos(4y^2)\,dx - 8xy\sin(4y^2)\,dy = 0$ 13. $\sin^2(x+y)\,dx - \cos^2(x+y)\,dy = 0$

14. Show that the integrating factor for $(x - y)y' + \alpha y(1 - y) = 0$ is $\mu(y) = y^a/(1 - y)^{a+2}$, $a + 1 = 1/\alpha$. Then show that the solution is

$$\alpha x\,\frac{y^{a+1}}{(1 - y)^{a+1}} - \int_0^y \frac{\xi^{a+1}}{(1 - \xi)^{a+2}}\,d\xi = C.$$

2.5 LINEAR EQUATIONS

In the case of first-order ordinary differential equations, any differential equation of the form

$$a_1(x)\frac{dy}{dx} + a_0(x)y = f(x) \tag{2.5.1}$$

is said to be linear.

Consider now the linear ordinary differential equation

$$x\frac{dy}{dx} - 4y = x^6 e^x \tag{2.5.2}$$

or

$$\frac{dy}{dx} - \frac{4}{x}y = x^5 e^x. \tag{2.5.3}$$

Let us now multiply (2.5.3) by x^{-4}. (How we knew that it should be x^{-4} and not something else will be addressed shortly.) This magical factor is called an *integrating factor* because (2.5.3) can be rewritten

$$\frac{1}{x^4}\frac{dy}{dx} - \frac{4}{x^5}y = xe^x, \tag{2.5.4}$$

or

$$\frac{d}{dx}\left(\frac{y}{x^4}\right) = xe^x. \tag{2.5.5}$$

Thus, our introduction of the integrating factor x^{-4} allows us to use the differentiation product rule in reverse and collapse the right side of (2.5.4) into a single x derivative of a function of x times y. If we had selected the incorrect integrating factor, the right side would not have collapsed into this useful form.

With (2.5.5), we may integrate both sides and find that

$$\frac{y}{x^4} = \int xe^x \, dx + C, \tag{2.5.6}$$

or

$$\frac{y}{x^4} = (x-1)e^x + C, \tag{2.5.7}$$

or

$$y = x^4(x-1)e^x + Cx^4. \tag{2.5.8}$$

From this example, it is clear that finding the integrating factor is crucial to solving first-order, linear, ordinary differential equations. To do this, let us first rewrite (2.5.1) by dividing through by $a_1(x)$ so that it becomes

$$\frac{dy}{dx} + P(x)y = Q(x), \tag{2.5.9}$$

or

$$dy + [P(x)y - Q(x)] \, dx = 0. \tag{2.5.10}$$

If we denote the integrating factor by $\mu(x)$, then

$$\mu(x)dy + \mu(x)[P(x)y - Q(x)] \, dx = 0. \tag{2.5.11}$$

Clearly, we can solve (2.5.11) by direct integration if it is an exact equation. If this is true, then

$$\frac{\partial \mu}{\partial x} = \frac{\partial}{\partial y} \{\mu(x)[P(x)y - Q(x)]\}, \tag{2.5.12}$$

or

$$\frac{d\mu}{dx} = \mu(x)P(x), \qquad \text{and} \qquad \frac{d\mu}{\mu} = P(x) \, dx. \tag{2.5.13}$$

Integrating (2.5.13),

$$\mu(x) = \exp\left[\int^x P(\xi) \, d\xi \right]. \tag{2.5.14}$$

Note that we do not need a constant of integration in (2.5.14) because (2.5.11) is unaffected by a constant multiple. It is also interesting that the integrating factor only depends on $P(x)$ and not $Q(x)$.

We can summarize our findings in the following theorem.

Theorem: Linear First-Order Equation

If the functions $P(x)$ and $Q(x)$ are continuous on the open interval I containing the point x_0, then the initial-value problem

$$\frac{dy}{dx} + P(x)y = Q(x), \qquad y(x_0) = y_0,$$

has a unique solution $y(x)$ on I, given by

$$y(x) = \frac{C}{\mu(x)} + \frac{1}{\mu(x)} \int^x Q(\xi)\mu(\xi)\,d\xi$$

with an appropriate value of C and $\mu(x)$ is defined by (2.5.14).

The procedure for implementing this theorem is as follows:

- **Step 1:** If necessary, divide the differential equation by the coefficient of dy/dx. This gives an equation of the form (2.5.9) and we can find $P(x)$ by inspection.

- **Step 2:** Find the integrating factor by (2.5.14).

- **Step 3:** Multiply the equation created in Step 1 by the integrating factor.

- **Step 4:** Run the derivative product rule in reverse, collapsing the left side of the differential equation into the form $d[\mu(x)y]/dx$. If you are unable to do this, you have made a mistake.

- **Step 5:** Integrate both sides of the differential equation to find the solution.

The following examples illustrate the technique.

- **Example 2.5.1**

Let us solve the linear, first-order ordinary differential equation

$$xy' - y = 4x\ln(x). \tag{2.5.15}$$

We begin by dividing through by x to convert (2.5.15) into its canonical form. This yields

$$y' - \frac{1}{x}y = 4\ln(x). \tag{2.5.16}$$

From (2.5.16), we see that $P(x) = 1/x$. Consequently, from (2.5.14), we have that

$$\mu(x) = \exp\left[\int^x P(\xi)\,d\xi\right] = \exp\left(-\int^x \frac{d\xi}{\xi}\right) = \frac{1}{x}. \tag{2.5.17}$$

Multiplying (2.5.16) by the integrating factor, we find that

$$\frac{y'}{x} - \frac{y}{x^2} = \frac{4\ln(x)}{x}, \tag{2.5.18}$$

or

$$\frac{d}{dx}\left(\frac{y}{x}\right) = \frac{4\ln(x)}{x}. \tag{2.5.19}$$

Integrating both sides of (2.5.19),

$$\frac{y}{x} = 4\int \frac{\ln(x)}{x}dx = 2\ln^2(x) + C. \tag{2.5.20}$$

Multiplying (2.5.20) through by x yields the general solution

$$y = 2x\ln^2(x) + Cx. \tag{2.5.21}$$

Although it is nice to have a closed form solution, considerable insight can be gained by graphing the solution for a wide variety of initial conditions. To illustrate this, consider the MATLAB script

```
clear
% use symbolic toolbox to solve (2.5.15)
y = dsolve('x*Dy-y=4*x*log(x)','y(1) = c','x');
% take the symbolic version of the solution
%     and convert it into executable code
solution = inline(vectorize(y),'x','c');
close all; axes; hold on
% now plot the solution for a wide variety of initial conditions
x = 0.1:0.1:2;
for c = -2:4
if (c==-2) plot(x,solution(x,c),'.'); end
if (c==-1) plot(x,solution(x,c),'o'); end
if (c== 0) plot(x,solution(x,c),'x'); end
if (c== 1) plot(x,solution(x,c),'+'); end
if (c== 2) plot(x,solution(x,c),'*'); end
if (c== 3) plot(x,solution(x,c),'s'); end
if (c== 4) plot(x,solution(x,c),'d'); end
end
axis tight
xlabel('x','Fontsize',20); ylabel('y','Fontsize',20)
legend('c = -2','c = -1','c = 0','c = 1',...
       'c = 2','c = 3','c = 4'); legend boxoff
```

This script does two things. First, it uses MATLAB's symbolic toolbox to solve (2.5.15). Alternatively we could have used (2.5.21) and introduced it as a function. The second portion of this script plots this solution for $y(1) = C$ where $C = -2, -1, 0, 1, 2, 3, 4$. Figure 2.5.1 shows the results. As $x \to 0$, we note how all of the solutions behave like $2x\ln^2(x)$.

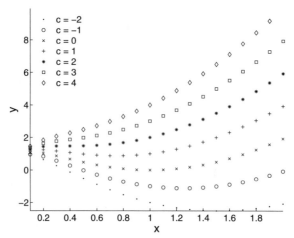

Figure 2.5.1: The solution to (2.5.15) when the initial condition is $y(1) = c$.

• **Example 2.5.2**

Let us solve the first-order ordinary differential equation

$$\frac{dy}{dx} = \frac{y}{y - x} \tag{2.5.22}$$

subject to the initial condition $y(2) = 6$.

Beginning as before, we rewrite (2.5.22) in the canonical form

$$(y - x)y' - y = 0. \tag{2.5.23}$$

Examining (2.5.23) more closely, we see that it is a nonlinear equation in y. On the other hand, if we treat x as the *dependent* variable and y as the *independent variable*, we can write (2.5.23) as the *linear* equation

$$\frac{dx}{dy} + \frac{x}{y} = 1. \tag{2.5.24}$$

Proceeding as before, we have that $P(y) = 1/y$ and $\mu(y) = y$ so that (2.5.24) can be rewritten

$$\frac{d}{dy}(yx) = y \tag{2.5.25}$$

or

$$yx = \tfrac{1}{2}y^2 + C. \tag{2.5.26}$$

Introducing the initial condition, we find that $C = -6$. Solving for y, we obtain

$$y = x \pm \sqrt{x^2 + 12}. \tag{2.5.27}$$

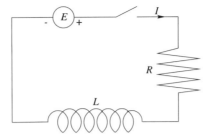

Figure 2.5.2: Schematic diagram for an electric circuit that contains a resistor of resistance R and an inductor of inductance L.

We must take the positive sign in order that $y(2) = 6$ and

$$y = x + \sqrt{x^2 + 12}. \qquad (2.5.28)$$

• Example 2.5.3: Electric circuits

A rich source of first-order differential equations is the analysis of simple electrical circuits. These electrical circuits are constructed from three fundamental components: the resistor, the inductor, and the capacitor. Each of these devices gives the following voltage drop: In the case of a resistor, the voltage drop equals the product of the resistance R times the current I. For the inductor, the voltage drop is $L\, dI/dt$, where L is called the inductance, while the voltage drop for a capacitor equals Q/C, where Q is the instantaneous charge and C is called the capacitance.

How are these voltage drops applied to mathematically describe an electrical circuit? This question leads to one of the fundamental laws in physics, **Kirchhoff's law**: *The algebraic sum of all the voltage drops around an electric loop or circuit is zero.*

To illustrate Kirchhoff's law, consider the electrical circuit shown in Figure 2.5.2. By Kirchhoff's law, the electromotive force E, provided by a battery, for example, equals the sum of the voltage drops across the resistor RI and $L\, dI/dt$. Thus the (differential) equation that governs this circuit is

$$L\frac{dI}{dt} + RI = E. \qquad (2.5.29)$$

Assuming that E, I, and R are constant, we can rewrite (2.5.29) as

$$\frac{d}{dt}\left[e^{Rt/L} I(t)\right] = \frac{E}{L} e^{Rt/L}. \qquad (2.5.30)$$

Integrating both sides of (2.5.30),

$$e^{Rt/L} I(t) = \frac{E}{R} e^{Rt/L} + C_1, \qquad (2.5.31)$$

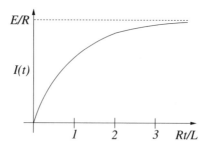

Figure 2.5.3: The temporal evolution of current $I(t)$ inside an electrical circuit shown in Figure 2.5.2 with a constant electromotive force E.

or

$$I(t) = \frac{E}{R} + C_1 e^{-Rt/L}. \tag{2.5.32}$$

To determine C_1, we apply the initial condition. Because the circuit is initially dead, $I(0) = 0$, and

$$I(t) = \frac{E}{R}\left(1 - e^{-Rt/L}\right). \tag{2.5.33}$$

Figure 2.5.3 illustrates (2.5.33) as a function of time. Initially the current increases rapidly but the growth slows with time. Note that we could also have solved this problem by separation of variables.

Quite often, the solution is separated into two parts: the *steady-state solution* and the *transient solution*. The steady-state solution is that portion of the solution which remains as $t \to \infty$. It can equal zero. Presently it equals the constant value, E/R. The transient solution is that portion of the solution that vanishes as time increases. Here it equals $-Ee^{-Rt/L}/R$.

Although our analysis is a useful approximation to the real world, a more realistic one would include the nonlinear properties of the resistor.[9] To illustrate this, consider the case of a RL circuit without any electromotive source $(E = 0)$ where the initial value for the current is I_0. Equation (2.5.29) now reads

$$L\frac{dI}{dt} + RI(1 - aI) = 0, \qquad I(0) = I_0. \tag{2.5.34}$$

Separating the variables,

$$\frac{dI}{I(aI-1)} = \frac{dI}{I - 1/a} - \frac{dI}{I} = \frac{R}{L}dt. \tag{2.5.35}$$

---

[9] For the analysis of

$$L\frac{dI}{dt} + RI + KI^\beta = 0,$$

see Fairweather, A., and J. Ingham, 1941: Subsidence transients in circuits containing a non-linear resistor, with reference to the problem of spark-quenching. *J. IEE, Part 1*, **88**, 330–339.

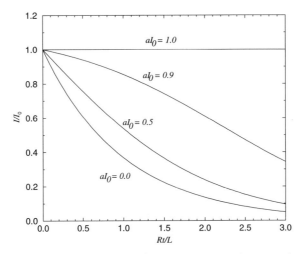

Figure 2.5.4: The variation of current I/I_0 as a function of time Rt/L with different values of aI_0.

Upon integrating and applying the initial condition, we have that

$$I = \frac{I_0 e^{-Rt/L}}{1 - aI_0 + aI_0 e^{-Rt/L}}. \tag{2.5.36}$$

Figure 2.5.4 shows $I(t)$ for various values of a. As the nonlinearity reduces resistance, the decay in the current is reduced. If $aI_0 > 1$, (2.5.36) predicts that the current would grow with time. The point here is that nonlinearity can have a dramatic influence on a physical system.

Consider now the electrical circuit shown in Figure 2.5.5 which contains a resistor with resistance R and a capacitor with capacitance C. Here the voltage drop across the resistor is still RI while the voltage drop across the capacitor is Q/C. Therefore, by Kirchhoff's law,

$$RI + \frac{Q}{C} = E. \tag{2.5.37}$$

Equation (2.5.37) is *not* a differential equation. However, because current is the time rate of change in charge $I = dQ/dt$, (2.5.37) becomes

$$R\frac{dQ}{dt} + \frac{Q}{C} = E, \tag{2.5.38}$$

which is the differential equation for the instantaneous charge.

Let us solve (2.5.38) when the resistance and capacitance is constant but the electromotive force equals $E_0 \cos(\omega t)$. The corresponding differential equation is now

$$R\frac{dQ}{dt} + \frac{Q}{C} = E_0 \cos(\omega t). \tag{2.5.39}$$

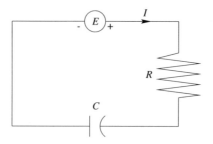

Figure 2.5.5: Schematic diagram for an electric circuit that contains a resistor of resistance R and a capacitor of capacitance C.

The differential equation has the integrating factor $e^{t/(RC)}$ so that it can be rewritten

$$\frac{d}{dt}\left[e^{t/(RC)}Q(t)\right] = \frac{E_0}{R}e^{t/(RC)}\cos(\omega t). \qquad (2.5.40)$$

Integrating (2.5.40),

$$e^{t/(RC)}Q(t) = \frac{CE_0}{1+R^2C^2\omega^2}e^{t/(RC)}\left[\cos(\omega t) + RC\omega\sin(\omega t)\right] + C_1 \qquad (2.5.41)$$

or

$$Q(t) = \frac{CE_0}{1+R^2C^2\omega^2}\left[\cos(\omega t) + RC\omega\sin(\omega t)\right] + C_1e^{-t/(RC)}. \qquad (2.5.42)$$

If we take the initial condition as $Q(0) = 0$, then the final solution is

$$Q(t) = \frac{CE_0}{1+R^2C^2\omega^2}\left[\cos(\omega t) - e^{-t/(RC)} + RC\omega\sin(\omega t)\right]. \qquad (2.5.43)$$

Figure 2.5.6 illustrates (2.5.43). Note how the circuit eventually supports a purely oscillatory solution (the steady-state solution) as the exponential term decays to zero (the transient solution). Indeed the purpose of the transient solution is to allow the system to adjust from its initial condition to the final steady state.

• **Example 2.5.4: Terminal velocity**

When an object passes through a fluid, the viscosity of the fluid resists the motion by exerting a force on the object proportional to its velocity. Let us find the motion of a mass m that is initially thrown upward with the speed v_0.

If we choose the coordinate system so that it increases in the vertical direction, then the equation of motion is

$$m\frac{dv}{dt} = -kv - mg \qquad (2.5.44)$$

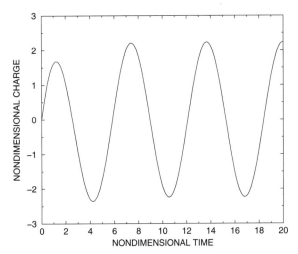

Figure 2.5.6: The temporal evolution of the nondimensional charge $(1 + R^2 C^2 \omega^2) Q(t)$ $/(C E_0)$ in the electric circuit shown in Figure 2.5.4 as a function of nondimensional time ωt when the circuit is driven by the electromotive force $E_0 \cos(\omega t)$ and $RC\omega = 2$.

with $v(0) = v_0$ and $k > 0$. Rewriting (2.5.44), we obtain the first-order linear differential equation

$$\frac{dv}{dt} + \frac{k}{m} v = -g. \tag{2.5.45}$$

Its solution in *nondimensional* form is

$$\frac{kv(t)}{mg} = -1 + \left(1 + \frac{kv_0}{mg}\right) e^{-kt/m}. \tag{2.5.46}$$

The displacement from its initial position is

$$\frac{k^2 x(t)}{m^2 g} = \frac{k^2 x_0}{m^2 g} - \frac{kt}{m} + \left(1 + \frac{kv_0}{mg}\right)\left(1 - e^{-kt/m}\right). \tag{2.5.47}$$

As $t \to \infty$, the velocity tends to a constant downward value, $-mg/k$, the so-called "terminal velocity," where the aerodynamic drag balances the gravitational acceleration. This is the steady-state solution.

Why have we written (2.5.46)–(2.5.47) in this nondimensional form? There are two reasons. First, the solution reduces to three fundamental variables, a *nondimensional* displacement $x_* = k^2 x(t)/(m^2 g)$, velocity $v_* = kv(t)/(mg)$, and time $t_* = kt/m$, rather than the six original parameters and variables: g, k, m, t, v, and x. Indeed, if we had substituted t_*, v_*, and x_* into (2.5.45), we would have obtained the following simplified initial-value problem:

$$\frac{dv_*}{dt_*} + v_* = -1, \quad \frac{dx_*}{dt_*} = v_*, \quad v_*(0) = \frac{kv_0}{mg}, \quad x_*(0) = \frac{k^2 x_0}{m^2 g} \tag{2.5.48}$$

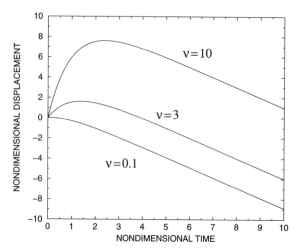

Figure 2.5.7: The nondimensional displacement $k^2 x(t)/(m^2 g)$ as a function of nondimensional time kt/m of an object of mass m thrown upward at the initial nondimensional speed $\nu = kv_0/(mg)$ in a fluid which retards its motion as $-kv$.

right from the start. The second advantage of the nondimensional form is the compact manner in which the results can be displayed as Figure 2.5.7 shows.

From (2.5.46)–(2.5.47), the trajectory of the ball is as follows: If we define the coordinate system so that $x_0 = 0$, then the object will initially rise to the height H given by

$$\frac{k^2 H}{m^2 g} = \frac{kv_0}{mg} - \ln\left(1 + \frac{kv_0}{mg}\right) \tag{2.5.49}$$

at the time

$$\frac{kt_{max}}{m} = \ln\left(1 + \frac{kv_0}{mg}\right), \tag{2.5.50}$$

when $v(t_{max}) = 0$. It will then fall toward the earth. Given sufficient time $kt/m \gg 1$, it would achieve terminal velocity.

• Example 2.5.5: The Bernoulli equation

Bernoulli's equation,

$$\frac{dy}{dx} + p(x)y = q(x)y^n, \qquad n \neq 0, 1, \tag{2.5.51}$$

is a first-order, nonlinear differential equation. This equation can be transformed into a first-order, linear differential equation by introducing the change of variable $z = y^{1-n}$. Because

$$\frac{dz}{dx} = (1 - n)y^{-n}\frac{dy}{dx}, \tag{2.5.52}$$

the transformed Bernoulli equation becomes

$$\frac{dz}{dx} + (1 - n)p(x)z = (1 - n)q(x). \qquad (2.5.53)$$

This is now a first-order linear differential equation for z and can be solved using the methods introduced in this section. Once z is known, the solution is found by transforming back from z to y.

To illustrate this procedure, consider the nonlinear ordinary differential equation

$$x^2 y \frac{dy}{dx} - xy^2 = 1, \qquad (2.5.54)$$

or

$$\frac{dy}{dx} - \frac{y}{x} = \frac{y^{-1}}{x^2}. \qquad (2.5.55)$$

Equation (2.5.55) is a Bernoulli equation with $p(x) = -1/x$, $q(x) = 1/x^2$, and $n = -1$. Introducing $z = y^2$, it becomes

$$\frac{dz}{dx} - \frac{2z}{x} = \frac{2}{x^2}. \qquad (2.5.56)$$

This first-order linear differential equation has the integrating factor $\mu(x) = 1/x^2$ and

$$\frac{d}{dx}\left(\frac{z}{x^2}\right) = \frac{2}{x^4}. \qquad (2.5.57)$$

Integration gives

$$\frac{z}{x^2} = C - \frac{2}{3x^3}. \qquad (2.5.58)$$

Therefore, the general solution is

$$y^2 = z = Cx^2 - \frac{2}{3x}. \qquad (2.5.59)$$

Problems

Find the solution for the following differential equations. State the interval on which the general solution is valid. Then use MATLAB to examine their behavior for a wide class of initial conditions.

1. $y' + y = e^x$

2. $y' + 2xy = x$

3. $x^2 y' + xy = 1$

4. $(2y + x^2)\,dx = x\,dy$

5. $y' - 3y/x = 2x^2$

6. $y' + 2y = 2\sin(x)$

7. $y' + 2\cos(2x)y = 0$

8. $xy' + y = \ln(x)$

9. $y' + 3y = 4, \quad y(0) = 5$ 10. $y' - y = e^x/x, \quad y(e) = 0$

11. $\sin(x)y' + \cos(x)y = 1$ 12. $[1 - \cos(x)]y' + 2\sin(x)y = \tan(x)$

13. $y' + [a\tan(x) + b\sec(x)]y = c\sec(x)$

14. $(xy + y - 1)\,dx + x\,dy = 0$

15. $y' + 2ay = \dfrac{x}{2} - \dfrac{\sin(2\omega x)}{4\omega}, \qquad y(0) = 0.$

16. $y' + \dfrac{2k}{x^3}y = \ln\left(\dfrac{x+1}{x}\right), \qquad k > 0, \qquad y(1) = 0.$

17. Solve the following initial-value problem:

$$kxy\frac{dy}{dx} = y^2 - x^2, \qquad y(1) = 0.$$

Hint: Introduce the new dependent variable $p = y^2$.

18. If $x(t)$ denotes the equity capital of a company, then under certain assumptions[10] $x(t)$ is governed by

$$\frac{dx}{dt} = (1 - N)rx + S,$$

where N is the dividend payout ratio, r is the rate of return of equity, and S is the rate of net new stock financing. If the initial value of $x(t)$ is $x(0)$, find $x(t)$.

19. The assimilation[11] of a drug into a body can be modeled by the chemical reaction A $\xrightarrow{k_1}$ B $\xrightarrow{k_2}$ C which is governed by the chemical kinetics equations

$$\frac{d[A]}{dt} = -k_1[A], \qquad \frac{d[B]}{dt} = k_1[A] - k_2[B], \qquad \frac{d[C]}{dt} = k_2[B],$$

where [A] denotes the concentration of the drug in the gastrointestinal tract or in the site of injection, [B] is the concentration of the drug in the body, and [C] is either the amount of drug eliminated by various metabolic functions or the amount of the drug utilized by various action sites in the body. If $[A]_0$

---

[10] See Lebowitz, J. L., C. O. Lee, and P. B. Linhart, 1976: Some effects of inflation on a firm with original cost depreciation. *Bell J. Economics*, **7**, 463–477. Reprinted by permission of RAND. Copyright ©1976.

[11] See Calder, G. V., 1974: The time evolution of drugs in the body: An application of the principle of chemical kinetics. *J. Chem. Educ.*, **51**, 19–22.

denotes the initial concentration of A, find [A], [B], and [C] as a function of time t.

20. Find the current in a RL circuit when the electromotive source equals $E_0 \cos^2(\omega t)$. Initially the circuit is dead.

Find the general solution for the following Bernoulli equations:

21. $\dfrac{dy}{dx} + \dfrac{y}{x} = -y^2$

22. $x^2 \dfrac{dy}{dx} = xy + y^2$

23. $\dfrac{dy}{dx} - \dfrac{4y}{x} = x\sqrt{y}$

24. $\dfrac{dy}{dx} + \dfrac{y}{x} = -xy^2$

25. $2xy\dfrac{dy}{dx} - y^2 + x = 0$

26. $x\dfrac{dy}{dx} + y = \frac{1}{2}xy^3$

2.6 GRAPHICAL SOLUTIONS

In spite of the many techniques developed for their solution, many ordinary differential equations cannot be solved analytically. In the next two sections, we highlight two alternative methods when analytical methods fail. Graphical methods seek to understand the nature of the solution by examining the differential equations at various points and infer the complete solution from these results. In the last section, we highlight the numerical techniques that are now commonly used to solve ordinary differential equations on the computer.

• *Direction fields*

One of the simplest numerical methods for solving first-order ordinary differential equations follows from the fundamental concept that the derivative gives the *slope* of a straight line that is tangent to a curve at a given point. Consider the first-order differential equation

$$y' = f(x, y) \tag{2.6.1}$$

which has the initial value $y(x_0) = y_0$. For any (x, y) it is possible to draw a short line segment whose slope equals $f(x, y)$. This graphical representation is known as the *direction field* or *slope field* of (2.6.1). Starting with the initial point (x_0, y_0), we can then construct the *solution curve* by extending the initial line segment in such a manner that the tangent of the solution curve parallels the direction field at each point through which the curve passes.

Before the days of computers, it was common to first draw lines of constant slope (*isoclines*) or $f(x, y) = c$. Because along any isocline all of the line segments had the same slope, considerable computational savings were realized. Today, computer software exists which perform these graphical computations with great speed.

To illustrate this technique, consider the ordinary differential equation

$$\frac{dx}{dt} = x - t^2. \tag{2.6.2}$$

Its exact solution is

$$x(t) = Ce^t + t^2 + 2t + 2, \tag{2.6.3}$$

where C is an arbitrary constant. Using the MATLAB script

```
clear
% create grid points in t and x
[t,x] = meshgrid(-2:0.2:3,-1:0.2:2);
% load in the slope
slope = x - t.*t;
% find the length of the vector (1,slope)
length = sqrt(1 + slope .* slope);
% create and plot the vector arrows
quiver(t,x,1./length,slope./length,0.5)
axis equal tight
hold on
% plot the exact solution for various initial conditions
tt = [-2:0.2:3];
for cval = -10:1:10
x_exact = cval * exp(tt) + tt.*tt + 2*tt + 2;
plot(tt,x_exact)
xlabel('t','Fontsize',20)
ylabel('x','Fontsize',20)
end
```

we show in Figure 2.6.1 the directional field associated with (2.6.2) along with some of the particular solutions. Clearly the vectors are parallel to the various particular solutions. Therefore, without knowing the solution, we could choose an arbitrary initial condition and sketch its behavior at subsequent times. The same holds true for nonlinear equations.

• *Rest points and autonomous equations*

In the case of autonomous differential equations (equations where the independent variable does not explicitly appear in the equation), considerable information can be gleaned from a graphical analysis of the equation.

Consider the nonlinear ordinary differential equation

$$x' = \frac{dx}{dt} = x(x^2 - 1). \tag{2.6.4}$$

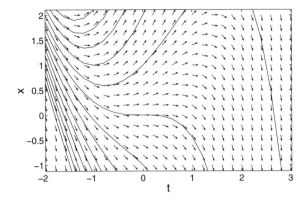

Figure 2.6.1: The direction field for (2.6.2). The solid lines are plots of the solution with various initial conditions.

The time derivative x' vanishes at $x = -1, 0, 1$. Consequently, if $x(0) = 0$, $x(t)$ will remain zero forever. Similarly, if $x(0) = 1$ or $x(0) = -1$, then $x(t)$ will equal 1 or -1 for all time. For this reason, values of x for which the derivative x' is zero are called *rest points, equilibrium points*, or *critical points* of the differential equation.

The behavior of solutions near rest points is often of considerable interest. For example, what happens to the solution when x is near one of the rest points $x = -1, 0, 1$?

Consider the point $x = 0$. For x slightly greater than zero, $x' < 0$. For x slightly less than 0, $x' > 0$. Therefore, for any initial value of x near $x = 0$, x will tend to zero. In this case, the point $x = 0$ is an asymptotically *stable critical point* because whenever x is perturbed away from the critical point, it tends to return there again.

Turning to the point $x = 1$, for x slightly greater than 1, $x' > 0$; for x slightly less than 1, $x' < 0$. Because any x near $x = 1$, but not equal to 1, will move away from $x = 1$, the point $x = 1$ is called an *unstable critical point*. A similar analysis applies at the point $x = -1$. This procedure of determining the behavior of an ordinary differential equation near its critical points is called a *graphical stability analysis*.

- *Phase line*

A graphical representation of the results of our graphical stability analysis is the *phase line*. On a phase line, the equilibrium points are denoted by circles. See Figure 2.6.2. Also on the phase line we identify the sign of x' for all values of x. From the sign of x', we then indicate whether x is increasing or deceasing by an appropriate arrow. If the arrow points toward the right, x is increasing; toward the left x decreases. Then, by knowing the *sign* of the derivative for all values of x, together with the starting value of x, we can determine what happens as $t \to \infty$. Any solution that is approached

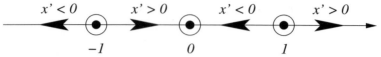

Figure 2.6.2: The phase line diagram for the ordinary differential equation (2.6.4).

asymptotically as $t \to \infty$ is called a *steady-state output*. In our present example, $x = 0$ is a steady-state output.

Problems

In previous sections, you used various techniques to solve first-order ordinary differential equations. Now check your work by using MATLAB to draw the direction field and plot your analytic solution for the following problems taken from previous sections:

1. §2.2, Problem 5   2. §2.3, Problem 1

3. §2.4, Problem 5   4. §2.5, Problem 3

For the following autonomous ordinary differential equations, draw the phase line. Then classify each equilibrium solution as either stable or unstable.

5. $x' = \alpha x(1 - x)(x - \frac{1}{2})$ 6. $x' = (x^2 - 1)(x^2 - 4)$

7. $x' = -4x - x^3$ 8. $x' = 4x - x^3$

2.7 NUMERICAL METHODS

By now you have seen most of the exact methods for finding solutions to first-order ordinary differential equations. The methods have also given you a view of the general behavior and properties of solutions to differential equations. However, it must be admitted that in many instances exact solutions cannot be found and we must resort to numerical solutions.

In this section we present the two most commonly used methods for solving differential equations: Euler and Runge-Kutta methods. There are many more methods and the interested student is referred to one of countless numerical methods books. A straightforward extension of these techniques can be applied to systems of first-order and higher-order differential equations.

• *Euler and modified Euler methods*

Consider the following first-order differential equation and initial condition:

$$\frac{dy}{dx} = f(x, y), \qquad y(x_0) = y_0. \tag{2.7.1}$$

Euler's method is based on a Taylor series expansion of the solution about x_0 or

$$y(x_0 + h) = y(x_0) + hy'(x_0) + \tfrac{1}{2}y''(\xi)h^2, \qquad x_0 < \xi < x_0 + h, \qquad (2.7.2)$$

where h is the step size. Euler's method consists of taking a sufficiently small h so that only the first two terms of this Taylor expansion are significant.

Let us now replace $y'(x_0)$ by $f(x_0, y_0)$. Using subscript notation, we have that

$$y_{i+1} = y_i + hf(x_i, y_i) + O(h^2). \qquad (2.7.3)$$

Equation (2.7.3) states that if we know the values of y_i and $f(x_i, y_i)$ at the position x_i, then the solution at x_{i+1} can be obtained with an error[12] $O(h^2)$.

The trouble with Euler's method is its lack of accuracy, often requiring an extremely small time step. How might we improve this method with little additional effort?

One possible method would retain the first three terms of the Taylor expansion rather than the first two. This scheme, known as the *modified Euler method*, is

$$y_{i+1} = y_i + hy'(x_i) + \tfrac{1}{2}h^2 y_i'' + O(h^3). \qquad (2.7.4)$$

This is clearly more accurate than (2.7.3).

An obvious question is how do we evaluate y_i'' because we do not have any information on its value? Using the forward derivative approximation, we find that

$$y_i'' = \frac{y_{i+1}' - y_i'}{h}. \qquad (2.7.5)$$

Substituting (2.7.5) into (2.7.4) and simplifying

$$y_{i+1} = y_i + \frac{h}{2}\left(y_i' + y_{i+1}'\right) + O(h^3). \qquad (2.7.6)$$

Using the differential equation,

$$y_{i+1} = y_i + \frac{h}{2}\left[f(x_i, y_i) + f(x_{i+1}, y_{i+1})\right] + O(h^3). \qquad (2.7.7)$$

Although $f(x_i, y_i)$ at (x_i, y_i) are easily calculated, how do we compute $f(x_{i+1}, y_{i+1})$ at (x_{i+1}, y_{i+1})? For this we compute a first guess via the Euler method (2.7.3); equation (2.7.7) then provides a refinement on the value of y_{i+1}.

In summary then, the simple Euler scheme is

$$y_{i+1} = y_i + k_1 + O(h^2), \qquad k_1 = hf(x_i, y_i). \qquad (2.7.8)$$

---

[12] The symbol O is a mathematical notation indicating relative magnitude of terms, namely that $f(\epsilon) = O(\epsilon^n)$ provided $\lim_{\epsilon \to 0} |f(\epsilon)/\epsilon^n| < \infty$. For example, as $\epsilon \to 0$, $\sin(\epsilon) = O(\epsilon)$, $\sin(\epsilon^2) = O(\epsilon^2)$, and $\cos(\epsilon) = O(1)$.

while the modified Euler method is

$$y_{i+1} = y_i + \tfrac{1}{2}(k_1 + k_2) + O(h^3), \ \ k_1 = hf(x_i, y_i), \ \ k_2 = hf(x_i + h, y_i + k_1).$$
$$\textbf{(2.7.9)}$$

• **Example 2.7.1**

Let us illustrate Euler's method by numerically solving

$$x' = x + t, \qquad x(0) = 1. \tag{2.7.10}$$

A quick check shows that (2.7.10) has the exact solution $x_{\text{exact}}(t) = 2e^t - t - 1$. Using the MATLAB script

```
clear
for i = 1:3
% set up time step increment and number of time steps
h = 1/10^i; n = 10/h;
% set up initial conditions
t=zeros(n+1,1); t(1) = 0;
x_euler=zeros(n+1,1); x_euler(1) = 1;
x_modified=zeros(n+1,1); x_modified(1) = 1;
x_exact=zeros(n+1,1); x_exact(1) = 1;
% set up difference arrays for plotting purposes
diff1 = zeros(n,1); diff2 = zeros(n,1); tplot = zeros(n,1);
% define right side of differential equation (2.7.10)
f = inline('xx+tt','tt','xx');
for k = 1:n
t(k+1) = t(k) + h;
% compute exact solution
x_exact(k+1) = 2*exp(t(k+1)) - t(k+1) - 1;
% compute solution via Euler's method
k1 = h * f(t(k),x_euler(k));
x_euler(k+1) = x_euler(k) + k1;
tplot(k) = t(k+1);
diff1(k) = x_euler(k+1) - x_exact(k+1);
diff1(k) = abs(diff1(k) / x_exact(k+1));
% compute solution via modified Euler method
k1 = h * f(t(k),x_modified(k));
k2 = h * f(t(k+1),x_modified(k)+k1);
x_modified(k+1) = x_modified(k) + 0.5 * (k1+k2);
diff2(k) = x_modified(k+1) - x_exact(k+1);
diff2(k) = abs(diff2(k) / x_exact(k+1));
end
% plot relative errors
semilogy(tplot,diff1,'-',tplot,diff2,':')
```

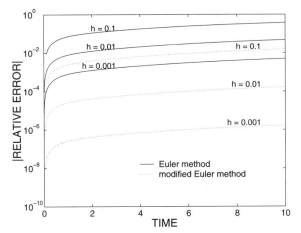

Figure 2.7.1: The relative error $[x(t) - x_{\text{exact}}(t)]/x_{\text{exact}}(t)$ of the numerical solution of (2.7.10) using Euler's method (the solid line) and modified Euler's method (the dotted line) with different time steps h.

```
hold on
xlabel('TIME','Fontsize',20)
ylabel('|RELATIVE ERROR|','Fontsize',20)
legend('Euler method','modified Euler method')
legend boxoff;
num1 = 0.2*n; num2 = 0.8*n;
text(3,diff1(num1),['h = ',num2str(h)],'Fontsize',15,...
    'HorizontalAlignment','right',...
    'VerticalAlignment','bottom')
text(9,diff2(num2),['h = ',num2str(h)],'Fontsize',15,...
    'HorizontalAlignment','right',...
    'VerticalAlignment','bottom')
end
```

Both the Euler and modified Euler methods have been used to numerically integrate (2.7.10) and the absolute value of the relative error is plotted in Figure 2.7.1 as a function of time for various time steps. In general, the error grows with time. The decrease of error with smaller time steps, as predicted in our analysis, is quite apparent. Furthermore, the superiority of the modified Euler method over the original Euler method is clearly seen.

• *Runge-Kutta method*

As we have just shown, the accuracy of numerical solutions of ordinary differential equations can be improved by adding more terms to the Taylor expansion. The Runge-Kutta method[13] builds upon this idea, just as the

---

[13] Runge, C., 1895: Ueber die numerische Auflösung von Differentialgleichungen. *Math.*

Figure 2.7.2: Although Carl David Tolmé Runge (1856–1927) began his studies in Munich, his friendship with Max Planck led him to Berlin and pure mathematics with Kronecker and Weierstrass. It was during his professorship at Hanover beginning in 1886 and subsequent work in spectroscopy that led him to his celebrated paper on the numerical integration of ordinary differential equations. Runge's final years were spent in Göttingen as a professor in applied mathematics. (Protrait taken with permission from Reid, C., 1976: *Courant in Göttingen and New York: The Story of an Improbable Mathematician.* Springer-Verlag, 314 pp. ©1976, by Springer-Verlag New York Inc.)

modified Euler method did.

Let us assume that the numerical solution can be approximated by

$$y_{i+1} = y_i + ak_1 + bk_2, \qquad (2.7.11)$$

where

$$k_1 = hf(x_i, y_i) \quad \text{and} \quad k_2 = hf(x_i + A_1h, y_i + B_1k_1). \qquad (2.7.12)$$

Here a, b, A_1, and B_1 are four unknowns. Equation (2.7.11) was suggested by the modified Euler method that we just presented. In that case, the truncated Taylor series had an error of $O(h^3)$. We anticipate such an error in the present case.

Ann., **46**, 167–178; Kutta, W., 1901: Beitrag zur Näherungsweisen Integration totaler Differentialgleichungen. *Zeit. Math. Phys.*, **46**, 435–453. For a historical review, see Butcher, J. C., 1996: A history of Runge-Kutta methods. *Appl. Numer. Math.*, **20**, 247–260 and Butcher, J. C., and G. Wanner, 1996: Runge-Kutta methods: Some historical notes. *Appl. Numer. Math.*, **22**, 113–151.

Because the Taylor series expansion of $f(x+h, y+k)$ about (x, y) is

$$f(x+h, y+k) = f(x, y) + (hf_x + kf_y) + \tfrac{1}{2}\left(h^2 f_{xx} + 2hk f_{xy} + k^2 f_{yy}\right)$$
$$+ \tfrac{1}{6}\left(h^3 f_{xxx} + 3h^2 k f_{xxy} + 3hk^2 f_{xyy} + k^3 f_{yyy}\right) + \cdots,$$

$$(2.7.13)$$

k_2 can be rewritten

$$k_2 = hf[x_i + A_1 h, y_i + Bhf(x_i, y_i)] \tag{2.7.14}$$
$$= h\left[f(x_i, y_i) + (A_1 h f_x + B_1 h f f_y)\right] \tag{2.7.15}$$
$$= hf + A_1 h^2 f_x + B_1 h^2 f f_y, \tag{2.7.16}$$

where we have retained only terms up to $O(h^2)$ and neglected all higher-order terms. Finally, substituting (2.7.16) into (2.7.11) gives

$$y_{i+1} = y_i + (a+b)hf + (A_1 b f_x + B_1 b f f_y)h^2. \tag{2.7.17}$$

This equation corresponds to the second-order Taylor expansion:

$$y_{i+1} = y_i + hy_i' + \tfrac{1}{2}h^2 y_i''. \tag{2.7.18}$$

Therefore, if we wish to solve the differential equation $y' = f(x, y)$, then

$$y'' = f_x + f_y y' = f_x + f f_y. \tag{2.7.19}$$

Substituting (2.7.19) into (2.7.18), we have that

$$y_{i+1} = y_i + hf + \tfrac{1}{2}h^2 (f_x + f f_y). \tag{2.7.20}$$

A direct comparison of (2.7.17) and (2.7.20) yields

$$a+b = 1, \qquad A_1 b = \tfrac{1}{2}, \quad \text{and} \quad B_1 b = \tfrac{1}{2}. \tag{2.7.21}$$

These three equations have four unknowns. If we choose $a = \tfrac{1}{2}$, we immediately calculate $b = \tfrac{1}{2}$ and $A_1 = B_1 = 1$. Hence the second-order Runge-Kutta scheme is

$$y_{i+1} = y_i + \tfrac{1}{2}(k_1 + k_2), \tag{2.7.22}$$

where $k_1 = hf(x_i, y_i)$ and $k_2 = hf(x_i + h, y_i + k_1)$. Thus, second-order Runge-Kutta scheme is identical to the modified Euler method.

Although the derivation of the second-order Runge-Kutta scheme yields the modified Euler scheme, it does provide a framework for computing higher-order and more accurate schemes. A particularly popular one is the fourth-order Runge-Kutta scheme

$$y_{i+1} = y_i + \tfrac{1}{6}(k_1 + 2k_2 + 2k_3 + k_4), \tag{2.7.23}$$

where

$$k_1 = hf(x_i, y_i), \qquad (2.7.24)$$

$$k_2 = hf(x_i + \tfrac{1}{2}h, y_i + \tfrac{1}{2}k_1), \qquad (2.7.25)$$

$$k_3 = hf(x_i + \tfrac{1}{2}h, y_i + \tfrac{1}{2}k_2), \qquad (2.7.26)$$

and

$$k_4 = hf(x_i + h, y_i + k_3). \qquad (2.7.27)$$

• Example 2.7.2

Let us illustrate the fourth-order Runge-Kutta by redoing the previous example using the MATLAB script

```
clear
% test out different time steps
for i = 1:4
% set up time step increment and number of time steps
if i==1 h = 0.50; end; if i==2 h = 0.10; end;
if i==3 h = 0.05; end; if i==4 h = 0.01; end;
n = 10/h;
% set up initial conditions
t=zeros(n+1,1); t(1) = 0;
x_rk=zeros(n+1,1); x_rk(1) = 1;
x_exact=zeros(n+1,1); x_exact(1) = 1;
% set up difference arrays for plotting purposes
diff = zeros(n,1); tplot = zeros(n,1);
% define right side of differential equation
f = inline('xx+tt','tt','xx');
for k = 1:n
x_local = x_rk(k); t_local = t(k);
k1 = h * f(t_local,x_local);
k2 = h * f(t_local + h/2,x_local + k1/2);
k3 = h * f(t_local + h/2,x_local + k2/2);
k4 = h * f(t_local + h,x_local + k3);
t(k+1) = t_local + h;
x_rk(k+1) = x_local + (k1+2*k2+2*k3+k4) / 6;
x_exact(k+1) = 2*exp(t(k+1)) - t(k+1) - 1;
tplot(k) = t(k);
diff(k) = x_rk(k+1) - x_exact(k+1);
diff(k) = abs(diff(k) / x_exact(k+1));
end
% plot relative errors
semilogy(tplot,diff,'-')
hold on
```

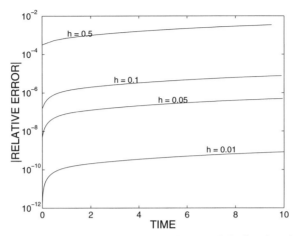

Figure 2.7.3: Same as Figure 2.7.1 except that we have used the fourth-order Runge-Kutta method.

```
xlabel('TIME','Fontsize',20)
ylabel('|RELATIVE ERROR|','Fontsize',20)
num1 = 2*i; num2 = 0.2*n;
text(num1,diff(num2),['h = ',num2str(h)],'Fontsize',15,...
    'HorizontalAlignment','right',...
    'VerticalAlignment','bottom')
end
```

The error growth with time is shown in Figure 2.7.3. Although this script could be used for any first-order ordinary differential equation, the people at MATLAB have an alternative called `ode45` which combines a fourth-order and a fifth-order method which are similar to our fourth-order Runge-Kutta method. Their scheme is more efficient because it varies the step size, choosing a new time step at each step in an attempt to achieve a given desired accuracy.

Problems

Using Euler's method for various values of $h = 10^{-n}$, find the numerical solution for the following initial-value problems. Check your answer by finding the exact solution:

1. $x' = x - t$, $\quad x(0) = 2$ $\qquad\qquad$ 2. $x' = tx$, $\quad x(0) = 1$

3. $x' = x^2/(t+1)$, $\quad x(0) = 1$ $\qquad$ 4. $x' = x + e^{-t}$, $\quad x(1) = 0$

5. Consider the integro-differential equation

$$\frac{dx}{dt} + \int_0^t x(\tau)\,d\tau + B\,\mathrm{sgn}(x)|x|^\beta = 1, \qquad B, \beta \geq 0,$$

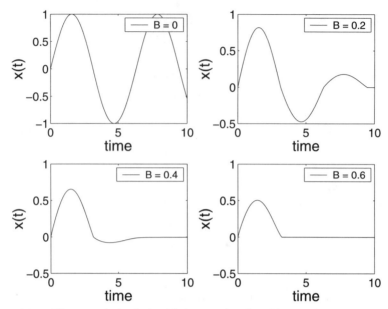

Figure 2.7.4: The numerical solution of the equation describing an electrical circuit with a nonlinear resistor. Here $\beta = 0.2$ and $\Delta t = 0.01$.

where the signum function is defined by (5.2.11). This equation describes the (nondimensional) current,[14] $x(t)$, within an electrical circuit that contains a capacitor, inductor, and nonlinear resistor. Assuming that the circuit is initially dead, x(0) = 0, write a MATLAB script that uses Euler's method to compute $x(t)$. Use a simple Riemann sum to approximate the integral. Examine the solution for various values of B and β as well as time step Δt.

[14] Monahan, T. F., 1960: Calculation of the current in non-linear surge-current-generation circuits. *Proc. IEE, Part C*, **107**, 288–291.

Chapter 3
Higher-Order Ordinary
Differential Equations

Although first-order ordinary differential equations exhibit most of the properties of differential equations, higher-order ordinary differential equations are more ubiquitous in the sciences and engineering. This chapter is devoted to the most commonly employed techniques for their solution.

A *linear* nth-order ordinary differential equation is a differential equation of the form

$$a_n(x)\frac{d^n y}{dx^n} + a_{n-1}(x)\frac{d^{n-1} y}{dx^{n-1}} + \cdots + a_1(x)\frac{dy}{dx} + a_0(x)y = f(x). \qquad (3.0.1)$$

If $f(x) = 0$, then (3.0.1) is said to be *homogeneous*; otherwise, it is *nonhomogeneous*. A linear differential equation is *normal* on an interval I if its coefficients and $f(x)$ are continuous, and the value of $a_n(x)$ is never zero on I.

Solutions to (3.0.1) generally must satisfy not only the differential equations but also certain specified conditions at one or more points. *Initial-value problems* are problems where *all* of the conditions are specified at a single point $x = a$ and have the form: $y(a) = b_0$, $y'(a) = b_1$, $y''(a) = b_2$, ..., $y^{(n-1)}(a) = b_{n-1}$, where b_0, b_1, b_2, ..., b_{n-1} are arbitrary constants. A quick check shows that if (3.0.1) is homogeneous and normal on an interval I and *all* of the initial conditions equal zero at the point $x = a$ which lies in I, then $y(x) \equiv 0$ on I. This follows because $y = 0$ is a solution of (3.0.1) and satisfies the initial conditions.

At this point a natural question would be whether the solution exists for this initial-value problem and, if so, how many? From a detailed study of this question,[1] we have the following useful theorem.

Theorem: Existence and Uniqueness

Suppose that the differential equation (3.0.1) is normal on the open interval I containing the point $x = a$. Then, given n numbers $b_0, b_1, \ldots, b_{n-1}$, the nth-order linear equation (3.0.1) has a unique solution on the entire interval I that satisfies the n initial conditions $y(a) = b_0, y'(a) = b_1, \ldots, y^{(n-1)}(a) = b_{n-1}$. □

• Example 3.0.1

The solution $y(x) = \frac{4}{3}e^x - \frac{1}{3}e^{-2x}$ to the ordinary differential equation $y''' + 2y'' - y' - 2y = 0$ satisfies the initial conditions $y(0) = 1$, $y'(0) = 2$, and $y''(0) = 0$ at $x = 0$. Our theorem guarantees us that this is the *only* solution with *these* initial values.

Another class of problems, commonly called (two-point) *boundary-value problems*, occurs when conditions are specified at two *different* points $x = a$ and $x = b$ with $b > a$. An important example, in the case of second-order ordinary differential equations, is the Sturm-Liouville problem where the boundary conditions are $\alpha_1 y(a) + \beta_1 y'(a) = 0$ at $x = a$ and $\alpha_2 y(b) + \beta_2 y'(b) = 0$ at $x = b$. The Sturm-Liouville problem is treated in Chapter 9.

Having introduced some of the terms associated with higher-order ordinary linear differential equations, how do we solve them? One way is to recognize that these equations are really a set of linear, first-order ordinary differential equations. For example, the linear second-order linear differential equation

$$y'' - 3y' + 2y = 3x \tag{3.0.2}$$

can be rewritten as the following system of first-order ordinary differential equations:

$$y' - y = v, \qquad \text{and} \qquad v' - 2v = 3x \tag{3.0.3}$$

because

$$y'' - y' = v' = 2v + 3x = 2y' - 2y + 3x, \tag{3.0.4}$$

---

[1] The proof of the existence and uniqueness of solutions to (3.0.1) is beyond the scope of this book. See Ince, E. L., 1956: *Ordinary Differential Equations.* Dover Publications, Inc., §3.32.

which is the same as (3.0.2). This suggests that (3.0.2) can be solved by applying the techniques from the previous chapter. Proceeding along this line, we first find that

$$v(x) = C_1 e^{2x} - \tfrac{3}{2}x - \tfrac{3}{4}. \tag{3.0.5}$$

Therefore,

$$y' - y = C_1 e^{2x} - \tfrac{3}{2}x - \tfrac{3}{4}. \tag{3.0.6}$$

Again, applying the techniques from the previous chapter, we have that

$$y = C_1 e^{2x} + C_2 e^x + \tfrac{3}{2}x + \tfrac{9}{4}. \tag{3.0.7}$$

Note that the solution to this second-order ordinary differential equation contains *two* arbitrary constants.

- **Example 3.0.2**

In the case of linear, second-order ordinary differential equations, a similar technique, called *reduction in order*, provides a method for solving differential equations if we know one of its solutions.

Consider the second-order ordinary differential equation

$$x^2 y'' - 5xy' + 9y = 0. \tag{3.0.8}$$

A quick check shows that $y_1(x) = x^3 \ln(x)$ is a solution of (3.0.8). Let us now assume that the *general* solution can be written $y(x) = u(x)x^3 \ln(x)$. Then

$$y' = u'(x)x^3 \ln(x) + u(x)\left[3x^2 \ln(x) + x^2\right], \tag{3.0.9}$$

and

$$y'' = u''(x)x^3 \ln(x) + 2u'(x)\left[3x^2 \ln(x) + x^2\right] + u(x)\left[6x\ln(x) + 5x\right]. \tag{3.0.10}$$

Substitution of $y(x)$, $y'(x)$, and $y''(x)$ into (3.0.8) yields

$$x^5 \ln(x)u'' + \left[x^4 \ln(x) + 2x^4\right]u' = 0. \tag{3.0.11}$$

Setting $u' = w$, separation of variables leads to

$$\frac{w'}{w} = -\frac{1}{x} - \frac{2}{x\ln(x)}. \tag{3.0.12}$$

Note how our replacement of $u'(x)$ with $w(x)$ has reduced the second-order ordinary differential equation to a first-order one. Solving (3.0.12), we find that

$$w(x) = u'(x) = -\frac{C_1}{x\ln^2(x)}, \tag{3.0.13}$$

and

$$u(x) = \frac{C_1}{\ln(x)} + C_2. \tag{3.0.14}$$

Because $y(x) = u(x)x^3 \ln(x)$, the complete solution is

$$y(x) = C_1 x^3 + C_2 x^3 \ln(x). \tag{3.0.15}$$

Substitution of (3.0.15) into (3.0.8) confirms that we have the correct solution.

We can verify our answer by using the symbolic toolbox in MATLAB. Typing the command:

```
dsolve('x*x*D2y-5*x*Dy+9*y=0','x')
```

yields

```
ans =
C1*x^3+C2*x^3*log(x)
```

In summary, we can reduce (in principle) any higher-order, linear ordinary differential equations into a system of first-order ordinary differential equations. This system of differential equations can then be solved using techniques from the previous chapter. In Chapter 14 we will pursue this idea further. Right now, however, we will introduce methods that allow us to find the solution in a more direct manner.

• **Example 3.0.3**

An *autonomous differential equation* is one where the independent variable does not appear explicitly. In certain cases we can reduce the order of the differential equation and then solve it.

Consider the autonomous ordinary differential equation

$$y'' = 2y^3. \tag{3.0.16}$$

The trick here is to note that

$$y'' = \frac{dv}{dx} = v\frac{dv}{dy} = 2y^3, \tag{3.0.17}$$

where $v = dy/dx$. Integrating both sides of (3.0.17), we find that

$$v^2 = y^4 + C_1. \tag{3.0.18}$$

Solving for v,

$$\frac{dy}{dx} = v = \sqrt{C_1 + y^4}. \tag{3.0.19}$$

Integrating once more, we have the final result that

$$x + C_2 = \int \frac{dy}{\sqrt{C_1 + y^4}}. \tag{3.0.20}$$

Problems

For the following differential equations, use reduction of order to find a second solution. Can you obtain the general solution using `dsolve` in MATLAB?

1. $xy'' + 2y' = 0,$ $\quad y_1(x) = 1$ $\qquad$ 2. $y'' + y' - 2y = 0,$ $\quad y_1(x) = e^x$

3. $x^2 y'' + 4xy' - 4y = 0,$ $\quad y_1(x) = x$

4. $xy'' - (x + 1)y' + y = 0,$ $\quad y_1(x) = e^x$

5. $(2x - x^2)y'' + 2(x - 1)y' - 2y = 0,$ $\quad y_1(x) = x - 1$

6. $y'' + \tan(x)y' - 6\cot^2(x)y = 0,$ $\quad y_1(x) = \sin^3(x)$

7. $4x^2 y'' + 4xy' + (4x^2 - 1)y = 0,$ $\quad y_1(x) = \cos(x)/\sqrt{x}$

8. $y'' + ay' + b(1 + ax - bx^2)y = 0,$ $\quad y_1(x) = e^{-bx^2/2}$

Solve the following autonomous ordinary differential equations:

9. $yy'' = y'^2$ $\qquad\qquad\qquad$ 10. $y'' = 2yy',$ $\quad y(0) = y'(0) = 1$

11. $yy'' = y' + y'^2$ $\qquad\qquad\quad$ 12. $2yy'' = 1 + y'^2$

13. $y'' = e^{2y},$ $\quad y(0) = 0, y'(0) = 1$

14. $y''' = 3yy',$ $\quad y(0) = y'(0) = 1, y''(0) = \frac{3}{2}$

15. Solve the nonlinear second-order ordinary differential equation

$$\frac{d^2 y}{dx^2} - \frac{1}{x}\frac{dy}{dx} - \frac{1}{2}\left(\frac{dy}{dx}\right)^2 = 0$$

by (1) reducing it to the Bernoulli equation

$$\frac{dv}{dx} - \frac{v}{x} - \frac{v^2}{2} = 0, \qquad v(x) = u'(x),$$

(2) solving for $v(x)$, and finally (3) integrating $u' = v$ to find $u(x)$.

16. Consider the differential equation

$$a_2(x)y'' + a_1(x)y' + a_0(x)y = 0, \qquad a_2(x) \neq 0.$$

Show that this ordinary differential equation can be rewritten

$$u'' + f(x)u = 0, \qquad f(x) = \frac{a_0(x)}{a_2(x)} - \frac{1}{4}\left[\frac{a_1(x)}{a_2(x)}\right]^2 - \frac{1}{2}\frac{d}{dx}\left[\frac{a_1(x)}{a_2(x)}\right],$$

using the substitution

$$y(x) = u(x)\exp\left[-\frac{1}{2}\int^x \frac{a_1(\xi)}{a_2(\xi)}\,d\xi\right].$$

3.1 HOMOGENEOUS LINEAR EQUATIONS WITH CONSTANT COEFFICIENTS

In our drive for more efficient methods to solve higher-order, linear, ordinary differential equations, let us examine the simplest possible case of a homogeneous differential equation with constant coefficients:

$$a_n\frac{d^n y}{dx^n} + a_{n-1}\frac{d^{n-1}y}{dx^{n-1}} + \cdots + a_2 y'' + a_1\frac{dy}{dx} + a_0 y = 0. \qquad (3.1.1)$$

Although we could explore (3.1.1) in its most general form, we will begin by studying the second-order version, namely

$$ay'' + by' + cy = 0, \qquad (3.1.2)$$

since it is the next step up the ladder in complexity from first-order ordinary differential equations.

Motivated by the fact that the solution to the first-order ordinary differential equation $y' + ay = 0$ is $y(x) = C_1 e^{-ax}$, we make the educated guess that the solution to (3.1.2) is $y(x) = Ae^{mx}$. Direct substitution into (3.1.2) yields

$$\left(am^2 + bm + c\right)Ae^{mx} = 0. \qquad (3.1.3)$$

Because $A \neq 0$ or we would have a trivial solution and since $e^{mx} \neq 0$ for arbitrary x, (3.1.3) simplifies to

$$am^2 + bm + c = 0. \qquad (3.1.4)$$

Equation (3.1.4) is called the *auxiliary* or *characteristic equation*. At this point we must consider three separate cases.

• *Distinct real roots*

In this case the roots to (3.1.4) are real and unequal. Let us denote these roots by $m = m_1$, and $m = m_2$. Thus, we have the two solutions:

$$y_1(x) = C_1 e^{m_1 x}, \qquad \text{and} \qquad y_2(x) = C_2 e^{m_2 x}. \qquad (3.1.5)$$

We will now show that the most general solution to (3.1.2) is

$$y(x) = C_1 e^{m_1 x} + C_2 e^{m_2 x}. \tag{3.1.6}$$

This result follows from the *principle of (linear) superposition.*

Theorem: *Let $y_1, y_2, \ldots, y_k$ be solutions of the homogeneous equation (3.1.1) on an interval I. Then the linear combination*

$$y(x) = C_1 y_1(x) + C_2 y_2(x) + \cdots + C_k y_k(x), \tag{3.1.7}$$

where C_i, $i = 1, 2, \ldots, k$, are arbitrary constants, is also a solution on the interval I.

Proof: We will prove this theorem for second-order ordinary differential equations; it is easily extended to higher orders. By the superposition principle, $y(x) = C_1 y_1(x) + C_2 y_2(x)$. Upon substitution into (3.1.2), we have that

$$a \left(C_1 y_1'' + C_2 y_2'' \right) + b \left(C_1 y_1' + C_2 y_2' \right) + c \left(C_1 y_1 + C_2 y_2 \right) = 0. \tag{3.1.8}$$

Recombining the terms, we obtain

$$C_1 \left(a y_1'' + b y_1' + c y_1 \right) + C_2 \left(a y_2'' + b y_2' + c y_2 \right) = 0, \tag{3.1.9}$$

or

$$0 C_1 + 0 C_2 = 0. \tag{3.1.10}$$

$\square$

• Example 3.1.1

A quick check shows that $y_1(x) = e^x$ and $y_2(x) = e^{-x}$ are two solutions of $y'' - y = 0$. Our theorem tells us that *any* linear combination of these solutions, such as $y(x) = 5e^x - 3e^{-x}$, is also a solution.

How about the converse? Is *every* solution to $y'' - y = 0$ a linear combination of $y_1(x)$ and $y_2(x)$? We will address this question shortly.

• Example 3.1.2

Let us find the general solution to

$$y'' + 2y' - 15y = 0. \tag{3.1.11}$$

Assuming a solution of the form $y(x) = A e^{mx}$, we have that

$$(m^2 + 2m - 15) A e^{mx} = 0. \tag{3.1.12}$$

Because $A \neq 0$ and e^{mx} generally do not equal zero, we obtain the auxiliary or characteristic equation

$$m^2 + 2m - 15 = (m+5)(m-3) = 0. \tag{3.1.13}$$

Therefore, the general solution is

$$y(x) = C_1 e^{3x} + C_2 e^{-5x}. \tag{3.1.14}$$

• *Repeated real roots*

When $m = m_1 = m_2$, we have only the single exponential solution $y_1(x) = C_1 e^{m_1 x}$. To find the second solution we apply the reduction of order technique shown in Example 3.0.2. Performing the calculation, we find

$$y_2(x) = C_2 e^{m_1 x} \int \frac{e^{-bx/a}}{e^{2m_1 x}} \, dx. \tag{3.1.15}$$

Since $m_1 = -b/(2a)$, the integral simplifies to $\int dx$ and

$$y(x) = C_1 e^{m_1 x} + C_2 x e^{m_1 x}. \tag{3.1.16}$$

• **Example 3.1.3**

Let us find the general solution to

$$y'' + 4y' + 4y = 0. \tag{3.1.17}$$

Here the auxiliary or characteristic equation is

$$m^2 + 4m + 4 = (m+2)^2 = 0. \tag{3.1.18}$$

Therefore, the general solution is

$$y(x) = (C_1 + C_2 x)e^{-2x}. \tag{3.1.19}$$

• *Complex conjugate roots*

When $b^2 - 4ac < 0$, the roots become the complex pair $m_1 = \alpha + i\beta$ and $m_2 = \alpha - \beta i$, where α and β are real and $i^2 = -1$. Therefore, the general solution is

$$y(x) = C_1 e^{(\alpha + i\beta)x} + C_2 e^{(\alpha - \beta i)x}. \tag{3.1.20}$$

Although (3.1.20) is quite correct, most engineers prefer to work with real functions rather than complex exponentials. To this end, we apply Euler's formula[2] to eliminate $e^{i\beta x}$ and $e^{-i\beta x}$ since

$$e^{i\beta x} = \cos(\beta x) + i\sin(\beta x), \qquad (3.1.21)$$

and

$$e^{-i\beta x} = \cos(\beta x) - i\sin(\beta x). \qquad (3.1.22)$$

Therefore,

$$y(x) = C_1 e^{\alpha x}\left[\cos(\beta x) + i\sin(\beta x)\right] + C_2 e^{\alpha x}\left[\cos(\beta x) - i\sin(\beta x)\right] \quad (3.1.23)$$
$$= C_3 e^{\alpha x}\cos(\beta x) + C_4 e^{\alpha x}\sin(\beta x), \qquad (3.1.24)$$

where $C_3 = C_1 + C_2$, and $C_4 = iC_1 - iC_2$.

• Example 3.1.4

Let us find the general solution to

$$y'' + 4y' + 5y = 0. \qquad (3.1.25)$$

Here the auxiliary or characteristic equation is

$$m^2 + 4m + 5 = (m+2)^2 + 1 = 0, \qquad (3.1.26)$$

or $m = -2 \pm i$. Therefore, the general solution is

$$y(x) = e^{-2x}[C_1\cos(x) + C_2\sin(x)]. \qquad (3.1.27)$$

So far we have only dealt with second-order differential equations. When we turn to higher-order ordinary differential equations, similar considerations hold. In place of (3.1.4), we now have the nth-degree polynomial equation

$$a_n m^n + a_{n-1} m^{n-1} + \cdots + a_2 m^2 + a_1 m + a_0 = 0 \qquad (3.1.28)$$

for its auxiliary equation.

When we treated second-order ordinary differential equations we were able to classify the roots to the auxiliary equation as distinct real roots, repeated roots, and complex roots. In the case of higher-order differential equations, such classifications are again useful although all three types may occur with the same equation. For example, the auxiliary equation

$$m^6 - m^5 + 2m^4 - 2m^3 + m^2 - m = 0 \qquad (3.1.29)$$

---

[2] If you are unfamiliar with Euler's formula, see §1.1.

has the distinct roots $m = 0$ and $m = 1$ with the twice repeated, complex roots $m = \pm i$.

Although the possible combinations increase with higher-order differential equations, the solution technique remains the same. For each distinct real root $m = m_1$, we have a corresponding homogeneous solution $e^{m_1 x}$. For each complex pair $m = \alpha \pm \beta i$, we have the corresponding pair of homogeneous solutions $e^{\alpha x} \cos(\beta x)$ and $e^{\alpha x} \sin(\beta x)$. For a repeated root $m = m_1$ of multiplicity k, regardless of whether it is real or complex, we have either $e^{m_1 x}, x e^{m_1 x}, x^2 e^{m_1 x}, \ldots, x^k e^{m_1 x}$ in the case of real m_1 or

$$e^{\alpha x} \cos(\beta x), e^{\alpha x} \sin(\beta x), x e^{\alpha x} \cos(\beta x), x e^{\alpha x} \sin(\beta x),$$

$$x^2 e^{\alpha x} \cos(\beta x), x^2 e^{\alpha x} \sin(\beta x), \ldots, x^k e^{\alpha x} \cos(\beta x), x^k e^{\alpha x} \sin(\beta x)$$

in the case of complex roots $\alpha \pm \beta i$. For example, the general solution for the roots to (3.1.29) is

$$y(x) = C_1 + C_2 e^x + C_3 \cos(x) + C_4 \sin(x) + C_5 x \cos(x) + C_6 x \sin(x). \quad \textbf{(3.1.30)}$$

- **Example 3.1.5**

Let us find the general solution to

$$y''' + y' - 10y = 0. \quad \textbf{(3.1.31)}$$

Here the auxiliary or characteristic equation is

$$m^3 + m - 10 = (m - 2)(m^2 + 2m + 5) = (m - 2)[(m + 1)^2 + 4] = 0, \quad \textbf{(3.1.32)}$$

or $m = -2$ and $m = -1 \pm 2i$. Therefore, the general solution is

$$y(x) = C_1 e^{-2x} + e^{-x}[C_2 \cos(2x) + C_3 \sin(2x)]. \quad \textbf{(3.1.33)}$$

Having presented the technique for solving constant coefficient, linear, ordinary differential equations, an obvious question is: How do we know that we have captured all of the solutions? Before we can answer this question, we must introduce the concept of linear dependence.

A set of functions $f_1(x), f_2(x), \ldots, f_n(x)$ is said to be *linearly dependent* on an interval I if there exists constants $C_1, C_2, \ldots, C_n$, not all zero, such that

$$C_1 f_1(x) + C_2 f_2(x) + C_3 f_3(x) + \cdots + C_n f_n(x) = 0 \quad \textbf{(3.1.34)}$$

for each x in the interval; otherwise, the set of functions is said to be *linearly independent*. This concept is easily understood when we have only two functions $f_1(x)$ and $f_2(x)$. If the functions are linearly dependent on an interval, then there exists constants C_1 and C_2 that are not both zero where

$$C_1 f_1(x) + C_2 f_2(x) = 0 \quad \textbf{(3.1.35)}$$

for every x in the interval. If $C_1 \neq 0$, then

$$f_1(x) = -\frac{C_2}{C_1} f_2(x). \qquad (3.1.36)$$

In other words, if two functions are linearly dependent, then one is a constant multiple of the other. Conversely, two functions are linearly independent when neither is a constant multiple of the other on an interval.

• **Example 3.1.6**

Let us show that $f(x) = 2x$, $g(x) = 3x^2$, and $h(x) = 5x - 8x^2$ are linearly dependent on the real line.

To show this, we must choose three constants, C_1, C_2, and C_3, such that

$$C_1 f(x) + C_2 g(x) + C_3 h(x) = 0, \qquad (3.1.37)$$

where not all of these constants are nonzero. A quick check shows that

$$15 f(x) - 16 g(x) - 6 h(x) = 0. \qquad (3.1.38)$$

Clearly, $f(x)$, $g(x)$, and $h(x)$ are linearly dependent.

• **Example 3.1.7**

This example shows the importance of defining the interval on which a function is linearly dependent or independent. Consider the two functions $f(x) = x$ and $g(x) = |x|$. They are linearly dependent on the interval $(0, \infty)$ since $C_1 x + C_2 |x| = C_1 x + C_2 x = 0$ is satisfied for any nonzero choice of C_1 and C_2 where $C_1 = -C_2$. What happens on the interval $(-\infty, 0)$? They are still linearly dependent but now $C_1 = C_2$.

Although we could use the fundamental concept of linear independence to check and see whether a set of functions is linearly independent or not, the following theorem introduces a procedure that is very straightforward.

Theorem: Wronskian Test of Linear Independence

Suppose $f_1(x), f_2(x), \ldots, f_n(x)$ possess at least $n - 1$ derivatives. If the determinant[3]

$$\begin{vmatrix} f_1 & f_2 & \cdots & f_n \\ f_1' & f_2' & \cdots & f_n' \\ \vdots & \vdots & & \vdots \\ f_1^{(n-1)} & f_2^{(n-1)} & \cdots & f_n^{(n-1)} \end{vmatrix}$$

---

[3] If you are unfamiliar with determinants, see §14.2.

is not zero for at least one point in the interval I, then the functions $f_1(x)$, $f_2(x)$, ..., $f_n(x)$ are linearly independent on the interval. The determinant in this theorem is denoted by $W[f_1(x), f_2(x), ..., f_n(x)]$ and is called the *Wronskian* of the functions.

Proof: We prove this theorem by contradiction when $n = 2$. Let us assume that $W[f_1(x_0), f_2(x_0)] \neq 0$ for some fixed x_0 in the interval I and that $f_1(x)$ and $f_2(x)$ are linearly dependent on the interval. Since the functions are linearly dependent, there exists C_1 and C_2, both not zero, for which

$$C_1 f_1(x) + C_2 f_2(x) = 0 \tag{3.1.39}$$

for every x in I. Differentiating (3.1.39) gives

$$C_1 f_1'(x) + C_2 f_2'(x) = 0. \tag{3.1.40}$$

We may view (3.1.39)–(3.1.40) as a system of equations with C_1 and C_2 as the unknowns. Because the linear dependence of f_1 and f_2 implies that $C_1 \neq 0$ and/or $C_2 \neq 0$ for each x in the interval,

$$W[f_1(x), f_2(x)] = \begin{vmatrix} f_1 & f_2 \\ f_1' & f_2' \end{vmatrix} = 0 \tag{3.1.41}$$

for every x in I. This contradicts the assumption that $W[f_1(x_0), f_2(x_0)] \neq 0$ and f_1 and f_2 are linearly independent. $\qquad\qquad\square$

• Example 3.1.8

Are the functions $f(x) = x$, $g(x) = xe^x$, and $h(x) = x^2 e^x$ linearly dependent on the real line? To find out, we compute the Wronskian or

$$W[f(x), g(x), h(x)] = \begin{vmatrix} e^x & xe^x & x^2 e^x \\ e^x & (x+1)e^x & (x^2 + 2x)e^x \\ e^x & (x+2)e^x & (x^2 + 4x + 2)e^x \end{vmatrix} \tag{3.1.42}$$

$$= e^{3x} \begin{vmatrix} 1 & x & x^2 \\ 0 & 1 & 2x \\ 0 & 0 & 2 \end{vmatrix} = 2e^{3x} \neq 0. \tag{3.1.43}$$

Therefore, x, xe^x, and $x^2 e^x$ are linearly *independent*.

Having introduced this concept of linear independence, we are now ready to address the question of how many linearly independent solutions a homogeneous linear equation has.

Theorem:

On any interval I over which an n-th order homogeneous linear differential equation is normal, the equation has n linearly independent solutions $y_1(x), y_2(x), \ldots, y_n(x)$ and any particular solution of the equation on I can be expressed as a linear combination of these linearly independent solutions.

Proof: Again for convenience and clarity we prove this theorem for the special case of $n = 2$. Let $y_1(x)$ and $y_2(x)$ denote solutions on I of (3.1.2). We know that these solutions exist by the existence theorem and have the following values:

$$y_1(a) = 1, \quad y_2(a) = 0, \quad y_1'(a) = 0, \quad y_2'(a) = 1 \qquad (3.1.44)$$

at some point a on I. To establish the linear independence of y_1 and y_2 we note that, if $C_1 y_1(x) + C_2 y_2(x) = 0$ holds identically on I, then $C_1 y_1'(x) + C_2 y_2'(x) = 0$ there too. Because $x = a$ lies in I, we have that

$$C_1 y_1(a) + C_2 y_2(a) = 0, \qquad (3.1.45)$$

and

$$C_1 y_1'(a) + C_2 y_2'(a) = 0, \qquad (3.1.46)$$

which yields $C_1 = C_2 = 0$ after substituting (3.1.44). Hence, the solutions y_1 and y_2 are linearly independent.

To complete the proof we must now show that any particular solution of (3.1.2) can be expressed as a linear combination of y_1 and y_2. Because y, y_1, and y_2 are all solutions of (3.1.2) on I, so is the function

$$Y(x) = y(x) - y(a)y_1(x) - y'(a)y_2(x), \qquad (3.1.47)$$

where $y(a)$ and $y'(a)$ are the values of the solution y and its derivative at $x = a$. Evaluating Y and Y' at $x = a$, we have that

$$Y(a) = y(a) - y(a)y_1(a) - y'(a)y_2(a) = y(a) - y(a) = 0, \qquad (3.1.48)$$

and

$$Y'(a) = y'(a) - y(a)y_1'(a) - y'(a)y_2'(a) = y'(a) - y'(a) = 0. \qquad (3.1.49)$$

Thus, Y is the trivial solution to (3.1.2). Hence, for every x in I,

$$y(x) - y(a)y_1(x) - y'(a)y_2(x) = 0. \qquad (3.1.50)$$

Solving (3.1.50) for $y(x)$, we see that y is expressible as the linear combination

$$y(x) = y(a)y_1(x) + y'(a)y_2(x) \qquad (3.1.51)$$

of y_1 and y_2, and the proof is complete for $n = 2$.   □

Problems

Find the general solution to the following differential equations. Check your general solution by using dsolve in MATLAB.

1. $y'' + 6y' + 5y = 0$ 2. $y'' - 6y' + 10y = 0$

3. $y'' - 2y' + y = 0$ 4. $y'' - 3y' + 2y = 0$

5. $y'' - 4y' + 8y = 0$ 6. $y'' + 6y' + 9y = 0$

7. $y'' + 6y' - 40y = 0$ 8. $y'' + 4y' + 5y = 0$

9. $y'' + 8y' + 25y = 0$ 10. $4y'' - 12y' + 9y = 0$

11. $y'' + 8y' + 16y = 0$ 12. $y''' + 4y'' = 0$

13. $y'''' + 4y'' = 0$ 14. $y'''' + 2y''' + y'' = 0$

15. $y''' - 8y = 0$ 16. $y'''' - 3y''' + 3y'' - y' = 0$

17. The simplest differential equation with "memory" — its past behavior affects the present — is

$$y' = -\frac{A}{2\tau} \int_{-\infty}^{t} e^{-(t-x)/\tau} y(x)\, dx.$$

Solve this integro-differential equation by differentiating it with respect to t to eliminate the integral.

3.2 SIMPLE HARMONIC MOTION

Second-order, linear, ordinary differential equations often arise in mechanical or electrical problems. The purpose of this section is to illustrate how the techniques that we just derived may be applied to these problems.

We begin by considering the mass-spring system illustrated in Figure 3.2.1 where a mass m is attached to a flexible spring suspended from a rigid support. If there were no spring, then the mass would simply fall downward due to the gravitational force mg. Because there is no motion, the gravitational force must be balanced by an upward force due to the presence of the spring. This upward force is usually assumed to obey Hooke's law which states that the restoring force is opposite to the direction of elongation and proportional to

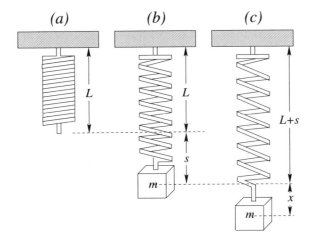

Figure 3.2.1: Various configurations of a mass/spring system. The spring alone has a length L which increases to $L + s$ when the mass is attached. During simple harmonic motion, the length of the mass/spring system varies as $L + s + x$.

the amount of elongation. Mathematically the equilibrium condition can be expressed $mg = ks$.

Consider now what happens when we disturb this equilibrium. This may occur in one of two ways: We could move the mass either upward or downward and then release it. Another method would be to impart an initial velocity to the mass. In either case, the motion of the mass/spring system would be governed by Newton's second law which states that the acceleration of the mass equals the imbalance of the forces. If we denote the downward displacement of the mass from its equilibrium position by positive x, then

$$m\frac{d^2x}{dt^2} = -k(s + x) + mg = -kx, \tag{3.2.1}$$

since $ks = mg$. After dividing (3.2.1) by the mass, we obtain the second-order differential equation

$$\frac{d^2x}{dt^2} + \frac{k}{m}x = 0, \tag{3.2.2}$$

or

$$\frac{d^2x}{dt^2} + \omega^2 x = 0, \tag{3.2.3}$$

where $\omega^2 = k/m$ and ω is the *circular frequency*. Equation (3.2.3) describes *simple harmonic motion* or *free undamped motion*. The two initial conditions associated with this differential equation are

$$x(0) = \alpha, \qquad x'(0) = \beta. \tag{3.2.4}$$

The first condition gives the initial amount of displacement while the second condition specifies the initial velocity. If $\alpha > 0$ while $\beta < 0$, then the mass starts from a point below the equilibrium position with an initial upward velocity. On the other hand, if $\alpha < 0$ with $\beta = 0$ the mass is at rest when it is released $|\alpha|$ units above the equilibrium position. Similar considerations hold for other values of α and β.

To solve (3.2.3), we note that the solutions of the auxiliary equation $m^2 + \omega^2 = 0$ are the complex numbers $m_1 = \omega i$, and $m_2 = -\omega i$. Therefore, the general solution is

$$x(t) = A\cos(\omega t) + B\sin(\omega t). \tag{3.2.5}$$

The (natural) *period* of free vibrations is $T = 2\pi/\omega$ while the (natural) frequency is $f = 1/T = \omega/(2\pi)$.

● **Example 3.2.1**

Let us solve the initial-value problem

$$\frac{d^2x}{dt^2} + 4x = 0, \qquad x(0) = 10, \qquad x'(0) = 0. \tag{3.2.6}$$

The physical interpretation is that we have pulled the mass on a spring down 10 units *below* the equilibrium position and then release it from rest at $t = 0$. Here, $\omega = 2$ so that

$$x(t) = A\cos(2t) + B\sin(2t) \tag{3.2.7}$$

from (3.2.5).

Because $x(0) = 10$, we find that

$$x(0) = 10 = A \cdot 1 + B \cdot 0 \tag{3.2.8}$$

so that $A = 10$. Next, we note that

$$\frac{dx}{dt} = -20\sin(2t) + 2B\cos(2t). \tag{3.2.9}$$

Therefore, at $t = 0$,

$$x'(0) = 0 = -20 \cdot 0 + 2B \cdot 1 \tag{3.2.10}$$

and $B = 0$. Thus, the equation of motion is $x(t) = 10\cos(2t)$.

What is the physical interpretation of our equation of motion? Once the system is set into motion, it stays in motion with the mass oscillating back and forth 10 units above and below the equilibrium position $x = 0$. The period of oscillation is $2\pi/2 = \pi$ units of time.

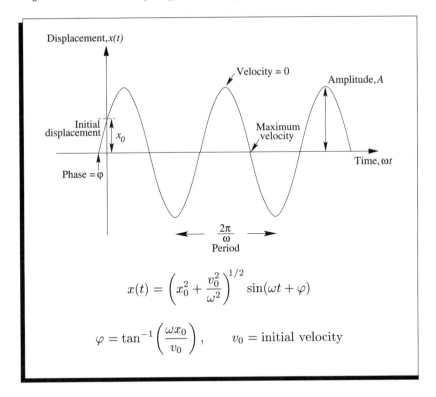

$$x(t) = \left(x_0^2 + \frac{v_0^2}{\omega^2}\right)^{1/2} \sin(\omega t + \varphi)$$

$$\varphi = \tan^{-1}\left(\frac{\omega x_0}{v_0}\right), \qquad v_0 = \text{initial velocity}$$

- **Example 3.2.2**

A weight of 45 N stretches a spring 5 cm. At time $t = 0$, the weight is released from its equilibrium position with an upward velocity of 28 cm s^{-1}. Determine the displacement $x(t)$ that describes the subsequent free motion.

From Hooke's law,

$$F = mg = 45\,\text{N} = k \times 5\,\text{cm} \tag{3.2.11}$$

so that $k = 9$ N cm^{-1}. Therefore, the differential equation is

$$\frac{d^2x}{dt^2} + 196\,\text{s}^{-2}x = 0. \tag{3.2.12}$$

The initial displacement and initial velocity are $x(0) = 0$ cm and $x'(0) = -28$ cm s^{-1}. The negative sign in the initial velocity reflects the fact that the weight has an initial velocity in the negative or upward direction.

Because $\omega^2 = 196$ s^{-2} or $\omega = 14$ s^{-1}, the general solution to the differential equation is

$$x(t) = A\cos(14\,\text{s}^{-1}t) + B\sin(14\,\text{s}^{-1}t). \tag{3.2.13}$$

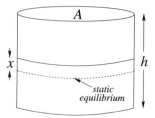

Figure 3.2.2: Schematic of a floating body partially submerged in pure water.

Substituting for the initial displacement $x(0)$ in (3.2.13), we find that

$$x(0) = 0\,\text{cm} = A \cdot 1 + B \cdot 0, \qquad (3.2.14)$$

and $A = 0$ cm. Therefore,

$$x(t) = B \sin(14\,\text{s}^{-1}t) \qquad (3.2.15)$$

and

$$x'(t) = 14\,\text{s}^{-1} B \cos(14\,\text{s}^{-1}t). \qquad (3.2.16)$$

Substituting for the initial velocity,

$$x'(0) = -28\,\text{cm}\,\text{s}^{-1} = 14\,\text{s}^{-1}B, \qquad (3.2.17)$$

and $B = -2$ cm. Thus the equation of motion is

$$x(t) = -2\,\text{cm}\ \sin(14\,\text{s}^{-1}t). \qquad (3.2.18)$$

• **Example 3.2.3: Vibration of floating bodies**

Consider a solid cylinder of radius a that is partially submerged in a bath of pure water as shown in Figure 3.2.2. Let us find the motion of this cylinder in the vertical direction assuming that it remains in an upright position.

If the displacement of the cylinder from its static equilibrium position is x, the weight of water displaced equals $Ag\rho_w x$, where ρ_w is the density of the water and g is the gravitational acceleration. This is the restoring force according to the Archimedes principle. The mass of the cylinder is $Ah\rho$, where ρ is the density of cylinder. From second Newton's law, the equation of motion is

$$\rho Ah x'' + Ag\rho_w x = 0, \qquad (3.2.19)$$

or

$$x'' + \frac{\rho_w g}{\rho h} x = 0. \qquad (3.2.20)$$

From (3.2.20) we see that the cylinder will oscillate about its static equilibrium position $x = 0$ with a frequency of

$$\omega = \left(\frac{\rho_w g}{\rho h}\right)^{1/2}. \tag{3.2.21}$$

When both A and B are both nonzero, it is often useful to rewrite the homogeneous solution (3.2.5) as

$$x(t) = C \sin(\omega t + \varphi) \tag{3.2.22}$$

to highlight the amplitude and phase of the oscillation. Upon employing the trigonometric angle-sum formula, (3.2.22) can be rewritten

$$x(t) = C \sin(\omega t) \cos(\varphi) + C \cos(\omega t) \sin(\varphi) = A \cos(\omega t) + B \sin(\omega t). \tag{3.2.23}$$

From (3.2.23), we see that $A = C \sin(\varphi)$ and $B = C \cos(\varphi)$. Therefore,

$$A^2 + B^2 = C^2 \sin^2(\varphi) + C^2 \cos^2(\varphi) = C^2, \tag{3.2.24}$$

and $C = \sqrt{A^2 + B^2}$. Similarly, $\tan(\varphi) = A/B$. Because the tangent is positive in both the first and third quadrants and negative in both the second and fourth quadrants, there are two possible choices for φ. The proper choice of φ satisfies the equations $A = C \sin(\varphi)$ and $B = C \cos(\varphi)$.

If we prefer the amplitude/phase solution

$$x(t) = C \cos(\omega t - \varphi), \tag{3.2.25}$$

we now have

$$x(t) = C \cos(\omega t) \cos(\varphi) + C \sin(\omega t) \sin(\varphi) = A \cos(\omega t) + B \sin(\omega t). \tag{3.2.26}$$

Consequently, $A = C \cos(\varphi)$ and $B = C \sin(\varphi)$. Once again, we obtain $C = \sqrt{A^2 + B^2}$. On the other hand, $\tan(\varphi) = B/A$.

Problems

Solve the following initial-value problems and write their solutions in terms of amplitude and phase:

1. $x'' + 25x = 0$, $\qquad x(0) = 10$, $\qquad x'(0) = -10$

2. $4x'' + 9x = 0$, $\qquad x(0) = 2\pi$, $\qquad x'(0) = 3\pi$

3. $x'' + \pi^2 x = 0$, $\qquad x(0) = 1$, $\qquad x'(0) = \pi\sqrt{3}$

4. A 4-kg mass is suspended from a 100 N/m spring. The mass is set in motion by giving it an initial downward velocity of 5 m/s from its equilibrium position. Find the displacement as a function of time.

5. A spring hangs vertically. A weight of mass M kg stretches it L m. This weight is removed. A body weighing m kg is then attached and allowed to come to rest. It is then pulled down s_0 m and released with a velocity v_0. Find the displacement of the body from its point of rest and its velocity at any time t.

6. A particle of mass m moving in a straight line is *repelled* from the origin by a force F. (a) If the force is proportional to the distance from the origin, find the position of the particle as a function of time. (b) If the initial velocity of the particle is $a\sqrt{k}$, where k is the proportionality constant and a is the distance from the origin, find the position of the particle as a function of time. What happens if $m < 1$ and $m = 1$?

3.3 DAMPED HARMONIC MOTION

Free harmonic motion is unrealistic because there are always frictional forces which act to retard motion. In mechanics, the drag is often modeled as a resistance that is proportional to the instantaneous velocity. Adopting this resistance law, it follows from Newton's second law that the harmonic oscillator is governed by

$$m\frac{d^2x}{dt^2} = -kx - \beta\frac{dx}{dt}, \tag{3.3.1}$$

where β is a positive *damping constant*. The negative sign is necessary since this resistance acts in a direction opposite to the motion.

Dividing (3.3.1) by the mass m, we obtain the differential equation of *free damped motion*,

$$\frac{d^2x}{dt^2} + \frac{\beta}{m}\frac{dx}{dt} + \frac{k}{m}x = 0, \tag{3.3.2}$$

or

$$\frac{d^2x}{dt^2} + 2\lambda\frac{dx}{dt} + \omega^2 x = 0. \tag{3.3.3}$$

We have written 2λ rather than just λ because it simplifies future computations. The auxiliary equation is $m^2 + 2\lambda m + \omega^2 = 0$ which has the roots

$$m_1 = -\lambda + \sqrt{\lambda^2 - \omega^2}, \qquad m_2 = -\lambda - \sqrt{\lambda^2 - \omega^2}. \tag{3.3.4}$$

From (3.3.4) we see that there are three possible cases which depend on the algebraic sign of $\lambda^2 - \omega^2$. Because all of the solutions contain the damping factor $e^{-\lambda t}$, $\lambda > 0$, $x(t)$ vanishes as $t \to \infty$.

- *Case I:* $\lambda > \omega$

Here the system is *overdamped* because the damping coefficient β is large compared to the spring constant k. The corresponding solution is

$$x(t) = Ae^{m_1 t} + Be^{m_2 t}, \tag{3.3.5}$$

or

$$x(t) = e^{-\lambda t}\left(Ae^{t\sqrt{\lambda^2 - \omega^2}} + Be^{-t\sqrt{\lambda^2 - \omega^2}}\right). \tag{3.3.6}$$

In this case the motion is smooth and nonoscillatory.

- *Case II:* $\lambda = \omega$

The system is *critically damped* because any slight decrease in the damping force would result in oscillatory motion. The general solution is

$$x(t) = Ae^{m_1 t} + Bte^{m_1 t}, \tag{3.3.7}$$

or

$$x(t) = e^{-\lambda t}(A + Bt). \tag{3.3.8}$$

The motion is quite similar to that of an overdamped system.

- *Case III:* $\lambda < \omega$

In this case the system is *underdamped* because the damping coefficient is small compared to the spring constant. The roots m_1 and m_2 are complex:

$$m_1 = -\lambda + i\sqrt{\omega^2 - \lambda^2}, \qquad m_2 = -\lambda - i\sqrt{\omega^2 - \lambda^2}. \tag{3.3.9}$$

The general solution now becomes

$$x(t) = e^{-\lambda t}\left[A\cos\left(t\sqrt{\omega^2 - \lambda^2}\right) + B\sin\left(t\sqrt{\omega^2 - \lambda^2}\right)\right]. \tag{3.3.10}$$

Equation (3.3.10) describes oscillatory motion which decays as $e^{-\lambda t}$. Equations (3.3.6), (3.3.8), and (3.3.10) are illustrated in Figure 3.3.1 when the initial conditions are $x(0) = 1$ and $x'(0) = 0$.

Just as we could write the solution for the simple harmonic motion in the amplitude/phase format, we can write any damped solution (3.3.10) in the alternative form

$$x(t) = Ce^{-\lambda t}\sin\left(t\sqrt{\omega^2 - \lambda^2} + \varphi\right), \tag{3.3.11}$$

where $C = \sqrt{A^2 + B^2}$ and the phase angle φ is given by $\tan(\varphi) = A/B$ such that $A = C\sin(\varphi)$ and $B = C\cos(\varphi)$. The coefficient $Ce^{-\lambda t}$ is sometimes

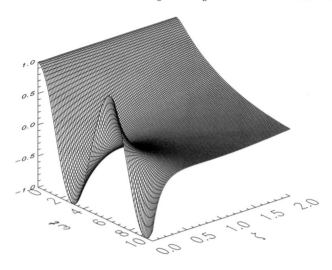

Figure 3.3.1: The displacement $x(t)$ of a damped harmonic oscillator as a function of time and $\zeta = \lambda/\omega$.

called the *damped coefficient* of vibrations. Because (3.3.11) is *not* a periodic function, the quantity $2\pi/\sqrt{\omega^2 - \lambda^2}$ is called the *quasi period* and $\sqrt{\omega^2 - \lambda^2}$ is the *quasi frequency*. The quasi period is the time interval between two successive maxima of $x(t)$.

● **Example 3.3.1**

A body with mass $m = \frac{1}{2}$ kg is attached to the end of a spring that is stretched 2 m by a force of 100 N. Furthermore, there is also attached a dashpot[4] that provides 6 N of resistance for each m/s of velocity. If the mass is set in motion by further stretching the spring $\frac{1}{2}$ m and giving it an upward velocity of 10 m/s, let us find the subsequent motion.

We begin by first computing the constants. The spring constant is $k = (100 \text{ N})/(2 \text{ m}) = 50$ N/m. Therefore, the differential equation is

$$\tfrac{1}{2}x'' + 6x' + 50x = 0 \qquad (3.3.12)$$

with $x(0) = \frac{1}{2}$ m and $x'(0) = -10$ m/s. Here the units of $x(t)$ are meters. The characteristic or auxiliary equation is

$$m^2 + 12m + 100 = (m + 6)^2 + 64 = 0, \qquad (3.3.13)$$

or $m = -6 \pm 8i$. Therefore, we have an underdamped harmonic oscillator and the general solution is

$$x(t) = e^{-6t}\left[A\cos(8t) + B\sin(8t)\right]. \qquad (3.3.14)$$

---

[4] A mechanical device — usually a piston that slides within a liquid-filled cylinder — used to damp the vibration or control the motion of a mechanism to which is attached.

**Review of the Solution of the
Underdamped Homogeneous Oscillator Problem**

$mx'' + \beta x' + kx = 0$ subject to $x(0) = x_0$, $x'(0) = v_0$ has the solution

$$x(t) = Ae^{-\lambda t} \sin(\omega_d t + \varphi),$$

where

$\omega = \sqrt{k/m}$ is the undamped natural frequency,

$\lambda = \beta/(2m)$ is the damping factor,

$\omega_d = \sqrt{\omega^2 - \lambda^2}$ is the damped natural frequency,

and the constants A and φ are determined by

$$A = \sqrt{x_0^2 + \left(\frac{v_0 + \lambda x_0}{\omega_d}\right)^2}$$

and

$$\varphi = \tan^{-1}\left(\frac{x_0 \omega_d}{v_0 + \lambda x_0}\right).$$

Consequently, each cycle takes $2\pi/8 = 0.79$ second. This is longer than the 0.63 second that would occur if the system were undamped.

From the initial conditions,

$$x(0) = A = \tfrac{1}{2}, \quad \text{and} \quad x'(0) = -10 = -6A + 8B. \tag{3.3.15}$$

Therefore, $A = \tfrac{1}{2}$ and $B = -\tfrac{7}{8}$. Consequently,

$$x(t) = e^{-6t}\left[\tfrac{1}{2}\cos(8t) - \tfrac{7}{8}\sin(8t)\right] = \frac{\sqrt{65}}{8}e^{-6t}\cos(8t + 2.62244). \tag{3.3.16}$$

• **Example 3.3.2: Design of a wind vane**

In its simplest form a wind vane is a flat plate or airfoil that can rotate about a vertical shaft. See Figure 3.3.2. In static equilibrium it points into the wind. There is usually a counterweight to balance the vane about the vertical shaft.

A vane uses a combination of the lift and drag forces on the vane to align itself with the wind. As the wind shifts direction from θ_0 to the new

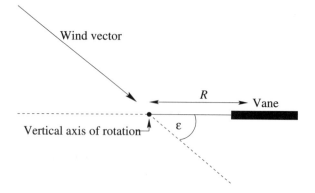

Figure 3.3.2: Schematic of a wind vane. The counterbalance is not shown.

direction θ_i, the direction θ in which the vane currently points is governed by the equation of motion[5]

$$I\frac{d^2\theta}{dt^2} + \frac{NR}{V}\frac{d\theta}{dt} = N(\theta_i - \theta), \qquad (3.3.17)$$

where I is the vane's moment of inertia, N is the aerodynamic torque per unit angle, and R is the distance from the axis of rotation to the effective center of the aerodynamic force on the vane. The aerodynamic torque is given by

$$N = \tfrac{1}{2}C_L\rho A V^2 R, \qquad (3.3.18)$$

where C_L is the lift coefficient, ρ is the air density, A is the vane area, and V is the wind speed.

Dividing (3.3.17) by I, we obtain the second-order ordinary differential equation

$$\frac{d^2(\theta - \theta_i)}{dt^2} + \frac{NR}{IV}\frac{d(\theta - \theta_i)}{dt} + \frac{N}{I}(\theta - \theta_i) = 0. \qquad (3.3.19)$$

The solution to (3.3.19) is

$$\theta - \theta_i = A\exp\left(-\frac{NRt}{2IV}\right)\cos(\omega t + \varphi), \qquad (3.3.20)$$

where

$$\omega^2 = \frac{N}{I} - \frac{N^2 R^2}{4I^2 V^2} \qquad (3.3.21)$$

---

[5] For a derivation of (3.3.12) and (3.3.13), see subsection 2 of §3 in Barthelt, H. P., and G. H. Ruppersberg, 1957: Die mechanische Windfahne, eine theoretische und experimentelle Untersuchung. *Beitr. Phys. Atmos.*, **29**, 154–185.

and A and φ are the two arbitrary constants which would be determined by presently unspecified initial conditions. Consequently an ideal wind vane is a damped harmonic oscillator where the wind torque should be large and its moment of inertia should be small.

Problems

For the following values of m, β, and k, find the position $x(t)$ of a damped oscillator for the given initial conditions:

1. $m = \frac{1}{2}$, $\beta = 3$, $k = 4$, $x(0) = 2$, $x'(0) = 0$

2. $m = 1$, $\beta = 10$, $k = 125$, $x(0) = 3$, $x'(0) = 25$

3. $m = 4$, $\beta = 20$, $k = 169$, $x(0) = 4$, $x'(0) = 16$

4. For a fixed value of λ/ω, what is the minimum number of cycles required to produce a reduction of at least 50% in the maxima of a underdamped oscillator?

5. For what values of c does $x'' + cx' + 4x = 0$ have critically damped solutions?

6. For what values of c are the motions governed by $4x'' + cx' + 9x = 0$ (a) overdamped, (b) underdamped, and (c) critically damped?

7. For an overdamped mass-spring system, prove that the mass can pass through its equilibrium position $x = 0$ at most once.

3.4 METHOD OF UNDETERMINED COEFFICIENTS

Homogeneous ordinary differential equations become nonhomogeneous when the right side of (3.0.1) is nonzero. How does this case differ from the homogeneous one that we have treated so far?

To answer this question, let us begin by introducing a function $y_p(x)$ — called a *particular solution* — whose only requirement is that it satisfies the differential equation

$$a_n(x)\frac{d^n y_p}{dx^n} + a_{n-1}(x)\frac{d^{n-1}y_p}{dx^{n-1}} + \cdots + a_1(x)\frac{dy_p}{dx} + a_0(x)y_p = f(x). \quad (3.4.1)$$

Then, by direct substitution, it can be seen that the general solution to any nonhomogeneous, linear, ordinary differential equation is

$$y(x) = y_H(x) + y_p(x), \qquad (3.4.2)$$

where $y_H(x)$ — the *homogeneous* or *complementary solution* — satisfies

$$a_n(x)\frac{d^n y_H}{dx^n} + a_{n-1}(x)\frac{d^{n-1}y_H}{dx^{n-1}} + \cdots + a_1(x)\frac{dy_H}{dx} + a_0(x)y_H = 0. \quad (3.4.3)$$

Why have we introduced this complementary solution because the particular solution already satisfies the ordinary differential equation? The purpose of the complementary solution is to introduce the arbitrary constants that any general solution of an ordinary differential equation must have. Thus, because we already know how to find $y_H(x)$, we must only invent a method for finding the particular solution to have our general solution.

• **Example 3.4.1**

Let us illustrate this technique with the second-order, linear, nonhomogeneous ordinary differential equation

$$y'' - 4y' + 4y = 2e^{2x} + 4x - 12. \quad (3.4.4)$$

Taking $y(x) = y_H(x) + y_p(x)$, direction substitution yields

$$y_H'' + y_p'' - 4(y_H' + y_p') + 4(y_H + y_p) = 2e^{2x} + 4x - 12. \quad (3.4.5)$$

If we now require that the particular solution $y_p(x)$ satisfies the differential equation

$$y_p'' - 4y_p' + 4y_p = 2e^{2x} + 4x - 12, \quad (3.4.6)$$

(3.4.5) simplifies to the homogeneous ordinary differential equation

$$y_H'' - 4y_H' + 4y_H = 0. \quad (3.4.7)$$

A quick check[6] shows that the particular solution to (3.4.6) is $y_p(x) = x^2 e^{2x} + x - 2$. Using techniques from the previous section, the complementary solution is $y_H(x) = C_1 e^{2x} + C_2 x e^{2x}$.

In general, finding $y_p(x)$ is a formidable task. In the case of constant coefficients, several techniques have been developed. The most commonly employed technique is called the *method of undetermined coefficients* which is used with linear, constant coefficient, ordinary differential equations when $f(x)$ is a constant, a polynomial, an exponential function $e^{\alpha x}$, $\sin(\beta x)$, $\cos(\beta x)$, or finite sum and products of these functions. Thus, this technique applies when the function $f(x)$ equals $e^x \sin(x) - (3x - 2)e^{-2x}$ but not when it equals $\ln(x)$.

---

[6] We will show how $y_p(x)$ was obtained momentarily.

Why does this technique work? The reason lies in the set of functions that we have allowed to be included in $f(x)$. They enjoy the remarkable property that derivatives of their sums and products yield sums and products that are also constants, polynomials, exponentials, sines, and cosines. Because a linear combination of derivatives such as $ay_p'' + by_p' + cy_p$ must equal $f(x)$, it seems reasonable to assume that $y_p(x)$ has the same form as $f(x)$. The following examples show that our conjecture is correct.

- **Example 3.4.2**

Let us illustrate the method of undetermined coefficients by finding the particular solution to

$$y'' - 2y' + y = x + \sin(x) \tag{3.4.8}$$

by the method of undetermined coefficients.

From the form of the right side of (3.4.8), we guess the particular solution

$$y_p(x) = Ax + B + C\sin(x) + D\cos(x). \tag{3.4.9}$$

Therefore,

$$y_p'(x) = A + C\cos(x) - D\sin(x), \tag{3.4.10}$$

and

$$y_p''(x) = -C\sin(x) - D\cos(x). \tag{3.4.11}$$

Substituting into (3.4.8), we find that

$$y_p'' - 2y_p' + y_p = Ax + B - 2A - 2C\cos(x) + 2D\sin(x) = x + \sin(x). \tag{3.4.12}$$

Since (3.4.12) must be true for all x, the constant terms must sum to zero or $B - 2A = 0$. Similarly, all of the terms involving the polynomial x must balance, yielding $A = 1$ and $B = 2A = 2$. Turning to the trigonometric terms, the coefficients of $\sin(x)$ and $\cos(x)$ give $2D = 1$ and $-2C = 0$, respectively. Therefore, the particular solution is

$$y_p(x) = x + 2 + \tfrac{1}{2}\cos(x), \tag{3.4.13}$$

and the general solution is

$$y(x) = y_H(x) + y_p(x) = C_1 e^x + C_2 x e^x + x + 2 + \tfrac{1}{2}\cos(x). \tag{3.4.14}$$

We can verify our result by using the symbolic toolbox in MATLAB. Typing the command:

```
dsolve('D2y-2*Dy+y=x+sin(x)','x')
```

yields

```
ans =
x+2+1/2*cos(x)+C1*exp(x)+C2*exp(x)*x
```

• **Example 3.4.3**

Let us find the particular solution to

$$y'' + y' - 2y = xe^x \qquad (3.4.15)$$

by the method of undetermined coefficients.

From the form of the right side of (3.4.15), we guess the particular solution

$$y_p(x) = Axe^x + Be^x. \qquad (3.4.16)$$

Therefore,

$$y_p'(x) = Axe^x + Ae^x + Be^x, \qquad (3.4.17)$$

and

$$y_p''(x) = Axe^x + 2Ae^x + Be^x. \qquad (3.4.18)$$

Substituting into (3.4.15), we find that

$$3Ae^x = xe^x. \qquad (3.4.19)$$

Clearly we cannot choose a constant A such that (3.4.19) is satisfied. What went wrong?

To understand why, let us find the homogeneous or complementary solution to (3.4.15); it is

$$y_H(x) = C_1 e^{-2x} + C_2 e^x. \qquad (3.4.20)$$

Therefore, one of the assumed particular solutions, Be^x, is also a homogeneous solution and cannot possibly give a nonzero left side when substituted into the differential equation. Consequently, it would appear that the method of undetermined coefficients does not work when one of the terms on the right side is also a homogeneous solution.

Before we give up, let us recall that we had a similar situation in the case of linear homogeneous second-order ordinary differential equations when the roots from the auxiliary equation were equal. There we found one of the homogeneous solution was $e^{m_1 x}$. We eventually found that the second solution was $xe^{m_1 x}$. Could such a solution work here? Let us try.

We begin by modifying (3.4.16) by multiplying it by x. Thus, our new guess for the particular solution reads

$$y_p(x) = Ax^2 e^x + Bxe^x. \qquad (3.4.21)$$

Then,

$$y_p' = Ax^2 e^x + 2Axe^x + Bxe^x + Be^x, \qquad (3.4.22)$$

and

$$y_p'' = Ax^2 e^x + 4Axe^x + 2Ae^x + Bxe^x + 2Be^x. \qquad (3.4.23)$$

Substituting (3.4.21) into (3.4.15) gives

$$y_p'' + y_p' - 2y_p = 6Axe^x + 2Ae^x + 3Be^x = xe^x. \tag{3.4.24}$$

Grouping together terms that vary as xe^x, we find that $6A = 1$. Similarly, terms that vary as e^x yield $2A + 3B = 0$. Therefore,

$$y_p(x) = \tfrac{1}{6}x^2 e^x - \tfrac{1}{9}xe^x, \tag{3.4.25}$$

so that the general solution is

$$y(x) = y_H(x) + y_p(x) = C_1 e^{-2x} + C_2 e^x + \tfrac{1}{6}x^2 e^x - \tfrac{1}{9}xe^x. \tag{3.4.26}$$

In summary, the method of finding particular solutions to higher-order ordinary differential equations by the method of undetermined coefficients is as follows:

- **Step 1:** Find the homogeneous solution to the differential equation.

- **Step 2:** Make an initial guess at the particular solution. The form of $y_p(x)$ is a linear combination of all linearly independent functions that are generated by repeated differentiations of $f(x)$.

- **Step 3:** If any of the terms in $y_p(x)$ given in Step 2 duplicate any of the homogeneous solutions, then that particular term in $y_p(x)$ must be multiplied by x^n, where n is the smallest positive integer that eliminates the duplication.

- **Example 3.4.4**

Let us apply the method of undetermined coefficients to solve

$$y'' + y = \sin(x) - e^{3x}\cos(5x). \tag{3.4.27}$$

We begin by first finding the solution to the homogeneous version of (3.4.27):

$$y_H'' + y_H = 0. \tag{3.4.28}$$

Its solution is

$$y_H(x) = A\cos(x) + B\sin(x). \tag{3.4.29}$$

To find the particular solution we examine the right side of (3.4.27) or

$$f(x) = \sin(x) - e^{3x}\cos(5x). \tag{3.4.30}$$

Taking a few derivatives of $f(x)$, we find that

$$f'(x) = \cos(x) - 3e^{3x}\cos(5x) + 5e^{3x}\sin(5x), \qquad (3.4.31)$$

$$f''(x) = -\sin(x) - 9e^{3x}\cos(5x) + 30e^{3x}\sin(5x) + 25e^{3x}\cos(5x), \qquad (3.4.32)$$

and so forth. Therefore, our guess at the particular solution is

$$y_p(x) = Cx\sin(x) + Dx\cos(x) + Ee^{3x}\cos(5x) + Fe^{3x}\sin(5x). \qquad (3.4.33)$$

Why have we chosen $x\sin(x)$ and $x\cos(x)$ rather than $\sin(x)$ and $\cos(x)$? Because $\sin(x)$ and $\cos(x)$ are homogeneous solutions to (3.4.27), we must multiply them by a power of x.

Since

$$y_p''(x) = 2C\cos(x) - Cx\sin(x) - 2D\sin(x) - Dx\cos(x)$$
$$+ (30F - 16E)e^{3x}\cos(5x) - (30E + 16F)e^{3x}\sin(5x), \qquad (3.4.34)$$

$$y_p'' + y_p = 2C\cos(x) - 2D\sin(x)$$
$$+ (30F - 15E)e^{3x}\cos(5x) - (30E + 15F)e^{3x}\sin(5x) \qquad (3.4.35)$$
$$= \sin(x) - e^{3x}\cos(5x). \qquad (3.4.36)$$

Therefore, $2C = 0$, $-2D = 1$, $30F - 15E = -1$, and $30E + 15F = 0$. Solving this system of equations yields $C = 0$, $D = -\frac{1}{2}$, $E = \frac{1}{75}$, and $F = -\frac{2}{75}$. Thus, the general solution is

$$y(x) = A\cos(x) + B\sin(x) - \tfrac{1}{2}x\cos(x) + \tfrac{1}{75}e^{3x}[\cos(5x) - 2\sin(5x)]. \qquad (3.4.37)$$

Problems

Use the method of undetermined coefficients to find the general solution of the following differential equations. Verify your solution by using `dsolve` in MATLAB.

1. $y'' + 4y' + 3y = x + 1$

2. $y'' - y = e^x - 2e^{-2x}$

3. $y'' + 2y' + 2y = 2x^2 + 2x + 4$

4. $y'' + y' = x^2 + x$

5. $y'' + 2y' = 2x + 5 - e^{-2x}$

6. $y'' - 4y' + 4y = (x + 1)e^{2x}$

7. $y'' + 4y' + 4y = xe^x$

8. $y'' - 4y = 4\sinh(2x)$

9. $y'' + 9y = x\cos(3x)$

10. $y'' + y = \sin(x) + x\cos(x)$

11. Solve

$$y'' + 2ay' = \sin^2(\omega x), \qquad y(0) = y'(0) = 0,$$

by (a) the method of undetermined coefficients and (b) integrating the ordinary differential equation so that it reduces to

$$y' + 2ay = \frac{x}{2} - \frac{\sin(2ax)}{4a}$$

and then using the techniques from the previous chapter to solve this first-order ordinary differential equation.

3.5 FORCED HARMONIC MOTION

Let us now consider the situation when an external force $f(t)$ acts on a vibrating mass on a spring. For example, $f(t)$ could represent a driving force that periodically raises and lowers the support of the spring. The inclusion of $f(t)$ in the formulation of Newton's second law yields the differential equation

$$m\frac{d^2x}{dt^2} = -kx - \beta\frac{dx}{dt} + f(t), \qquad (3.5.1)$$

$$\frac{d^2x}{dt^2} + \frac{\beta}{m}\frac{dx}{dt} + \frac{k}{m}x = \frac{f(t)}{m}, \qquad (3.5.2)$$

or

$$\frac{d^2x}{dt^2} + 2\lambda\frac{dx}{dt} + \omega^2 x = F(t), \qquad (3.5.3)$$

where $F(t) = f(t)/m$, $2\lambda = \beta/m$, and $\omega^2 = k/m$. To solve this nonhomogeneous equation we will use the method of undetermined coefficients.

● **Example 3.5.1**

Let us find the solution to the nonhomogeneous differential equation

$$y'' + 2y' + y = 2\sin(t), \qquad (3.5.4)$$

subject to the initial conditions $y(0) = 2$ and $y'(0) = 1$.
    The homogeneous solution is easily found and equals

$$y_H(t) = Ae^{-t} + Bte^{-t}. \qquad (3.5.5)$$

From the method of undetermined coefficients, we guess that the particular solution is

$$y_p(t) = C\cos(t) + D\sin(t), \qquad (3.5.6)$$

so that

$$y_p'(t) = -C\sin(t) + D\cos(t), \qquad (3.5.7)$$

and

$$y_p''(t) = -C\cos(t) - D\sin(t). \tag{3.5.8}$$

Substituting $y_p(t)$, $y_p'(t)$, and $y_p''(t)$ into (3.5.4) and simplifying, we find that

$$-2C\sin(t) + 2D\cos(t) = 2\sin(t) \tag{3.5.9}$$

or $D = 0$ and $C = -1$.

To find A and B, we now apply the initial conditions on the general solution

$$y(t) = Ae^{-t} + Bte^{-t} - \cos(t). \tag{3.5.10}$$

The initial condition $y(0) = 2$ yields

$$y(0) = A + 0 - 1 = 2, \tag{3.5.11}$$

or $A = 3$. The initial condition $y'(0) = 1$ gives

$$y'(0) = -A + B = 1, \tag{3.5.12}$$

or $B = 4$, since

$$y'(t) = -Ae^{-t} + Be^{-t} - Bte^{-t} + \sin(t). \tag{3.5.13}$$

Therefore, the solution which satisfies the differential equation and initial conditions is

$$y(t) = 3e^{-t} + 4te^{-t} - \cos(t). \tag{3.5.14}$$

• **Example 3.5.2**

Let us solve the differential equation for a weakly damped harmonic oscillator when the constant forcing F_0 "turns on" at $t = t_0$. The initial conditions are that $x(0) = x_0$ and $x'(0) = v_0$. Mathematically, the problem is

$$x'' + 2\lambda x' + \omega^2 x = \begin{cases} 0, & 0 < t < t_0, \\ F_0, & t_0 < t, \end{cases} \tag{3.5.15}$$

with $x(0) = x_0$ and $x'(0) = v_0$.

To solve (3.5.15), we first divide the time domain into two regions: $0 < t < t_0$ and $t_0 < t$. For $0 < t < t_0$,

$$x(t) = Ae^{-\lambda t}\cos(\omega_d t) + Be^{-\lambda t}\sin(\omega_d t), \tag{3.5.16}$$

where $\omega_d^2 = \omega^2 - \lambda^2$. Upon applying the initial conditions,

$$x(t) = x_0 e^{-\lambda t}\cos(\omega_d t) + \frac{v_0 + \lambda x_0}{\omega_d}e^{-\lambda t}\sin(\omega_d t), \tag{3.5.17}$$

as before.

For the region $t_0 < t$, we write the general solution as

$$x(t) = Ae^{-\lambda t}\cos(\omega_d t) + Be^{-\lambda t}\sin(\omega_d t) + \frac{F_0}{\omega^2}$$
$$+ Ce^{-\lambda(t-t_0)}\cos[\omega_d(t-t_0)] + De^{-\lambda(t-t_0)}\sin[\omega_d(t-t_0)]. \quad (3.5.18)$$

Why have we written our solution in this particular form rather than the simpler

$$x(t) = Ce^{-\lambda t}\cos(\omega_d t) + De^{-\lambda t}\sin(\omega_d t) + \frac{F_0}{\omega^2}? \quad (3.5.19)$$

Both solutions satisfy the differential equation as direct substitution verifies. However, the algebra is greatly simplified when (3.5.18) rather than (3.5.19) is used in matching the solution from each region at $t = t_0$. There both the solution and its first derivative must be continuous or

$$x(t_0^-) = x(t_0^+), \quad \text{and} \quad x'(t_0^-) = x'(t_0^+), \quad (3.5.20)$$

where t_0^- and t_0^+ are points just below and above t_0, respectively. When (3.5.17) and (3.5.18) are substituted, we find that $C = -F_0/\omega^2$, and $\omega_d D = \lambda C$. Thus, the solution for the region $t_0 < t$ is

$$x(t) = x_0 e^{-\lambda t}\cos(\omega_d t) + \frac{v_0 + \lambda x_0}{\omega_d}e^{-\lambda t}\sin(\omega_d t) + \frac{F_0}{\omega^2}$$
$$- \frac{F_0}{\omega^2}e^{-\lambda(t-t_0)}\cos[\omega_d(t-t_0)] - \frac{\lambda F_0}{\omega_d\omega^2}e^{-\lambda(t-t_0)}\sin[\omega_d(t-t_0)].$$
$$(3.5.21)$$

As we will see in Chapter 6, the technique of Laplace transforms is particularly well suited for this type of problem when the forcing function changes abruptly at one or more times.

As noted earlier, nonhomogeneous solutions consist of the homogeneous solution plus a particular solution. In the case of a damped harmonic oscillator, another, more physical, way of describing the solution involves its behavior at large time. That portion of the solution that eventually becomes negligible as $t \to \infty$ is often referred to as the *transient term*, or *transient solution*. In (3.5.14) the transient solution equals $3e^{-t} + 4te^{-t}$. On the other hand, the portion of the solution that remains as $t \to \infty$ is called the *steady-state solution*. In (3.5.14) the steady-state solution equals $-\cos(t)$.

One of the most interesting forced oscillator problems occurs when $\beta = 0$ and the forcing function equals $F_0 \sin(\omega_0 t)$, where F_0 is a constant. Then the initial-value problem becomes

$$\frac{d^2 x}{dt^2} + \omega^2 x = F_0 \sin(\omega_0 t). \quad (3.5.22)$$

Review of the Solution of the
Forced Harmonic Oscillator Problem

The undamped system $mx'' + kx = F_0 \cos(\omega_0 t)$ subject to the initial conditions $x(0) = x_0$ and $x'(0) = v_0$ has the solution

$$x(t) = \frac{v_0}{\omega} \sin(\omega t) + \left(x_0 - \frac{f_0}{\omega^2 - \omega_0^2} \right) \cos(\omega t) + \frac{f_0}{\omega^2 - \omega_0^2} \cos(\omega_0 t),$$

where $f_0 = F_0/m$ and $\omega = \sqrt{k/m}$. The underdamped system $mx'' + \beta x' + kx = F_0 \cos(\omega_0 t)$ has the *steady-state* solution

$$x(t) = \frac{f_0}{\sqrt{(\omega^2 - \omega_0^2)^2 + (2\lambda\omega_0)^2}} \cos\left[\omega_0 t - \tan^{-1} \left(\frac{2\lambda\omega_0}{\omega^2 - \omega_0^2} \right) \right],$$

where $2\lambda = \beta/m$.

Let us solve this problem when $x(0) = x'(0) = 0$.

The homogeneous solution to (3.5.22) is

$$x_H(t) = A \cos(\omega t) + B \sin(\omega t). \tag{3.5.23}$$

To obtain the particular solution, we assume that

$$x_p(t) = C \cos(\omega_0 t) + D \sin(\omega_0 t). \tag{3.5.24}$$

This leads to

$$x_p'(t) = -C\omega_0 \sin(\omega_0 t) + D\omega_0 \cos(\omega_0 t), \tag{3.5.25}$$

$$x_p''(t) = -C\omega_0^2 \cos(\omega_0 t) + D\omega_0^2 \sin(\omega_0 t), \tag{3.5.26}$$

and

$$x_p'' + \omega^2 x_p = C(\omega^2 - \omega_0^2) \cos(\omega_0 t) + D(\omega^2 - \omega_0^2) \sin(\omega_0 t) = F_0 \sin(\omega_0 t). \tag{3.5.27}$$

We immediately conclude that $C(\omega^2 - \omega_0^2) = 0$, and $D(\omega^2 - \omega_0^2) = F_0$. Therefore,

$$C = 0, \quad \text{and} \quad D = \frac{F_0}{\omega^2 - \omega_0^2}, \tag{3.5.28}$$

provided that $\omega \neq \omega_0$. Thus,

$$x_p(t) = \frac{F_0}{\omega^2 - \omega_0^2} \sin(\omega_0 t). \tag{3.5.29}$$

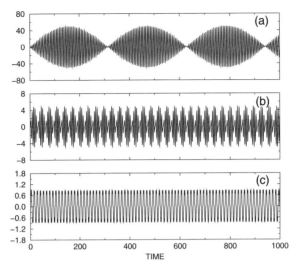

Figure 3.5.1: The solution (3.5.31) as a function of time when $\omega = 1$ and ω_0 equals (a) 1.02, (b) 1.2, and (c) 2.

To finish the problem, we must apply the initial conditions to the general solution

$$x(t) = A \cos(\omega t) + B \sin(\omega t) + \frac{F_0}{\omega^2 - \omega_0^2} \sin(\omega_0 t). \qquad (3.5.30)$$

From $x(0) = 0$, we find that $A = 0$. On the other hand, $x'(0) = 0$ yields $B = -\omega_0 F_0 / [\omega(\omega^2 - \omega_0^2)]$. Thus, the final result is

$$x(t) = \frac{F_0}{\omega(\omega^2 - \omega_0^2)} \left[\omega \sin(\omega_0 t) - \omega_0 \sin(\omega t) \right]. \qquad (3.5.31)$$

Equation (3.5.31) is illustrated in Figure 3.5.1 as a function of time.

The most arresting feature in Figure 3.5.1 is the evolution of the uniform amplitude of the oscillation shown in frame (c) into the one shown in frame (a) where the amplitude exhibits a sinusoidal variation as $\omega_0 \to \omega$. In acoustics these fluctuations in the amplitude are called *beats*, the loud sounds corresponding to the larger amplitudes.

As our analysis indicates, (3.5.31) does not apply when $\omega = \omega_0$. As we shall shortly see, this is probably the most interesting configuration. We can use (3.5.31) to examine this case by applying L'Hôpital's rule in the limiting case of $\omega_0 \to \omega$. This limiting process is analogous to "tuning in" the frequency of the driving frequency $[\omega_0/(2\pi)]$ to the frequency of free vibrations $[\omega/(2\pi)]$. From experience, we expect that given enough time we should be able to substantially increase the amplitudes of vibrations. Mathematical confirmation of our physical intuition is as follows:

$$x(t) = \lim_{\omega_0 \to \omega} F_0 \frac{\omega \sin(\omega_0 t) - \omega_0 \sin(\omega t)}{\omega(\omega^2 - \omega_0^2)} \qquad (3.5.32)$$

$$x(t) = F_0 \lim_{\omega_0 \to \omega} \frac{d[\omega \sin(\omega_0 t) - \omega_0 \sin(\omega t)]/d\omega_0}{d[\omega(\omega^2 - \omega_0^2)]/d\omega_0} \tag{3.5.33}$$

$$= F_0 \lim_{\omega_0 \to \omega} \frac{\omega t \cos(\omega_0 t) - \sin(\omega t)}{-2\omega_0 \omega} \tag{3.5.34}$$

$$= F_0 \frac{\omega t \cos(\omega t) - \sin(\omega t)}{-2\omega^2} \tag{3.5.35}$$

$$= \frac{F_0}{2\omega^2} \sin(\omega t) - \frac{F_0 t}{2\omega} \cos(\omega t). \tag{3.5.36}$$

As we suspected, as $t \to \infty$, the displacement grows without bounds. This phenomenon is known as *pure resonance*. We could also have obtained (3.5.36) directly using the method of undetermined coefficients involving the initial value problem

$$\frac{d^2 x}{dt^2} + \omega^2 x = F_0 \sin(\omega t), \qquad x(0) = x'(0) = 0. \tag{3.5.37}$$

Because there is almost always some friction, pure resonance rarely occurs and the more realistic differential equation is

$$\frac{d^2 x}{dt^2} + 2\lambda \frac{dx}{dt} + \omega^2 x = F_0 \sin(\omega_0 t). \tag{3.5.38}$$

Its solution is

$$x(t) = Ce^{-\lambda t} \sin\left(t\sqrt{\omega^2 - \omega_0^2} + \varphi\right)$$

$$+ \frac{F_0}{\sqrt{(\omega^2 - \omega_0^2)^2 + 4\lambda^2 \omega_0^2}} \sin(\omega_0 t - \theta), \tag{3.5.39}$$

where

$$\sin(\theta) = \frac{2\lambda \omega_0}{\sqrt{(\omega^2 - \omega_0^2)^2 + 4\lambda^2 \omega_0^2}}, \tag{3.5.40}$$

$$\cos(\theta) = \frac{\omega^2 - \omega_0^2}{\sqrt{(\omega^2 - \omega_0^2)^2 + 4\lambda^2 \omega_0^2}}, \tag{3.5.41}$$

and C and φ are determined by the initial conditions. To illustrate (3.5.39) we rewrite the amplitude and phase of the particular solution as

$$\frac{F_0}{\sqrt{(\omega^2 - \omega_0^2)^2 + 4\lambda^2 \omega_0^2}} = \frac{F_0}{\omega^2 \sqrt{(1 - r^2)^2 + 4\beta^2 r^2}} \tag{3.5.42}$$

and

$$\tan(\theta) = \frac{2\beta r}{1 - r^2}, \tag{3.5.43}$$

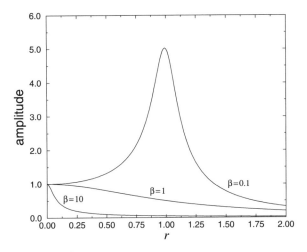

Figure 3.5.2: The amplitude of the particular solution (3.5.39) for a forced, damped simple harmonic oscillator (normalized with F_0/ω^2) as a function of $r = \omega_0/\omega$.

where $r = \omega_0/\omega$ and $\beta = \lambda/\omega$. Figures 3.5.2 and 3.5.3 graph (3.5.42) and (3.5.43) as functions of r for various values of β.

● **Example 3.5.3: Electrical circuits**

In the previous chapter, we saw how the mathematical analysis of electrical circuits yields first-order linear differential equations. In those cases we only had a resistor and capacitor or a resistor and inductor. One of the fundamental problems of electrical circuits is a circuit where a resistor, capacitor, and inductor are connected in series, as shown in Figure 3.5.4.

In this RCL circuit, an instantaneous current flows when the key or switch K is closed. If $Q(t)$ denotes the instantaneous charge on the capacitor, Kirchhoff's law yields the differential equation

$$L\frac{dI}{dt} + RI + \frac{Q}{C} = E(t), \tag{3.5.44}$$

where $E(t)$, the electromotive force, may depend on time, but where L, R, and C are constant. Because $I = dQ/dt$, (3.5.44) becomes

$$L\frac{d^2Q}{dt^2} + R\frac{dQ}{dt} + \frac{Q}{C} = E(t). \tag{3.5.45}$$

Consider now the case when resistance is negligibly small. Equation (3.5.45) will become identical to the differential equation for the forced simple harmonic oscillator, (3.5.3), with $\lambda = 0$. Similarly, the general case yields various analogs to the damped harmonic oscillator:

Case 1	Overdamped	$R^2 > 4L/C$
Case 2	Critically damped	$R^2 = 4L/C$
Case 3	Underdamped	$R^2 < 4L/C$

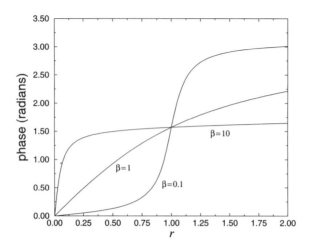

Figure 3.5.3: The phase of the particular solution (3.5.39) for a forced, damped simple harmonic oscillator as a function of $r = \omega_0/\omega$.

In each of these three cases, $Q(t) \to 0$ as $t \to \infty$. (See Problem 6.) Therefore, an RLC electrical circuit behaves like a damped mass-spring mechanical system, where inductance acts like mass, resistance is the damping coefficient, and $1/C$ is the spring constant.

Problems

1. Find the values of γ so that $x'' + 6x' + 18 = \cos(\gamma t)$ is in resonance.

2. The differential equation

$$x'' + 2x' + 2x = 10\sin(2t)$$

describes a damped, forced oscillator. If the initial conditions are $x(0) = x_0$ and $x'(0) = 0$, find its solution by hand and by using MATLAB. Plot the solution when $x_0 = -10, -9, \ldots, 9, 10$. Give a physical interpretation to what you observe.

3. At time $t = 0$, a mass m is suddenly attached to the end of a hanging spring with a spring constant k. Neglecting friction, find the subsequent motion if the coordinate system is chosen so that $x(0) = 0$.

Step 1: Show that the differential equation is

$$m\frac{d^2x}{dt^2} + kx = mg,$$

with the initial conditions $x(0) = x'(0) = 0$.

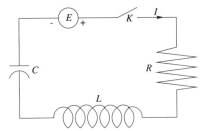

Figure 3.5.4: A simple electrical circuit containing a resistor of constant resistance R, capacitor of constant capacitance C, and inductor of constant inductance L driven by a time-dependent electromotive force $E(t)$.

Step 2: Show that the solution to Step 1 is

$$x(t) = mg\left[1 - \cos(\omega t)\right]/k, \qquad \omega^2 = k/m.$$

4. Consider the electrical circuit shown in Figure 3.5.4 which now possesses negligible resistance and has an applied voltage $E(t) = E_0[1 - \cos(\omega t)]$. Find the *current* if the circuit is initially dead.

5. Find the general solution to the differential equation governing a forced, damped harmonic equation

$$mx'' + cx' + kx = F_0 \sin(\omega t),$$

where m, c, k, F_0, and ω are constants. Write the particular solution in amplitude/phase format.

6. Prove that the *transient* solution to (3.5.45) tends to zero as $t \to \infty$ if R, C, and L are greater than zero.

3.6 VARIATION OF PARAMETERS

As the previous section has shown, the method of undetermined coefficients can be used when the right side of the differential equation contains constants, polynomials, exponentials, sines, and cosines. On the other hand, when the right side contains terms other than these, variation of parameters provides a method for finding the particular solution.

To understand this technique, let us return to our solution of the first-order ordinary differential equation

$$\frac{dy}{dx} + P(x)y = f(x). \tag{3.6.1}$$

Its solution is

$$y(x) = C_1 e^{-\int P(x)\,dx} + e^{-\int P(x)\,dx}\int e^{\int P(x)\,dx} f(x)\,dx. \tag{3.6.2}$$

The solution (3.6.2) consists of two parts: The first term is the homogeneous solution and can be written $y_H(x) = C_1 y_1(x)$, where $y_1(x) = e^{-\int P(x)\,dx}$. The second term is the particular solution and equals the product of some function of x, say $u_1(x)$, times $y_1(x)$:

$$y_p(x) = e^{-\int P(x)\,dx} \int e^{\int P(x)\,dx} f(x)\,dx = u_1(x) y_1(x). \qquad (3.6.3)$$

This particular solution bears a striking resemblance to the homogeneous solution if we replace $u_1(x)$ with C_1.

Variation of parameters builds upon this observation by using the homogeneous solution $y_1(x)$ to construct a guess for the particular solution $y_p(x) = u_1(x) y_1(x)$. Upon substituting this guessed $y_p(x)$ into (3.6.1), we have that

$$\frac{d}{dx}(u_1 y_1) + P(x) u_1 y_1 = f(x), \qquad (3.6.4)$$

$$u_1 \frac{dy_1}{dx} + y_1 \frac{du_1}{dx} + P(x) u_1 y_1 = f(x), \qquad (3.6.5)$$

or

$$y_1 \frac{du_1}{dx} = f(x), \qquad (3.6.6)$$

since $y_1' + P(x) y_1 = 0$.

Using the technique of separating the variables, we have that

$$du_1 = \frac{f(x)}{y_1(x)}\,dx, \quad \text{and} \quad u_1(x) = \int \frac{f(x)}{y_1(x)}\,dx. \qquad (3.6.7)$$

Consequently, the particular solution equals

$$y_p(x) = u_1(x) y_1(x) = y_1(x) \int \frac{f(x)}{y_1(x)}\,dx. \qquad (3.6.8)$$

Upon substituting for $y_1(x)$, we obtain (3.6.3).

How do we apply this method to the linear second-order differential equation

$$a_2(x) y'' + a_1 y'(x) + a_0(x) y = g(x), \qquad (3.6.9)$$

or

$$y'' + P(x) y' + Q(x) y = f(x), \qquad (3.6.10)$$

where $P(x)$, $Q(x)$, and $f(x)$ are continuous on some interval I?

Let $y_1(x)$ and $y_2(x)$ denote the homogeneous solutions of (3.6.10). That is, $y_1(x)$ and $y_2(x)$ satisfy

$$y_1'' + P(x) y_1' + Q(x) y_1 = 0, \qquad (3.6.11)$$

and

$$y_2'' + P(x)y_2' + Q(x)y_2 = 0. \tag{3.6.12}$$

Following our previous example, we now seek two function $u_1(x)$ and $u_2(x)$ such that

$$y_p(x) = u_1(x)y_1(x) + u_2(x)y_2(x) \tag{3.6.13}$$

is a particular solution of (3.6.10). Once again, we replaced our arbitrary constants C_1 and C_2 by the "variable parameters" $u_1(x)$ and $u_2(x)$. Because we have two unknown functions, we require two equations to solve for $u_1(x)$ and $u_2(x)$. One of them follows from substituting $y_p(x) = u_1(x)y_1(x) + u_2(x)y_2(x)$ into (3.6.10). The other equation is

$$y_1(x)u_1'(x) + y_2(x)u_2'(x) = 0. \tag{3.6.14}$$

This equation is an assumption that is made to simplify the first and second derivative which is clearly seen by computing

$$y_p' = u_1y_1' + y_1u_1' + u_2y_2' + y_2u_2' = u_1y_1' + u_2y_2', \tag{3.6.15}$$

after applying (3.6.14). Continuing to the second derivative,

$$y_p'' = u_1y_1'' + y_1'u_1' + u_2y_2'' + y_2'u_2'. \tag{3.6.16}$$

Substituting these results into (3.6.10), we obtain

$$\begin{aligned}
y_p'' + P(x)y_p' + Q(x)y_p &= u_1y_1'' + y_1'u_1' + u_2y_2'' + y_2'u_2' \\
&\quad + Pu_1y_1' + Pu_2y_2' + Qu_1y_1 + Qu_2y_2, \tag{3.6.17} \\
&= u_1\left[y_1'' + P(x)y_1' + Q(x)y_1\right] \\
&\quad + u_2\left[y_2'' + P(x)y_2' + Q(x)y_2\right] \\
&\quad + y_1'u_1' + y_2'u_2' = f(x). \tag{3.6.18}
\end{aligned}$$

Hence, $u_1(x)$ and $u_2(x)$ must be functions that also satisfy the condition

$$y_1'u_1' + y_2'u_2' = f(x). \tag{3.6.19}$$

It is important to note that the differential equation must be written so that it conforms to (3.6.10). This may require the division of the differential equation by $a_2(x)$ so that you have the correct $f(x)$.

Equation (3.6.14) and (3.6.19) constitute a linear system of equations for determining the unknown derivatives u_1' and u_2'. By Cramer's rule,[7] the solutions of (3.6.14) and (3.6.19) equal

$$u_1'(x) = \frac{W_1}{W}, \quad \text{and} \quad u_2'(x) = \frac{W_2}{W}, \tag{3.6.20}$$

---

[7] If you are unfamiliar with Cramer's rule, see §14.3.

where

$$W = \begin{vmatrix} y_1 & y_2 \\ y_1' & y_2' \end{vmatrix}, \quad W_1 = \begin{vmatrix} 0 & y_2 \\ f(x) & y_2' \end{vmatrix}, \quad \text{and} \quad W_2 = \begin{vmatrix} y_1 & 0 \\ y_1' & f(x) \end{vmatrix}. \quad (3.6.21)$$

The determinant W is the Wronskian of y_1 and y_2. Because y_1 and y_2 are linearly independent on I, the Wronskian will never equal to zero for every x in the interval.

These results can be generalized to any nonhomogeneous, nth-order, linear equation of the form

$$y^{(n)} + P_{n-1}(x)y^{(n-1)} + P_1(x)y' + P_0(x) = f(x). \quad (3.6.22)$$

If $y_H(x) = C_1 y_1(x) + C_2 y_2(x) + \cdots + C_n y_n(x)$ is the complementary function for (3.6.22), then a particular solution is

$$y_p(x) = u_1(x)y_1(x) + u_2(x)y_2(x) + \cdots + u_n(x)y_n(x), \quad (3.6.23)$$

where the u_k', $k = 1, 2, \ldots, n$, are determined by the n equations:

$$y_1 u_1' + y_2 u_2' + \cdots + y_n u_n' = 0, \quad (3.6.24)$$

$$y_1' u_1' + y_2' u_2' + \cdots + y_n' u_n' = 0, \quad (3.6.25)$$

$$\vdots$$

$$y_1^{(n-1)} u_1' + y_2^{(n-1)} u_2' + \cdots + y_n^{(n-1)} u_n' = f(x). \quad (3.6.26)$$

The first $n - 1$ equations in this system, like (3.6.14), are assumptions made to simplify the first $n - 1$ derivatives of $y_p(x)$. The last equation of the system results from substituting the n derivative of $y_p(x)$ and the simplified lower derivatives into (3.6.22). Then, by Cramer's rule, we find that

$$u_k' = \frac{W_k}{W}, \quad k = 1, 2, \ldots, n, \quad (3.6.27)$$

where W is the Wronskian of y_1, y_2, ..., y_n, and W_k is the determinant obtained by replacing the kth column of the Wronskian by the column $[0, 0, 0, \cdots, f(x)]^T$.

• **Example 3.6.1**

Let us apply variation of parameters to find the general solution to

$$y'' + y' - 2y = xe^x. \quad (3.6.28)$$

We begin by first finding the homogeneous solution which satisfies the differential equation

$$y_H'' + y_H' - 2y_H = 0. \quad (3.6.29)$$

Applying the techniques from §3.1, the homogeneous solution is

$$y_H(x) = Ae^x + Be^{-2x}, \qquad (3.6.30)$$

yielding the two independent solutions $y_1(x) = e^x$, and $y_2(x) = e^{-2x}$. Thus, the method of variation of parameters yields the particular solution

$$y_p(x) = e^x u_1(x) + e^{-2x} u_2(x). \qquad (3.6.31)$$

From (3.6.14), we have that

$$e^x u_1'(x) + e^{-2x} u_2'(x) = 0, \qquad (3.6.32)$$

while

$$e^x u_1'(x) - 2e^{-2x} u_2'(x) = xe^x. \qquad (3.6.33)$$

Solving for $u_1'(x)$ and $u_2'(x)$, we find that

$$u_1'(x) = \tfrac{1}{3}x, \qquad (3.6.34)$$

or

$$u_1(x) = \tfrac{1}{6}x^2, \qquad (3.6.35)$$

and

$$u_2'(x) = -\tfrac{1}{3}xe^{3x}, \qquad (3.6.36)$$

or

$$u_2(x) = \tfrac{1}{27}(1 - 3x)e^{3x}. \qquad (3.6.37)$$

Therefore, the general solution is

$$
\begin{align}
y(x) &= Ae^x + Be^{-2x} + e^x u_1(x) + e^{-2x} u_2(x) && (3.6.38)\\
&= Ae^x + Be^{-2x} + \tfrac{1}{6}x^2 e^x + \tfrac{1}{27}(1 - 3x)e^x && (3.6.39)\\
&= Ce^x + Be^{-2x} + \left(\tfrac{1}{6}x^2 - \tfrac{1}{9}x\right)e^x. && (3.6.40)
\end{align}
$$

• Example 3.6.2

Let us find the general solution to

$$y'' + 2y' + y = e^{-x}\ln(x) \qquad (3.6.41)$$

by variation of parameters on the interval $(0, \infty)$.

We start by finding the homogeneous solution which satisfies the differential equation

$$y_H'' + 2y_H' + y_H = 0. \qquad (3.6.42)$$

Applying the techniques from §3.1, the homogeneous solution is

$$y_H(x) = Ae^{-x} + Bxe^{-x}, \qquad (3.6.43)$$

yielding the two independent solutions $y_1(x) = e^{-x}$ and $y_2(x) = xe^{-x}$. Thus, the particular solution equals

$$y_p(x) = e^{-x}u_1(x) + xe^{-x}u_2(x). \qquad (3.6.44)$$

From (3.6.14), we have that

$$e^{-x}u_1'(x) + xe^{-x}u_2'(x) = 0, \qquad (3.6.45)$$

while

$$-e^{-x}u_1'(x) + (1-x)e^{-x}u_2'(x) = e^{-x}\ln(x). \qquad (3.6.46)$$

Solving for $u_1'(x)$ and $u_2'(x)$, we find that

$$u_1'(x) = -x\ln(x), \qquad (3.6.47)$$

or

$$u_1(x) = \tfrac{1}{4}x^2 - \tfrac{1}{2}x^2\ln(x), \qquad (3.6.48)$$

and

$$u_2'(x) = \ln(x), \qquad (3.6.49)$$

or

$$u_2(x) = x\ln(x) - x. \qquad (3.6.50)$$

Therefore, the general solution is

$$y(x) = Ae^{-x} + Bxe^{-x} + e^{-x}u_1(x) + xe^{-x}u_2(x) \qquad (3.6.51)$$
$$= Ae^{-x} + Bxe^{-x} + \tfrac{1}{2}x^2\ln(x)e^{-x} - \tfrac{3}{4}x^2e^{-x}. \qquad (3.6.52)$$

We can verify our result by using the symbolic toolbox in MATLAB. Typing the command:

```
dsolve('D2y+2*Dy+y=exp(-x)*log(x)','x')
```

yields

```
ans =
1/2*exp(-x)*x^2*log(x)-3/4*exp(-x)*x^2+C1*exp(-x)+C2*exp(-x)*x
```

• **Example 3.6.3**

So far, all of our examples have yielded closed form solutions. To show that this is not necessarily so, let us solve

$$y'' - 4y = e^{2x}/x \qquad (3.6.53)$$

by variation of parameters.

Again we begin by solving the homogeneous differential equation

$$y_H'' - 4y_H = 0, \tag{3.6.54}$$

which has the solution

$$y_H(x) = Ae^{2x} + Be^{-2x}. \tag{3.6.55}$$

Thus, our two independent solutions are $y_1(x) = e^{2x}$ and $y_2(x) = e^{-2x}$. Therefore, the particular solution equals

$$y_P(x) = e^{2x}u_1(x) + e^{-2x}u_2(x). \tag{3.6.56}$$

From (3.6.14), we have that

$$e^{2x}u_1'(x) + e^{-2x}u_2'(x) = 0, \tag{3.6.57}$$

while

$$2e^{2x}u_1'(x) - 2e^{-2x}u_2'(x) = e^{2x}/x. \tag{3.6.58}$$

Solving for $u_1'(x)$ and $u_2'(x)$, we find that

$$u_1'(x) = \frac{1}{4x}, \tag{3.6.59}$$

or

$$u_1(x) = \tfrac{1}{4}\ln|x|, \tag{3.6.60}$$

and

$$u_2'(x) = -\frac{e^{4x}}{4x}, \tag{3.6.61}$$

or

$$u_2(x) = -\tfrac{1}{4}\int_{x_0}^x \frac{e^{4t}}{t}\,dt. \tag{3.6.62}$$

Therefore, the general solution is

$$y(x) = Ae^{2x} + Be^{-2x} + e^{2x}u_1(x) + e^{-2x}u_2(x) \tag{3.6.63}$$

$$= Ae^{2x} + Be^{-2x} + \tfrac{1}{4}\ln|x|e^{2x} - \tfrac{1}{4}e^{-2x}\int_{x_0}^x \frac{e^{4t}}{t}\,dt. \tag{3.6.64}$$

Problems

Use variation of parameters to find the general solution for the following differential equations. Then see if you can obtain your solution by using `dsolve` in MATLAB.

1. $y'' - 4y' + 3y = e^{-x}$ 　　　　　　　2. $y'' - y' - 2y = x$

3. $y'' - 4y = xe^{x}$ 　　　　　　　　　4. $y'' + 9y = 2\sec(x)$

5. $y'' + 4y' + 4y = xe^{-2x}$ 　　　　　6. $y'' + 2ay' = \sin^2(\omega x)$

7. $y'' - 4y' + 4y = (x+1)e^{2x}$ 　　　8. $y'' - 4y = \sin^2(x)$

9. $y'' - 2y' + y = e^{x}/x$ 　　　　　　10. $y'' + y = \tan(x)$

3.7 EULER-CAUCHY EQUATION

The Euler-Cauchy or equidimensional equation is a linear differential equation of the form

$$a_n x^n \frac{d^n y}{dx^n} + a_{n-1} x^{n-1} \frac{d^{n-1} y}{dx^{n-1}} + \cdots + a_1 x \frac{dy}{dx} + a_0 y = f(x), \qquad (3.7.1)$$

where a_n, a_{n-1}, ..., a_0 are constants. The important point here is that in each term the power to which x is raised equals the *order* of differentiation.

To illustrate this equation, we will focus on the homogeneous, second-order, ordinary differential equation

$$ax^2 \frac{d^2 y}{dx^2} + bx \frac{dy}{dx} + cy = 0. \qquad (3.7.2)$$

The solution of higher-order ordinary differential equations follows by analog. If we wish to solve the nonhomogeneous equation

$$ax^2 \frac{d^2 y}{dx^2} + bx \frac{dy}{dx} + cy = f(x), \qquad (3.7.3)$$

we can do so by applying variation of parameters using the complementary solutions that satisfy (3.7.2).

Our analysis starts by trying a solution of the form $y = x^m$, where m is presently undetermined. The first and second derivatives are

$$\frac{dy}{dx} = mx^{m-1}, \quad \text{and} \quad \frac{d^2 y}{dx^2} = m(m-1)x^{m-2}, \qquad (3.7.4)$$

respectively. Consequently, substitution yields the differential equation

$$ax^2 \frac{d^2y}{dx^2} + bx\frac{dy}{dx} + cy = ax^2 \cdot m(m-1)x^{m-2} + bx \cdot mx^{m-1} + cx^m \quad \textbf{(3.7.5)}$$

$$= am(m-1)x^m + bmx^m + cx^m \quad \textbf{(3.7.6)}$$

$$= [am(m-1) + bm + c]\,x^m. \quad \textbf{(3.7.7)}$$

Thus, $y = x^m$ is a solution of the differential equation whenever m is a solution of the *auxiliary equation*

$$am(m-1) + bm + c = 0, \quad \text{or} \quad am^2 + (b-a)m + c = 0. \quad \textbf{(3.7.8)}$$

At this point we must consider three different cases which depend upon the values of a, b, and c.

- *Distinct real roots*

Let m_1 and m_2 denote the real roots of (3.7.8) such that $m_1 \neq m_2$. Then,

$$y_1(x) = x^{m_1} \quad \text{and} \quad y_2(x) = x^{m_2} \quad \textbf{(3.7.9)}$$

are homogeneous solutions to (3.7.2). Therefore, the general solution is

$$y(x) = C_1 x^{m_1} + C_2 x^{m_2}. \quad \textbf{(3.7.10)}$$

- *Repeated real roots*

If the roots of (3.7.8) are repeated $[m_1 = m_2 = -(b-a)/2]$, then we presently have only one solution, $y = x^{m_1}$. To construct the second solution y_2, we use reduction in order. We begin by first rewriting the Euler-Cauchy equation as

$$\frac{d^2y}{dx^2} + \frac{b}{ax}\frac{dy}{dx} + \frac{c}{ax^2}y = 0. \quad \textbf{(3.7.11)}$$

Letting $P(x) = b/(ax)$, we have

$$y_2(x) = x^{m_1} \int \frac{e^{-\int [b/(ax)]\,dx}}{(x^{m_1})^2}\,dx = x^{m_1} \int \frac{e^{-(b/a)\ln(x)}}{x^{2m_1}}\,dx \quad \textbf{(3.7.12)}$$

$$= x^{m_1} \int x^{-b/a}x^{-2m_1}\,dx = x^{m_1} \int x^{-b/a}x^{(b-a)/a}\,dx \quad \textbf{(3.7.13)}$$

$$= x^{m_1} \int \frac{dx}{x} = x^{m_1}\ln(x). \quad \textbf{(3.7.14)}$$

The general solution is then

$$y(x) = C_1 x^{m_1} + C_2 x^{m_1}\ln(x). \quad \textbf{(3.7.15)}$$

For higher-order equations, if m_1 is a root of multiplicity k, then it can be shown that

$$x^{m_1}, x^{m_1} \ln(x), x^{m_1} [\ln(x)]^2, \ldots, x^{m_1} [\ln(x)]^{k-1}$$

are the k linearly independent solutions. Therefore, the general solution of the differential equation equals a linear combination of these k solutions.

• *Conjugate complex roots*

If the roots of (3.7.8) are the complex conjugate pair $m_1 = \alpha + i\beta$, and $m_2 = \alpha - i\beta$, where α and β are real and $\beta > 0$, then a solution is

$$y(x) = C_1 x^{\alpha+i\beta} + C_2 x^{\alpha-i\beta}. \tag{3.7.16}$$

However, because $x^{i\theta} = [e^{\ln(x)}]^{i\theta} = e^{i\theta \ln(x)}$, we have by Euler's formula

$$x^{i\theta} = \cos[\theta \ln(x)] + i \sin[\theta \ln(x)], \tag{3.7.17}$$

and

$$x^{-i\theta} = \cos[\theta \ln(x)] - i \sin[\theta \ln(x)], \tag{3.7.18}$$

Substitution into (3.7.16) leads to

$$y(x) = C_3 x^\alpha \cos[\beta \ln(x)] + C_4 x^\alpha \sin[\beta \ln(x)], \tag{3.7.19}$$

where $C_3 = C_1 + C_2$, and $C_4 = iC_1 - iC_2$.

• **Example 3.7.1**

Let us find the general solution to

$$x^2 y'' + 5xy' - 12y = \ln(x) \tag{3.7.20}$$

by the method of undetermined coefficients and variation of parameters.

In the case of undetermined coefficients, we begin by letting $t = \ln(x)$ and $y(x) = Y(t)$. Substituting these variables into (3.7.20), we find that

$$Y'' + 4Y' - 12Y = t. \tag{3.7.21}$$

The homogeneous solution to (3.7.21) is

$$Y_H(t) = A'e^{-6t} + B'e^{2t}, \tag{3.7.22}$$

while the particular solution is

$$Y_p(t) = Ct + D \tag{3.7.23}$$

from the method of undetermined coefficients. Substituting (3.7.23) into (3.7.21) yields $C = -\frac{1}{12}$ and $D = -\frac{1}{36}$. Therefore,

$$Y(t) = A'e^{-6t} + B'e^{2t} - \frac{1}{12}t - \frac{1}{36}, \tag{3.7.24}$$

or

$$y(x) = \frac{A}{x^6} + Bx^2 - \frac{1}{12}\ln(x) - \frac{1}{36}. \tag{3.7.25}$$

To find the particular solution via variation of parameters, we use the homogeneous solution

$$y_H(x) = \frac{A}{x^6} + Bx^2 \tag{3.7.26}$$

to obtain $y_1(x) = x^{-6}$ and $y_2(x) = x^2$. Therefore,

$$y_p(x) = x^{-6}u_1(x) + x^2u_2(x). \tag{3.7.27}$$

Substitution of (3.7.27) in (3.7.20) yields the system of equations:

$$x^{-6}u_1'(x) + x^2u_2'(x) = 0, \tag{3.7.28}$$

and

$$-6x^{-7}u_1'(x) + 2xu_2'(x) = \ln(x)/x^2. \tag{3.7.29}$$

Solving for $u_1'(x)$ and $u_2'(x)$,

$$u_1'(x) = -\frac{x^5 \ln(x)}{8}, \quad \text{and} \quad u_2'(x) = -\frac{\ln(x)}{8x^3}. \tag{3.7.30}$$

The solutions of these equations are

$$u_1(x) = -\frac{x^6 \ln(x)}{48} + \frac{x^6}{288}, \tag{3.7.31}$$

and

$$u_2(x) = -\frac{\ln(x)}{16x^2} - \frac{1}{32x^2}. \tag{3.7.32}$$

The general solution then equals

$$y(x) = \frac{A}{x^6} + Bx^2 + x^{-6}u_1(x) + x^2u_2(x) = \frac{A}{x^6} + Bx^2 - \frac{1}{12}\ln(x) - \frac{1}{36}. \tag{3.7.33}$$

We can verify this result by using the symbolic toolbox in MATLAB. Typing the command:

```
dsolve('x^2*D2y+5*x*Dy-12*y=log(x)','x')
```

yields

```
ans =
-1/12*log(x)-1/36+C1*x^2+C2/x^6
```

Problems

Find the general solution for the following Euler-Cauchy equations valid over the domain $(-\infty, \infty)$. Then check your answer by using dsolve in MATLAB.

1. $x^2 y'' + xy' - y = 0$

2. $x^2 y'' + 2xy' - 2y = 0$

3. $x^2 y'' - 2y = 0$

4. $x^2 y'' - xy' + y = 0$

5. $x^2 y'' + 3xy' + y = 0$

6. $x^2 y'' - 3xy' + 4y = 0$

7. $x^2 y'' - y' + 5y = 0$

8. $4x^2 y'' + 8xy' + 5y = 0$

9. $x^2 y'' + xy' + y = 0$

10. $x^2 y'' - 3xy' + 13y = 0$

11. $x^3 y''' - 2x^2 y'' - 2xy' + 8y = 0$　　12. $x^2 y'' - 2xy' - 4y = x$

3.8 PHASE DIAGRAMS

In §2.6 we showed how solutions to first-order ordinary differential equations could be *qualitatively* solved through the use of the phase line. This concept of qualitatively studying differential equations showed promise as a method for deducing many of the characteristics of the solution to a differential equation without actually solving it. In this section we extend these concepts to second-order ordinary differential equations by introducing the *phase plane*.

Consider the differential equation

$$x'' + \mathrm{sgn}(x) = 0, \tag{3.8.1}$$

where the signum function is defined by (5.2.11). Equation (3.8.1) describes, for example, the motion of an infinitesimal ball rolling in a "V" shaped trough in a constant gravitational field.[8]

Our analysis begins by introducing the new dependent variable $v = x'$ so that (3.8.1) can be written

$$v\frac{dv}{dx} + \mathrm{sgn}(x) = 0, \tag{3.8.2}$$

since

$$x'' = \frac{d^2 x}{dt^2} = \frac{dv}{dt} = \frac{dx}{dt}\frac{dv}{dx} = v\frac{dv}{dx}. \tag{3.8.3}$$

---

[8] See Lipscomb, T., and R. E. Mickens, 1994: Exact solution to the axisymmetric, constant force oscillator equation. *J. Sound Vibr.*, **169**, 138–140.

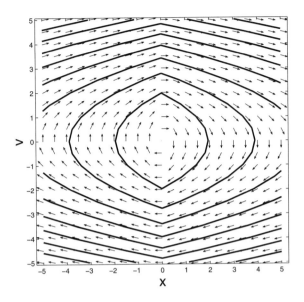

Figure 3.8.1: Phase diagram for the differential equation (3.8.1).

Equation (3.8.2) relates v to x and t has disappeared explicitly from the problem. Integrating (3.8.2) with respect to x, we obtain

$$\int v \, dv + \int \operatorname{sgn}(x) \, dx = C, \tag{3.8.4}$$

or

$$\tfrac{1}{2}v^2 + |x| = C. \tag{3.8.5}$$

Equation (3.8.5) expresses conservation of energy because the first term on the left side of (3.8.5) is kinetic energy while the second term is the potential energy. The value of C depends upon the initial condition $x(0)$ and $v(0)$. Thus, for a specific initial condition, (3.8.5) gives the relationship between x and v for the motion corresponding to the initial condition.

Although there is a closed-form solution for (3.8.1), let us imagine that there is none. What could we learn from (3.8.5)?

Equation (3.8.5) can be represented in a diagram, called a *phase plane*, where x and v are its axes. A given pair of (x, v) is called a *state* of the system. A given state determines all subsequent states because it serves as initial conditions for any subsequent motion.

For each different value of C, we will obtain a curve, commonly known as *phase paths*, *trajectories*, or *integral curves*, on the phase plane. In Figure 3.8.1, we used the MATLAB script

```
clear
% set up grid points in the (x,v) plane
```

```
[x,v] = meshgrid(-5:0.5:5,-5:0.5:5);
% compute slopes
dxdt = v; dvdt = -sign(x);
% find magnitude of vector [dxdt,dydt]
L = sqrt(dxdt.*dxdt + dvdt.*dvdt);
% plot scaled vectors
quiver(x,v,dxdt./L,dvdt./L,0.5); axis equal tight
hold
% contour trajectories
contour(x,v,v.*v/2 + abs(x),8)
h = findobj('Type','patch'); set(h,'Linewidth',2);
xlabel('x','Fontsize',20); ylabel('v','Fontsize',20)
```

to graph the phase plane for (3.8.1). Here the phase paths are simply closed, oval shaped curves which are symmetric with respect to both the x and v *phase space axes*. Each phase path corresponds to a particular possible motion of the system. Associated with each path is a direction, indicated by an arrow, showing how the state of the system changes as time increases.

An interesting feature on Figure 3.8.1 is the point $(0,0)$. What is happening there? In our discussion of phase line, we sought to determine whether there were any *equilibrium* or *critical points*. Recall that at an equilibrium or critical point the solution is constant and was given by $x' = 0$. In the case of second-order differential equations, we again have the condition $x' = v = 0$. For this reason equilibrium points are always situated on the abscissa of the phase diagram.

The condition $x' = 0$ is insufficient for determining critical points. For example, when a ball is thrown upward, its velocity equals zero at the peak height. However, this is clearly not a point of equilibrium. Consequently, we must impose the additional constraint that $x'' = v' = 0$. In the present example, equilibrium points occur where $x' = v = 0$ and $v' = -\text{sgn}(x) = 0$ or $x = 0$. Therefore, the point $(0,0)$ is the critical point for (3.8.1).

The closed curves immediately surrounding the origin in Figure 3.8.1 show that we have periodic solutions there because on completing a circuit, the original state returns and the motion simply repeats itself indefinitely.

Once we have found an equilibrium point, an obvious question is whether it is stable or not. To determine this, consider what happens if the initial state is displaced slightly from the origin. It lands on one of the nearby closed curves and the particle oscillates with small amplitude about the origin. Thus, this critical point is *stable*.

In the following examples, we further illustrate the details that may be gleaned from a phase diagram.

- **Example 3.8.1**

The equation describing a simple pendulum is

$$ma^2\theta'' + mga\sin(\theta) = 0, \tag{3.8.6}$$

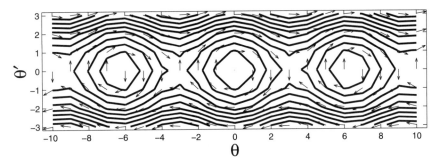

Figure 3.8.2: Phase diagram for a simple pendulum.

where m denotes the mass of the bob, a is the length of the rod or light string, and g is the acceleration due to gravity. Here the conservation of energy equation is

$$\tfrac{1}{2}ma^2\theta'^2 - mga\cos(\theta) = C. \tag{3.8.7}$$

Figure 3.8.2 is the phase diagram for the simple pendulum. Some of the critical points are located at $\theta = \pm 2n\pi$, $n = 0, 1, 2, \ldots$, and $\theta' = 0$. Near these critical points, we have closed patterns surrounding these critical points, just as we did in the earlier case of an infinitesimal ball rolling in a "V" shaped trough. Once again, these critical points are *stable* and the region around these equilibrium points corresponds to a pendulum swinging to and fro about the vertical. On the other hand, there is a new type of critical point at $\theta = \pm(2n - 1)\pi$, $n = 0, 1, 2, \ldots$ and $\theta' = 0$. Here the trajectories form hyperbolas near these equilibrium points. Thus, for any initial state that is near these critical points we have solutions that move away from the equilibrium point. This is an example of an *unstable* critical point. Physically these critical points correspond to a pendulum that is balanced on end. Any displacement from the equilibrium results in the bob falling from the inverted position.

Finally, we have a wavy line as $\theta' \to \pm\infty$. This corresponds to whirling motions of the pendulum where θ' has the same sign and θ continuously increases or decreases.

● **Example 3.8.2: Damped harmonic oscillator**

Consider the ordinary differential equation

$$x'' + 2x' + 5x = 0. \tag{3.8.8}$$

The exact solution to this differential equation is

$$x(t) = e^{-t}\left[A\cos(2t) + B\sin(2t)\right], \tag{3.8.9}$$

and

$$x'(t) = 2e^{-t}\left[B\cos(2t) - A\sin(2t)\right] - e^{-t}\left[A\cos(2t) + B\sin(2t)\right]. \tag{3.8.10}$$

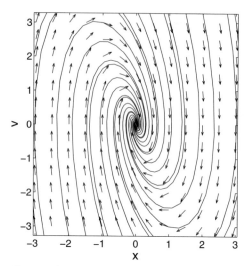

Figure 3.8.3: Phase diagram for the damped harmonic oscillator (3.8.8).

To construct its phase diagram, we again define $v = x'$ and replace (3.8.8) with $v' = -2v - 5x$. The MATLAB script

```
clear
% set up grid points in the x,x' plane
[x,v] = meshgrid(-3:0.5:3,-3:0.5:3);
% compute slopes
dxdt = v; dvdt = -2*v - 5*x;
% find length of vector
L = sqrt(dxdt.*dxdt + dvdt.*dvdt);
% plot direction field
quiver(x,v,dxdt./L,dvdt./L,0.5); axis equal tight
hold
% compute x(t) and v(t) at various times and a's and b's
for b = -3:2:3; for a = -3:2:3;
t = [-5:0.1:5];
xx = exp(-t) .* (a*cos(2*t) + b*sin(2*t));
vv = 2 * exp(-t) .* (b*cos(2*t) - a*sin(2*t)) - xx;
% plot these values
plot(xx,vv)
end; end;
xlabel('x','Fontsize',20); ylabel('v','Fontsize',20)
```

was used to construct the phase diagram for (3.8.8) and is shown in Figure 3.8.3. Here the equilibrium point is at $x = v = 0$. This is a new type of critical point. It is called a *stable node* because all slight displacements from this critical point eventually return to this equilibrium point.

Problems

1. Using MATLAB, construct the phase diagram for $x'' - 3x' + 2x = 0$. What happens around the point $x = v = 0$?

2. Consider the nonlinear differential equation $x'' = x^3 - x$. This equation arises in the study of simple pendulums with swings of moderate amplitude.

(a) Show that the conservation law is

$$\tfrac{1}{2}v^2 - \tfrac{1}{4}x^4 + \tfrac{1}{2}x^2 = C.$$

What is special about $C = 0$ and $C = \tfrac{1}{4}$?

(b) Show that there are three critical points: $x = 0$ and $x = \pm 1$ with $v = 0$.

(c) Using MATLAB, graph the phase diagram with axes x and v.

For the following ordinary differential equations, find the equilibrium points and then classify them. Use MATLAB to draw the phase diagrams.

3. $x'' = 2x'$

4. $x'' + \mathrm{sgn}(x)x = 0$

5. $x'' = \begin{cases} 1, & |x| > 2, \\ 0, & |x| < 2. \end{cases}$

3.9 NUMERICAL METHODS

When differential equations cannot be integrated in closed form, numerical methods must be employed. In the finite difference method, the discrete variable x_i or t_i replaces the continuous variable x or t and the differential equation is solved progressively in increments h starting from known initial conditions. The solution is approximate, but with a sufficiently small increment, you can obtain a solution of acceptable accuracy.

Although there are many different finite difference schemes available, we consider here only two methods that are chosen for their simplicity. The interested student may read any number of texts on numerical analysis if he or she wishes a wider view of other possible schemes.

Let us focus on second-order differential equations; the solution of higher-order differential equations follows by analog. In the case of second-order ordinary differential equations, the differential equation can be rewritten as

$$x'' = f(x, x', t), \qquad x_0 = x(0), \quad x_0' = x'(0), \qquad (3.9.1)$$

where the initial conditions x_0 and x_0' are assumed to be known.

For the present moment, let us treat the second-order ordinary differential equation

$$x'' = f(x, t), \qquad x_0 = x(0), \quad x_0' = x'(0). \tag{3.9.2}$$

The following scheme, known as the *central difference method*, computes the solution from Taylor expansions at x_{i+1} and x_{i-1}:

$$x_{i+1} = x_i + hx_i' + \tfrac{1}{2}h^2 x_i'' + \tfrac{1}{6}h^3 x_i''' + O(h^4) \tag{3.9.3}$$

and

$$x_{i-1} = x_i - hx_i' + \tfrac{1}{2}h^2 x_i'' - \tfrac{1}{6}h^3 x_i''' + O(h^4), \tag{3.9.4}$$

where h denotes the time interval Δt. Subtracting and ignoring higher-order terms, we obtain

$$x_i' = \frac{x_{i+1} - x_{i-1}}{2h}. \tag{3.9.5}$$

Adding (3.9.3) and (3.9.4) yields

$$x_i'' = \frac{x_{i+1} - 2x_i + x_{i-1}}{h^2}. \tag{3.9.6}$$

In both (3.9.5) and (3.9.6) we ignored terms of $O(h^2)$. After substituting into the differential equation (3.9.2), (3.9.6) can be rearranged to

$$x_{i+1} = 2x_i - x_{i-1} + h^2 f(x_i, t_i), \qquad i \geq 1, \tag{3.9.7}$$

which is known as the *recurrence formula*.

Consider now the situation when $i = 0$. We note that although we have x_0 we do not have x_{-1}. Thus, to start the computation, we need another equation for x_1. This is supplied by (3.9.3) which gives

$$x_1 = x_0 + hx_0' + \tfrac{1}{2}h^2 x_0'' = x_0 + hx_0' + \tfrac{1}{2}h^2 f(x_0, t_0). \tag{3.9.8}$$

Once we have computed x_1, then we can switch to (3.9.6) for all subsequent calculations.

In this development we have ignored higher-order terms that introduce what is known as *truncation errors*. Other errors, such as *round-off errors*, are introduced due to loss of significant figures. These errors are all related to the time increment h in a rather complicated manner that are investigated in numerical analysis books. In general, better accuracy is obtained by choosing a smaller h, but the number of computations will then increase together with errors.

• Example 3.9.1

Let us solve $x'' - 4x = 2t$ subject to $x(0) = x'(0) = 1$. The exact solution is

$$x(t) = \tfrac{7}{8}e^{2t} + \tfrac{1}{8}e^{-2t} - \tfrac{1}{2}t. \tag{3.9.9}$$

The MATLAB script

```
clear
% test out different time steps
for i = 1:3
% set up time step increment and number of time steps
h = 1/10^i; n = 10/h;
% set up initial conditions
t=zeros(n+1,1); t(1) = 0; x(1) = 1; x_exact(1) = 1;
% define right side of differential equation
f = inline('4*xx+2*tt','tt','xx');
% set up difference arrays for plotting purposes
diff = zeros(n,1); t_plot = zeros(n,1);
% compute first time step
t(2) = t(1) + h; x(2) = x(1) + h + 0.5*h*h*f(t(1),x(1));
x_exact(2) = (7/8)*exp(2*t(2))+(1/8)*exp(-2*t(2))-t(2)/2;
t_plot(1) = t(2);
diff(1) = x(2) - x_exact(2); diff(1) = abs(diff(1)/x_exact(2));
% compute the remaining time steps
for k = 2:n
t(k+1) = t(k) + h; t_plot(k) = t(k+1);
x(k+1) = 2*x(k) - x(k-1) + h*h*f(t(k),x(k));
x_exact(k+1) = (7/8)*exp(2*t(k+1)) + (1/8)*exp(-2*t(k+1)) ...
            - t(k+1)/2;
diff(k) = x(k+1) - x_exact(k+1);
diff(k) = abs(diff(k) / x_exact(k+1));
end
% plot the relative error
semilogy(t_plot,diff,'-')
hold on
num = 0.2*n;
text(3*i,diff(num),['h = ',num2str(h)],'Fontsize',15,...
    'HorizontalAlignment','right','VerticalAlignment','bottom')
xlabel('TIME','Fontsize',20);
ylabel('|RELATIVE ERROR|','Fontsize',20);
end
```

implements our simple finite difference method of solving a second-order ordinary differential equation. In Figure 3.9.1 we have plotted results for three different values of the time step. As our analysis suggests, the relative error is related to h^2.

An alternative method for integrating higher-order ordinary differential equations is Runge-Kutta. It is popular because it is self-starting and the results are very accurate.

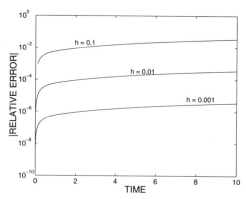

Figure 3.9.1: The numerical solution of $x'' - 4x = 2t$ when $x(0) = x'(0) = 1$ using a simple finite difference approach.

For second-order ordinary differential equations this method first reduces the differential equation into two first-order equations. For example, the differential equation

$$x'' = \frac{f(t) - kx - cx'}{m} = F(x, x', t) \tag{3.9.10}$$

becomes the first-order differential equations

$$x' = y, \qquad y' = F(x, y, t). \tag{3.9.11}$$

The Runge-Kutta procedure can then be applied to each of these equations. Using a fourth-order scheme, the procedure is as follows:

$$x_{i+1} = x_i + \tfrac{1}{6}h(k_1 + 2k_2 + 2k_3 + k_4), \tag{3.9.12}$$

and

$$y_{i+1} = y_i + \tfrac{1}{6}h(K_1 + 2K_2 + 2K_3 + K_4), \tag{3.9.13}$$

where

$$k_1 = y_i, \qquad K_1 = F(x_i, y_i, t_i), \tag{3.9.14}$$

$$k_2 = y_i + \tfrac{h}{2}K_1, \qquad K_2 = F(x_i + \tfrac{h}{2}k_1, k_2, t_i + \tfrac{h}{2}), \tag{3.9.15}$$

$$k_3 = y_i + \tfrac{h}{2}K_2, \qquad K_3 = F(x_i + \tfrac{h}{2}k_2, k_3, t_i + \tfrac{h}{2}), \tag{3.9.16}$$

and

$$k_4 = y_i + K_3 h, \qquad K_4 = F(x_i + hk_3, k_4, t_i + h). \tag{3.9.17}$$

- **Example 3.9.2**

    The MATLAB script

```
clear
% test out different time steps
for i = 1:4
% set up time step increment and number of time steps
if i==1 h = 0.50; end; if i==2 h = 0.10; end;
if i==3 h = 0.05; end; if i==4 h = 0.01; end;
n = 10/h;
% set up initial conditions
t=zeros(n+1,1); t(1) = 0;
x_rk=zeros(n+1,1); x_rk(1) = 1;
y_rk=zeros(n+1,1); y_rk(1) = 1;
x_exact=zeros(n+1,1); x_exact(1) = 1;
% set up difference arrays for plotting purposes
t_plot = zeros(n,1); diff = zeros(n,1);
% define right side of differential equation
f = inline('4*xx+2*tt','tt','xx','yy');
for k = 1:n
t_local = t(k); x_local = x_rk(k); y_local = y_rk(k);
k1 = y_local; K1 = f(t_local,x_local,y_local);
k2 = y_local + h*K1/2;
K2 = f(t_local + h/2,x_local + h*k1/2,k2);
k3 = y_local + h*K2/2;
K3 = f(t_local + h/2,x_local + h*k2/2,k3);
k4 = y_local + h*K3; K4 = f(t_local + h,x_local + h*k3,k4);
t(k+1) = t_local + h;
x_rk(k+1) = x_local + (h/6) * (k1+2*k2+2*k3+k4);
y_rk(k+1) = y_local + (h/6) * (K1+2*K2+2*K3+K4);
x_exact(k+1) = (7/8)*exp(2*t(k+1)) + (1/8)*exp(-2*t(k+1)) ...
            - t(k+1)/2;
t_plot(k) = t(k);
diff(k) = x_rk(k+1) - x_exact(k+1);
diff(k) = abs(diff(k) / x_exact(k+1));
end
% plot the relative errors
semilogy(t_plot,diff,'-')
hold on
xlabel('TIME','Fontsize',20);
ylabel('|RELATIVE ERROR|','Fontsize',20);
text(2*i,diff(0.2*n),['h = ',num2str(h)],'Fontsize',15,...
    'HorizontalAlignment','right','VerticalAlignment','bottom')
end
```

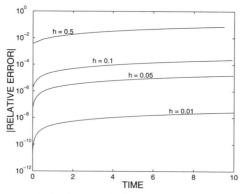

Figure 3.9.2: The numerical solution of $x'' - 4x = 2t$ when $x(0) = x'(0) = 1$ using Runge-Kutta method.

was used to resolve Example 3.9.1 using the Runge-Kutta approach. Figure 3.9.2 illustrates the results for time steps of various size.

Problems

In previous sections, you found exact solutions to second-order ordinary differential equations. Confirm these earlier results by using MATLAB and the Runge-Kutta scheme to find the numerical solution to the following problems drawn from previous sections.

1. §3.1, Problem 1   2. §3.1, Problem 5

3. §3.4, Problem 1   4. §3.4, Problem 5

5. §3.6, Problem 1   6. §3.6, Problem 5

Chapter 4

Fourier Series

Fourier series arose during the eighteenth century as a formal solution to the classic wave equation. Later on, it was used to describe physical processes in which events recur in a regular pattern. For example, a musical note usually consists of a simple note, called the fundamental, and a series of auxiliary vibrations, called overtones. Fourier's theorem provides the mathematical language which allows us to precisely describe this complex structure.

4.1 FOURIER SERIES

One of the crowning glories[1] of nineteenth century mathematics was the discovery that the infinite series

$$f(t) = \frac{a_0}{2} + \sum_{n=1}^{\infty} a_n \cos\left(\frac{n\pi t}{L}\right) + b_n \sin\left(\frac{n\pi t}{L}\right) \tag{4.1.1}$$

---

[1] "Fourier's Theorem ... is not only one of the most beautiful results of modern analysis, but may be said to furnish an indispensable instrument in the treatment of nearly every recondite question in modern physics. To mention only sonorous vibrations, the propagation of electric signals along a telegraph wire, and the conduction of heat by the earth's crust, as subjects in their generality intractable without it, is to give but a feeble idea of its importance." (Quote taken from Thomson, W., and P. G. Tait, 1879: *Treatise on Natural Philosophy, Part 1*. Cambridge University Press, §75.)

can represent a function $f(t)$ under certain general conditions. This series, called a *Fourier series*, converges to the value of the function $f(t)$ at every point in the interval $[-L, L]$ with the possible exceptions of the points at any discontinuities and the endpoints of the interval. Because each term has a period of $2L$, the sum of the series also has the same period. The *fundamental* of the periodic function $f(t)$ is the $n = 1$ term while the *harmonics* are the remaining terms whose frequencies are integer multiples of the fundamental.

We must now find some easy method for computing the coefficients a_n and b_n for a given function $f(t)$. As a first attempt, we integrate (4.1.1) term by term[2] from $-L$ to L. On the right side, all of the integrals multiplied by a_n and b_n vanish because the average of $\cos(n\pi t/L)$ and $\sin(n\pi t/L)$ is zero. Therefore, we are left with

$$a_0 = \frac{1}{L} \int_{-L}^{L} f(t)\, dt. \qquad (4.1.2)$$

Consequently a_0 is twice the mean value of $f(t)$ over one period.

We next multiply each side of (4.1.1) by $\cos(m\pi t/L)$, where m is a fixed integer. Integrating from $-L$ to L,

$$\int_{-L}^{L} f(t) \cos\left(\frac{m\pi t}{L}\right) dt = \frac{a_0}{2} \int_{-L}^{L} \cos\left(\frac{m\pi t}{L}\right) dt$$

$$+ \sum_{n=1}^{\infty} a_n \int_{-L}^{L} \cos\left(\frac{n\pi t}{L}\right) \cos\left(\frac{m\pi t}{L}\right) dt$$

$$+ \sum_{n=1}^{\infty} b_n \int_{-L}^{L} \sin\left(\frac{n\pi t}{L}\right) \cos\left(\frac{m\pi t}{L}\right) dt. \qquad (4.1.3)$$

The a_0 and b_n terms vanish by direct integration. Finally all of the a_n integrals vanish when $n \neq m$. Consequently, (4.1.3) simplifies to

$$a_n = \frac{1}{L} \int_{-L}^{L} f(t) \cos\left(\frac{n\pi t}{L}\right) dt, \qquad (4.1.4)$$

because $\int_{-L}^{L} \cos^2(n\pi t/L)\, dt = L$. Finally, by multiplying both sides of (4.1.1) by $\sin(m\pi t/L)$ (m is again a fixed integer) and integrating from $-L$ to L,

$$b_n = \frac{1}{L} \int_{-L}^{L} f(t) \sin\left(\frac{n\pi t}{L}\right) dt. \qquad (4.1.5)$$

---

[2] We assume that the integration of the series can be carried out term by term. This is sometimes difficult to justify but we do it anyway.

Although (4.1.2), (4.1.4), and (4.1.5) give us a_0, a_n, and b_n for periodic functions over the interval $[-L, L]$, in certain situations it is convenient to use the interval $[\tau, \tau + 2L]$, where τ is any real number. In that case, (4.1.1) still gives the Fourier series of $f(t)$ and

$$
\begin{aligned}
a_0 &= \frac{1}{L} \int_\tau^{\tau+2L} f(t)\, dt, \\
a_n &= \frac{1}{L} \int_\tau^{\tau+2L} f(t) \cos\left(\frac{n\pi t}{L}\right) dt, \\
b_n &= \frac{1}{L} \int_\tau^{\tau+2L} f(t) \sin\left(\frac{n\pi t}{L}\right) dt.
\end{aligned}
\qquad \textbf{(4.1.6)}
$$

These results follow when we recall that the function $f(t)$ is a periodic function that extends from minus infinity to plus infinity. The results must remain unchanged, therefore, when we shift from the interval $[-L, L]$ to the new interval $[\tau, \tau + 2L]$.

We now ask the question: what types of functions have Fourier series? Secondly, if a function is discontinuous at a point, what value will the Fourier series give? Dirichlet[3,4] answered these questions in the first half of the nineteenth century. His results may be summarized as follows.

Dirichlet's Theorem: *If for the interval $[-L, L]$ the function $f(t)$ (1) is single-valued, (2) is bounded, (3) has at most a finite number of maxima and minima, and (4) has only a finite number of discontinuities (piecewise continuous), and if (5) $f(t + 2L) = f(t)$ for values of t outside of $[-L, L]$, then*

$$
f(t) = \frac{a_0}{2} + \sum_{n=1}^{N} a_n \cos\left(\frac{n\pi t}{L}\right) + b_n \sin\left(\frac{n\pi t}{L}\right) \qquad \textbf{(4.1.7)}
$$

converges to $f(t)$ as $N \to \infty$ at values of t for which $f(t)$ is continuous and to $\frac{1}{2}[f(t^+) + f(t^-)]$ at points of discontinuity.

The quantities t^+ and t^- denote points infinitesimally to the right and left of t. The coefficients in (4.1.7) are given by (4.1.2), (4.1.4), and (4.1.5). A function $f(t)$ is bounded if the inequality $|f(t)| \leq M$ holds for some constant M for all values of t. Because the *Dirichlet's conditions* (1)–(4) are very mild,

---

[3] Dirichlet, P. G. L., 1829: Sur la convergence des séries trigonométriques qui servent à représenter une fonction arbitraire entre des limites données. *J. Reine Angew. Math.*, **4**, 157–169.

[4] Dirichlet, P. G. L., 1837: Sur l'usage des intégrales définies dans la sommation des séries finies ou infinies. *J. Reine Angew. Math.*, **17**, 57–67.

Figure 4.1.1: A product of the French Revolution, (Jean Baptiste) Joseph Fourier (1768–1830) held positions within the Napoleonic Empire during his early career. After Napoleon's fall from power, Fourier devoted his talents exclusively to science. Although he won the Institut de France prize in 1811 for his work on heat diffusion, criticism of its mathematical rigor and generality led him to publish the classic book *Théorie analytique de la chaleur* in 1823. Within this book he introduced the world to the series that bears his name. (Portrait courtesy of the Archives de l'Académie des sciences, Paris.)

it is very rare that a convergent Fourier series does not exist for a function that appears in an engineering or scientific problem.

- **Example 4.1.1**

Let us find the Fourier series for the function

$$f(t) = \begin{cases} 0, & -\pi < t \le 0, \\ t, & 0 \le t < \pi. \end{cases} \tag{4.1.8}$$

We compute the Fourier coefficients a_n and b_n using (4.1.6) by letting $L = \pi$ and $\tau = -\pi$. We then find that

$$a_0 = \frac{1}{\pi} \int_{-\pi}^{\pi} f(t)\,dt = \frac{1}{\pi} \int_0^{\pi} t\,dt = \frac{\pi}{2}, \tag{4.1.9}$$

$$a_n = \frac{1}{\pi} \int_0^{\pi} t \cos(nt)\,dt = \frac{1}{\pi} \left[\frac{t\sin(nt)}{n} + \frac{\cos(nt)}{n^2} \right] \Bigg|_0^{\pi} \tag{4.1.10}$$

$$= \frac{\cos(n\pi) - 1}{n^2 \pi} = \frac{(-1)^n - 1}{n^2 \pi} \tag{4.1.11}$$

Figure 4.1.2: Second to Gauss, Peter Gustav Lejeune Dirichlet (1805–1859) was Germany's leading mathematician during the first half of the nineteenth century. Initially drawn to number theory, his later studies in analysis and applied mathematics led him to consider the convergence of Fourier series. These studies eventually produced the modern concept of a function as a correspondence that associates with each real x in an interval some unique value denoted by $f(x)$. (Taken from the frontispiece of Dirichlet, P. G. L., 1889: *Werke*. Druck und Verlag von Georg Reimer, 644 pp.)

because $\cos(n\pi) = (-1)^n$, and

$$b_n = \frac{1}{\pi} \int_0^\pi t \sin(nt)\, dt = \frac{1}{\pi} \left[\frac{-t\cos(nt)}{n} + \frac{\sin(nt)}{n^2} \right]\Big|_0^\pi \qquad (4.1.12)$$

$$= -\frac{\cos(n\pi)}{n} = \frac{(-1)^{n+1}}{n} \qquad (4.1.13)$$

for $n = 1, 2, 3, \ldots$. Thus, the Fourier series for $f(t)$ is

$$f(t) = \frac{\pi}{4} + \sum_{n=1}^\infty \frac{(-1)^n - 1}{n^2 \pi} \cos(nt) + \frac{(-1)^{n+1}}{n} \sin(nt) \qquad (4.1.14)$$

$$= \frac{\pi}{4} - \frac{2}{\pi} \sum_{m=1}^\infty \frac{\cos[(2m-1)t]}{(2m-1)^2} - \sum_{n=1}^\infty \frac{(-1)^n}{n} \sin(nt). \qquad (4.1.15)$$

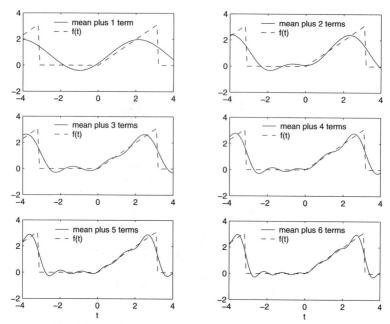

Figure 4.1.3: Partial sum of the Fourier series for (4.1.8).

We note that at the points $t = \pm(2n-1)\pi$, where $n = 1, 2, 3, \ldots$, the function jumps from zero to π. To what value does the Fourier series converge at these points? From Dirichlet's theorem, the series converges to the average of the values of the function just to the right and left of the point of discontinuity, i.e., $(\pi + 0)/2 = \pi/2$. At the remaining points the series converges to $f(t)$.

Figure 4.1.3 shows how well (4.1.14) approximates the function by graphing various partial sums of (4.1.14) as we include more and more terms (harmonics). The MATLAB script that created this figure is:

```
clear;
t = [-4:0.1:4]; % create time points in plot
f = zeros(size(t)); % initialize function f(t)
for k = 1:length(t) % construct function f(t)
    if t(k) < 0; f(k) = 0; else f(k) = t(k); end;
    if t(k) < -pi; f(k) = t(k) + 2*pi; end;
    if t(k) > pi ; f(k) = 0; end;
end
% initialize fourier series with the mean term
fs = (pi/4) * ones(size(t));
clf % clear any figures
for n = 1:6
% create plot of truncated FS with only n harmonic
```

```
fs = fs - (2/pi) * cos((2*n-1)*t) / (2*n-1)^2;
fs = fs - (-1)^n * sin(n*t) / n;
subplot(3,2,n), plot(t,fs,t,f,'--')
if n==1
    legend('mean plus 1 term','f(t)'); legend boxoff;
else
    legend(['mean plus ',num2str(n),' terms'],'f(t)')
    legend boxoff
end
    if n >= 5; xlabel('t'); end;
end
```

As the figure shows, successive corrections are made to the mean value of the series, $\pi/2$. As each harmonic is added, the Fourier series fits the function better in the sense of least squares:

$$\int_{\tau}^{\tau+2L} [f(x) - f_N(x)]^2 \, dx = \text{minimum}, \tag{4.1.16}$$

where $f_N(x)$ is the truncated Fourier series of N terms.

- **Example 4.1.2**

Let us calculate the Fourier series of the function $f(t) = |t|$ which is defined over the range $-\pi \le t \le \pi$.

From the definition of the Fourier coefficients,

$$a_0 = \frac{1}{\pi} \left[\int_{-\pi}^{0} -t \, dt + \int_{0}^{\pi} t \, dt \right] = \frac{\pi}{2} + \frac{\pi}{2} = \pi, \tag{4.1.17}$$

$$a_n = \frac{1}{\pi} \left[\int_{-\pi}^{0} -t \cos(nt) \, dt + \int_{0}^{\pi} t \cos(nt) \, dt \right] \tag{4.1.18}$$

$$= -\frac{nt \sin(nt) + \cos(nt)}{n^2 \pi} \bigg|_{-\pi}^{0} + \frac{nt \sin(nt) + \cos(nt)}{n^2 \pi} \bigg|_{0}^{\pi} \tag{4.1.19}$$

$$= \frac{2}{n^2 \pi} [(-1)^n - 1] \tag{4.1.20}$$

and

$$b_n = \frac{1}{\pi} \left[\int_{-\pi}^{0} -t \sin(nt) \, dt + \int_{0}^{\pi} t \sin(nt) \, dt \right] \tag{4.1.21}$$

$$= \frac{nt \cos(nt) - \sin(nt)}{n^2 \pi} \bigg|_{-\pi}^{0} - \frac{nt \cos(nt) - \sin(nt)}{n^2 \pi} \bigg|_{0}^{\pi} = 0 \tag{4.1.22}$$

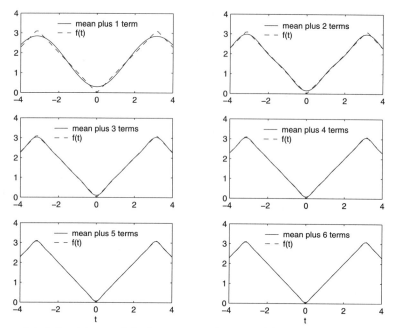

Figure 4.1.4: Partial sum of the Fourier series for $f(t) = |t|$.

for $n = 1, 2, 3, \ldots$. Therefore,

$$|t| = \frac{\pi}{2} + \frac{2}{\pi} \sum_{n=1}^{\infty} \frac{[(-1)^n - 1]}{n^2} \cos(nt) = \frac{\pi}{2} - \frac{4}{\pi} \sum_{m=1}^{\infty} \frac{\cos[(2m-1)t]}{(2m-1)^2} \quad (4.1.23)$$

for $-\pi \leq t \leq \pi$.

In Figure 4.1.4 we show how well (4.1.23) approximates the function by graphing various partial sums of (4.1.23). As the figure shows, the Fourier series does very well even when we use very few terms. The reason for this rapid convergence is the nature of the function: it does not possess any jump discontinuities.

• **Example 4.1.3**

Sometimes the function $f(t)$ is an even or odd function.[5] Can we use this property to simplify our work? The answer is yes.

Let $f(t)$ be an even function. Then

$$a_0 = \frac{1}{L} \int_{-L}^{L} f(t)\, dt = \frac{2}{L} \int_0^L f(t)\, dt, \quad (4.1.24)$$

---

[5] An even function $f_e(t)$ has the property that $f_e(-t) = f_e(t)$; an odd function $f_o(t)$ has the property that $f_o(-t) = -f_o(t)$.

and

$$a_n = \frac{1}{L} \int_{-L}^{L} f(t) \cos\left(\frac{n\pi t}{L}\right) dt = \frac{2}{L} \int_{0}^{L} f(t) \cos\left(\frac{n\pi t}{L}\right) dt, \qquad (4.1.25)$$

whereas

$$b_n = \frac{1}{L} \int_{-L}^{L} f(t) \sin\left(\frac{n\pi t}{L}\right) dt = 0. \qquad (4.1.26)$$

Here we used the properties that $\int_{-L}^{L} f_e(x)\, dx = 2 \int_{0}^{L} f_e(x)\, dx$ and $\int_{-L}^{L} f_o(x)\,dx = 0$. Thus, if we have an even function, we merely compute a_0 and a_n via (4.1.24)–(4.1.25) and $b_n = 0$. Because the corresponding series contains only cosine terms, it is often called a *Fourier cosine series*.

Similarly, if $f(t)$ is odd, then

$$a_0 = a_n = 0, \qquad \text{and} \qquad b_n = \frac{2}{L} \int_{0}^{L} f(t) \sin\left(\frac{n\pi t}{L}\right) dt. \qquad (4.1.27)$$

Thus, if we have an odd function, we merely compute b_n via (4.1.27) and $a_0 = a_n = 0$. Because the corresponding series contains only sine terms, it is often called a *Fourier sine series*.

• **Example 4.1.4**

In the case when $f(x)$ consists of a constant and/or trigonometric functions, it is much easier to find the corresponding Fourier series by inspection rather than by using (4.1.6). For example, let us find the Fourier series for $f(x) = \sin^2(x)$ defined over the range $-\pi \le x \le \pi$.

We begin by rewriting $f(x) = \sin^2(x)$ as $f(x) = \frac{1}{2}[1 - \cos(2x)]$. Next, we note that any function defined over the range $-\pi < x < \pi$ has the Fourier series

$$f(x) = \frac{a_0}{2} + \sum_{n=1}^{\infty} a_n \cos(nx) + b_n \sin(nx) \qquad (4.1.28)$$

$$= \frac{a_0}{2} + a_1 \cos(x) + b_1 \sin(x) + a_2 \cos(2x) + b_2 \sin(2x) + \cdots. \qquad (4.1.29)$$

On the other hand,

$$f(x) = \frac{1}{2} - \frac{1}{2} \cos(2x) \qquad (4.1.30)$$

$$= \frac{1}{2} + 0 \cos(x) + 0 \sin(x) - \frac{1}{2} \cos(2x) + 0 \sin(2x) + \cdots. \qquad (4.1.31)$$

Consequently, by inspection, we can immediately write that

$$a_0 = 1, a_1 = b_1 = 0, a_2 = -\tfrac{1}{2}, b_2 = 0, a_n = b_n = 0, n \ge 3. \qquad (4.1.32)$$

Thus, instead of the usual expansion involving an infinite number of sine and cosine terms, our Fourier series contains only two terms and is simply

$$f(x) = \tfrac{1}{2} - \tfrac{1}{2}\cos(2x), \qquad -\pi \le x \le \pi. \qquad (4.1.33)$$

• Example 4.1.5: Quieting snow tires

An application of Fourier series to a problem in industry occurred several years ago, when drivers found that snow tires produced a loud whine[6] on dry pavement. Tire sounds are produced primarily by the dynamic interaction of the tread elements with the road surface.[7] As each tread element passes through the contact patch, it contributes a pulse of acoustic energy to the total sound field radiated by the tire.

For evenly spaced treads we envision that the release of acoustic energy resembles the top of Figure 4.1.5. If we perform a Fourier analysis of this distribution, we find that

$$a_0 = \frac{1}{\pi}\left[\int_{-\pi/2-\epsilon}^{-\pi/2+\epsilon} 1\,dt + \int_{\pi/2-\epsilon}^{\pi/2+\epsilon} 1\,dt\right] = \frac{4\epsilon}{\pi}, \qquad (4.1.34)$$

where ϵ is half of the width of the tread and

$$a_n = \frac{1}{\pi}\left[\int_{-\pi/2-\epsilon}^{-\pi/2+\epsilon} \cos(nt)\,dt + \int_{\pi/2-\epsilon}^{\pi/2+\epsilon} \cos(nt)\,dt\right] \qquad (4.1.35)$$

$$= \frac{1}{n\pi}\left[\sin(nt)\big|_{-\pi/2-\epsilon}^{-\pi/2+\epsilon} + \sin(nt)\big|_{\pi/2-\epsilon}^{\pi/2+\epsilon}\right] \qquad (4.1.36)$$

$$= \frac{1}{n\pi}\left[\sin\left(-\frac{n\pi}{2}+n\epsilon\right) - \sin\left(-\frac{n\pi}{2}-n\epsilon\right)\right.$$
$$\left. + \sin\left(\frac{n\pi}{2}+n\epsilon\right) - \sin\left(\frac{n\pi}{2}-n\epsilon\right)\right] \qquad (4.1.37)$$

$$= \frac{1}{n\pi}\left[2\cos\left(-\frac{n\pi}{2}\right) + 2\cos\left(\frac{n\pi}{2}\right)\right]\sin(n\epsilon) \qquad (4.1.38)$$

$$= \frac{4}{n\pi}\cos\left(\frac{n\pi}{2}\right)\sin(n\epsilon). \qquad (4.1.39)$$

Because $f(t)$ is an even function, $b_n = 0$.

The question now arises of how to best illustrate our Fourier coefficients. In §4.4 we will show that any harmonic can be represented as a

---

[6] Information based on Varterasian, J. H., 1969: Math quiets rotating machines. *SAE J.*, **77(10)**, 53.

[7] Willett, P. R., 1975: Tire tread pattern sound generation. *Tire Sci. Tech.*, **3**, 252–266.

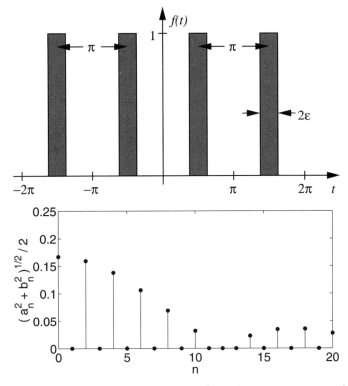

Figure 4.1.5: Temporal spacing (over two periods) and frequency spectrum of uniformly spaced snow tire treads.

single wave $A_n \cos(n\pi t/L + \varphi_n)$ or $A_n \sin(n\pi t/L + \psi_n)$, where the amplitude $A_n = \sqrt{a_n^2 + b_n^2}$. In the bottom frame of Figure 4.1.5, MATLAB was used to plot this amplitude, usually called the *amplitude* or *frequency spectrum* $\frac{1}{2}\sqrt{a_n^2 + b_n^2}$, as a function of n for an arbitrarily chosen $\epsilon = \pi/12$. Although the value of ϵ will affect the exact shape of the spectrum, the qualitative arguments that we will present remain unchanged. We have added the factor $\frac{1}{2}$ so that our definition of the frequency spectrum is consistent with that for a complex Fourier series stated after (4.5.13). The amplitude spectrum in Figure 4.1.5 shows that the spectrum for periodically placed tire treads has its largest amplitude at small n. This produces one loud tone plus strong harmonic overtones because the fundamental and its overtones are the dominant terms in the Fourier series representation.

Clearly this loud, monotone whine is undesirable. How might we avoid it? Just as soldiers marching in step produce a loud uniform sound, we suspect that our uniform tread pattern is the problem. Therefore, let us now vary the interval between the treads so that the distance between any tread and its nearest neighbor is not equal as illustrated in Figure 4.1.6. Again we perform

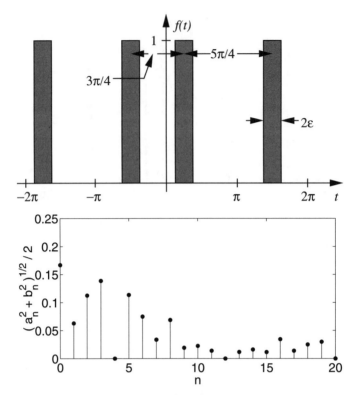

Figure 4.1.6: Temporal spacing and frequency spectrum of nonuniformly spaced snow tire treads.

its Fourier analysis and obtain that

$$a_0 = \frac{1}{\pi} \left[\int_{-\pi/2-\epsilon}^{-\pi/2+\epsilon} 1 \, dt + \int_{\pi/4-\epsilon}^{\pi/4+\epsilon} 1 \, dt \right] = \frac{4\epsilon}{\pi}, \tag{4.1.40}$$

$$a_n = \frac{1}{\pi} \left[\int_{-\pi/2-\epsilon}^{-\pi/2+\epsilon} \cos(nt) \, dt + \int_{\pi/4-\epsilon}^{\pi/4+\epsilon} \cos(nt) \, dt \right] \tag{4.1.41}$$

$$a_n = \frac{1}{n\pi} \sin(nt) \Big|_{-\pi/2-\epsilon}^{-\pi/2+\epsilon} + \frac{1}{n\pi} \sin(nt) \Big|_{\pi/4-\epsilon}^{\pi/4+\epsilon} \tag{4.1.42}$$

$$= -\frac{1}{n\pi} \left[\sin\left(\frac{n\pi}{2} - n\epsilon\right) - \sin\left(\frac{n\pi}{2} + n\epsilon\right) \right]$$

$$+ \frac{1}{n\pi} \left[\sin\left(\frac{n\pi}{4} + n\epsilon\right) - \sin\left(\frac{n\pi}{4} - n\epsilon\right) \right] \tag{4.1.43}$$

$$= \frac{2}{n\pi} \left[\cos\left(\frac{n\pi}{2}\right) + \cos\left(\frac{n\pi}{4}\right) \right] \sin(n\epsilon), \tag{4.1.44}$$

and

$$b_n = \frac{1}{\pi}\left[\int_{-\pi/2-\epsilon}^{-\pi/2+\epsilon} \sin(nt)\,dt + \int_{\pi/4-\epsilon}^{\pi/4+\epsilon} \sin(nt)\,dt\right] \qquad (4.1.45)$$

$$= -\frac{1}{n\pi}\left[\cos\left(\frac{n\pi}{2} - n\epsilon\right) - \cos\left(\frac{n\pi}{2} + n\epsilon\right)\right]$$

$$\quad - \frac{1}{n\pi}\left[\cos\left(\frac{n\pi}{4} + n\epsilon\right) - \cos\left(\frac{n\pi}{4} - n\epsilon\right)\right] \qquad (4.1.46)$$

$$b_n = \frac{2}{n\pi}\left[\sin\left(\frac{n\pi}{4}\right) - \sin\left(\frac{n\pi}{2}\right)\right]\sin(n\epsilon). \qquad (4.1.47)$$

The MATLAB script

```
epsilon = pi/12; % set up parameter for fs coefficient
n = 1:20; % number of harmonics
arg1 = (pi/2)*n; arg2 = (pi/4)*n; arg3 = epsilon*n;
% compute the fourier coefficient a_n
an = (cos(arg1) + cos(arg2)).*sin(arg3);
an = (2/pi) * an./n;
% compute the fourier coefficient b_n
bn = (sin(arg2) - sin(arg1)).*sin(arg3);
bn = (2/pi) * bn./n;
% compute the magnitude
cn = 0.5 * sqrt(an.*an + bn.*bn);
% add in the a_0 term
cn = [2*epsilon/pi,cn];
n = [0,n];
clf % clear any figures
axes('FontSize',20) % set font size
stem(n,cn,'filled') % plot spectrum
set (gca,'PlotBoxAspectRatio',[8 4 1]) % set aspect ratio
xlabel('n') % label x-axis
ylabel('( a_n^2 + b_n^2 )^{1/2}/2') % label y-axis,
```

was used to compute the amplitude of each harmonic as a function of n and the results were plotted. See Figure 4.1.6. The important point is that our new choice for the spacing of the treads has reduced or eliminated some of the harmonics compared to the case of equally spaced treads. On the negative side we have excited some of the harmonics that were previously absent. However, the net effect is advantageous because the treads produce less noise at more frequencies rather than a lot of noise at a few select frequencies.

If we were to extend this technique so that the treads occurred at completely random positions, then the treads would produce very little noise at many frequencies and the total noise would be comparable to that generated by other sources within the car. To find the distribution of treads with the whitest noise[8] is a process of trial and error. Assuming a distribution, we can perform a Fourier analysis to obtain its frequency spectrum. If annoying peaks are present in the spectrum, we can then adjust the elements in the tread distribution that may contribute to the peak and analyze the revised distribution. You are finished when no peaks appear.

Problems

Find the Fourier series for the following functions. Using MATLAB, plot the Fourier spectrum. Then plot various partial sums and compare them against the exact function.

1. $f(t) = \begin{cases} 1, & -\pi < t < 0 \\ 0, & 0 < t < \pi \end{cases}$

2. $f(t) = \begin{cases} t, & -\pi < t \leq 0 \\ 0, & 0 \leq t < \pi \end{cases}$

3. $f(t) = \begin{cases} -\pi, & -\pi < t < 0 \\ t, & 0 < t < \pi \end{cases}$

4. $f(t) = \begin{cases} \frac{1}{2} + t, & -1 \leq t \leq 0 \\ \frac{1}{2} - t, & 0 \leq t \leq 1 \end{cases}$

5. $f(t) = \begin{cases} 0, & -\pi \leq t \leq 0 \\ t, & 0 \leq t \leq \pi/2 \\ \pi - t, & \pi/2 \leq t \leq \pi \end{cases}$

6. $f(t) = \begin{cases} 0, & -\pi \leq t \leq -\pi/2 \\ \sin(2t), & -\pi/2 \leq t \leq \pi/2 \\ 0, & \pi/2 \leq t \leq \pi \end{cases}$

7. $f(t) = e^{at}, \quad -L < t < L$

8. $f(t) = t + t^2, \quad -L < t < L$

9. $f(t) = \begin{cases} 0, & -\pi \leq t \leq 0 \\ \sin(t), & 0 \leq t \leq \pi \end{cases}$

10. $f(t) = \begin{cases} t, & -\frac{1}{2} \leq t \leq \frac{1}{2} \\ 1 - t, & \frac{1}{2} \leq t \leq \frac{3}{2} \end{cases}$

11. $f(t) = \begin{cases} 0, & -a < t < 0 \\ 2t, & 0 < t < a \end{cases}$

12. $f(t) = \begin{cases} 0, & -\pi < t \leq 0 \\ t^2, & 0 \leq t < \pi \end{cases}$

13. $f(t) = \dfrac{\pi - t}{2}, \quad 0 < t < 2$

14. $f(t) = t\cos\left(\dfrac{\pi t}{L}\right), \quad -L < t < L$

15.
$$f(t) = \sinh\left[a\left(\frac{\pi}{2} - |t|\right)\right], \quad -\pi \leq t \leq \pi$$

16.
$$f(t) = \begin{cases} x(2L - x), & 0 \leq t \leq 2L \\ x^2 - 6Lx + 8L^2, & 2L \leq t \leq 4L \end{cases}$$

---

[8] White noise is sound that is analogous to white light in that it is uniformly distributed throughout the complete audible sound spectrum.

4.2 PROPERTIES OF FOURIER SERIES

In the previous section we introduced the Fourier series and showed how to compute one given the function $f(t)$. In this section we examine some particular properties of these series.

> Differentiation of a Fourier series

In certain instances we only have the Fourier series representation of a function $f(t)$. Can we find the derivative or the integral of $f(t)$ merely by differentiating or integrating the Fourier series term by term? Is this permitted? Let us consider the case of differentiation first.

Consider a function $f(t)$ of period $2L$ which has the derivative $f'(t)$. Let us assume that we can expand $f'(t)$ as a Fourier series. This implies that $f'(t)$ is continuous except for a finite number of discontinuities and $f(t)$ is continuous over an interval that starts at $t = \tau$ and ends at $t = \tau + 2L$. Then

$$f'(t) = \frac{a_0'}{2} + \sum_{n=1}^{\infty} a_n' \cos\left(\frac{n\pi t}{L}\right) + b_n' \sin\left(\frac{n\pi t}{L}\right), \qquad (\textbf{4.2.1})$$

where we denoted the Fourier coefficients of $f'(t)$ with a prime. Computing the Fourier coefficients,

$$a_0' = \frac{1}{L}\int_{\tau}^{\tau+2L} f'(t)\, dt = \frac{1}{L}[f(\tau + 2L) - f(\tau)] = 0, \qquad (\textbf{4.2.2})$$

if $f(\tau + 2L) = f(\tau)$. Similarly, by integrating by parts,

$$a_n' = \frac{1}{L}\int_{\tau}^{\tau+2L} f'(t) \cos\left(\frac{n\pi t}{L}\right)\, dt \qquad (\textbf{4.2.3})$$

$$= \frac{1}{L}\left[f(t)\cos\left(\frac{n\pi t}{L}\right)\right]\Big|_{\tau}^{\tau+2L} + \frac{n\pi}{L^2}\int_{\tau}^{\tau+2L} f(t) \sin\left(\frac{n\pi t}{L}\right)\, dt \qquad (\textbf{4.2.4})$$

$$= \frac{n\pi b_n}{L}, \qquad (\textbf{4.2.5})$$

and

$$b_n' = \frac{1}{L}\int_{\tau}^{\tau+2L} f'(t) \sin\left(\frac{n\pi t}{L}\right)\, dt \qquad (\textbf{4.2.6})$$

$$= \frac{1}{L}\left[f(t)\sin\left(\frac{n\pi t}{L}\right)\right]\Big|_{\tau}^{\tau+2L} - \frac{n\pi}{L^2}\int_{\tau}^{\tau+2L} f(t) \cos\left(\frac{n\pi t}{L}\right)\, dt \qquad (\textbf{4.2.7})$$

$$= -\frac{n\pi a_n}{L}. \qquad (\textbf{4.2.8})$$

Consequently, if we have a function $f(t)$ whose derivative $f'(t)$ is continuous except for a finite number of discontinuities and $f(\tau) = f(\tau + 2L)$, then

$$f'(t) = \sum_{n=1}^{\infty} \frac{n\pi}{L} \left[b_n \cos\left(\frac{n\pi t}{L}\right) - a_n \sin\left(\frac{n\pi t}{L}\right) \right]. \qquad (4.2.9)$$

That is, the derivative of $f(t)$ is given by a term-by-term differentiation of the Fourier series of $f(t)$.

• **Example 4.2.1**

The Fourier series for the periodic function

$$f(t) = \begin{cases} 0, & -\pi \le t \le 0, \\ t, & 0 \le t \le \pi/2, \\ \pi - t, & \pi/2 \le t \le \pi, \end{cases} \qquad f(t) = f(t + 2\pi), \qquad (4.2.10)$$

is

$$f(t) = \frac{\pi}{8} - \frac{1}{\pi} \sum_{n=1}^{\infty} \frac{\cos[2(2n-1)t]}{(2n-1)^2} - \frac{2}{\pi} \sum_{n=1}^{\infty} \frac{(-1)^n}{(2n-1)^2} \sin[(2n-1)t]. \quad (4.2.11)$$

Because $f(t)$ is continuous over the entire interval $(-\pi, \pi)$ and $f(-\pi) = f(\pi) = 0$, we can find $f'(t)$ by taking the derivative of (4.2.11) term by term:

$$f'(t) = \frac{2}{\pi} \sum_{n=1}^{\infty} \frac{\sin[2(2n-1)t]}{2n-1} - \frac{2}{\pi} \sum_{n=1}^{\infty} \frac{(-1)^n}{2n-1} \cos[(2n-1)t]. \qquad (4.2.12)$$

This is the same Fourier series that we would obtain by computing the Fourier series for

$$f'(t) = \begin{cases} 0, & -\pi < t < 0, \\ 1, & 0 < t < \pi/2, \\ -1, & \pi/2 < t < \pi. \end{cases} \qquad (4.2.13)$$

> **Integration of a Fourier series**

To determine whether we can find the integral of $f(t)$ by term-by-term integration of its Fourier series, consider a form of the antiderivative of $f(t)$:

$$F(t) = \int_0^t \left[f(\tau) - \frac{a_0}{2} \right] d\tau. \qquad (4.2.14)$$

Now

$$F(t + 2L) = \int_0^t \left[f(\tau) - \frac{a_0}{2} \right] d\tau + \int_t^{t+2L} \left[f(\tau) - \frac{a_0}{2} \right] d\tau \qquad (4.2.15)$$

$$= F(t) + \int_{-L}^L \left[f(\tau) - \frac{a_0}{2} \right] d\tau \qquad (4.2.16)$$

$$= F(t) + \int_{-L}^L f(\tau)\, d\tau - La_0 = F(t), \qquad (4.2.17)$$

so that $F(t)$ has a period of $2L$. Consequently we may expand $F(t)$ as the Fourier series

$$F(t) = \frac{A_0}{2} + \sum_{n=1}^{\infty} A_n \cos\left(\frac{n\pi t}{L}\right) + B_n \sin\left(\frac{n\pi t}{L}\right). \qquad (4.2.18)$$

For A_n,

$$A_n = \frac{1}{L} \int_{-L}^L F(t) \cos\left(\frac{n\pi t}{L}\right) dt \qquad (4.2.19)$$

$$= \frac{1}{L} \left[F(t) \frac{\sin(n\pi t/L)}{n\pi/L} \right] \Big|_{-L}^L - \frac{1}{n\pi} \int_{-L}^L \left[f(t) - \frac{a_0}{2} \right] \sin\left(\frac{n\pi t}{L}\right) dt \quad (4.2.20)$$

$$= -\frac{b_n}{n\pi/L}. \qquad (4.2.21)$$

Similarly,

$$B_n = \frac{a_n}{n\pi/L}. \qquad (4.2.22)$$

Therefore,

$$\int_0^t f(\tau)\, d\tau = \frac{a_0 t}{2} + \frac{A_0}{2} + \sum_{n=1}^{\infty} \frac{a_n \sin(n\pi t/L) - b_n \cos(n\pi t/L)}{n\pi/L}. \qquad (4.2.23)$$

This is identical to a term-by-term integration of the Fourier series for $f(t)$. Thus, we can always find the integral of $f(t)$ by a term-by-term integration of its Fourier series.

- **Example 4.2.2**

The Fourier series for $f(t) = t$ for $-\pi < t < \pi$ is

$$f(t) = -2 \sum_{n=1}^{\infty} \frac{(-1)^n}{n} \sin(nt). \qquad (4.2.24)$$

To find the Fourier series for $f(t) = t^2$, we integrate (4.2.24) term by term and find that

$$\frac{\tau^2}{2}\bigg|_0^t = 2\sum_{n=1}^{\infty} \frac{(-1)^n}{n^2} \cos(nt) - 2\sum_{n=1}^{\infty} \frac{(-1)^n}{n^2}. \qquad (4.2.25)$$

But $\sum_{n=1}^{\infty}(-1)^n/n^2 = -\pi^2/12$. Substituting and multiplying by 2, we obtain the final result that

$$t^2 = \frac{\pi^2}{3} + 4\sum_{n=1}^{\infty} \frac{(-1)^n}{n^2} \cos(nt). \qquad (4.2.26)$$

Parseval's equality

One of the fundamental quantities in engineering is power. The *power content* of a periodic signal $f(t)$ of period $2L$ is $\int_{\tau}^{\tau+2L} f^2(t)\, dt/L$. This mathematical definition mirrors the power dissipation I^2R that occurs in a resistor of resistance R where I is the root mean square (RMS) of the current. We would like to compute this power content as simply as possible given the coefficients of its Fourier series.

Assume that $f(t)$ has the Fourier series

$$f(t) = \frac{a_0}{2} + \sum_{n=1}^{\infty} a_n \cos\left(\frac{n\pi t}{L}\right) + b_n \sin\left(\frac{n\pi t}{L}\right). \qquad (4.2.27)$$

Then,

$$\frac{1}{L}\int_{\tau}^{\tau+2L} f^2(t)\, dt = \frac{a_0}{2L}\int_{\tau}^{\tau+2L} f(t)\, dt + \sum_{n=1}^{\infty} \frac{a_n}{L}\int_{\tau}^{\tau+2L} f(t)\cos\left(\frac{n\pi t}{L}\right) dt$$

$$+ \sum_{n=1}^{\infty} \frac{b_n}{L}\int_{\tau}^{\tau+2L} f(t)\sin\left(\frac{n\pi t}{L}\right) dt \qquad (4.2.28)$$

$$= \frac{a_0^2}{2} + \sum_{n=1}^{\infty}(a_n^2 + b_n^2). \qquad (4.2.29)$$

Equation (4.2.29) is *Parseval's equality*.[9] It allows us to sum squares of Fourier coefficients (which we have already computed) rather than performing the integration $\int_{\tau}^{\tau+2L} f^2(t)\, dt$ analytically or numerically.

---

[9] Parseval, M.-A., 1805: Mémoire sur les séries et sur l'intégration complète d'une équation aux différences partielles linéaires du second ordre, à coefficients constants. *Mémoires présentés a l'Institut des sciences, lettres et arts, par divers savans, et lus dans ses assemblées: Sciences mathématiques et Physiques*, **1**, 638–648.

- **Example 4.2.3**

The Fourier series for $f(t) = t^2$ over the interval $[-\pi, \pi]$ is

$$t^2 = \frac{\pi^2}{3} + 4 \sum_{n=1}^{\infty} \frac{(-1)^n}{n^2} \cos(nt). \qquad (4.2.30)$$

Then, by Parseval's equality,

$$\frac{1}{\pi} \int_{-\pi}^{\pi} t^4 \, dt = \left. \frac{2t^5}{5\pi} \right|_0^{\pi} = \frac{4\pi^4}{18} + 16 \sum_{n=1}^{\infty} \frac{1}{n^4} \qquad (4.2.31)$$

$$\left(\frac{2}{5} - \frac{4}{18} \right) \pi^4 = 16 \sum_{n=1}^{\infty} \frac{1}{n^4} \qquad (4.2.32)$$

$$\frac{\pi^4}{90} = \sum_{n=1}^{\infty} \frac{1}{n^4}. \qquad (4.2.33)$$

Gibbs phenomena

In the actual application of Fourier series, we cannot sum an infinite number of terms but must be content with N terms. If we denote this partial sum of the Fourier series by $S_N(t)$, we have from the definition of the Fourier series:

$$S_N(t) = \tfrac{1}{2}a_0 + \sum_{n=1}^{N} a_n \cos(nt) + b_n \sin(nt) \qquad (4.2.34)$$

$$= \frac{1}{2\pi} \int_0^{2\pi} f(x) \, dx$$

$$+ \frac{1}{\pi} \int_0^{2\pi} f(x) \left[\sum_{n=1}^{N} \cos(nt) \cos(nx) + \sin(nt) \sin(nx) \right] dx \qquad (4.2.35)$$

$$= \frac{1}{\pi} \int_0^{2\pi} f(x) \left\{ \frac{1}{2} + \sum_{n=1}^{N} \cos[n(t-x)] \right\} dx \qquad (4.2.36)$$

$$= \frac{1}{2\pi} \int_0^{2\pi} f(x) \frac{\sin[(N + \frac{1}{2})(x - t)]}{\sin[\frac{1}{2}(x - t)]} \, dx. \qquad (4.2.37)$$

The quantity $\sin[(N + \frac{1}{2})(x - t)] / \sin[\frac{1}{2}(x - t)]$ is called a *scanning function*. Over the range $0 \le x \le 2\pi$ it has a very large peak at $x = t$ where the amplitude equals $2N + 1$. See Figure 4.2.1. On either side of this peak

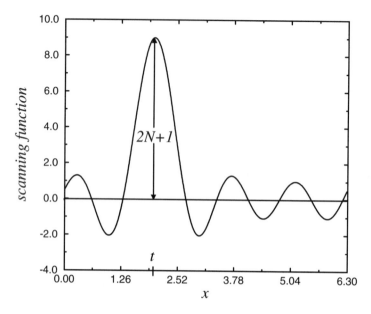

Figure 4.2.1: The scanning function over $0 \leq x \leq 2\pi$ for $N = 5$.

there are oscillations which decrease rapidly with distance from the peak. Consequently, as $N \to \infty$, the scanning function becomes essentially a long narrow slit corresponding to the area under the large peak at $x = t$. If we neglect for the moment the small area under the minor ripples adjacent to this slit, then the integral (4.2.37) essentially equals $f(t)$ times the area of the slit divided by 2π. If $1/2\pi$ times the area of the slit equals unity, then the value of $S_N(t) \approx f(t)$ to a good approximation for large N.

For relatively small values of N, the scanning function deviates considerably from its ideal form, and the partial sum $S_N(t)$ only crudely approximates $f(t)$. As the partial sum includes more terms and N becomes relatively large, the form of the scanning function improves and so does the agreement between $S_N(t)$ and $f(t)$. The improvement in the scanning function is due to the large hump becoming taller and narrower. At the same time, the adjacent ripples become more numerous as well as narrower in the same proportion as the large hump does.

The reason why $S_N(t)$ and $f(t)$ will never become identical, even in the limit of $N \to \infty$, is the presence of the positive and negative side lobes near the large peak. Because

$$\frac{\sin[(N + \frac{1}{2})(x - t)]}{\sin[\frac{1}{2}(x - t)]} = 1 + 2 \sum_{n=1}^{N} \cos[n(t - x)], \qquad (4.2.38)$$

an integration of the scanning function over the interval 0 to 2π shows that the total area under the scanning function equals 2π. However, from Figure

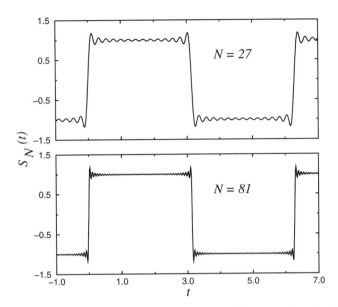

Figure 4.2.2: The finite Fourier series representation $S_N(t)$ for the function (4.2.39) for the range $-1 \leq t \leq 7$ for $N = 27$ and $N = 81$.

4.2.1 the net area contributed by the ripples is numerically negative so that the area under the large peak must exceed 2π if the total area equals 2π. Although the exact value depends upon N, it is important to note that this excess does not become zero as $N \to \infty$.

Thus, the presence of these negative side lobes explains the departure of our scanning function from the idealized slit of area 2π. To illustrate this departure, consider the function:

$$f(t) = \begin{cases} 1, & 0 < t < \pi, \\ -1, & \pi < t < 2\pi. \end{cases} \tag{4.2.39}$$

Then,

$$S_N(t) = \frac{1}{2\pi} \int_0^\pi \frac{\sin[(N + \frac{1}{2})(x - t)]}{\sin[\frac{1}{2}(x - t)]} \, dx - \frac{1}{2\pi} \int_\pi^{2\pi} \frac{\sin[(N + \frac{1}{2})(x - t)]}{\sin[\frac{1}{2}(x - t)]} \, dx \tag{4.2.40}$$

$$= \frac{1}{2\pi} \int_0^\pi \left\{ \frac{\sin[(N + \frac{1}{2})(x - t)]}{\sin[\frac{1}{2}(x - t)]} \, dx + \frac{\sin[(N + \frac{1}{2})(x + t)]}{\sin[\frac{1}{2}(x + t)]} \, dx \right\} \tag{4.2.41}$$

$$= \frac{1}{2\pi} \int_{-t}^{\pi - t} \frac{\sin[(N + \frac{1}{2})\theta]}{\sin(\frac{1}{2}\theta)} \, d\theta - \frac{1}{2\pi} \int_t^{\pi + t} \frac{\sin[(N + \frac{1}{2})\theta]}{\sin(\frac{1}{2}\theta)} \, d\theta. \tag{4.2.42}$$

The first integral in (4.2.42) gives the contribution to $S_N(t)$ from the jump discontinuity at $t = 0$ while the second integral gives the contribution from $t = \pi$. In Figure 4.2.2 we have plotted $S_N(t)$ when $N = 27$ and $N = 81$. Residual discrepancies remain even for very large values of N. Indeed, as N

increases, this figure changes only in that the ripples in the vicinity of the discontinuity of $f(t)$ proportionally increase their rate of oscillation as a function of t while their relative magnitude remains the same. As $N \to \infty$ these ripples compress into a single vertical line at the point of discontinuity. True, these oscillations occupy smaller and smaller spaces but they still remain. Thus, we can never approximate a function in the vicinity of a discontinuity by a finite Fourier series without suffering from this over- and undershooting of the series. This peculiarity of Fourier series is called the *Gibbs phenomena*.[10] Gibbs phenomena can only be eliminated by removing the discontinuity.[11]

Problems

Additional Fourier series representations can be generated by differentiating or integrating known Fourier series. Work out the following two examples.

1. Given

$$\frac{\pi^2 - 2\pi x}{8} = \sum_{n=0}^{\infty} \frac{\cos[(2n+1)x]}{(2n+1)^2}, \qquad 0 \le x \le \pi,$$

obtain

$$\frac{\pi^2 x - \pi x^2}{8} = \sum_{n=0}^{\infty} \frac{\sin[(2n+1)x]}{(2n+1)^3}, \qquad 0 \le x \le \pi,$$

by term-by-term integration. Could we go the other way, i.e., take the derivative of the second equation to obtain the first? Explain.

2. Given

$$\frac{\pi^2 - 3x^2}{12} = \sum_{n=1}^{\infty} (-1)^{n+1} \frac{\cos(nx)}{n^2}, \qquad -\pi \le x \le \pi,$$

obtain

$$\frac{\pi^2 x - x^3}{12} = \sum_{n=1}^{\infty} (-1)^{n+1} \frac{\sin(nx)}{n^3}, \qquad -\pi \le x \le \pi,$$

by term-by-term integration. Could we go the other way, i.e., take the derivative of the second equation to obtain the first? Explain.

---

[10] Gibbs, J. W., 1898: Fourier's series. *Nature*, **59**, 200; Gibbs, J. W., 1899: Fourier's series. *Nature*, **59**, 606. For the historical development, see Hewitt, E., and R. E. Hewitt, 1979: The Gibbs-Wilbraham phenomenon: An episode in Fourier analysis. *Arch. Hist. Exact Sci.*, **21**, 129–160.

[11] For a particularly clever method for improving the convergence of a trigonometric series, see Kantorovich, L. V., and V. I. Krylov, 1964: *Approximate Methods of Higher Analysis*. Interscience, pp. 77–88.

3. (a) Show that the Fourier series for the odd function:

$$f(t) = \begin{cases} 2t + t^2, & -2 \le t \le 0, \\ 2t - t^2, & 0 \le t \le 2, \end{cases}$$

is

$$f(t) = \frac{32}{\pi^3} \sum_{n=1}^{\infty} \frac{1}{(2n-1)^3} \sin\left[\frac{(2n-1)\pi t}{2}\right].$$

(b) Use Parseval's equality to show that

$$\frac{\pi^6}{960} = \sum_{n=1}^{\infty} \frac{1}{(2n-1)^6}.$$

This series converges very rapidly to $\pi^6/960$ and provides a convenient method for computing π^6.

4.3 HALF-RANGE EXPANSIONS

In certain applications, we will find that we need a Fourier series representation for a function $f(x)$ that applies over the interval $(0, L)$ rather than $(-L, L)$. Because we are completely free to define the function over the interval $(-L, 0)$, it is simplest to have a series that consists only of sines or cosines. In this section we shall show how we can obtain these so-called *half-range expansions*.

Recall in Example 4.1.3 how we saw that if $f(x)$ is an even function, then $b_n = 0$ for all n. Similarly, if $f(x)$ is an odd function, then $a_0 = a_n = 0$ for all n. We now use these results to find a Fourier half-range expansion by extending the function defined over the interval $(0, L)$ as either an even or odd function into the interval $(-L, 0)$. If we extend $f(x)$ as an even function, we will get a half-range cosine series; if we extend $f(x)$ as an odd function, we obtain a half-range sine series.

It is important to remember that half-range expansions are a special case of the general Fourier series. For any $f(x)$ we can construct either a Fourier sine or cosine series over the interval $(-L, L)$. Both of these series will give the correct answer over the interval of $(0, L)$. Which one we choose to use depends upon whether we wish to deal with a cosine or sine series.

- **Example 4.3.1**

Let us find the half-range sine expansion of

$$f(x) = 1, \qquad 0 < x < \pi. \tag{4.3.1}$$

We begin by defining the periodic odd function

$$\widetilde{f}(x) = \begin{cases} -1, & -\pi < x < 0, \\ 1, & 0 < x < \pi, \end{cases} \qquad (4.3.2)$$

with $\widetilde{f}(x + 2\pi) = \widetilde{f}(x)$. Because $\widetilde{f}(x)$ is odd, $a_0 = a_n = 0$ and

$$b_n = \frac{2}{\pi} \int_0^\pi 1 \sin(nx)\, dx = -\frac{2}{n\pi} \cos(nx) \Big|_0^\pi \qquad (4.3.3)$$

$$= -\frac{2}{n\pi} [\cos(n\pi) - 1] = -\frac{2}{n\pi} [(-1)^n - 1]. \qquad (4.3.4)$$

The Fourier half-range sine series expansion of $f(x)$ is therefore

$$f(x) = \frac{2}{\pi} \sum_{n=1}^\infty \frac{[1 - (-1)^n]}{n} \sin(nx) = \frac{4}{\pi} \sum_{m=1}^\infty \frac{\sin[(2m-1)x]}{2m-1}. \qquad (4.3.5)$$

As counterpoint, let us find the half-range cosine expansion of $f(x) = 1$, $0 < x < \pi$. Now, we have that $b_n = 0$,

$$a_0 = \frac{2}{\pi} \int_0^\pi 1\, dx = 2, \qquad (4.3.6)$$

and

$$a_n = \frac{2}{\pi} \int_0^\pi \cos(nx)\, dx = \frac{2}{n\pi} \sin(nx) \Big|_0^\pi = 0. \qquad (4.3.7)$$

Thus, the Fourier half-range cosine expansion equals the single term:

$$f(x) = 1, \qquad 0 < x < \pi. \qquad (4.3.8)$$

This is perfectly reasonable. To form a half-range cosine expansion we extend $f(x)$ as an even function into the interval $(-\pi, 0)$. In this case, we would obtain $\widetilde{f}(x) = 1$ for $-\pi < x < \pi$. Finally, we note that the Fourier series of a constant is simply that constant.

In practice it is impossible to sum (4.3.5) exactly and we actually sum only the first N terms. Figure 4.3.1 illustrates $f(x)$ when the Fourier series (4.3.5) contains N terms. As seen from the figure, the truncated series tries to achieve the infinite slope at $x = 0$, but in the attempt, it *overshoots* the discontinuity by a certain amount (in this particular case, by 17.9%). This is another example of the Gibbs phenomena. Increasing the number of terms does not remove this peculiarity; it merely shifts it nearer to the discontinuity.

• Example 4.3.2: Inertial supercharging of an engine

An important aspect of designing any gasoline engine involves the motion of the fuel, air, and exhaust gas mixture through the engine. Ordinarily

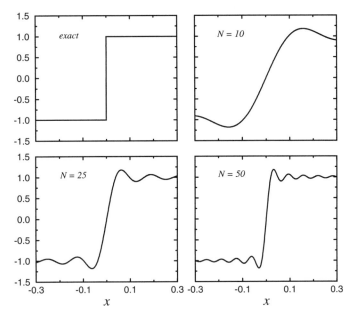

Figure 4.3.1: Partial sum of N terms in the Fourier half-range sine representation of a square wave.

an engineer would consider the motion as steady flow; but in the case of a four-stroke, single-cylinder gasoline engine, the closing of the intake valve interrupts the steady flow of the gasoline-air mixture for nearly three quarters of the engine cycle. This periodic interruption sets up standing waves in the intake pipe—waves which can build up an appreciable pressure amplitude just outside the input value.

When one of the harmonics of the engine frequency equals one of the resonance frequencies of the intake pipe, then the pressure fluctuations at the valve will be large. If the intake valve closes during that portion of the cycle when the pressure is less than average, then the waves will reduce the power output. However, if the intake valve closes when the pressure is greater than atmospheric, then the waves will have a supercharging effect and will produce an increase of power. This effect is called *inertia supercharging*.

While studying this problem, Morse et al.[12] found it necessary to express the velocity of the air-gas mixture in the valve, given by

$$f(t) = \begin{cases} 0, & -\pi < \omega t < -\pi/4, \\ \pi \cos(2\omega t)/2, & -\pi/4 < \omega t < \pi/4, \\ 0, & \pi/4 < \omega t < \pi, \end{cases} \qquad (\mathbf{4.3.9})$$

[12] Morse, P. M., R. H. Boden, and H. Schecter, 1938: Acoustic vibrations and internal combustion engine performance. I. Standing waves in the intake pipe system. *J. Appl. Phys.*, **9**, 16–23.

in terms of a Fourier expansion. The advantage of working with the Fourier series rather than the function itself lies in the ability to write the velocity as a periodic forcing function that highlights the various harmonics that might resonate with the structure comprising the fuel line.

Clearly $f(t)$ is an even function and its Fourier representation will be a cosine series. In this problem $\tau = -\pi/\omega$, and $L = \pi/\omega$. Therefore,

$$a_0 = \frac{2\omega}{\pi} \int_{-\pi/4\omega}^{\pi/4\omega} \frac{\pi}{2} \cos(2\omega t)\, dt = \tfrac{1}{2} \sin(2\omega t) \Big|_{-\pi/4\omega}^{\pi/4\omega} = 1, \qquad (\mathbf{4.3.10})$$

and

$$a_n = \frac{2\omega}{\pi} \int_{-\pi/4\omega}^{\pi/4\omega} \frac{\pi}{2} \cos(2\omega t) \cos\left(\frac{n\pi t}{\pi/\omega}\right)\, dt \qquad (\mathbf{4.3.11})$$

$$= \frac{\omega}{2} \int_{-\pi/4\omega}^{\pi/4\omega} \{\cos[(n+2)\omega t] + \cos[(n-2)\omega t]\}\, dt \qquad (\mathbf{4.3.12})$$

$$= \begin{cases} \frac{\sin[(n+2)\omega t]}{2(n+2)} + \frac{\sin[(n-2)\omega t]}{2(n-2)} \Big|_{-\pi/4\omega}^{\pi/4\omega}, & n \neq 2, \\[2ex] \frac{\omega t}{2} + \frac{\sin(4\omega t)}{4} \Big|_{-\pi/4\omega}^{\pi/4\omega}, & n = 2, \end{cases} \qquad (\mathbf{4.3.13})$$

$$= \begin{cases} -\frac{4}{n^2-4} \cos\left(\frac{n\pi}{4}\right), & n \neq 2, \\[1.5ex] \frac{\pi}{4}, & n = 2. \end{cases} \qquad (\mathbf{4.3.14})$$

Plotting these Fourier coefficients using the MATLAB script:

```
for m = 1:21;
n = m-1;
% compute the fourier coefficients a_n
if n == 2; an(m) = pi/4; else;
an(m) = 4.*cos(pi*n/4)/(4-n*n); end;
end
nn=0:20; % create indices for x-axis
fzero=zeros(size(nn)); % create the zero line
clf % clear any figures
axes('FontSize',20) % set font size
stem(nn,an,'filled') % plot spectrum
hold on
plot(nn,fzero,'-') % plot the zero line
set (gca,'PlotBoxAspectRatio',[8 4 1]) % set aspect ratio
xlabel('n') % label x-axis
ylabel('a_n') % label y-axis,
```

we see that these Fourier coefficients become small rapidly (see Figure 4.3.2). For that reason, Morse et al. showed that there are only about three resonances where the acoustic properties of the intake pipe can enhance engine

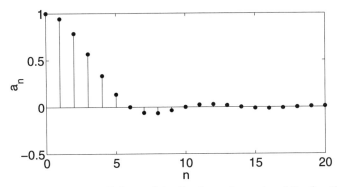

Figure 4.3.2: The spectral coefficients of the Fourier cosine series of the function (4.3.9).

performance. These peaks occur when $q = 30c/NL = 3, 4$, or 5, where c is the velocity of sound in the air-gas mixture, L is the effective length of the intake pipe, and N is the engine speed in rpm. See Figure 4.3.3. Subsequent experiments[13] verified these results.

Such analyses are valuable to automotive engineers. Engineers are always seeking ways to optimize a system with little or no additional cost. Our analysis shows that by tuning the length of the intake pipe so that it falls on one of the resonance peaks, we could obtain higher performance from the engine with little or no extra work. Of course, the problem is that no car always performs at some optimal condition.

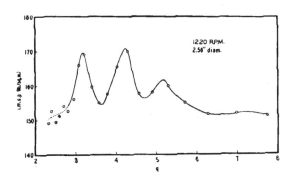

Figure 4.3.3: Experimental verification of inertial supercharging within a gasoline engine resulting from the resonance of the air-gas mixture and the intake pipe system. The peaks correspond to the $n = 3, 4$, and 5 harmonics of the Fourier representation (4.3.14) and the parameter q is defined in the text. (From Morse, P., R. H. Boden, and H. Schecter, 1938: Acoustic vibrations and internal combustion engine performance. *J. Appl. Phys.*, **9**, 17 with permission.)

---

[13] Boden, R. H., and H. Schecter, 1944: Dynamics of the inlet system of a four-stroke engine. *NACA Tech. Note 935*.

Problems

Find the Fourier cosine and sine series for the following functions. Then, use MATLAB to plot the Fourier coefficients.

1. $f(t) = t, \quad 0 < t < \pi$

2. $f(t) = \pi - t, \quad 0 < t < \pi$

3. $f(t) = t(a - t), \quad 0 < t < a$

4. $f(t) = e^{kt}, \quad 0 < t < a$

5. $f(t) = \begin{cases} t, & 0 \le t \le \frac{1}{2} \\ 1 - t, & \frac{1}{2} \le t \le 1 \end{cases}$

6. $f(t) = \begin{cases} t, & 0 < t \le 1 \\ 1, & 1 \le t < 2 \end{cases}$

7. $f(t) = \pi^2 - t^2, \quad 0 < t < \pi$

8. $f(t) = \begin{cases} 0, & 0 < t < \dfrac{a}{2} \\ 1, & \dfrac{a}{2} < t < a \end{cases}$

9. $f(t) = \begin{cases} 0, & 0 < t \le \dfrac{a}{3} \\ t - \dfrac{a}{3}, & \dfrac{a}{3} \le t \le \dfrac{2a}{3} \\ \dfrac{a}{3}, & \dfrac{2a}{3} \le t < a \end{cases}$

10. $f(t) = \begin{cases} 0, & 0 < t < \dfrac{a}{4} \\ 1, & \dfrac{a}{4} < t < \dfrac{3a}{4} \\ 0, & \dfrac{3a}{4} < t < a \end{cases}$

11. $f(t) = \begin{cases} \dfrac{1}{2}, & 0 < t < \dfrac{a}{2} \\ 1, & \dfrac{a}{2} < t < a \end{cases}$

12. $f(t) = \begin{cases} \dfrac{2t}{a}, & 0 < t \le \dfrac{a}{2} \\ \dfrac{3a - 2t}{2a}, & \dfrac{a}{2} \le t < a \end{cases}$

13. $f(t) = \begin{cases} t, & 0 < t \le \dfrac{a}{2} \\ \dfrac{a}{2}, & \dfrac{a}{2} \le t < a \end{cases}$

14. $f(t) = \dfrac{a - t}{a}, \quad 0 < t < a$

15. Using the relationships[14] that

$$\int_0^1 \frac{\cos(ax)}{\sqrt{1 - x^2}}\, dx = \frac{\pi}{2} J_0(a),$$

and

$$\int_0^u (u^2 - x^2)^{\nu - \frac{1}{2}} \cos(ax)\, dx = \frac{\sqrt{\pi}}{2} \left(\frac{2u}{a}\right)^{\nu} \Gamma\left(\nu + \tfrac{1}{2}\right) J_\nu(au),$$

---

[14] Gradshteyn, I. S., and I. M. Ryzhik, 1965: *Table of Integrals, Series, and Products.* Academic Press, §3.753, Formula 2 and §3.771, Formula 8.

with $a > 0$, $u > 0$, $\mathrm{Re}(\nu) > -\frac{1}{2}$, obtain the following half-range expansions:

$$\frac{1}{\sqrt{1-x^2}} = \frac{\pi}{2} + \pi \sum_{n=1}^{\infty} J_0(n\pi) \cos(n\pi x), \qquad 0 < x < 1,$$

and

$$\sqrt{1-x^2} = 2 \sum_{n=1}^{\infty} \frac{J_1[(2n-1)\pi/2]}{2n-1} \cos[(2n-1)\pi x/2], \qquad 0 < x < 1.$$

Here $J_\nu(\ )$ denotes the Bessel function of the first kind and order ν (see §9.5) and $\Gamma(\ )$ is the gamma function.[15]

16. The function

$$f(t) = 1 - (1+a)\frac{t}{\pi} + (a-1)\frac{t^2}{\pi^2} + (a+1)\frac{t^3}{\pi^3} - a\frac{t^4}{\pi^4}, \qquad 0 < t < \pi,$$

is a curve fit to the observed pressure trace of an explosion wave in the atmosphere. Because the observed transmission of atmospheric waves depends on the five-fourths power of the frequency, Reed[16] had to re-express this curve fit as a Fourier sine series before he could use the transmission law. He found that

$$f(t) = \frac{1}{\pi} \sum_{n=1}^{\infty} \frac{1}{n} \left[1 - \frac{3(a-1)}{2\pi^2 n^2}\right] \sin(2nt)$$

$$+ \frac{1}{\pi} \sum_{n=1}^{\infty} \frac{2}{2n-1} \left[1 + \frac{2(a-1)}{\pi^2(2n-1)^2} - \frac{48a}{\pi^4(2n-1)^4}\right] \sin[(2n-1)t].$$

Confirm his result.

4.4 FOURIER SERIES WITH PHASE ANGLES

Sometimes it is desirable to rewrite a general Fourier series as a purely cosine or purely sine series with a phase angle. Engineers often speak of some quantity leading or lagging another quantity. Re-expressing a Fourier series in terms of amplitude and phase provides a convenient method for determining these phase relationships.

---

[15] Gradshteyn and Ryzhik, *op. cit.*, §6.41.

[16] From Reed, J. W., 1977: Atmospheric attenuation of explosion waves. *J. Acoust. Soc. Am.*, **61**, 39–47 with permission.

Suppose, for example, that we have a function $f(t)$ of period $2L$, given in the interval $[-L, L]$, whose Fourier series expansion is

$$f(t) = \frac{a_0}{2} + \sum_{n=1}^{\infty} a_n \cos\left(\frac{n\pi t}{L}\right) + b_n \sin\left(\frac{n\pi t}{L}\right). \qquad (4.4.1)$$

We wish to replace (4.4.1) by the series:

$$f(t) = \frac{a_0}{2} + \sum_{n=1}^{\infty} B_n \sin\left(\frac{n\pi t}{L} + \varphi_n\right). \qquad (4.4.2)$$

To do this we note that

$$B_n \sin\left(\frac{n\pi t}{L} + \varphi_n\right) = a_n \cos\left(\frac{n\pi t}{L}\right) + b_n \sin\left(\frac{n\pi t}{L}\right) \qquad (4.4.3)$$

$$= B_n \sin\left(\frac{n\pi t}{L}\right)\cos(\varphi_n) + B_n \sin(\varphi_n)\cos\left(\frac{n\pi t}{L}\right). \qquad (4.4.4)$$

We equate coefficients of $\sin(n\pi t/L)$ and $\cos(n\pi t/L)$ on both sides and obtain

$$a_n = B_n \sin(\varphi_n), \quad \text{and} \quad b_n = B_n \cos(\varphi_n). \qquad (4.4.5)$$

Hence, upon squaring and adding,

$$B_n = \sqrt{a_n^2 + b_n^2}, \qquad (4.4.6)$$

while taking the ratio gives

$$\varphi_n = \tan^{-1}(a_n/b_n). \qquad (4.4.7)$$

Similarly we could rewrite (4.4.1) as

$$f(t) = \frac{a_0}{2} + \sum_{n=1}^{\infty} A_n \cos\left(\frac{n\pi t}{L} + \varphi_n\right), \qquad (4.4.8)$$

where

$$A_n = \sqrt{a_n^2 + b_n^2}, \quad \text{and} \quad \varphi_n = \tan^{-1}(-b_n/a_n), \qquad (4.4.9)$$

and

$$a_n = A_n \cos(\varphi_n), \quad \text{and} \quad b_n = -A_n \sin(\varphi_n). \qquad (4.4.10)$$

In both cases, we must be careful in computing φ_n because there are two possible values of φ_n which satisfy (4.4.7) or (4.4.9). These angles φ_n must give the correct a_n and b_n using either (4.4.5) or (4.4.10).

• **Example 4.4.1**

The Fourier series for $f(t) = e^t$ over the interval $-L < t < L$ is

$$f(t) = \frac{\sinh(aL)}{aL} + 2\sinh(aL) \sum_{n=1}^{\infty} \frac{aL(-1)^n}{a^2L^2 + n^2\pi^2} \cos\left(\frac{n\pi t}{L}\right)$$

$$- 2\sinh(aL) \sum_{n=1}^{\infty} \frac{n\pi(-1)^n}{a^2L^2 + n^2\pi^2} \sin\left(\frac{n\pi t}{L}\right). \qquad (4.4.11)$$

Let us rewrite (4.4.11) as a Fourier series with a phase angle. Regardless of whether we want the new series to contain $\cos(n\pi t/L + \varphi_n)$ or $\sin(n\pi t/L + \varphi_n)$, the amplitude A_n or B_n is the same in both series:

$$A_n = B_n = \sqrt{a_n^2 + b_n^2} = \frac{2\sinh(aL)}{\sqrt{a^2L^2 + n^2\pi^2}}. \qquad (4.4.12)$$

If we want our Fourier series to read

$$f(t) = \frac{\sinh(aL)}{aL} + 2\sinh(aL) \sum_{n=1}^{\infty} \frac{\cos(n\pi t/L + \varphi_n)}{\sqrt{a^2L^2 + n^2\pi^2}}, \qquad (4.4.13)$$

then

$$\varphi_n = \tan^{-1}\left(-\frac{b_n}{a_n}\right) = \tan^{-1}\left(\frac{n\pi}{aL}\right), \qquad (4.4.14)$$

where φ_n lies in the first quadrant if n is even and in the third quadrant if n is odd. This ensures that the sign from the $(-1)^n$ is correct.

On the other hand, if we prefer

$$f(t) = \frac{\sinh(aL)}{aL} + 2\sinh(aL) \sum_{n=1}^{\infty} \frac{\sin(n\pi t/L + \varphi_n)}{\sqrt{a^2L^2 + n^2\pi^2}}, \qquad (4.4.15)$$

then

$$\varphi_n = \tan^{-1}\left(\frac{a_n}{b_n}\right) = -\tan^{-1}\left(\frac{aL}{n\pi}\right), \qquad (4.4.16)$$

where φ_n lies in the fourth quadrant if n is odd and in the second quadrant if n is even.

Problems

Write the following Fourier series in both the cosine and sine phase angle form:

1.

$$f(t) = \frac{1}{2} + \frac{2}{\pi} \sum_{n=1}^{\infty} \frac{\sin[(2n-1)\pi t]}{2n-1}$$

2.

$$f(t) = \frac{3}{2} + \frac{2}{\pi} \sum_{n=1}^{\infty} \frac{(-1)^n}{2n-1} \cos\left[\frac{(2n-1)\pi t}{2}\right]$$

3.

$$f(t) = -2 \sum_{n=1}^{\infty} \frac{(-1)^n}{n} \sin(nt)$$

4.

$$f(t) = \frac{\pi}{2} - \frac{4}{\pi} \sum_{n=1}^{\infty} \frac{\cos[(2n-1)t]}{(2n-1)^2}$$

4.5 COMPLEX FOURIER SERIES

So far in our discussion, we expressed Fourier series in terms of sines and cosines. We are now ready to re-express a Fourier series as a series of complex exponentials. There are two reasons for this. First, in certain engineering and scientific applications of Fourier series, the expansion of a function in terms of complex exponentials results in coefficients of considerable simplicity and clarity. Second, these complex Fourier series point the way to the development of the Fourier transform in the next chapter.

We begin by introducing the variable $\omega_n = n\pi/L$, where $n = 0, \pm 1, \pm 2, \ldots$ Using Euler's formula we can replace the sine and cosine in the Fourier series by exponentials and find that

$$f(t) = \frac{a_0}{2} + \sum_{n=1}^{\infty} \frac{a_n}{2}\left(e^{i\omega_n t} + e^{-i\omega_n t}\right) + \frac{b_n}{2i}\left(e^{i\omega_n t} - e^{-i\omega_n t}\right) \qquad (4.5.1)$$

$$= \frac{a_0}{2} + \sum_{n=1}^{\infty} \left(\frac{a_n}{2} - \frac{b_n i}{2}\right)e^{i\omega_n t} + \left(\frac{a_n}{2} + \frac{b_n i}{2}\right)e^{-i\omega_n t}. \qquad (4.5.2)$$

If we define $c_n = \frac{1}{2}(a_n - ib_n)$, then

$$c_n = \frac{1}{2}(a_n - ib_n) = \frac{1}{2L} \int_{\tau}^{\tau+2L} f(t)[\cos(\omega_n t) - i\sin(\omega_n t)]\, dt \qquad (4.5.3)$$

$$= \frac{1}{2L} \int_{\tau}^{\tau+2L} f(t)e^{-i\omega_n t}\, dt. \qquad (4.5.4)$$

Similarly, the complex conjugate of c_n, c_n^*, equals

$$c_n^* = \frac{1}{2}(a_n + ib_n) = \frac{1}{2L} \int_{\tau}^{\tau+2L} f(t)e^{i\omega_n t}\, dt. \qquad (4.5.5)$$

To simplify (4.5.2) we note that

$$\omega_{-n} = \frac{(-n)\pi}{L} = -\frac{n\pi}{L} = -\omega_n, \tag{4.5.6}$$

which yields the result that

$$c_{-n} = \frac{1}{2L} \int_\tau^{\tau+2L} f(t)e^{-i\omega_{-n}t} dt = \frac{1}{2L} \int_\tau^{\tau+2L} f(t)e^{i\omega_n t} dt = c_n^* \tag{4.5.7}$$

so that we can write (4.5.2) as

$$f(t) = \frac{a_0}{2} + \sum_{n=1}^\infty c_n e^{i\omega_n t} + c_n^* e^{-i\omega_n t} = \frac{a_0}{2} + \sum_{n=1}^\infty c_n e^{i\omega_n t} + c_{-n} e^{-i\omega_n t}. \tag{4.5.8}$$

Letting $n = -m$ in the second summation on the right side of (4.5.8),

$$\sum_{n=1}^\infty c_{-n} e^{-i\omega_n t} = \sum_{m=-1}^{-\infty} c_m e^{-i\omega_{-m}t} = \sum_{m=-\infty}^{-1} c_m e^{i\omega_m t} = \sum_{n=-\infty}^{-1} c_n e^{i\omega_n t}, \tag{4.5.9}$$

where we introduced $m = n$ into the last summation in (4.5.9). Therefore,

$$f(t) = \frac{a_0}{2} + \sum_{n=1}^\infty c_n e^{i\omega_n t} + \sum_{n=-\infty}^{-1} c_n e^{i\omega_n t}. \tag{4.5.10}$$

On the other hand,

$$\frac{a_0}{2} = \frac{1}{2L} \int_\tau^{\tau+2L} f(t) \, dt = c_0 = c_0 e^{i\omega_0 t}, \tag{4.5.11}$$

because $\omega_0 = 0\pi/L = 0$. Thus, our final result is

$$f(t) = \sum_{n=-\infty}^\infty c_n e^{i\omega_n t}, \tag{4.5.12}$$

where

$$c_n = \frac{1}{2L} \int_\tau^{\tau+2L} f(t)e^{-i\omega_n t} \, dt \tag{4.5.13}$$

and $n = 0, \pm 1, \pm 2, \ldots$. Note that even though c_n is generally complex, the summation (4.5.12) always gives a *real-valued* function $f(t)$.

Just as we can represent the function $f(t)$ graphically by a plot of t against $f(t)$, we can plot c_n as a function of n, commonly called the frequency *spectrum*. Because c_n is generally complex, it is necessary to make two plots. Typically the plotted quantities are the amplitude spectra $|c_n|$ and the phase spectra φ_n, where φ_n is the phase of c_n. However, we could just as well plot the real and imaginary parts of c_n. Because n is an integer, these plots consist merely of a series of vertical lines representing the ordinates of the quantity $|c_n|$ or φ_n for each n. For this reason we refer to these plots as the *line spectra*.

Because $2c_n = a_n - ib_n$, the coefficients c_n for an even function will be purely real; the coefficients c_n for an odd function are purely imaginary. It is important to note that we lose the advantage of even and odd functions in the sense that we cannot just integrate over the interval 0 to L and then double the result. In the present case we have a line integral of a complex function along the real axis.

• Example 4.5.1

Let us find the complex Fourier series for

$$f(t) = \begin{cases} 1, & 0 < t < \pi, \\ -1, & -\pi < t < 0, \end{cases} \qquad (4.5.14)$$

which has the periodicity $f(t + 2\pi) = f(t)$.

With $L = \pi$ and $\tau = -\pi$, $\omega_n = n\pi/L = n$. Therefore,

$$c_n = \frac{1}{2\pi} \int_{-\pi}^{0} (-1)e^{-int} \, dt + \frac{1}{2\pi} \int_{0}^{\pi} (1)e^{-int} \, dt \qquad (4.5.15)$$

$$= \frac{1}{2n\pi i} e^{-int} \Big|_{-\pi}^{0} - \frac{1}{2n\pi i} e^{-int} \Big|_{0}^{\pi} \qquad (4.5.16)$$

$$= -\frac{i}{2n\pi} \left(1 - e^{n\pi i}\right) + \frac{i}{2n\pi} \left(e^{-n\pi i} - 1\right), \qquad (4.5.17)$$

if $n \neq 0$. Because $e^{n\pi i} = \cos(n\pi) + i\sin(n\pi) = (-1)^n$ and $e^{-n\pi i} = \cos(-n\pi) + i\sin(-n\pi) = (-1)^n$, then

$$c_n = -\frac{i}{n\pi}[1 - (-1)^n] = \begin{cases} 0, & n \text{ even}, \\ -\frac{2i}{n\pi}, & n \text{ odd}, \end{cases} \qquad (4.5.18)$$

with

$$f(t) = \sum_{n=-\infty}^{\infty} c_n e^{int}. \qquad (4.5.19)$$

In this particular problem we must treat the case $n = 0$ specially because (4.5.16) is undefined for $n = 0$. In that case,

$$c_0 = \frac{1}{2\pi} \int_{-\pi}^{0} (-1)\, dt + \frac{1}{2\pi} \int_{0}^{\pi} (1)\, dt = \frac{1}{2\pi}(-t)\Big|_{-\pi}^{0} + \frac{1}{2\pi}(t)\Big|_{0}^{\pi} = 0. \quad \textbf{(4.5.20)}$$

Because $c_0 = 0$, we can write the expansion:

$$f(t) = -\frac{2i}{\pi} \sum_{m=-\infty}^{\infty} \frac{e^{(2m-1)it}}{2m-1}, \quad \textbf{(4.5.21)}$$

since we can write all odd integers as $2m - 1$, where $m = 0, \pm 1, \pm 2, \pm 3, \ldots$ Using the MATLAB script

```
max = 31; % total number of harmonics
mid = (max+1)/2; % in the array, location of c_0
for m = 1:max;
n = m - mid; % compute value of harmonic
% compute complex Fourier coefficient c_n = (cnr,cni)
if mod(n,2) == 0; cnr(m) = 0; cni(m) = 0; else;
cnr(m) = 0; cni(m) = - 2/(pi*n); end;
end
nn=(1-mid):(max-mid); % create indices for x-axis
fzero=zeros(size(nn)); % create the zero line
clf % clear any figures
amplitude = sqrt(cnr.*cnr+cni.*cni);
phase = atan2(cni,cnr);
% plot amplitude of c_n
subplot(2,1,1), stem(nn,amplitude,'filled')
% label amplitude plot
text(6,0.75,'amplitude','FontSize',20)
subplot(2,1,2), stem(nn,phase,'filled') % plot phases of c_n
text(7,1,'phase','FontSize',20) % label phase plot
xlabel('n','Fontsize',20) % label x-axis,
```

we plot the amplitude and phase spectra for the function (4.5.14) as a function of n in Figure 4.5.1.

• **Example 4.5.2**

The concept of Fourier series can be generalized to multivariable functions. Consider the function $f(x,y)$ defined over $0 < x < L$ and $0 < y < H$. Taking y constant, we have that

$$c_n(y) = \frac{1}{L} \int_{0}^{L} f(x,y) e^{-i\xi_n x}\, dx, \qquad \xi_n = \frac{2\pi n}{L}. \quad \textbf{(4.5.22)}$$

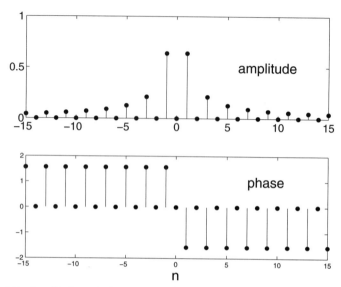

Figure 4.5.1: Amplitude and phase spectra for the function (4.5.14).

Similarly, holding ξ_n constant,

$$c_{nm} = \frac{1}{H} \int_0^H c_n(y)e^{-i\eta_m y} \, dy, \qquad \eta_m = \frac{2\pi m}{H}. \tag{4.5.23}$$

Therefore, the (complex) Fourier coefficient for the two-dimensional function $f(x,y)$ is

$$c_{nm} = \frac{1}{LH} \int_0^L \int_0^H f(x,y)e^{-i(\xi_n x + \eta_m y)} \, dx \, dy, \tag{4.5.24}$$

assuming that the integral exists.

To recover $f(x,y)$ given c_{nm}, we reverse the process of deriving c_{nm}. Starting with

$$c_n(y) = \sum_{m=-\infty}^{\infty} c_{nm} e^{i\eta_m y}, \tag{4.5.25}$$

we find that

$$f(x,y) = \sum_{n=-\infty}^{\infty} c_n(y)e^{i\xi_n x}. \tag{4.5.26}$$

Therefore,

$$f(x,y) = \sum_{n=-\infty}^{\infty} \sum_{m=-\infty}^{\infty} c_{nm} e^{i(\xi_n x + \eta_m y)}. \tag{4.5.27}$$

Problems

Find the complex Fourier series for the following functions. Then use MAT-LAB to plot the corresponding spectra.

1. $f(t) = |t|, \quad -\pi \leq t \leq \pi$

2. $f(t) = e^t, \quad 0 < t < 2$

3. $f(t) = t, \quad 0 < t < 2$

4. $f(t) = t^2, \quad -\pi \leq t \leq \pi$

5. $f(t) = \begin{cases} 0, & -\pi/2 < t < 0 \\ 1, & 0 < t < \pi/2 \end{cases}$

6. $f(t) = t, \quad -1 < t < 1$

4.6 THE USE OF FOURIER SERIES IN THE SOLUTION OF ORDINARY DIFFERENTIAL EQUATIONS

An important application of Fourier series is the solution of ordinary differential equations. Structural engineers especially use this technique because the occupants of buildings and bridges often subject these structures to forcings that are periodic in nature.[17]

• Example 4.6.1

Let us find the general solution to the ordinary differential equation

$$y'' + 9y = f(t), \tag{4.6.1}$$

where the forcing is

$$f(t) = |t|, \qquad -\pi \leq t \leq \pi, \qquad f(t + 2\pi) = f(t). \tag{4.6.2}$$

This equation represents an oscillator forced by a driver whose displacement is the saw-tooth function.

We begin by replacing the function $f(t)$ by its Fourier series representation because the forcing function is periodic. The advantage of expressing $f(t)$ as a Fourier series is its validity for any time t. The alternative would be to construct a solution over each interval $n\pi < t < (n + 1)\pi$ and then piece together the final solution assuming that the solution and its first derivative is continuous at each junction $t = n\pi$. Because the function is an even function, all of the sine terms vanish and the Fourier series is

$$|t| = \frac{\pi}{2} - \frac{4}{\pi} \sum_{n=1}^{\infty} \frac{\cos[(2n-1)t]}{(2n-1)^2}. \tag{4.6.3}$$

---

[17] Timoshenko, S. P., 1943: Theory of suspension bridges. Part II. *J. Franklin Inst.*, **235**, 327–349; Inglis, C. E., 1934: *A Mathematical Treatise on Vibrations in Railway Bridges.* Cambridge University Press, 203 pp.

Next, we note that the general solution consists of the complementary solution, which equals

$$y_H(t) = A\cos(3t) + B\sin(3t), \tag{4.6.4}$$

and the particular solution $y_p(t)$ which satisfies the differential equation

$$y_p'' + 9y_p = \frac{\pi}{2} - \frac{4}{\pi} \sum_{n=1}^{\infty} \frac{\cos[(2n-1)t]}{(2n-1)^2}. \tag{4.6.5}$$

To determine this particular solution, we write (4.6.5) as

$$y_p'' + 9y_p = \frac{\pi}{2} - \frac{4}{\pi}\cos(t) - \frac{4}{9\pi}\cos(3t) - \frac{4}{25\pi}\cos(5t) - \cdots. \tag{4.6.6}$$

By the method of undetermined coefficients, we guess the particular solution:

$$y_p(t) = \frac{a_0}{2} + a_1\cos(t) + b_1\sin(t) + a_2\cos(3t) + b_2\sin(3t) + \cdots \tag{4.6.7}$$

or

$$y_p(t) = \tfrac{1}{2}a_0 + \sum_{n=1}^{\infty} a_n\cos[(2n-1)t] + b_n\sin[(2n-1)t]. \tag{4.6.8}$$

Because

$$y_p''(t) = \sum_{n=1}^{\infty} -(2n-1)^2\{a_n\cos[(2n-1)t] + b_n\sin[(2n-1)t]\}, \tag{4.6.9}$$

$$\sum_{n=1}^{\infty} -(2n-1)^2\{a_n\cos[(2n-1)t] + b_n\sin[(2n-1)t]\}$$

$$+ \tfrac{9}{2}a_0 + 9\sum_{n=1}^{\infty} a_n\cos[(2n-1)t] + b_n\sin[(2n-1)t]$$

$$= \frac{\pi}{2} - \frac{4}{\pi} \sum_{n=1}^{\infty} \frac{\cos[(2n-1)t]}{(2n-1)^2}, \tag{4.6.10}$$

or

$$\frac{9a_0}{2} - \frac{\pi}{2} + \sum_{n=1}^{\infty} \left\{ [9 - (2n-1)^2]a_n + \frac{4}{\pi(2n-1)^2} \right\} \cos[(2n-1)t]$$

$$+ \sum_{n=1}^{\infty} [9 - (2n-1)^2]b_n\sin[(2n-1)t] = 0. \tag{4.6.11}$$

Because (4.6.11) must hold true for any time, each harmonic must vanish separately and

$$a_0 = \frac{\pi}{9}, \qquad a_n = -\frac{4}{\pi(2n-1)^2[9-(2n-1)^2]} \qquad (4.6.12)$$

and $b_n = 0$. All of the coefficients a_n are finite except for $n = 2$, where a_2 becomes undefined. This coefficient is undefined because the harmonic $\cos(3t)$ in the forcing function resonates with the natural mode of the system.

Let us review our analysis to date. We found that each harmonic in the forcing function yields a corresponding harmonic in the particular solution (4.6.8). The only difficulty arises with the harmonic $n = 2$. Although our particular solution is not correct because it contains $\cos(3t)$, we suspect that if we remove that term then the remaining harmonic solutions are correct. The problem is linear, and difficulties with one harmonic term should not affect other harmonics. But how shall we deal with the $\cos(3t)$ term in the forcing function? Let us denote that particular solution by $Y(t)$ and modify our particular solution as follows:

$$y_p(t) = \tfrac{1}{2}a_0 + a_1\cos(t) + Y(t) + a_3\cos(5t) + \cdots. \qquad (4.6.13)$$

Substituting this solution into the differential equation and simplifying, everything cancels except

$$Y'' + 9Y = -\frac{4}{9\pi}\cos(3t). \qquad (4.6.14)$$

The solution of this equation by the method of undetermined coefficients is

$$Y(t) = -\frac{2}{27\pi}t\sin(3t). \qquad (4.6.15)$$

This term, called a *secular term*, is the most important one in the solution. While the other terms merely represent simple oscillatory motion, the term $t\sin(3t)$ grows linearly with time and eventually becomes the dominant term in the series. Consequently, the general solution equals the complementary plus the particular solution or

$$\begin{aligned} y(t) = {} & A\cos(3t) + B\sin(3t) \\ & + \frac{\pi}{18} - \frac{2}{27\pi}t\sin(3t) - \frac{4}{\pi}\sum_{\substack{n=1 \\ n\neq 2}}^{\infty}\frac{\cos[(2n-1)t]}{(2n-1)^2[9-(2n-1)^2]}. \end{aligned} \qquad (4.6.16)$$

● **Example 4.6.2**

Let us redo the previous problem only using complex Fourier series. That is, let us find the general solution to the ordinary differential equation

$$y'' + 9y = \frac{\pi}{2} - \frac{2}{\pi} \sum_{n=-\infty}^{\infty} \frac{e^{i(2n-1)t}}{(2n-1)^2}. \tag{4.6.17}$$

From the method of undetermined coefficients we guess the particular solution for (4.6.17) to be

$$y_p(t) = c_0 + \sum_{n=-\infty}^{\infty} c_n e^{i(2n-1)t}. \tag{4.6.18}$$

Then

$$y_p''(t) = \sum_{n=-\infty}^{\infty} -(2n-1)^2 c_n e^{i(2n-1)t}. \tag{4.6.19}$$

Substituting (4.6.18) and (4.6.19) into (4.6.17),

$$9c_0 + \sum_{n=-\infty}^{\infty} [9 - (2n-1)^2] c_n e^{i(2n-1)t} = \frac{\pi}{2} - \frac{2}{\pi} \sum_{n=-\infty}^{\infty} \frac{e^{i(2n-1)t}}{(2n-1)^2}. \tag{4.6.20}$$

Because (4.6.20) must be true for any t,

$$c_0 = \frac{\pi}{18}, \quad \text{and} \quad c_n = \frac{2}{\pi(2n-1)^2[(2n-1)^2 - 9]}. \tag{4.6.21}$$

Therefore,

$$y_p(t) = \frac{\pi}{18} + \frac{2}{\pi} \sum_{n=-\infty}^{\infty} \frac{e^{i(2n-1)t}}{(2n-1)^2[(2n-1)^2 - 9]} e^{i(2n-1)t}. \tag{4.6.22}$$

However, there is a problem when $n = -1$ and $n = 2$. Therefore, we modify (4.6.22) to read

$$y_p(t) = \frac{\pi}{18} + c_2 t e^{3it} + c_{-1} t e^{-3it}$$

$$+ \frac{2}{\pi} \sum_{\substack{n=-\infty \\ n \neq -1,2}}^{\infty} \frac{e^{i(2n-1)t}}{(2n-1)^2[(2n-1)^2 - 9]} e^{i(2n-1)t}. \tag{4.6.23}$$

Substituting (4.6.23) into (4.6.17) and simplifying,

$$c_2 = -\frac{1}{27\pi i}, \quad \text{and} \quad c_{-1} = -\frac{1}{27\pi i}. \tag{4.6.24}$$

The general solution is then

$$y(t) = Ae^{3it} + Be^{-3it} + \frac{\pi}{18} - \frac{te^{3it}}{27\pi i} + \frac{te^{-3it}}{27\pi i} + \frac{2}{\pi} \sum_{\substack{n=-\infty \\ n \neq -1,2}}^{\infty} \frac{e^{i(2n-1)t}}{(2n-1)^2[(2n-1)^2 - 9]}.$$

$$(4.6.25)$$

The first two terms on the right side of (4.6.25) represent the complementary solution. Although (4.6.25) is equivalent to (4.6.16), we have all of the advantages of dealing with exponentials rather than sines and cosines. These advantages include ease of differentiation and integration, and writing the series in terms of amplitude and phase.

• Example 4.6.3: Temperature within a spinning satellite

In the design of artificial satellites, it is important to determine the temperature distribution on the spacecraft's surface. An interesting special case is the temperature fluctuation in the skin due to the spinning of the vehicle. If the craft is thin-walled so that there is no radial dependence, Hrycak[18] showed that he could approximate the nondimensional temperature field at the equator of the rotating satellite by

$$\frac{d^2T}{d\eta^2} + b\frac{dT}{d\eta} - c\left(T - \frac{3}{4}\right) = -\frac{\pi c}{4} \frac{F(\eta) + \beta/4}{1 + \pi\beta/4}, \tag{4.6.26}$$

where

$$b = 4\pi^2 r^2 f/a, \quad c = \frac{16\pi S}{\gamma T_\infty}\left(1 + \frac{\pi\beta}{4}\right), \quad T_\infty = \left(\frac{S}{\pi\sigma\epsilon}\right)^{1/4}\left(\frac{1 + \pi\beta/4}{1 + \beta}\right)^{1/4}, \tag{4.6.27}$$

$$F(\eta) = \begin{cases} \cos(2\pi\eta), & 0 \leq \eta \leq \frac{1}{4}, \\ 0, & \frac{1}{4} \leq \eta \leq \frac{3}{4}, \\ \cos(2\pi\eta), & \frac{3}{4} \leq \eta \leq 1, \end{cases} \tag{4.6.28}$$

a is the thermal diffusivity of the shell, f is the rate of spin, r is the radius of the spacecraft, S is the net direct solar heating, β is the ratio of the emissivity of the interior shell to the emissivity of the exterior surface, ϵ is the overall emissivity of the exterior surface, γ is the satellite's skin conductance, and σ is the Stefan-Boltzmann constant. The independent variable η is the longitude along the equator with the effect of rotation subtracted out ($2\pi\eta = \varphi - 2\pi ft$). The reference temperature T_∞ equals the temperature that the spacecraft would have if it spun with infinite angular speed so that the solar heating would be uniform around the craft. We nondimensionalized the temperature with respect to T_∞.

---

[18] Hrycak, P., 1963: Temperature distribution in a spinning spherical space vehicle. *AIAA J.*, **1**, 96–99.

We begin by introducing the new variables

$$y = T - \frac{3}{4} - \frac{\pi\beta}{16 + 4\pi\beta}, \quad \nu_0 = \frac{2\pi^2 r^2 f}{a\rho_0}, \quad A_0 = -\frac{\pi\rho^2}{4 + \pi\beta} \qquad (4.6.29)$$

and $\rho_0^2 = c$ so that (4.6.26) becomes

$$\frac{d^2y}{d\eta^2} + 2\rho_0\nu_0\frac{dy}{d\eta} - \rho_0^2 y = A_0 F(\eta). \qquad (4.6.30)$$

Next, we expand $F(\eta)$ as a Fourier series because it is a periodic function of period 1. Because it is an even function,

$$f(\eta) = \tfrac{1}{2}a_0 + \sum_{n=1}^{\infty} a_n \cos(2n\pi\eta), \qquad (4.6.31)$$

where

$$a_0 = \frac{1}{1/2}\int_0^{1/4} \cos(2\pi x)\,dx + \frac{1}{1/2}\int_{3/4}^1 \cos(2\pi x)\,dx = \frac{2}{\pi}, \qquad (4.6.32)$$

$$a_1 = \frac{1}{1/2}\int_0^{1/4} \cos^2(2\pi x)\,dx + \frac{1}{1/2}\int_{3/4}^1 \cos^2(2\pi x)\,dx = \frac{1}{2} \qquad (4.6.33)$$

and

$$a_n = \frac{1}{1/2}\int_0^{1/4} \cos(2\pi x)\cos(2n\pi x)\,dx + \frac{1}{1/2}\int_{3/4}^1 \cos(2\pi x)\cos(2n\pi x)\,dx$$

$$\hspace{9cm} (4.6.34)$$

$$= -\frac{2(-1)^n}{\pi(n^2 - 1)}\cos\left(\frac{n\pi}{2}\right), \qquad (4.6.35)$$

if $n \geq 2$. Therefore,

$$f(\eta) = \frac{1}{\pi} + \frac{1}{2}\cos(2\pi\eta) - \frac{2}{\pi}\sum_{n=1}^{\infty} \frac{(-1)^n}{4n^2 - 1}\cos(4n\pi\eta). \qquad (4.6.36)$$

From the method of undetermined coefficients, the particular solution is

$$y_p(\eta) = \tfrac{1}{2}a_0 + a_1\cos(2\pi\eta) + b_1\sin(2\pi\eta) + \sum_{n=1}^{\infty} a_{2n}\cos(4n\pi\eta) + b_{2n}\sin(4n\pi\eta),$$

$$\hspace{9cm} (4.6.37)$$

which yields

$$y_p'(\eta) = -2\pi a_1\sin(2\pi\eta) + 2\pi b_1\cos(2\pi\eta)$$

$$+ \sum_{n=1}^{\infty} [-4n\pi a_{2n}\sin(4n\pi\eta) + 4n\pi b_{2n}\cos(4n\pi\eta)], \qquad (4.6.38)$$

and

$$y_p''(\eta) = -4\pi^2 a_1 \cos(2\pi\eta) - 4\pi^2 b_1 \sin(2\pi\eta)$$
$$+ \sum_{n=1}^{\infty}[-16n^2\pi^2 a_{2n}\cos(4n\pi\eta) - 16n^2\pi^2 b_{2n}\sin(4n\pi\eta)]. \qquad \textbf{(4.6.39)}$$

Substituting into (4.6.30),

$$-\frac{1}{2}\rho_0^2 a_0 - \frac{A_0}{\pi} + \left(-4\pi^2 a_1 + 4\pi\rho_0\nu_0 b_1 - \rho_0^2 a_1 - \frac{A_0}{2}\right)\cos(2\pi\eta)$$
$$+ \left(-4\pi^2 b_1 - 4\pi\rho_0\nu_0 a_1 - \rho_0^2 b_1\right)\sin(2\pi\eta)$$
$$+ \sum_{n=1}^{\infty}\left[-16n^2\pi^2 a_{2n} + 8n\pi\rho_0\nu_0 b_{2n} - \rho_0^2 a_{2n} + \frac{2A_0(-1)^n}{\pi(4n^2-1)}\right]\cos(4n\pi\eta)$$
$$+ \sum_{n=1}^{\infty}\left(-16n^2\pi^2 b_{2n} - 8n\pi\rho_0\nu_0 a_{2n} - \rho_0^2 b_{2n}\right)\sin(4n\pi\eta) = 0. \qquad \textbf{(4.6.40)}$$

To satisfy (4.6.40) for any η, we set

$$a_0 = -\frac{2A_0}{\pi\rho_0^2}, \qquad \textbf{(4.6.41)}$$

$$-(4\pi^2 + \rho_0^2)a_1 + 4\pi\rho_0\nu_0 b_1 = \frac{A_0}{2}, \qquad \textbf{(4.6.42)}$$

$$4\pi\rho_0\nu_0 a_1 + (4\pi^2 + \rho_0^2)b_1 = 0, \qquad \textbf{(4.6.43)}$$

$$(16n^2\pi^2 + \rho_0^2)a_{2n} - 8n\pi\rho_0\nu_0 b_{2n} = \frac{2A_0(-1)^n}{\pi(4n^2-1)}, \qquad \textbf{(4.6.44)}$$

and

$$8n\pi\rho_0\nu_0 a_{2n} + (16n^2\pi^2 + \rho_0^2)b_{2n} = 0, \qquad \textbf{(4.6.45)}$$

or

$$[16\pi^2\rho_0^2\nu_0^2 + (4\pi^2 + \rho_0^2)^2]a_1 = -\frac{(4\pi^2 + \rho_0^2)A_0}{2}, \qquad \textbf{(4.6.46)}$$

$$[16\pi^2\rho_0^2\nu_0^2 + (4\pi^2 + \rho_0^2)^2]b_1 = 2\pi\rho_0\nu_0 A_0, \qquad \textbf{(4.6.47)}$$

$$[64n^2\pi^2\rho_0^2\nu_0^2 + (16n^2\pi^2 + \rho_0^2)^2]a_{2n} = \frac{2A_0(-1)^n(16n^2\pi^2 + \rho_0^2)}{\pi(4n^2-1)}, \qquad \textbf{(4.6.48)}$$

and

$$[64n^2\pi^2\rho_0^2\nu_0^2 + (16n^2\pi^2 + \rho_0^2)^2]b_{2n} = -\frac{16(-1)^n\rho_0\nu_0 n A_0}{4n^2-1}. \qquad \textbf{(4.6.49)}$$

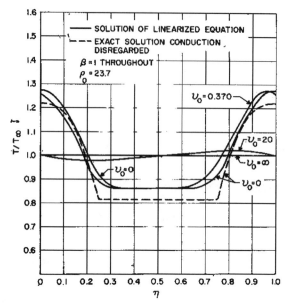

Figure 4.6.1: Temperature distribution along the equator of a spinning spherical satellite. (From Hrycak, P., 1963: Temperature distribution in a spinning spherical space vehicle. *AIAA J.*, **1**, 97. ©1963 AIAA, reprinted with permission.)

Substituting for a_0, a_1, b_1, a_{2n}, and b_{2n}, the particular solution is

$$
\begin{aligned}
y_p(\eta) = {}& -\frac{A_0}{\pi\rho_0^2} - \frac{(4\pi^2 + \rho_0^2)A_0 \cos(2\pi\eta)}{2[(4\pi^2 + \rho_0^2)^2 + 16\pi^2\rho_0^2\nu_0^2]} + \frac{2\pi\rho_0\nu_0 A_0 \sin(2\pi\eta)}{(4\pi^2 + \rho_0^2)^2 + 16\pi^2\rho_0^2\nu_0^2} \\
& + \frac{2A_0}{\pi} \sum_{n=1}^{\infty} \frac{(-1)^n(16n^2\pi^2 + \rho_0^2)\cos(2n\pi\eta)}{(4n^2 - 1)[64n^2\pi^2\rho_0^2\nu_0^2 + (16n^2\pi^2 + \rho_0^2)^2]} \\
& - 16\rho_0\nu_0 A_0 \sum_{n=1}^{\infty} \frac{(-1)^n n \sin(2n\pi\eta)}{(4n^2 - 1)[64n^2\pi^2\rho_0^2\nu_0^2 + (16n^2\pi^2 + \rho_0^2)^2]}. \quad (4.6.50)
\end{aligned}
$$

Figure 4.6.1 is from Hrycak's paper and shows the variation of the nondimensional temperature as a function of η for the spinning rate ν_0. The other parameters are typical of a satellite with aluminum skin and fully covered with glass-protected solar cells. As a check on the solution, we show the temperature field (the dashed line) of a nonrotating satellite where we neglect the effects of conduction and only radiation occurs. The difference between the $\nu_0 = 0$ solid and dashed lines arises primarily due to the *linearization* of the nonlinear radiation boundary condition during the derivation of the governing equations.

Problems

Solve the following ordinary differential equations by Fourier series if the forcing is given by the periodic function

$$f(t) = \begin{cases} 1, & 0 < t < \pi, \\ 0, & \pi < t < 2\pi, \end{cases}$$

and $f(t) = f(t + 2\pi)$:

1. $y'' - y = f(t)$,   2. $y'' + y = f(t)$,   3. $y'' - 3y' + 2y = f(t)$.

Solve the following ordinary differential equations by *complex* Fourier series if the forcing is given by the periodic function

$$f(t) = |t|, \qquad -\pi \le t \le \pi,$$

and $f(t) = f(t + 2\pi)$:

4. $y'' - y = f(t)$,   5. $y'' + 4y = f(t)$.

6. An object radiating into its nocturnal surrounding has a temperature $y(t)$ governed by the equation[19]

$$\frac{dy}{dt} + ay = A_0 + \sum_{n=1}^{\infty} A_n \cos(n\omega t) + B_n \sin(n\omega t),$$

where the constant a is the heat loss coefficient and the Fourier series describes the temporal variation of the atmospheric air temperature and the effective sky temperature. If $y(0) = T_0$, find $y(t)$.

7. The equation that governs the charge q on the capacitor of an LRC electrical circuit is

$$q'' + 2\alpha q' + \omega^2 q = \omega^2 E,$$

where $\alpha = R/(2L)$, $\omega^2 = 1/(LC)$, R denotes resistance, C denotes capacitance, L denotes the inductance, and E is the electromotive force driving the circuit. If E is given by

$$E = \sum_{n=-\infty}^{\infty} \varphi_n e^{in\omega_0 t},$$

find $q(t)$.

---

[19] Reprinted from *Solar Energy*, **28**, M. S. Sodha, Transient radiative cooling, 541, ©1982, with the kind permission from Elsevier Science Ltd, The Boulevard, Langford Lane, Kidlington, OX5 1GB, UK.

4.7 FINITE FOURIER SERIES

In many applications we must construct a Fourier series from values given by data or a graph. Unlike the situation with analytic formulas where we have an infinite number of data points and, consequently, an infinite number of terms in the Fourier series, the Fourier series contain a finite number of sines and cosines where the number of coefficients equals the number of data points.

Assuming that these series are useful, the next question is how do we find the Fourier coefficients? We could compute them by numerically integrating (4.1.6). However, the results would suffer from the truncation errors that afflict all numerical schemes. On the other hand, we can avoid this problem if we again employ the orthogonality properties of sines and cosines, now in their discrete form. Just as in the case of conventional Fourier series, we can use these properties to derive formulas for computing the Fourier coefficients. These results will be *exact* except for roundoff errors.

We start by deriving some preliminary results. Let us define $x_m = mP/(2N)$. Then, if k is an integer,

$$\sum_{m=0}^{2N-1} \exp\left(\frac{2\pi i k x_m}{P}\right) = \sum_{m=0}^{2N-1} \exp\left(\frac{km\pi i}{N}\right) = \sum_{m=0}^{2N-1} r^m \qquad (4.7.1)$$

$$= \begin{cases} \frac{1-r^{2N}}{1-r} = 0, & r \neq 1, \\ \\ 2N, & r = 1, \end{cases} \qquad (4.7.2)$$

because $r^{2N} = \exp(2\pi ki) = 1$ if $r \neq 1$. If $r = 1$, then the sum consists of $2N$ terms, each of which equals one. The condition $r = 1$ corresponds to $k = 0, \pm 2N, \pm 4N, \ldots$. Taking the real and imaginary part of (4.7.2),

$$\sum_{m=0}^{2N-1} \cos\left(\frac{2\pi k x_m}{P}\right) = \begin{cases} 0, & k \neq 0, \pm 2N, \pm 4N, \ldots, \\ 2N, & k = 0, \pm 2N, \pm 4N, \ldots, \end{cases} \qquad (4.7.3)$$

and

$$\sum_{m=0}^{2N-1} \sin\left(\frac{2\pi k x_m}{P}\right) = 0 \qquad (4.7.4)$$

for all k.

Consider now the following sum:

$$\sum_{m=0}^{2N-1} \cos\left(\frac{2\pi k x_m}{P}\right) \cos\left(\frac{2\pi j x_m}{P}\right)$$

$$= \frac{1}{2} \sum_{m=0}^{2N-1} \left\{ \cos\left[\frac{2\pi(k+j)x_m}{P}\right] + \cos\left[\frac{2\pi(k-j)x_m}{P}\right] \right\} \qquad (4.7.5)$$

$$= \begin{cases} 0, & |k-j| \text{ and } |k+m| \neq 0, 2N, 4N, \ldots, \\ N, & |k-j| \text{ or } |k+m| \neq 0, 2N, 4N, \ldots, \\ 2N, & |k-j| \text{ and } |k+m| = 0, 2N, 4N, \ldots. \end{cases} \qquad (4.7.6)$$

Let us simplify the right side of (4.7.6) by restricting ourselves to $k + j$ lying between 0 to $2N$. This is permissible because of the periodic nature of (4.7.5). If $k + j = 0$, $k = j = 0$; if $k + j = 2N$, $k = j = N$. In either case, $k - j = 0$ and the right side of (4.7.6) equals $2N$. Consider now the case $k \neq j$. Then $k + j \neq 0$ or $2N$ and $k - j \neq 0$ or $2N$. The right side of (4.7.6) must equal 0. Finally, if $k = j \neq 0$ or N, then $k + j \neq 0$ or $2N$ but $k - j = 0$ and the right side of (4.7.6) equals N. In summary,

$$\sum_{m=0}^{2N-1} \cos\left(\frac{2\pi k x_m}{P}\right) \cos\left(\frac{2\pi j x_m}{P}\right) = \begin{cases} 0, & k \neq j \\ N, & k = j \neq 0, N \\ 2N, & k = j = 0, N. \end{cases} \quad \textbf{(4.7.7)}$$

In a similar manner,

$$\sum_{m=0}^{2N-1} \cos\left(\frac{2\pi k x_m}{P}\right) \sin\left(\frac{2\pi j x_m}{P}\right) = 0 \quad \textbf{(4.7.8)}$$

for all k and j and

$$\sum_{m=0}^{2N-1} \sin\left(\frac{2\pi k x_m}{P}\right) \sin\left(\frac{2\pi j x_m}{P}\right) = \begin{cases} 0, & k \neq j \\ N, & k = j \neq 0, N \\ 0, & k = j = 0, N. \end{cases} \quad \textbf{(4.7.9)}$$

Armed with (4.7.7)–(4.7.9) we are ready to find the coefficients A_n and B_n of the finite Fourier series,

$$f(x) = \frac{A_0}{2} + \sum_{k=1}^{N-1}\left[A_k \cos\left(\frac{2\pi k x}{P}\right) + B_k \sin\left(\frac{2\pi k x}{P}\right)\right] + \frac{A_N}{2}\cos\left(\frac{2\pi N x}{P}\right),$$
$$\textbf{(4.7.10)}$$

where we have $2N$ data points and now define P as the period of the function.

To find A_k we proceed as before and multiply (4.7.10) by $\cos(2\pi j x / P)$ (j may take on values from 0 to N) and sum from 0 to $2N - 1$. At the point $x = x_m$,

$$\sum_{m=0}^{2N-1} f(x_m) \cos\left(\frac{2\pi j}{P} x_m\right) = \frac{A_0}{2} \sum_{m=0}^{2N-1} \cos\left(\frac{2\pi j}{P} x_m\right)$$
$$+ \sum_{k=1}^{N-1} A_k \sum_{m=0}^{2N-1} \cos\left(\frac{2\pi k}{P} x_m\right) \cos\left(\frac{2\pi j}{P} x_m\right)$$
$$+ \sum_{k=1}^{N-1} B_k \sum_{m=0}^{2N-1} \sin\left(\frac{2\pi k}{P} x_m\right) \cos\left(\frac{2\pi j}{P} x_m\right)$$
$$+ \frac{A_N}{2} \sum_{m=0}^{2N-1} \cos\left(\frac{2\pi N}{P} x_m\right) \cos\left(\frac{2\pi j}{P} x_m\right).$$
$$\textbf{(4.7.11)}$$

If $j \neq 0$ or N, then the first summation on the right side vanishes by (4.7.3), the third by (4.7.9), and the fourth by (4.7.7). The second summation does *not* vanish if $k = j$ and equals N. Similar considerations lead to the formulas for the calculation of A_k and B_k:

$$A_k = \frac{1}{N} \sum_{m=0}^{2N-1} f(x_m) \cos\left(\frac{2\pi k}{P} x_m\right), \qquad k = 0, 1, 2, \ldots, N, \qquad (4.7.12)$$

and

$$B_k = \frac{1}{N} \sum_{m=0}^{2N-1} f(x_m) \sin\left(\frac{2\pi k}{P} x_m\right), \qquad k = 1, 2, \ldots, N-1. \qquad (4.7.13)$$

If there are $2N+1$ data points and $f(x_0) = f(x_{2N})$, then (4.7.12)–(4.7.13) is still valid and we need only consider the first $2N$ points. If $f(x_0) \neq f(x_{2N})$, we can still use our formulas if we require that the endpoints have the value of $[f(x_0) + f(x_{2N})]/2$. In this case the formulas for the coefficients A_k and B_k are

$$A_k = \frac{1}{N} \left[\frac{f(x_0) + f(x_{2N})}{2} + \sum_{m=1}^{2N-1} f(x_m) \cos\left(\frac{2\pi k}{P} x_m\right) \right], \qquad (4.7.14)$$

where $k = 0, 1, 2, \ldots, N$, and

$$B_k = \frac{1}{N} \sum_{m=1}^{2N-1} f(x_m) \sin\left(\frac{2\pi k}{P} x_m\right), \qquad (4.7.15)$$

where $k = 1, 2, \ldots, N-1$.

It is important to note that $2N$ data points yield $2N$ Fourier coefficients A_k and B_k. Consequently our sampling frequency will always limit the amount of information, whether in the form of data points or Fourier coefficients. It might be argued that from the Fourier series representation of $f(t)$ we could find the value of $f(t)$ for any given t, which is more than we can do with the data alone. This is not true. Although we can calculate $f(t)$ at any t using the finite Fourier series, the values may or may not be correct since the constraint on the finite Fourier series is that the series must fit the data in a least squared sense. Despite the limitations imposed by only having a finite number of Fourier coefficients, the Fourier analysis of finite data sets yields valuable physical insights into the processes governing many physical systems.

- **Example 4.7.1: Water depth at Buffalo, NY**

Each entry[20] in Table 4.7.1 gives the observed depth of water at Buffalo, NY (minus the low-water datum of 568.6 ft) on the 15th of the corresponding

---

[20] National Ocean Survey, 1977: *Great Lakes Water Level, 1977, Daily and Monthly Average Water Surface Elevations*. National Oceanic and Atmospheric Administration.

Table 4.7.1: The Depth of Water in the Harbor at Buffalo, NY (Minus the Low-Water Datum of 568.8 ft) on the 15th Day of Each Month During 1977

mo	n	depth	mo	n	depth	mo	n	depth
Jan	1	1.61	May	5	3.16	Sep	9	2.42
Feb	2	1.57	Jun	6	2.95	Oct	10	2.95
Mar	3	2.01	Jul	7	3.10	Nov	11	2.74
Apr	4	2.68	Aug	8	2.90	Dec	12	2.63

month during 1977. Assuming that the water level is a periodic function of 1 year, and that we took the observations at equal intervals, let us construct a finite Fourier series from these data. This corresponds to computing the Fourier coefficients $A_0, A_1, \ldots, A_6, B_1, \ldots, B_5$, which give the mean level and harmonic fluctuations of the depth of water, the harmonics having the periods 12 months, 6 months, 4 months, and so forth.

In this problem, P equals 12 months, $N = P/2 = 6$ mo and $x_m = mP/(2N) = m(12 \text{ mo})/12 \text{ mo} = m$. That is, there should be a data point for each month. From (4.7.12) and (4.7.13),

$$A_k = \frac{1}{6} \sum_{m=0}^{11} f(x_m) \cos\left(\frac{mk\pi}{6}\right), \quad k = 0,1,2,3,4,5,6, \qquad (4.7.16)$$

and

$$B_k = \frac{1}{6} \sum_{m=0}^{11} f(x_m) \sin\left(\frac{mk\pi}{6}\right), \quad k = 1,2,3,4,5. \qquad (4.7.17)$$

Substituting the data into (4.7.16)–(4.7.17) yields

A_0 = twice the mean level $\qquad\qquad\qquad\qquad\quad = +5.120$ ft
A_1 = harmonic component with a period of 12 mo $= -0.566$ ft
B_1 = harmonic component with a period of 12 mo $= -0.128$ ft
A_2 = harmonic component with a period of 6 mo $= -0.177$ ft
B_2 = harmonic component with a period of 6 mo $= -0.372$ ft
A_3 = harmonic component with a period of 4 mo $= -0.110$ ft
B_3 = harmonic component with a period of 4 mo $= -0.123$ ft
A_4 = harmonic component with a period of 3 mo $= +0.025$ ft
B_4 = harmonic component with a period of 3 mo $= +0.052$ ft
A_5 = harmonic component with a period of 2.4 mo $= -0.079$ ft
B_5 = harmonic component with a period of 2.4 mo $= -0.131$ ft
A_6 = harmonic component with a period of 2 mo $= -0.107$ ft

Figure 4.7.1 is a plot of our results using (4.7.10). Note that when we include all of the harmonic terms, the finite Fourier series fits the data points

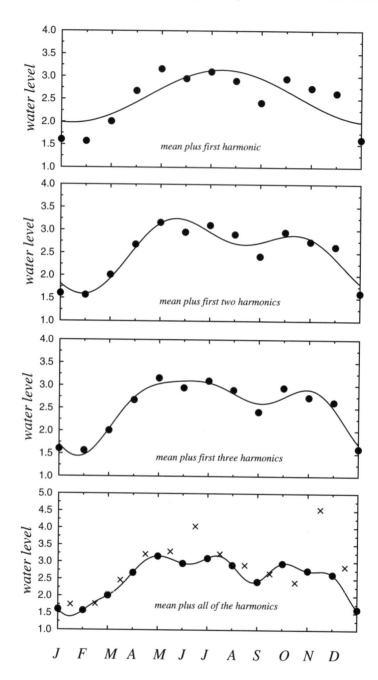

Figure 4.7.1: Partial sums of the finite Fourier series for the depth of water in the harbor of Buffalo, NY during 1977. Circles indicate observations on the 15th of the month; crosses are observations on the first.

exactly. The values given by the series at points between the data points may be right or they may not. To illustrate this, we also plotted the values for the first of each month. Sometimes the values given by the Fourier series and these intermediate data points are quite different.

Let us now examine our results in terms of various physical processes. In the long run the depth of water in the harbor at Buffalo, NY depends upon the three-way balance between precipitation, evaporation, and inflow-outflow of any rivers. Because the inflow and outflow of the rivers depends strongly upon precipitation, and evaporation is of secondary importance, the water level should correlate with the precipitation rate. It is well known that more precipitation falls during the warmer months rather than the colder months. The large amplitude of the Fourier coefficient A_1 and B_1, corresponding to the annual cycle ($k = 1$), reflects this.

Another important term in the harmonic analysis corresponds to the semiannual cycle ($k = 2$). During the winter months around Lake Ontario, precipitation falls as snow. Therefore, the inflow from rivers is greatly reduced. When spring comes, the snow and ice melt and a jump in the water level occurs. Because the second harmonic gives periodic variations associated with seasonal variations, this harmonic is absolutely necessary if we want to get the correct answer while the higher harmonics do not represent any specific physical process.

• Example 4.7.2: Numerical computation of Fourier coefficients

At the begining of this chapter, we showed how you could compute the Fourier coefficients a_0, a_n, and b_n from (4.1.6) given a function $f(t)$. All of this assumed that you could carry out the integrations. What do you do if you cannot perform the integrations? The obvious solution is perform it numerically. In this section we showed that the best approximation to (4.1.6) is given by (4.7.12)–(4.7.13). In the case when we have $f(t)$ this is still true but we may choose N as large as necessary to obtain the desired number of Fourier coefficients.

To illustrate this we have redone Example 4.1.1 and ploted the exact (analytic) and numerically computed Fourier coefficients in Figure 4.7.2. This figure was created using the MATLAB script

```
clear;
N = 15, M = 2*N; dt = 2*pi/M; % number of points in interval
% create time points assuming x(t) = x(t+period)
t = [-pi:dt:pi-dt];
%
f = zeros(size(t)); % initialize function f(t)
for k = 1:length(t) % construct function f(t)
    if t(k) < 0; f(k) = 0; else f(k) = t(k); end; end;
%
% compute Fourier coefficients using fast Fourier transform
```

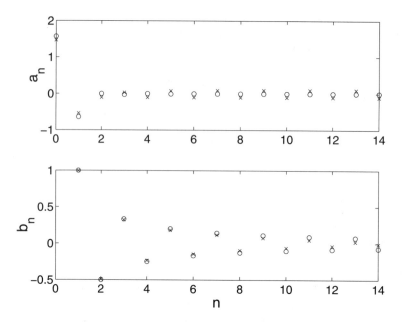

Figure 4.7.2: The computation of Fourier coefficients using a finite Fourier series when $f(t)$ is given by (4.1.8). The circles give a_n and b_n as computed from (4.1.9), (4.1.11), and (4.1.13). The crosses give the corresponding Fourier coefficients given by the finite Fourier series with $N = 15$.

```
%
fourier = fft(f) / N;
a_0_comp = real(fourier(1)); sign = 1;
for n = 2:N;
a_n_comp(n-1) = - sign * real(fourier(n));
b_n_comp(n-1) = sign * imag(fourier(n));
sign = - sign;
end
%
% plot comparisons
%
NN = linspace(0,N-1,N);
exact_coeff(1) = pi/2;
numerical_coeff(1) = a_0_comp;
for n = 1:N-1;
exact_coeff(n+1) = ((-1)^n-1) / (pi*(2*n-1)^2);
numerical_coeff(n+1) = a_n_comp(n);
end;
subplot(2,1,1), plot(NN,exact_coeff,'o',NN,numerical_coeff,'kx')
ylabel('a_n','Fontsize',20)
clear exact_coeff numerical_coeff
```

```
NN = linspace(1,N-1,N-1);
for n = 1:N-1;
exact_coeff(n) = - (-1)^n / n; numerical_coeff(n) = b_n_comp(n);
end;
subplot(2,1,2), plot(NN,exact_coeff,'o',NN,numerical_coeff,'kx')
xlabel('n','Fontsize',20); ylabel('b_n','Fontsize',20);
```

It shows that a relative few data points can yield quite reasonable answers.

Let us examine this script a little closer. One of the first things that you will note is that there is no explicit reference to (4.7.12)–(4.7.13). How did we get the correct answer?

Although we could have coded (4.7.12)–(4.7.13), no one does that any more. In the 1960s, J. W. Cooley and J. W. Tukey[21] devised an incredibly clever method of performing (4.7.12)–(4.7.13). This method, commonly called a fast Fourier transform or FFT, is so popular that all computational packages contain it as an intrinsic function and MATLAB is no exception, calling it `fft`. This is what has been used here.

Although we now have a `fft` to compute the coefficients, this routine does not directly give the coefficients a_n and b_n but rather some mysterious (complex) number that is related to $a_n + ib_n$. This is a common problem in using a package's FFT rather than your own and why the script divides by N and we keep changing the sign. The best method for discovering how to extract the coefficients a_n and b_n is to test it with a dataset created by a simple, finite series such as

$$f(x) = 20 + \cos(t) + 3\sin(t) + 6\cos(2t) - 20\sin(2t) - 10\cos(3t) - 30\sin(3t).$$
$$(4.7.18)$$

If the code is correct, it must give back the coefficient in (4.7.18) to within round-off. Otherwise, something is wrong.

Finally, most FFTs assume that the dataset will start repeating after the final data point. Therefore, when reading in the dataset, the point corresponding to $x = L$ must be excluded.

• Example 4.7.3: Aliasing

In the previous example, we could only resolve phenomena with a period of 2 months or greater although we had data for each of the 12 months. This is an example of *Nyquist's sampling criteria*[22]: At least two samples are required to resolve the highest frequency in a periodically sampled record.

Figure 4.7.3 will help explain this phenomenon. In case (a) we have quite a few data points over one cycle. Consequently our picture, constructed from

[21] Cooley, J. W., and J. W. Tukey, 1965: An algorithm for machine calculation of complex Fourier series. *Math. Comput.*, **19**, 297–301.

[22] Nyquist, H., 1928: Certain topics in telegraph transmission theory. *AIEE Trans.*, **47**, 617–644.

real world   *instrument data points*

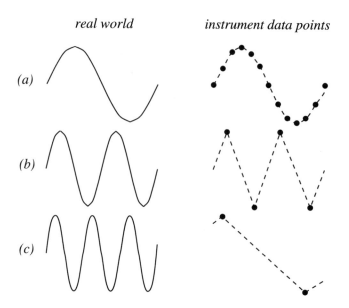

(a)

(b)

(c)

Figure 4.7.3: The effect of sampling in the representation of periodic functions.

data, is fairly good. In case (b), we took only samples at the ridges and troughs of the wave. Although our picture of the real phenomenon is poor, at least we know that there is a wave. From this picture we see that even if we are lucky enough to take our observations at the ridges and troughs of a wave, we need at least two data points per cycle (one for the ridge, the other for the trough) to resolve the highest-frequency wave.

In case (c) we have made a big mistake. We have taken a wave of frequency N Hz and misrepresented it as a wave of frequency $N/2$ Hz. This misrepresentation of a high-frequency wave by a lower-frequency wave is called *aliasing*. It arises because we are sampling a continuous signal at equal intervals. By comparing cases (b) and (c), we see that there is a cutoff between aliased and nonaliased frequencies. This frequency is called the *Nyquist* or *folding* frequency. It corresponds to the highest frequency resolved by our finite Fourier analysis.

Because most periodic functions require an infinite number of harmonics for their representation, aliasing of signals is a common problem. Thus the question is not "can I avoid aliasing?" but "can I live with it?" Quite often, we can construct our experiments to say yes. An example where aliasing is unavoidable occurs in a Western at the movies when we see the rapidly rotating spokes of the stagecoach's wheel. A movie is a sampling of continuous motion where we present the data as a succession of pictures. Consequently, a film aliases the high rate of revolution of the stagecoach's wheel in such a manner so that it appears to be stationary or rotating very slowly.

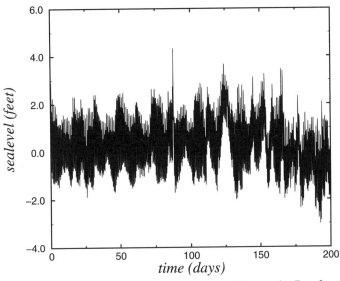

Figure 4.7.4: The sea elevation at the mouth of the Chesapeake Bay from its average depth as a function of time after 1 July 1985.

● **Example 4.7.4: Spectrum of the Chesapeake Bay**

For our final example, we perform a Fourier analysis of hourly sea-level measurements taken at the mouth of the Chesapeake Bay during the 2000 days from 9 April 1985 to 29 June 1990. Figure 4.7.4 shows 200 days of this record, starting from 1 July 1985. As this figure shows, the measurements contain a wide range of oscillations. In particular, note the large peak near day 90 which corresponds to the passage of hurricane Gloria during the early hours of 27 September 1985.

Utilizing the entire 2000 days, we plotted the amplitude of the Fourier coefficients as a function of period in Figure 4.7.5. We see a general rise of the amplitude as the period increases. Especially noteworthy are the sharp peaks near periods of 12 and 24 hours. The largest peak is at 12.417 hours and corresponds to the semidiurnal tide. Thus, our Fourier analysis shows that the dominant oscillations at the mouth of the Chesapeake Bay are the tides. A similar situation occurs in Baltimore harbor. Furthermore, with this spectral information we could predict high and low tides very accurately.

Although the tides are of great interest to some, they are a nuisance to others because they mask other physical processes that might be occurring. For that reason we would like to remove them from the tidal gauge history and see what is left. One way would be to zero out the Fourier coefficients corresponding to the tidal components and then plot the resulting Fourier series. Another method is to replace each hourly report with an average of hourly reports that occurred 24 hours ahead and behind of a particular report. We construct this average in such a manner that waves with periods of the

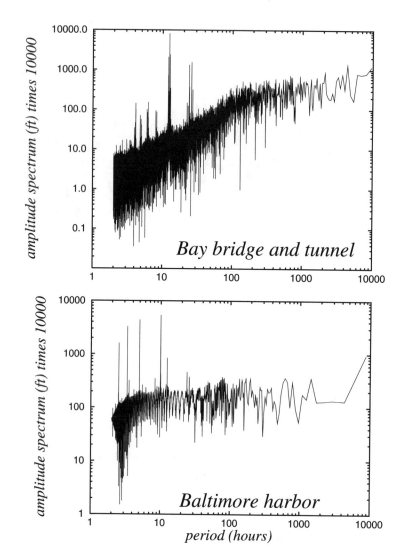

Figure 4.7.5: The amplitude of the Fourier coefficients for the sea elevation at the Chesapeake Bay bridge and tunnel (top) and Baltimore harbor (bottom) as a function of period.

tides sum to zero.[23] Such a *filter* is a popular method for eliminating unwanted waves from a record. Filters play an important role in the analysis of data. We plotted the filtered sea level data in Figure 4.7.6. Note that summertime (0–50 days) produces little variation in the sea level compared to wintertime (100–150 days) when intense coastal storms occur.

---

[23] See Godin, G., 1972: *The Analysis of Tides.* University of Toronto Press, §2.1.

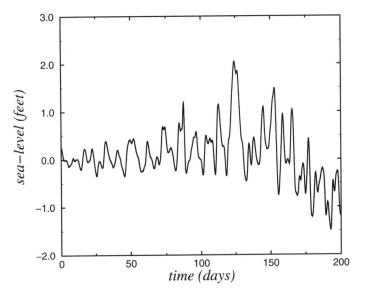

Figure 4.7.6: Same as Figure 4.7.4 but with the tides removed.

Problems

Find the finite Fourier series for the following pieces of data:

1. $f(0) = 0$, $f(1) = 1$, $f(2) = 2$, $f(3) = 3$, and $N = 2$.

2. $f(0) = 1$, $f(1) = 1$, $f(2) = -1$, $f(3) = -1$, and $N = 2$.

Project: Spectrum of the Earth's Orography

Table 4.7.2 gives the orographic height of the earth's surface used in an atmospheric general circulation model (GCM) at a resolution of 2.5° longitude along the latitude belts of 28°S, 36°N, and 66°N. In this project you will find the spectrum of this orographic field along the various latitude belts.

Step 1: Write a MATLAB script that reads in the data and find A_n and B_n and then construct the amplitude spectra for this data.

Step 2: Construct several spectra by using every data point, every other data point, etc. How do the magnitudes of the Fourier coefficient change? You might like to read about *leakage* from a book on harmonic analysis.[24]

---

[24] For example, Bloomfield, P., 1976: *Fourier Analysis of Time Series: An Introduction.* John Wiley & Sons, 258 pp.

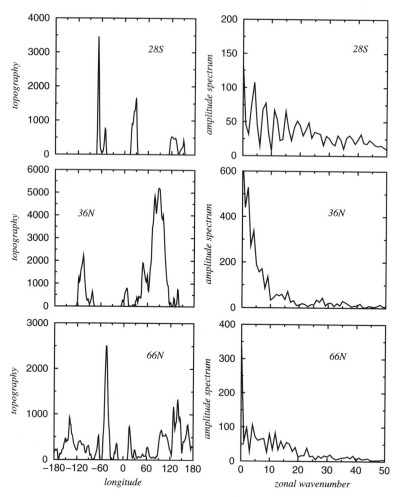

Figure 4.7.7: The orography of the earth and its spectrum in meters along three latitude belts using a topography dataset with a resolution of 1.25° longitude.

Step 3: Compare and contrast the spectra from the various latitude belts. How do the magnitudes of the Fourier coefficients decrease with n? Why are there these differences?

Step 4: You may have noted that some of the heights are negative, even in the middle of the ocean! Take the original data (for any latitude belt) and zero out all of the negative heights. Find the spectra for this new data set. How have the spectra changed? Is there a reason why the negative heights were introduced?

Table 4.7.2: Orographic Heights (in m) Times the Gravitational Acceleration Constant ($g = 9.81$ m/s^2) Along Three Latitude Belts

Longitude	28°S	36°N	66°N	Longitude	28°S	36°N	66°N
−180.0	4.	3.	2532.	−82.5	36.	4047.	737.
−177.5	1.	−2.	1665.	−80.0	−64.	3938.	185.
−175.0	1.	2.	1432.	−77.5	138.	1669.	71.
−172.5	1.	−3.	1213.	−75.0	−363.	236.	160.
−170.0	1.	1.	501.	−72.5	4692.	31.	823.
−167.5	1.	−3.	367.	−70.0	19317.	−8.	1830.
−165.0	1.	1.	963.	−67.ᴗ	21681.	0.	3000.
−162.5	0.	0.	1814.	−65.0	9222.	−2.	3668.
−160.0	−1.	6.	2562.	−62.5	1949.	−2.	2147.
−157.5	0.	1.	3150.	−60.0	774.	0.	391.
−155.0	0.	3.	4008.	−57.5	955.	5.	−77.
−152.5	1.	−2.	4980.	−55.0	2268.	6.	601.
−150.0	−1.	4.	6011.	−52.5	4636.	−1.	3266.
−147.5	6.	−1.	6273.	−50.0	4621.	2.	9128.
−145.0	14.	3.	5928.	−47.5	1300.	−4.	17808.
−142.5	6.	−1.	6509.	−45.0	−91.	1.	22960.
−140.0	−2.	6.	7865.	−42.5	57.	−1.	20559.
−137.5	0.	3.	7752.	−40.0	−25.	4.	14296.
−135.0	−2.	5.	6817.	−37.5	13.	−1.	9783.
−132.5	1.	−2.	6272.	−35.0	−10.	6.	5969.
−130.0	−2.	0.	5582.	−32.5	8.	2.	1972.
−127.5	0.	5.	4412.	−30.0	−4.	22.	640.
−125.0	−2.	423.	3206.	−27.5	6.	33.	379.
−122.5	1.	3688.	2653.	−25.0	−2.	39.	286.
−120.0	−3.	10919.	2702.	−22.5	3.	2.	981.
−117.5	2.	16148.	3062.	−20.0	−3.	11.	1971.
−115.0	−3.	17624.	3344.	−17.5	1.	−6.	2576.
−112.5	7.	18132.	3444.	−15.0	−1.	19.	1692.
−110.0	12.	19511.	3262.	−12.5	0.	−18.	357.
−107.5	9.	22619.	3001.	−10.0	−1.	490.	−21.
−105.0	−5.	20273.	2931.	−7.5	0.	2164.	−5.
−102.5	3.	12914.	2633.	−5.0	1.	4728.	−10.
−100.0	−5.	7434.	1933.	−2.5	0.	5347.	0.
−97.5	6.	4311.	1473.	0.0	4.	2667.	−6.
−95.0	−8.	2933.	1689.	2.5	−5.	1213.	−1.
−92.5	8.	2404.	2318.	5.0	7.	1612.	−31.
−90.0	−12.	1721.	2285.	7.5	−13.	1744.	−58.
−87.5	18.	1681.	1561.	10.0	28.	1153.	381.
−85.0	−23.	2666.	1199.	12.5	107.	838.	2472.

Table 4.7.2, contd.: Orographic Heights (in m) Times the Gravitational Acceleration Constant ($g = 9.81$ m/s^2) Along Three Latitude Belts

Longitude	28°S	36°N	66°N	Longitude	28°S	36°N	66°N
15.0	2208.	1313.	5263.	97.5	0.	35538.	6222.
17.5	6566.	862.	5646.	100.0	−2.	31985.	5523.
20.0	9091.	1509.	3672.	102.5	0.	23246.	4823.
22.5	10690.	2483.	1628.	105.0	−4.	17363.	4689.
25.0	12715.	1697.	889.	107.5	2.	14315.	4698.
27.5	14583.	3377.	1366.	110.0	−17.	12639.	4674.
30.0	11351.	7682.	1857.	112.5	302.	10543.	4435.
32.5	3370.	9663.	1534.	115.0	1874.	4967.	3646.
35.0	15.	10197.	993.	117.5	4005.	1119.	2655.
37.5	49.	10792.	863.	120.0	4989.	696.	2065.
40.0	−31.	11322.	756.	122.5	4887.	475.	1583.
42.5	20.	13321.	620.	125.0	4445.	1631.	3072.
45.0	−17.	15414.	626.	127.5	4362.	2933.	7290.
47.5	−19.	12873.	836.	130.0	4368.	1329.	8541.
50.0	−18.	6114.	1029.	132.5	3485.	88.	7078.
52.5	6.	2962.	946.	135.0	1921.	598.	7322.
55.0	−2.	4913.	828.	137.5	670.	1983.	9445.
57.5	3.	6600.	1247.	140.0	666.	2511.	10692.
60.0	−3.	4885.	2091.	142.5	1275.	866.	9280.
62.5	2.	3380.	2276.	145.0	1865.	13.	8372.
65.0	−1.	5842.	1870.	147.5	2452.	11.	6624.
67.5	2.	12106.	1215.	150.0	3160.	−4.	3617.
70.0	0.	23032.	680.	152.5	2676.	−1.	2717.
72.5	2.	35376.	531.	155.0	697.	0.	3474.
75.0	−1.	36415.	539.	157.5	−67.	−3.	4337.
77.5	1.	26544.	579.	160.0	25.	3.	4824.
80.0	0.	19363.	554.	162.5	−12.	−1.	5525.
82.5	1.	17915.	632.	165.0	10.	4.	6323.
85.0	−2.	22260.	791.	167.5	−5.	−2.	5899.
87.5	−1.	30442.	1455.	170.0	0.	1.	4330.
90.0	−3.	33601.	3194.	172.5	0.	−4.	3338.
92.5	−1.	30873.	4878.	175.0	4.	3.	3408.
95.0	0.	31865.	5903.	177.5	3.	−1.	3407.

Chapter 5
The Fourier Transform

In the previous chapter we showed how we could expand a periodic function in terms of an infinite sum of sines and cosines. However, most functions encountered in engineering are aperiodic. As we shall see, the extension of Fourier series to these functions leads to the Fourier transform.

5.1 FOURIER TRANSFORMS

The Fourier transform is the natural extension of Fourier series to a function $f(t)$ of infinite period. To show this, consider a periodic function $f(t)$ of period $2T$ that satisfies the so-called Dirichlet's conditions.[1] If the integral $\int_a^b |f(t)|\, dt$ exists, this function has the complex Fourier series

$$f(t) = \sum_{n=-\infty}^{\infty} c_n e^{in\pi t/T}, \tag{5.1.1}$$

where

$$c_n = \frac{1}{2T} \int_{-T}^{T} f(t) e^{-in\pi t/T} dt. \tag{5.1.2}$$

---

[1] A function $f(t)$ satisfies Dirichlet's conditions in the interval (a, b) if (1) it is bounded in (a, b), and (2) it has at most a finite number of discontinuities and a finite number of maxima and minima in that interval.

227

Equation (5.1.1) applies only if $f(t)$ is continuous at t; if $f(t)$ suffers from a jump discontinuity at t, then the left side of (5.1.1) equals $\frac{1}{2}[f(t^+) + f(t^-)]$, where $f(t^+) = \lim_{x \to t+} f(x)$ and $f(t^-) = \lim_{x \to t-} f(x)$. Substituting (5.1.2) into (5.1.1),

$$f(t) = \frac{1}{2T} \sum_{n=-\infty}^{\infty} e^{in\pi t/T} \int_{-T}^{T} f(x)e^{-in\pi x/T}dx. \qquad (5.1.3)$$

Let us now introduce the notation $\omega_n = n\pi/T$ so that $\Delta\omega_n = \omega_{n+1} - \omega_n = \pi/T$. Then,

$$f(t) = \frac{1}{2\pi} \sum_{n=-\infty}^{\infty} F(\omega_n)e^{i\omega_n t}\Delta\omega_n, \qquad (5.1.4)$$

where

$$F(\omega_n) = \int_{-T}^{T} f(x)e^{-i\omega_n x}dx. \qquad (5.1.5)$$

As $T \to \infty$, ω_n approaches a continuous variable ω, and $\Delta\omega_n$ may be interpreted as the infinitesimal $d\omega$. Therefore, ignoring any possible difficulties,[2]

$$f(t) = \frac{1}{2\pi} \int_{-\infty}^{\infty} F(\omega)e^{i\omega t}d\omega, \qquad (5.1.6)$$

and

$$F(\omega) = \int_{-\infty}^{\infty} f(t)e^{-i\omega t}dt. \qquad (5.1.7)$$

Equation (5.1.7) is the *Fourier transform* of $f(t)$ while (5.1.6) is the *inverse Fourier transform* which converts a Fourier transform back to $f(t)$. Alternatively, we may combine (5.1.6)–(5.1.7) to yield the equivalent real form

$$f(t) = \frac{1}{\pi} \int_{0}^{\infty} \left\{ \int_{-\infty}^{\infty} f(x) \cos[\omega(t-x)] \, dx \right\} d\omega. \qquad (5.1.8)$$

---

[2] For a rigorous derivation, see Titchmarsh, E. C., 1948: *Introduction to the Theory of Fourier Integrals*. Oxford University Press, Chapter 1.

Hamming[3] suggested the following analog in understanding the Fourier transform. Let us imagine that $f(t)$ is a light beam. Then the Fourier transform, like a glass prism, breaks up the function into its component frequencies ω, each of intensity $F(\omega)$. In optics, the various frequencies are called colors; by analogy the Fourier transform gives us the color spectrum of a function. On the other hand, the inverse Fourier transform blends a function's spectrum to give back the original function.

Most signals encountered in practice have Fourier transforms because they are absolutely integrable since they are bounded and of finite duration. However, there are some notable exceptions. Examples include the trigonometric functions sine and cosine.

• Example 5.1.1

Let us find the Fourier transform for

$$f(t) = \begin{cases} 1, & |t| < a, \\ 0, & |t| > a. \end{cases} \tag{5.1.9}$$

From the definition of the Fourier transform,

$$F(\omega) = \int_{-\infty}^{-a} 0\,e^{-i\omega t}\,dt + \int_{-a}^{a} 1\,e^{-i\omega t}\,dt + \int_{a}^{\infty} 0\,e^{-i\omega t}\,dt \tag{5.1.10}$$

$$= \frac{e^{\omega a i} - e^{-\omega a i}}{\omega i} = \frac{2\sin(\omega a)}{\omega} = 2a\,\mathrm{sinc}(\omega a), \tag{5.1.11}$$

where $\mathrm{sinc}(x) = \sin(x)/x$ is the *sinc function*.

Although this particular example does not show it, the Fourier transform is, in general, a complex function. The most common method of displaying it is to plot its amplitude and phase on two separate graphs for all values of ω. Another problem here is ratio of $0/0$ when $\omega = 0$. Applying L'Hôpital's rule, we find that $F(0) = 2$. Thus, we can plot the amplitude and phase of $F(\omega)$ using the MATLAB script:

```
clear; % clear all previous computations
omegan = [-20:0.01:-0.01]; % set up negative frequencies
omegap = [0.01:0.01:20]; % set up positive frequencies
% compute Fourier transform for negative frequencies
f_omegan = 2.*sin(omegan)./omegan;
% compute Fourier transform for positive frequencies
f_omegap = 2.*sin(omegap)./omegap;
% concatenate all of the frequencies
omega = [omegan,0,omegap];
% bring together the Fourier transforms found
```

---

[3] Hamming, R. W., 1977: *Digital Filters*. Prentice-Hall, p. 136.

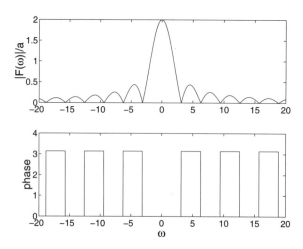

Figure 5.1.1: Graph of the Fourier transform for (5.1.9).

```
%       at positive and negative frequencies
f_omega = [f_omegan,2,f_omegap];
amplitude = abs(f_omega); % compute the amplitude
phase = atan2(0,f_omega); % compute the phase
clf; % clear all previous figures
% plot frequency spectrum
subplot(2,1,1), plot(omega,amplitude)
% label amplitude plot
ylabel('|F(\omega)|/a','FontSize',15)
subplot(2,1,2), plot(omega,phase) % plot phase of transform
ylabel('phase','FontSize',15) % label amplitude plot
xlabel('\omega','FontSize',15) % label x-axis.
```

Figure 5.1.1 shows the output from the MATLAB script. Of these two quantities, the amplitude is by far the more popular one and is given the special name of *frequency spectrum*.

From the definition of the inverse Fourier transform,

$$f(t) = \frac{1}{\pi} \int_{-\infty}^{\infty} \frac{\sin(\omega a)}{\omega} e^{i\omega t} \, d\omega = \begin{cases} 1, & |t| < a, \\ 0, & |t| > a. \end{cases} \tag{5.1.12}$$

An important question is what value does $f(t)$ converge to in the limit as $t \to a$ and $t \to -a$? Because Fourier transforms are an extension of Fourier series, the behavior at a jump is the same as that for a Fourier series. For that reason, $f(a) = \frac{1}{2}[f(a^+) + f(a^-)] = \frac{1}{2}$ and $f(-a) = \frac{1}{2}[f(-a^+) + f(-a^-)] = \frac{1}{2}$.

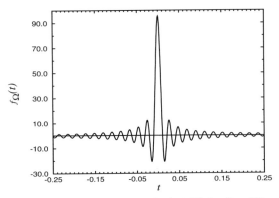

Figure 5.1.2: Graph of the function given in (5.1.15) for $\Omega = 300$.

- **Example 5.1.2: Dirac delta function**

Of the many functions that have a Fourier transform, a particularly important one is the *(Dirac) delta function*.[4] For example, in §5.6 we will use it to solve differential equations. We *define* it as the inverse of the Fourier transform $F(\omega) = 1$. Therefore,

$$\delta(t) = \frac{1}{2\pi} \int_{-\infty}^{\infty} e^{i\omega t} d\omega. \qquad (5.1.13)$$

To give some insight into the nature of the delta function, consider another band-limited transform

$$F_\Omega(\omega) = \begin{cases} 1, & |\omega| < \Omega, \\ 0, & |\omega| > \Omega, \end{cases} \qquad (5.1.14)$$

where Ω is real and positive. Then,

$$f_\Omega(t) = \frac{1}{2\pi} \int_{-\Omega}^{\Omega} e^{i\omega t} d\omega = \frac{\Omega}{\pi} \frac{\sin(\Omega t)}{\Omega t}. \qquad (5.1.15)$$

Figure 5.1.2 illustrates $f_\Omega(t)$ for a large value of Ω. We observe that as $\Omega \to \infty$, $f_\Omega(t)$ becomes very large near $t = 0$ as well as very narrow. On the other hand, $f_\Omega(t)$ rapidly approaches zero as $|t|$ increases. Therefore, the delta function is given by the limit

$$\delta(t) = \lim_{\Omega \to \infty} \frac{\sin(\Omega t)}{\pi t}, \qquad (5.1.16)$$

or

$$\delta(t) = \begin{cases} \infty, & t = 0, \\ 0, & t \neq 0. \end{cases} \qquad (5.1.17)$$

---

[4] Dirac, P. A. M., 1947: *The Principles of Quantum Mechanics.* Oxford University Press, §15.

Table 5.1.1: The Fourier Transforms of Some Commonly Encountered Functions

	$f(t),\ |t| < \infty$	$F(\omega)$
1.	$e^{-at}H(t),\quad a > 0$	$\dfrac{1}{a + \omega i}$
2.	$e^{at}H(-t),\quad a > 0$	$\dfrac{1}{a - \omega i}$
3.	$te^{-at}H(t),\quad a > 0$	$\dfrac{1}{(a + \omega i)^2}$
4.	$te^{at}H(-t),\quad a > 0$	$\dfrac{-1}{(a - \omega i)^2}$
5.	$t^n e^{-at}H(t),\ \mathrm{Re}(a) > 0,\ n = 1, 2, \ldots$	$\dfrac{n!}{(a + \omega i)^{n+1}}$
6.	$e^{-a|t|},\quad a > 0$	$\dfrac{2a}{\omega^2 + a^2}$
7.	$te^{-a|t|},\quad a > 0$	$\dfrac{-4a\omega i}{(\omega^2 + a^2)^2}$
8.	$\dfrac{1}{1 + a^2 t^2}$	$\dfrac{\pi}{|a|}e^{-|\omega/a|}$
9.	$\dfrac{\cos(at)}{1 + t^2}$	$\frac{\pi}{2}\left(e^{-|\omega - a|} + e^{-|\omega + a|}\right)$
10.	$\dfrac{\sin(at)}{1 + t^2}$	$\frac{\pi}{2i}\left(e^{-|\omega - a|} - e^{-|\omega + a|}\right)$
11.	$\begin{cases} 1, & |t| < a \\ 0, & |t| > a \end{cases}$	$\dfrac{2\sin(\omega a)}{\omega}$
12.	$\dfrac{\sin(at)}{at}$	$\begin{cases} \pi/a, & |\omega| < a \\ 0, & |\omega| > a \end{cases}$
13.	$e^{-at^2},\quad a > 0$	$\sqrt{\dfrac{\pi}{a}}\exp\left(-\dfrac{\omega^2}{4a}\right)$

Note: The Heaviside step function $H(t)$ is defined by (5.1.31).

Because the Fourier transform of the delta function equals one,

$$\int_{-\infty}^{\infty} \delta(t)e^{-i\omega t}\,dt = 1. \tag{5.1.18}$$

Since (5.1.18) must hold for any ω, we take $\omega = 0$ and find that

$$\int_{-\infty}^{\infty} \delta(t)\,dt = 1. \tag{5.1.19}$$

Thus, the area under the delta function equals unity. Taking (5.1.17) into account, we can also write (5.1.19) as

$$\int_{-a}^{b} \delta(t)\, dt = 1, \qquad a, b > 0. \tag{5.1.20}$$

Finally, from the law of the mean of integrals, we have the *sifting property* that

$$\int_{a}^{b} f(t)\delta(t - t_0)\, dt = f(t_0), \tag{5.1.21}$$

if $a < t_0 < b$. This property is given its name because $\delta(t - t_0)$ acts as a sieve, selecting from all possible values of $f(t)$ its value at $t = t_0$.

We can also use several other functions with equal validity to represent the delta function. These include the limiting case of the following rectangular or triangular distributions:

$$\delta(t) = \lim_{\epsilon \to 0} \begin{cases} \frac{1}{\epsilon}, & |t| < \frac{\epsilon}{2}, \\ 0, & |t| > \frac{\epsilon}{2}, \end{cases} \tag{5.1.22}$$

or

$$\delta(t) = \lim_{\epsilon \to 0} \begin{cases} \frac{1}{\epsilon}\left(1 - \frac{|t|}{\epsilon}\right), & |t| < \epsilon, \\ 0, & |t| > \epsilon, \end{cases} \tag{5.1.23}$$

and the Gaussian function:

$$\delta(t) = \lim_{\epsilon \to 0} \frac{\exp(-\pi t^2/\epsilon)}{\sqrt{\epsilon}}. \tag{5.1.24}$$

Note that the delta function is an even function.

• **Example 5.1.3: Multiple Fourier transforms**

The concept of Fourier transforms can be extended to multivariable functions. Consider a two-dimensional function $f(x, y)$. Then, holding y constant,

$$G(\xi, y) = \int_{-\infty}^{\infty} f(x, y)\, e^{-i\xi x}\, dx. \tag{5.1.25}$$

Then, holding ξ constant,

$$F(\xi, \eta) = \int_{-\infty}^{\infty} G(\xi, y)\, e^{-i\eta y}\, dy. \tag{5.1.26}$$

Therefore, the double Fourier transform of $f(x, y)$ is

$$F(\xi, \eta) = \int_{-\infty}^{\infty} \int_{-\infty}^{\infty} f(x, y)\, e^{-i(\xi x + \eta y)}\, dx\, dy, \tag{5.1.27}$$

assuming that the integral exists.

In a similar manner, we can compute $f(x, y)$ given $F(\xi, \eta)$ by reversing the process. Starting with

$$G(\xi, y) = \frac{1}{2\pi} \int_{-\infty}^{\infty} F(\xi, \eta) \, e^{i\eta y} \, d\eta, \qquad (5.1.28)$$

followed by

$$f(x, y) = \frac{1}{2\pi} \int_{-\infty}^{\infty} G(\xi, y) \, e^{i\xi x} \, d\xi, \qquad (5.1.29)$$

we find that

$$f(x, y) = \frac{1}{4\pi^2} \int_{-\infty}^{\infty} \int_{-\infty}^{\infty} F(\xi, \eta) \, e^{i(\xi x + \eta y)} \, d\xi \, d\eta. \qquad (5.1.30)$$

• **Example 5.1.4: Computation of Fourier transforms using MATLAB**

The Heaviside (unit) step function is a piecewise continuous function defined by

$$H(t - a) = \begin{cases} 1, & t > a, \\ 0, & t < a, \end{cases} \qquad (5.1.31)$$

where $a \geq 0$. We will have much to say about this very useful function in the chapter on Laplace transforms. Presently we will use it to express functions whose definition changes over different ranges of t. For example, the "top hat" function (5.1.9) can be rewritten $f(t) = H(t + a) - H(t - a)$. We can see that this is correct by considering various ranges of t. For example, if $t < -a$, both step functions equal zero and $f(t) = 0$. On the other hand, if $t > a$, both step functions equal one and again $f(t) = 0$. Finally, for $-a < t < a$, the first step function equals one while the second one equals zero. In this case, $f(t) = 1$. Therefore, $f(t) = H(t + a) - H(t - a)$ is equivalent to (5.1.9).

This ability to rewrite functions in terms of the step function is crucial if you want to use MATLAB to compute Fourier transform via the MATLAB routine `fourier`. For example, how would we compute the Fourier transform of the signum function? The MATLAB commands

```
>> syms omega t; syms a positive
>> fourier('Heaviside(t+a)-Heaviside(t-a)',t,omega)
>> simplify(ans)
```

yields

```
ans =
2*sin(a*omega)/omega
```

the correct answer.

Problems

1. (a) Show that the Fourier transform of

$$f(t) = e^{-a|t|}, \qquad a > 0,$$

is

$$F(\omega) = \frac{2a}{\omega^2 + a^2}.$$

Using MATLAB, plot the amplitude and phase spectra for this transform.

(b) Use MATLAB's `fourier` to find $F(\omega)$.

2. (a) Show that the Fourier transform of

$$f(t) = te^{-a|t|}, \qquad a > 0,$$

is

$$F(\omega) = -\frac{4a\omega i}{(\omega^2 + a^2)^2}.$$

Using MATLAB, plot the amplitude and phase spectra for this transform.

(b) Use MATLAB's `fourier` to find $F(\omega)$.

3. (a) Show that the Fourier transform of

$$f(t) = e^{-at^2}, \qquad a > 0,$$

is

$$F(\omega) = \sqrt{\frac{\pi}{a}} \exp\left(-\frac{\omega^2}{4a}\right).$$

Using MATLAB, plot the amplitude and phase spectra for this transform.

(b) Use MATLAB's `fourier` to find $F(\omega)$.

4. (a) Show that the Fourier transform of

$$f(t) = \begin{cases} e^{2t}, & t < 0, \\ e^{-t}, & t > 0, \end{cases}$$

is

$$F(\omega) = \frac{3}{(2 - i\omega)(1 + i\omega)}.$$

Using MATLAB, plot the amplitude and phase spectra for this transform.

(b) Rewrite $f(t)$ in terms of step functions. Then use MATLAB's `fourier` to find $F(\omega)$.

5. (a) Show that the Fourier transform of

$$f(t) = \begin{cases} e^{-(1+i)t}, & t > 0, \\ -e^{(1-i)t}, & t < 0, \end{cases}$$

is

$$F(\omega) = \frac{-2i(\omega + 1)}{(\omega + 1)^2 + 1}.$$

Using MATLAB, plot the amplitude and phase spectra for this transform.

(b) Rewrite $f(t)$ in terms of step functions. Then use MATLAB's `fourier` to find $F(\omega)$.

6. (a) Show that the Fourier transform of

$$f(t) = \begin{cases} \cos(at), & |t| < 1, \\ 0, & |t| > 1, \end{cases}$$

is

$$F(\omega) = \frac{\sin(\omega - a)}{\omega - a} + \frac{\sin(\omega + a)}{\omega + a}.$$

Using MATLAB, plot the amplitude and phase spectra for this transform.

(b) Rewrite $f(t)$ in terms of step functions. Then use MATLAB's `fourier` to find $F(\omega)$.

7. (a) Show that the Fourier transform of

$$f(t) = \begin{cases} \sin(t), & 0 \le t < 1, \\ 0, & \text{otherwise}, \end{cases}$$

is

$$F(\omega) = -\frac{1}{2}\left[\frac{1 - \cos(\omega - 1)}{\omega - 1} + \frac{\cos(\omega + 1) - 1}{\omega + 1}\right]$$
$$- \frac{i}{2}\left[\frac{\sin(\omega - 1)}{\omega - 1} - \frac{\sin(\omega + 1)}{\omega + 1}\right].$$

Using MATLAB, plot the amplitude and phase spectra for this transform.

(b) Rewrite $f(t)$ in terms of step functions. Then use MATLAB's `fourier` to find $F(\omega)$.

8. (a) Show that the Fourier transform of

$$f(t) = \begin{cases} t/a, & |t| < a, \\ 0, & |t| > a, \end{cases}$$

is

$$F(\omega) = \frac{2i\cos(\omega a)}{\omega} - \frac{2i\sin(\omega a)}{\omega^2 a}.$$

Using MATLAB, plot the amplitude and phase spectra for this transform.

(b) Rewrite $f(t)$ in terms of step functions. Then use MATLAB's `fourier` to find $F(\omega)$.

9. (a) Show that the Fourier transform of

$$f(t) = \begin{cases} (t/a)^2, & |t| < a, \\ 0, & |t| > a, \end{cases}$$

is

$$F(\omega) = \frac{4\cos(\omega a)}{\omega^2 a} - \frac{4\sin(\omega a)}{\omega^3 a^2} + \frac{2\sin(\omega a)}{\omega}.$$

Using MATLAB, plot the amplitude and phase spectra for this transform.

(b) Rewrite $f(t)$ in terms of step functions. Then use MATLAB's `fourier` to find $F(\omega)$.

10. (a) Show that the Fourier transform of

$$f(t) = \begin{cases} 1 - t/\tau, & 0 \le t < 2\tau, \\ 0, & \text{otherwise}, \end{cases}$$

is

$$F(\omega) = \frac{2e^{-i\omega\tau}}{i\omega} \left[\frac{\sin(\omega\tau)}{\omega\tau} - \cos(\omega\tau) \right].$$

Using MATLAB, plot the amplitude and phase spectra for this transform.

(b) Rewrite $f(t)$ in terms of step functions. Then use MATLAB's `fourier` to find $F(\omega)$.

11. (a) Show that the Fourier transform of

$$f(t) = \begin{cases} 1 - (t/a)^2, & |t| \le a, \\ 0, & |t| \ge a, \end{cases}$$

is

$$F(\omega) = \frac{4\sin(\omega a) - 4a\omega\cos(\omega a)}{a^2\omega^3}.$$

Using MATLAB, plot the amplitude and phase spectra for this transform.

(b) Rewrite $f(t)$ in terms of step functions. Then use MATLAB's `fourier` to find $F(\omega)$.

12. The integral representation[5] of the modified Bessel function $K_\nu(\ )$ is

$$K_\nu(a|\omega|) = \frac{\Gamma\left(\nu + \frac{1}{2}\right)(2a)^\nu}{|\omega|^\nu \Gamma\left(\frac{1}{2}\right)} \int_0^\infty \frac{\cos(\omega t)}{(t^2 + a^2)^{\nu+1/2}}\, dt,$$

where $\Gamma(\ )$ is the gamma function, $\nu \geq 0$ and $a > 0$. Use this relationship to show that

$$\mathcal{F}\left[\frac{1}{(t^2 + a^2)^{\nu+1/2}}\right] = \frac{2|\omega|^\nu \Gamma\left(\frac{1}{2}\right) K_\nu(a|\omega|)}{\Gamma\left(\nu + \frac{1}{2}\right)(2a)^\nu}.$$

13. Show that the Fourier transform of a constant K is $2\pi\delta(\omega)K$.

14. Show that

$$\int_a^b \tau\,\delta(t - \tau)\, d\tau = t\left[H(t - a) - H(t - b)\right].$$

Hint: Use integration by parts.

15. For the real function $f(t)$ with Fourier transform $F(\omega)$, prove that $|F(\omega)| = |F(-\omega)|$ and the phase of $F(\omega)$ is an odd function of ω.

5.2 FOURIER TRANSFORMS CONTAINING THE DELTA FUNCTION

In the previous section we stressed the fact that such simple functions as cosine and sine are not absolutely integrable. Does this mean that these functions do not possess a Fourier transform? In this section we shall show that certain functions can still have a Fourier transform even though we cannot compute them directly.

The reason why we can find the Fourier transform of certain functions that are not absolutely integrable lies with the introduction of the delta function because

$$\int_{-\infty}^\infty \delta(\omega - \omega_0)e^{it\omega}\, d\omega = e^{i\omega_0 t} \tag{5.2.1}$$

for all t. Thus, the inverse of the Fourier transform $\delta(\omega - \omega_0)$ is the complex exponential $e^{i\omega_0 t}/2\pi$ or

$$\mathcal{F}\left(e^{i\omega_0 t}\right) = 2\pi\delta(\omega - \omega_0). \tag{5.2.2}$$

[5] Watson, G. N., 1966: *A Treatise on the Theory of Bessel Functions.* Cambridge University Press, p. 185.

This yields immediately the result that

$$\mathcal{F}(1) = 2\pi\delta(\omega), \qquad (5.2.3)$$

if we set $\omega_0 = 0$. Thus, the Fourier transform of 1 is an impulse at $\omega = 0$ with weight 2π. Because the Fourier transform equals zero for all $\omega \neq 0$, $f(t) = 1$ does not contain a nonzero frequency and is consequently a DC signal.

Another set of transforms arises from Euler's formula because we have that

$$\mathcal{F}[\sin(\omega_0 t)] = \left[\mathcal{F}\left(e^{i\omega_0 t}\right) - \mathcal{F}\left(e^{-i\omega_0 t}\right)\right] / (2i) \qquad (5.2.4)$$
$$= \pi\left[\delta(\omega - \omega_0) - \delta(\omega + \omega_0)\right] / i \qquad (5.2.5)$$
$$= -\pi i\delta(\omega - \omega_0) + \pi i\delta(\omega + \omega_0) \qquad (5.2.6)$$

and

$$\mathcal{F}[\cos(\omega_0 t)] = \tfrac{1}{2}\left[\mathcal{F}\left(e^{i\omega_0 t}\right) + \mathcal{F}\left(e^{-i\omega_0 t}\right)\right] \qquad (5.2.7)$$
$$= \pi\left[\delta(\omega - \omega_0) + \delta(\omega + \omega_0)\right]. \qquad (5.2.8)$$

Note that although the amplitude spectra of $\sin(\omega_0 t)$ and $\cos(\omega_0 t)$ are the same, their phase spectra are different.

Let us consider the Fourier transform of any arbitrary periodic function. Recall that any such function $f(t)$ with period $2L$ can be rewritten as the complex Fourier series

$$f(t) = \sum_{n=-\infty}^{\infty} c_n e^{in\omega_0 t}, \qquad (5.2.9)$$

where $\omega_0 = \pi/L$. The Fourier transform of $f(t)$ is

$$F(\omega) = \mathcal{F}[f(t)] = \sum_{n=-\infty}^{\infty} 2\pi c_n \delta(\omega - n\omega_0). \qquad (5.2.10)$$

Therefore, the Fourier transform of any arbitrary periodic function is a sequence of impulses with weight $2\pi c_n$ located at $\omega = n\omega_0$ with $n = 0, \pm 1, \pm 2,$ .... Thus, the Fourier series and transform of a periodic function are closely related.

• **Example 5.2.1: Fourier transform of the sign function**

Consider the sign function

$$\text{sgn}(t) = \begin{cases} 1, & t > 0, \\ 0, & t = 0, \\ -1, & t < 0. \end{cases} \qquad (5.2.11)$$

The function is not absolutely integrable. However, let us approximate it by $e^{-\epsilon|t|}\text{sgn}(t)$, where ϵ is a small positive number. This new function is absolutely integrable and we have that

$$\mathcal{F}[\text{sgn}(t)] = \lim_{\epsilon \to 0} \left[-\int_{-\infty}^{0} e^{\epsilon t} e^{-i\omega t}\, dt + \int_{0}^{\infty} e^{-\epsilon t} e^{-i\omega t}\, dt \right] \qquad (5.2.12)$$

$$= \lim_{\epsilon \to 0} \left(\frac{-1}{\epsilon - i\omega} + \frac{1}{\epsilon + i\omega} \right). \qquad (5.2.13)$$

If $\omega \neq 0$, (5.2.13) equals $2/i\omega$. If $\omega = 0$, (5.2.13) equals 0 because

$$\lim_{\epsilon \to 0} \left(\frac{-1}{\epsilon} + \frac{1}{\epsilon} \right) = 0. \qquad (5.2.14)$$

Thus, we conclude that

$$\mathcal{F}[\text{sgn}(t)] = \begin{cases} 2/i\omega, & \omega \neq 0, \\ 0, & \omega = 0. \end{cases} \qquad (5.2.15)$$

• **Example 5.2.2: Fourier transform of the step function**

An important function in transform methods is the *(Heaviside) step function*

$$H(t) = \begin{cases} 1, & t > 0, \\ 0, & t < 0. \end{cases} \qquad (5.2.16)$$

In terms of the sign function it can be written

$$H(t) = \tfrac{1}{2} + \tfrac{1}{2}\text{sgn}(t). \qquad (5.2.17)$$

Because the Fourier transforms of 1 and $\text{sgn}(t)$ are $2\pi\delta(\omega)$ and $2/i\omega$, respectively, we have that

$$\mathcal{F}[H(t)] = \pi\delta(\omega) + \frac{1}{i\omega}. \qquad (5.2.18)$$

These transforms are used in engineering but the presence of the delta function requires extra care to ensure their proper use.

Problems

1. Verify that

$$\mathcal{F}[\sin(\omega_0 t) H(t)] = \frac{\omega_0}{\omega_0^2 - \omega^2} + \frac{\pi i}{2}[\delta(\omega + \omega_0) - \delta(\omega - \omega_0)].$$

2. Verify that

$$\mathcal{F}[\cos(\omega_0 t)H(t)] = \frac{i\omega}{\omega_0^2 - \omega^2} + \frac{\pi}{2}[\delta(\omega + \omega_0) + \delta(\omega - \omega_0)].$$

3. Using the definition of Fourier transforms and (5.2.18), show that

$$\int_0^\infty e^{-i\omega t}\, dt = \pi\delta(\omega) - \frac{i}{\omega}, \qquad \text{or} \qquad \int_0^\infty e^{i\omega t}\, dt = \pi\delta(\omega) + \frac{i}{\omega}.$$

4. Following Example 5.2.1, show that

$$\mathcal{F}[\operatorname{sgn}(t)\sin(\omega_0 t)] = \frac{2\omega_0}{\omega_0^2 - \omega^2}$$

and

$$\mathcal{F}[\operatorname{sgn}(t)\cos(\omega_0 t)] = \frac{2\omega i}{\omega_0^2 - \omega^2}.$$

5.3 PROPERTIES OF FOURIER TRANSFORMS

In principle we can compute any Fourier transform from its definition. However, it is far more efficient to derive some simple relationships that relate transforms to each other. This is the purpose of this section.

Linearity

If $f(t)$ and $g(t)$ are functions with Fourier transforms $F(\omega)$ and $G(\omega)$, respectively, then

$$\mathcal{F}[c_1 f(t) + c_2 g(t)] = c_1 F(\omega) + c_2 G(\omega), \qquad (5.3.1)$$

where c_1 and c_2 are (real or complex) constants.
This result follows from the integral definition

$$\mathcal{F}[c_1 f(t) + c_2 g(t)] = \int_{-\infty}^{\infty} [c_1 f(t) + c_2 g(t)]e^{-i\omega t}\, dt \qquad (5.3.2)$$

$$= c_1 \int_{-\infty}^{\infty} f(t)e^{-i\omega t}\, dt + c_2 \int_{-\infty}^{\infty} g(t)e^{-i\omega t}\, dt \qquad (5.3.3)$$

$$= c_1 F(\omega) + c_2 G(\omega). \qquad (5.3.4)$$

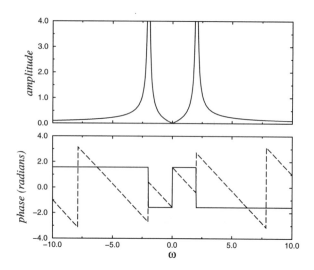

Figure 5.3.1: The amplitude and phase spectra of the Fourier transform for $\cos(2t)\,H(t)$ (solid line) and $\cos[2(t-1)]H(t-1)$ (dashed line). The amplitude becomes infinite at $\omega = \pm 2$.

Time shifting

If $f(t)$ is a function with a Fourier transform $F(\omega)$, then $\mathcal{F}[f(t-\tau)] = e^{-i\omega\tau}F(\omega)$.

This follows from the definition of the Fourier transform

$$\mathcal{F}[f(t-\tau)] = \int_{-\infty}^{\infty} f(t-\tau)e^{-i\omega t}\,dt = \int_{-\infty}^{\infty} f(x)e^{-i\omega(x+\tau)}\,dx \qquad (5.3.5)$$

$$= e^{-i\omega\tau}\int_{-\infty}^{\infty} f(x)e^{-i\omega x}\,dx = e^{-i\omega\tau}F(\omega). \qquad (5.3.6)$$

• **Example 5.3.1**

The Fourier transform of $f(t) = \cos(at)H(t)$ is $F(\omega) = i\omega/(a^2 - \omega^2) + \pi[\delta(\omega + a) + \delta(\omega - a)]/2$. Therefore,

$$\mathcal{F}\{\cos[a(t-k)]H(t-k)\} = e^{-ik\omega}\mathcal{F}[\cos(at)H(t)], \qquad (5.3.7)$$

or

$$\mathcal{F}\{\cos[a(t-k)]H(t-k)\} = \frac{i\omega e^{-ik\omega}}{a^2 - \omega^2} + \frac{\pi}{2}e^{-ik\omega}[\delta(\omega + a) + \delta(\omega - a)]. \quad (5.3.8)$$

In Figure 5.3.1 we present the amplitude and phase spectra for $\cos(2t)\,H(t)$ (the solid line) while the dashed line gives these spectra for $\cos[2(t-1)]H(t-1)$.

This figure shows that the amplitude spectra are identical (why?) while the phase spectra are considerably different.

Scaling factor

Let $f(t)$ be a function with a Fourier transform $F(\omega)$ and k be a real, nonzero constant. Then $\mathcal{F}[f(kt)] = F(\omega/k)/|k|$.

From the definition of the Fourier transform:

$$\mathcal{F}[f(kt)] = \int_{-\infty}^{\infty} f(kt)e^{-i\omega t}\,dt = \frac{1}{|k|}\int_{-\infty}^{\infty} f(x)e^{-i(\omega/k)x}\,dx = \frac{1}{|k|}F\left(\frac{\omega}{k}\right).$$

$$(5.3.9)$$

• **Example 5.3.2**

The Fourier transform of $f(t) = e^{-t}H(t)$ is $F(\omega) = 1/(1+\omega i)$. Therefore, the Fourier transform for $f(at) = e^{-at}H(t)$, $a > 0$, is

$$\mathcal{F}[f(at)] = \left(\frac{1}{a}\right)\left(\frac{1}{1+i\omega/a}\right) = \frac{1}{a+\omega i}. \qquad (5.3.10)$$

To illustrate this scaling property we use the MATLAB script

```
clear; % clear all previous computations
omega = [-10:0.01:10]; % set up frequencies
% real part of transform with a = 1
f1r_omega = 1./(1+omega.*omega);
% imaginary part of transform with a = 1
f1i_omega = - omega./(1+omega.*omega);
% real part of transform with a = 2
f2r_omega = 2./(4+omega.*omega);
% imaginary part of transform with a = 2
f2i_omega = - omega./(4+omega.*omega);
% compute the amplitude of the first transform
ampl1 = sqrt(f1r_omega.*f1r_omega + f1i_omega.*f1i_omega);
% compute the amplitude of the second transform
ampl2 = sqrt(f2r_omega.*f2r_omega + f2i_omega.*f2i_omega);
% compute phase of first transform
phase1 = atan2(f1i_omega,f1r_omega);
% compute phase of second transform
phase2 = atan2(f2i_omega,f2r_omega);
clf; % clear all previous figures
% plot amplitudes of Fourier transforms
subplot(2,1,1), plot(omega,ampl1,omega,ampl2,'--')
ylabel('|F(\omega)|','FontSize',15) % label amplitude plot
```

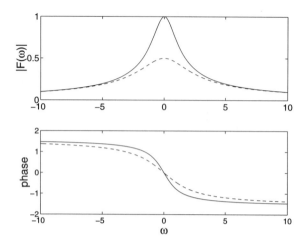

Figure 5.3.2: The amplitude and phase spectra of the Fourier transform for $e^{-t}H(t)$ (solid line) and $e^{-2t}H(t)$ (dashed line).

```
% plot phases of Fourier transforms
subplot(2,1,2), plot(omega,phase1,omega,phase2,'--')
ylabel('phase','FontSize',15) % label amplitude plot
xlabel('\omega','FontSize',15) % label x-axis
```

to plot the amplitude and phase when $a = 1$ and $a = 2$. Figure 5.3.2 shows the results from the MATLAB script: The amplitude spectra decreased by a factor of two for $e^{-2t}H(t)$ compared to $e^{-t}H(t)$ while the differences in the phase are smaller.

Symmetry

If the function $f(t)$ has the Fourier transform $F(\omega)$, then $\mathcal{F}[F(t)] = 2\pi f(-\omega)$.

From the definition of the inverse Fourier transform,

$$f(t) = \frac{1}{2\pi} \int_{-\infty}^{\infty} F(\omega)e^{i\omega t}\,d\omega = \frac{1}{2\pi} \int_{-\infty}^{\infty} F(x)e^{ixt}\,dx. \qquad (5.3.11)$$

Then

$$2\pi f(-\omega) = \int_{-\infty}^{\infty} F(x)e^{-i\omega x}\,dx = \int_{-\infty}^{\infty} F(t)e^{-i\omega t}\,dt = \mathcal{F}[F(t)]. \qquad (5.3.12)$$

● **Example 5.3.3**

The Fourier transform of $1/(1 + t^2)$ is $\pi e^{-|\omega|}$. Therefore,

$$\mathcal{F}\left(\pi e^{-|t|}\right) = \frac{2\pi}{1 + \omega^2} \tag{5.3.13}$$

or

$$\mathcal{F}\left(e^{-|t|}\right) = \frac{2}{1 + \omega^2}. \tag{5.3.14}$$

Derivatives of functions

Let $f^{(k)}(t), k = 0, 1, 2, \ldots, n - 1$, be continuous and $f^{(n)}(t)$ be piecewise continuous. Let $|f^{(k)}(t)| \leq Ke^{-bt}, b > 0, 0 \leq t < \infty; |f^{(k)}(t)| \leq Me^{at}, a > 0, -\infty < t \leq 0, k = 0, 1, \ldots, n$. Then, $\mathcal{F}[f^{(n)}(t)] = (i\omega)^n F(\omega)$.

We begin by noting that if the transform $\mathcal{F}[f'(t)]$ exists, then

$$\mathcal{F}[f'(t)] = \int_{-\infty}^{\infty} f'(t)e^{-i\omega t}\,dt \tag{5.3.15}$$

$$= \int_{-\infty}^{\infty} f'(t)e^{\omega_i t}[\cos(\omega_r t) - i\sin(\omega_r t)]\,dt \tag{5.3.16}$$

$$= (-\omega_i + i\omega_r)\int_{-\infty}^{\infty} f(t)e^{\omega_i t}[\cos(\omega_r t) - i\sin(\omega_r t)]\,dt \tag{5.3.17}$$

$$= i\omega \int_{-\infty}^{\infty} f(t)e^{-i\omega t}\,dt = i\omega F(\omega). \tag{5.3.18}$$

Finally,

$$\mathcal{F}[f^{(n)}(t)] = i\omega\mathcal{F}[f^{(n-1)}(t)] = (i\omega)^2\mathcal{F}[f^{(n-2)}(t)] = \cdots = (i\omega)^n F(\omega). \tag{5.3.19}$$

● **Example 5.3.4**

The Fourier transform of $f(t) = 1/(1 + t^2)$ is $F(\omega) = \pi e^{-|\omega|}$. Therefore,

$$\mathcal{F}\left[-\frac{2t}{(1 + t^2)^2}\right] = i\omega\pi e^{-|\omega|} \tag{5.3.20}$$

or

$$\mathcal{F}\left[\frac{t}{(1 + t^2)^2}\right] = -\frac{i\omega\pi}{2}e^{-|\omega|}. \tag{5.3.21}$$

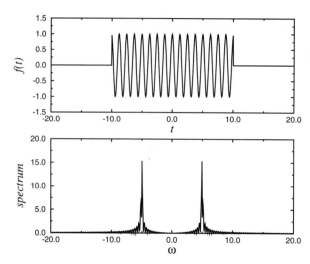

Figure 5.3.3: The (amplitude) spectrum of a rectangular pulse (5.1.9) with a half width $a = 10$ that has been modulated with $\cos(5t)$.

Modulation

In communications a popular method of transmitting information is by *amplitude modulation* (AM). In this process the signal is carried according to the expression $f(t)e^{i\omega_0 t}$, where ω_0 is the *carrier frequency* and $f(t)$ is an arbitrary function of time whose amplitude spectrum peaks at some frequency that is usually small compared to ω_0. We now show that the Fourier transform of $f(t)e^{i\omega_0 t}$ is $F(\omega - \omega_0)$, where $F(\omega)$ is the Fourier transform of $f(t)$.

We begin by using the definition of the Fourier transform or

$$\mathcal{F}[f(t)e^{i\omega_0 t}] = \int_{-\infty}^{\infty} f(t)e^{i\omega_0 t}e^{-i\omega t}dt = \int_{-\infty}^{\infty} f(t)e^{-i(\omega - \omega_0)t}dt \qquad (5.3.22)$$

$$= F(\omega - \omega_0). \qquad (5.3.23)$$

Therefore, if we have the spectrum of a particular function $f(t)$, then the Fourier transform of the modulated function $f(t)e^{i\omega_0 t}$ is the same as that for $f(t)$ except that it is now centered on the frequency ω_0 rather than on the zero frequency.

● **Example 5.3.5**

Let us determine the Fourier transform of a square pulse modulated by a cosine wave as shown in Figures 5.3.3 and 5.3.4. Because $\cos(\omega_0 t) = \frac{1}{2}[e^{i\omega_0 t} + e^{-i\omega_0 t}]$ and the Fourier transform of a square pulse is $F(\omega) = 2\sin(\omega a)/\omega$,

$$\mathcal{F}[f(t)\cos(\omega_0 t)] = \frac{\sin[(\omega - \omega_0)a]}{\omega - \omega_0} + \frac{\sin[(\omega + \omega_0)a]}{\omega + \omega_0}. \qquad (5.3.24)$$

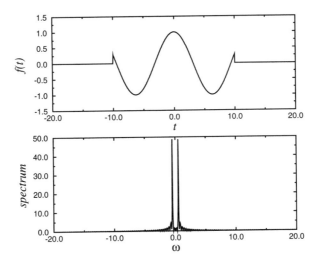

Figure 5.3.4: The (amplitude) spectrum of a rectangular pulse (5.1.9) with a half width $a = 10$ that has been modulated with $\cos(t/2)$.

Therefore, the Fourier transform of the modulated pulse equals one half of the sum of the Fourier transform of the pulse centered on ω_0 and $-\omega_0$. See Figures 5.3.3 and 5.3.4.

In many practical situations, $\omega_0 \gg \pi/a$. In this case we may treat each term as completely independent from the other; the contribution from the peak at $\omega = \omega_0$ has a negligible effect on the peak at $\omega = -\omega_0$.

- **Example 5.3.6**

The Fourier transform of $f(t) = e^{-bt}H(t)$ is $F(\omega) = 1/(b+i\omega)$. Therefore,

$$\mathcal{F}[e^{-bt}\cos(at)H(t)] = \tfrac{1}{2}\mathcal{F}\left(e^{iat}e^{-bt} + e^{-iat}e^{-bt}\right) \tag{5.3.25}$$

$$= \frac{1}{2}\left(\frac{1}{b+i\omega'}\bigg|_{\omega'=\omega-a} + \frac{1}{b+i\omega'}\bigg|_{\omega'=\omega+a}\right) \tag{5.3.26}$$

$$= \frac{1}{2}\left[\frac{1}{(b+i\omega)-ai} + \frac{1}{(b+i\omega)+ai}\right] \tag{5.3.27}$$

$$= \frac{b+i\omega}{(b+i\omega)^2 + a^2}. \tag{5.3.28}$$

We illustrate this result using $e^{-2t}H(t)$ and $e^{-2t}\cos(4t)H(t)$ in Figure 5.3.5.

- **Example 5.3.7: Frequency modulation**

In contrast to amplitude modulation, *frequency modulation* (FM) transmits information by instantaneous variations of the carrier frequency. It can be

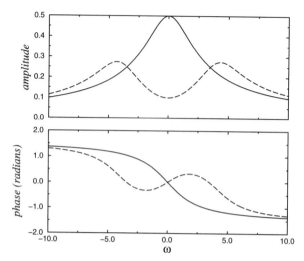

Figure 5.3.5: The amplitude and phase spectra of the Fourier transform for $e^{-2t} H(t)$ (solid line) and $e^{-2t} \cos(4t) H(t)$ (dashed line).

expressed mathematically as $\exp\left[i \int_{-\infty}^{t} f(\tau)\,d\tau + iC\right] e^{i\omega_0 t}$, where C is a constant. To illustrate this concept, let us find the Fourier transform of a simple frequency modulation

$$f(t) = \begin{cases} \omega_1, & |t| < T/2, \\ 0, & |t| > T/2, \end{cases} \tag{5.3.29}$$

and $C = -\omega_1 T/2$. In this case, the signal in the time domain is

$$g(t) = \exp\left[i \int_{-\infty}^{t} f(\tau)\,d\tau + iC\right] e^{i\omega_0 t} \tag{5.3.30}$$

$$= \begin{cases} e^{-i\omega_1 T/2} e^{i\omega_0 t}, & t < -T/2, \\ e^{i\omega_1 t} e^{i\omega_0 t}, & -T/2 < t < T/2, \\ e^{i\omega_1 T/2} e^{i\omega_0 t}, & T/2 < t. \end{cases} \tag{5.3.31}$$

We illustrate this signal in Figures 5.3.6 and 5.3.7.

The Fourier transform of the signal $G(\omega)$ equals

$$G(\omega) = e^{-i\omega_1 T/2} \int_{-\infty}^{-T/2} e^{i(\omega_0 - \omega)t}\,dt + \int_{-T/2}^{T/2} e^{i(\omega_0 + \omega_1 - \omega)t}\,dt$$

$$+ e^{i\omega_1 T/2} \int_{T/2}^{\infty} e^{i(\omega_0 - \omega)t}\,dt \tag{5.3.32}$$

$$= e^{-i\omega_1 T/2} \int_{-\infty}^{0} e^{i(\omega_0 - \omega)t}\,dt + e^{i\omega_1 T/2} \int_{0}^{\infty} e^{i(\omega_0 - \omega)t}\,dt$$

$$- e^{-i\omega_1 T/2} \int_{-T/2}^{0} e^{i(\omega_0 - \omega)t}\,dt + \int_{-T/2}^{T/2} e^{i(\omega_0 + \omega_1 - \omega)t}\,dt$$

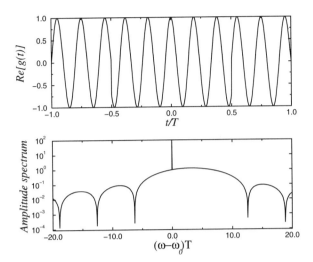

Figure 5.3.6: The (amplitude) spectrum $|G(\omega)|/T$ of a frequency-modulated signal (shown top) when $\omega_1 T = 2\pi$ and $\omega_0 T = 10\pi$. The transform becomes undefined at $\omega = \omega_0$.

$$- e^{i\omega_1 T/2} \int_0^{T/2} e^{i(\omega_0 - \omega)t}\, dt. \tag{5.3.33}$$

Applying the fact that

$$\int_0^{\infty} e^{\pm i\alpha t}\, dt = \pi \delta(\alpha) \pm \frac{i}{\alpha}, \tag{5.3.34}$$

$$G(\omega) = \pi \delta(\omega - \omega_0) \left[e^{i\omega_1 T/2} + e^{-i\omega_1 T/2} \right]$$
$$+ \frac{\left[e^{i(\omega_0 + \omega_1 - \omega)T/2} - e^{-i(\omega_0 + \omega_1 - \omega)T/2} \right]}{i(\omega_0 + \omega_1 - \omega)}$$
$$- \frac{\left[e^{i(\omega_0 + \omega_1 - \omega)T/2} - e^{-i(\omega_0 + \omega_1 - \omega)T/2} \right]}{i(\omega_0 - \omega)} \tag{5.3.35}$$

$$= 2\pi \delta(\omega - \omega_0) \cos(\omega_1 T/2) + \frac{2\omega_1 \sin[(\omega - \omega_0 - \omega_1)T/2]}{(\omega - \omega_0)(\omega - \omega_0 - \omega_1)}. \tag{5.3.36}$$

Figures 5.3.6 and 5.3.7 illustrate the amplitude spectrum for various parameters. In general, the transform is not symmetric, with an increasing number of humped curves as $\omega_1 T$ increases.

> **Parseval's equality**

In applying Fourier methods to practical problems we may encounter a situation where we are interested in computing the energy of a system.

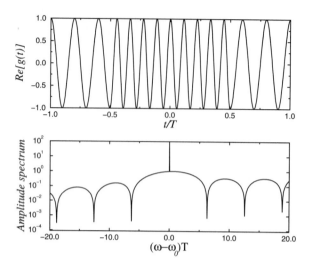

Figure 5.3.7: The (amplitude) spectrum $|G(\omega)|/T$ of a frequency-modulated signal (shown top) when $\omega_1 T = 8\pi$ and $\omega_0 T = 10\pi$. The transform becomes undefined at $\omega = \omega_0$.

Energy is usually expressed by the integral $\int_{-\infty}^{\infty} |f(t)|^2\, dt$. Can we compute this integral if we only have the Fourier transform of $F(\omega)$?

From the definition of the inverse Fourier transform

$$f(t) = \frac{1}{2\pi} \int_{-\infty}^{\infty} F(\omega) e^{i\omega t}\, d\omega, \tag{5.3.37}$$

we have that

$$\int_{-\infty}^{\infty} |f(t)|^2\, dt = \frac{1}{2\pi} \int_{-\infty}^{\infty} f(t) \left[\int_{-\infty}^{\infty} F(\omega) e^{i\omega t}\, d\omega \right] dt. \tag{5.3.38}$$

Interchanging the order of integration on the right side of (5.3.38),

$$\int_{-\infty}^{\infty} |f(t)|^2\, dt = \frac{1}{2\pi} \int_{-\infty}^{\infty} F(\omega) \left[\int_{-\infty}^{\infty} f(t) e^{i\omega t}\, dt \right] d\omega. \tag{5.3.39}$$

However,

$$F^*(\omega) = \int_{-\infty}^{\infty} f(t) e^{i\omega t}\, dt. \tag{5.3.40}$$

Therefore,

$$\int_{-\infty}^{\infty} |f(t)|^2\, dt = \frac{1}{2\pi} \int_{-\infty}^{\infty} |F(\omega)|^2\, d\omega. \tag{5.3.41}$$

Table 5.3.1: Some General Properties of Fourier Transforms

	function, f(t)	Fourier transform, F(ω)
1. Linearity	$c_1 f(t) + c_2 g(t)$	$c_1 F(\omega) + c_2 G(\omega)$
2. Complex conjugate	$f^*(t)$	$F^*(-\omega)$
3. Scaling	$f(\alpha t)$	$F(\omega/\alpha)/|\alpha|$
4. Delay	$f(t - \tau)$	$e^{-i\omega\tau} F(\omega)$
5. Frequency translation	$e^{i\omega_0 t} f(t)$	$F(\omega - \omega_0)$
6. Duality-time frequency	$F(t)$	$2\pi f(-\omega)$
7. Time differentiation	$f'(t)$	$i\omega F(\omega)$

This is *Parseval's equality*[6] as it applies to Fourier transforms. The quantity $|F(\omega)|^2$ is called the *power spectrum*.

• **Example 5.3.8**

In Example 5.1.1, we showed that the Fourier transform for a unit rectangular pulse between $-a < t < a$ is $2\sin(\omega a)/\omega$. Therefore, by Parseval's equality,

$$\frac{2}{\pi} \int_{-\infty}^{\infty} \frac{\sin^2(\omega a)}{\omega^2} \, d\omega = \int_{-a}^{a} 1^2 \, dt = 2a \qquad (5.3.42)$$

or

$$\int_{-\infty}^{\infty} \frac{\sin^2(\omega a)}{\omega^2} \, d\omega = \pi a. \qquad (5.3.43)$$

---

[6] Apparently first derived by Rayleigh, J. W., 1889: On the character of the complete radiation at a given temperature. *Philos. Mag., Ser. 5*, **27**, 460–469.

| Poisson's summation formula |

If $f(x)$ is integrable over $(-\infty, \infty)$, there exists a relationship between the function and its Fourier transform, commonly called *Poisson's summation formula*.[7]

We begin by inventing a periodic function $g(x)$ defined by

$$g(x) = \sum_{k=-\infty}^{\infty} f(x + 2\pi k). \qquad (5.3.44)$$

Because $g(x)$ is a periodic function of 2π, it can be represented by the complex Fourier series:

$$g(x) = \sum_{n=-\infty}^{\infty} c_n e^{inx}, \qquad (5.3.45)$$

or

$$g(0) = \sum_{k=-\infty}^{\infty} f(2\pi k) = \sum_{n=-\infty}^{\infty} c_n. \qquad (5.3.46)$$

Computing c_n, we find that

$$c_n = \frac{1}{2\pi} \int_{-\pi}^{\pi} g(x) e^{-inx}\, dx = \frac{1}{2\pi} \int_{-\pi}^{\pi} \sum_{k=-\infty}^{\infty} f(x + 2k\pi) e^{-inx}\, dx \qquad (5.3.47)$$

$$= \frac{1}{2\pi} \sum_{k=-\infty}^{\infty} \int_{-\pi}^{\pi} f(x + 2k\pi) e^{-inx}\, dx = \frac{1}{2\pi} \int_{-\infty}^{\infty} f(x) e^{-inx}\, dx \qquad (5.3.48)$$

$$= \frac{F(n)}{2\pi}, \qquad (5.3.49)$$

where $F(\omega)$ is the Fourier transform of $f(x)$. Substituting (5.3.49) into (5.3.46), we obtain

$$\sum_{k=-\infty}^{\infty} f(2\pi k) = \frac{1}{2\pi} \sum_{n=-\infty}^{\infty} F(n) \qquad (5.3.50)$$

or

$$\boxed{\sum_{k=-\infty}^{\infty} f(\alpha k) = \frac{1}{\alpha} \sum_{n=-\infty}^{\infty} F\left(\frac{2\pi n}{\alpha}\right).} \qquad (5.3.51)$$

---

[7] Poisson, S. D., 1823: Suite du mémoire sur les intégrales définies et sur la sommation des séries. *J. École Polytech.*, **19**, 404–509. See page 451.

• Example 5.3.9

One of the popular uses of Poisson's summation formula is the evaluation of infinite series. For example, let $f(x) = 1/(a^2 + x^2)$ with a real and nonzero. Then, $F(\omega) = \pi e^{-|a\omega|}/|a|$ and

$$\sum_{k=-\infty}^{\infty} \frac{1}{a^2 + (2\pi k)^2} = \frac{1}{2} \sum_{n=-\infty}^{\infty} \frac{1}{|a|} e^{-|an|} = \frac{1}{2|a|} \left(1 + 2 \sum_{n=1}^{\infty} e^{-|a|n} \right) \quad (5.3.52)$$

$$= \frac{1}{2|a|} \left(-1 + \frac{2}{1 - e^{-|a|}} \right) = \frac{1}{2|a|} \coth\left(\frac{|a|}{2} \right). \quad (5.3.53)$$

Problems

1. Find the Fourier transform of $1/(1 + a^2 t^2)$, where a is real, given that $\mathcal{F}[1/(1 + t^2)] = \pi e^{-|\omega|}$.

2. Find the Fourier transform of $\cos(at)/(1 + t^2)$, where a is real, given that $\mathcal{F}[1/(1 + t^2)] = \pi e^{-|\omega|}$.

3. Use the fact that $\mathcal{F}[e^{-at} H(t)] = 1/(a + i\omega)$ with $a > 0$ and Parseval's equality to show that

$$\int_{-\infty}^{\infty} \frac{dx}{x^2 + a^2} = \frac{\pi}{a}.$$

4. Use the fact that $\mathcal{F}[1/(1 + t^2)] = \pi e^{-|\omega|}$ and Parseval's equality to show that

$$\int_{-\infty}^{\infty} \frac{dx}{(x^2 + 1)^2} = \frac{\pi}{2}.$$

5. Use the function $f(t) = e^{-at} \sin(bt) H(t)$ with $a > 0$ and Parseval's equality to show that

$$2 \int_0^{\infty} \frac{dx}{(x^2 + a^2 - b^2)^2 + 4a^2 b^2} = \int_{-\infty}^{\infty} \frac{dx}{(x^2 + a^2 - b^2)^2 + 4a^2 b^2} = \frac{\pi}{2a(a^2 + b^2)}.$$

6. Using the modulation property and $\mathcal{F}[e^{-bt} H(t)] = 1/(b + i\omega)$, show that

$$\mathcal{F}\left[e^{-bt} \sin(at) H(t) \right] = \frac{a}{(b + i\omega)^2 + a^2}.$$

Use MATLAB to plot and compare the amplitude and phase spectra for $e^{-t} H(t)$ and $e^{-t} \sin(2t) H(t)$.

7. Use Poisson's summation formula with $f(t) = e^{-|t|}$ to show that

$$\sum_{n=-\infty}^{\infty} \frac{1}{n^2 + 1} = \pi \frac{1 + e^{-2\pi}}{1 - e^{-2\pi}}.$$

8. Use Poisson's summation formula to prove[8] that

$$\sum_{n=-\infty}^{\infty} e^{-a(n+c)^2 + 2b(n+c)} = \sqrt{\frac{\pi}{a}} e^{b^2/a} \sum_{n=-\infty}^{\infty} e^{-n^2\pi^2/a - 2n\pi i(b/a - c)}.$$

9. Use Poisson's summation formula to prove that

$$\sum_{n=-\infty}^{\infty} e^{-ianT} = \frac{2\pi}{T} \sum_{n=-\infty}^{\infty} \delta\left(\frac{2\pi n}{T} - a\right),$$

where $\delta(\ )$ is the Dirac delta function.

10. Prove the two-dimensional form[9] of Poisson's summation formula:

$$\sum_{k_1=-\infty}^{\infty} \sum_{k_2=-\infty}^{\infty} f(\alpha_1 k_1, \alpha_2 k_2) = \frac{1}{\alpha_1 \alpha_2} \sum_{n_1=-\infty}^{\infty} \sum_{n_2=-\infty}^{\infty} F\left(\frac{2\pi n_1}{\alpha_1}, \frac{2\pi n_2}{\alpha_2}\right),$$

where

$$F(\omega_1, \omega_2) = \int_{-\infty}^{\infty} \int_{-\infty}^{\infty} f(x, y) e^{-i\omega_1 x - i\omega_2 y} \, dx \, dy.$$

5.4 INVERSION OF FOURIER TRANSFORMS

Having focused on the Fourier transform in the previous sections, we now consider the inverse Fourier transform. Recall that the improper integral (5.1.6) defines the inverse. Consequently one method of inversion is direct integration.

- **Example 5.4.1**

Let us find the inverse of $F(\omega) = \pi e^{-|\omega|}$.

---

[8] First proved by Ewald, P. P., 1921: Die Berechnung optischer und elektrostatischer Gitterpotentiale. *Ann. Phys., 4te Folge,* **64**, 253–287.

[9] Taken from Lucas, S. K., R. Sipcic, and H. A. Stone, 1997: An integral equation solution for the steady-state current at a periodic array of surface microelectrodes. *SIAM J. Appl. Math.,* **57**, 1615–1638.

From the definition of the inverse Fourier transform,

$$f(t) = \frac{1}{2\pi} \int_{-\infty}^{\infty} \pi e^{-|\omega|} e^{i\omega t} d\omega = \frac{1}{2} \int_{-\infty}^{0} e^{(1+it)\omega} d\omega + \frac{1}{2} \int_{0}^{\infty} e^{(-1+it)\omega} d\omega$$

$$(5.4.1)$$

$$= \frac{1}{2} \left[\frac{e^{(1+it)\omega}}{1+it} \Big|_{-\infty}^{0} + \frac{e^{(-1+it)\omega}}{-1+it} \Big|_{0}^{\infty} \right] = \frac{1}{2} \left[\frac{1}{1+it} - \frac{1}{-1+it} \right] = \frac{1}{1+t^2}.$$

$$(5.4.2)$$

An alternative to direct integration is the MATLAB function ifourier. For example, to invert $F(\omega) = \pi e^{-|\omega|}$, we type in the commands:

```
>> syms pi omega t
>> ifourier('pi*exp(-abs(omega))',omega,t)
```

This yields

```
ans =
1/(1+t^2)
```

Another method for inverting Fourier transforms is rewriting the Fourier transform using partial fractions so that we can use transform tables. The following example illustrates this technique.

• **Example 5.4.2**

Let us invert the transform

$$F(\omega) = \frac{1}{(1+i\omega)(1-2i\omega)^2}.$$ $$(5.4.3)$$

We begin by rewriting (5.4.3) as

$$F(\omega) = \frac{1}{9} \left[\frac{1}{1+i\omega} + \frac{2}{1-2i\omega} + \frac{6}{(1-2i\omega)^2} \right]$$ $$(5.4.4)$$

$$= \frac{1}{9(1+i\omega)} + \frac{1}{9(\frac{1}{2}-i\omega)} + \frac{1}{6(\frac{1}{2}-i\omega)^2}.$$ $$(5.4.5)$$

Using Table 5.1.1, we invert (5.4.5) term by term and find that

$$f(t) = \tfrac{1}{9}e^{-t}H(t) + \tfrac{1}{9}e^{t/2}H(-t) - \tfrac{1}{6}te^{t/2}H(-t).$$ $$(5.4.6)$$

To check our answer, we type the following commands into MATLAB:

```
>> syms omega t
>> ifourier(1/((1+i*omega)*(1-2*i*omega)^2),omega,t)
```

which yields

```
ans =
1/9*exp(-t)*Heaviside(t)-1/6*exp(1/2*t)*t*Heaviside(-t)
   +1/9*exp(1/2*t)*Heaviside(-t)
```

Although we may find the inverse by direct integration or partial fractions, in many instances the Fourier transform does not lend itself to these techniques. On the other hand, if we view the inverse Fourier transform as a line integral along the real axis in the complex ω-plane, then perhaps some of the techniques that we developed in Chapter 1 might be applicable to this problem. To this end, we rewrite the inversion integral (5.1.6) as

$$f(t) = \frac{1}{2\pi} \int_{-\infty}^{\infty} F(\omega)e^{it\omega} \, d\omega = \frac{1}{2\pi} \oint_C F(z)e^{itz} \, dz - \frac{1}{2\pi} \int_{C_R} F(z)e^{itz} \, dz,$$

(5.4.7)

where C denotes a closed contour consisting of the entire real axis plus a new contour C_R that joins the point $(\infty, 0)$ to $(-\infty, 0)$. There are countless possibilities for C_R. For example, it could be the loop $(\infty, 0)$ to (∞, R) to $(-\infty, R)$ to $(-\infty, 0)$ with $R > 0$. However, any choice of C_R must be such that we can compute $\int_{C_R} F(z)e^{itz} \, dz$. When we take that constraint into account, the number of acceptable contours decreases to just a few. The best is given by *Jordan's lemma*.[10]

Jordan's lemma: *Suppose that, on a circular arc C_R with radius R and center at the origin, $f(z) \to 0$ uniformly as $R \to \infty$. Then*

$$(1) \qquad \lim_{R \to \infty} \int_{C_R} f(z)e^{imz} \, dz = 0, \qquad (m > 0) \qquad (5.4.8)$$

if C_R lies in the first and/or second quadrant;

$$(2) \qquad \lim_{R \to \infty} \int_{C_R} f(z)e^{-imz} \, dz = 0, \qquad (m > 0) \qquad (5.4.9)$$

if C_R lies in the third and/or fourth quadrant;

$$(3) \qquad \lim_{R \to \infty} \int_{C_R} f(z)e^{mz} \, dz = 0, \qquad (m > 0) \qquad (5.4.10)$$

if C_R lies in the second and/or third quadrant; and

$$(4) \qquad \lim_{R \to \infty} \int_{C_R} f(z)e^{-mz} \, dz = 0, \qquad (m > 0) \qquad (5.4.11)$$

---

[10] Jordan, C., 1894: *Cours D'Analyse de l'École Polytechnique. Vol. 2.* Gauthier-Villars, pp. 285–286. See also Whittaker, E. T., and G. N. Watson, 1963: *A Course of Modern Analysis.* Cambridge University Press, p. 115.

if C_R lies in the first and/or fourth quadrant.

Technically, only (1) is actually Jordan's lemma while the remaining points are variations.

Proof: We shall prove the first part; the remaining portions follow by analog. We begin by noting that

$$|I_R| = \left| \int_{C_R} f(z) e^{imz} \, dz \right| \leq \int_{C_R} |f(z)| \left| e^{imz} \right| |dz|. \qquad (5.4.12)$$

Now

$$|dz| = R \, d\theta, \quad |f(z)| \leq M_R, \qquad (5.4.13)$$

$$\left| e^{imz} \right| = \left| \exp(imRe^{\theta i}) \right| = \left| \exp\{imR[\cos(\theta) + i \sin(\theta)]\} \right| = e^{-mR \sin(\theta)}. \qquad (5.4.14)$$

Therefore,

$$|I_R| \leq R M_R \int_{\theta_0}^{\theta_1} \exp[-mR \sin(\theta)] \, d\theta, \qquad (5.4.15)$$

where $0 \leq \theta_0 < \theta_1 \leq \pi$. Because the integrand is positive, the right side of (5.4.15) is largest if we take $\theta_0 = 0$ and $\theta_1 = \pi$. Then

$$|I_R| \leq R M_R \int_0^{\pi} e^{-mR \sin(\theta)} \, d\theta = 2R M_R \int_0^{\pi/2} e^{-mR \sin(\theta)} \, d\theta. \qquad (5.4.16)$$

We cannot evaluate the integrals in (5.4.16) as they stand. However, because $\sin(\theta) \geq 2\theta/\pi$ if $0 \leq \theta \leq \pi/2$, we can bound the value of the integral by

$$|I_R| \leq 2R M_R \int_0^{\pi/2} e^{-2mR\theta/\pi} \, d\theta = \frac{\pi}{m} M_R \left(1 - e^{-mR} \right). \qquad (5.4.17)$$

If $m > 0$, $|I_R|$ tends to zero with M_R as $R \to \infty$. □

Consider now the following inversions of Fourier transforms:

• **Example 5.4.3**

For our first example we find the inverse for

$$F(\omega) = \frac{1}{\omega^2 - 2ib\omega - a^2 - b^2}, \quad a, b > 0. \qquad (5.4.18)$$

From the inversion integral,

$$f(t) = \frac{1}{2\pi} \int_{-\infty}^{\infty} \frac{e^{it\omega}}{\omega^2 - 2ib\omega - a^2 - b^2} \, d\omega, \qquad (5.4.19)$$

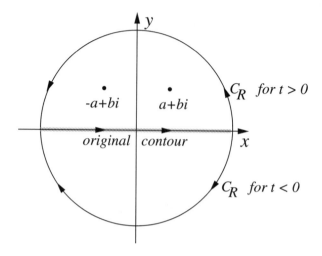

Figure 5.4.1: Contour used to find the inverse of the Fourier transform (5.4.18). The contour C consists of the line integral along the real axis plus C_R.

or

$$f(t) = \frac{1}{2\pi} \oint_C \frac{e^{itz}}{z^2 - 2ibz - a^2 - b^2}\, dz - \frac{1}{2\pi} \int_{C_R} \frac{e^{itz}}{z^2 - 2ibz - a^2 - b^2}\, dz,$$

(5.4.20)

where C denotes a closed contour consisting of the entire real axis plus C_R. Because $f(z) = 1/(z^2 - 2ibz - a^2 - b^2)$ tends to zero uniformly as $|z| \to \infty$ and $m = t$, the second integral in (5.4.20) vanishes by Jordan's lemma if C_R is a semicircle of infinite radius in the upper half of the z-plane when $t > 0$ and a semicircle in the lower half of the z-plane when $t < 0$.

Next we must find the location and nature of the singularities. They are located at

$$z^2 - 2ibz - a^2 - b^2 = 0,$$

(5.4.21)

or

$$z = \pm a + bi.$$

(5.4.22)

Therefore we can rewrite (5.4.20) as

$$f(t) = \frac{1}{2\pi} \oint_C \frac{e^{itz}}{(z - a - bi)(z + a - bi)}\, dz.$$

(5.4.23)

Thus, all of the singularities are simple poles.

Consider now $t > 0$. As stated earlier, we close the line integral with an infinite semicircle in the upper half-plane. See Figure 5.4.1. Inside this closed contour there are two singularities: $z = \pm a + bi$. For these poles,

$$\text{Res}\left(\frac{e^{itz}}{z^2 - 2ibz - a^2 - b^2}; a + bi\right) = \lim_{z \to a+bi} \frac{(z - a - bi)e^{itz}}{(z - a - bi)(z + a - bi)} \quad (5.4.24)$$

$$= \frac{e^{iat}e^{-bt}}{2a} = \frac{e^{-bt}}{2a}[\cos(at) + i\sin(at)], \quad (5.4.25)$$

where we used Euler's formula to eliminate e^{iat}. Similarly,

$$\text{Res}\left(\frac{e^{itz}}{z^2 - 2ibz - a^2 - b^2}; -a + bi\right) = -\frac{e^{-bt}}{2a}[\cos(at) - i\sin(at)]. \quad (5.4.26)$$

Consequently the inverse Fourier transform follows from (5.4.23) after applying the residue theorem and equals

$$f(t) = -\frac{e^{-bt}}{2a}\sin(at) \quad (5.4.27)$$

for $t > 0$.

For $t < 0$ the semicircle is in the lower half-plane because the contribution from the semicircle vanishes as $R \to \infty$. Because there are no singularities within the closed contour, $f(t) = 0$. Therefore, we can write in general that

$$f(t) = -\frac{e^{-bt}}{2a}\sin(at)H(t). \quad (5.4.28)$$

• Example 5.4.4

Let us find the inverse of the Fourier transform

$$F(\omega) = \frac{e^{-\omega i}}{\omega^2 + a^2}, \quad (5.4.29)$$

where a is real and positive.

From the inversion integral,

$$f(t) = \frac{1}{2\pi}\int_{-\infty}^{\infty} \frac{e^{i(t-1)\omega}}{\omega^2 + a^2}\,d\omega \quad (5.4.30)$$

$$= \frac{1}{2\pi}\oint_C \frac{e^{i(t-1)z}}{z^2 + a^2}\,dz - \frac{1}{2\pi}\int_{C_R} \frac{e^{i(t-1)z}}{z^2 + a^2}\,dz, \quad (5.4.31)$$

where C denotes a closed contour consisting of the entire real axis plus C_R. The contour C_R is determined by Jordan's lemma because $1/(z^2 + a^2) \to 0$ uniformly as $|z| \to \infty$. Since $m = t - 1$, the semicircle C_R of infinite radius lies in the upper half-plane if $t > 1$ and in the lower half-plane if $t < 1$. Thus, if $t > 1$,

$$f(t) = \frac{1}{2\pi}(2\pi i)\mathrm{Res}\left[\frac{e^{i(t-1)z}}{z^2 + a^2}; ai\right] = \frac{e^{-a(t-1)}}{2a}, \tag{5.4.32}$$

whereas for $t < 1$,

$$f(t) = \frac{1}{2\pi}(-2\pi i)\mathrm{Res}\left[\frac{e^{i(t-1)z}}{z^2 + a^2}; -ai\right] = \frac{e^{a(t-1)}}{2a}. \tag{5.4.33}$$

The minus sign in front of the $2\pi i$ arises from the clockwise direction or negative sense of the contour. We can write the inverse as the single expression

$$f(t) = \frac{e^{-a|t-1|}}{2a}. \tag{5.4.34}$$

• Example 5.4.5

Let us evaluate the integral

$$\int_{-\infty}^{\infty} \frac{\cos(kx)}{x^2 + a^2}\, dx, \tag{5.4.35}$$

where $a, k > 0$.

We begin by noting that

$$\int_{-\infty}^{\infty} \frac{\cos(kx)}{x^2 + a^2}\, dx = \mathrm{Re}\left(\int_{-\infty}^{\infty} \frac{e^{ikx}}{x^2 + a^2}\, dx\right) = \mathrm{Re}\left(\int_{C_1} \frac{e^{ikz}}{z^2 + a^2}\, dz\right), \tag{5.4.36}$$

where C_1 denotes a line integral along the real axis from $-\infty$ to ∞. A quick check shows that the integrand of the right side of (5.4.36) satisfies Jordan's lemma. Therefore,

$$\int_{-\infty}^{\infty} \frac{e^{ikx}}{x^2 + a^2}\, dx = \oint_C \frac{e^{ikz}}{z^2 + a^2}\, dz = 2\pi i\, \mathrm{Res}\left(\frac{e^{ikz}}{z^2 + a^2}; ai\right) \tag{5.4.37}$$

$$= 2\pi i\, \lim_{z \to ai} \frac{(z - ai)e^{ikz}}{z^2 + a^2} = \frac{\pi}{a}e^{-ka}, \tag{5.4.38}$$

where C denotes the closed infinite semicircle in the upper half-plane. Taking the real and imaginary parts of (5.4.38),

$$\int_{-\infty}^{\infty} \frac{\cos(kx)}{x^2 + a^2}\, dx = \frac{\pi}{a}e^{-ka} \tag{5.4.39}$$

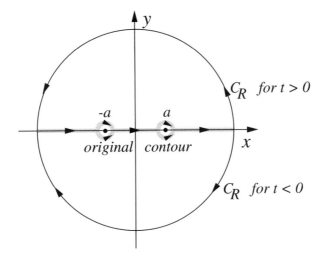

Figure 5.4.2: Contour used in Example 5.4.6.

and

$$\int_{-\infty}^{\infty} \frac{\sin(kx)}{x^2 + a^2}\, dx = 0. \tag{5.4.40}$$

● **Example 5.4.6**

Let us now invert the Fourier transform $F(\omega) = 2a/(a^2 - \omega^2)$, where a is real. The interesting aspect of this problem is the presence of singularities at $\omega = \pm a$ which lie *along* the contour of integration. How do we use contour integration to compute

$$f(t) = \frac{a}{\pi} \int_{-\infty}^{\infty} \frac{e^{it\omega}}{a^2 - \omega^2}\, d\omega? \tag{5.4.41}$$

The answer to this question involves the concept of Cauchy principal value integrals which allows us to extend the conventional definition of integrals to include integrands that become infinite at a finite number of points. See §1.10. Thus, by treating (5.4.41) as a Cauchy principal value integral, we again convert (5.4.41) into a closed contour integration by closing the line integration along the real axis as shown in Figure 5.4.2. The semicircles at infinity vanish by Jordan's lemma and

$$f(t) = \frac{a}{\pi} \oint_C \frac{e^{itz}}{a^2 - z^2}\, dz. \tag{5.4.42}$$

For $t > 0$,

$$f(t) = -\frac{2\pi i a}{\pi} \frac{1}{2} \mathrm{Res}\left[\frac{e^{itz}}{z^2 - a^2}; -a\right] - \frac{2\pi i a}{\pi} \frac{1}{2} \mathrm{Res}\left[\frac{e^{itz}}{z^2 - a^2}; a\right]. \tag{5.4.43}$$

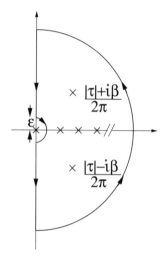

Figure 5.4.3: Contour used in Example 5.4.7.

We have the factor $\frac{1}{2}$ because we are only passing over the "top" of the singularity at $z = a$ and $z = -a$. Computing the residues and simiplifying the results, we obtain

$$f(t) = \sin(at). \tag{5.4.44}$$

Similarly, when $t < 0$,

$$f(t) = \frac{2\pi i a}{\pi} \frac{1}{2} \operatorname{Res}\left[\frac{e^{itz}}{z^2 - a^2}; -a\right] + \frac{2\pi i a}{\pi} \frac{1}{2} \operatorname{Res}\left[\frac{e^{itz}}{z^2 - a^2}; a\right] = -\sin(at).$$
$$\tag{5.4.45}$$

These results can be collapsed down to the single expression

$$f(t) = \operatorname{sgn}(t) \sin(at). \tag{5.4.46}$$

● **Example 5.4.7**

So far, we used only the first two points of Jordan's lemma. In this example[11] we illustrate how the remaining two points may be applied.

Consider the contour integral

$$\oint_C \cot(\pi z)\left[\frac{e^{-cz}}{(\tau + 2\pi z)^2 + \beta^2} + \frac{e^{-cz}}{(\tau - 2\pi z)^2 + \beta^2}\right] dz,$$

where $c > 0$ and β, τ are real. Let us evaluate this contour integral where the contour is shown in Figure 5.4.3.

─────────────────

[11] Reprinted from *Int. J. Heat Mass Transfer*, **15**, T. C. Hsieh and R. Greif, Theoretical determination of the absorption coefficient and the total band absorptance including a specific application to carbon monoxide, 1477–1487, ©1972, with kind permission from Elsevier Science Ltd., The Boulevard, Langford Lane, Kidlington OX5 1GB, UK.

From the residue theorem,

$$\oint_C \cot(\pi z)\left[\frac{e^{-cz}}{(\tau + 2\pi z)^2 + \beta^2} + \frac{e^{-cz}}{(\tau - 2\pi z)^2 + \beta^2}\right] dz$$

$$= 2\pi i \sum_{n=1}^{\infty} \text{Res}\left\{\cot(\pi z)\left[\frac{e^{-cz}}{(\tau + 2\pi z)^2 + \beta^2} + \frac{e^{-cz}}{(\tau - 2\pi z)^2 + \beta^2}\right]; n\right\}$$

$$+ 2\pi i \, \text{Res}\left\{\cot(\pi z)\left[\frac{e^{-cz}}{(\tau + 2\pi z)^2 + \beta^2} + \frac{e^{-cz}}{(\tau - 2\pi z)^2 + \beta^2}\right]; \frac{|\tau| + \beta i}{2\pi}\right\}$$

$$+ 2\pi i \, \text{Res}\left\{\cot(\pi z)\left[\frac{e^{-cz}}{(\tau + 2\pi z)^2 + \beta^2} + \frac{e^{-cz}}{(\tau - 2\pi z)^2 + \beta^2}\right]; \frac{|\tau| - \beta i}{2\pi}\right\}.$$

$$(5.4.47)$$

Now

$$\text{Res}\left\{\cot(\pi z)\left[\frac{e^{-cz}}{(\tau + 2\pi z)^2 + \beta^2} + \frac{e^{-cz}}{(\tau - 2\pi z)^2 + \beta^2}\right]; n\right\}$$

$$= \lim_{z \to n} \frac{(z - n)\cos(\pi z)}{\sin(\pi z)} \lim_{z \to n}\left[\frac{e^{-cz}}{(\tau + 2\pi z)^2 + \beta^2} + \frac{e^{-cz}}{(\tau - 2\pi z)^2 + \beta^2}\right]$$

$$(5.4.48)$$

$$= \frac{1}{\pi}\left[\frac{e^{-nc}}{(\tau + 2n\pi)^2 + \beta^2} + \frac{e^{-nc}}{(\tau - 2n\pi)^2 + \beta^2}\right], \qquad (5.4.49)$$

$$\text{Res}\left\{\cot(\pi z)\left[\frac{e^{-cz}}{(\tau + 2\pi z)^2 + \beta^2} + \frac{e^{-cz}}{(\tau - 2\pi z)^2 + \beta^2}\right]; \frac{|\tau| + \beta i}{2\pi}\right\}$$

$$= \lim_{z \to (|\tau| + \beta i)/2\pi} \frac{\cot(\pi z)}{4\pi^2}$$

$$\times \left[\frac{(z - |\tau| - \beta i)e^{-cz}}{(z + \tau/2\pi)^2 + \beta^2/4\pi^2} + \frac{(z - |\tau| - \beta i)e^{-cz}}{(z - \tau/2\pi)^2 + \beta^2/4\pi^2}\right] \qquad (5.4.50)$$

$$= \frac{\cot(|\tau|/2 + \beta i/2)\exp(-c|\tau|/2\pi)[\cos(c\beta/2\pi) - i\sin(c\beta/2\pi)]}{4\pi\beta i},$$

$$(5.4.51)$$

and

$$\text{Res}\left\{\cot(\pi z)\left[\frac{e^{-cz}}{(\tau + 2\pi z)^2 + \beta^2} + \frac{e^{-cz}}{(\tau - 2\pi z)^2 + \beta^2}\right]; \frac{|\tau| - \beta i}{2\pi}\right\}$$

$$= \lim_{z \to (|\tau| - \beta i)/2\pi} \frac{\cot(\pi z)}{4\pi^2}$$

$$\times \left[\frac{(z - |\tau| + \beta i)e^{-cz}}{(z + \tau/2\pi)^2 + \beta^2/4\pi^2} + \frac{(z - |\tau| + \beta i)e^{-cz}}{(z - \tau/2\pi)^2 + \beta^2/4\pi^2}\right] \qquad (5.4.52)$$

$$= \frac{\cot(|\tau|/2 - \beta i/2)\exp(-c|\tau|/2\pi)[\cos(c\beta/2\pi) + i\sin(c\beta/2\pi)]}{-4\pi\beta i}.$$

$$(5.4.53)$$

Therefore,

$$\oint_C \cot(\pi z) \left[\frac{e^{-cz}}{(\tau + 2\pi z)^2 + \beta^2} + \frac{e^{-cz}}{(\tau - 2\pi z)^2 + \beta^2} \right] dz$$

$$= 2i \sum_{n=1}^{\infty} \left[\frac{e^{-nc}}{(\tau + 2n\pi)^2 + \beta^2} + \frac{e^{-nc}}{(\tau - 2n\pi)^2 + \beta^2} \right]$$

$$+ \frac{i}{2\beta} \frac{e^{i|\tau|} + e^{\beta}}{e^{i|\tau|} - e^{\beta}} e^{-c|\tau|/2\pi} [\cos(c\beta/2\pi) - i\sin(c\beta/2\pi)]$$

$$- \frac{i}{2\beta} \frac{e^{i|\tau|} + e^{-\beta}}{e^{i|\tau|} - e^{-\beta}} e^{-c|\tau|/2\pi} [\cos(c\beta/2\pi) + i\sin(c\beta/2\pi)] \qquad (5.4.54)$$

$$= 2i \sum_{n=1}^{\infty} \left[\frac{e^{-nc}}{(\tau + 2n\pi)^2 + \beta^2} + \frac{e^{-nc}}{(\tau - 2n\pi)^2 + \beta^2} \right]$$

$$- \frac{i}{\beta} \frac{\sinh(\beta)\cos(c\beta/2\pi) + \sin(|\tau|)\sin(c\beta/2\pi)}{\cosh(\beta) - \cos(\tau)} e^{-c|\tau|/2\pi}, \qquad (5.4.55)$$

where $\cot(\alpha) = i(e^{2i\alpha} + 1)/(e^{2i\alpha} - 1)$ and we made extensive use of Euler's formula.

Let us now evaluate the contour integral by direct integration. The contribution from the integration along the semicircle at infinity vanishes according to Jordan's lemma. Indeed that is why this particular contour was chosen. Therefore,

$$\oint_C \cot(\pi z) \left[\frac{e^{-cz}}{(\tau + 2\pi z)^2 + \beta^2} + \frac{e^{-cz}}{(\tau - 2\pi z)^2 + \beta^2} \right] dz$$

$$= \int_{i\infty}^{i\epsilon} \cot(\pi z) \left[\frac{e^{-cz}}{(\tau + 2\pi z)^2 + \beta^2} + \frac{e^{-cz}}{(\tau - 2\pi z)^2 + \beta^2} \right] dz$$

$$+ \int_{C_\epsilon} \cot(\pi z) \left[\frac{e^{-cz}}{(\tau + 2\pi z)^2 + \beta^2} + \frac{e^{-cz}}{(\tau - 2\pi z)^2 + \beta^2} \right] dz$$

$$+ \int_{-i\epsilon}^{-i\infty} \cot(\pi z) \left[\frac{e^{-cz}}{(\tau + 2\pi z)^2 + \beta^2} + \frac{e^{-cz}}{(\tau - 2\pi z)^2 + \beta^2} \right] dz. (5.4.56)$$

Now, because $z = iy$,

$$\int_{i\infty}^{i\epsilon} \cot(\pi z) \left[\frac{e^{-cz}}{(\tau + 2\pi z)^2 + \beta^2} + \frac{e^{-cz}}{(\tau - 2\pi z)^2 + \beta^2} \right] dz$$

$$= \int_{\infty}^{\epsilon} \coth(\pi y) \left[\frac{e^{-icy}}{(\tau + 2\pi iy)^2 + \beta^2} + \frac{e^{-icy}}{(\tau - 2\pi iy)^2 + \beta^2} \right] dy \qquad (5.4.57)$$

$$= -2 \int_{\epsilon}^{\infty} \frac{\coth(\pi y)(\tau^2 + \beta^2 - 4\pi^2 y^2)e^{-icy}}{(\tau^2 + \beta^2 - 4\pi^2 y^2)^2 + 16\pi^2 \tau^2 y^2} dy, \qquad (5.4.58)$$

$$\int_{-i\epsilon}^{-i\infty} \cot(\pi z) \left[\frac{e^{-cz}}{(\tau + 2\pi z)^2 + \beta^2} + \frac{e^{-cz}}{(\tau - 2\pi z)^2 + \beta^2} \right] dz$$

$$= \int_{-\epsilon}^{-\infty} \coth(\pi y) \left[\frac{e^{-icy}}{(\tau + 2\pi iy)^2 + \beta^2} + \frac{e^{-icy}}{(\tau - 2\pi iy)^2 + \beta^2} \right] dy \tag{5.4.59}$$

$$= 2 \int_{\epsilon}^{\infty} \frac{\coth(\pi y)(\tau^2 + \beta^2 - 4\pi^2 y^2) e^{icy}}{(\tau^2 + \beta^2 - 4\pi^2 y^2)^2 + 16\pi^2 \tau^2 y^2} dy, \tag{5.4.60}$$

and

$$\int_{C_\epsilon} \cot(\pi z) \left[\frac{e^{-cz}}{(\tau + 2\pi z)^2 + \beta^2} + \frac{e^{-cz}}{(\tau - 2\pi z)^2 + \beta^2} \right] dz$$

$$= \int_{\pi/2}^{-\pi/2} \left[\frac{1}{\pi \epsilon e^{\theta i}} - \frac{\pi \epsilon e^{\theta i}}{3} - \cdots \right] \epsilon i e^{\theta i} \, d\theta$$

$$\times \left[\frac{\exp(-c\epsilon e^{\theta i})}{(\tau + 2\pi \epsilon e^{\theta i})^2 + \beta^2} + \frac{\exp(-c\epsilon e^{\theta i})}{(\tau - 2\pi \epsilon e^{\theta i})^2 + \beta^2} \right]. \tag{5.4.61}$$

In the limit of $\epsilon \to 0$,

$$\oint_C \cot(\pi z) \left[\frac{e^{-cz}}{(\tau + 2\pi z)^2 + \beta^2} + \frac{e^{-cz}}{(\tau - 2\pi z)^2 + \beta^2} \right] dz$$

$$= 4i \int_0^{\infty} \frac{\coth(\pi y)(\tau^2 + \beta^2 - 4\pi^2 y^2) \sin(cy)}{(\tau^2 + \beta^2 - 4\pi^2 y^2)^2 + 16\pi^2 \tau^2 y^2} dy - \frac{2i}{\tau^2 + \beta^2} \tag{5.4.62}$$

$$= 2i \sum_{n=1}^{\infty} \left[\frac{e^{-nc}}{(\tau + 2n\pi)^2 + \beta^2} + \frac{e^{-nc}}{(\tau - 2n\pi)^2 + \beta^2} \right]$$

$$- \frac{i}{\beta} \frac{\sinh(\beta) \cos(c\beta/2\pi) + \sin(|\tau|) \sin(c\beta/2\pi)}{\cosh(\beta) - \cos(\tau)} e^{-c|\tau|/2\pi}, \tag{5.4.63}$$

or

$$4 \int_0^{\infty} \frac{\coth(\pi y)(\tau^2 + \beta^2 - 4\pi^2 y^2) \sin(cy)}{(\tau^2 + \beta^2 - 4\pi^2 y^2)^2 + 16\pi^2 \tau^2 y^2} dy$$

$$= 2 \sum_{n=1}^{\infty} \left[\frac{e^{-nc}}{(\tau + 2n\pi)^2 + \beta^2} + \frac{e^{-nc}}{(\tau - 2n\pi)^2 + \beta^2} \right]$$

$$- \frac{1}{\beta} \frac{\sinh(\beta) \cos(c\beta/2\pi) + \sin(|\tau|) \sin(c\beta/2\pi)}{\cosh(\beta) - \cos(\tau)} e^{-c|\tau|/2\pi} + \frac{2}{\tau^2 + \beta^2}. \tag{5.4.64}$$

If we let $y = x/2\pi$,

$$\frac{\beta}{\pi} \int_0^{\infty} \frac{\coth(x/2)(\tau^2 + \beta^2 - x^2) \sin(cx/2\pi)}{(\tau^2 + \beta^2 - x^2)^2 + 4\tau^2 x^2} dx$$

$$= 2\beta \sum_{n=1}^{\infty} \left[\frac{e^{-nc}}{(\tau + 2n\pi)^2 + \beta^2} + \frac{e^{-nc}}{(\tau - 2n\pi)^2 + \beta^2} \right]$$

$$- \frac{\sinh(\beta) \cos(c\beta/2\pi) + \sin(|\tau|) \sin(c\beta/2\pi)}{\cosh(\beta) - \cos(\tau)} e^{-c|\tau|/2\pi} + \frac{2\beta}{\tau^2 + \beta^2}. \tag{5.4.65}$$

• **Example 5.4.8**

An additional benefit of understanding inversion by the residue method is the ability to qualitatively anticipate the inverse by knowing the location of the poles of $F(\omega)$. This intuition is important because many engineering analyses discuss stability and performance entirely in terms of the properties of the system's Fourier transform. In Figure 5.4.4 we graphed the location of the poles of $F(\omega)$ and the corresponding $f(t)$. The student should go through the mental exercise of connecting the two pictures.

Problems

1. Use direct integration to find the inverse of the Fourier transform

$$F(\omega) = \frac{i\omega\pi}{2} e^{-|\omega|}.$$

Check your answer using MATLAB.

Use partial fractions to invert the following Fourier transforms:

2. $\dfrac{1}{(1 + i\omega)(1 + 2i\omega)}$

3. $\dfrac{1}{(1 + i\omega)(1 - i\omega)}$

4. $\dfrac{i\omega}{(1 + i\omega)(1 + 2i\omega)}$

5. $\dfrac{1}{(1 + i\omega)(1 + 2i\omega)^2}$

Then check your answer using MATLAB.

By taking the appropriate closed contour, find the inverse of the following Fourier transforms by contour integration. The parameter a is real and positive.

6. $\dfrac{1}{\omega^2 + a^2}$

7. $\dfrac{\omega}{\omega^2 + a^2}$

8. $\dfrac{\omega}{(\omega^2 + a^2)^2}$

9. $\dfrac{\omega^2}{(\omega^2 + a^2)^2}$

10. $\dfrac{1}{\omega^2 - 3i\omega - 3}$

11. $\dfrac{1}{(\omega - ia)^{2n+2}}$

12. $\dfrac{\omega^2}{(\omega^2 - 1)^2 + 4a^2\omega^2}$

13. $\dfrac{3}{(2 - \omega i)(1 + \omega i)}$

Then check your answer using MATLAB.

14. Find the inverse of $F(\omega) = \cos(\omega)/(\omega^2 + a^2)$, $a > 0$, by first rewriting the transform as

$$F(\omega) = \frac{e^{i\omega}}{2(\omega^2 + a^2)} + \frac{e^{-i\omega}}{2(\omega^2 + a^2)}$$

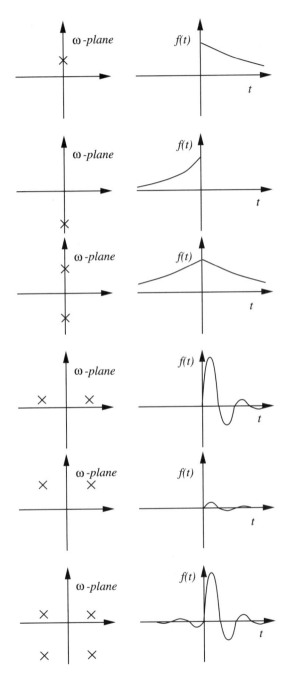

Figure 5.4.4: The correspondence between the location of the simple poles of the Fourier transform $F(\omega)$ and the behavior of $f(t)$.

and then using the residue theorem on each term.

15. Find[12] the inverse Fourier transform for

$$
F_\pm(\omega) = \frac{e^{\pm i\omega}}{(\omega - ai)\,(R^2 e^{\omega i} - e^{-\omega i})} = \frac{e^{\pm i\omega - i\omega}}{(\omega - ai)\,(R^2 - e^{-2\omega i})},
$$

where $a > 0$ and $R > 1$. Hint: You must find separate inverses for different time intervals. For example, in the case of $F_+(\omega)$, you must examine the special cases of $t < 0$ and $t > 0$.

16. As we shall show shortly, Fourier transforms can be used to solve differential equations. During the solution of the heat equation, Taitel *et al.*[13] inverted the Fourier transform

$$
F(\omega) = \frac{\cosh(y\sqrt{\omega^2 + 1}\,)}{\sqrt{\omega^2 + 1}\,\sinh(p\sqrt{\omega^2 + 1}/2)},
$$

where y and p are real. Show that they should have found

$$
f(t) = \frac{e^{-|t|}}{p} + \frac{2}{p}\sum_{n=1}^{\infty}\frac{(-1)^n}{\sqrt{1 + 4n^2\pi^2/p^2}}\cos\left(\frac{2n\pi y}{p}\right)e^{-\sqrt{1 + 4n^2\pi^2/p^2}\,|t|}.
$$

In this case, our time variable t was their spatial variable $x - \xi$.

17. Find the inverse of the Fourier transform

$$
F(\omega) = \left[\cos\left\{\frac{\omega L}{\beta[1 + i\gamma\,\text{sgn}(\omega)]}\right\}\right]^{-1},
$$

where L, β, and γ are real and positive and $\text{sgn}(z) = 1$ if $\text{Re}(z) > 0$ and -1 if $\text{Re}(z) < 0$.

Use the residue theorem to verify the following integrals:

18. $\displaystyle\int_{-\infty}^{\infty}\frac{\sin(x)}{x^2 + 4x + 5}\,dx = -\frac{\pi}{e}\sin(2)$ 19. $\displaystyle\int_{0}^{\infty}\frac{\cos(x)}{(x^2 + 1)^2}\,dx = \frac{\pi}{2e}$

---

[12] Taken from Scharstein, R. W., 1992: Transient electromagnetic plane wave reflection from a dielectric slab. *IEEE Trans. Educ.*, **35**, 170–175.

[13] Reprinted from *Int. J. Heat Mass Transfer*, **16**, Y. Taitel, M. Bentwich, and A. Tamir, Effects of upstream and downstream boundary conditions on heat (mass) transfer with axial diffusion, 359–369, ©1973, with kind permission from Elsevier Science Ltd., The Boulevard, Langford Lane, Kidlington OX5 1GB, UK.

20.
$$\int_{-\infty}^{\infty} \frac{x \sin(ax)}{x^2 + 4} \, dx = \pi e^{-2a}, \qquad a > 0$$

21.
$$\int_{0}^{\infty} \frac{x^2 \cos(ax)}{(x^2 + b^2)^2} \, dx = \frac{\pi}{4b}(1 - ab)e^{-ab}, \qquad a, b > 0$$

22. The concept of forced convection is normally associated with heat streaming through a duct or past an obstacle. Bentwich[14] showed that a similar transport can exist when convection results from a wave traveling through an essentially stagnant fluid. In the process of computing the amount of heating he proved the following identity:

$$\int_{-\infty}^{\infty} \frac{\cosh(hx) - 1}{x \sinh(hx)} \cos(ax) \, dx = \ln[\coth(|a|\pi/h)], \qquad h > 0.$$

Confirm his result.

5.5 CONVOLUTION

The most important property of Fourier transforms is convolution. We shall use it extensively in the solution of differential equations and the design of filters because it yields in time or space the effect of multiplying two transforms together.

The convolution operation is

$$f(t) * g(t) = \int_{-\infty}^{\infty} f(x)g(t - x) \, dx = \int_{-\infty}^{\infty} f(t - x)g(x) \, dx. \qquad (5.5.1)$$

Then,

$$\mathcal{F}[f(t) * g(t)] = \int_{-\infty}^{\infty} f(x)e^{-i\omega x} \left[\int_{-\infty}^{\infty} g(t - x)e^{-i\omega(t-x)} \, dt \right] dx \qquad (5.5.2)$$

$$= \int_{-\infty}^{\infty} f(x)G(\omega)e^{-i\omega x} \, dx = F(\omega)G(\omega). \qquad (5.5.3)$$

Thus, the Fourier transform of the convolution of two functions equals the product of the Fourier transforms of each of the functions.

---

[14] Reprinted from *Int. J. Heat Mass Transfer*, **9**, M. Bentwich, Convection enforced by surface and tidal waves, 663–670, ©1966, with kind permission from Elsevier Science Ltd., The Boulevard, Langford Lane, Kidlington OX5 1GB, UK.

• **Example 5.5.1**

Let us verify the convolution theorem using the functions $f(t) = H(t + a) - H(t - a)$ and $g(t) = e^{-t}H(t)$, where $a > 0$.

The convolution of $f(t)$ with $g(t)$ is

$$f(t) * g(t) = \int_{-\infty}^{\infty} e^{-(t-x)} H(t - x) \left[H(x + a) - H(x - a) \right] dx \qquad (5.5.4)$$

$$= e^{-t} \int_{-a}^{a} e^x H(t - x) \, dx. \qquad (5.5.5)$$

If $t < -a$, then the integrand of (5.5.5) is always zero and $f(t) * g(t) = 0$. If $t > a$,

$$f(t) * g(t) = e^{-t} \int_{-a}^{a} e^x dx = e^{-(t-a)} - e^{-(t+a)}. \qquad (5.5.6)$$

Finally, for $-a < t < a$,

$$f(t) * g(t) = e^{-t} \int_{-a}^{t} e^x dx = 1 - e^{-(t+a)}. \qquad (5.5.7)$$

In summary,

$$f(t) * g(t) = \begin{cases} 0, & t \leq -a, \\ 1 - e^{-(t+a)}, & -a \leq t \leq a, \\ e^{-(t-a)} - e^{-(t+a)}, & a \leq t. \end{cases} \qquad (5.5.8)$$

An alternative to examining various cases involving the value of t, we could have used MATLAB to evaluate (5.5.5). The MATLAB instructions are as follows:

```
>> syms f t x
>> syms a positive
>> f = 'exp(x-t)*Heaviside(t-x)*(Heaviside(x+a)-Heaviside(x-a))'
>> int(f,x,-inf,inf)
```

This yields

```
ans =
Heaviside(t+a)-Heaviside(t+a)*exp(-a-t)
    -Heaviside(t-a)+Heaviside(t-a)*exp(a-t)
```

The Fourier transform of $f(t) * g(t)$ is

$$\mathcal{F}[f(t) * g(t)] = \int_{-a}^{a} \left[1 - e^{-(t+a)} \right] e^{-i\omega t} dt + \int_{a}^{\infty} \left[e^{-(t-a)} - e^{-(t+a)} \right] e^{-i\omega t} dt \qquad (5.5.9)$$

$$= \frac{2 \sin(\omega a)}{\omega} - \frac{2i \sin(\omega a)}{1 + \omega i} \qquad (5.5.10)$$

$$= \frac{2 \sin(\omega a)}{\omega} \left(\frac{1}{1 + \omega i} \right) = F(\omega)G(\omega) \qquad (5.5.11)$$

and the convolution theorem is true for this special case. The Fourier transform (5.5.11) could also be obtained by substituting our earlier MATLAB result into `fourier` and then using `simplify(ans)`.

● **Example 5.5.2**

Let us consider the convolution of $f(t) = f_+(t)H(t)$ with $g(t) = g_+H(t)$. Note that both of the functions are nonzero only for $t > 0$.

From the definition of convolution,

$$f(t) * g(t) = \int_{-\infty}^{\infty} f_+(t - x)H(t - x)g_+(x)H(x)\, dx \qquad (5.5.12)$$

$$= \int_0^{\infty} f_+(t - x)H(t - x)g_+(x)\, dx. \qquad (5.5.13)$$

For $t < 0$, the integrand is always zero and $f(t) * g(t) = 0$. For $t > 0$,

$$f(t) * g(t) = \int_0^t f_+(t - x)g_+(x)\, dx. \qquad (5.5.14)$$

Therefore, in general,

$$f(t) * g(t) = \left[\int_0^t f_+(t - x)g_+(x)\, dx \right] H(t). \qquad (5.5.15)$$

This is the definition of convolution that we will use for Laplace transforms where all of the functions equal zero for $t < 0$.

The convolution operation also applies to Fourier transforms, in what is commonly known as *frequency convolution*. We now prove that

$$\mathcal{F}[f(t)g(t)] = \frac{F(\omega) * G(\omega)}{2\pi}, \qquad (5.5.16)$$

where

$$F(\omega) * G(\omega) = \int_{-\infty}^{\infty} F(\tau)G(\omega - \tau)\, d\tau, \qquad (5.5.17)$$

where $F(\omega)$ and $G(\omega)$ are the Fourier transforms of $f(t)$ and $g(t)$, respectively.

Proof: Starting with

$$f(t) = \frac{1}{2\pi} \int_{-\infty}^{\infty} F(\tau)e^{i\tau t}\, d\tau, \qquad (5.5.18)$$

we can multiply the inverse of $F(\tau)$ by $g(t)$ so that we obtain

$$f(t)g(t) = \frac{1}{2\pi} \int_{-\infty}^{\infty} F(\tau)g(t)e^{i\tau t}\, d\tau. \qquad (5.5.19)$$

Then, taking the Fourier transform of (5.5.19), we find that

$$\mathcal{F}[f(t)g(t)] = \int_{-\infty}^{\infty} \left[\frac{1}{2\pi} \int_{-\infty}^{\infty} F(\tau)g(t)e^{i\tau t}\,d\tau \right] e^{-i\omega t}\,dt \qquad (5.5.20)$$

$$= \frac{1}{2\pi} \int_{-\infty}^{\infty} F(\tau) \left[\int_{-\infty}^{\infty} g(t)e^{-i(\omega-\tau)t}\,dt \right] d\tau \qquad (5.5.21)$$

$$= \frac{1}{2\pi} \int_{-\infty}^{\infty} F(\tau)G(\omega-\tau)\,d\tau = \frac{F(\omega) * G(\omega)}{2\pi}. \qquad (5.5.22)$$

Thus, the multiplication of two functions in the time domain is equivalent to the convolution of their spectral densities in the frequency domain. □

Problems

1. Show that

$$e^{-t}H(t) * e^{-t}H(t) = te^{-t}H(t).$$

Then verify your result using MATLAB.

2. Show that

$$e^{-t}H(t) * e^{t}H(-t) = \tfrac{1}{2}e^{-|t|}.$$

Then verify your result using MATLAB.

3. Show that

$$e^{-t}H(t) * e^{-2t}H(t) = \left(e^{-t} - e^{-2t}\right)H(t).$$

Then verify your result using MATLAB.

4. Show that

$$e^{t}H(-t) * [H(t) - H(t-2)] = \begin{cases} e^t - e^{t-2}, & t \leq 0, \\ 1 - e^{t-2}, & 0 \leq t \leq 2, \\ 0, & 2 \leq t. \end{cases}$$

Then verify your result using MATLAB.

5. Show that

$$[H(t) - H(t-2)] * [H(t) - H(t-2)] = \begin{cases} 0, & t \leq 0, \\ t, & 0 \leq t \leq 2, \\ 4 - t, & 2 \leq t \leq 4, \\ 0, & 4 \leq t. \end{cases}$$

Then try and verify your result using MATLAB. What do you have to do to make it work?

6. Show that

$$e^{-|t|} * e^{-|t|} = (1 + |t|)e^{-|t|}.$$

7. Prove that the convolution of two Dirac delta functions is a Dirac delta function.

5.6 SOLUTION OF ORDINARY DIFFERENTIAL EQUATIONS BY FOURIER TRANSFORMS

As with Laplace transforms, we may use Fourier transforms to solve ordinary differential equations. However, this method gives only the particular solution and we must find the complementary solution separately.

Consider the differential equation

$$y' + y = \tfrac{1}{2}e^{-|t|}, \qquad -\infty < t < \infty. \tag{5.6.1}$$

Taking the Fourier transform of both sides of (5.6.1),

$$i\omega Y(\omega) + Y(\omega) = \frac{1}{\omega^2 + 1}, \tag{5.6.2}$$

where we used the derivative rule (5.3.19) to obtain the transform of y' and $Y(\omega) = \mathcal{F}[y(t)]$. Therefore,

$$Y(\omega) = \frac{1}{(\omega^2 + 1)(1 + \omega i)}. \tag{5.6.3}$$

Applying the inversion integral to (5.6.3),

$$y(t) = \frac{1}{2\pi} \int_{-\infty}^{\infty} \frac{e^{it\omega}}{(\omega^2 + 1)(1 + \omega i)} \, d\omega. \tag{5.6.4}$$

We evaluate (5.6.4) by contour integration. For $t > 0$ we close the line integral with an infinite semicircle in the upper half of the ω-plane. The integration along this arc equals zero by Jordan's lemma. Within this closed contour we have a second-order pole at $z = i$. Therefore,

$$\text{Res}\left[\frac{e^{itz}}{(z^2 + 1)(1 + zi)}; i\right] = \lim_{z \to i} \frac{d}{dz}\left[(z - i)^2 \frac{e^{itz}}{i(z - i)^2(z + i)}\right] \tag{5.6.5}$$

$$= \frac{te^{-t}}{2i} + \frac{e^{-t}}{4i} \tag{5.6.6}$$

and

$$y(t) = \frac{1}{2\pi}(2\pi i)\left[\frac{te^{-t}}{2i} + \frac{e^{-t}}{4i}\right] = \frac{e^{-t}}{4}(2t + 1). \tag{5.6.7}$$

For $t < 0$, we again close the line integral with an infinite semicircle but this time it is in the lower half of the ω-plane. The contribution from the line integral along the arc vanishes by Jordan's lemma. Within the contour, we have a simple pole at $z = -i$. Therefore,

$$\text{Res}\left[\frac{e^{itz}}{(z^2+1)(1+zi)}; -i\right] = \lim_{z \to -i}(z+i)\frac{e^{itz}}{i(z+i)(z-i)^2} = -\frac{e^t}{4i}, \qquad (5.6.8)$$

and

$$y(t) = \frac{1}{2\pi}(-2\pi i)\left(-\frac{e^t}{4i}\right) = \frac{e^t}{4}. \qquad (5.6.9)$$

The minus sign in front of the $2\pi i$ results from the contour being taken in the clockwise direction or negative sense. Using the step function, we can combine (5.6.7) and (5.6.9) into the single expression

$$y(t) = \tfrac{1}{4}e^{-|t|} + \tfrac{1}{2}te^{-t}H(t). \qquad (5.6.10)$$

Note that we only found the particular or forced solution to (5.6.1). The most general solution therefore requires that we add the complementary solution Ae^{-t}, yielding

$$y(t) = Ae^{-t} + \tfrac{1}{4}e^{-|t|} + \tfrac{1}{2}te^{-t}H(t). \qquad (5.6.11)$$

The arbitrary constant A would be determined by the initial condition which we have not specified.

We could also have solved this problem using MATLAB. The MATLAB script

```
clear
% define symbolic variables
syms omega t Y
% take Fourier transform of left side of differential equation
LHS = fourier(diff(sym('y(t)'))+sym('y(t)'),t,omega);
% take Fourier transform of right side of differential equation
RHS = fourier(1/2*exp(-abs(t)),t,omega);
% set Y for Fourier transform of y
%     and introduce initial conditions
newLHS = subs(LHS,'fourier(y(t),t,omega)',Y);
% solve for Y
Y = solve(newLHS-RHS,Y);
% invert Fourier transform and find y(t)
y = ifourier(Y,omega,t)
```

yields

```
y =
1/4*exp(t)*Heaviside(-t)+1/2*exp(-t)*t*Heaviside(t)
    +1/4*exp(-t)*Heaviside(t)
```

which is equivalent to (5.6.10).

Consider now a more general problem of

$$y' + y = f(t), \qquad -\infty < t < \infty, \qquad (5.6.12)$$

where we assume that $f(t)$ has the Fourier transform $F(\omega)$. Then the Fourier-transformed solution to (5.6.12) is

$$Y(\omega) = \frac{1}{1 + \omega i} F(\omega) = G(\omega) F(\omega) \qquad (5.6.13)$$

or

$$y(t) = g(t) * f(t), \qquad (5.6.14)$$

where $g(t) = \mathcal{F}^{-1}[1/(1 + \omega i)] = e^{-t} H(t)$. Thus, we can obtain our solution in one of two ways. First, we can take the Fourier transform of $f(t)$, multiply this transform by $G(\omega)$, and finally compute the inverse. The second method requires a convolution of $f(t)$ with $g(t)$. Which method is easiest depends upon $f(t)$ and $g(t)$.

The function $g(t)$ can also be viewed as the particular solution of (5.6.12) resulting from the forcing function $\delta(t)$, the Dirac delta function, because $\mathcal{F}[\delta(t)] = 1$. Traditionally this forced solution $g(t)$ is called the *Green's function* and $G(\omega)$ is called the *frequency response* or *steady-state transfer function* of our system. Engineers often extensively study the frequency response in their analysis rather than the Green's function because the frequency response is easier to obtain experimentally and the output from a linear system is just the product of two transforms [see (5.6.13)] rather than an integration.

In summary, we can use Fourier transforms to find particular solutions to differential equations. The complete solution consists of this particular solution plus any homogeneous solution that we need to satisfy the initial conditions. Convolution of the Green's function with the forcing function also gives the particular solution.

• Example 5.6.1: Spectrum of a damped harmonic oscillator

Second-order differential equations are ubiquitous in engineering. In electrical engineering many electrical circuits are governed by second-order, linear ordinary differential equations. In mechanical engineering they arise during the application of Newton's second law. For example, in mechanics the damped oscillations of a mass m attached to a spring with a spring constant k and damped with a velocity dependent resistance are governed by the equation

$$my'' + cy' + ky = f(t), \qquad (5.6.15)$$

where $y(t)$ denotes the displacement of the oscillator from its equilibrium position, c denotes the damping coefficient, and $f(t)$ denotes the forcing.

Assuming that both $f(t)$ and $y(t)$ have Fourier transforms, let us analyze this system by finding its frequency response. We begin by solving for the Green's function $g(t)$ which is given by

$$mg'' + cg' + kg = \delta(t), \qquad (5.6.16)$$

because the Green's function is the response of a system to a delta function forcing. Taking the Fourier transform of both sides of (5.6.16), the frequency response is

$$G(\omega) = \frac{1}{k + ic\omega - m\omega^2} = \frac{1/m}{\omega_0^2 + ic\omega/m - \omega^2}, \qquad (5.6.17)$$

where $\omega_0^2 = k/m$ is the natural frequency of the system. The most useful quantity to plot is the frequency response or

$$|G(\omega)| = \frac{\omega_0^2}{k\sqrt{(\omega^2 - \omega_0^2)^2 + \omega^2\omega_0^2(c^2/km)}} \qquad (5.6.18)$$

$$= \frac{1}{k\sqrt{[(\omega/\omega_0)^2 - 1]^2 + (c^2/km)(\omega/\omega_0)^2}}. \qquad (5.6.19)$$

In Figure 5.6.1 we plotted the frequency response as a function of $c^2/(km)$. Note that as the damping becomes larger, the sharp peak at $\omega = \omega_0$ essentially vanishes. As $c^2/(km) \to 0$, we obtain a very finely tuned response curve.

Let us now find the Green's function. From the definition of the inverse Fourier transform,

$$mg(t) = -\frac{1}{2\pi}\int_{-\infty}^{\infty}\frac{e^{i\omega t}}{\omega^2 - ic\omega/m - \omega_0^2}\,d\omega = -\frac{1}{2\pi}\int_{-\infty}^{\infty}\frac{e^{i\omega t}}{(\omega - \omega_1)(\omega - \omega_2)}\,d\omega, \qquad (5.6.20)$$

where

$$\omega_{1,2} = \pm\sqrt{\omega_0^2 - \gamma^2} + \gamma i, \qquad (5.6.21)$$

and $\gamma = c/2m > 0$. We can evaluate (5.6.20) by residues. Clearly the poles always lie in the upper half of the ω-plane. Thus, if $t < 0$ in (5.6.20) we can close the line integration along the real axis with a semicircle of infinite radius in the lower half of the ω-plane by Jordan's lemma. Because the integrand is analytic within the closed contour, $g(t) = 0$ for $t < 0$. This is simply the causality condition,[15] the impulse forcing being the cause of the excitation. Clearly, causality is closely connected with the analyticity of the frequency response in the lower half of the ω-plane.

---

[15] The principle stating that an event cannot precede its cause.

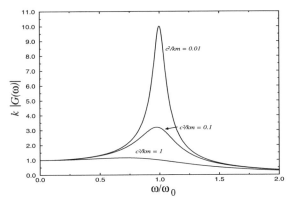

Figure 5.6.1: The variation of the frequency response for a damped harmonic oscillator as a function of driving frequency ω. See the text for the definition of the parameters.

If $t > 0$, we close the line integration along the real axis with a semicircle of infinite radius in the upper half of the ω-plane and obtain

$$mg(t) = 2\pi i \left(-\frac{1}{2\pi}\right) \left\{ \text{Res}\left[\frac{e^{izt}}{(z - \omega_1)(z - \omega_2)}; \omega_1\right] \right.$$

$$\left. + \text{Res}\left[\frac{e^{izt}}{(z - \omega_1)(z - \omega_2)}; \omega_2\right]\right\} \qquad (5.6.22)$$

$$= \frac{-i}{\omega_1 - \omega_2}\left(e^{i\omega_1 t} - e^{i\omega_2 t}\right) = \frac{e^{-\gamma t}\sin\left(t\sqrt{\omega_0^2 - \gamma^2}\right)}{\sqrt{\omega_0^2 - \gamma^2}}H(t). \quad (5.6.23)$$

Let us now examine the damped harmonic oscillator by describing the migration of the poles $\omega_{1,2}$ in the complex ω-plane as γ increases from 0 to ∞. See Figure 5.6.2. For $\gamma \ll \omega_0$ (weak damping), the poles $\omega_{1,2}$ are very near to the real axis, above the points $\pm\omega_0$, respectively. This corresponds to the narrow resonance band discussed earlier and we have an underdamped harmonic oscillator. As γ increases from 0 to ω_0, the poles approach the positive imaginary axis, moving along a semicircle of radius ω_0 centered at the origin. They coalesce at the point $i\omega_0$ for $\gamma = \omega_0$, yielding repeated roots, and we have a critically damped oscillator. For $\gamma > \omega_0$, the poles move in opposite directions along the positive imaginary axis; one of them approaches the origin, while the other tends to $i\infty$ as $\gamma \to \infty$. The solution then has two purely decaying, overdamped solutions.

During the early 1950s, a similar diagram was invented by Evans[16] where the movement of closed-loop poles is plotted for all values of a system parameter, usually the gain. This *root-locus method* is very popular in system control

---

[16] Evans, W. R., 1948: Graphical analysis of control systems. *Trans. AIEE*, **67**, 547–551; Evans, W. R., 1954: *Control-System Dynamics*. McGraw-Hill, 282 pp.

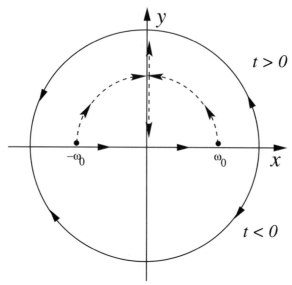

Figure 5.6.2: The migration of the poles of the frequency response of a damped harmonic oscillator as a function of γ.

theory for two reasons. First, the investigator can easily determine the contribution of a particular closed-loop pole to the transient response. Second, he can determine the manner in which open-loop poles or zeros should be introduced or their location modified so that he will achieve a desired performance characteristic for his system.

• **Example 5.6.2: Low frequency filter**

Consider the ordinary differential equation

$$Ry' + \frac{y}{C} = f(t), \tag{5.6.24}$$

where R and C are real, positive constants. If $y(t)$ denotes current, then (5.6.24) would be the equation that gives the voltage across a capacitor in a RC circuit. Let us find the frequency response and Green's function for this system.

We begin by writing (5.6.24) as

$$Rg' + \frac{g}{C} = \delta(t), \tag{5.6.25}$$

where $g(t)$ denotes the Green's function. If the Fourier transform of $g(t)$ is $G(\omega)$, the frequency response $G(\omega)$ is given by

$$i\omega RG(\omega) + \frac{G(\omega)}{C} = 1, \tag{5.6.26}$$

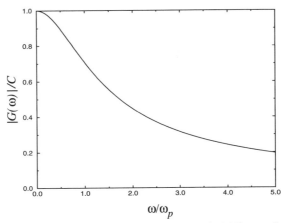

Figure 5.6.3: The variation of the frequency response (5.6.28) as a function of driving frequency ω. See the text for the definition of the parameters.

or

$$G(\omega) = \frac{1}{i\omega R + 1/C} = \frac{C}{1 + i\omega RC}, \tag{5.6.27}$$

and

$$|G(\omega)| = \frac{C}{\sqrt{1 + \omega^2 R^2 C^2}} = \frac{C}{\sqrt{1 + \omega^2/\omega_p^2}}, \tag{5.6.28}$$

where $\omega_p = 1/(RC)$ is an intrinsic constant of the system. In Figure 5.6.3 we plotted $|G(\omega)|$ as a function of ω. From this figure, we see that the response is largest for small ω and decreases as ω increases.

This is an example of a *low frequency filter* because relatively more signal passes through at lower frequencies than at higher frequencies. To understand this, let us drive the system with a forcing function that has the Fourier transform $F(\omega)$. The response of the system will be $G(\omega)F(\omega)$. Thus, that portion of the forcing function's spectrum at the lower frequencies is relatively unaffected because $|G(\omega)|$ is near unity. However, at higher frequencies where $|G(\omega)|$ is smaller, the magnitude of the output is greatly reduced.

- **Example 5.6.3**

During his study of tumor growth, Adam[17] found the particular solution to an ordinary differential equation which, in its simplest form, is

$$y'' - \alpha^2 y = \begin{cases} |x|/L - 1, & |x| < L, \\ 0, & |x| > L, \end{cases} \tag{5.6.29}$$

---

[17] Reprinted from *Math. Biosci.*, **81**, J. A. Adam, A simplified mathematical model of tumor growth, 229–244, ©1986, with permission from Elsevier Science.

by the method of Green's functions. Let us retrace his steps and see how he did it.

The first step is finding the Green's function. We do this by solving

$$g'' - \alpha^2 g = \delta(x), \tag{5.6.30}$$

subject to the boundary conditions $\lim_{|x| \to \infty} g(x) \to 0$. Taking the Fourier transform of (5.6.30), we obtain

$$G(\omega) = -\frac{1}{\omega^2 + \alpha^2}. \tag{5.6.31}$$

The function $G(\omega)$ is the frequency response for our problem. Straightforward inversion yields the Green's function

$$g(x) = -\frac{e^{-\alpha|x|}}{2\alpha}. \tag{5.6.32}$$

Therefore, by the convolution integral (5.6.14),

$$y(x) = \int_{-L}^{L} g(x - \xi)\, (|\xi|/L - 1)\, d\xi = \frac{1}{2\alpha} \int_{-L}^{L} (1 - |\xi|/L)\, e^{-\alpha|x-\xi|}\, d\xi. \tag{5.6.33}$$

To evaluate (5.6.33) we must consider four separate cases: $-\infty < x < -L$, $-L < x < 0$, $0 < x < L$, and $L < x < \infty$. Turning to the $-\infty < x < -L$ case first, we have

$$y(x) = \frac{1}{2\alpha} \int_{-L}^{L} (1 - |\xi|/L)\, e^{\alpha(x-\xi)}\, d\xi \tag{5.6.34}$$

$$= \frac{e^{\alpha x}}{2\alpha} \int_{-L}^{0} (1 + \xi/L)\, e^{-\alpha\xi}\, d\xi + \frac{e^{\alpha x}}{2\alpha} \int_{0}^{L} (1 - \xi/L)\, e^{-\alpha\xi}\, d\xi \tag{5.6.35}$$

$$= \frac{e^{\alpha x}}{2\alpha^3 L} \left(e^{\alpha L} + e^{-\alpha L} - 2 \right). \tag{5.6.36}$$

Similarly, for $x > L$,

$$y(x) = \frac{1}{2\alpha} \int_{-L}^{L} (1 - |\xi|/L)\, e^{-\alpha(x-\xi)}\, d\xi \tag{5.6.37}$$

$$= \frac{e^{-\alpha x}}{2\alpha} \int_{-L}^{0} (1 + \xi/L)\, e^{\alpha\xi}\, d\xi + \frac{e^{-\alpha x}}{2\alpha} \int_{0}^{L} (1 - \xi/L)\, e^{\alpha\xi}\, d\xi \tag{5.6.38}$$

$$= \frac{e^{-\alpha x}}{2\alpha^3 L} \left(e^{\alpha L} + e^{-\alpha L} - 2 \right). \tag{5.6.39}$$

On the other hand, for $-L < x < 0$, we find that

$$y(x) = \frac{1}{2\alpha} \int_{-L}^{x} \left(1 - |\xi|/L\right) e^{-\alpha(x-\xi)} \, d\xi + \frac{1}{2\alpha} \int_{x}^{L} \left(1 - |\xi|/L\right) e^{\alpha(x-\xi)} \, d\xi$$

(5.6.40)

$$= \frac{e^{-\alpha x}}{2\alpha} \int_{-L}^{x} \left(1 + \xi/L\right) e^{\alpha\xi} \, d\xi + \frac{e^{\alpha x}}{2\alpha} \int_{x}^{0} \left(1 + \xi/L\right) e^{-\alpha\xi} \, d\xi$$

$$+ \frac{e^{\alpha x}}{2\alpha} \int_{0}^{L} \left(1 - \xi/L\right) e^{-\alpha\xi} \, d\xi$$

(5.6.41)

$$= \frac{1}{\alpha^3 L} \left[e^{-\alpha L} \cosh(\alpha x) + \alpha(x + L) - e^{\alpha x} \right].$$

(5.6.42)

Finally, for $0 < x < L$, we have that

$$y(x) = \frac{1}{2\alpha} \int_{-L}^{x} \left(1 - |\xi|/L\right) e^{-\alpha(x-\xi)} \, d\xi + \frac{1}{2\alpha} \int_{x}^{L} \left(1 - |\xi|/L\right) e^{\alpha(x-\xi)} \, d\xi$$

(5.6.43)

$$= \frac{e^{-\alpha x}}{2\alpha} \int_{-L}^{0} \left(1 + \xi/L\right) e^{\alpha\xi} \, d\xi + \frac{e^{-\alpha x}}{2\alpha} \int_{0}^{x} \left(1 - \xi/L\right) e^{\alpha\xi} \, d\xi$$

$$+ \frac{e^{\alpha x}}{2\alpha} \int_{x}^{L} \left(1 - \xi/L\right) e^{-\alpha\xi} \, d\xi$$

(5.6.44)

$$= \frac{1}{\alpha^3 L} \left[e^{-\alpha L} \cosh(\alpha x) + \alpha(L - x) - e^{-\alpha x} \right].$$

(5.6.45)

These results can be collapsed down into

$$y(x) = \frac{1}{\alpha^3 L} \left[e^{-\alpha L} \cosh(\alpha x) + \alpha(L - |x|) - e^{-\alpha|x|} \right]$$

(5.6.46)

if $|x| < L$ and

$$y(x) = \frac{e^{-\alpha|x|}}{2\alpha^3 L} \left(e^{\alpha L} + e^{-\alpha L} - 2 \right)$$

(5.6.47)

if $|x| > L$.

Problems

Find the particular solutions for the following differential equations. For Problems 1–3, verify your solution using MATLAB.

1. $y'' + 3y' + 2y = e^{-t} H(t)$ 2. $y'' + 4y' + 4y = \frac{1}{2} e^{-|t|}$

3. $y'' - 4y' + 4y = e^{-t} H(t)$ 4. $y^{iv} - \lambda^4 y = \delta(x)$,

where λ has a positive real part and a negative imaginary part.

Chapter 6
The Laplace Transform

The previous chapter introduced the concept of the Fourier integral. If the function is nonzero only when $t > 0$, a similar transform, the *Laplace transform*,[1] exists. It is particularly useful in solving initial-value problems involving linear, constant coefficient, ordinary and partial differential equations. The present chapter develops the general properties and techniques of Laplace transforms.

6.1 DEFINITION AND ELEMENTARY PROPERTIES

Consider a function $f(t)$ such that $f(t) = 0$ for $t < 0$. Then the *Laplace integral*:

$$\mathcal{L}[f(t)] = F(s) = \int_0^\infty f(t)e^{-st}\,dt \qquad (6.1.1)$$

defines the Laplace transform of $f(t)$, which we shall write $\mathcal{L}[f(t)]$ or $F(s)$. The Laplace transform converts a function of t into a function of the transform variable s.

---

[1] The standard reference for Laplace transforms is Doetsch, G., 1950: *Handbuch der Laplace-Transformation. Band 1. Theorie der Laplace-Transformation.* Birkhäuser Verlag, 581 pp.; Doetsch, G., 1955: *Handbuch der Laplace-Transformation. Band 2. Anwendungen der Laplace-Transformation. 1. Abteilung.* Birkhäuser Verlag, 433 pp.; Doetsch, G., 1956: *Handbuch der Laplace-Transformation. Band 3. Anwendungen der Laplace-Transformation. 2. Abteilung.* Birkhäuser Verlag, 298 pp.

Not all functions have a Laplace transform because the integral (6.1.1) may fail to exist. For example, the function may have infinite discontinuities. For this reason, $f(t) = \tan(t)$ does *not* have a Laplace transform. We can avoid this difficulty by requiring that $f(t)$ be *piece-wise continuous*. That is, we can divide a finite range into a finite number of intervals in such a manner that $f(t)$ is continuous inside each interval and approaches finite values as we approach either end of any interval from the interior.

Another unacceptable function is $f(t) = 1/t$ because the integral (6.1.1) fails to exist. This leads to the requirement that the product $t^n|f(t)|$ is bounded near $t = 0$ for some number $n < 1$.

Finally $|f(t)|$ cannot grow too rapidly or it could overwhelm the e^{-st} term. To express this, we introduce the concept of functions of *exponential order*. By exponential order we mean that there exist some constants, M and k, for which

$$|f(t)| \leq Me^{kt} \qquad (6.1.2)$$

for all $t > 0$. Then, the Laplace transform of $f(t)$ exists if s, or just the real part of s, is greater than k.

In summary, the Laplace transform of $f(t)$ exists, for sufficiently large s, provided $f(t)$ satisfies the following conditions:

- $f(t) = 0$ for $t < 0$,
- $f(t)$ is continuous or piece-wise continuous in every interval,
- $t^n|f(t)| < \infty$ as $t \to 0$ for some number n, where $n < 1$,
- $e^{-s_0 t}|f(t)| < \infty$ as $t \to \infty$, for some number s_0. The quantity s_0 is called the *abscissa of convergence*.

- **Example 6.1.1**

Let us find the Laplace transform of 1, e^{at}, $\sin(at)$, $\cos(at)$, and t^n from the definition of the Laplace transform. From (6.1.1), direct integration yields

$$\mathcal{L}(1) = \int_0^\infty e^{-st}\,dt = -\left.\frac{e^{-st}}{s}\right|_0^\infty = \frac{1}{s}, \qquad s > 0, \qquad (6.1.3)$$

$$\mathcal{L}(e^{at}) = \int_0^\infty e^{at}e^{-st}\,dt = \int_0^\infty e^{-(s-a)t}\,dt \qquad (6.1.4)$$

$$= -\left.\frac{e^{-(s-a)t}}{s-a}\right|_0^\infty = \frac{1}{s-a}, \qquad s > a, \qquad (6.1.5)$$

$$\mathcal{L}[\sin(at)] = \int_0^\infty \sin(at)e^{-st}\,dt = -\left.\frac{e^{-st}}{s^2+a^2}[s\sin(at) + a\cos(at)]\right|_0^\infty \quad (6.1.6)$$

$$= \frac{a}{s^2+a^2}, \qquad s > 0, \qquad (6.1.7)$$

$$\mathcal{L}[\cos(at)] = \int_0^\infty \cos(at)e^{-st}dt = \frac{e^{-st}}{s^2+a^2}[-s\cos(at)+a\sin(at)]\Big|_0^\infty \quad \textbf{(6.1.8)}$$

$$= \frac{s}{s^2+a^2}, \qquad s > 0, \tag{6.1.9}$$

and

$$\mathcal{L}(t^n) = \int_0^\infty t^n e^{-st}dt = n!e^{-st}\sum_{m=0}^n \frac{t^{n-m}}{(n-m)!s^{m+1}}\Big|_0^\infty = \frac{n!}{s^{n+1}}, \quad s > 0,$$
$$\textbf{(6.1.10)}$$

where n is a positive integer.

MATLAB provides the routine `laplace` to compute the Laplace transform for a given function. For example,

```
>> syms a n s t
>> laplace(1,t,s)
ans =
1/s
>> laplace(exp(a*t),t,s)
ans =
1/(s-a)
>> laplace(sin(a*t),t,s)
ans =
a/(s^2+a^2)
>> laplace(cos(a*t),t,s)
ans =
s/(s^2+a^2)
>> laplace(t^5,t,s)
ans =
120/s^6
```

The Laplace transform inherits two important properties from its integral definition. First, the transform of a sum equals the sum of the transforms or

$$\mathcal{L}[c_1 f(t) + c_2 g(t)] = c_1 \mathcal{L}[f(t)] + c_2 \mathcal{L}[g(t)]. \tag{6.1.11}$$

This linearity property holds with complex numbers and functions as well.

● **Example 6.1.2**

Success with Laplace transforms often rests with the ability to manipulate a given transform into a form which you can invert by inspection. Consider the following examples.

Given $F(s) = 4/s^3$, then

$$F(s) = 2 \times \frac{2}{s^3}, \quad \text{and} \quad f(t) = 2t^2 \tag{6.1.12}$$

from (6.1.10).

Given

$$F(s) = \frac{s+2}{s^2+1} = \frac{s}{s^2+1} + \frac{2}{s^2+1}, \tag{6.1.13}$$

then

$$f(t) = \cos(t) + 2\sin(t) \tag{6.1.14}$$

by (6.1.7), (6.1.9), and (6.1.11).

Because

$$F(s) = \frac{1}{s(s-1)} = \frac{1}{s-1} - \frac{1}{s} \tag{6.1.15}$$

by partial fractions, then

$$f(t) = e^t - 1 \tag{6.1.16}$$

by (6.1.3), (6.1.5), and (6.1.11).

MATLAB also provides the routine `ilaplace` to compute the inverse Laplace transform for a given function. For example,

```
>> syms s t
>> ilaplace(4/s^3,s,t)
ans =
2*t^2
>> ilaplace((s+2)/(s^2+1),s,t)
ans =
cos(t)+2*sin(t)
>> ilaplace(1/(s*(s-1)),s,t)
ans =
-1+exp(t)
```

The second important property deals with derivatives. Suppose $f(t)$ is continuous and has a piece-wise continuous derivative $f'(t)$. Then

$$\mathcal{L}[f'(t)] = \int_0^\infty f'(t)e^{-st}\,dt = e^{-st}f(t)\big|_0^\infty + s\int_0^\infty f(t)e^{-st}\,dt \tag{6.1.17}$$

by integration by parts. If $f(t)$ is of exponential order, $e^{-st}f(t)$ tends to zero as $t \to \infty$, for large enough s, so that

$$\mathcal{L}[f'(t)] = sF(s) - f(0). \tag{6.1.18}$$

Similarly, if $f(t)$ and $f'(t)$ are continuous, $f''(t)$ is piece-wise continuous, and all three functions are of exponential order, then

$$\mathcal{L}[f''(t)] = s\mathcal{L}[f'(t)] - f'(0) = s^2F(s) - sf(0) - f'(0). \tag{6.1.19}$$

Table 6.1.1: The Laplace Transforms of Some Commonly Encountered Functions

	$f(t), \ t \geq 0$	$F(s)$
1.	1	$\dfrac{1}{s}$
2.	e^{-at}	$\dfrac{1}{s+a}$
3.	$\frac{1}{a}\left(1 - e^{-at}\right)$	$\dfrac{1}{s(s+a)}$
4.	$\frac{1}{a-b}\left(e^{-bt} - e^{-at}\right)$	$\dfrac{1}{(s+a)(s+b)}$
5.	$\frac{1}{b-a}\left(be^{-bt} - ae^{-at}\right)$	$\dfrac{s}{(s+a)(s+b)}$
6.	$\sin(at)$	$\dfrac{a}{s^2 + a^2}$
7.	$\cos(at)$	$\dfrac{s}{s^2 + a^2}$
8.	$\sinh(at)$	$\dfrac{a}{s^2 - a^2}$
9.	$\cosh(at)$	$\dfrac{s}{s^2 - a^2}$
10.	$t\sin(at)$	$\dfrac{2as}{(s^2 + a^2)^2}$
11.	$1 - \cos(at)$	$\dfrac{a^2}{s(s^2 + a^2)}$
12.	$at - \sin(at)$	$\dfrac{a^3}{s^2(s^2 + a^2)}$
13.	$t\cos(at)$	$\dfrac{s^2 - a^2}{(s^2 + a^2)^2}$
14.	$\sin(at) - at\cos(at)$	$\dfrac{2a^3}{(s^2 + a^2)^2}$
15.	$t\sinh(at)$	$\dfrac{2as}{(s^2 - a^2)^2}$
16.	$t\cosh(at)$	$\dfrac{s^2 + a^2}{(s^2 - a^2)^2}$
17.	$at\cosh(at) - \sinh(at)$	$\dfrac{2a^3}{(s^2 - a^2)^2}$
18.	$e^{-bt}\sin(at)$	$\dfrac{a}{(s+b)^2 + a^2}$

Table 6.1.1 (contd.): The Laplace Transforms of Some Commonly Encountered Functions

	$f(t)$, $t \geq 0$	$F(s)$
19.	$e^{-bt}\cos(at)$	$\dfrac{s+b}{(s+b)^2+a^2}$
20.	$(1+a^2t^2)\sin(at) - at\cos(at)$	$\dfrac{8a^3s^2}{(s^2+a^2)^3}$
21.	$\sin(at)\cosh(at) - \cos(at)\sinh(at)$	$\dfrac{4a^3}{s^4+4a^4}$
22.	$\sin(at)\sinh(at)$	$\dfrac{2a^2s}{s^4+4a^4}$
23.	$\sinh(at) - \sin(at)$	$\dfrac{2a^3}{s^4-a^4}$
24.	$\cosh(at) - \cos(at)$	$\dfrac{2a^2s}{s^4-a^4}$
25.	$\dfrac{a\sin(at)-b\sin(bt)}{a^2-b^2}, a^2 \neq b^2$	$\dfrac{s^2}{(s^2+a^2)(s^2+b^2)}$
26.	$\dfrac{b\sin(at)-a\sin(bt)}{ab(b^2-a^2)}, a^2 \neq b^2$	$\dfrac{1}{(s^2+a^2)(s^2+b^2)}$
27.	$\dfrac{\cos(at)-\cos(bt)}{b^2-a^2}, a^2 \neq b^2$	$\dfrac{s}{(s^2+a^2)(s^2+b^2)}$
28.	$t^n, n \geq 0$	$\dfrac{n!}{s^{n+1}}$
29.	$\dfrac{t^{n-1}e^{-at}}{(n-1)!}, n > 0$	$\dfrac{1}{(s+a)^n}$
30.	$\dfrac{(n-1)-at}{(n-1)!}t^{n-2}e^{-at}, n > 1$	$\dfrac{s}{(s+a)^n}$
31.	$t^n e^{-at}, n \geq 0$	$\dfrac{n!}{(s+a)^{n+1}}$
32.	$\dfrac{2^n t^{n-(1/2)}}{1\cdot3\cdot5\cdots(2n-1)\sqrt{\pi}}, n \geq 1$	$s^{-[n+(1/2)]}$
33.	$J_0(at)$	$\dfrac{1}{\sqrt{s^2+a^2}}$
34.	$I_0(at)$	$\dfrac{1}{\sqrt{s^2-a^2}}$
35.	$\dfrac{1}{\sqrt{a}}\,\text{erf}(\sqrt{at})$	$\dfrac{1}{s\sqrt{s+a}}$

Table 6.1.1 (contd.): The Laplace Transforms of Some Commonly Encountered Functions

	$f(t), \ t \geq 0$	$F(s)$
36.	$\dfrac{1}{\sqrt{\pi t}}e^{-at} + \sqrt{a}\,\text{erf}(\sqrt{at})$	$\dfrac{\sqrt{s+a}}{s}$
37.	$\dfrac{1}{\sqrt{\pi t}} - ae^{a^2 t}\text{erfc}(a\sqrt{t})$	$\dfrac{1}{a+\sqrt{s}}$
38.	$e^{at}\text{erfc}(\sqrt{at})$	$\dfrac{1}{s+\sqrt{as}}$
39.	$\dfrac{1}{2\sqrt{\pi t^3}}\left(e^{bt} - e^{at}\right)$	$\sqrt{s-a} - \sqrt{s-b}$
40.	$\dfrac{1}{\sqrt{\pi t}} + ae^{a^2 t}\text{erf}(a\sqrt{t})$	$\dfrac{\sqrt{s}}{s-a^2}$
41.	$\dfrac{1}{\sqrt{\pi t}}e^{at}(1 + 2at)$	$\dfrac{s}{(s-a)\sqrt{s-a}}$
42.	$\dfrac{1}{a}e^{a^2 t}\text{erf}(a\sqrt{t})$	$\dfrac{1}{(s-a^2)\sqrt{s}}$
43.	$\sqrt{\dfrac{a}{\pi t^3}}\,e^{-a/t}, a > 0$	$e^{-2\sqrt{as}}$
44.	$\dfrac{1}{\sqrt{\pi t}}e^{-a/t}, a \geq 0$	$\dfrac{1}{\sqrt{s}}e^{-2\sqrt{as}}$
45.	$\text{erfc}\left(\sqrt{\dfrac{a}{t}}\right), a \geq 0$	$\dfrac{1}{s}e^{-2\sqrt{as}}$
46.	$2\sqrt{\dfrac{t}{\pi}}\exp\left(-\dfrac{a^2}{4t}\right) - a\,\text{erfc}\left(\dfrac{a}{2\sqrt{t}}\right), \ a \geq 0$	$\dfrac{e^{-a\sqrt{s}}}{s\sqrt{s}}$
47.	$-e^{b^2 t+ab}\text{erfc}\left(b\sqrt{t} + \dfrac{a}{2\sqrt{t}}\right) + \text{erfc}\left(\dfrac{a}{2\sqrt{t}}\right), a \geq 0$	$\dfrac{be^{-a\sqrt{s}}}{s(b+\sqrt{s})}$
48.	$e^{ab}e^{b^2 t}\text{erfc}\left(b\sqrt{t} + \dfrac{a}{2\sqrt{t}}\right), a \geq 0$	$\dfrac{e^{-a\sqrt{s}}}{\sqrt{s}\,(b+\sqrt{s})}$

Notes:

$$\text{Error function: erf}(x) = \frac{2}{\sqrt{\pi}}\int_0^x e^{-y^2}\,dy$$

$$\text{Complementary error function: erfc}(x) = 1 - \text{erf}(x)$$

In general,

$$\mathcal{L}[f^{(n)}(t)] = s^n F(s) - s^{n-1} f(0) - \cdots - s f^{(n-2)}(0) - f^{(n-1)}(0) \quad \textbf{(6.1.20)}$$

on the assumption that $f(t)$ and its first $n-1$ derivatives are continuous, $f^{(n)}(t)$ is piece-wise continuous, and all are of exponential order so that the Laplace transform exists.

The converse of (6.1.20) is also of some importance. If

$$u(t) = \int_0^t f(\tau)\, d\tau, \qquad (6.1.21)$$

then

$$\mathcal{L}[u(t)] = \int_0^\infty e^{-st} \left[\int_0^t f(\tau)\, d\tau \right] dt \qquad (6.1.22)$$

$$= -\frac{e^{-st}}{s} \int_0^t f(\tau)\, d\tau \bigg|_0^\infty + \frac{1}{s} \int_0^\infty f(t) e^{-st}\, dt, \qquad (6.1.23)$$

and

$$\mathcal{L}\left[\int_0^t f(\tau)\, d\tau \right] = \frac{F(s)}{s}, \qquad (6.1.24)$$

where $u(0) = 0$.

Problems

Using the definition of the Laplace transform, find the Laplace transform of the following functions. For Problems 1–4, check your answer using MATLAB.

1. $f(t) = \cosh(at)$

2. $f(t) = \cos^2(at)$

3. $f(t) = (t+1)^2$

4. $f(t) = (t+1)e^{-at}$

5. $f(t) = \begin{cases} e^t, & 0 < t < 2 \\ 0, & 2 < t \end{cases}$

6. $f(t) = \begin{cases} \sin(t), & 0 \le t \le \pi \\ 0, & \pi \le t \end{cases}$

Using your knowledge of the transform for 1, e^{at}, $\sin(at)$, $\cos(at)$, and t^n, find the Laplace transform of

7. $f(t) = 2\sin(t) - \cos(2t) + \cos(3) - t$

8. $f(t) = t - 2 + e^{-5t} - \sin(5t) + \cos(2)$.

Find the inverse of the following transforms. Verify your result using MAT-LAB.

9. $F(s) = 1/(s + 3)$ 10. $F(s) = 1/s^4$

11. $F(s) = 1/(s^2 + 9)$ 12. $F(s) = (2s + 3)/(s^2 + 9)$

13. $F(s) = 2/(s^2 + 1) - 15/s^3 + 2/(s + 1) - 6s/(s^2 + 4)$

14. $F(s) = 3/s + 15/s^3 + (s + 5)/(s^2 + 1) - 6/(s - 2)$.

15. Verify the derivative rule for Laplace transforms using the function $f(t) = \sin(at)$.

16. Show that $\mathcal{L}[f(at)] = F(s/a)/a$, where $F(s) = \mathcal{L}[f(t)]$.

17. Using the trigonometric identity $\sin^2(x) = [1 - \cos(2x)]/2$, find the Laplace transform of $f(t) = \sin^2[\pi t/(2T)]$.

6.2 THE HEAVISIDE STEP AND DIRAC DELTA FUNCTIONS

Change can occur abruptly. We throw a switch and electricity suddenly flows. In this section we introduce two functions, the Heaviside step and Dirac delta, that will give us the ability to construct complicated discontinuous functions to express these changes.

> Heaviside step function

We define the *Heaviside step function* as

$$H(t - a) = \begin{cases} 1, & t > a, \\ 0, & t < a, \end{cases} \tag{6.2.1}$$

where $a \geq 0$. From this definition,

$$\mathcal{L}[H(t - a)] = \int_a^\infty e^{-st}\,dt = \frac{e^{-as}}{s}, \qquad s > 0. \tag{6.2.2}$$

Figure 6.2.1: Largely a self-educated man, Oliver Heaviside (1850–1925) lived the life of a recluse. It was during his studies of the implications of Maxwell's theory of electricity and magnetism that he re-invented Laplace transforms. Initially rejected, it would require the work of Bromwich to justify its use. (Portrait courtesy of the Institution of Electrical Engineers, London.)

Note that this transform is identical to that for $f(t) = 1$ if $a = 0$. This should not surprise us. As pointed out earlier, the function $f(t)$ is zero for all $t < 0$ by definition. Thus, when dealing with Laplace transforms $f(t) = 1$ and $H(t)$ are identical. Generally we will take 1 rather than $H(t)$ as the inverse of $1/s$.

The Heaviside step function is essentially a bookkeeping device that gives us the ability to "switch on" and "switch off" a given function. For example, if we want a function $f(t)$ to become nonzero at time $t = a$, we represent this process by the product $f(t)H(t-a)$. On the other hand, if we only want the function to be "turned on" when $a < t < b$, the desired expression is then $f(t)[H(t-a) - H(t-b)]$. For $t < a$, both step functions in the brackets have the value of zero. For $a < t < b$, the first step function has the value of unity and the second step function has the value of zero, so that we have $f(t)$. For $t > b$, both step functions equal unity so that their difference is zero.

- **Example 6.2.1**

Quite often we need to express the graphical representation of a function by a mathematical equation. We can conveniently do this through the use of

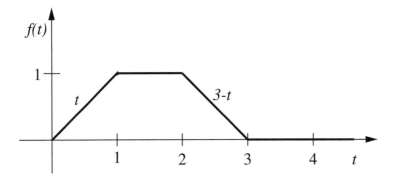

Figure 6.2.2: Graphical representation of (6.2.5).

step functions in a two-step procedure. The following example illustrates this procedure.

Consider Figure 6.2.2. We would like to express this graph in terms of Heaviside step functions. We begin by introducing step functions at each point where there is a kink (discontinuity in the first derivative) or jump in the graph—in the present case at $t = 0$, $t = 1$, $t = 2$, and $t = 3$. These are the points of abrupt change. Thus,

$$f(t) = a_0(t)H(t) + a_1(t)H(t-1) + a_2(t)H(t-2) + a_3(t)H(t-3), \quad (6.2.3)$$

where the coefficients $a_0(t), a_1(t), \ldots$ are yet to be determined. Proceeding from left to right in Figure 6.2.2, the coefficient of each step function equals the mathematical expression that we want after the kink or jump minus the expression before the kink or jump. As each Heaviside turns on, we need to add in the new t behavior and subtract out the old t behavior. Thus, in the present example,

$$f(t) = (t-0)H(t) + (1-t)H(t-1) + [(3-t)-1]H(t-2) + [0-(3-t)]H(t-3) \quad (6.2.4)$$

or

$$f(t) = tH(t) - (t-1)H(t-1) - (t-2)H(t-2) + (t-3)H(t-3). \quad (6.2.5)$$

We can easily find the Laplace transform of (6.2.5) by the "second shifting" theorem introduced in the next section.

● **Example 6.2.2**

Laplace transforms are particularly useful in solving initial-value problems involving linear, constant coefficient, ordinary differential equations where the nonhomogeneous term is discontinuous. As we shall show in the next section, we must first rewrite the nonhomogeneous term using the Heaviside step function before we can use Laplace transforms. For example, given the nonhomogeneous ordinary differential equation:

$$y'' + 3y' + 2y = \begin{cases} t, & 0 < t < 1 \\ 0, & 1 < t, \end{cases} \qquad \textbf{(6.2.6)}$$

we can rewrite the right side of (6.2.6) as

$$y'' + 3y' + 2y = t - tH(t-1) = t - (t-1)H(t-1) - H(t-1). \qquad \textbf{(6.2.7)}$$

In §6.8 we will show how to solve this type of ordinary differential equation using Laplace transforms.

Dirac delta function

The second special function is the *Dirac delta function* or *impulse function*. We define it by

$$\delta(t-a) = \begin{cases} \infty, & t = a, \\ 0, & t \neq a, \end{cases} \qquad \int_0^\infty \delta(t-a)\,dt = 1, \qquad \textbf{(6.2.8)}$$

where $a \geq 0$.

A popular way of visualizing the delta function is as a very narrow rectangular pulse:

$$\delta(t-a) = \lim_{\epsilon \to 0} \begin{cases} 1/\epsilon, & 0 < |t-a| < \epsilon/2, \\ 0, & |t-a| > \epsilon/2, \end{cases} \qquad \textbf{(6.2.9)}$$

where $\epsilon > 0$ is some small number and $a > 0$. See Figure 6.2.3. This pulse has a width ϵ, height $1/\epsilon$, and its center at $t = a$ so that its area is unity. Now as this pulse shrinks in width ($\epsilon \to 0$), its height increases so that it remains centered at $t = a$ and its area equals unity. If we continue this process, always keeping the area unity and the pulse symmetric about $t = a$, eventually we obtain an extremely narrow, very large amplitude pulse at $t = a$. If we proceed to the limit, where the width approaches zero and the height approaches infinity (but still with unit area), we obtain the delta function $\delta(t-a)$.

The delta function was introduced earlier during our study of Fourier transforms. So what is the difference between the delta function introduced then and the delta function now? Simply put, the delta function can now only be used on the interval $[0, \infty)$. Outside of that, we shall use it very much as we did with Fourier transforms.

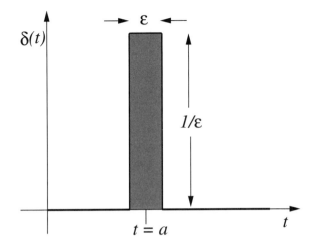

Figure 6.2.3: The Dirac delta function.

Using (6.2.9), the Laplace transform of the delta function is

$$\mathcal{L}[\delta(t-a)] = \int_0^\infty \delta(t-a)e^{-st}\,dt = \lim_{\epsilon \to 0} \frac{1}{\epsilon} \int_{a-\epsilon/2}^{a+\epsilon/2} e^{-st}\,dt \qquad (6.2.10)$$

$$= \lim_{\epsilon \to 0} \frac{1}{\epsilon s}\left(e^{-as+\epsilon s/2} - e^{-as-\epsilon s/2}\right) \qquad (6.2.11)$$

$$= \lim_{\epsilon \to 0} \frac{e^{-as}}{\epsilon s}\left(1 + \frac{\epsilon s}{2} + \frac{\epsilon^2 s^2}{8} + \cdots - 1 + \frac{\epsilon s}{2} - \frac{\epsilon^2 s^2}{8} + \cdots\right) \qquad (6.2.12)$$

$$= e^{-as}. \qquad (6.2.13)$$

In the special case when $a = 0$, $\mathcal{L}[\delta(t)] = 1$, a property that we will use in §6.9. Note that this is exactly the result that we obtained for the Fourier transform of the delta function.

If we integrate the impulse function,

$$\int_0^t \delta(\tau - a)\,d\tau = \begin{cases} 0, & t < a, \\ 1, & t > a, \end{cases} \qquad (6.2.14)$$

according to whether the impulse does or does not come within the range of integration. This integral gives a result that is precisely the definition of the Heaviside step function so that we can rewrite (6.2.14)

$$\int_0^t \delta(\tau - a)\,d\tau = H(t - a). \qquad (6.2.15)$$

Consequently the delta function behaves like the derivative of the step function or

$$\frac{d}{dt}\left[H(t - a)\right] = \delta(t - a). \qquad (6.2.16)$$

Because the conventional derivative does not exist at a point of discontinuity, we can only make sense of (6.2.16) if we extend the definition of the derivative. Here we extended the definition formally, but a richer and deeper understanding arises from the theory of generalized functions.[2]

- **Example 6.2.3**

Let us find the (generalized) derivative of

$$f(t) = 3t^2 \left[H(t) - H(t-1)\right]. \tag{6.2.17}$$

Proceeding formally,

$$
\begin{align}
f'(t) &= 6t\left[H(t) - H(t-1)\right] + 3t^2\left[\delta(t) - \delta(t-1)\right] \tag{6.2.18}\\
&= 6t\left[H(t) - H(t-1)\right] + 0 - 3\delta(t-1) \tag{6.2.19}\\
&= 6t\left[H(t) - H(t-1)\right] - 3\delta(t-1), \tag{6.2.20}
\end{align}
$$

because $f(t)\delta(t - t_0) = f(t_0)\delta(t - t_0)$.

- **Example 6.2.4**

MATLAB also includes the step and Dirac delta functions among its intrinsic functions. There are two types of step functions. In symbolic calculations, the function is Heaviside while stepfunction is used in numerical calculations. For example, the Laplace transform of (6.2.5) is

```
>>syms s,t
>>laplace('t-(t-1)*Heaviside(t-1)-(t-2)*Heaviside(t-2)'...
    '+(t-3)*Heaviside(t-3)',t,s)
ans =
1/s^2-exp(-s)/s^2-exp(-2*s)/s^2+exp(-3*s)/s^2
```

In a similar manner, the symbolic function for the Dirac delta function is Dirac. Therefore, the Laplace transform of $(t - 1)\delta(t - 2)$ is

```
>>syms s,t
>>laplace('(t-1)*Dirac(t-2)',t,s)
ans =
exp(-2*s)
```

---

[2] The generalization of the definition of a function so that it can express in a mathematically correct form such idealized concepts as the density of a material point, a point charge or point dipole, the space charge of a simple or double layer, the intensity of an instantaneous source, etc.

Problems

Sketch the following functions and express them in terms of the Heaviside step functions:

1. $f(t) = \begin{cases} 0, & 0 \le t \le 2 \\ t - 2, & 2 \le t < 3 \\ 0, & 3 < t \end{cases}$

2. $f(t) = \begin{cases} 0, & 0 < t < a \\ 1, & a < t < 2a \\ -1, & 2a < t < 3a \\ 0, & 3a < t \end{cases}$

Rewrite the following nonhomogeneous ordinary differential equations using the Heaviside step functions:

3. $y'' + 3y' + 2y = \begin{cases} 0, & 0 < t < 1 \\ 1, & 1 < t \end{cases}$

4. $y'' + 4y = \begin{cases} 0, & 0 < t < 4 \\ 3, & 4 < t \end{cases}$

5. $y'' + 4y' + 4y = \begin{cases} 0, & 0 < t < 2 \\ t, & 2 < t \end{cases}$

6. $y'' + 3y' + 2y = \begin{cases} 0, & 0 < t < 1 \\ e^t, & 1 < t \end{cases}$

7. $y'' - 3y' + 2y = \begin{cases} 0, & 0 < t < 2 \\ e^{-t}, & 2 < t \end{cases}$

8. $y'' - 3y' + 2y = \begin{cases} 0, & 0 < t < 1 \\ t^2, & 1 < t \end{cases}$

9. $y'' + y = \begin{cases} \sin(t), & 0 \le t \le \pi \\ 0, & \pi \le t \end{cases}$

10. $y'' + 3y' + 2y = \begin{cases} t, & 0 \le t \le a \\ ae^{-(t-a)}, & a \le t \end{cases}$

6.3 SOME USEFUL THEOREMS

Although at first sight there would appear to be a bewildering number of transforms to either memorize or tabulate, there are several useful theorems which can extend the applicability of a given transform.

> First shifting theorem

Consider the transform of the function $e^{-at} f(t)$, where a is any real number. Then, by definition,

$$\mathcal{L}\left[e^{-at} f(t)\right] = \int_0^\infty e^{-st} e^{-at} f(t)\, dt = \int_0^\infty e^{-(s+a)t} f(t)\, dt, \qquad (6.3.1)$$

or

$$\mathcal{L}\left[e^{-at} f(t)\right] = F(s + a). \qquad (6.3.2)$$

That is, if $F(s)$ is the transform of $f(t)$ and a is a constant, then $F(s+a)$ is the transform of $e^{-at}f(t)$.

• **Example 6.3.1**

Let us find the Laplace transform of $f(t) = e^{-at}\sin(bt)$. Because the Laplace transform of $\sin(bt)$ is $b/(s^2 + b^2)$,

$$\mathcal{L}\left[e^{-at}\sin(bt)\right] = \frac{b}{(s+a)^2 + b^2}, \tag{6.3.3}$$

where we simply replaced s by $s + a$ in the transform for $\sin(bt)$.

• **Example 6.3.2**

Let us find the inverse of the Laplace transform

$$F(s) = \frac{s+2}{s^2 + 6s + 1}. \tag{6.3.4}$$

Rearranging terms,

$$F(s) = \frac{s+2}{s^2 + 6s + 1} = \frac{s+2}{(s+3)^2 - 8} \tag{6.3.5}$$

$$= \frac{s+3}{(s+3)^2 - 8} - \frac{1}{2\sqrt{2}}\frac{2\sqrt{2}}{(s+3)^2 - 8}. \tag{6.3.6}$$

Immediately, from the first shifting theorem,

$$f(t) = e^{-3t}\cosh(2\sqrt{2}t) - \frac{e^{-3t}}{2\sqrt{2}}\sinh(2\sqrt{2}t). \tag{6.3.7}$$

Second shifting theorem

The *second shifting theorem* states that if $F(s)$ is the transform of $f(t)$, then $e^{-bs}F(s)$ is the transform of $f(t-b)H(t-b)$, where b is real and positive. To show this, consider the Laplace transform of $f(t-b)H(t-b)$. Then, from the definition,

$$\mathcal{L}[f(t-b)H(t-b)] = \int_0^\infty f(t-b)H(t-b)e^{-st}\,dt \tag{6.3.8}$$

$$= \int_b^\infty f(t-b)e^{-st}\,dt = \int_0^\infty e^{-bs}e^{-sx}f(x)\,dx \tag{6.3.9}$$

$$= e^{-bs}\int_0^\infty e^{-sx}f(x)\,dx, \tag{6.3.10}$$

or

$$\mathcal{L}[f(t-b)H(t-b)] = e^{-bs}F(s), \qquad (\textbf{6.3.11})$$

where we set $x = t - b$. This theorem is of fundamental importance because it allows us to write down the transforms for "delayed" time functions. That is, functions which "turn on" b units after the initial time.

- **Example 6.3.3**

Let us find the inverse of the transform $(1 - e^{-s})/s$. Since

$$\frac{1 - e^{-s}}{s} = \frac{1}{s} - \frac{e^{-s}}{s}, \qquad (\textbf{6.3.12})$$

$$\mathcal{L}^{-1}\left(\frac{1}{s} - \frac{e^{-s}}{s}\right) = \mathcal{L}^{-1}\left(\frac{1}{s}\right) - \mathcal{L}^{-1}\left(\frac{e^{-s}}{s}\right) = H(t) - H(t-1), \quad (\textbf{6.3.13})$$

because $\mathcal{L}^{-1}(1/s) = f(t) = 1$, and $f(t-1) = 1$.

- **Example 6.3.4**

Let us find the Laplace transform of $f(t) = (t^2 - 1)H(t-1)$.
We begin by noting that

$$(t^2 - 1)H(t-1) = [(t-1+1)^2 - 1]H(t-1) \qquad (\textbf{6.3.14})$$
$$= [(t-1)^2 + 2(t-1)]H(t-1) \qquad (\textbf{6.3.15})$$
$$= (t-1)^2 H(t-1) + 2(t-1)H(t-1). \qquad (\textbf{6.3.16})$$

A direct application of the second shifting theorem leads then to

$$\mathcal{L}[(t^2 - 1)H(t-1)] = \frac{2e^{-s}}{s^3} + \frac{2e^{-s}}{s^2}. \qquad (\textbf{6.3.17})$$

- **Example 6.3.5**

In Example 6.2.2 we discussed the use of Laplace transforms in solving ordinary differential equations. One further step along the road consists of finding $Y(s) = \mathcal{L}[y(t)]$. Now that we have the second shifting theorem, let us do this.

Continuing Example 6.2.2 with $y(0) = 0$ and $y'(0) = 1$, let us take the Laplace transform of (6.2.8). Employing the second shifting theorem and (6.1.20), we find that

$$s^2 Y(s) - sy(0) - y'(0) + 3sY(s) - 3y(0) + 2Y(s) = \frac{1}{s^2} - \frac{e^{-s}}{s^2} - \frac{e^{-s}}{s}. \qquad (\textbf{6.3.18})$$

Substituting in the initial conditions and solving for $Y(s)$, we finally obtain

$$Y(s) = \frac{1}{(s+2)(s+1)} + \frac{1}{s^2(s+2)(s+1)} + \frac{e^{-s}}{s^2(s+2)(s+1)} + \frac{e^{-s}}{s(s+2)(s+1)}. \tag{6.3.19}$$

> **Laplace transform of $t^n f(t)$**

In addition to the shifting theorems, there are two other particularly useful theorems that involve the derivative and integral of the transform $F(s)$. For example, if we write

$$F(s) = \mathcal{L}[f(t)] = \int_0^\infty f(t)e^{-st}\,dt \tag{6.3.20}$$

and differentiate with respect to s, then

$$F'(s) = \int_0^\infty -tf(t)e^{-st}\,dt = -\mathcal{L}[tf(t)]. \tag{6.3.21}$$

In general, we have that

$$F^{(n)}(s) = (-1)^n \mathcal{L}[t^n f(t)]. \tag{6.3.22}$$

> **Laplace transform of $f(t)/t$**

Consider the following integration of the Laplace transform $F(s)$:

$$\int_s^\infty F(z)\,dz = \int_s^\infty \left[\int_0^\infty f(t)e^{-zt}\,dt\right]\,dz. \tag{6.3.23}$$

Upon interchanging the order of integration, we find that

$$\int_s^\infty F(z)\,dz = \int_0^\infty f(t)\left[\int_s^\infty e^{-zt}\,dz\right]\,dt \tag{6.3.24}$$

$$= -\int_0^\infty f(t)\left.\frac{e^{-zt}}{t}\right|_s^\infty\,dt = \int_0^\infty \frac{f(t)}{t}e^{-st}\,dt. \tag{6.3.25}$$

Therefore,

$$\int_s^\infty F(z)\,dz = \mathcal{L}\left[\frac{f(t)}{t}\right]. \tag{6.3.26}$$

• **Example 6.3.6**

Let us find the transform of $t\sin(at)$. From (6.3.21),

$$\mathcal{L}[t\sin(at)] = -\frac{d}{ds}\left\{\mathcal{L}[\sin(at)]\right\} = -\frac{d}{ds}\left[\frac{a}{s^2+a^2}\right] = \frac{2as}{(s^2+a^2)^2}. \quad \textbf{(6.3.27)}$$

• **Example 6.3.7**

Let us find the transform of $[1 - \cos(at)]/t$. To solve this problem, we apply (6.3.26) and find that

$$\mathcal{L}\left[\frac{1-\cos(at)}{t}\right] = \int_s^\infty \mathcal{L}[1-\cos(at)]\Big|_{s=z} dz = \int_s^\infty \left(\frac{1}{z} - \frac{z}{z^2+a^2}\right) dz \tag*{(6.3.28)}$$

$$= \ln(z) - \tfrac{1}{2}\ln(z^2+a^2)\Big|_s^\infty = \ln\left(\frac{z}{\sqrt{z^2+a^2}}\right)\Big|_s^\infty \tag*{(6.3.29)}$$

$$= \ln(1) - \ln\left(\frac{s}{\sqrt{s^2+a^2}}\right) = -\ln\left(\frac{s}{\sqrt{s^2+a^2}}\right). \tag*{(6.3.30)}$$

Initial-value theorem

Let $f(t)$ and $f'(t)$ possess Laplace transforms. Then, from the definition of the Laplace transform,

$$\int_0^\infty f'(t)e^{-st}\, dt = sF(s) - f(0). \tag*{(6.3.31)}$$

Because s is a parameter in (6.3.31) and the existence of the integral is implied by the derivative rule, we can let $s \to \infty$ before we integrate. In that case, the left side of (6.3.31) vanishes to zero, which leads to

$$\lim_{s\to\infty} sF(s) = f(0). \tag*{(6.3.32)}$$

This is the *initial-value theorem*.

• **Example 6.3.8**

Let us verify the initial-value theorem using $f(t) = e^{3t}$. Because $F(s) = 1/(s-3)$, $\lim_{s\to\infty} s/(s-3) = 1$. This agrees with $f(0) = 1$.

In the common case when the Laplace transform is ratio to two polynomials, we can use MATLAB to find the initial value. This consists of two steps. First, we construct $sF(s)$ by creating vectors which describe the numerator and denominator of $sF(s)$ and then evaluate the numerator and denominator using very large values of s. For example, in the previous example,

```
>>num = [1 0];
>>den = [1 -3];
>>initialvalue = polyval(num,1e20) / polyval(den,1e20)
initialvalue =
    1
```

| Final-value theorem |

Let $f(t)$ and $f'(t)$ possess Laplace transforms. Then, in the limit of $s \to 0$, (6.3.31) becomes

$$\int_0^\infty f'(t)\,dt = \lim_{t\to\infty}\int_0^t f'(\tau)\,d\tau = \lim_{t\to\infty} f(t) - f(0) = \lim_{s\to 0} sF(s) - f(0).$$
(6.3.33)

Because $f(0)$ is not a function of t or s, the quantity $f(0)$ cancels from the (6.3.33), leaving

$$\lim_{t\to\infty} f(t) = \lim_{s\to 0} sF(s).$$
(6.3.34)

Equation (6.3.34) is the *final-value theorem.* It should be noted that this theorem assumes that $\lim_{t\to\infty} f(t)$ exists. For example, it does not apply to sinusoidal functions. Thus, we must restrict ourselves to Laplace transforms that have singularities in the left half of the s-plane unless they occur at the origin.

● **Example 6.3.9**

Let us verify the final-value theorem using $f(t) = t$. Because $F(s) = 1/s^2$,

$$\lim_{s\to 0} sF(s) = \lim_{s\to 0} 1/s = \infty.$$
(6.3.35)

The limit of $f(t)$ as $t \to \infty$ is also undefined.

Just as we can use MATLAB to find the initial value of a Laplace transform in the case when $F(s)$ is a ratio of two polynomials, we can do the same here for the final value. Again we define vectors num and den that give $sF(s)$ and then evaluate them at $s = 0$. Using the previous example, the MATLAB commands are:

```
>>num = [0 1 0];
```

```
>>den = [1 0 0];
>>finalvalue = polyval(num,0) / polyval(den,0)
Warning: Divide by zero.
finalvalue =
   NaN
```

This agrees with the result from a hand calculation and shows what happens when the denominator has a zero.

• **Example 6.3.10**

Looking ahead, we will shortly need to find the Laplace transform of $y(t)$ which is defined by a differential equation. For example, we will want $Y(s)$ where $y(t)$ is governed by

$$y'' + 2y' + 2y = \cos(t) + \delta(t - \pi/2), \qquad y(0) = y'(0) = 0. \qquad (6.3.36)$$

Applying Laplace transforms to both sides of (6.3.36), we have that

$$\mathcal{L}(y'') + 2\mathcal{L}(y') + 2\mathcal{L}(y) = \mathcal{L}[\cos(t)] + \mathcal{L}[\delta(t - \pi/2)], \qquad (6.3.37)$$

or

$$s^2 Y(s) - sy(0) - y'(0) + 2sY(s) - 2y(0) + 2Y(s) = \frac{s}{s^2 + 1} + e^{-s\pi/2}. \qquad (6.3.38)$$

Substituting for $y(0)$ and $y'(0)$ and solving for $Y(s)$, we find that

$$Y(s) = \frac{s}{(s^2 + 1)(s^2 + 2s + 2)} + \frac{e^{-s\pi/2}}{s^2 + 2s + 2}. \qquad (6.3.39)$$

Presently this is as far as we can go.

How would we use MATLAB to find $Y(s)$? The following MATLAB script shows you how:

```
clear
% define symbolic variables
syms pi s t Y
% take Laplace transform of left side of differential equation
LHS = laplace(diff(diff(sym('y(t)')))+2*diff(sym('y(t)')))...
    +2*sym('y(t)'));
% take Laplace transform of right side of differential equation
RHS = laplace(cos(t)+'Dirac(t-pi/2)',t,s);
% set Y for Laplace transform of y
%      and introduce initial conditions
newLHS = subs(LHS,'laplace(y(t),t,s)','y(0)','D(y)(0)',Y,0,0);
% solve for Y
Y = solve(newLHS-RHS,Y)
```

It yields

```
Y =
(s+exp(-1/2*pi*s)*s^2+exp(-1/2*pi*s))/(s^4+3*s^2+2*s^3+2*s+2)
```

Problems

Find the Laplace transform of the following functions and then check your work using MATLAB.

1. $f(t) = e^{-t} \sin(2t)$ 2. $f(t) = e^{-2t} \cos(2t)$

3. $f(t) = t^2 H(t-1)$ 4. $f(t) = e^{2t} H(t-3)$

5. $f(t) = te^t + \sin(3t)e^t + \cos(5t)e^{2t}$

6. $f(t) = t^4 e^{-2t} + \sin(3t)e^t + \cos(4t)e^{2t}$

7. $f(t) = t^2 e^{-t} + \sin(2t)e^t + \cos(3t)e^{-3t}$

8. $f(t) = t^2 H(t-1) + e^t H(t-2)$

9. $f(t) = (t^2 + 2)H(t-1) + H(t-2)$

10. $f(t) = (t+1)^2 H(t-1) + e^t H(t-2)$

11. $f(t) = \begin{cases} \sin(t), & 0 \le t \le \pi \\ 0, & \pi \le t \end{cases}$

12. $f(t) = \begin{cases} t, & 0 \le t \le 2 \\ 2, & 2 \le t \end{cases}$

13. $f(t) = te^{-3t} \sin(2t)$

Find the inverse of the following Laplace transforms by hand and using MATLAB:

14. $F(s) = \dfrac{1}{(s+2)^4}$ 15. $F(s) = \dfrac{s}{(s+2)^4}$

16. $F(s) = \dfrac{s}{s^2 + 2s + 2}$ 17. $F(s) = \dfrac{s+3}{s^2 + 2s + 2}$

18. $F(s) = \dfrac{s}{(s+1)^3} + \dfrac{s+1}{s^2 + 2s + 2}$ 19. $F(s) = \dfrac{s}{(s+2)^2} + \dfrac{s+2}{s^2 + 2s + 2}$

20. $F(s) = \dfrac{s}{(s+2)^3} + \dfrac{s+4}{s^2 + 4s + 5}$ 21. $F(s) = \dfrac{e^{-3s}}{s-1}$

22. $F(s) = \dfrac{e^{-2s}}{(s+1)^2}$

23. $F(s) = \dfrac{s\,e^{-s}}{s^2 + 2s + 2}$

24. $F(s) = \dfrac{e^{-4s}}{s^2 + 4s + 5}$

25. $F(s) = \dfrac{s\,e^{-s}}{s^2 + 4} + \dfrac{e^{-3s}}{(s-2)^4}$

26. $F(s) = \dfrac{e^{-s}}{s^2 + 4} + \dfrac{(s-1)\,e^{-3s}}{s^4}$

27. $F(s) = \dfrac{(s+1)\,e^{-s}}{s^2 + 4} + \dfrac{e^{-3s}}{s^4}$

28. Find the Laplace transform of $f(t) = te^t[H(t-1) - H(t-2)]$ by using (a) the definition of the Laplace transform, and (b) a joint application of the first and second shifting theorems.

29. Write the function

$$f(t) = \begin{cases} t, & 0 < t < a, \\ 0, & a < t, \end{cases}$$

in terms of Heaviside's step functions. Then find its transform using (a) the definition of the Laplace transform, and (b) the second shifting theorem.

In Problems 30–33, write the function $f(t)$ in terms of the Heaviside step functions and then find its transform using the second shifting theorem. Check your answer using MATLAB.

30. $f(t) = \begin{cases} t/2, & 0 \le t < 2 \\ 0, & 2 < t \end{cases}$

31. $f(t) = \begin{cases} t, & 0 \le t \le 1 \\ 1, & 1 \le t < 2 \\ 0, & 2 < t \end{cases}$

32. $f(t) = \begin{cases} t, & 0 \le t \le 2 \\ 4 - t, & 2 \le t \le 4 \\ 0, & 4 \le t \end{cases}$

33. $f(t) = \begin{cases} 0, & 0 \le t \le 1 \\ t - 1, & 1 \le t \le 2 \\ 1, & 2 \le t < 3 \\ 0, & 3 < t \end{cases}$

Find $Y(s)$ for the following ordinary differential equations and then use MATLAB to check your work.

34. $y'' + 3y' + 2y = H(t-1);$ $y(0) = y'(0) = 0$

35. $y'' + 4y = 3H(t-4);$ $y(0) = 1,\ y'(0) = 0$

36. $y'' + 4y' + 4y = tH(t-2);$ $y(0) = 0,\ y'(0) = 2$

37. $y'' + 3y' + 2y = e^t H(t-1);$ $y(0) = y'(0) = 0$

38. $y'' - 3y' + 2y = e^{-t}H(t-2);$ $y(0) = 2,\ y'(0) = 0$

39. $y'' - 3y' + 2y = t^2 H(t-1);$ $y(0) = 0,\ y'(0) = 5$

40. $y'' + y = \sin(t)[1 - H(t - \pi)];$ $y(0) = y'(0) = 0$

41. $y'' + 3y' + 2y = t + \left[ae^{-(t-a)} - t\right] H(t - a);$ $y(0) = y'(0) = 0.$

For each of the following functions, find its value at $t = 0$. Then check your answer using the initial-value theorem by hand and using MATLAB.

42. $f(t) = t$ 43. $f(t) = \cos(at)$

44. $f(t) = te^{-t}$ 45. $f(t) = e^t \sin(3t)$

For each of the following Laplace transforms, state whether you can or cannot apply the final-value theorem. If you can, find the final value by hand and using MATLAB. Check your result by finding the inverse and finding the limit as $t \to \infty$.

46. $F(s) = \dfrac{1}{s - 1}$ 47. $F(s) = \dfrac{1}{s}$

48. $F(s) = \dfrac{1}{s + 1}$ 49. $F(s) = \dfrac{s}{s^2 + 1}$

50. $F(s) = \dfrac{2}{s(s^2 + 3s + 2)}$ 51. $F(s) = \dfrac{2}{s(s^2 - 3s + 2)}$

6.4 THE LAPLACE TRANSFORM OF A PERIODIC FUNCTION

Periodic functions frequently occur in engineering problems and we shall now show how to calculate their transform. They possess the property that $f(t + T) = f(t)$ for $t > 0$ and equal zero for $t < 0$, where T is the period of the function.

For convenience let us define a function $x(t)$ which equals zero except over the interval $(0, T)$ where it equals $f(t)$:

$$x(t) = \begin{cases} f(t), & 0 < t < T \\ 0, & T < t. \end{cases} \tag{6.4.1}$$

By definition

$$F(s) = \int_0^\infty f(t)e^{-st}\,dt \tag{6.4.2}$$

$$= \int_0^T f(t)e^{-st}\,dt + \int_T^{2T} f(t)e^{-st}\,dt + \cdots + \int_{kT}^{(k+1)T} f(t)e^{-st}\,dt + \cdots. \tag{6.4.3}$$

Now let $z = t - kT$, where $k = 0, 1, 2, \ldots$, in the kth integral and $F(s)$ becomes

$$F(s) = \int_0^T f(z)e^{-sz}\, dz + \int_0^T f(z+T)e^{-s(z+T)}\, dz + \cdots$$

$$+ \int_0^T f(z+kT)e^{-s(z+kT)}\, dz + \cdots. \qquad (6.4.4)$$

However,

$$x(z) = f(z) = f(z+T) = \ldots = f(z+kT) = \ldots, \qquad (6.4.5)$$

because the range of integration in each integral is from 0 to T. Thus, $F(s)$ becomes

$$F(s) = \int_0^T x(z)e^{-sz}\, dz + e^{-sT} \int_0^T x(z)e^{-sz}\, dz + \cdots$$

$$+ e^{-ksT} \int_0^T x(z)e^{-sz}\, dz + \cdots \qquad (6.4.6)$$

or

$$F(s) = \left(1 + e^{-sT} + e^{-2sT} + \cdots + e^{-ksT} + \cdots\right)X(s). \qquad (6.4.7)$$

The first term on the right side of (6.4.7) is a geometric series with common ratio e^{-sT}. If $|e^{-sT}| < 1$, then the series converges and

$$\boxed{F(s) = \frac{X(s)}{1 - e^{-sT}}.} \qquad (6.4.8)$$

• Example 6.4.1

Let us find the Laplace transform of the square wave with period T:

$$f(t) = \begin{cases} h, & 0 < t < T/2, \\ -h, & T/2 < t < T. \end{cases} \qquad (6.4.9)$$

By definition $x(t)$ is

$$x(t) = \begin{cases} h, & 0 < t < T/2, \\ -h, & T/2 < t < T, \\ 0, & T < t. \end{cases} \qquad (6.4.10)$$

Then

$$X(s) = \int_0^\infty x(t)e^{-st}\, dt = \int_0^{T/2} h\,e^{-st}\, dt + \int_{T/2}^T (-h)\,e^{-st}\, dt \qquad (6.4.11)$$

$$= \frac{h}{s}\left(1 - 2e^{-sT/2} + e^{-sT}\right) = \frac{h}{s}\left(1 - e^{-sT/2}\right)^2, \qquad (6.4.12)$$

and

$$F(s) = \frac{h\left(1 - e^{-sT/2}\right)^2}{s\left(1 - e^{-sT}\right)} = \frac{h\left(1 - e^{-sT/2}\right)}{s\left(1 + e^{-sT/2}\right)}.$$ (6.4.13)

If we multiply numerator and denominator by $\exp(sT/4)$ and recall that $\tanh(u) = (e^u - e^{-u})/(e^u + e^{-u})$, we have that

$$F(s) = \frac{h}{s}\tanh\left(\frac{sT}{4}\right).$$ (6.4.14)

• Example 6.4.2

Let us find the Laplace transform of the periodic function

$$f(t) = \begin{cases} \sin(2\pi t/T), & 0 \le t \le T/2, \\ 0, & T/2 \le t \le T. \end{cases}$$ (6.4.15)

By definition $x(t)$ is

$$x(t) = \begin{cases} \sin(2\pi t/T), & 0 \le t \le T/2, \\ 0, & T/2 \le t. \end{cases}$$ (6.4.16)

Then

$$X(s) = \int_0^{T/2} \sin\left(\frac{2\pi t}{T}\right) e^{-st}\, dt = \frac{2\pi T}{s^2 T^2 + 4\pi^2}\left(1 + e^{-sT/2}\right).$$ (6.4.17)

Hence,

$$F(s) = \frac{X(s)}{1 - e^{-sT}} = \frac{2\pi T}{s^2 T^2 + 4\pi^2} \times \frac{1 + e^{-sT/2}}{1 - e^{-sT}}$$ (6.4.18)

$$= \frac{2\pi T}{s^2 T^2 + 4\pi^2} \times \frac{1}{1 - e^{-sT/2}}.$$ (6.4.19)

Problems

Find the Laplace transform for the following periodic functions:

1. $f(t) = \sin(t),\qquad 0 \le t \le \pi,\qquad\qquad f(t) = f(t + \pi)$

2. $f(t) = \begin{cases} \sin(t), & 0 \le t \le \pi, \\ 0, & \pi \le t \le 2\pi, \end{cases}\qquad f(t) = f(t + 2\pi)$

3. $f(t) = \begin{cases} t, & 0 \le t < a, \\ 0, & a < t \le 2a, \end{cases}\qquad f(t) = f(t + 2a)$

4. $f(t) = \begin{cases} 1, & 0 < t < a, \\ 0, & a < t < 2a, \\ -1, & 2a < t < 3a, \\ 0, & 3a < t < 4a, \end{cases} \qquad f(t) = f(t + 4a)$

6.5 INVERSION BY PARTIAL FRACTIONS: HEAVISIDE'S EXPANSION THEOREM

In the previous sections, we devoted our efforts to calculating the Laplace transform of a given function. Obviously we must have a method for going the other way. Given a transform, we must find the corresponding function. This is often a very formidable task. In the next few sections we shall present some general techniques for the inversion of a Laplace transform.

The first technique involves transforms that we can express as the ratio of two polynomials: $F(s) = q(s)/p(s)$. We shall assume that the order of $q(s)$ is *less* than $p(s)$ and we have divided out any common factor between them. In principle we know that $p(s)$ has n zeros, where n is the order of the $p(s)$ polynomial. Some of the zeros may be complex, some of them may be real, and some of them may be duplicates of other zeros. In the case when $p(s)$ has n simple zeros (nonrepeating roots), a simple method exists for inverting the transform.

We want to rewrite $F(s)$ in the form:

$$F(s) = \frac{a_1}{s - s_1} + \frac{a_2}{s - s_2} + \cdots + \frac{a_n}{s - s_n} = \frac{q(s)}{p(s)}, \qquad (6.5.1)$$

where $s_1, s_2, \ldots, s_n$ are the n simple zeros of $p(s)$. We now multiply both sides of (6.5.1) by $s - s_1$ so that

$$\frac{(s - s_1)q(s)}{p(s)} = a_1 + \frac{(s - s_1)a_2}{s - s_2} + \cdots + \frac{(s - s_1)a_n}{s - s_n}. \qquad (6.5.2)$$

If we set $s = s_1$, the right side of (6.5.2) becomes simply a_1. The left side takes the form $0/0$ and there are two cases. If $p(s) = (s - s_1)g(s)$, then $a_1 = q(s_1)/g(s_1)$. If we cannot explicitly factor out $s - s_1$, l'Hôspital's rule gives

$$a_1 = \lim_{s \to s_1} \frac{(s - s_1)q(s)}{p(s)} = \lim_{s \to s_1} \frac{(s - s_1)q'(s) + q(s)}{p'(s)} = \frac{q(s_1)}{p'(s_1)}. \qquad (6.5.3)$$

In a similar manner, we can compute all of the coefficients a_k, where $k = 1, 2, \ldots, n$. Therefore,

$$\mathcal{L}^{-1}[F(s)] = \mathcal{L}^{-1}\left[\frac{q(s)}{p(s)}\right] = \mathcal{L}^{-1}\left(\frac{a_1}{s - s_1} + \frac{a_2}{s - s_2} + \cdots + \frac{a_n}{s - s_n}\right)$$

$$\qquad (6.5.4)$$

$$= a_1 e^{s_1 t} + a_2 e^{s_2 t} + \cdots + a_n e^{s_n t}. \qquad (6.5.5)$$

This is *Heaviside's expansion theorem*, applicable when $p(s)$ has only simple poles.

● **Example 6.5.1**

Let us invert the transform $s/[(s+2)(s^2+1)]$. It has three simple poles at $s = -2$ and $s = \pm i$. From our earlier discussion, $q(s) = s$, $p(s) = (s+2)(s^2+1)$, and $p'(s) = 3s^2 + 4s + 1$. Therefore,

$$\mathcal{L}^{-1}\left[\frac{s}{(s+2)(s^2+1)}\right] = \frac{-2}{12 - 8 + 1}e^{-2t} + \frac{i}{-3 + 4i + 1}e^{it} + \frac{-i}{-3 - 4i + 1}e^{-it} \tag{6.5.6}$$

$$= -\frac{2}{5}e^{-2t} + \frac{i}{-2 + 4i}e^{it} - \frac{i}{-2 - 4i}e^{-it} \tag{6.5.7}$$

$$= -\frac{2}{5}e^{-2t} + i\frac{-2 - 4i}{4 + 16}e^{it} - i\frac{-2 + 4i}{4 + 16}e^{-it} \tag{6.5.8}$$

$$= -\frac{2}{5}e^{-2t} + \frac{1}{5}\sin(t) + \frac{2}{5}\cos(t), \tag{6.5.9}$$

where we used $\sin(t) = \frac{1}{2i}(e^{it} - e^{-it})$, and $\cos(t) = \frac{1}{2}(e^{it} + e^{-it})$.

● **Example 6.5.2**

Let us invert the transform $1/[(s-1)(s-2)(s-3)]$. There are three simple poles at $s_1 = 1$, $s_2 = 2$, and $s_3 = 3$. In this case, the easiest method for computing a_1, a_2, and a_3 is

$$a_1 = \lim_{s \to 1} \frac{s-1}{(s-1)(s-2)(s-3)} = \frac{1}{2}, \tag{6.5.10}$$

$$a_2 = \lim_{s \to 2} \frac{s-2}{(s-1)(s-2)(s-3)} = -1 \tag{6.5.11}$$

and

$$a_3 = \lim_{s \to 3} \frac{s-3}{(s-1)(s-2)(s-3)} = \frac{1}{2}. \tag{6.5.12}$$

Therefore,

$$\mathcal{L}^{-1}\left[\frac{1}{(s-1)(s-2)(s-3)}\right] = \mathcal{L}^{-1}\left[\frac{a_1}{s-1} + \frac{a_2}{s-2} + \frac{a_3}{s-3}\right] \tag{6.5.13}$$

$$= \tfrac{1}{2}e^t - e^{2t} + \tfrac{1}{2}e^{3t}. \tag{6.5.14}$$

Note that for inverting transforms of the form $F(s)e^{-as}$ with $a > 0$, you should use Heaviside's expansion theorem to first invert $F(s)$ and then apply the second shifting theorem.

Let us now find the expansion when we have multiple roots, namely

$$F(s) = \frac{q(s)}{p(s)} = \frac{q(s)}{(s-s_1)^{m_1}(s-s_2)^{m_2}\cdots(s-s_n)^{m_n}}, \tag{6.5.15}$$

where the order of the denominator, $m_1 + m_2 + \cdots + m_n$, is greater than that for the numerator. Once again we eliminated any common factor between the numerator and denominator. Now we can write $F(s)$ as

$$F(s) = \sum_{k=1}^{n}\sum_{j=1}^{m_k}\frac{a_{kj}}{(s-s_k)^{m_k-j+1}}. \tag{6.5.16}$$

Multiplying (6.5.16) by $(s-s_k)^{m_k}$,

$$\frac{(s-s_k)^{m_k}q(s)}{p(s)} = a_{k1} + a_{k2}(s-s_k) + \cdots + a_{km_k}(s-s_k)^{m_k-1}$$

$$+ (s-s_k)^{m_k}\left[\frac{a_{11}}{(s-s_1)^{m_1}} + \cdots + \frac{a_{nm_n}}{s-s_n}\right], \tag{6.5.17}$$

where we grouped together into the square-bracketed term all of the terms except for those with a_{kj} coefficients. Taking the limit as $s \to s_k$,

$$a_{k1} = \lim_{s\to s_k}\frac{(s-s_k)^{m_k}q(s)}{p(s)}. \tag{6.5.18}$$

Let us now take the derivative of (6.5.17),

$$\frac{d}{ds}\left[\frac{(s-s_k)^{m_k}q(s)}{p(s)}\right] = a_{k2} + 2a_{k3}(s-s_k) + \cdots + (m_k-1)a_{km_k}(s-s_k)^{m_k-2}$$

$$+ \frac{d}{ds}\left\{(s-s_k)^{m_k}\left[\frac{a_{11}}{(s-s_1)^{m_1}} + \cdots + \frac{a_{nm_n}}{s-s_n}\right]\right\}. \tag{6.5.19}$$

Taking the limit as $s \to s_k$,

$$a_{k2} = \lim_{s\to s_k}\frac{d}{ds}\left[\frac{(s-s_k)^{m_k}q(s)}{p(s)}\right]. \tag{6.5.20}$$

In general,

$$a_{kj} = \lim_{s\to s_k}\frac{1}{(j-1)!}\frac{d^{j-1}}{ds^{j-1}}\left[\frac{(s-s_k)^{m_k}q(s)}{p(s)}\right], \tag{6.5.21}$$

and by direct inversion,

$$f(t) = \sum_{k=1}^{n}\sum_{j=1}^{m_k}\frac{a_{kj}}{(m_k-j)!}t^{m_k-j}e^{s_k t}. \tag{6.5.22}$$

• **Example 6.5.3**

Let us find the inverse of

$$F(s) = \frac{s}{(s+2)^2(s^2+1)}. \tag{6.5.23}$$

We first note that the denominator has simple zeros at $s = \pm i$ and a repeated root at $s = -2$. Therefore,

$$F(s) = \frac{A}{s-i} + \frac{B}{s+i} + \frac{C}{s+2} + \frac{D}{(s+2)^2}, \tag{6.5.24}$$

where

$$A = \lim_{s \to i} (s-i)F(s) = \tfrac{1}{6+8i}, \tag{6.5.25}$$

$$B = \lim_{s \to -i} (s+i)F(s) = \tfrac{1}{6-8i}, \tag{6.5.26}$$

$$C = \lim_{s \to -2} \frac{d}{ds}\left[(s+2)^2 F(s)\right] = \lim_{s \to -2} \frac{d}{ds}\left(\frac{s}{s^2+1}\right) = -\tfrac{3}{25}, \tag{6.5.27}$$

and

$$D = \lim_{s \to -2} (s+2)^2 F(s) = -\tfrac{2}{5}. \tag{6.5.28}$$

Thus,

$$f(t) = \tfrac{1}{6+8i}e^{it} + \tfrac{1}{6-8i}e^{-it} - \tfrac{3}{25}e^{-2t} - \tfrac{2}{5}te^{-2t} \tag{6.5.29}$$

$$= \tfrac{3}{25}\cos(t) + \tfrac{4}{25}\sin(t) - \tfrac{3}{25}e^{-2t} - \tfrac{10}{25}te^{-2t}. \tag{6.5.30}$$

In §6.10 we shall see that we can invert transforms just as easily with the residue theorem.

Let us now find the inverse of

$$F(s) = \frac{cs + (ca - \omega d)}{(s+a)^2 + \omega^2} = \frac{cs + (ca - \omega d)}{(s+a-\omega i)(s+a+\omega i)} \tag{6.5.31}$$

by Heaviside's expansion theorem. Then

$$F(s) = \frac{c+di}{2(s+a-\omega i)} + \frac{c-di}{2(s+a+\omega i)} \tag{6.5.32}$$

$$= \frac{\sqrt{c^2+d^2}e^{\theta i}}{2(s+a-\omega i)} + \frac{\sqrt{c^2+d^2}e^{-\theta i}}{2(s+a+\omega i)}, \tag{6.5.33}$$

where $\theta = \tan^{-1}(d/c)$. Note that we must choose θ so that it gives the correct sign for c and d.

Taking the inverse of (6.5.33),

$$f(t) = \tfrac{1}{2}\sqrt{c^2 + d^2}\,e^{-at+\omega ti+\theta i} + \tfrac{1}{2}\sqrt{c^2 + d^2}\,e^{-at-\omega ti-\theta i} \qquad (6.5.34)$$
$$= \sqrt{c^2 + d^2}\,e^{-at}\cos(\omega t + \theta). \qquad (6.5.35)$$

Equation (6.5.35) is the amplitude/phase form of the inverse of (6.5.31). It is particularly popular with electrical engineers.

● **Example 6.5.4**

Let us express the inverse of

$$F(s) = \frac{8s - 3}{s^2 + 4s + 13} \qquad (6.5.36)$$

in the amplitude/phase form.

Starting with

$$F(s) = \frac{8s - 3}{(s + 2 - 3i)(s + 2 + 3i)} = \frac{4 + 19i/6}{s + 2 - 3i} + \frac{4 - 19i/6}{s + 2 + 3i} \qquad (6.5.37)$$

$$= \frac{5.1017e^{38.3675°i}}{s + 2 - 3i} + \frac{5.1017e^{-38.3675°i}}{s + 2 + 3i}, \qquad (6.5.38)$$

or

$$f(t) = 5.1017e^{-2t+3it+38.3675°i} + 5.1017e^{-2t-3it-38.3675°i} \qquad (6.5.39)$$
$$= 10.2034e^{-2t}\cos(3t + 38.3675°). \qquad (6.5.40)$$

● **Example 6.5.5: The design of film projectors**

For our final example we anticipate future work. The primary use of Laplace transforms is the solution of differential equations. In this example we illustrate this technique that includes Heaviside's expansion theorem in the form of amplitude and phase.

This problem[3] arose in the design of projectors for motion pictures. An early problem was ensuring that the speed at which the film passed the electric eye remained essentially constant; otherwise, a frequency modulation of the reproduced sound resulted. Figure 6.5.1(A) shows a diagram of the projector. Many will remember this design from their days as a school projectionist. In this section we shall show that this particular design filters out variations in the film speed caused by irregularities either in the driving-gear trains or in the engagement of the sprocket teeth with the holes in the film.

---

3 Cook, E. D., 1935: The technical aspects of the high-fidelity reproducer. *J. Soc. Motion Pict. Eng.*, **25**, 289–312.

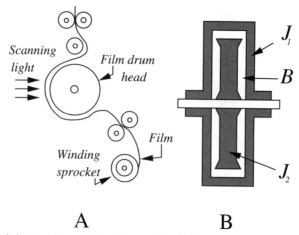

Figure 6.5.1: (A) The schematic for the scanning light in a motion-picture projector and (B) interior of the film drum head.

Let us now focus on the film head—a hollow drum of small moment of inertia J_1. See Figure 6.5.1(B). Within it there is a concentric inner flywheel of moment of inertia J_2, where $J_2 \gg J_1$. The remainder of the space within the drum is filled with oil. The inner flywheel rotates on precision ball bearings on the drum shaft. The only coupling between the drum and flywheel is through fluid friction and the very small friction in the ball bearings. The flection of the film loops between the drum head and idler pulleys provides the spring restoring force for the system as the film runs rapidly through the system.

From Figure 6.5.1 the dynamical equations governing the outer case and inner flywheel are (1) the rate of change of the outer casing of the film head equals the frictional torque given to the casing from the inner flywheel plus the restoring torque due to the flection of the film, and (2) the rate of change of the inner flywheel equals the negative of the frictional torque given to the outer casing by the inner flywheel.

Assuming that the frictional torque between the two flywheels is proportional to the difference in their angular velocities, the frictional torque given to the casing from the inner flywheel is $B(\omega_2 - \omega_1)$, where B is the frictional resistance, ω_1 and ω_2 are the deviations of the drum and inner flywheel from their normal angular velocities, respectively. If r is the ratio of the diameter of the winding sprocket to the diameter of the drum, the restoring torque due to the flection of the film and its corresponding angular twist equals $K \int_0^t (r\omega_0 - \omega_1) \, d\tau$, where K is the rotational stiffness and ω_0 is the deviation of the winding sprocket from its normal angular velocity. The quantity $r\omega_0$ gives the angular velocity at which the film is running through the projector because the winding sprocket is the mechanism that pulls the film. Consequently the equations governing this mechanical system are

$$J_1 \frac{d\omega_1}{dt} = K \int_0^t (r\omega_0 - \omega_1) \, d\tau + B(\omega_2 - \omega_1), \qquad (6.5.41)$$

and

$$J_2 \frac{d\omega_2}{dt} = -B(\omega_2 - \omega_1). \tag{6.5.42}$$

With the winding sprocket, the drum, and the flywheel running at their normal uniform angular velocities, let us assume that the winding sprocket introduces a disturbance equivalent to an unit increase in its angular velocity for 0.15 second, followed by the resumption of its normal velocity. It is assumed that the film in contact with the drum cannot slip. The initial conditions are $\omega_1(0) = \omega_2(0) = 0$.

Taking the Laplace transform of (6.5.41)–(6.5.42) using (6.1.18),

$$\left(J_1 s + B + \frac{K}{s}\right)\Omega_1(s) - B\Omega_2(s) = \frac{rK}{s}\Omega_0(s) = rK\mathcal{L}\left[\int_0^t \omega_0(\tau)\,d\tau\right], \tag{6.5.43}$$

and

$$-B\Omega_1(s) + (J_2 s + B)\Omega_2(s) = 0. \tag{6.5.44}$$

The solution of (6.5.43)–(6.5.44) for $\Omega_1(s)$ is

$$\Omega_1(s) = \frac{rK}{J_1}\,\frac{(s + a_0)\Omega_0(s)}{s^3 + b_2 s^2 + b_1 s + b_0}, \tag{6.5.45}$$

where typical values[4] are

$$\frac{rK}{J_1} = 90.8, \quad a_0 = \frac{B}{J_2} = 1.47, \quad b_0 = \frac{BK}{J_1 J_2} = 231, \tag{6.5.46}$$

$$b_1 = \frac{K}{J_1} = 157, \quad \text{and} \quad b_2 = \frac{B(J_1 + J_2)}{J_1 J_2} = 8.20. \tag{6.5.47}$$

The transform $\Omega_1(s)$ has three simple poles located at $s_1 = -1.58, s_2 = -3.32 + 11.6i$, and $s_3 = -3.32 - 11.6i$.

Because the sprocket angular velocity deviation $\omega_0(t)$ is a pulse of unit amplitude and 0.15 second duration, we express it as the difference of two Heaviside step functions

$$\omega_0(t) = H(t) - H(t - 0.15). \tag{6.5.48}$$

Its Laplace transform is

$$\Omega_0(s) = \frac{1}{s} - \frac{1}{s}e^{-0.15s} \tag{6.5.49}$$

---

[4] $J_1 = 1.84 \times 10^4$ dyne cm sec^2 per radian, $J_2 = 8.43 \times 10^4$ dyne cm sec^2 per radian, $B = 12.4 \times 10^4$ dyne cm sec per radian, $K = 2.89 \times 10^6$ dyne cm per radian, and $r = 0.578$.

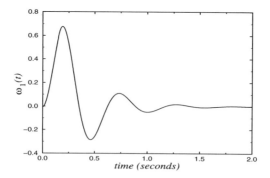

Figure 6.5.2: The deviation $\omega_1(t)$ of a film drum head from its uniform angular velocity when the sprocket angular velocity is perturbed by a unit amount for the duration of 0.15 second.

so that (6.5.45) becomes

$$\Omega_1(s) = \frac{rK}{J_1}\frac{(s+a_0)}{s(s-s_1)(s-s_2)(s-s_3)}\left(1 - e^{-0.15s}\right). \tag{6.5.50}$$

The inversion of (6.5.50) follows directly from the second shifting theorem and Heaviside's expansion theorem or

$$\omega_1(t) = K_0 + K_1 e^{s_1 t} + K_2 e^{s_2 t} + K_3 e^{s_3 t} \tag{6.5.51}$$
$$- [K_0 + K_1 e^{s_1(t-0.15)} + K_2 e^{s_2(t-0.15)} + K_3 e^{s_3(t-0.15)}]H(t - 0.15),$$

where

$$K_0 = \frac{rK}{J_1}\left.\frac{s+a_0}{(s-s_1)(s-s_2)(s-s_3)}\right|_{s=0} = 0.578, \tag{6.5.52}$$

$$K_1 = \frac{rK}{J_1}\left.\frac{s+a_0}{s(s-s_2)(s-s_3)}\right|_{s=s_1} = 0.046, \tag{6.5.53}$$

$$K_2 = \frac{rK}{J_1}\left.\frac{s+a_0}{s(s-s_1)(s-s_3)}\right|_{s=s_2} = 0.326e^{165°i}, \tag{6.5.54}$$

and

$$K_3 = \frac{rK}{J_1}\left.\frac{s+a_0}{s(s-s_1)(s-s_2)}\right|_{s=s_3} = 0.326e^{-165°i}. \tag{6.5.55}$$

Using Euler's identity $\cos(t) = (e^{it} + e^{-it})/2$, we can write (6.5.51) as

$$\omega_1(t) = 0.578 + 0.046e^{-1.58t} + 0.652e^{-3.32t}\cos(11.6t + 165°)$$
$$- \{0.578 + 0.046e^{-1.58(t-0.15)} + 0.652e^{-3.32(t-0.15)}$$
$$\times \cos[11.6(t - 0.15) + 165°]\}H(t - 0.15). \tag{6.5.56}$$

Equation (6.5.56) is plotted in Figure 6.5.2. Note that fluctuations in $\omega_1(t)$ are damped out by the particular design of this film projector. Because this mechanical device dampens unwanted fluctuations (or noise) in the motion-picture projector, this particular device is an example of a *mechanical filter*.

Problems

Use Heaviside's expansion theorem to find the inverse of the following Laplace transforms:

1. $F(s) = \dfrac{1}{s^2 + 3s + 2}$

2. $F(s) = \dfrac{s + 3}{(s + 4)(s - 2)}$

3. $F(s) = \dfrac{s - 4}{(s + 2)(s + 1)(s - 3)}$

4. $F(s) = \dfrac{s - 3}{(s^2 + 4)(s + 1)}$.

Find the inverse of the following transforms and express them in amplitude/phase form:

5. $F(s) = \dfrac{1}{s^2 + 4s + 5}$

6. $F(s) = \dfrac{1}{s^2 + 6s + 13}$

7. $F(s) = \dfrac{2s - 5}{s^2 + 16}$

8. $F(s) = \dfrac{1}{s(s^2 + 2s + 2)}$

9. $F(s) = \dfrac{s + 2}{s(s^2 + 4)}$

6.6 CONVOLUTION

In this section we turn to a fundamental concept in Laplace transforms: convolution. We shall restrict ourselves to its use in finding the inverse of a transform when that transform consists of the *product* of two simpler transforms. In subsequent sections we will use it to solve ordinary differential equations.

We begin by formally introducing the mathematical operation of the *convolution product*

$$f(t) * g(t) = \int_0^t f(t - x)g(x)\, dx = \int_0^t f(x)g(t - x)\, dx. \qquad (6.6.1)$$

In most cases the operations required by (6.6.1) are straightforward.

• Example 6.6.1

Let us find the convolution between $\cos(t)$ and $\sin(t)$.

$$\cos(t) * \sin(t) = \int_0^t \sin(t - x)\cos(x)\, dx = \tfrac{1}{2}\int_0^t [\sin(t) + \sin(t - 2x)]\, dx \qquad (6.6.2)$$

$$= \tfrac{1}{2}\int_0^t \sin(t)\, dx + \tfrac{1}{2}\int_0^t \sin(t - 2x)\, dx \qquad (6.6.3)$$

$$= \tfrac{1}{2}\sin(t)\, x\Big|_0^t + \tfrac{1}{4}\cos(t - 2x)\Big|_0^t = \tfrac{1}{2}t\sin(t). \qquad (6.6.4)$$

- **Example 6.6.2**

Similarly, the convolution between t^2 and $\sin(t)$ is

$$t^2 * \sin(t) = \int_0^t (t-x)^2 \sin(x)\, dx \tag{6.6.5}$$

$$= -(t-x)^2 \cos(x)\big|_0^t - 2\int_0^t (t-x)\cos(x)\, dx \tag{6.6.6}$$

$$= t^2 - 2(t-x)\sin(x)\big|_0^t - 2\int_0^t \sin(x)\, dx \tag{6.6.7}$$

$$= t^2 + 2\cos(t) - 2 \tag{6.6.8}$$

by integration by parts.

- **Example 6.6.3**

Consider now the convolution between e^t and the discontinuous function $H(t-1) - H(t-2)$:

$$e^t * [H(t-1) - H(t-2)] = \int_0^t e^{t-x}[H(x-1) - H(x-2)]\, dx \tag{6.6.9}$$

$$= e^t \int_0^t e^{-x}[H(x-1) - H(x-2)]\, dx. \tag{6.6.10}$$

In order to evaluate the integral (6.6.10) we must examine various cases. If $t < 1$, then both of the step functions equal zero and the convolution equals zero. However, when $1 < t < 2$, the first step function equals one while the second equals zero as the dummy variable x runs between 1 and t. Therefore,

$$e^t * [H(t-1) - H(t-2)] = e^t \int_1^t e^{-x}\, dx = e^{t-1} - 1, \tag{6.6.11}$$

because the portion of the integral from zero to one equals zero. Finally, when $t > 2$, the integrand is only nonzero for that portion of the integration when $1 < x < 2$. Consequently,

$$e^t * [H(t-1) - H(t-2)] = e^t \int_1^2 e^{-x}\, dx = e^{t-1} - e^{t-2}. \tag{6.6.12}$$

Thus, the convolution of e^t with the pulse $H(t-1) - H(t-2)$ is

$$e^t * [H(t-1) - H(t-2)] = \begin{cases} 0, & 0 \le t \le 1, \\ e^{t-1} - 1, & 1 \le t \le 2, \\ e^{t-1} - e^{t-2}, & 2 \le t. \end{cases} \tag{6.6.13}$$

MATLAB can also be used to find the convolution of two functions. For example, in the present case the commands

```
syms x t positive
int('exp(t-x)*(Heaviside(x-1)-Heaviside(x-2))',x,0,t)
```

yields

```
ans =
-Heaviside(t-1)+Heaviside(t-1)*exp(t-1)+Heaviside(t-2)
    -Heaviside(t-2)*exp(t-2)
```

The reason why we introduced convolution stems from the following fundamental theorem (often called *Borel's theorem*[5]). If

$$w(t) = u(t) * v(t) \tag{6.6.14}$$

then

$$W(s) = U(s)V(s). \tag{6.6.15}$$

In other words, we can invert a complicated transform by convoluting the inverses to two simpler functions. The proof is as follows:

$$W(s) = \int_0^\infty \left[\int_0^t u(x)v(t-x)\,dx \right] e^{-st}dt \tag{6.6.16}$$

$$= \int_0^\infty \left[\int_x^\infty u(x)v(t-x)e^{-st}dt \right] dx \tag{6.6.17}$$

$$= \int_0^\infty u(x) \left[\int_0^\infty v(r)e^{-s(r+x)}dr \right] dx \tag{6.6.18}$$

$$= \left[\int_0^\infty u(x)e^{-sx}dx \right] \left[\int_0^\infty v(r)e^{-sr}dr \right] = U(s)V(s), \tag{6.6.19}$$

where $t = r + x$.   □

• **Example 6.6.4**

Let us find the inverse of the transform

$$\frac{s}{(s^2+1)^2} = \frac{s}{s^2+1} \times \frac{1}{s^2+1} = \mathcal{L}[\cos(t)]\mathcal{L}[\sin(t)] \tag{6.6.20}$$

$$= \mathcal{L}[\cos(t) * \sin(t)] = \mathcal{L}[\tfrac{1}{2}t\sin(t)] \tag{6.6.21}$$

from Example 6.6.1.

---

[5] Borel, É., 1901: *Leçons sur les séries divergentes*. Gauthier-Villars, p. 104.

- **Example 6.6.5**

Let us find the inverse of the transform

$$\frac{1}{(s^2 + a^2)^2} = \frac{1}{a^2}\left(\frac{a}{s^2 + a^2} \times \frac{a}{s^2 + a^2}\right) = \frac{1}{a^2}\mathcal{L}[\sin(at)]\mathcal{L}[\sin(at)]. \quad (\textbf{6.6.22})$$

Therefore,

$$\mathcal{L}^{-1}\left[\frac{1}{(s^2 + a^2)^2}\right] = \frac{1}{a^2}\int_0^t \sin[a(t - x)]\sin(ax)\,dx \qquad (\textbf{6.6.23})$$

$$= \frac{1}{2a^2}\int_0^t \cos[a(t - 2x)]\,dx - \frac{1}{2a^2}\int_0^t \cos(at)\,dx \quad (\textbf{6.6.24})$$

$$= -\frac{1}{4a^3}\sin[a(t - 2x)]\Big|_0^t - \frac{1}{2a^2}\cos(at)\,x\Big|_0^t \qquad (\textbf{6.6.25})$$

$$= \frac{1}{2a^3}[\sin(at) - at\cos(at)]. \qquad (\textbf{6.6.26})$$

- **Example 6.6.6**

Let us use the results from Example 6.6.3 to verify the convolution theorem.

We begin by rewriting (6.6.13) in terms of the Heaviside step functions. Using the method outline in Example 6.2.1,

$$f(t) * g(t) = \left(e^{t-1} - 1\right)H(t - 1) + \left(1 - e^{t-2}\right)H(t - 2). \qquad (\textbf{6.6.27})$$

Employing the second shifting theorem,

$$\mathcal{L}[f * g] = \frac{e^{-s}}{s - 1} - \frac{e^{-s}}{s} + \frac{e^{-2s}}{s} - \frac{e^{-2s}}{s - 1} \qquad (\textbf{6.6.28})$$

$$= \frac{e^{-s}}{s(s - 1)} - \frac{e^{-2s}}{s(s - 1)} = \frac{1}{s - 1}\left(\frac{e^{-s}}{s} - \frac{e^{-2s}}{s}\right) \qquad (\textbf{6.6.29})$$

$$= \mathcal{L}[e^t]\mathcal{L}[H(t - 1) - H(t - 2)] \qquad (\textbf{6.6.30})$$

and the convolution theorem holds true. If we had not rewritten (6.6.13) in terms of step functions, we could still have found $\mathcal{L}[f * g]$ from the definition of the Laplace transform.

Problems

Verify the following convolutions and then show that the convolution theorem is true. Use MATLAB to check your answer.

1. $1 * 1 = t$

2. $1 * \cos(at) = \sin(at)/a$

3. $1 * e^t = e^t - 1$

4. $t * t = t^3/6$

5. $t * \sin(t) = t - \sin(t)$

6. $t * e^t = e^t - t - 1$

7. $t^2 * \sin(at) = \dfrac{t^2}{a} - \dfrac{4}{a^3}\sin^2\left(\dfrac{at}{2}\right)$

8. $t * H(t-1) = \frac{1}{2}(t-1)^2 H(t-1)$

9. $H(t-a) * H(t-b) = (t-a-b)H(t-a-b)$

10. $t * [H(t) - H(t-2)] = \dfrac{t^2}{2} - \dfrac{(t-2)^2}{2}H(t-2)$

Use the convolution theorem to invert the following functions:

11. $F(s) = \dfrac{1}{s^2(s-1)}$

12. $F(s) = \dfrac{1}{s^2(s+a)^2}$

13. Prove that the convolution of two Dirac delta functions is a Dirac delta function.

6.7 INTEGRAL EQUATIONS

An *integral equation* contains the dependent variable under an integral sign. The convolution theorem provides an excellent tool for solving a very special class of these equations, *Volterra equation of the second kind* :[6]

$$f(t) - \int_0^t K[t, x, f(x)]\, dx = g(t), \qquad 0 \le t \le T. \tag{6.7.1}$$

These equations appear in history-dependent problems, such as epidemics,[7] vibration problems,[8] and viscoelasticity.[9]

---

[6] Fock, V., 1924: Über eine Klasse von Integralgleichungen. *Math. Zeit.*, **21**, 161–173; Koizumi, S., 1931: On Heaviside's operational solution of a Volterra's integral equation when its nucleus is a function of $(x - \xi)$. *Philos. Mag.*, Ser. 7, **11**, 432–441.

[7] Wang, F. J. S., 1978: Asymptotic behavior of some deterministic epidemic models. *SIAM J. Math. Anal.*, **9**, 529–534.

[8] Lin, S. P., 1975: Damped vibration of a string. *J. Fluid Mech.*, **72**, 787–797.

[9] Rogers, T. G., and E. H. Lee, 1964: The cylinder problem in viscoelastic stress analysis. *Q. Appl. Math.*, **22**, 117–131.

● **Example 6.7.1**

Let us find $f(t)$ from the integral equation

$$f(t) = 4t - 3 \int_0^t f(x) \sin(t - x)\, dx. \tag{6.7.2}$$

The integral in (6.7.2) is such that we can use the convolution theorem to find its Laplace transform. Then, because $\mathcal{L}[\sin(t)] = 1/(s^2 + 1)$, the convolution theorem yields

$$\mathcal{L}\left[\int_0^t f(x) \sin(t - x)\, dx\right] = \frac{F(s)}{s^2 + 1}. \tag{6.7.3}$$

Therefore, the Laplace transform converts (6.7.2) into

$$F(s) = \frac{4}{s^2} - \frac{3F(s)}{s^2 + 1}. \tag{6.7.4}$$

Solving for $F(s)$,

$$F(s) = \frac{4(s^2 + 1)}{s^2(s^2 + 4)}. \tag{6.7.5}$$

By partial fractions, or by inspection,

$$F(s) = \frac{1}{s^2} + \frac{3}{s^2 + 4}. \tag{6.7.6}$$

Therefore, inverting term by term,

$$f(t) = t + \tfrac{3}{2} \sin(2t). \tag{6.7.7}$$

Note that the integral equation

$$f(t) = 4t - 3 \int_0^t f(t - x) \sin(x)\, dx \tag{6.7.8}$$

also has the same solution.

● **Example 6.7.2**

Let us solve the equation

$$f'(t) + \alpha^2 \int_0^t f(\tau)\, d\tau = B - C \cos(\omega t), \qquad f(0) = 0. \tag{6.7.9}$$

Again the integral is one of the convolution type; it differs from the previous example in that it includes a derivative. Taking the Laplace transform of (6.7.9),

$$sF(s) - f(0) + \frac{a^2 F(s)}{s} = \frac{B}{s} - \frac{sC}{s^2 + \omega^2}. \tag{6.7.10}$$

Because $f(0) = 0$, (6.7.10) simplifies to

$$(s^2 + a^2)F(s) = B - \frac{Cs^2}{s^2 + \omega^2}. \tag{6.7.11}$$

Solving for $F(s)$,

$$F(s) = \frac{B}{s^2 + a^2} - \frac{Cs^2}{(s^2 + a^2)(s^2 + \omega^2)}. \tag{6.7.12}$$

Using partial fractions to invert (6.7.12),

$$f(t) = \left(\frac{B}{\alpha} + \frac{\alpha C}{\omega^2 - \alpha^2} \right) \sin(\alpha t) - \frac{\omega C}{\omega^2 - \alpha^2} \sin(\omega t). \tag{6.7.13}$$

Problems

Solve the following integral equations:

1. $f(t) = 1 + 2 \int_0^t f(t - x) e^{-2x} \, dx$

2. $f(t) = 1 + \int_0^t f(x) \sin(t - x) \, dx$

3. $f(t) = t + \int_0^t f(t - x) e^{-x} \, dx$

4. $f(t) = 4t^2 - \int_0^t f(t - x) e^{-x} \, dx$

5. $f(t) = t^3 + \int_0^t f(x) \sin(t - x) \, dx$

6. $f(t) = 8t^2 - 3 \int_0^t f(x) \sin(t - x) \, dx$

7. $f(t) = t^2 - 2 \int_0^t f(t - x) \sinh(2x) \, dx$

8. $f(t) = 1 + 2 \int_0^t f(t - x) \cos(x) \, dx$

9. $f(t) = e^{2t} - 2 \int_0^t f(t - x) \cos(x) \, dx$

10. $f(t) = t^2 + \int_0^t f(x) \sin(t - x) \, dx$

11. $f(t) = e^{-t} - 2 \int_0^t f(x) \cos(t - x)\, dx$

12. $f(t) = 6t + 4 \int_0^t f(x)(x - t)^2\, dx$

13. $f(t) = a\sqrt{t} - \int_0^t \dfrac{f(t - x)}{\sqrt{x}}\, dx$

14. Solve the following equation for $f(t)$ with the condition that $f(0) = 4$:

$$f'(t) = t + \int_0^t f(t - x) \cos(x)\, dx.$$

15. Solve the following equation for $f(t)$ with the condition that $f(0) = 0$:

$$f'(t) = \sin(t) + \int_0^t f(t - x) \cos(x)\, dx.$$

16. During a study of nucleation involving idealized active sites along a boiling surface, Marto and Rohsenow[10] solved the integral equation

$$A = B\sqrt{t} + C \int_0^t \frac{x'(\tau)}{\sqrt{t - \tau}}\, d\tau$$

to find the position $x(t)$ of the liquid/vapor interface. If A, B, and C are constants and $x(0) = 0$, find the solution for them.

17. Solve the following equation for $x(t)$ with the condition that $x(0) = 0$:

$$x(t) + t = \frac{1}{c\sqrt{\pi}} \int_0^t \frac{x'(\tau)}{\sqrt{t - \tau}}\, d\tau,$$

where c is constant.

18. During a study of the temperature $f(t)$ of a heat reservoir attached to a semi-infinite heat-conducting rod, Huber[11] solved the integral equation

$$f'(t) = \alpha - \frac{\beta}{\sqrt{\pi}} \int_0^t \frac{f'(\tau)}{\sqrt{t - \tau}}\, d\tau,$$

[10] From Marto, P. J., and W. M. Rohsenow, 1966: Nucleate boiling instability of alkali metals. *J. Heat Transfer*, **88**, 183–193 with permission.

[11] From Huber, A., 1934: Eine Methode zur Bestimmung der Wärme- und Temperatur-leitfähigkeit. *Monatsh. Math. Phys.*, **41**, 35–42.

where α and β are constants and $f(0) = 0$. Find $f(t)$ for him. Hint:

$$\frac{\alpha}{s^{3/2}(s^{1/2} + \beta)} = \frac{\alpha}{s(s - \beta^2)} - \frac{\alpha\beta}{s^{3/2}(s - \beta^2)}.$$

19. During the solution of a diffusion problem, Zhdanov, Chikhachev, and Yavlinskii[12] solved an integral equation similar to

$$\int_0^t f(\tau) \left[1 - \operatorname{erf}\left(a\sqrt{t - \tau}\right)\right] d\tau = at,$$

where $\operatorname{erf}(x) = \dfrac{2}{\sqrt{\pi}} \displaystyle\int_0^x e^{-y^2}\, dy$ is the error function. What should they have found? Hint: You will need to prove that

$$\mathcal{L}\left[t\,\operatorname{erf}(a\sqrt{t}) - \frac{1}{2a^2}\operatorname{erf}(a\sqrt{t}) + \frac{\sqrt{t}}{a\sqrt{\pi}}e^{-a^2 t}\right] = \frac{a}{s^2\sqrt{s + a^2}}.$$

20. The *Laguerre polynomial*[13]

$$y(t) = L_n(t) = \frac{e^t}{n!}\frac{d^n}{dt^n}\left(t^n e^{-t}\right), \qquad n = 0, 1, 2, 3, \ldots$$

satisfies the ordinary differential equation

$$ty'' + (1 - t)y' + ny = (ty')' - ty' + ny = 0,$$

with $y(0) = 1$ and $y'(0) = -n$.

Step 1: Using (6.1.20) and (6.3.22), show that the Laplace transformed version of this differential equation is

$$Y'(s) = \frac{n + 1 - s}{s(s - 1)}Y(s) = \frac{n}{s - 1}Y(s) - \frac{n + 1}{s}Y(s),$$

where $Y(s)$ is the Laplace transform of $y(t)$.

Step 2: Using (6.3.22) and the convolution theorem, show that Laguerre polynomials are the solution to the integral equation

$$ty(t) = (n + 1)\int_0^t y(\tau)\, d\tau - ne^t \int_0^t y(\tau)\, e^{-\tau}\, d\tau.$$

---

[12] Zhdanov, S. K., A. S. Chikhachev, and Yu. N. Yavlinskii, 1976: Diffusion boundary-value problem for regions with moving boundaries and conservation of particles. *Sov. Phys. Tech. Phys.*, **21**, 883–884.

[13] See §5.3 in Andrews, L. C., 1985: *Special Functions for Engineers and Applied Mathematicians*. MacMillian, 357 pp.

6.8 SOLUTION OF LINEAR DIFFERENTIAL EQUATIONS WITH CONSTANT COEFFICIENTS

For the engineer, as it was for Oliver Heaviside, the primary use of Laplace transforms is the solution of ordinary, constant coefficient, linear differential equations. These equations are important not only because they appear in many engineering problems but also because they may serve as approximations, even if locally, to ordinary differential equations with nonconstant coefficients or to nonlinear ordinary differential equations.

For all of these reasons, we wish to solve the *initial-value problem*

$$\frac{d^n y}{dt^n} + a_1 \frac{d^{n-1} y}{dt^{n-1}} + \cdots + a_{n-1} \frac{dy}{dt} + a_n y = f(t), \quad t > 0, \tag{6.8.1}$$

by Laplace transforms, where $a_1, a_2, \ldots$ are constants and we know the value of $y, y', \ldots, y^{(n-1)}$ at $t = 0$. The procedure is as follows. Applying the derivative rule (6.1.20) to (6.8.1), we reduce the *differential* equation to an *algebraic* one involving the constants $a_1, a_2, \ldots, a_n$, the parameter s, the Laplace transform of $f(t)$, and the values of the initial conditions. We then solve for the Laplace transform of $y(t)$, $Y(s)$. Finally, we apply one of the many techniques of inverting a Laplace transform to find $y(t)$.

Similar considerations hold with *systems* of ordinary differential equations. The Laplace transform of the system of ordinary differential equations results in an algebraic set of equations containing $Y_1(s), Y_2(s), \ldots, Y_n(s)$. By some method we solve this set of equations and invert each transform $Y_1(s), Y_2(s), \ldots, Y_n(s)$ in turn to give $y_1(t), y_2(t), \ldots, y_n(t)$.

The following examples will illustrate the details of the process.

• **Example 6.8.1**

Let us solve the ordinary differential equation

$$y'' + 2y' = 8t, \tag{6.8.2}$$

subject to the initial conditions that $y'(0) = y(0) = 0$. Taking the Laplace transform of both sides of (6.8.2),

$$\mathcal{L}(y'') + 2\mathcal{L}(y') = 8\mathcal{L}(t), \tag{6.8.3}$$

or

$$s^2 Y(s) - s y(0) - y'(0) + 2s Y(s) - 2y(0) = \frac{8}{s^2}, \tag{6.8.4}$$

where $Y(s) = \mathcal{L}[y(t)]$. Substituting the initial conditions into (6.8.4) and solving for $Y(s)$,

$$Y(s) = \frac{8}{s^3(s+2)} = \frac{A}{s^3} + \frac{B}{s^2} + \frac{C}{s} + \frac{D}{s+2} \tag{6.8.5}$$

$$= \frac{8}{s^3(s+2)} = \frac{(s+2)A + s(s+2)B + s^2(s+2)C + s^3 D}{s^3(s+2)}. \tag{6.8.6}$$

Matching powers of s in the numerators of (6.8.6), $C + D = 0$, $B + 2C = 0$, $A + 2B = 0$, and $2A = 8$ or $A = 4$, $B = -2$, $C = 1$, and $D = -1$. Therefore,

$$Y(s) = \frac{4}{s^3} - \frac{2}{s^2} + \frac{1}{s} - \frac{1}{s+2}. \tag{6.8.7}$$

Finally, performing term-by-term inversion of (6.8.7), the final solution is

$$y(t) = 2t^2 - 2t + 1 - e^{-2t}. \tag{6.8.8}$$

We could have performed the same operations using the symbolic toolbox with MATLAB. The MATLAB script

```
clear
% define symbolic variables
syms s t Y
% take Laplace transform of left side of differential equation
LHS = laplace(diff(diff(sym('y(t)')))+2*diff(sym('y(t)')));
% take Laplace transform of right side of differential equation
RHS = laplace(8*t);
% set Y for Laplace transform of y
%      and introduce initial conditions
newLHS = subs(LHS,'laplace(y(t),t,s)','y(0)','D(y)(0)',Y,0,0);
% solve for Y
Y = solve(newLHS-RHS,Y);
% invert Laplace transform and find y(t)
y = ilaplace(Y,s,t)
```

yields the result

```
y =
1-exp(-2*t)-2*t+2*t^2
```

which agrees with (6.8.8).

- **Example 6.8.2**

Let us solve the ordinary differential equation

$$y'' + y = H(t) - H(t-1) \tag{6.8.9}$$

with the initial conditions that $y'(0) = y(0) = 0$. Taking the Laplace transform of both sides of (6.8.9),

$$s^2 Y(s) - sy(0) - y'(0) + Y(s) = \frac{1}{s} - \frac{e^{-s}}{s}, \tag{6.8.10}$$

where $Y(s) = \mathcal{L}[y(t)]$. Substituting the initial conditions into (6.8.10) and solving for $Y(s)$,

$$Y(s) = \left(\frac{1}{s} - \frac{s}{s^2+1}\right) - \left(\frac{1}{s} - \frac{s}{s^2+1}\right) e^{-s}. \tag{6.8.11}$$

Using the second shifting theorem, the final solution is

$$y(t) = 1 - \cos(t) - [1 - \cos(t-1)]H(t-1). \qquad (6.8.12)$$

We can check our results using the MATLAB script

```
clear
% define symbolic variables
syms s t Y
% take Laplace transform of left side of differential equation
LHS = laplace(diff(diff(sym('y(t)')))+sym('y(t)'));
% take Laplace transform of right side of differential equation
RHS = laplace('Heaviside(t) - Heaviside(t-1)',t,s);
% set Y for Laplace transform of y
%      and introduce initial conditions
newLHS = subs(LHS,'laplace(y(t),t,s)','y(0)','D(y)(0)',Y,0,0);
% solve for Y
Y = solve(newLHS-RHS,Y);
% invert Laplace transform and find y(t)
y = ilaplace(Y,s,t)
```

which yields

```
y =
1-cos(t)-Heaviside(t-1)+Heaviside(t-1)*cos(t-1)
```

• **Example 6.8.3**

Let us solve the ordinary differential equation

$$y'' + 2y' + y = f(t) \qquad (6.8.13)$$

with the initial conditions that $y'(0) = y(0) = 0$, where $f(t)$ is an unknown function whose Laplace transform exists. Taking the Laplace transform of both sides of (6.8.13),

$$s^2 Y(s) - sy(0) - y'(0) + 2sY(s) - 2y(0) + Y(s) = F(s), \qquad (6.8.14)$$

where $Y(s) = \mathcal{L}[y(t)]$. Substituting the initial conditions into (6.8.14) and solving for $Y(s)$,

$$Y(s) = \frac{1}{(s+1)^2} F(s). \qquad (6.8.15)$$

We wrote (6.8.15) in this form because the transform $Y(s)$ equals the product of two transforms $1/(s+1)^2$ and $F(s)$. Therefore, by the convolution theorem we can immediately write

$$y(t) = te^{-t} * f(t) = \int_0^t x e^{-x} f(t-x) \, dx. \qquad (6.8.16)$$

Without knowing $f(t)$, this is as far as we can go.

• Example 6.8.4: Forced harmonic oscillator

Let us solve the *simple harmonic oscillator* forced by a harmonic forcing

$$y'' + \omega^2 y = \cos(\omega t), \tag{6.8.17}$$

subject to the initial conditions that $y'(0) = y(0) = 0$. Although the complete solution could be found by summing the complementary solution and a particular solution obtained, say, from the method of undetermined coefficients, we now illustrate how we can use Laplace transforms to solve this problem.

Taking the Laplace transform of both sides of (6.8.17), substituting in the initial conditions, and solving for $Y(s)$,

$$Y(s) = \frac{s}{(s^2 + \omega^2)^2}, \tag{6.8.18}$$

and

$$y(t) = \frac{1}{\omega} \sin(\omega t) * \cos(\omega t) = \frac{t}{2\omega} \sin(\omega t). \tag{6.8.19}$$

Equation (6.8.19) gives an oscillation that grows linearly with time although the forcing function is simply periodic. Why does this occur? Recall that our simple harmonic oscillator has the natural frequency ω. But that is exactly the frequency at which we drive the system. Consequently, our choice of forcing has resulted in *resonance* where energy continuously feeds into the oscillator.

• Example 6.8.5

Let us solve the *system* of ordinary differential equations:

$$2x' + y = \cos(t), \tag{6.8.20}$$

and

$$y' - 2x = \sin(t), \tag{6.8.21}$$

subject to the initial conditions that $x(0) = 0$, and $y(0) = 1$. Taking the Laplace transform of (6.8.20) and (6.8.21),

$$2sX(s) + Y(s) = \frac{s}{s^2 + 1}, \tag{6.8.22}$$

and

$$-2X(s) + sY(s) = 1 + \frac{1}{s^2 + 1}, \tag{6.8.23}$$

after introducing the initial conditions. Solving for $X(s)$ and $Y(s)$,

$$X(s) = -\frac{1}{(s^2 + 1)^2}, \tag{6.8.24}$$

and

$$Y(s) = \frac{s}{s^2 + 1} + \frac{2s}{(s^2 + 1)^2}. \tag{6.8.25}$$

Taking the inverse of (6.8.24)–(6.8.25) term by term,

$$x(t) = \tfrac{1}{2}[t\cos(t) - \sin(t)], \tag{6.8.26}$$

and

$$y(t) = t\sin(t) + \cos(t). \tag{6.8.27}$$

The MATLAB script

```
clear
% define symbolic variables
syms s t X Y
% take Laplace transform of left side of differential equations
LHS1 = laplace(2*diff(sym('x(t)'))+sym('y(t)'));
LHS2 = laplace(diff(sym('y(t)'))-2*sym('x(t)'));
% take Laplace transform of right side of differential equations
RHS1 = laplace(cos(t)); RHS2 = laplace(sin(t));
% set X and Y for Laplace transforms of x and y
%      and introduce initial conditions
newLHS1 = subs(LHS1,'laplace(x(t),t,s)','laplace(y(t),t,s)',...
    'x(0)','y(0)',X,Y,0,1);
newLHS2 = subs(LHS2,'laplace(x(t),t,s)','laplace(y(t),t,s)',...
    'x(0)','y(0)',X,Y,0,1);
% solve for X and Y
[X,Y] = solve(newLHS1-RHS1,newLHS2-RHS2,X,Y);
% invert Laplace transform and find x(t) and y(t)
x = ilaplace(X,s,t); y = ilaplace(Y,s,t)
```

uses the symbolic toolbox to solve (6.8.20)–(6.8.21). MATLAB finally gives

```
x =
1/2*t*cos(t)-1/2*sin(t)
y =
t*sin(t)+cos(t)
```

• Example 6.8.6

Let us determine the displacement of a mass m attached to a spring and excited by the driving force

$$F(t) = mA\left(1 - \frac{t}{T}\right)e^{-t/T}. \tag{6.8.28}$$

The dynamical equation governing this system is

$$y'' + \omega^2 y = A\left(1 - \frac{t}{T}\right)e^{-t/T}, \tag{6.8.29}$$

where $\omega^2 = k/m$ and k is the spring constant. Assuming that the system is initially at rest, the Laplace transform of the dynamical system is

$$(s^2 + \omega^2)Y(s) = \frac{A}{s + 1/T} - \frac{A}{T(s + 1/T)^2}, \qquad (6.8.30)$$

or

$$Y(s) = \frac{A}{(s^2 + \omega^2)(s + 1/T)} - \frac{A}{T(s^2 + \omega^2)(s + 1/T)^2}. \qquad (6.8.31)$$

Partial fractions yield

$$Y(s) = \frac{A}{\omega^2 + 1/T^2}\left(\frac{1}{s + 1/T} - \frac{s - 1/T}{s^2 + \omega^2}\right) - \frac{A}{T(\omega^2 + 1/T^2)^2}$$

$$\times \left[\frac{1/T^2 - \omega^2}{s^2 + \omega^2} - \frac{2s/T}{s^2 + \omega^2} + \frac{\omega^2 + 1/T^2}{(s + 1/T)^2} + \frac{2/T}{s + 1/T}\right]. \qquad (6.8.32)$$

Inverting (6.8.32) term by term,

$$y(t) = \frac{AT^2}{1 + \omega^2 T^2}\left[e^{-t/T} - \cos(\omega t) + \frac{\sin(\omega t)}{\omega T}\right]$$

$$- \frac{AT^2}{(1 + \omega^2 T^2)^2}\left\{(1 - \omega^2 T^2)\frac{\sin(\omega t)}{\omega T}\right.$$

$$\left. + 2\left[e^{-t/T} - \cos(\omega t)\right] + (1 + \omega^2 T^2)(t/T)e^{-t/T}\right\}. \qquad (6.8.33)$$

The solution to this problem consists of two parts. The exponential terms result from the forcing and will die away with time. This is the *transient* portion of the solution. The sinusoidal terms are those natural oscillations that are necessary so that the solution satisfies the initial conditions. They are the *steady-state* portion of the solution and endure forever. Figure 6.8.1 illustrates the solution when $\omega T = 0.1$, 1, and 2. Note that the displacement decreases in magnitude as the nondimensional frequency of the oscillator increases.

• Example 6.8.7

Let us solve the equation

$$y'' + 16y = \delta(t - \pi/4) \qquad (6.8.34)$$

with the initial conditions that $y(0) = 1$, and $y'(0) = 0$.

Taking the Laplace transform of (6.8.34) and inserting the initial conditions,

$$(s^2 + 16)Y(s) = s + e^{-s\pi/4}, \qquad (6.8.35)$$

or

$$Y(s) = \frac{s}{s^2 + 16} + \frac{e^{-s\pi/4}}{s^2 + 16}. \qquad (6.8.36)$$

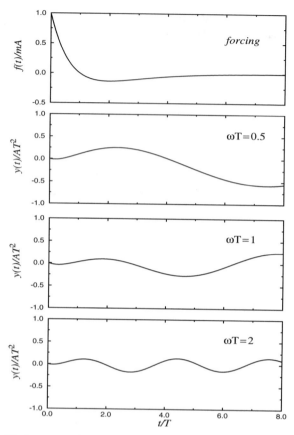

Figure 6.8.1: Displacement of a simple harmonic oscillator with nondimensional frequency ωT as a function of time t/T. The top frame shows the forcing function.

Applying the second shifting theorem,

$$y(t) = \cos(4t) + \tfrac{1}{4}\sin[4(t - \pi/4)]H(t - \pi/4) \qquad (\mathbf{6.8.37})$$
$$= \cos(4t) - \tfrac{1}{4}\sin(4t)H(t - \pi/4). \qquad (\mathbf{6.8.38})$$

We can check our results using the MATLAB script

```
clear
% define symbolic variables
syms pi s t Y
% take Laplace transform of left side of differential equation
LHS = laplace(diff(diff(sym('y(t)')))+16*sym('y(t)'));
% take Laplace transform of right side of differential equation
RHS = laplace('Dirac(t-pi/4)',t,s);
% set Y for Laplace transform of y
%      and introduce initial conditions
```

```
newLHS = subs(LHS,'laplace(y(t),t,s)','y(0)','D(y)(0)',Y,1,0);
% solve for Y
Y = solve(newLHS-RHS,Y);
% invert Laplace transform and find y(t)
y = ilaplace(Y,s,t)
```

which yields

```
y =
cos(4*t)-1/4*Heaviside(t-1/4*pi)*sin(4*t)
```

We can also verify that (6.8.38) is the solution to our initial-value problem by computing the (generalized) derivative of (6.8.38) or

$$y'(t) = -4\sin(4t) - \cos(4t)H(t - \pi/4) - \tfrac{1}{4}\sin(4t)\delta(t - \pi/4) \qquad (6.8.39)$$
$$= -4\sin(4t) - \cos(4t)H(t - \pi/4) - \tfrac{1}{4}\sin(\pi)\delta(t - \pi/4) \qquad (6.8.40)$$
$$= -4\sin(4t) - \cos(4t)H(t - \pi/4), \qquad (6.8.41)$$

since $f(t)\delta(t - t_0) = f(t_0)\delta(t - t_0)$. Similarly,

$$y''(t) = -16\cos(4t) + 4\sin(4t)H(t - \pi/4) - \cos(4t)\delta(t - \pi/4) \qquad (6.8.42)$$
$$= -16\cos(4t) + 4\sin(4t)H(t - \pi/4) - \cos(\pi)\delta(t - \pi/4) \qquad (6.8.43)$$
$$= -16\cos(4t) + 4\sin(4t)H(t - \pi/4) + \delta(t - \pi/4). \qquad (6.8.44)$$

Substituting (6.8.38) and (6.8.44) into (6.8.34) completes the verification. A quick check of $y(0)$ and $y'(0)$ also shows that we have the correct solution.

• Example 6.8.8: Oscillations in electric circuits

During the middle of the nineteenth century, Lord Kelvin[14] analyzed the LCR electrical circuit shown in Figure 6.8.2 which contains resistance R, capacitance C, and inductance L. For reasons that we shall shortly show, this LCR circuit has become one of the quintessential circuits for electrical engineers. In this example, we shall solve the problem by Laplace transforms.

Because we can add the potential differences across the elements, the equation governing the LCR circuit is

$$L\frac{dI}{dt} + RI + \frac{1}{C}\int_0^t I\,d\tau = E(t), \qquad (6.8.45)$$

where I denotes the current in the circuit. Let us solve (6.8.45) when we close the circuit and the initial conditions are $I(0) = 0$ and $Q(0) = -Q_0$. Taking the Laplace transform of (6.8.45),

$$\left(Ls + R + \frac{1}{Cs}\right)\overline{I}(s) = LI(0) - \frac{Q(0)}{Cs}. \qquad (6.8.46)$$

---

[14] Thomson, W., 1853: On transient electric currents. *Philos. Mag., Ser. 4*, **5**, 393–405.

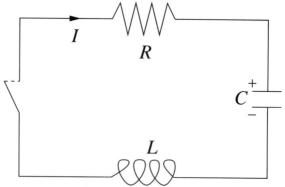

Figure 6.8.2: Schematic of a LCR circuit.

Solving for $\bar{I}(s)$,

$$\bar{I}(s) = \frac{Q_0}{Cs(Ls + R + 1/Cs)} = \frac{\omega_0^2 Q_0}{s^2 + 2\alpha s + \omega_0^2} = \frac{\omega_0^2 Q_0}{(s+\alpha)^2 + \omega_0^2 - \alpha^2},$$

(6.8.47)

where $\alpha = R/(2L)$, and $\omega_0^2 = 1/(LC)$. From the first shifting theorem,

$$I(t) = \frac{\omega_0^2 Q_0}{\omega} e^{-\alpha t} \sin(\omega t),$$

(6.8.48)

where $\omega^2 = \omega_0^2 - \alpha^2 > 0$. The quantity ω is the natural frequency of the circuit, which is lower than the free frequency ω_0 of a circuit formed by a condenser and coil. Most importantly, the solution decays in amplitude with time.

Although Kelvin's solution was of academic interest when he originally published it, this radically changed with the advent of radio telegraphy[15] because the LCR circuit described the fundamental physical properties of wireless transmitters and receivers.[16] The inescapable conclusion from numerous analyses was that no matter how cleverly the receiver was designed, eventually the resistance in the circuit would dampen the electrical oscillations and thus limit the strength of the received signal.

This technical problem was overcome by Armstrong[17] who invented an electrical circuit that used De Forest's audion (the first vacuum tube) for generating electrical oscillations and for amplifying externally impressed oscillations by "regenerative action." The effect of adding the "thermionic am-

[15] Stone, J S., 1914: The resistance of the spark and its effect on the oscillations of electrical oscillators. *Proc. IRE*, **2**, 307–324.

[16] See Hogan, J. L., 1916: Physical aspects of radio telegraphy. *Proc. IRE*, **4**, 397–420.

[17] Armstrong, E. H., 1915: Some recent developments in the audion receiver. *Proc. IRE*, **3**, 215–247.

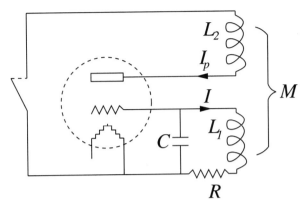

Figure 6.8.3: Schematic of a LCR circuit with the addition of a thermionic amplifier. [From Ballantine, S., 1919: The operational characteristics of thermionic amplifiers. *Proc. IRE*, **7**, 155. ©IRE (now IEEE).]

plifier" is seen by again considering the LRC circuit as shown in Figure 6.8.3 with the modification suggested by Armstrong.[18]

The governing equations of this new circuit are

$$L_1 \frac{dI}{dt} + RI + \frac{1}{C} \int_0^t I \, d\tau + M \frac{dI_p}{dt} = 0, \tag{6.8.49}$$

and

$$L_2 \frac{dI_p}{dt} + R_0 I_p + M \frac{dI}{dt} + \frac{\mu}{C} \int_0^t I \, d\tau = 0, \tag{6.8.50}$$

where the plate circuit has the current I_p, the resistance R_0, the inductance L_2, and the electromotive force (emf) of $\mu \int_0^t I \, d\tau / C$. The mutual inductance between the two circuits is given by M. Taking the Laplace transform of (6.8.49)–(6.8.50),

$$L_1 s \bar{I}(s) + R \bar{I}(s) + \frac{\bar{I}(s)}{sC} + M s \bar{I}_p(s) = \frac{Q_0}{sC}, \tag{6.8.51}$$

and

$$L_2 s \bar{I}_p(s) + R_0 \bar{I}_p(s) + M s \bar{I}(s) + \frac{\mu}{sC} \bar{I}(s) = 0. \tag{6.8.52}$$

Eliminating $\bar{I}_p(s)$ between (6.8.51)–(6.8.52) and solving for $\bar{I}(s)$,

$$\bar{I}(s) = \frac{(L_2 s + R_0) Q_0}{(L_1 L_2 - M^2) C s^3 + (R L_2 + R_0 L_1) C s^2 + (L_2 + C R R_0 - \mu M) s + R_0}. \tag{6.8.53}$$

For high-frequency radio circuits, we can approximate the roots of the denominator of (6.8.53) as

$$s_1 \approx -\frac{R_0}{L_2 + CRR_0 - \mu M}, \tag{6.8.54}$$

and

$$s_{2,3} \approx \frac{R_0}{2(L_2 + CRR_0 - \mu M)} - \frac{R_0 L_1 + RL_2}{2(L_1 L_2 - M^2)} \pm i\omega. \tag{6.8.55}$$

In the limit of M and R_0 vanishing, we recover our previous result for the LRC circuit. However, in reality, R_0 is very large and our solution has three terms. The term associated with s_1 is a rapidly decaying transient while the s_2 and s_3 roots yield oscillatory solutions with a *slight* amount of damping. Thus, our analysis shows that in the ordinary regenerative circuit, the tube effectively introduces sufficient "negative" resistance so that the resultant positive resistance of the equivalent LCR circuit is relatively low, and the response of an applied signal voltage at the resonant frequency of the circuit is therefore relatively great. Later, Armstrong[19] extended his work on regeneration by introducing an electrical circuit—the superregenerative circuit—where the regeneration is made large enough so that the resultant resistance is negative, and self-sustained oscillations can occur.[20] It was this circuit[21] which led to the explosive development of radio in the 1920s and 1930s.

• Example 6.8.9: Resonance transformer circuit

One of the fundamental electrical circuits of early radio telegraphy[22] is the resonance transformer circuit shown in Figure 6.8.4. Its development gave transmitters and receivers the ability to tune to each other.

The governing equations follow from Kirchhoff's law and are

$$L_1 \frac{dI_1}{dt} + M \frac{dI_2}{dt} + \frac{1}{C_1} \int_0^t I_1 \, d\tau = E(t), \tag{6.8.56}$$

and

$$M \frac{dI_1}{dt} + L_2 \frac{dI_2}{dt} + RI_2 + \frac{1}{C_2} \int_0^t I_2 \, d\tau = 0. \tag{6.8.57}$$

Let us examine the oscillations generated if initially the system has no currents or charges and the forcing function is $E(t) = \delta(t)$.

[19] Armstrong, E. H., 1922: Some recent developments of regenerative circuits. *Proc. IRE*, **10**, 244–260.

[20] See Frink, F. W., 1938: The basic principles of superregenerative reception. *Proc. IRE*, **26**, 76–106.

[21] Lewis, T., 1991: *Empire of the Air: The Men Who Made Radio*. HarperCollins Publishers, 421 pp.

[22] Fleming, J. A., 1919: *The Principles of Electric Wave Telegraphy and Telephony*. Longmans, Green, 911 pp.

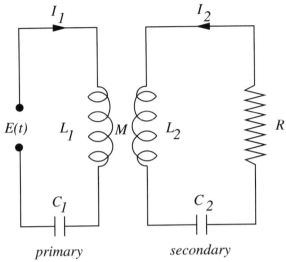

Figure 6.8.4: Schematic of a resonance transformer circuit.

Taking the Laplace transform of (6.8.56)–(6.8.57),

$$L_1 s \overline{I}_1 + M s \overline{I}_2 + \frac{\overline{I}_1}{sC_1} = 1, \qquad (6.8.58)$$

and

$$M s \overline{I}_1 + L_2 s \overline{I}_2 + R \overline{I}_2 + \frac{\overline{I}_2}{sC_2} = 0. \qquad (6.8.59)$$

Because the current in the second circuit is of greater interest, we solve for $\overline{I}_2$ and find that

$$\overline{I}_2(s) = -\frac{M s^3}{L_1 L_2 [(1 - k^2)s^4 + 2\alpha \omega_2^2 s^3 + (\omega_1^2 + \omega_2^2)s^2 + 2\alpha \omega_1^2 s + \omega_1^2 \omega_2^2]}, \qquad (6.8.60)$$

where $\alpha = R/(2L_2)$, $\omega_1^2 = 1/(L_1 C_1)$, $\omega_2^2 = 1/(L_2 C_2)$, and $k^2 = M^2/(L_1 L_2)$, the so-called coefficient of coupling.

We can obtain analytic solutions if we assume that the coupling is weak ($k^2 \ll 1$). Equation (6.8.60) becomes

$$\overline{I}_2 = -\frac{M s^3}{L_1 L_2 (s^2 + \omega_1^2)(s^2 + 2\alpha s + \omega_2^2)}. \qquad (6.8.61)$$

Using partial fractions and inverting term by term, we find that

$$I_2(t) = \frac{M}{L_1 L_2} \left[\frac{2\alpha \omega_1^3 \sin(\omega_1 t)}{(\omega_2^2 - \omega_1^2)^2 + 4\alpha^2 \omega_1^2} + \frac{\omega_1^2(\omega_2^2 - \omega_1^2)\cos(\omega_1 t)}{(\omega_2^2 - \omega_1^2)^2 + 4\alpha^2 \omega_1^2} \right.$$
$$+ \frac{\alpha \omega_2^4 - 3\alpha \omega_1^2 \omega_2^2 + 4\alpha^3 \omega_1^2}{(\omega_2^2 - \omega_1^2)^2 + 4\alpha^2 \omega_1^2} e^{-\alpha t} \frac{\sin(\omega t)}{\omega}$$
$$\left. - \frac{\omega_2^2(\omega_2^2 - \omega_1^2) + 4\alpha^2 \omega_1^2}{(\omega_2^2 - \omega_1^2)^2 + 4\alpha^2 \omega_1^2} e^{-\alpha t} \cos(\omega t) \right], \qquad (6.8.62)$$

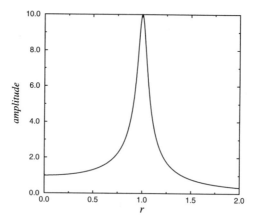

Figure 6.8.5: The resonance curve $1/\sqrt{(r^2-1)^2+0.01}$ for a resonance transformer circuit with $r = \omega_2/\omega_1$.

where $\omega^2 = \omega_2^2 - \alpha^2$.

The exponentially damped solutions will eventually disappear, leaving only the steady-state oscillations which vibrate with the angular frequency ω_1, the natural frequency of the primary circuit. If we rewrite this steady-state solution in amplitude/phase form, the amplitude is

$$\frac{M}{L_1 L_2 \sqrt{(r^2-1)^2 + 4\alpha^2/\omega_1^2}}, \tag{6.8.63}$$

where $r = \omega_2/\omega_1$. As Figure 6.8.5 shows, as r increases from zero to two, the amplitude rises until a very sharp peak occurs at $r = 1$ and then decreases just as rapidly as we approach $r = 2$. Thus, the resonance transformer circuit provides a convenient way to tune a transmitter or receiver to the frequency ω_1.

• **Example 6.8.10: Delay differential equation**

Laplace transforms provide a valuable tool in solving a general class of ordinary differential equations called *delay differential equations*. These equations arise in such diverse fields as chemical kinetics[23] and population dynamics.[24]

---

[23] See Roussel, M. R., 1996: The use of delay differential equations in chemical kinetics. *J. Phys. Chem.*, **100**, 8323–8330.

[24] See the first chapter of MacDonald, N., 1989: *Biological Delay Systems: Linear Stability Theory*. Cambridge University Press, 235 pp.

To illustrate the technique,[25] consider the differential equation

$$x' = -ax(t-1) \tag{6.8.64}$$

with $x(t) = 1 - at$ for $0 < t < 1$. Clearly, $x(0) = 1$.

Multiplying (6.8.64) by e^{-st} and integrating from 1 to ∞,

$$\int_1^\infty x'(t)e^{-st}\,dt = -a\int_1^\infty x(t-1)e^{-st}\,dt \tag{6.8.65}$$

$$\int_0^\infty x'(t)e^{-st}\,dt - \int_0^1 x'(t)e^{-st}\,dt = -a\int_0^\infty x(\tau)e^{-s(\tau+1)}\,d\tau \tag{6.8.66}$$

$$sX(s) - 1 + a\int_0^1 e^{-st}\,dt = -ae^{-s}X(s) \tag{6.8.67}$$

$$sX(s) - 1 - \frac{a}{s}\,e^{-st}\Big|_0^1 = -ae^{-s}X(s) \tag{6.8.68}$$

since $x'(t) = -a$ for $0 < t < 1$. Solving for $X(s)$,

$$X(s) = (1 + ae^{-s}/s - a/s)/[s(1 + ae^{-s}/s)]. \tag{6.8.69}$$

To facilitate the inversion of (6.8.69), we expand its denominator in terms of a geometric series and find that

$$X(s) = \sum_{n=0}^\infty (-a)^n e^{-ns}/s^{n+1} + \sum_{n=0}^\infty (-a)^{n+1} e^{-ns}/s^{n+2}$$

$$- \sum_{n=0}^\infty (-a)^{n+1} e^{-(n+1)s}/s^{n+2}. \tag{6.8.70}$$

The first and third sums cancel, except for the $n = 0$ term in the first sum. Therefore,

$$X(s) = \frac{1}{s} + \sum_{n=0}^\infty (-a)^{n+1} e^{-ns}/s^{n+2} \tag{6.8.71}$$

and

$$x(t) = 1 + \sum_{n=0}^\infty \frac{(-a)^{n+1}}{(n+1)!} H(t-n)(t-n)^{n+1}. \tag{6.8.72}$$

Figure 6.8.6 illustrates (6.8.72) as a function of time for various values of a. For $0 < a < e^{-1}$, $x(t)$ decays monotonically from 1 to an asymptotic limit of zero. For $e^{-1} < a < \pi/2$, the solution is a damped oscillatory function. If

---

[25] Reprinted with permission from Epstein, I. R., 1990: Differential delay equations in chemical kinetics: Some simple linear model systems. *J. Chem. Phys.*, **92**, 1702–1712. ©1990, American Institute of Physics.

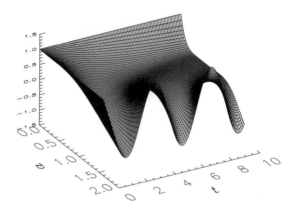

Figure 6.8.6: The solution to the differential delay equation (6.8.59) at various times t and values of a.

$\pi/2 < a$, then $x(t)$ is oscillatory with an exponentially increasing envelope. When $a = \pi/2$, $x(t)$ oscillates periodically.

● **Example 6.8.11**

Laplace transforms can sometimes be used to solve ordinary differential equations where the coefficients are powers of t. To illustrate this, let us solve

$$y'' + 2ty' - 4y = 0, \qquad y(0) = 1, \qquad \lim_{t \to \infty} y(t) \to 0. \qquad (6.8.73)$$

We begin by taking the Laplace transform of (6.8.73) and find that

$$s^2 Y(s) - sy(0) - y'(0) - 2\frac{d}{ds}[sY(s) - y(0)] - 4Y(s) = 0 \qquad (6.8.74)$$

An interesting aspect of this problem is the fact that we not know $y'(0)$. To circumvent this difficulty, let us temporarily set $y'(0) = -A$ so that (6.8.74) becomes

$$\frac{dY}{ds} + \left(\frac{3}{s} - \frac{s}{2}\right)Y = \frac{A}{2s} - \frac{1}{2}. \qquad (6.8.75)$$

Later on, we will find A.

Equation (6.8.75) is a first-order, linear, ordinary differential equation with s as its independent variable. To find $Y(s)$, we use the standard technique of multiplying it by its integrating factor, here $\mu(s) = s^3 e^{-s^2/4}$, and rewriting it as

$$\frac{d}{ds}\left[s^3 e^{-s^2/4} Y(s)\right] = \tfrac{1}{2} A s^2 e^{-s^2/4} - \tfrac{1}{2} s^3 e^{-s^2/4}. \qquad (6.8.76)$$

Integrating (6.8.76) from s to ∞, we obtain

$$s^3 e^{-s^2/4} Y(s) = (s^2 + 4)e^{-s^2/4} - A\left[se^{-s^2/4} + \sqrt{\pi}\,\mathrm{erfc}(s/2)\right], \qquad (6.8.77)$$

or

$$Y(s) = \frac{4}{s^3} + \frac{1}{s} - \frac{A}{s^2} - \frac{A\sqrt{\pi}}{s^3} e^{s^2/4} \operatorname{erfc}(s/2). \qquad (6.8.78)$$

We must now evaluate A. From the final-value theorem, $\lim_{t\to\infty} y(t) = \lim_{s\to 0} sY(s) = 0$. Therefore, multiplying (6.8.78) by s and using the expansion for the complementary error function for small s, we have that

$$sY(s) = \frac{4}{s^2} + 1 - \frac{A}{s} - \frac{A\sqrt{\pi}}{s^2}\left[1 + \frac{s^2}{4} - \frac{s}{\sqrt{\pi}} + \cdots\right]. \qquad (6.8.79)$$

In order that $\lim_{s\to 0} sY(s) = 0$, $A = 4/\sqrt{\pi}$. Therefore,

$$Y(s) = \frac{4}{s^3} + \frac{1}{s} - \frac{4}{\sqrt{\pi}\,s^2} - \frac{4}{s^3} e^{s^2/4} \operatorname{erfc}(s/2). \qquad (6.8.80)$$

The final step is to invert (6.8.80). Applying tables and the convolution theorem,

$$y(t) = 2t^2 + 1 - \frac{4t}{\sqrt{\pi}} - \frac{4}{\sqrt{\pi}} \int_0^t (t-x)^2 e^{-x^2}\, dx \qquad (6.8.81)$$

$$= (2t^2 + 1)[1 - \operatorname{erf}(t)] - \frac{2t}{\sqrt{\pi}} e^{-t^2}. \qquad (6.8.82)$$

Problems

Solve the following ordinary differential equations by Laplace transforms. Then use MATLAB to verify your solution.

1. $y' - 2y = 1 - t; \quad y(0) = 1$

2. $y'' - 4y' + 3y = e^t; \quad y(0) = 0, y'(0) = 0$

3. $y'' - 4y' + 3y = e^{2t}; \quad y(0) = 0, y'(0) = 1$

4. $y'' - 6y' + 8y = e^t; \quad y(0) = 3, y'(0) = 9$

5. $y'' + 4y' + 3y = e^{-t}; \quad y(0) = 1, y'(0) = 1$

6. $y'' + y = t; \quad y(0) = 1, y'(0) = 0$

7. $y'' + 4y' + 3y = e^t; \quad y(0) = 0, y'(0) = 2$

8. $y'' - 4y' + 5y = 0; \quad y(0) = 2, y'(0) = 4$

9. $y' + y = tH(t-1); \quad y(0) = 0$

10.　　$y'' + 3y' + 2y = H(t - 1);$　　$y(0) = 0, y'(0) = 1$

11.　　$y'' - 3y' + 2y = H(t - 1);$　　$y(0) = 0, y'(0) = 1$

12.　　$y'' + 4y = 3H(t - 4);$　　$y(0) = 1, y'(0) = 0$

13.　　$y'' + 4y' + 4y = 4H(t - 2);$　　$y(0) = 0, y'(0) = 0$

14.　　$y'' + 3y' + 2y = e^{t-1} H(t - 1);$　　$y(0) = 0, y'(0) = 1$

15.　　$y'' - 3y' + 2y = e^{-(t-2)} H(t - 2);$　　$y(0) = 0, y'(0) = 0$

16.　　$y'' - 3y' + 2y = H(t - 1) - H(t - 2);$　　$y(0) = 0, y'(0) = 0$

17.　　$y'' + y = 1 - H(t - T);$　　$y(0) = 0, y'(0) = 0$

18.　　$y'' + y = \begin{cases} \sin(t), & 0 \le t \le \pi, \\ 0, & \pi \le t; \end{cases}$　　$y(0) = 0, y'(0) = 0$

19.　　$y'' + 3y' + 2y = \begin{cases} t, & 0 \le t \le a, \\ ae^{-(t-a)}, & a \le t; \end{cases}$　　$y(0) = 0, y'(0) = 0$

20.　　$y'' + \omega^2 y = \begin{cases} t/a, & 0 \le t \le a, \\ 1 - (t - a)/(b - a), & a \le t \le b, \\ 0, & b \le t; \end{cases}$　　$y(0) = 0, y'(0) = 0$

21.　　$y'' - 2y' + y = 3\delta(t - 2);$　　$y(0) = 0, y'(0) = 1$

22.　　$y'' - 5y' + 4y = \delta(t - 1);$　　$y(0) = 0, y'(0) = 0$

23.　　$y'' + 5y' + 6y = 3\delta(t - 2) - 4\delta(t - 5);$　　$y(0) = y'(0) = 0$

24.　　$y'' + \omega y' = A\delta(t - \tau) - BH(t - \tau);$　　$y(0) = y'(0) = 0$

25.　　$x' - 2x + y = 0,$　　$y' - 3x - 4y = 0;$　　$x(0) = 1, y(0) = 0$

26.　　$x' - 2y' = 1,$　　$x' + y - x = 0;$　　$x(0) = y(0) = 0$

27.　　$x' + 2x - y' = 0,$　　$x' + y + x = t^2;$　　$x(0) = y(0) = 0$

28.　　$x' + 3x - y = 1,$　　$x' + y' + 3x = 0;$　　$x(0) = 2, y(0) = 0$

29.   Forster, Escobal, and Lieske[26] used Laplace transforms to solve the linearized equations of motion of a vehicle in a gravitational field created by two other bodies. A simplified form of this problem involves solving the following system of ordinary differential equations:

$$x'' - 2y' = F_1 + x + 2y, \qquad 2x' + y'' = F_2 + 2x + 3y,$$

subject to the initial conditions that $x(0) = y(0) = x'(0) = y'(0) = 0$. Find the solution to this system.

Use Laplace transforms to find the solution for the following ordinary differential equations:

30. $y'' + 2ty' - 8y = 0, \qquad y(0) = 1, \quad y'(0) = 0$

31. $y'' - ty' + 2y = 0, \qquad y(0) = -1, \quad y'(0) = 0$

Step 1: Show that the Laplace transform for these differential equations is

30. $2sY'(s) + (10 - s^2)Y(s) = -s$

31. $sY'(s) + (s^2 + 3)Y(s) = -s$

Step 2: Solve these first-order ordinary differential equations and show that

30. $Y(s) = 1/s + 8/s^3 + 32/s^5 + Ae^{s^2/4}/s^5$

31. $Y(s) = (A - 2)e^{-s^2/2}/s^3 + 2/s^3 - 1/s$

Step 3: Invert $Y(s)$ and show that the general solutions are

30. $y(t) = 1 + 4t^2 + 4t^4/3$ 31. $y(t) = t^2 - 1$

Use Laplace transforms to find the general solutions for the following ordinary differential equations:

32. $ty'' - (2 - t)y' - y = 0$

33. $ty'' - 2(a + bt)y' + b(2a + bt)y = 0, \qquad a \geq 0$

Step 1: Show that the Laplace transform for these differential equations is

---

[26] Reprinted from *Astronaut. Acta*, **14**, K. Forster, P. R. Escobal, and H. A. Lieske, Motion of a vehicle in the transition region of the three-body problem, 1–10, ©1968, with kind permission from Elsevier Science Ltd, The Boulevard, Langford Lane, Kidlington OX5 1GB, UK.

32. $s(s+1)Y'(s) + 2(2s+1)Y(s) = 3y(0)$

33. $(s-b)^2Y'(s) + 2(1+a)(s-b)Y(s) = (1+2a)y(0)$

Step 2: Solve these first-order ordinary differential equations and show that

32. $Y(s) = y(0)/(s+1) + y(0)/[2(s+1)^2] + A/[s^2(s+1)^2]$

33. $Y(s) = y(0)/(s-b) + A/(s-b)^{2+2a}$

Step 3: Invert $Y(s)$ and show that the general solutions are

32. $y(t) = C_1(t+2)e^{-t} + C_2(t-2)$ 33. $y(t) = C_1 e^{bt} + C_2 t^{2a+1} e^{bt}$

6.9 TRANSFER FUNCTIONS, GREEN'S FUNCTION, AND INDICIAL ADMITTANCE

One of the drawbacks of using Laplace transforms to solve ordinary differential equations with a forcing term is its lack of generality. Each new forcing function requires a repetition of the entire process. In this section we give some methods for finding the solution in a somewhat more general manner for stationary systems where the forcing, not any initially stored energy (i.e., nonzero initial conditions), produces the total output. Unfortunately, the solution must be written as an integral.

In Example 6.8.3 we solved the linear differential equation

$$y'' + 2y' + y = f(t), \tag{6.9.1}$$

subject to the initial conditions $y(0) = y'(0) = 0$. At that time we wrote the Laplace transform of $y(t)$, $Y(s)$, as the product of two Laplace transforms:

$$Y(s) = \frac{1}{(s+1)^2}F(s). \tag{6.9.2}$$

One drawback in using (6.9.2) is its dependence upon an unspecified Laplace transform $F(s)$. Is there a way to eliminate this dependence and yet retain the essence of the solution?

One way of obtaining a quantity that is independent of the forcing is to consider the ratio:

$$\frac{Y(s)}{F(s)} = G(s) = \frac{1}{(s+1)^2}. \tag{6.9.3}$$

This ratio is called the *transfer function* because we can transfer the input $F(s)$ into the output $Y(s)$ by multiplying $F(s)$ by $G(s)$. It depends only upon the properties of the system.

Let us now consider a related problem to (6.9.1), namely

$$g'' + 2g' + g = \delta(t), \qquad t > 0, \qquad (6.9.4)$$

with $g(0) = g'(0) = 0$. Because the forcing equals the Dirac delta function, $g(t)$ is called the *impulse response* or *Green's function*.[27] Computing $G(s)$,

$$G(s) = \frac{1}{(s+1)^2}. \qquad (6.9.5)$$

From (6.9.3) we see that $G(s)$ is also the transfer function. Thus, an alternative method for computing the transfer function is to subject the system to impulse forcing and the Laplace transform of the response is the transfer function.

From (6.9.3),

$$Y(s) = G(s)F(s), \qquad (6.9.6)$$

or

$$y(t) = g(t) * f(t). \qquad (6.9.7)$$

That is, the convolution of the impulse response with the particular forcing gives the response of the system. Thus, we may describe a stationary system in one of two ways: (1) in the transform domain we have the transfer function, and (2) in the time domain there is the impulse response.

Despite the fundamental importance of the impulse response or Green's function for a given linear system, it is often quite difficult to determine, especially experimentally, and a more convenient practice is to deal with the response to the unit step $H(t)$. This response is called the *indicial admittance* or *step response*, which we shall denote by $a(t)$. Because $\mathcal{L}[H(t)] = 1/s$, we can determine the transfer function from the indicial admittance because $\mathcal{L}[a(t)] = G(s)\mathcal{L}[H(t)]$ or $sA(s) = G(s)$. Furthermore, because

$$\mathcal{L}[g(t)] = G(s) = \frac{\mathcal{L}[a(t)]}{\mathcal{L}[H(t)]}, \qquad (6.9.8)$$

then

$$g(t) = \frac{da(t)}{dt} \qquad (6.9.9)$$

from (6.1.18).

---

[27] For the origin of the Green's function, see Farina, J. E. G., 1976: The work and significance of George Green, the miller mathematician, 1793–1841. *Bull. Inst. Math. Appl.*, **12**, 98–105.

• **Example 6.9.1**

Let us find the transfer function, impulse response, and step response for the system

$$y'' - 3y' + 2y = f(t), \qquad (6.9.10)$$

with $y(0) = y'(0) = 0$. To find the impulse response, we solve

$$g'' - 3g' + 2g = \delta(t), \qquad (6.9.11)$$

with $g(0) = g'(0) = 0$. Taking the Laplace transform of (6.9.11), we find that

$$G(s) = \frac{1}{s^2 - 3s + 2}, \qquad (6.9.12)$$

which is the transfer function for this system. The impulse response equals the inverse of $G(s)$ or

$$g(t) = e^{2t} - e^t. \qquad (6.9.13)$$

To find the step response, we solve

$$a'' - 3a' + 2a = H(t), \qquad (6.9.14)$$

with $a(0) = a'(0) = 0$. Taking the Laplace transform of (6.9.14),

$$A(s) = \frac{1}{s(s-1)(s-2)}, \qquad (6.9.15)$$

or

$$a(t) = \tfrac{1}{2} + \tfrac{1}{2}e^{2t} - e^t. \qquad (6.9.16)$$

Note that $a'(t) = g(t)$.

• **Example 6.9.2**

MATLAB's control toolbox contains several routines for the numerical computation of impulse and step responses if the transfer function can be written as the ratio of two polynomials. To illustrate this capacity, let us redo the previous example where the transfer function is given by (6.9.12). The transfer function is introduced by loading in the polynomial in the numerator num and in the denominator den followed by calling tf. The MATLAB script

```
clear
% load in coefficients of the numerator and denominator
%      of the transfer function
num = [0 0 1]; den = [1 -3 2];
% create the transfer function
sys = tf(num,den);
% find the step response, a
```

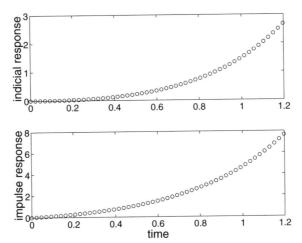

Figure 6.9.1: The impulse and step responses corresponding to the transfer function (6.9.12).

```
[a,t] = step(sys);
% plot the indicial admittance
subplot(2,1,1), plot(t, a, 'o')
ylabel('indicial response','Fontsize',20)
% find the impulse response, g
[g,t] = impulse(sys);
% plot the impulse response
subplot(2,1,2), plot(t, g, 'o')
ylabel('impulse response','Fontsize',20)
xlabel('time','Fontsize',20)
```

shows how the impulse and step responses are found. Both of them are shown in Figure 6.9.1.

• **Example 6.9.3**

There is an old joke about a man who took his car into a garage because of a terrible knocking sound. Upon his arrival the mechanic took one look at it and gave it a hefty kick.[28] Then, without a moment's hesitation he opened the hood, bent over, and tightened up a loose bolt. Turning to the owner, he said, "Your car is fine. That'll be $50." The owner felt that the charge was somewhat excessive, and demanded an itemized account. The mechanic said, "The kicking of the car and tightening one bolt, cost you a buck. The remaining $49 comes from knowing where to kick the car and finding the loose bolt."

---

[28] This is obviously a very old joke.

Although the moral of the story may be about expertise as a marketable commodity, it also illustrates the concept of transfer function.[29] Let us model the car as a linear system where the equation

$$a_n \frac{d^n y}{dt^n} + a_{n-1} \frac{d^{n-1} y}{dt^{n-1}} + \cdots + a_1 \frac{dy}{dt} + a_0 y = f(t) \qquad (6.9.17)$$

governs the response $y(t)$ to a forcing $f(t)$. Assuming that the car has been sitting still, the initial conditions are zero and the Laplace transform of (6.9.17) is

$$K(s)Y(s) = F(s), \qquad (6.9.18)$$

where

$$K(s) = a_n s^n + a_{n-1} s^{n-1} + \cdots + a_1 s + a_0. \qquad (6.9.19)$$

Hence

$$Y(s) = \frac{F(s)}{K(s)} = G(s)F(s), \qquad (6.9.20)$$

where the transfer function $G(s)$ clearly depends only on the internal workings of the car. So if we know the transfer function, we understand how the car vibrates because

$$y(t) = \int_0^t g(t-x)f(x)\,dx. \qquad (6.9.21)$$

But what does this have to do with our mechanic? He realized that a short sharp kick mimics an impulse forcing with $f(t) = \delta(t)$ and $y(t) = g(t)$. Therefore, by observing the response of the car to his kick, he diagnosed the loose bolt and fixed the car.

In this section we showed how the response of any system can be expressed in terms of its Green's function and the arbitrary forcing. Can we also determine the response using the indicial admittance $a(t)$?

Consider first a system that is dormant until a certain time $t = \tau_1$. At that instant we subject the system to a forcing $H(t - \tau_1)$. Then the response will be zero if $t < \tau_1$ and will equal the indicial admittance $a(t - \tau_1)$ when $t > \tau_1$ because the indicial admittance is the response of a system to the step function. Here $t - \tau_1$ is the time measured from the instant of change.

Next, suppose that we now force the system with the value $f(0)$ when $t = 0$ and hold that value until $t = \tau_1$. We then abruptly change the forcing by an amount $f(\tau_1) - f(0)$ to the value $f(\tau_1)$ at the time τ_1 and hold it at that value until $t = \tau_2$. Then we again abruptly change the forcing by an amount $f(\tau_2) - f(\tau_1)$ at the time τ_2, and so forth (see Figure 6.9.2). From

---

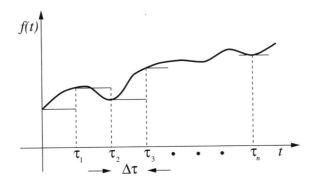

Figure 6.9.2: Diagram used in the derivation of Duhamel's integral.

the *linearity* of the problem the response after the instant $t = \tau_n$ equals the sum

$$y(t) = f(0)a(t) + [f(\tau_1) - f(0)]a(t - \tau_1) + [f(\tau_2) - f(\tau_1)]a(t - \tau_2)$$
$$+ \cdots + [f(\tau_n) - f(\tau_{n-1})]a(t - \tau_n). \qquad (6.9.22)$$

If we write $f(\tau_k) - f(\tau_{k-1}) = \Delta f_k$ and $\tau_k - \tau_{k-1} = \Delta \tau_k$, (6.9.22) becomes

$$y(t) = f(0)a(t) + \sum_{k=1}^{n} a(t - \tau_k) \frac{\Delta f_k}{\Delta \tau_k} \Delta \tau_k. \qquad (6.9.23)$$

Finally, proceeding to the limit as the number n of jumps becomes infinite, in such a manner that all jumps and intervals between successive jumps tend to zero, this sum has the limit

$$y(t) = f(0)a(t) + \int_0^t f'(\tau)a(t - \tau) \, d\tau. \qquad (6.9.24)$$

Because the total response of the system equals the weighted sum [the weights being $a(t)$] of the forcing from the initial moment up to the time t, we refer to (6.9.24) as the *superposition integral*, or *Duhamel's integral*,[30] named after the French mathematical physicist Jean-Marie-Constant Duhamel (1797–1872) who first derived it in conjunction with heat conduction.

We can also express (6.9.24) in several different forms. Integration by parts yields

$$y(t) = f(t)a(0) + \int_0^t f(\tau)a'(t - \tau) \, d\tau = \frac{d}{dt} \left[\int_0^t f(\tau)a(t - \tau) \, d\tau \right]. \qquad (6.9.25)$$

---

[30] Duhamel, J.-M.-C., 1833: Mémoire sur la méthode générale relative au mouvement de la chaleur dans les corps solides plongés dans des milieux dont la température varie avec le temps. *J. École Polytech.*, **22**, 20–77.

• **Example 6.9.4**

Suppose that a system has the step response of $a(t) = A[1 - e^{-t/T}]$, where A and T are positive constants. Let us find the response if we force this system by $f(t) = kt$, where k is a constant.

From the superposition integral (6.9.24),

$$y(t) = 0 + \int_0^t kA[1 - e^{-(t-\tau)/T}]\,d\tau = kA[t - T(1 - e^{-t/T})]. \qquad (\textbf{6.9.26})$$

Problems

For the following nonhomogeneous differential equations, find the transfer function, impulse response, and step response. Assume that all of the necessary initial conditions are zero. If you have MATLAB's control toolbox, use MATLAB to check your work.

1. $y' + ky = f(t)$ 　　　　　　　　　　2. $y'' - 2y' - 3y = f(t)$

3. $y'' + 4y' + 3y = f(t)$ 　　　　　　　4. $y'' - 2y' + 5y = f(t)$

5. $y'' - 3y' + 2y = f(t)$ 　　　　　　　6. $y'' + 4y' + 4y = f(t)$

7. $y'' - 9y = f(t)$ 　　　　　　　　　　8. $y'' + y = f(t)$

9. $y'' - y' = f(t)$

6.10 INVERSION BY CONTOUR INTEGRATION

In §6.5 and 6.6 we showed how we can use partial fractions and convolution to find the inverse of the Laplace transform $F(s)$. In many instances these methods fail simply because of the complexity of the transform to be inverted. In this section we shall show how we can invert transforms through the powerful method of contour integration. Of course, the student must be proficient in the use of complex variables.

Consider the piece-wise differentiable function $f(x)$ which vanishes for $x < 0$. We can express the function $e^{-cx}f(x)$ by the complex Fourier representation of

$$f(x)e^{-cx} = \frac{1}{2\pi} \int_{-\infty}^{\infty} e^{i\omega x}\left[\int_0^{\infty} e^{-ct}f(t)e^{-i\omega t}\,dt\right]d\omega, \qquad (\textbf{6.10.1})$$

for any value of the real constant c, where the integral

$$I = \int_0^{\infty} e^{-ct}|f(t)|\,dt \qquad (\textbf{6.10.2})$$

exists. By multiplying both sides of (6.10.1) by e^{cx} and bringing it inside the first integral,

$$f(x) = \frac{1}{2\pi} \int_{-\infty}^{\infty} e^{(c+\omega i)x} \left[\int_0^{\infty} f(t) e^{-(c+\omega i)t} \, dt \right] d\omega. \qquad (6.10.3)$$

With the substitution $z = c + \omega i$, where z is a new, complex variable of integration,

$$f(x) = \frac{1}{2\pi i} \int_{c-\infty i}^{c+\infty i} e^{zx} \left[\int_0^{\infty} f(t) e^{-zt} \, dt \right] dz. \qquad (6.10.4)$$

The quantity inside the square brackets is the Laplace transform $F(z)$. Therefore, we can express $f(t)$ in terms of its transform by the complex contour integral

$$f(t) = \frac{1}{2\pi i} \int_{c-\infty i}^{c+\infty i} F(z) e^{tz} \, dz. \qquad (6.10.5)$$

This line integral, *Bromwich's integral*,[31] runs along the line $x = c$ parallel to the imaginary axis and c units to the right of it, the so-called *Bromwich contour*. We select the value of c sufficiently large so that the integral (6.10.2) exists; subsequent analysis shows that this occurs when c is larger than the real part of any of the singularities of $F(z)$.

We must now evaluate the contour integral. Because of the power of the *residue* theorem in complex variables, the contour integral is usually transformed into a closed contour through the use of *Jordan's lemma*. See §5.4, Equations (5.4.12) and (5.4.13). The following examples will illustrate the proper use of (6.10.5).

- **Example 6.10.1**

Let us invert

$$F(s) = \frac{e^{-3s}}{s^2(s-1)}. \qquad (6.10.6)$$

From Bromwich's integral,

$$f(t) = \frac{1}{2\pi i} \int_{c-\infty i}^{c+\infty i} \frac{e^{(t-3)z}}{z^2(z-1)} \, dz \qquad (6.10.7)$$

$$= \frac{1}{2\pi i} \oint_C \frac{e^{(t-3)z}}{z^2(z-1)} \, dz - \frac{1}{2\pi i} \int_{C_R} \frac{e^{(t-3)z}}{z^2(z-1)} \, dz, \qquad (6.10.8)$$

---

[31] Bromwich, T. J. I'A., 1916: Normal coordinates in dynamical systems. *Proc. London Math. Soc., Ser. 2*, **15**, 401–448.

Figure 6.10.1: An outstanding mathematician at Cambridge University at the turn of the twentieth century, Thomas John l'Anson Bromwich (1875–1929) came to Heaviside's operational calculus through his interest in divergent series. Beginning a correspondence with Heaviside, Bromwich was able to justify operational calculus through the use of contour integrals by 1915. After his premature death, individuals such as J. R. Carson and Sir H. Jeffreys brought Laplace transforms to the increasing attention of scientists and engineers. (Portrait courtesy of the Royal Society of London.)

where C_R is a semicircle of infinite radius in either the right or left half of the z-plane and C is the closed contour that includes C_R and Bromwich's contour. See Figure 6.10.2.

Our first task is to choose an appropriate contour so that the integral along C_R vanishes. By Jordan's lemma this requires a semicircle in the right half-plane if $t - 3 < 0$ and a semicircle in the left half-plane if $t - 3 > 0$. Consequently, by considering these two separate cases, we force the second integral in (6.10.8) to zero and the inversion simply equals the closed contour.

Consider the case $t < 3$ first. Because Bromwich's contour lies to the right of any singularities, there are no singularities within the closed contour and $f(t) = 0$.

Consider now the case $t > 3$. Within the closed contour in the left half-plane, there is a second-order pole at $z = 0$ and a simple pole at $z = 1$.

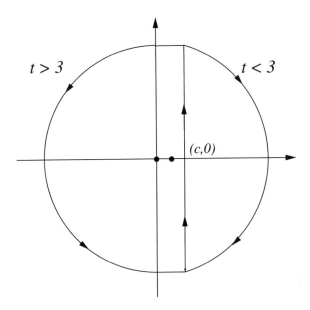

Figure 6.10.2: Contours used in the inversion of (6.10.6).

Therefore,

$$f(t) = \text{Res}\left[\frac{e^{(t-3)z}}{z^2(z-1)};0\right] + \text{Res}\left[\frac{e^{(t-3)z}}{z^2(z-1)};1\right], \qquad (6.10.9)$$

where

$$\text{Res}\left[\frac{e^{(t-3)z}}{z^2(z-1)};0\right] = \lim_{z \to 0} \frac{d}{dz}\left[z^2\frac{e^{(t-3)z}}{z^2(z-1)}\right] \qquad (6.10.10)$$

$$= \lim_{z \to 0}\left[\frac{(t-3)e^{(t-3)z}}{z-1} - \frac{e^{(t-3)z}}{(z-1)^2}\right] \qquad (6.10.11)$$

$$= 2 - t, \qquad (6.10.12)$$

and

$$\text{Res}\left[\frac{e^{(t-3)z}}{z^2(z-1)};1\right] = \lim_{z \to 1}(z-1)\frac{e^{(t-3)z}}{z^2(z-1)} = e^{t-3}. \qquad (6.10.13)$$

Taking our earlier results into account, the inverse equals

$$f(t) = \left[e^{t-3} - (t-3) - 1\right]H(t-3) \qquad (6.10.14)$$

which we would have obtained from the second shifting theorem and tables.

• **Example 6.10.2**

For our second example of the inversion of Laplace transforms by complex integration, let us find the inverse of

$$F(s) = \frac{1}{s \sinh(as)}, \tag{6.10.15}$$

where a is real. From Bromwich's integral,

$$f(t) = \frac{1}{2\pi i} \int_{c-\infty i}^{c+\infty i} \frac{e^{tz}}{z \sinh(az)} \, dz. \tag{6.10.16}$$

Here c is greater than the real part of any of the singularities in (6.10.15). Using the infinite product for the hyperbolic sine,[32]

$$\frac{e^{tz}}{z \sinh(az)} = \frac{e^{tz}}{az^2[1 + a^2z^2/\pi^2][1 + a^2z^2/(4\pi^2)][1 + a^2z^2/(9\pi^2)] \cdots}. \tag{6.10.17}$$

Thus, we have a second-order pole at $z = 0$ and simple poles at $z_n = \pm n\pi i/a$, where $n = 1, 2, 3, \ldots$.

We can convert the line integral (6.10.16), with the Bromwich contour lying parallel and slightly to the right of the imaginary axis, into a closed contour using Jordan's lemma through the addition of an infinite semicircle joining $i\infty$ to $-i\infty$ as shown in Figure 6.10.3. We now apply the residue theorem. For the second-order pole at $z = 0$,

$$\text{Res}\left[\frac{e^{tz}}{z \sinh(az)}; 0\right] = \frac{1}{1!} \lim_{z \to 0} \frac{d}{dz}\left[\frac{(z-0)^2 e^{tz}}{z \sinh(az)}\right] \tag{6.10.18}$$

$$= \lim_{z \to 0} \frac{d}{dz}\left[\frac{z e^{tz}}{\sinh(az)}\right] \tag{6.10.19}$$

$$= \lim_{z \to 0}\left[\frac{e^{tz}}{\sinh(az)} + \frac{zt e^{tz}}{\sinh(az)} - \frac{az \cosh(az) e^{tz}}{\sinh^2(az)}\right] \tag{6.10.20}$$

$$= \frac{t}{a} \tag{6.10.21}$$

after using $\sinh(az) = az + O(z^3)$. For the simple poles $z_n = \pm n\pi i/a$,

$$\text{Res}\left[\frac{e^{tz}}{z \sinh(az)}; z_n\right] = \lim_{z \to z_n} \frac{(z - z_n) e^{tz}}{z \sinh(az)} \tag{6.10.22}$$

$$= \lim_{z \to z_n} \frac{e^{tz}}{\sinh(az) + az \cosh(az)} \tag{6.10.23}$$

$$= \frac{\exp(\pm n\pi it/a)}{(-1)^n(\pm n\pi i)}, \tag{6.10.24}$$

---

[32] Gradshteyn, I. S., and I. M. Ryzhik, 1965: *Table of Integrals, Series and Products.* Academic Press, §1.431, Formula 2.

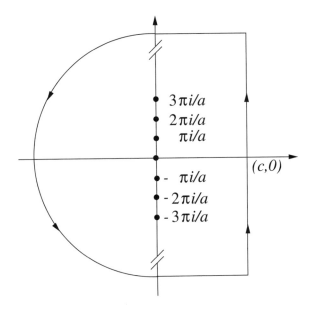

Figure 6.10.3: Contours used in the inversion of (6.10.15).

because $\cosh(\pm n\pi i) = \cos(n\pi) = (-1)^n$. Thus, summing up all of the residues gives

$$f(t) = \frac{t}{a} + \sum_{n=1}^{\infty} \frac{(-1)^n \exp(n\pi it/a)}{n\pi i} - \sum_{n=1}^{\infty} \frac{(-1)^n \exp(-n\pi it/a)}{n\pi i} \quad \text{(6.10.25)}$$

$$= \frac{t}{a} + \frac{2}{\pi} \sum_{n=1}^{\infty} \frac{(-1)^n}{n} \sin(n\pi t/a). \quad \text{(6.10.26)}$$

In addition to computing the inverse of Laplace transforms, Bromwich's integral places certain restrictions on $F(s)$ in order that an inverse exists. If α denotes the minimum value that c may possess, the restrictions are threefold.[33] First, $F(z)$ must be analytic in the half-plane $x \geq \alpha$, where $z = x + iy$. Second, in the same half-plane it must behave as z^{-k}, where $k > 1$. Finally, $F(x)$ must be real when $x \geq \alpha$.

• **Example 6.10.3**

Is the function $\sin(s)/(s^2 + 4)$ a proper Laplace transform? Although the function satisfies the first and third criteria listed in the previous paragraph on the half-plane $x > 2$, the function becomes unbounded as $y \to \pm\infty$ for any fixed $x > 2$. Thus, $\sin(s)/(s^2 + 4)$ cannot be a Laplace transform.

---

[33] For the proof, see Churchill, R. V., 1972: *Operational Mathematics*. McGraw-Hill, §67.

• Example 6.10.4

An additional benefit of understanding inversion by the residue method is the ability to qualitatively anticipate the inverse by knowing the location of the poles of $F(s)$. This intuition is important because many engineering analyses discuss stability and performance entirely in terms of the properties of the system's Laplace transform. In Figure 6.10.4 we have graphed the location of the poles of $F(s)$ and the corresponding $f(t)$. The student should go through the mental exercise of connecting the two pictures.

Problems

Use Bromwich's integral to invert the following Laplace transforms:

1. $F(s) = \dfrac{s+1}{(s+2)^2(s+3)}$

2. $F(s) = \dfrac{1}{s^2(s+a)^2}$

3. $F(s) = \dfrac{1}{s(s-2)^3}$

4. $F(s) = \dfrac{1}{s(s+a)^2(s^2+b^2)}$

5. $F(s) = \dfrac{e^{-s}}{s^2(s+2)}$

6. $F(s) = \dfrac{1}{s(1+e^{-as})}$

7. $F(s) = \dfrac{1}{(s+b)\cosh(as)}$

8. $F(s) = \dfrac{1}{s(1-e^{-as})}$

9. Consider a function $f(t)$ which has the Laplace transform $F(z)$ which is analytic in the half-plane $\text{Re}(z) > s_0$. Can we use this knowledge to find $g(t)$ whose Laplace transform $G(z)$ equals $F[\varphi(z)]$, where $\varphi(z)$ is also analytic for $\text{Re}(z) > s_0$? The answer to this question leads to the Schouten[34]–Van der Pol[35] theorem.

Step 1: Show that the following relationships hold true:

$$G(z) = F[\varphi(z)] = \int_0^\infty f(\tau)e^{-\varphi(z)\tau}\,d\tau,$$

and

$$g(t) = \frac{1}{2\pi i}\int_{c-\infty i}^{c+\infty i} F[\varphi(z)]e^{tz}\,dz.$$

---

[34] Schouten, J. P., 1935: A new theorem in operational calculus together with an application of it. *Physica*, **2**, 75–80.

[35] Van der Pol, B., 1934: A theorem on electrical networks with applications to filters. *Physica*, **1**, 521–530.

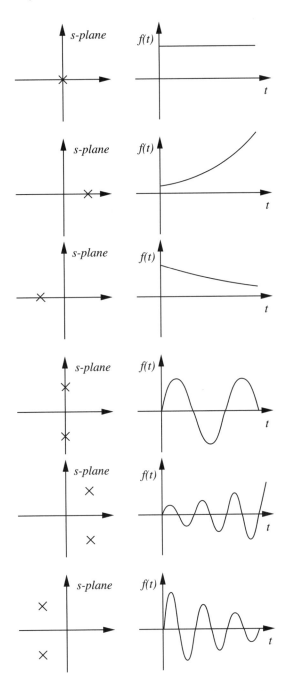

Figure 6.10.4: The correspondence between the location of the simple poles of the Laplace transform $F(s)$ and the behavior of $f(t)$.

Step 2: Using the results from Step 1, show that

$$g(t) = \int_0^\infty f(\tau) \left[\frac{1}{2\pi i} \int_{c-\infty i}^{c+\infty i} e^{-\varphi(z)\tau} e^{tz} \, dz \right] d\tau.$$

This is the Schouten-Van der Pol theorem.

Step 3: If $G(z) = F(\sqrt{z})$ show that

$$g(t) = \frac{1}{2\sqrt{\pi t^3}} \int_0^\infty \tau f(\tau) \exp\left(-\frac{\tau^2}{4t} \right) d\tau.$$

Hint: Do not evaluate the contour integral. Instead, ask yourself: What function of time has a Laplace transform that equals $e^{-\varphi(z)\tau}$, where τ is a parameter? Then use tables.

Chapter 7
The Z-Transform

Since the Second World War, the rise of digital technology has resulted in a corresponding demand for designing and understanding discrete-time (data sampled) systems. These systems are governed by *difference equations* in which members of the sequence y_n are coupled to each other.

One source of difference equations is the numerical evaluation of integrals on a digital computer. Because we can only have values at discrete time points $t_k = kT$ for $k = 0, 1, 2, \ldots$, the value of the integral $y(t) = \int_0^t f(\tau)\, d\tau$ is

$$y(kT) = \int_0^{kT} f(\tau)\, d\tau = \int_0^{(k-1)T} f(\tau)\, d\tau + \int_{(k-1)T}^{kT} f(\tau)\, d\tau \qquad (7.0.1)$$

$$= y[(k-1)T] + \int_{(k-1)T}^{kT} f(\tau)\, d\tau \qquad (7.0.2)$$

$$= y[(k-1)T] + Tf(kT), \qquad (7.0.3)$$

because $\int_{(k-1)T}^{kT} f(\tau)\, d\tau \approx Tf(kT)$. Equation (7.0.3) is an example of a first-order difference equation because the numerical scheme couples the sequence value $y(kT)$ directly to the previous sequence value $y[(k-1)T]$. If (7.0.3) had contained $y[(k-2)T]$, then it would have been a second-order difference equation, and so forth.

Although we could use the conventional Laplace transform to solve these difference equations, the use of z-transforms can greatly facilitate the analysis,

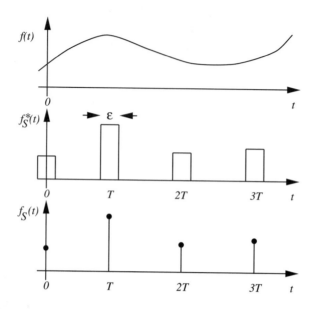

Figure 7.1.1: Schematic of how a continuous function $f(t)$ is sampled by a narrow-width pulse sampler $f_S^*(t)$ and an ideal sampler $f_S(t)$.

especially when we only desire responses at the sampling instants. Often the entire analysis can be done using only the transforms and the analyst does not actually find the sequence $y(kT)$.

In this chapter we will first define the z-transform and discuss its properties. Then we will show how to find its inverse. Finally we shall use them to solve difference equations.

7.1 THE RELATIONSHIP OF THE Z-TRANSFORM TO THE LAPLACE TRANSFORM[1]

Let $f(t)$ be a continuous function that an instrument samples every T units of time. We denote this data-sampled function by $f_S^*(t)$. See Figure 7.1.1. Taking ϵ, the duration of an individual sampling event, to be small, we may approximate the narrow-width pulse in Figure 7.1.1 by flat-topped pulses. Then $f_S^*(t)$ approximately equals

$$f_S^*(t) \approx \frac{1}{\epsilon} \sum_{n=0}^{\infty} f(nT) \left[H(t - nT + \epsilon/2) - H(t - nT - \epsilon/2) \right], \qquad (7.1.1)$$

if $\epsilon \ll T$.

---

[1] Gera [Gera, A. E., 1999: The relationship between the z-transform and the discrete-time Fourier transform. *IEEE Trans. Auto. Control*, **AC-44**, 370–371] has explored the general relationship between the one-sided discrete-time Fourier transform and the one-sided z-transform. See also Naumović, M. B., 2001: Interrelationship between the one-sided discrete-time Fourier transform and one-sided delta transform. *Electr. Engng.*, **83**, 99–101.

Clearly the presence of ϵ is troublesome in (7.1.1); it adds one more parameter to our problem. For this reason we introduce the concept of the *ideal sampler*, where the sampling time becomes infinitesimally small so that

$$f_S(t) = \lim_{\epsilon \to 0} \sum_{n=0}^{\infty} f(nT) \left[\frac{H(t - nT + \epsilon/2) - H(t - nT - \epsilon/2)}{\epsilon} \right] \qquad (7.1.2)$$

$$= \sum_{n=0}^{\infty} f(nT)\delta(t - nT). \qquad (7.1.3)$$

Let us now find the Laplace transform of this data-sampled function. From the linearity property of Laplace transforms,

$$F_S(s) = \mathcal{L}[f_S(t)] = \mathcal{L}\left[\sum_{n=0}^{\infty} f(nT)\delta(t - nT) \right] \qquad (7.1.4)$$

$$= \sum_{n=0}^{\infty} f(nT)\mathcal{L}[\delta(t - nT)]. \qquad (7.1.5)$$

Because $\mathcal{L}[\delta(t - nT)] = e^{-nsT}$, (7.1.5) simplifies to

$$F_S(s) = \sum_{n=0}^{\infty} f(nT)e^{-nsT}. \qquad (7.1.6)$$

If we now make the substitution that $z = e^{sT}$, then $F_S(s)$ becomes

$$F(z) = \mathcal{Z}(f_n) = \sum_{n=0}^{\infty} f_n z^{-n}, \qquad (7.1.7)$$

where $F(z)$ is the one-sided z-transform[2] of the sequence $f(nT)$, which we shall now denote by f_n. Here $\mathcal{Z}$ denotes the operation of taking the z-transform while $\mathcal{Z}^{-1}$ represents the inverse z-transformation. We will consider methods for finding the inverse z-transform in §7.3.

Just as the Laplace transform was defined by an integration in t, the z-transform is defined by a power series (Laurent series) in z. Consequently, every z-transform has a region of convergence which must be implicitly understood if not explicitly stated. Furthermore, just as the Laplace integral diverged for certain functions, there are sequences where the associated power series diverges and its z-transform does not exist.

---

[2] The standard reference is Jury, E. I., 1964: *Theory and Application of the z-Transform Method.* John Wiley & Sons, 330 pp.

Consider now the following examples of how to find the z-transform.

• **Example 7.1.1**

Given the unit sequence $f_n = 1$, $n \geq 0$, let us find $F(z)$. Substituting f_n into the definition of the z-transform leads to

$$F(z) = \sum_{n=0}^{\infty} z^{-n} = \frac{z}{z-1}, \qquad (7.1.8)$$

because $\sum_{n=0}^{\infty} z^{-n}$ is a complex-valued *geometric series* with common ratio z^{-1}. This series converges if $|z^{-1}| < 1$ or $|z| > 1$, which gives the region of convergence of $F(z)$.

MATLAB's symbolic toolbox provides an alternative to the hand computation of the z-transform. In the present case, the command

```
>> syms z; syms n positive
>> ztrans(1,n,z)
```

yields

```
ans =
z/(z-1)
```

• **Example 7.1.2**

Let us find the z-transform of the sequence

$$f_n = e^{-anT}, \qquad n \geq 0, \qquad (7.1.9)$$

for a real and a imaginary.

For a real, substitution of the sequence into the definition of the z-transform yields

$$F(z) = \sum_{n=0}^{\infty} e^{-anT} z^{-n} = \sum_{n=0}^{\infty} \left(e^{-aT} z^{-1} \right)^n. \qquad (7.1.10)$$

If $u = e^{-aT} z^{-1}$, then (7.1.10) is a geometric series so that

$$F(z) = \sum_{n=0}^{\infty} u^n = \frac{1}{1-u}. \qquad (7.1.11)$$

Because $|u| = e^{-aT} |z^{-1}|$, the condition for convergence is that $|z| > e^{-aT}$. Thus,

$$F(z) = \frac{z}{z - e^{-aT}}, \qquad |z| > e^{-aT}. \qquad (7.1.12)$$

For imaginary a, the infinite series in (7.1.10) converges if $|z| > 1$, because $|u| = |z^{-1}|$ when a is imaginary. Thus,

$$F(z) = \frac{z}{z - e^{-aT}}, \qquad |z| > 1. \tag{7.1.13}$$

Although the z-transforms in (7.1.12) and (7.1.13) are the same in these two cases, the corresponding regions of convergence are different. If a is a complex number, then

$$F(z) = \frac{z}{z - e^{-aT}}, \qquad |z| > |e^{-aT}|. \tag{7.1.14}$$

Checking our work using MATLAB, we type the commands:

```
>> syms a z; syms n T positive
>> ztrans(exp(-a*n*T),n,z);
>> simplify(ans)
```

which yields

```
ans =
z*exp(a*T)/(z*exp(a*T)-1)
```

• **Example 7.1.3**

Let us find the z-transform of the sinusoidal sequence

$$f_n = \cos(n\omega T), \qquad n \geq 0. \tag{7.1.15}$$

Substituting (7.1.15) into the definition of the z-transform results in

$$F(z) = \sum_{n=0}^{\infty} \cos(n\omega T) z^{-n}. \tag{7.1.16}$$

From Euler's formula,

$$\cos(n\omega T) = \tfrac{1}{2}(e^{in\omega T} + e^{-in\omega T}), \tag{7.1.17}$$

so that (7.1.16) becomes

$$F(z) = \frac{1}{2} \sum_{n=0}^{\infty} \left(e^{in\omega T} z^{-n} + e^{-in\omega T} z^{-n} \right), \tag{7.1.18}$$

or

$$F(z) = \tfrac{1}{2}\left[\mathcal{Z}(e^{in\omega T}) + \mathcal{Z}(e^{-in\omega T}) \right]. \tag{7.1.19}$$

From (7.1.13),

$$\mathcal{Z}(e^{\pm in\omega T}) = \frac{z}{z - e^{\pm i\omega T}}, \qquad |z| > 1. \tag{7.1.20}$$

Substituting (7.1.20) into (7.1.19) and simplifying yields

$$F(z) = \frac{z[z - \cos(\omega T)]}{z^2 - 2z\cos(\omega T) + 1}, \qquad |z| > 1. \qquad (\textbf{7.1.21})$$

• **Example 7.1.4**

Let us find the z-transform for the sequence

$$f_n = \begin{cases} 1, & 0 \le n \le 5, \\ (\frac{1}{2})^n, & 6 \le n. \end{cases} \qquad (\textbf{7.1.22})$$

From the definition of the z-transform,

$$\mathcal{Z}(f_n) = F(z) = \sum_{n=0}^{5} z^{-n} + \sum_{n=6}^{\infty} \left(\frac{1}{2z}\right)^n. \qquad (\textbf{7.1.23})$$

$$= 1 + \frac{1}{z} + \frac{1}{z^2} + \frac{1}{z^3} + \frac{1}{z^4} + \frac{1}{z^5}$$

$$+ \frac{2z}{2z - 1} - 1 - \frac{1}{2z} - \frac{1}{4z^2} - \frac{1}{8z^3} - \frac{1}{16z^4} - \frac{1}{32z^5} \qquad (\textbf{7.1.24})$$

$$= \frac{2z}{2z - 1} + \frac{1}{2z} + \frac{3}{4z^2} + \frac{7}{8z^3} + \frac{15}{16z^4} + \frac{31}{32z^5}. \qquad (\textbf{7.1.25})$$

We could also have obtained (7.1.25) via MATLAB by typing the commands:

```
>> syms z; syms n positive
>> ztrans('1+((1/2)^n-1)*Heaviside(n-6)',n,z)
```
which yields

```
ans =
2*z/(2*z-1)+1/2/z+3/4/z^2+7/8/z^3+15/16/z^4+31/32/z^5
```

We summarize some of the more commonly encountered sequences and their transforms in Table 7.1.1 along with their regions of convergence.

• **Example 7.1.5**

In many engineering studies, the analysis is done entirely using transforms without actually finding any inverses. Consequently, it is useful to compare and contrast how various transforms behave in very simple test problems.

Consider the time function $f(t) = ae^{-at}H(t)$, $a > 0$. Its Laplace and Fourier transform are identical, namely $a/(a + i\omega)$, if we set $s = i\omega$. In Figure 7.1.2 we illustrate its behavior as a function of positive ω.

Let us now generate the sequence of observations that we would measure if we sampled $f(t)$ every T units of time apart: $f_n = ae^{-anT}$. Taking the

Table 7.1.1: Z-Transforms of Some Commonly Used Sequences

$f_n, \ n \geq 0$	$F(z)$	Region of convergence
1. $\quad f_0 = k = \text{const.}$ $\quad f_n = 0, \ n \geq 1$	k	$|z| > 0$
2. $\quad f_m = k = \text{const.}$ $\quad f_n = 0$, for all values $\quad\quad$ of $n \neq m$	kz^{-m}	$|z| > 0$
3. $\quad k = \text{constant}$	$kz/(z-1)$	$|z| > 1$
4. $\quad kn$	$kz/(z-1)^2$	$|z| > 1$
5. $\quad kn^2$	$kz(z+1)/(z-1)^3$	$|z| > 1$
6. ke^{-anT}, a complex	$kz/\left(z - e^{-aT}\right)$	$|z| > |e^{-aT}|$
7. kne^{-anT}, a complex	$\frac{kze^{-aT}}{(z-e^{-aT})^2}$	$|z| > |e^{-aT}|$
8. $\quad \sin(\omega_0 nT)$	$\frac{z \sin(\omega_0 T)}{z^2 - 2z\cos(\omega_0 T)+1}$	$|z| > 1$
9. $\quad \cos(\omega_0 nT)$	$\frac{z[z-\cos(\omega_0 T)]}{z^2 - 2z\cos(\omega_0 T)+1}$	$|z| > 1$
10. $e^{-anT}\sin(\omega_0 nT)$	$\frac{ze^{-aT}\sin(\omega_0 T)}{z^2 - 2ze^{-aT}\cos(\omega_0 T)+e^{-2aT}}$	$|z| > e^{-aT}$
11. $e^{-anT}\cos(\omega_0 nT)$	$\frac{ze^{-aT}[ze^{aT}-\cos(\omega_0 T)]}{z^2 - 2ze^{-aT}\cos(\omega_0 T)+e^{-2aT}}$	$|z| > e^{-aT}$
12. $\quad \alpha^n$, α constant	$z/(z-\alpha)$	$|z| > |\alpha|$
13. $\quad n\alpha^n$	$\alpha z/(z-\alpha)^2$	$|z| > |\alpha|$
14. $\quad n^2\alpha^n$	$\alpha z(z+\alpha)/(z-\alpha)^3$	$|z| > |\alpha|$
15. $\quad \sinh(\omega_0 nT)$	$\frac{z \sinh(\omega_0 T)}{z^2 - 2z\cosh(\omega_0 T)+1}$	$|z| > \cosh(\omega_0 T)$
16. $\quad \cosh(\omega_0 nT)$	$\frac{z[z-\cosh(\omega_0 T)]}{z^2 - 2z\cosh(\omega_0 T)+1}$	$|z| > \sinh(\omega_0 T)$
17. $\quad a^n/n!$	$e^{a/z}$	$|z| > 0$
18. $\quad [\ln(a)]^n/n!$	$a^{1/z}$	$|z| > 0$

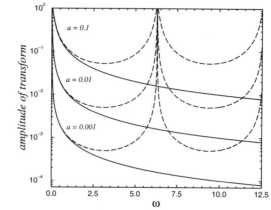

Figure 7.1.2: The amplitude of the Laplace or Fourier transform (solid line) for $ae^{-at}H(t)$ and the z-transform (dashed line) for $f_n = ae^{-anT}$ as a function of frequency ω for various positive values of a and $T = 1$.

z-transform of this sequence, it equals $az/\left(z - e^{-aT}\right)$. Recalling that $z = e^{sT} = e^{i\omega T}$, we can also plot this transform as a function of positive ω. For small ω, the transforms agree, but as ω becomes larger they diverge markedly. Why does this occur?

Recall that the z-transform is computed from a sequence comprised of samples from a continuous signal. One very important flaw in sampled data is the possible misrepresentation of high-frequency effects as lower-frequency phenomena. It is this *aliasing* or *folding* effect that we are observing here. Consequently, the z-transform of a sampled record can differ markedly from the corresponding Laplace or Fourier transforms of the continuous record at frequencies above one half of the sampling frequency. This also suggests that care should be exercised in interpolating between sampling instants. Indeed, in those applications where the output between sampling instants is very important, such as in a hybrid mixture of digital and analog systems, we must apply the so-called "modified z-transform."

Problems

From the fundamental definition of the z-transform, find the transform of the following sequences, where $n \geq 0$. Then check your answer using MATLAB.

1. $f_n = \left(\frac{1}{2}\right)^n$

2. $f_n = e^{in\theta}$

3. $f_n = \begin{cases} 1, & 0 \leq n \leq 5 \\ 0, & 5 < n \end{cases}$

4. $f_n = \begin{cases} \left(\frac{1}{2}\right)^n, & n = 0, 1, \ldots, 10 \\ \left(\frac{1}{4}\right)^n, & n \geq 11 \end{cases}$

5. $f_n = \begin{cases} 0, & n = 0 \\ -1, & n = 1 \\ a^n, & n \geq 2 \end{cases}$

7.2 SOME USEFUL PROPERTIES

In principle we could construct any desired transform from the definition of the z-transform. However, there are several general theorems that are much more effective in finding new transforms.

> Linearity

From the definition of the z-transform, it immediately follows that

$$\text{if} \quad h_n = c_1 f_n + c_2 g_n, \quad \text{then} \quad H(z) = c_1 F(z) + c_2 G(z), \qquad (7.2.1)$$

where $F(z) = \mathcal{Z}(f_n)$, $G(z) = \mathcal{Z}(g_n)$, $H(z) = \mathcal{Z}(h_n)$, and c_1, c_2 are arbitrary constants.

> Multiplication by an exponential sequence

If $\quad g_n = e^{-anT} f_n, \quad n \geq 0, \quad$ then $\quad G(z) = F(ze^{aT}).$ $\qquad (7.2.2)$

This follows from

$$G(z) = \mathcal{Z}(g_n) = \sum_{n=0}^{\infty} g_n z^{-n} = \sum_{n=0}^{\infty} e^{-anT} f_n z^{-n} \qquad (7.2.3)$$

$$= \sum_{n=0}^{\infty} f_n (ze^{aT})^{-n} = F(ze^{aT}). \qquad (7.2.4)$$

This is the z-transform analog to the first shifting theorem in Laplace transforms.

> Shifting

The effect of shifting depends upon whether it is to the right or to the left, as Table 7.2.1 illustrates. For the sequence f_{n-2}, no values from the sequence f_n are lost; thus, we anticipate that the z-transform of f_{n-2} only involves $F(z)$. However, in forming the sequence f_{n+2}, the first two values of f_n are lost, and we anticipate that the z-transform of f_{n+2} cannot be expressed solely in terms of $F(z)$ but must include those two lost pieces of information.

Table 7.2.1: Examples of Shifting Involving Sequences

n	f_n	f_{n-2}	f_{n+2}
0	1	0	4
1	2	0	8
2	4	1	16
3	8	2	64
4	16	4	128
$\vdots$	$\vdots$	$\vdots$	$\vdots$

Let us now confirm these conjectures by finding the z-transform of f_{n+1} which is a sequence that has been shifted one step to the left. From the definition of the z-transform, it follows that

$$\mathcal{Z}(f_{n+1}) = \sum_{n=0}^{\infty} f_{n+1} z^{-n} = z \sum_{n=0}^{\infty} f_{n+1} z^{-(n+1)} \qquad (7.2.5)$$

or

$$\mathcal{Z}(f_{n+1}) = z \sum_{k=1}^{\infty} f_k z^{-k} + z f_0 - z f_0, \qquad (7.2.6)$$

where we added zero in (7.2.6). This algebraic trick allows us to collapse the first two terms on the right side of (7.2.6) into one and

$$\mathcal{Z}(f_{n+1}) = z F(z) - z f_0. \qquad (7.2.7)$$

In a similar manner, repeated applications of (7.2.7) yield

$$\mathcal{Z}(f_{n+m}) = z^m F(z) - z^m f_0 - z^{m-1} f_1 - \ldots - z f_{m-1}, \qquad (7.2.8)$$

where $m > 0$. This shifting operation transforms f_{n+m} into an algebraic expression involving m. Furthermore, we introduced initial sequence values, just as we introduced initial conditions when we took the Laplace transform of the nth derivative of $f(t)$. We will make frequent use of this property in solving difference equations in §7.4.

Consider now shifting to the right by the positive integer k,

$$g_n = f_{n-k} H_{n-k}, \quad n \geq 0, \qquad (7.2.9)$$

where $H_{n-k} = 0$ for $n < k$ and 1 for $n \geq k$. Then the z-transform of (7.2.9) is

$$G(z) = z^{-k} F(z), \qquad (7.2.10)$$

where $G(z) = \mathcal{Z}(g_n)$, and $F(z) = \mathcal{Z}(f_n)$. This follows from

$$G(z) = \sum_{n=0}^{\infty} g_n z^{-n} = \sum_{n=0}^{\infty} f_{n-k} H_{n-k} z^{-n} = z^{-k} \sum_{n=k}^{\infty} f_{n-k} z^{-(n-k)} \quad \textbf{(7.2.11)}$$

$$= z^{-k} \sum_{m=0}^{\infty} f_m z^{-m} = z^{-k} F(z). \quad \textbf{(7.2.12)}$$

This result is the z-transform analog to the second shifting theorem in Laplace transforms.

In symbolic calculations involving MATLAB, the operator H_{n-k} can be expressed by Heaviside(n-k).

Initial-value theorem

The initial value of the sequence f_n, f_0, can be computed from $F(z)$ using the initial-value theorem:

$$f_0 = \lim_{z \to \infty} F(z). \quad \textbf{(7.2.13)}$$

From the definition of the z-transform,

$$F(z) = \sum_{n=0}^{\infty} f_n z^{-n} = f_0 + f_1 z^{-1} + f_2 z^{-2} + \dots. \quad \textbf{(7.2.14)}$$

In the limit of $z \to \infty$, we obtain the desired result.

Final-value theorem

The value of f_n, as $n \to \infty$, is given by the final-value theorem:

$$f_\infty = \lim_{z \to 1} (z - 1) F(z), \quad \textbf{(7.2.15)}$$

where $F(z)$ is the z-transform of f_n.

We begin by noting that

$$\mathcal{Z}(f_{n+1} - f_n) = \lim_{n \to \infty} \sum_{k=0}^{n} (f_{k+1} - f_k) z^{-k}. \quad \textbf{(7.2.16)}$$

Using the shifting theorem on the left side of (7.2.16),

$$z F(z) - z f_0 - F(z) = \lim_{n \to \infty} \sum_{k=0}^{n} (f_{k+1} - f_k) z^{-k}. \quad \textbf{(7.2.17)}$$

Applying the limit as z approaches 1 to both sides of (7.2.17):

$$\lim_{z \to 1} (z-1)F(z) - f_0 = \lim_{n \to \infty} \sum_{k=0}^{n} (f_{k+1} - f_k) \tag{7.2.18}$$

$$= \lim_{n \to \infty} \left[(f_1 - f_0) + (f_2 - f_1) + \dots \right.$$
$$\left. + (f_n - f_{n-1}) + (f_{n+1} - f_n) + \dots \right] \tag{7.2.19}$$

$$= \lim_{n \to \infty} (-f_0 + f_{n+1}) \tag{7.2.20}$$

$$= -f_0 + f_\infty. \tag{7.2.21}$$

Consequently,

$$f_\infty = \lim_{z \to 1} (z-1)F(z). \tag{7.2.22}$$

Note that this limit has meaning only if f_∞ exists. This occurs if $F(z)$ has no second-order or higher poles on the unit circle and no poles outside the unit circle.

> **Multiplication by n**

Given

$$g_n = n f_n, \qquad n \geq 0, \tag{7.2.23}$$

this theorem states that

$$G(z) = -z \frac{dF(z)}{dz}, \tag{7.2.24}$$

where $G(z) = \mathcal{Z}(g_n)$, and $F(z) = \mathcal{Z}(f_n)$.

This follows from

$$G(z) = \sum_{n=0}^{\infty} g_n z^{-n} = \sum_{n=0}^{\infty} n f_n z^{-n} = z \sum_{n=0}^{\infty} n f_n z^{-n-1} = -z \frac{dF(z)}{dz}. \tag{7.2.25}$$

> **Periodic sequence theorem**

Consider the N-periodic sequence:

$$f_n = \{ \underbrace{f_0 f_1 f_2 \dots f_{N-1}}_{\text{first period}} f_0 f_1 \dots \}, \tag{7.2.26}$$

and the related sequence:

$$x_n = \begin{cases} f_n, & 0 \le n \le N - 1, \\ 0, & N \le n. \end{cases} \quad (7.2.27)$$

This theorem allows us to find the z-transform of f_n if we can find the z-transform of x_n via the relationship

$$F(z) = \frac{X(z)}{1 - z^{-N}}, \quad |z^N| > 1, \quad (7.2.28)$$

where $X(z) = \mathscr{Z}(x_n)$.

This follows from

$$F(z) = \sum_{n=0}^{\infty} f_n z^{-n} \quad (7.2.29)$$

$$= \sum_{n=0}^{N-1} x_n z^{-n} + \sum_{n=N}^{2N-1} x_{n-N} z^{-n} + \sum_{n=2N}^{3N-1} x_{n-2N} z^{-n} + \cdots. \quad (7.2.30)$$

Application of the shifting theorem in (7.2.30) leads to

$$F(z) = X(z) + z^{-N} X(z) + z^{-2N} X(z) + \cdots \quad (7.2.31)$$

$$= X(z) \left[1 + z^{-N} + z^{-2N} + \cdots \right]. \quad (7.2.32)$$

Equation (7.2.32) contains an infinite geometric series with common ratio z^{-N}, which converges if $|z^{-N}| < 1$. Thus,

$$F(z) = \frac{X(z)}{1 - z^{-N}}, \quad |z^N| > 1. \quad (7.2.33)$$

Convolution

Given the sequences f_n and g_n, the convolution product of these two sequences is

$$w_n = f_n * g_n = \sum_{k=0}^{n} f_k g_{n-k} = \sum_{k=0}^{n} f_{n-k} g_k. \quad (7.2.34)$$

Given $F(z)$ and $G(z)$, we then have that $W(z) = F(z)G(z)$.

This follows from

$$W(z) = \sum_{n=0}^{\infty} \left[\sum_{k=0}^{n} f_k g_{n-k} \right] z^{-n} = \sum_{n=0}^{\infty} \sum_{k=0}^{\infty} f_k g_{n-k} z^{-n}, \quad (7.2.35)$$

because $g_{n-k} = 0$ for $k > n$. Reversing the order of summation and letting $m = n - k$,

$$W(z) = \sum_{k=0}^{\infty} \sum_{m=-k}^{\infty} f_k g_m z^{-(m+k)} \tag{7.2.36}$$

$$= \left[\sum_{k=0}^{\infty} f_k z^{-k} \right] \left[\sum_{m=0}^{\infty} g_m z^{-m} \right] = F(z)G(z). \tag{7.2.37}$$

We can use MATLAB's command conv( ) which multiplies two polynomials to perform discrete convolution as follows:

```
>>x = [1 1 1 1 1 1 1];
>>y = [1 2 4 8 16 32 64];
>>z = conv(x,y)
```

produces

```
z =
    1 3 7 15 31 63 127 126 124 120 112 96 64
```

The first seven values of z contains the convolution of the sequence x with the sequence y.

Consider now the following examples of the properties discussed in this section.

• **Example 7.2.1**

From

$$\mathcal{Z}\left(a^n\right) = \frac{1}{1 - az^{-1}}, \tag{7.2.38}$$

for $n \geq 0$ and $|z| > |a|$, we have that

$$\mathcal{Z}\left(e^{inx}\right) = \frac{1}{1 - e^{ix}z^{-1}}, \tag{7.2.39}$$

and

$$\mathcal{Z}\left(e^{-inx}\right) = \frac{1}{1 - e^{-ix}z^{-1}}, \tag{7.2.40}$$

if $n \geq 0$ and $|z| > 1$. Therefore, the sequence $f_n = \cos(nx)$ has the z-transform

$$F(z) = \mathcal{Z}[\cos(nx)] = \tfrac{1}{2}\mathcal{Z}\left(e^{inx}\right) + \tfrac{1}{2}\mathcal{Z}\left(e^{-inx}\right) \tag{7.2.41}$$

$$= \frac{1}{2}\frac{1}{1 - e^{ix}z^{-1}} + \frac{1}{2}\frac{1}{1 - e^{-ix}z^{-1}} = \frac{1 - \cos(x)z^{-1}}{1 - 2\cos(x)z^{-1} + z^{-2}}. \tag{7.2.42}$$

• **Example 7.2.2**

Using the z-transform,

$$\mathcal{Z}\left(a^n\right) = \frac{1}{1 - az^{-1}}, \qquad n \geq 0, \tag{7.2.43}$$

we find that

$$\mathcal{Z}\left(na^n\right) = -z\frac{d}{dz}\left[\left(1 - az^{-1}\right)^{-1}\right] \tag{7.2.44}$$

$$= (-z)(-1)\left(1 - az^{-1}\right)^{-2}(-a)(-1)z^{-2} \tag{7.2.45}$$

$$= \frac{az^{-1}}{\left(1 - az^{-1}\right)^2} = \frac{az}{(z - a)^2}. \tag{7.2.46}$$

• **Example 7.2.3**

Consider $F(z) = 2az^{-1}/(1 - az^{-1})^3$, where $|a| < |z|$ and $|a| < 1$. Here we have that

$$f_0 = \lim_{z \to \infty} F(z) = \lim_{z \to \infty} \frac{2az^{-1}}{(1 - az^{-1})^3} = 0 \tag{7.2.47}$$

from the initial-value theorem. This agrees with the inverse of $F(z)$:

$$f_n = n(n + 1)a^n, \qquad n \geq 0. \tag{7.2.48}$$

If the z-transform consists of the ratio of two polynomials, we can use MATLAB to find f_0. For example, if $F(z) = 2z^2/(z - 1)^3$, we can find f_0 as follows:

```
>>num = [2 0 0];
>>den = conv([1 -1],[1 -1]);
>>den = conv(den,[1 -1]);
>>initialvalue = polyval(num,1e20) / polyval(den,1e20)
initialvalue =
    2.0000e-20
```

Therefore, $f_0 = 0$.

• **Example 7.2.4**

Given the z-transform $F(z) = (1 - a)z/[(z - 1)(z - a)]$, where $|z| > 1 > a > 0$, then from the final-value theorem we have that

$$\lim_{n \to \infty} f_n = \lim_{z \to 1}(z - 1)F(z) = \lim_{z \to 1}\frac{1 - a}{1 - az^{-1}} = 1. \tag{7.2.49}$$

This is consistent with the inverse transform $f_n = 1 - a^n$ with $n \geq 0$.

• Example 7.2.5

Using the sequences $f_n = 1$ and $g_n = a^n$, where a is real, verify the convolution theorem.

We first compute the convolution of f_n with g_n, namely

$$w_n = f_n * g_n = \sum_{k=0}^{n} a^k = \frac{1}{1-a} - \frac{a^{n+1}}{1-a}. \tag{7.2.50}$$

Taking the z-transform of w_n,

$$W(z) = \frac{z}{(1-a)(z-1)} - \frac{az}{(1-a)(z-a)} = \frac{z^2}{(z-1)(z-a)} = F(z)G(z) \tag{7.2.51}$$

and convolution theorem holds true for this special case.

Problems

Use the properties of z-transforms and Table 7.1.1 to find the z-transform of the following sequences. Then check your answer using MATLAB.

1. $f_n = nTe^{-anT}$

2. $f_n = \begin{cases} 0, & n = 0 \\ na^{n-1}, & n \geq 1 \end{cases}$

3. $f_n = \begin{cases} 0, & n = 0 \\ n^2 a^{n-1}, & n \geq 1 \end{cases}$

4. $f_n = a^n \cos(n)$

[Use $\cos(n) = \frac{1}{2}(e^{in} + e^{-in})$]

5. $f_n = \cos(n-2)H_{n-2}$

6. $f_n = 3 + e^{-2nT}$

7. $f_n = \sin(n\omega_0 T + \theta)$

8. $f_n = \begin{cases} 0, & n = 0 \\ 1, & n = 1 \\ 2, & n = 2 \\ 1, & n = 3, \end{cases} \quad f_{n+4} = f_n$

9. $f_n = (-1)^n$

(Hint: f_n is a periodic sequence.)

10. Using the property stated in (7.2.23)–(7.2.24) *twice*, find the z-transform of $n^2 = n[n(1)^n]$. Then verify your result using MATLAB.

11. Verify the convolution theorem using the sequences $f_n = g_n = 1$. Then check your results using MATLAB.

12. Verify the convolution theorem using the sequences $f_n = 1$ and $g_n = n$. Then check your results using MATLAB.

13. Verify the convolution theorem using the sequences $f_n = g_n = 1/(n!)$. [Hint: Use the binomial theorem with $x = 1$ to evaluate the summation.] Then check your results using MATLAB.

14. If a is a real number, show that $\mathcal{Z}(a^n f_n) = F(z/a)$, where $\mathcal{Z}(f_n) = F(z)$.

7.3 INVERSE Z-TRANSFORMS

In the previous two sections we dealt with finding the z-transform. In this section we find f_n by inverting the z-transform $F(z)$. There are four methods for finding the inverse: (1) power series, (2) recursion, (3) partial fractions, and (4) the residue method. We will discuss each technique individually. The first three apply only to those functions $F(z)$ that are *rational* functions while the residue method is more general. For symbolic computations with MATLAB, you can use `iztrans`.

> Power series

By means of the long-division process, we can always rewrite $F(z)$ as the Laurent expansion:

$$F(z) = a_0 + a_1 z^{-1} + a_2 z^{-2} + \cdots . \tag{7.3.1}$$

From the definition of the z-transform,

$$F(z) = \sum_{n=0}^{\infty} f_n z^{-n} = f_0 + f_1 z^{-1} + f_2 z^{-2} + \cdots , \tag{7.3.2}$$

the desired sequence f_n is given by a_n.

- **Example 7.3.1**

Let

$$F(z) = \frac{z+1}{2z-2} = \frac{N(z)}{D(z)}. \tag{7.3.3}$$

Using long division, $N(z)$ is divided by $D(z)$ and we obtain

$$F(z) = \tfrac{1}{2} + z^{-1} + z^{-2} + z^{-3} + z^{-4} + \cdots . \tag{7.3.4}$$

Therefore,

$$a_0 = \tfrac{1}{2}, \ a_1 = 1, \ a_2 = 1, \ a_3 = 1, \ a_4 = 1, \ \text{etc.}, \tag{7.3.5}$$

which suggests that $f_0 = \frac{1}{2}$ and $f_n = 1$ for $n \geq 1$ is the inverse of $F(z)$.

• **Example 7.3.2**

Let us find the inverse of the z-transform:

$$F(z) = \frac{2z^2 - 1.5z}{z^2 - 1.5z + 0.5}. \tag{7.3.6}$$

By the long-division process, we have that

$$
\begin{array}{r}
2 \;+\; 1.5z^{-1} \;+\; 1.25z^{-2} \;+\; 1.125z^{-3} \;+\; \cdots \\
z^2 - 1.5z + 0.5 \;\overline{\smash{\big)}\; 2z^2 \;-\; 1.5z } \\
\end{array}
$$

$$
\begin{array}{r}
2z^2 \;-\;\; 3z \;\;+\;\; 1 \\
\hline
1.5z \;\;-\;\; 1 \\
1.5z \;\;-\;\; 2.25 \;\;+\; 0.75z^{-1} \\
\hline
1.25 \;\;-\; 0.75z^{-1} \\
1.25 \;\;-\; 1.87z^{-1} \;+\; \cdots \\
\hline
1.125z^{-1} \;+\; \cdots
\end{array}
$$

Thus, $f_0 = 2$, $f_1 = 1.5$, $f_2 = 1.25$, $f_3 = 1.125$, and so forth, or $f_n = 1 + (\frac{1}{2})^n$. In general, this technique only produces numerical values for some of the elements of the sequence. Note also that our long division must always yield the power series (7.3.1) in order for this method to be of any use.

To check our answer using MATLAB, we type the commands:

```
syms z; syms n positive
iztrans((2*z^2 - 1.5*z)/(z^2 - 1.5*z + 0.5),z,n)
```

which yields

```
ans =
1 + (1/2)^n
```

Recursive method

An alternative to long division was suggested[3] several years ago. It obtains the inverse recursively.

We begin by assuming that the z-transform is of the form

$$F(z) = \frac{a_0 z^m + a_1 z^{m-1} + a_2 z^{m-2} + \cdots + a_{m-1}z + a_m}{b_0 z^m + b_1 z^{m-1} + b_2 z^{m-2} + \cdots + b_{m-1}z + b_m}, \tag{7.3.7}$$

---

[3] Jury, E. I., 1964: *Theory and Application of the z-Transform Method.* John Wiley & Sons, p. 41; Pierre, D. A., 1963: A tabular algorithm for z-transform inversion. *Control Engng.*, **10(9)**, 110–111. The present derivation is by Jenkins, L. B., 1967: A useful recursive form for obtaining inverse z-transforms. *Proc. IEEE*, **55**, 574–575. ©IEEE.

where some of the coefficients a_i and b_i may be zero and $b_0 \neq 0$. Applying the initial-value theorem,

$$f_0 = \lim_{z \to \infty} F(z) = a_0/b_0. \tag{7.3.8}$$

Next, we apply the initial-value theorem to $z[F(z) - f_0]$ and find that

$$f_1 = \lim_{z \to \infty} z[F(z) - f_0] \tag{7.3.9}$$

$$= \lim_{z \to \infty} z \frac{(a_0 - b_0 f_0)z^m + (a_1 - b_1 f_0)z^{m-1} + \cdots + (a_m - b_m f_0)}{b_0 z^m + b_1 z^{m-1} + b_2 z^{m-2} + \cdots + b_{m-1}z + b_m} \tag{7.3.10}$$

$$= (a_1 - b_1 f_0)/b_0. \tag{7.3.11}$$

Note that the coefficient $a_0 - b_0 f_0 = 0$ from (7.3.8). Similarly,

$$f_2 = \lim_{z \to \infty} z[zF(z) - zf_0 - f_1] \tag{7.3.12}$$

$$= \lim_{z \to \infty} z \frac{(a_0 - b_0 f_0)z^{m+1} + (a_1 - b_1 f_0 - b_0 f_1)z^m + (a_2 - b_2 f_0 - b_1 f_1)z^{m-1} + \cdots - b_m f_1}{b_0 z^m + b_1 z^{m-1} + b_2 z^{m-2} + \cdots + b_{m-1}z + b_m} \tag{7.3.13}$$

$$= (a_2 - b_2 f_0 - b_1 f_1)/b_0 \tag{7.3.14}$$

because $a_0 - b_0 f_0 = a_1 - b_1 f_0 - f_1 b_0 = 0$. Continuing this process, we finally have that

$$f_n = (a_n - b_n f_0 - b_{n-1}f_1 - \cdots - b_1 f_{n-1})/b_0, \tag{7.3.15}$$

where $a_n = b_n \equiv 0$ for $n > m$.

● **Example 7.3.3**

Let us redo Example 7.3.2 using the recursive method. Comparing (7.3.7) to (7.3.6), $a_0 = 2$, $a_1 = -1.5$, $a_2 = 0$, $b_0 = 1$, $b_1 = -1.5$, $b_2 = 0.5$, and $a_n = b_n = 0$ if $n \geq 3$. From (7.3.15),

$$f_0 = a_0/b_0 = 2/1 = 2, \tag{7.3.16}$$

$$f_1 = (a_1 - b_1 f_0)/b_0 = [-1.5 - (-1.5)(2)]/1 = 1.5, \tag{7.3.17}$$

$$f_2 = (a_2 - b_2 f_0 - b_1 f_1)/b_0 \tag{7.3.18}$$
$$= [0 - (0.5)(2) - (-1.5)(1.5)]/1 = 1.25, \tag{7.3.19}$$

and

$$f_3 = (a_3 - b_3 f_0 - b_2 f_1 - b_1 f_2)/b_0 \tag{7.3.20}$$
$$= [0 - (0)(2) - (0.5)(1.5) - (-1.5)(1.25)]/1 = 1.125. \tag{7.3.21}$$

Partial fraction expansion

One of the popular methods for inverting Laplace transforms is partial fractions. A similar, but slightly different scheme works here.

• **Example 7.3.4**

Given $F(z) = z/(z^2 - 1)$, let us find f_n. The first step is to obtain the partial fraction expansion of $F(z)/z$. Why we want $F(z)/z$ rather than $F(z)$ will be made clear in a moment. Thus,

$$\frac{F(z)}{z} = \frac{1}{(z-1)(z+1)} = \frac{A}{z-1} + \frac{B}{z+1}, \qquad (7.3.22)$$

where

$$A = (z-1) \left. \frac{F(z)}{z} \right|_{z=1} = \frac{1}{2}, \qquad (7.3.23)$$

and

$$B = (z+1) \left. \frac{F(z)}{z} \right|_{z=-1} = -\frac{1}{2}. \qquad (7.3.24)$$

Multiplying (7.3.22) by z,

$$F(z) = \frac{1}{2} \left(\frac{z}{z-1} - \frac{z}{z+1} \right). \qquad (7.3.25)$$

Next, we find the inverse z-transform of $z/(z-1)$ and $z/(z+1)$ in Table 7.1.1. This yields

$$\mathcal{Z}^{-1} \left(\frac{z}{z-1} \right) = 1, \quad \text{and} \quad \mathcal{Z}^{-1} \left(\frac{z}{z+1} \right) = (-1)^n. \qquad (7.3.26)$$

Thus, the inverse is

$$f_n = \tfrac{1}{2} \left[1 - (-1)^n \right], \; n \geq 0. \qquad (7.3.27)$$

From this example it is clear that there are two steps: (1) obtain the partial fraction expansion of $F(z)/z$, and (2) find the inverse z-transform by referring to Table 7.1.1.

• **Example 7.3.5**

Given $F(z) = 2z^2/[(z+2)(z+1)^2]$, let us find f_n. We begin by expanding $F(z)/z$ as

$$\frac{F(z)}{z} = \frac{2z}{(z+2)(z+1)^2} = \frac{A}{z+2} + \frac{B}{z+1} + \frac{C}{(z+1)^2}, \qquad (7.3.28)$$

where

$$A = (z + 2) \left. \frac{F(z)}{z} \right|_{z=-2} = -4, \tag{7.3.29}$$

$$B = \frac{d}{dz} \left[(z+1)^2 \frac{F(z)}{z} \right] \Bigg|_{z=-1} = 4, \tag{7.3.30}$$

and

$$C = (z+1)^2 \left. \frac{F(z)}{z} \right|_{z=-1} = -2, \tag{7.3.31}$$

so that

$$F(z) = \frac{4z}{z+1} - \frac{4z}{z+2} - \frac{2z}{(z+1)^2}, \tag{7.3.32}$$

or

$$f_n = \mathcal{Z}^{-1} \left[\frac{4z}{z+1} \right] - \mathcal{Z}^{-1} \left[\frac{4z}{z+2} \right] - \mathcal{Z}^{-1} \left[\frac{2z}{(z+1)^2} \right]. \tag{7.3.33}$$

From Table 7.1.1,

$$\mathcal{Z}^{-1} \left(\frac{z}{z+1} \right) = (-1)^n, \tag{7.3.34}$$

$$\mathcal{Z}^{-1} \left(\frac{z}{z+2} \right) = (-2)^n, \tag{7.3.35}$$

and

$$\mathcal{Z}^{-1} \left[\frac{z}{(z+1)^2} \right] = - \, \mathcal{Z}^{-1} \left[\frac{-z}{(z+1)^2} \right] = -n(-1)^n = n(-1)^{n+1}. \tag{7.3.36}$$

Applying (7.3.34)–(7.3.36) to (7.3.33),

$$f_n = 4(-1)^n - 4(-2)^n + 2n(-1)^n, \quad n \geq 0. \tag{7.3.37}$$

• **Example 7.3.6**

Given $F(z) = (z^2 + z)/(z - 2)^2$, let us determine f_n. Because

$$\frac{F(z)}{z} = \frac{z+1}{(z-2)^2} = \frac{1}{z-2} + \frac{3}{(z-2)^2}, \tag{7.3.38}$$

$$f_n = \mathcal{Z}^{-1} \left[\frac{z}{z-2} \right] + \mathcal{Z}^{-1} \left[\frac{3z}{(z-2)^2} \right]. \tag{7.3.39}$$

Referring to Table 7.1.1,

$$\mathcal{Z}^{-1} \left(\frac{z}{z-2} \right) = 2^n, \quad \text{and} \quad \mathcal{Z}^{-1} \left[\frac{3z}{(z-2)^2} \right] = \tfrac{3}{2} n 2^n. \tag{7.3.40}$$

Substituting (7.3.40) into (7.3.39) yields

$$f_n = \left(\tfrac{3}{2}n + 1\right) 2^n, \quad n \geq 0. \tag{7.3.41}$$

Residue method

The power series, recursive, and partial fraction expansion methods are rather limited. We now prove that f_n may be computed from the following *inverse integral formula*:

$$f_n = \frac{1}{2\pi i} \oint_C z^{n-1} F(z) \, dz, \quad n \geq 0, \tag{7.3.42}$$

where C is any simple curve, taken in the positive sense, that encloses all of the singularities of $F(z)$. It is readily shown that the power series and partial fraction methods are *special cases* of the residue method.

Proof: Starting with the definition of the z-transform

$$F(z) = \sum_{n=0}^{\infty} f_n z^{-n}, \quad |z| > R_1, \tag{7.3.43}$$

we multiply (7.3.43) by z^{n-1} and integrating both sides around any contour C which includes all of the singularities,

$$\frac{1}{2\pi i} \oint_C z^{n-1} F(z) \, dz = \sum_{m=0}^{\infty} f_m \frac{1}{2\pi i} \oint_C z^{n-m} \frac{dz}{z}. \tag{7.3.44}$$

Let C be a circle of radius R, where $R > R_1$. Then, changing variables to $z = R e^{i\theta}$, and $dz = iz \, d\theta$,

$$\frac{1}{2\pi i} \oint_C z^{n-m} \frac{dz}{z} = \frac{R^{n-m}}{2\pi} \int_0^{2\pi} e^{i(n-m)\theta} d\theta = \begin{cases} 1, & m = n, \\ 0, & \text{otherwise.} \end{cases} \tag{7.3.45}$$

Substituting (7.3.45) into (7.3.44) yields the desired result that

$$\frac{1}{2\pi i} \oint_C z^{n-1} F(z) \, dz = f_n. \tag{7.3.46}$$

$\square$

We can easily evaluate the inversion integral (7.3.42) using Cauchy's residue theorem.

• **Example 7.3.7**

Let us find the inverse z-transform of

$$F(z) = \frac{1}{(z-1)(z-2)}. \tag{7.3.47}$$

From the inversion integral,

$$f_n = \frac{1}{2\pi i} \oint_C \frac{z^{n-1}}{(z-1)(z-2)} \, dz. \tag{7.3.48}$$

Clearly the integral has simple poles at $z = 1$ and $z = 2$. However, when $n = 0$ we also have a simple pole at $z = 0$. Thus the cases $n = 0$ and $n > 0$ must be considered separately.

Case 1: $n = 0$. The residue theorem yields

$$f_0 = \text{Res}\left[\frac{1}{z(z-1)(z-2)}; 0\right] + \text{Res}\left[\frac{1}{z(z-1)(z-2)}; 1\right]$$

$$+ \text{Res}\left[\frac{1}{z(z-1)(z-2)}; 2\right]. \tag{7.3.49}$$

Evaluating these residues,

$$\text{Res}\left[\frac{1}{z(z-1)(z-2)}; 0\right] = \frac{1}{(z-1)(z-2)}\bigg|_{z=0} = \frac{1}{2}, \tag{7.3.50}$$

$$\text{Res}\left[\frac{1}{z(z-1)(z-2)}; 1\right] = \frac{1}{z(z-2)}\bigg|_{z=1} = -1, \tag{7.3.51}$$

and

$$\text{Res}\left[\frac{1}{z(z-1)(z-2)}; 2\right] = \frac{1}{z(z-1)}\bigg|_{z=2} = \frac{1}{2}. \tag{7.3.52}$$

Substituting (7.3.50)–(7.3.52) into (7.3.49) yields $f_0 = 0$.

Case 2: $n > 0$. Here we only have contributions from $z = 1$ and $z = 2$.

$$f_n = \text{Res}\left[\frac{z^{n-1}}{(z-1)(z-2)}; 1\right] + \text{Res}\left[\frac{z^{n-1}}{(z-1)(z-2)}; 2\right], \quad n > 0, \tag{7.3.53}$$

where

$$\text{Res}\left[\frac{z^{n-1}}{(z-1)(z-2)}; 1\right] = \frac{z^{n-1}}{z-2}\bigg|_{z=1} = -1, \tag{7.3.54}$$

and

$$\text{Res}\left[\frac{z^{n-1}}{(z-1)(z-2)}; 2\right] = \frac{z^{n-1}}{z-1}\bigg|_{z=2} = 2^{n-1}, \quad n > 0. \tag{7.3.55}$$

Thus,

$$f_n = 2^{n-1} - 1, \quad n > 0. \tag{7.3.56}$$

Combining our results,

$$f_n = \begin{cases} 0, & n = 0, \\ \frac{1}{2}\left(2^n - 2\right), & n > 0. \end{cases} \tag{7.3.57}$$

• Example 7.3.8

Let us use the inversion integral to find the inverse of

$$F(z) = \frac{z^2 + 2z}{(z-1)^2}. \tag{7.3.58}$$

The inversion theorem gives

$$f_n = \frac{1}{2\pi i} \oint_C \frac{z^{n+1} + 2z^n}{(z-1)^2} \, dz = \text{Res}\left[\frac{z^{n+1} + 2z^n}{(z-1)^2}; 1\right], \tag{7.3.59}$$

where the pole at $z = 1$ is second order. Consequently, the corresponding residue is

$$\text{Res}\left[\frac{z^{n+1} + 2z^n}{(z-1)^2}; 1\right] = \frac{d}{dz}\left(z^{n+1} + 2z^n\right)\Big|_{z=1} = 3n + 1. \tag{7.3.60}$$

Thus, the inverse z-transform of (7.3.58) is

$$f_n = 3n + 1, \quad n \geq 0. \tag{7.3.61}$$

• Example 7.3.9

Let $F(z)$ be a z-transform whose poles lie within the unit circle $|z| = 1$. Then

$$F(z) = \sum_{n=0}^{\infty} f_n z^{-n}, \quad |z| > 1, \tag{7.3.62}$$

and

$$F(z)F(z^{-1}) = \sum_{n=0}^{\infty} f_n^2 + \sum_{\substack{n=0 \\ n \neq m}}^{\infty} \sum_{m=0}^{\infty} f_m f_n z^{m-n}. \tag{7.3.63}$$

We now multiply both sides of (7.3.63) by z^{-1} and integrate around the unit circle C. Therefore,

$$\oint_{|z|=1} F(z)F(z^{-1})z^{-1} \, dz = \sum_{n=0}^{\infty} \oint_{|z|=1} f_n^2 z^{-1} \, dz$$

$$+ \sum_{\substack{n=0 \\ n \neq m}}^{\infty} \sum_{m=0}^{\infty} f_m f_n \oint_{|z|=1} z^{m-n-1} \, dz, \tag{7.3.64}$$

after interchanging the order of integration and summation. Performing the integration,

$$\sum_{n=0}^{\infty} f_n^2 = \frac{1}{2\pi i} \oint_{|z|=1} F(z)F(z^{-1})z^{-1}\, dz, \qquad (7.3.65)$$

which is *Parseval's theorem* for one-sided z-transforms. Recall that there are similar theorems for Fourier series and transforms.

- **Example 7.3.10: Evaluation of partial summations**[4]

Consider the partial summation $S_N = \sum_{n=1}^{N} f_n$. We shall now show that z-transforms can be employed to compute S_N.

We begin by noting that

$$S_N = \sum_{n=1}^{N} f_n = \frac{1}{2\pi i} \oint_C F(z) \sum_{n=1}^{N} z^{n-1}\, dz. \qquad (7.3.66)$$

Here we employed the inversion integral to replace f_n and reversed the order of integration and summation. This interchange is permissible since we only have a partial summation. Because the summation in (7.3.66) is a geometric series, we have the final result that

$$S_N = \frac{1}{2\pi i} \oint_C \frac{F(z)(z^N - 1)}{z - 1}\, dz. \qquad (7.3.67)$$

Therefore, we can use the residue theorem and z-transforms to evaluate partial summations.

Let us find $S_N = \sum_{n=1}^{N} n^3$. Because $f_n = n^3$, $F(z) = z(z^2 + 4z + 1)/(z - 1)^4$. Consequently

$$S_N = \text{Res}\left[\frac{z(z^2 + 4z + 1)(z^N - 1)}{(z - 1)^5}; 1\right] \qquad (7.3.68)$$

$$= \frac{1}{4!} \frac{d^4}{dz^4}\left[z(z^2 + 4z + 1)(z^N - 1)\right]\Big|_{z=1} \qquad (7.3.69)$$

$$= \frac{1}{4!} \frac{d^4}{dz^4}\left(z^{N+3} + 4z^{N+2} + z^{N+1} - z^3 - 4z^2 - z\right)\Big|_{z=1} \qquad (7.3.70)$$

$$= \tfrac{1}{4}(N + 1)^2 N^2. \qquad (7.3.71)$$

---

[4] Taken from Bunch, K. J., W. N. Cain, and R. W. Grow, 1990: The z-transform method of evaluating partial summations in closed form. *J. Phys. A*, **23**, L1213–L1215. The material has been used with the permission of the authors and IOP Publishing Limited.

- **Example 7.3.11**

An additional benefit of understanding inversion by the residue method is the ability to *qualitatively* anticipate the inverse by knowing the location of the poles of $F(z)$. This intuition is important because many engineering analyses discuss stability and performance entirely in terms of the properties of the system's z-transform. In Figure 7.3.1 we graphed the location of the poles of $F(z)$ and the corresponding f_n. The student should go through the mental exercise of connecting the two pictures.

Problems

Use the power series or recursive method to compute the first few values of f_n of the following z-transforms. Then check your answers with MATLAB.

1. $F(z) = \dfrac{0.09z^2 + 0.9z + 0.09}{12.6z^2 - 24z + 11.4}$

2. $F(z) = \dfrac{z+1}{2z^4 - 2z^3 + 2z - 2}$

3. $F(z) = \dfrac{1.5z^2 + 1.5z}{15.25z^2 - 36.75z + 30.75}$

4. $F(z) = \dfrac{6z^2 + 6z}{19z^3 - 33z^2 + 21z - 7}$

Use partial fractions to find the inverse of the following z-transforms. Then verify your answers with MATLAB.

5. $F(z) = \dfrac{z(z+1)}{(z-1)(z^2 - z + 1/4)}$

6. $F(z) = \dfrac{(1 - e^{-aT})z}{(z-1)(z - e^{-aT})}$

7. $F(z) = \dfrac{z^2}{(z-1)(z-\alpha)}$

8. $F(z) = \dfrac{(2z - a - b)z}{(z-a)(z-b)}$

9. Using the property that the z-transform of $g_n = f_{n-k} H_{n-k}$ if $n \geq 0$ is $G(z) = z^{-k} F(z)$, find the inverse of

$$F(z) = \frac{z+1}{z^{10}(z - 1/2)}.$$

Then check your answer with MATLAB.

Use the residue method to find the inverse z-transform of the following z-transforms. Then verify your answer with MATLAB.

10. $F(z) = \dfrac{z^2 + 3z}{(z - 1/2)^3}$

11. $F(z) = \dfrac{z}{(z+1)^2(z-2)}$

12. $F(z) = \dfrac{z}{(z+1)^2(z-1)^2}$

13. $F(z) = e^{a/z}$

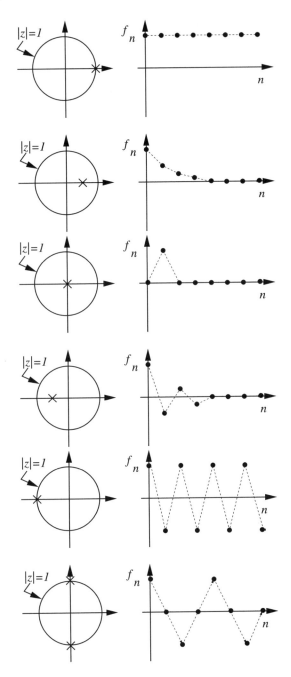

Figure 7.3.1: The correspondence between the location of the simple poles of the z-transform $F(z)$ and the behavior of f_n.

7.4 SOLUTION OF DIFFERENCE EQUATIONS

Having reached the point where we can take a z-transform and then find its inverse, we are ready to use it to solve difference equations. The procedure parallels that of solving ordinary differential equations by Laplace transforms. Essentially we reduce the difference equation to an algebraic problem. We then find the solution by inverting $Y(z)$.

- **Example 7.4.1**

Let us solve the second-order difference equation

$$2y_{n+2} - 3y_{n+1} + y_n = 5 \cdot 3^n, \quad n \geq 0, \tag{7.4.1}$$

where $y_0 = 0$ and $y_1 = 1$.

Taking the z-transform of both sides of (7.4.1), we obtain

$$2\mathcal{Z}(y_{n+2}) - 3\mathcal{Z}(y_{n+1}) + \mathcal{Z}(y_n) = 5\mathcal{Z}(3^n). \tag{7.4.2}$$

From the shifting theorem and Table 7.1.1,

$$2z^2 Y(z) - 2z^2 y_0 - 2zy_1 - 3[zY(z) - zy_0] + Y(z) = \frac{5z}{z-3}. \tag{7.4.3}$$

Substituting $y_0 = 0$ and $y_1 = 1$ into (7.4.3) and simplifying yields

$$(2z - 1)(z - 1)Y(z) = \frac{z(2z - 1)}{z - 3}, \tag{7.4.4}$$

or

$$Y(z) = \frac{z}{(z - 3)(z - 1)}. \tag{7.4.5}$$

To obtain y_n from $Y(z)$ we can employ partial fractions or the residue method. Applying partial fractions gives

$$\frac{Y(z)}{z} = \frac{A}{z - 1} + \frac{B}{z - 3}, \tag{7.4.6}$$

where

$$A = (z - 1) \left. \frac{Y(z)}{z} \right|_{z=1} = -\frac{1}{2}, \tag{7.4.7}$$

and

$$B = (z - 3) \left. \frac{Y(z)}{z} \right|_{z=3} = \frac{1}{2}. \tag{7.4.8}$$

Thus,

$$Y(z) = -\frac{1}{2} \frac{z}{z - 1} + \frac{1}{2} \frac{z}{z - 3}, \tag{7.4.9}$$

or

$$y_n = -\frac{1}{2}\mathcal{Z}^{-1}\left(\frac{z}{z-1}\right) + \frac{1}{2}\mathcal{Z}^{-1}\left(\frac{z}{z-3}\right). \tag{7.4.10}$$

From (7.4.10) and Table 7.1.1,

$$y_n = \tfrac{1}{2}\left(3^n - 1\right), \qquad n \geq 0. \tag{7.4.11}$$

An alternative to this hand calculation is to use MATLAB's ztrans and iztrans to solve difference equations. In the present case, the MATLAB script would read

```
clear
% define symbolic variables
syms z Y; syms n positive
% take z-transform of left side of difference equation
LHS = ztrans(2*sym('y(n+2)')-3*sym('y(n+1)')+sym('y(n)'),n,z);
% take z-transform of right side of difference equation
RHS = 5 * ztrans(3^n,n,z);
% set Y for z-transform of y and introduce initial conditions
newLHS = subs(LHS,'ztrans(y(n),n,z)','y(0)','y(1)',Y,0,1);
% solve for Y
Y = solve(newLHS-RHS,Y);
% invert z-transform and find y(n)
y = iztrans(Y,z,n)
```

This script produced

```
y =
-1/2+1/2*3^n
```

Two checks confirm that we have the *correct* solution. First, our solution must satisfy the initial values of the sequence. Computing y_0 and y_1,

$$y_0 = \tfrac{1}{2}(3^0 - 1) = \tfrac{1}{2}(1-1) = 0, \tag{7.4.12}$$

and

$$y_1 = \tfrac{1}{2}(3^1 - 1) = \tfrac{1}{2}(3-1) = 1. \tag{7.4.13}$$

Thus, our solution gives the correct initial values.

Our sequence y_n must also satisfy the difference equation. Now

$$y_{n+2} = \tfrac{1}{2}(3^{n+2} - 1) = \tfrac{1}{2}(9\,3^n - 1), \tag{7.4.14}$$

and

$$y_{n+1} = \tfrac{1}{2}(3^{n+1} - 1) = \tfrac{1}{2}(3\,3^n - 1). \tag{7.4.15}$$

Therefore,

$$2y_{n+2} - 3y_{n+1} + y_n = \left(9 - \tfrac{9}{2} + \tfrac{1}{2}\right)3^n - 1 + \tfrac{3}{2} - \tfrac{1}{2} = 5\,3^n \tag{7.4.16}$$

and our solution is correct.

Finally, we note that the term $3^n/2$ is necessary to give the right side of (7.4.1); it is the particular solution. The $-1/2$ term is necessary so that the sequence satisfies the initial values; it is the complementary solution.

- **Example 7.4.2**

Let us find the y_n in the difference equation

$$y_{n+2} - 2y_{n+1} + y_n = 1, \quad n \geq 0 \tag{7.4.17}$$

with the initial conditions $y_0 = 0$ and $y_1 = 3/2$.

From (7.4.17),

$$\mathcal{Z}(y_{n+2}) - 2\mathcal{Z}(y_{n+1}) + \mathcal{Z}(y_n) = \mathcal{Z}(1). \tag{7.4.18}$$

The z-transform of the left side of (7.4.18) is obtained from the shifting theorem and Table 7.1.1 yields $\mathcal{Z}(1)$. Thus,

$$z^2 Y(z) - z^2 y_0 - z y_1 - 2z Y(z) + 2z y_0 + Y(z) = \frac{z}{z-1}. \tag{7.4.19}$$

Substituting $y_0 = 0$ and $y_1 = 3/2$ in (7.4.19) and simplifying gives

$$Y(z) = \frac{3z^2 - z}{2(z-1)^3} \tag{7.4.20}$$

or

$$y_n = \mathcal{Z}^{-1}\left[\frac{3z^2 - z}{2(z-1)^3}\right]. \tag{7.4.21}$$

We find the inverse z-transform of (7.4.21) by the residue method or

$$y_n = \frac{1}{2\pi i} \oint_C \frac{3z^{n+1} - z^n}{2(z-1)^3}\, dz = \frac{1}{2!} \frac{d^2}{dz^2}\left[\frac{3z^{n+1}}{2} - \frac{z^n}{2}\right]\Bigg|_{z=1} \tag{7.4.22}$$

$$= \tfrac{1}{2}n^2 + n. \tag{7.4.23}$$

Thus,

$$y_n = \tfrac{1}{2}n^2 + n, \quad n \geq 0. \tag{7.4.24}$$

Note that $n^2/2$ gives the particular solution to (7.4.17), while n is there so that y_n satisfies the initial conditions. This problem is particularly interesting because our constant forcing produces a response that grows as n^2, just as in the case of resonance in a time-continuous system when a finite forcing such as $\sin(\omega_0 t)$ results in a response whose amplitude grows as t^m.

• **Example 7.4.3**

Let us solve the difference equation

$$b^2 y_n + y_{n+2} = 0, \qquad (7.4.25)$$

where the initial conditions are $y_0 = b^2$ and $y_1 = 0$.

We begin by taking the z-transform of each term in (7.4.25). This yields

$$b^2 \mathcal{Z}(y_n) + \mathcal{Z}(y_{n+2}) = 0. \qquad (7.4.26)$$

From the shifting theorem, it follows that

$$b^2 Y(z) + z^2 Y(z) - z^2 y_0 - z y_1 = 0. \qquad (7.4.27)$$

Substituting $y_0 = b^2$ and $y_1 = 0$ into (7.4.27),

$$b^2 Y(z) + z^2 Y(z) - b^2 z^2 = 0, \qquad (7.4.28)$$

or

$$Y(z) = \frac{b^2 z^2}{z^2 + b^2}. \qquad (7.4.29)$$

To find y_n we employ the residue method or

$$y_n = \frac{1}{2\pi i} \oint_C \frac{b^2 z^{n+1}}{(z - ib)(z + ib)} \, dz. \qquad (7.4.30)$$

Thus,

$$y_n = \left. \frac{b^2 z^{n+1}}{z + ib} \right|_{z=ib} + \left. \frac{b^2 z^{n+1}}{z - ib} \right|_{z=-ib} = \frac{b^{n+2} i^n}{2} + \frac{b^{n+2}(-i)^n}{2} \qquad (7.4.31)$$

$$= \frac{b^{n+2} e^{in\pi/2}}{2} + \frac{b^{n+2} e^{-in\pi/2}}{2} = b^{n+2} \cos\left(\frac{n\pi}{2}\right), \qquad (7.4.32)$$

because $\cos(x) = \frac{1}{2}\left(e^{ix} + e^{-ix}\right)$. Consequently, we obtain the desired result that

$$y_n = b^{n+2} \cos\left(\frac{n\pi}{2}\right) \quad \text{for } n \geq 0. \qquad (7.4.33)$$

• **Example 7.4.4: Compound interest**

Difference equations arise in finance because the increase or decrease in an account occurs in discrete steps. For example, the amount of money in a compound interest saving account after $n + 1$ conversion periods (the time period between interest payments) is

$$y_{n+1} = y_n + r y_n, \qquad (7.4.34)$$

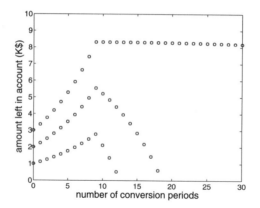

Figure 7.4.1: The amount in a saving account as a function of an annual conversion period when interest is compounded at the annual rate of 12% and a $1000 is taken from the account every period starting with period 10.

where r is the interest rate per conversion period. The second term on the right side of (7.4.34) is the amount of interest paid at the end of each period.

Let us ask a somewhat more difficult question of how much money we will have if we withdraw the amount A at the end of every period starting after the period ℓ. Now the difference equation reads

$$y_{n+1} = y_n + ry_n - AH_{n-\ell-1}. \tag{7.4.35}$$

Taking the z-transform of (7.4.35),

$$zY(z) - zy_0 = (1+r)Y(z) - \frac{Az^{2-\ell}}{z-1} \tag{7.4.36}$$

after using (7.2.10) or

$$Y(z) = \frac{y_0 z}{z-(1+r)} - \frac{Az^{2-\ell}}{(z-1)[z-(1+r)]}. \tag{7.4.37}$$

Taking the inverse of (7.4.37),

$$y_n = y_0(1+r)^n - \frac{A}{r}\left[(1+r)^{n-\ell+1} - 1\right]H_{n-\ell}. \tag{7.4.38}$$

The first term in (7.4.38) represents the growth of money by compound interest while the second term gives the depletion of the account by withdrawals.

Figure 7.4.1 gives the values of y_n for various starting amounts assuming an annual conversion period with $r = 0.12$, $\ell = 10$ years, and $A = \$1000$. These computations were done two ways using MATLAB as follows:

```
% load in parameters
```

```
clear; r = 0.12; A = 1; k = 0:30;
y = zeros(length(k),3); yanswer = zeros(length(k),3);
% set initial condition
for m=1:3
y(1,m) = m;
% compute other y values
for n = 1:30
y(n+1,m) = y(n,m)+r*y(n,m);
y(n+1,m) = y(n+1,m)-A*stepfun(n,11);
end
% now use (7.4.38)
for n = 1:31
yanswer(n,m) = y(1,m)*(1+r)^(n-1);
yanswer(n,m) = yanswer(n,m)-A*((1+r)^(n-10)-1)*stepfun(n,11)/r;
end; end;
plot(k,y,'o'); hold; plot(k,yanswer,'s');
axis([0 30 0 10])
xlabel('number of conversion periods','Fontsize',20)
ylabel('amount left in account (K$)','Fontsize',20)
```

Figure 7.4.1 shows that if an investor places an initial amount of $3000 in an account bearing 12% annually, after 10 years he can withdraw $1000 annually, essentially forever. This is because the amount that he removes every year is replaced by the interest on the funds that remain in the account.

● **Example 7.4.5**

Let us solve the following system of difference equations:

$$x_{n+1} = 4x_n + 2y_n, \qquad (7.4.39)$$

and

$$y_{n+1} = 3x_n + 3y_n, \qquad (7.4.40)$$

with the initial values of $x_0 = 0$ and $y_0 = 5$.
Taking the z-transform of (7.4.39)–(7.4.40),

$$zX(z) - x_0 z = 4X(z) + 2Y(z), \qquad (7.4.41)$$

$$zY(z) - y_0 z = 3X(z) + 3Y(z), \qquad (7.4.42)$$

or

$$(z - 4)X(z) - 2Y(z) = 0, \qquad (7.4.43)$$

$$3X(z) - (z - 3)Y(z) = -5z. \qquad (7.4.44)$$

Solving for $X(z)$ and $Y(z)$,

$$X(z) = \frac{10z}{(z-6)(z-1)} = \frac{2z}{z-6} - \frac{2z}{z-1},$$
(7.4.45)

and

$$Y(z) = \frac{5z(z-4)}{(z-6)(z-1)} = \frac{2z}{z-6} + \frac{3z}{z-1}.$$
(7.4.46)

Taking the inverse of (7.4.45)–(7.4.46) term by term,

$$x_n = -2 + 2\,6^n, \quad \text{and} \quad y_n = 3 + 2\,6^n.$$
(7.4.47)

We can also check our work using the MATLAB script

```
clear
% define symbolic variables
syms X Y z; syms n positive
% take z-transform of left side of differential equations
LHS1 = ztrans(sym('x(n+1)')-4*sym('x(n)')-2*sym('y(n)'),n,z);
LHS2 = ztrans(sym('y(n+1)')-3*sym('x(n)')-3*sym('y(n)'),n,z);
% set X and Y for the z-transform of x and y
%      and introduce initial conditions
newLHS1 = subs(LHS1,'ztrans(x(n),n,z)','ztrans(y(n),n,z)',...
    'x(0)','y(0)',X,Y,0,5);
newLHS2 = subs(LHS2,'ztrans(x(n),n,z)','ztrans(y(n),n,z)',...
    'x(0)','y(0)',X,Y,0,5);
% solve for X and Y
[X,Y] = solve(newLHS1,newLHS2,X,Y);
% invert z-transform and find x(n) and y(n)
x = iztrans(X,z,n)
y = iztrans(Y,z,n)
```

This script yields

```
x =
2*6^n-2
y =
2*6^n+3
```

Problems

Solve the following difference equations using z-transforms, where $n \geq 0$. Check your answer using MATLAB.

1. $y_{n+1} - y_n = n^2, \quad y_0 = 1.$

2. $y_{n+2} - 2y_{n+1} + y_n = 0, \quad y_0 = y_1 = 1.$

3. $y_{n+2} - 2y_{n+1} + y_n = 1, \quad y_0 = y_1 = 0.$

4. $y_{n+1} + 3y_n = n, \quad y_0 = 0.$

5. $y_{n+1} - 5y_n = \cos(n\pi), \quad y_0 = 0.$

6. $y_{n+2} - 4y_n = 1, \quad y_0 = 1, y_1 = 0.$

7. $y_{n+2} - \frac{1}{4}y_n = (\frac{1}{2})^n, \quad y_0 = y_1 = 0.$

8. $y_{n+2} - 5y_{n+1} + 6y_n = 0, \quad y_0 = y_1 = 1.$

9. $y_{n+2} - 3y_{n+1} + 2y_n = 1, \quad y_0 = y_1 = 0.$

10. $y_{n+2} - 2y_{n+1} + y_n = 2, \quad y_0 = 0, \quad y_1 = 2.$

11. $x_{n+1} = 3x_n - 4y_n, \ y_{n+1} = 2x_n - 3y_n, \quad x_0 = 3, \ y_0 = 2.$

12. $x_{n+1} = 2x_n - 10y_n, \ y_{n+1} = -x_n - y_n, \quad x_0 = 3, \ y_0 = -2.$

13. $x_{n+1} = x_n - 2y_n, \ y_{n+1} = -6y_n, \quad x_0 = -1, \ y_0 = -7.$

14. $x_{n+1} = 4x_n - 5y_n, \ y_{n+1} = x_n - 2y_n, \quad x_0 = 6, \ y_0 = 2.$

7.5 STABILITY OF DISCRETE-TIME SYSTEMS

When we discussed the solution of ordinary differential equations by Laplace transforms, we introduced the concept of transfer function and impulse response. In the case of discrete-time systems, similar considerations come into play.

Consider the recursive system

$$y_n = a_1 y_{n-1} H_{n-1} + a_2 y_{n-2} H_{n-2} + x_n, \quad n \geq 0, \qquad (7.5.1)$$

where H_{n-k} is the unit step function. It equals 0 for $n < k$ and 1 for $n \geq k$. Equation (7.5.1) is called a *recursive system* because future values of the sequence depend upon all of the previous values. At present, a_1 and a_2 are free parameters which we shall vary.

Using (7.2.10),

$$z^2 Y(z) - a_1 z Y(z) - a_2 Y(z) = z^2 X(z), \qquad (7.5.2)$$

or

$$G(z) = \frac{Y(z)}{X(z)} = \frac{z^2}{z^2 - a_1 z - a_2}. \qquad (7.5.3)$$

As in the case of Laplace transforms, the ratio $Y(z)/X(z)$ is the transfer function. The inverse of the transfer function gives the impulse response for our discrete-time system. This particular transfer function has two poles, namely

$$z_{1,2} = \frac{a_1}{2} \pm \sqrt{\frac{a_1^2}{4} + a_2}. \qquad (7.5.4)$$

At this point, we consider three cases.

Case 1: $a_1^2/4 + a_2 < 0$. In this case z_1 and z_2 are complex conjugates. Let us write them as $z_{1,2} = re^{\pm i\omega_0 T}$. Then

$$G(z) = \frac{z^2}{(z - re^{i\omega_0 T})(z - re^{-i\omega_0 T})} = \frac{z^2}{z^2 - 2r\cos(\omega_0 T)z + r^2}, \qquad (7.5.5)$$

where $r^2 = -a_2$, and $\omega_0 T = \cos^{-1}(a_1/2r)$. From the inversion integral,

$$g_n = \text{Res}\left[\frac{z^{n+1}}{z^2 - 2r\cos(\omega_0 T)z + r^2}; z_1\right]$$

$$+ \text{Res}\left[\frac{z^{n+1}}{z^2 - 2r\cos(\omega_0 T)z + r^2}; z_2\right], \qquad (7.5.6)$$

where g_n denotes the impulse response. Now

$$\text{Res}\left[\frac{z^{n+1}}{z^2 - 2r\cos(\omega_0 T)z + r^2}; z_1\right] = \lim_{z \to z_1} \frac{(z - z_1)z^{n+1}}{(z - z_1)(z - z_2)} \qquad (7.5.7)$$

$$= r^n \frac{\exp[i(n+1)\omega_0 T]}{e^{i\omega_0 T} - e^{-i\omega_0 T}} \qquad (7.5.8)$$

$$= \frac{r^n \exp[i(n+1)\omega_0 T]}{2i\sin(\omega_0 T)}. \qquad (7.5.9)$$

Similarly,

$$\text{Res}\left[\frac{z^{n+1}}{z^2 - 2r\cos(\omega_0 T)z + r^2}; z_2\right] = -\frac{r^n \exp[-i(n+1)\omega_0 T]}{2i\sin(\omega_0 T)}, \qquad (7.5.10)$$

and

$$g_n = \frac{r^n \sin[(n+1)\omega_0 T]}{\sin(\omega_0 T)}. \qquad (7.5.11)$$

A graph of $\sin[(n+1)\omega_0 T]/\sin(\omega_0 T)$ with respect to n gives a sinusoidal envelope. More importantly, if $|r| < 1$ these oscillations vanish as $n \to \infty$ and the system is stable. On the other hand, if $|r| > 1$ the oscillations grow without bound as $n \to \infty$ and the system is unstable.

Recall that $|r| > 1$ corresponds to poles that lie outside the unit circle while $|r| < 1$ is exactly the opposite. Our example suggests that for discrete-time systems to be stable, all of the poles of the transfer function must lie within the unit circle while an unstable system has at least one pole that lies outside of this circle.

Case 2: $a_1^2/4 + a_2 > 0$. This case leads to two real roots, z_1 and z_2. From the inversion integral, the sum of the residues gives the impulse response

$$g_n = \frac{z_1^{n+1} - z_2^{n+1}}{z_1 - z_2}. \tag{7.5.12}$$

Once again, if the poles lie within the unit circle, $|z_1| < 1$ and $|z_2| < 1$, the system is stable.

Case 3: $a_1^2/4 + a_2 = 0$. This case yields $z_1 = z_2$,

$$G(z) = \frac{z^2}{(z - a_1/2)^2} \tag{7.5.13}$$

and

$$g_n = \frac{1}{2\pi i} \oint_C \frac{z^{n+1}}{(z - a_1/2)^2} dz = \left(\frac{a_1}{2}\right)^n (n+1). \tag{7.5.14}$$

This system is obviously stable if $|a_1/2| < 1$ and the pole of the transfer function lies within the unit circle.

In summary, finding the transfer function of a discrete-time system is important in determining its stability. Because the location of the poles of $G(z)$ determines the response of the system, a stable system has all of its poles within the unit circle. Conversely, if any of the poles of $G(z)$ lie outside of the unit circle, the system is unstable. Finally, if $\lim_{n \to \infty} g_n = c$, the system is marginally stable. For example, if $G(z)$ has simple poles, some of the poles must lie *on* the unit circle.

• Example 7.5.1

Numerical methods of integration provide some of the simplest, yet most important, difference equations in the literature. In this example,[5] we show how z-transforms can be used to highlight the strengths and weaknesses of such schemes.

Consider the trapezoidal integration rule in numerical analysis. The integral y_n is updated by adding the latest trapezoidal approximation of the continuous curve. Thus, the integral is computed by

$$y_n = \tfrac{1}{2} T(x_n + x_{n-1} H_{n-1}) + y_{n-1} H_{n-1}, \tag{7.5.15}$$

[5] From Salzer, J. M., 1954: Frequency analysis of digital computers operating in real time. *Proc. IRE*, **42**, 457–466. ©IRE (now IEEE).

where T is the interval between evaluations of the integrand.

We first determine the stability of this rule because it is of little value if it is not stable. Using (7.2.10), the transfer function is

$$G(z) = \frac{Y(z)}{X(z)} = \frac{T}{2}\left(\frac{z+1}{z-1}\right). \qquad (7.5.16)$$

To find the impulse response, we use the inversion integral and find that

$$g_n = \frac{T}{4\pi i} \oint_C z^{n-1}\frac{z+1}{z-1}\, dz. \qquad (7.5.17)$$

At this point, we must consider two cases: $n = 0$ and $n > 0$. For $n = 0$,

$$g_0 = \frac{T}{2}\mathrm{Res}\left[\frac{z+1}{z(z-1)};0\right] + \frac{T}{2}\mathrm{Res}\left[\frac{z+1}{z(z-1)};1\right] = \frac{T}{2}. \qquad (7.5.18)$$

For $n > 0$,

$$g_0 = \frac{T}{2}\mathrm{Res}\left[\frac{z^{n-1}(z+1)}{z-1};1\right] = T. \qquad (7.5.19)$$

Therefore, the impulse response for this numerical scheme is $g_0 = \frac{T}{2}$ and $g_n = T$ for $n > 0$. Note that this is a marginally stable system (the solution neither grows nor decays with n) because the pole associated with the transfer function lies *on* the unit circle.

Having discovered that the system is not unstable, let us continue and explore some of its properties. Recall now that $z = e^{sT} = e^{i\omega T}$ if $s = i\omega$. Then the transfer function becomes

$$G(\omega) = \frac{T}{2}\frac{1+e^{-i\omega T}}{1-e^{-i\omega T}} = -\frac{iT}{2}\cot\left(\frac{\omega T}{2}\right). \qquad (7.5.20)$$

On the other hand, the transfer function of an ideal integrator is $1/s$ or $-i/\omega$. Thus, the trapezoidal rule has ideal phase but its shortcoming lies in its amplitude characteristic; it lies below the ideal integrator for $0 < \omega T < \pi$. We show this behavior, along with that for Simpson's one third rule and Simpson's three eighth rule, in Figure 7.5.1.

Figure 7.5.1 confirms the superiority of Simpson's one third rule over his three eighth rule. The figure also shows that certain schemes are better at suppressing noise at higher frequencies, an effect not generally emphasized in numerical calculus but often important in system design. For example, the trapezoidal rule is inferior to all others at low frequencies but only to Simpson's one third rule at higher frequencies. Furthermore, the trapezoidal rule might actually be preferred not only because of its simplicity but also because it attenuates at higher frequencies, thereby counteracting the effect of noise.

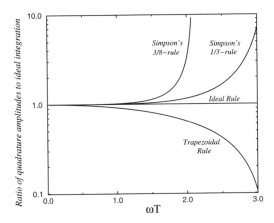

Figure 7.5.1: Comparison of various quadrature formulas by ratios of their amplitudes to that of an ideal integrator. [From Salzer, J. M., 1954: Frequency analysis of digital computers operating in real time. *Proc. IRE*, **42**, p. 463. ©IRE (now IEEE).]

• Example 7.5.2

Given the transfer function

$$G(z) = \frac{z^2}{(z-1)(z-1/2)}, \tag{7.5.21}$$

is this discrete-time system stable or marginally stable?

This transfer function has two simple poles. The pole at $z = 1/2$ gives rise to a term that varies as $\left(\frac{1}{2}\right)^n$ in the impulse response while the $z = 1$ pole gives rise to a constant. Because this constant neither grows nor decays with n, the system is marginally stable.

• Example 7.5.3

In most cases the transfer function consists of a ratio of two polynomials. In this case we can use the MATLAB function `filter` to compute the impulse response as follows: Consider the Kronecker delta sequence, $x_0 = 1$, and $x_n = 0$ for $n > 0$. From the definition of the z-transform, $X(z) = 1$. Therefore, if our input into `filter` is the Kronecker delta sequence, the output y_n will be the impulse response since $Y(z) = G(z)$. If the impulse response grows without bound as n increases, the system is unstable. If it goes to zero as n increases, the system is stable. If it remains constant, it is marginally stable.

To illustrate this concept, the following MATLAB script finds the impulse response corresponding to the transfer function (7.5.21):

```
% enter the coefficients of the numerator
%      of the transfer function (7.5.21)
```

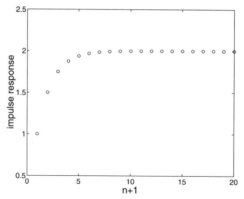

Figure 7.5.2: The impulse response for a discrete system with a transform function given by (7.5.21).

```
num = [1 0 0];
% enter the coefficients of the denominator
%     of the transfer function (7.5.21)
den = [1 -1.5 0.5];
% create the Kronecker delta sequence
x = [1 zeros(1,20)];
% find the impulse response
y = filter(num,den,x);
% plot impulse response
plot(y,'o'), axis([0 20 0.5 2.5])
xlabel('n+1','Fontsize',20)
ylabel('impulse response','Fontsize',20)
```

Figure 7.5.2 shows the computed impulse response. The asymptotic limit is two, so the system is marginally stable as we found before.

We note in closing that the same procedure can be used to find the inverse of *any* z-transform which consists of a ratio of two polynomials. Here we simply set $G(z)$ equal to the given z-transform and perform the same analysis.

Problems

For the following time-discrete systems, find the transfer function and determine whether the systems are unstable, marginally stable, or stable. Check your answer by graphing the impulse response using MATLAB.

1. $y_n = y_{n-1}H_{n-1} + x_n$ 2. $y_n = 2y_{n-1}H_{n-1} - y_{n-2}H_{n-2} + x_n$

3. $y_n = 3y_{n-1}H_{n-1} + x_n$ 4. $y_n = \frac{1}{4}y_{n-2}H_{n-2} + x_n$

Chapter 8

The Hilbert Transform

In addition to the Fourier, Laplace, and z-transforms, there are many other linear transforms which have their own special niche in engineering. Examples include Hankel, Walsh, Radon, and Hartley transforms. In this chapter we consider the *Hilbert transform* which is a commonly used technique for relating the real and imaginary parts of a spectral response, particularly in communication theory.

We begin our study of Hilbert transforms by first defining them and then exploring their properties. Next, we develop the concept of the analytic signal. Finally, we explore a property of Hilbert transforms that is frequently applied to data analysis: the Kramers-Kronig relationship.

8.1 DEFINITION

In Chapter 7 we motivated the development of z-transforms by exploring the concept of the ideal sampler. In the case of Hilbert transforms, we introduce another fundamental operation, namely *quadrature phase shifting* or the *ideal Hilbert transformer*. This procedure does nothing more than shift the phase of all input frequency components by $-\pi/2$. Hilbert transformers are frequently used in communication systems and signal processing; examples include the generation of single-sideband modulated signals and radar and speech signal processing.

Because a $-\pi/2$ phase shift is equivalent to multiplying the Fourier transform of a signal by $e^{-i\pi/2} = -i$, and because phase shifting must be an odd function of frequency,[1] the transfer function of the phase shifter is $G(\omega) = -i\operatorname{sgn}(\omega)$, where $\operatorname{sgn}(\ )$ is defined by (5.2.11). In other words, if $X(\omega)$ denotes the input spectrum to the phase shifter, the output spectrum must be $-i\operatorname{sgn}(\omega)X(\omega)$. If the process is repeated, the total phase shift is $-\pi$, a complete phase reversal of all frequency components. The output spectrum then equals $[-i\operatorname{sgn}(\omega)]^2X(\omega) = -X(\omega)$. This agrees with the notion of phase reversal because the output function is $-x(t)$.

Consider now the impulse response of the quadrature phase shifter, $g(t) = \mathcal{F}^{-1}[G(\omega)]$. From the definition of Fourier transforms,

$$\frac{dG}{d\omega} = -i\int_{-\infty}^{\infty} tg(t)e^{-i\omega t}\,dt, \tag{8.1.1}$$

and

$$g(t) = \frac{i}{t}\mathcal{F}^{-1}\left(\frac{dG}{d\omega}\right). \tag{8.1.2}$$

Since $G'(\omega) = -2i\delta(\omega)$, the corresponding impulse response is

$$g(t) = \frac{i}{t}\mathcal{F}^{-1}[-2i\delta(\omega)] = \frac{1}{\pi t}. \tag{8.1.3}$$

Consequently, if $x(t)$ is the input to a quadrature phase shifter, the superposition integral gives the output time function as

$$\widehat{x}(t) = x(t) * \frac{1}{\pi t} = \frac{1}{\pi}\int_{-\infty}^{\infty}\frac{x(\tau)}{t-\tau}\,d\tau. \tag{8.1.4}$$

We shall define $\widehat{x}(t)$ as the *Hilbert transform* of $x(t)$, although some authors use the negative of (8.1.4) corresponding to a $+\pi/2$ phase shift. The transform $\widehat{x}(t)$ is also called the *harmonic conjugate* of $x(t)$.

In similar fashion, $\widehat{\widehat{x}}(t)$ is the Hilbert transform of the Hilbert transform of $x(t)$ and corresponds to the output of two cascaded phase shifters. However, this output is known to be $-x(t)$, so $\widehat{\widehat{x}}(t) = -x(t)$, and we arrive at the *inverse Hilbert transform* relationship that

$$x(t) = -\widehat{x}(t) * \frac{1}{\pi t} = -\frac{1}{\pi}\int_{-\infty}^{\infty}\frac{\widehat{x}(\tau)}{t-\tau}\,d\tau. \tag{8.1.5}$$

---

[1] For a real function the phase of its Fourier transform must be an odd function of ω.

Figure 8.1.1: Descended from a Prussian middle-class family, David Hilbert (1862–1943) would make significant contributions in the fields of algebraic form, algebraic number theory, foundations of geometry, analysis, mathematical physics, and the foundations of mathematics. Hilbert transforms arose during his study of integral equations [Hilbert, D., 1912: *Grundzüge einer allgemeinen Theorie der linearen Integralgleichungen.* Teubner, p. 75]. (Portrait courtesy of Photo AKG, London.)

Taken together, $x(t)$ and $\widehat{x}(t)$ are called a *Hilbert pair*. Hilbert pairs enjoy the unique property that $x(t) + i\widehat{x}(t)$ is an *analytic function*.[2]

Because of the singularity at $\tau = t$, the integrals in (8.1.4) and (8.1.5) must be taken in the *Cauchy principal value* sense by approaching the singularity point from both sides, namely

$$\int_{-\infty}^{\infty} f(\tau)\, d\tau = \lim_{\epsilon \to 0} \left[\int_{-\infty}^{t-\epsilon} f(\tau)\, d\tau + \int_{t+\epsilon}^{\infty} f(\tau)\, d\tau \right], \qquad (\textbf{8.1.6})$$

so that the infinities to the right and left of $\tau = t$ cancel each other. See §1.10. We also note that the Hilbert transform is basically a convolution and

---

[2] For the proof, see Titchmarsh, E. C., 1948: *Introduction to the Theory of Fourier Integrals.* Oxford University Press, p. 125.

does not produce a change of domain; if x is a function of time, then $\hat{x}$ is also a function of time. This is quite different from what we encountered with Laplace or Fourier transforms.

From its origin in phase shifting, Hilbert transforms of sinusoidal functions are trivial. Some examples are

$$\widehat{\cos(\omega t + \varphi)} = \cos\left(\omega t + \varphi - \tfrac{\pi}{2}\right) = \operatorname{sgn}(\omega)\sin(\omega t + \varphi). \qquad (8.1.7)$$

Similarly,

$$\widehat{\sin(\omega t + \varphi)} = -\operatorname{sgn}(\omega)\cos(\omega t + \varphi), \qquad (8.1.8)$$

and

$$\widehat{e^{i\omega t + i\varphi}} = -i\,\operatorname{sgn}(\omega)e^{i\omega t + i\varphi}. \qquad (8.1.9)$$

Thus, Hilbert transformation does not change the amplitude of sine or cosine but does change their phase by $\pm\pi/2$.

• **Example 8.1.1**

Let us apply the integral definition of the Hilbert transform (8.1.4) to find the Hilbert transform of $\sin(\omega t)$, $\omega \neq 0$.

From the definition,

$$\mathcal{H}\left[\sin(\omega t)\right] = \frac{1}{\pi}\int_{-\infty}^{\infty} \frac{\sin(\omega \tau)}{t - \tau}\, d\tau. \qquad (8.1.10)$$

If $x = t - \tau$, then

$$\mathcal{H}\left[\sin(\omega t)\right] = -\frac{\cos(\omega t)}{\pi}\int_{-\infty}^{\infty} \frac{\sin(\omega x)}{x}\, dx = -\cos(\omega t)\,\operatorname{sgn}(\omega). \qquad (8.1.11)$$

• **Example 8.1.2**

Let us compute the Hilbert transform of $x(t) = \sin(t)/(t^2 + 1)$ from the definition of the Hilbert transform, (8.1.4).

From the definition,

$$\hat{x}(t) = \frac{1}{\pi}PV\int_{-\infty}^{\infty} \frac{\sin(\tau)}{(t - \tau)(\tau^2 + 1)}\, d\tau = \frac{1}{\pi}\operatorname{Im}\left[PV\int_{-\infty}^{\infty} \frac{e^{i\tau}}{(t - \tau)(\tau^2 + 1)}\, d\tau\right]. \qquad (8.1.12)$$

Because of the singularity on the real axis at $\tau = t$, we treat the integrals in (8.1.12) in the sense of Cauchy principal value.

To evaluate (8.1.12), we convert it into a closed contour integration by introducing a semicircle C_R of infinite radius in the upper half-plane. This

Table 8.1.1: The Hilbert Transform of Some Common Functions

	function, $x(t)$	Hilbert transform, $\hat{x}(t)$
1.	$\begin{cases} 1, & a < t < b \\ 0, & \text{otherwise} \end{cases}$	$\dfrac{1}{\pi} \ln \left\lvert \dfrac{t-a}{t-b} \right\rvert$
2.	$\sin(\omega t + \varphi)$	$-\operatorname{sgn}(\omega) \cos(\omega t + \varphi)$
3.	$\cos(\omega t + \varphi)$	$\operatorname{sgn}(\omega) \sin(\omega t + \varphi)$
4.	$e^{i\omega t + \varphi i}$	$-i\,\operatorname{sgn}(\omega) e^{i\omega t + \varphi i}$
5.	$\dfrac{1}{t}$	$-\pi\delta(t)$
6.	$\dfrac{1}{t^2 + a^2}, \quad 0 < \operatorname{Re}(a)$	$\dfrac{t}{a(t^2 + a^2)}$
7.	$\dfrac{\lambda t + \mu a}{t^2 + a^2}, \quad 0 < \operatorname{Re}(a)$	$\dfrac{\mu t - \lambda a}{t^2 + a^2}$
8.	$\dfrac{1}{1 + t^4}$	$\dfrac{t(1 + t^2)}{\sqrt{2}\,(1 + t^4)}$
9.	$\dfrac{\sin(at)}{t}, \quad 0 < a$	$\dfrac{1 - \cos(at)}{t}$
10.	$\dfrac{\sin(t)}{1 + t^2}$	$\dfrac{e^{-1} - \cos(t)}{1 + t^2}$
11.	$\sin(at) J_1(at), \quad 0 < a$	$-\cos(at) J_1(at)$
12.	$\sin(at) J_n(bt), \quad 0 < b < a$	$-\cos(at) J_n(bt)$
13.	$\cos(at) J_1(at), \quad 0 < a$	$\sin(at) J_1(at)$
14.	$\cos(at) J_n(bt), \quad 0 < b < a$	$\sin(at) J_n(at)$
15.	$\begin{cases} \sqrt{a^2 - t^2}, & -a < t < a \\ 0, & \text{otherwise} \end{cases}$	$\begin{cases} t + \sqrt{t^2 - a^2}, & -\infty < t < -a \\ t, & -a < t < a \\ t - \sqrt{t^2 - a^2}, & a < t < \infty \end{cases}$
16.	$\sin\left(a\sqrt{t}\,\right) H(t), \quad 0 < a$	$\begin{cases} -e^{-a\sqrt{\lvert t \rvert}}, & -\infty < t < 0 \\ -\cos\left(a\sqrt{t}\,\right), & 0 < t < \infty \end{cases}$

yields a closed contour C which consists of the real line plus this semicircle. Therefore, (8.1.12) can be rewritten

$$PV \int_{-\infty}^{\infty} \frac{e^{i\tau}}{(t-\tau)(\tau^2+1)} \, d\tau = PV \oint_C \frac{e^{iz}}{(t-z)(z^2+1)} \, dz$$

$$- \int_{C_R} \frac{e^{iz}}{(t-z)(z^2+1)} \, dz. \qquad (8.1.13)$$

The second integral on the right side of (8.1.13) vanishes by (1.9.7).

The evaluation of the closed integral in (8.1.13) follows from the residue theorem. We have that

$$\text{Res}\left[\frac{e^{iz}}{(t-z)(z^2+1)}; t\right] = \lim_{z \to t} \frac{(z-t)\, e^{iz}}{(t-z)(z^2+1)} = -\frac{e^{it}}{t^2+1}, \qquad (8.1.14)$$

and

$$\text{Res}\left[\frac{e^{iz}}{(t-z)(z^2+1)}; i\right] = \lim_{z \to i} \frac{(z-i)\, e^{iz}}{(t-z)(z^2+1)} = \frac{e^{-1}}{2i(t-i)}. \qquad (8.1.15)$$

We do not have a contribution from $z = -i$ because it lies *outside* of the closed contour.

Therefore,

$$PV \int_{-\infty}^{\infty} \frac{e^{i\tau}}{(t-\tau)(\tau^2+1)} \, d\tau = -\frac{\pi i\, e^{it}}{t^2+1} + \frac{\pi\, e^{-1}(t+i)}{t^2+1}. \qquad (8.1.16)$$

Only one half of the value of the residue at $z = t$ was included; this reflects the semicircular indentation around the singularity there. Substituting (8.1.16) into (8.1.12), we obtain the final result that

$$\mathcal{H}\left[\frac{\sin(t)}{t^2+1}\right] = \frac{e^{-1} - \cos(t)}{t^2+1}. \qquad (8.1.17)$$

• Example 8.1.3

Let us employ the relationship that the Fourier transform of $\hat{x}(t)$ equals $-i\,\text{sgn}(\omega)$ times the Fourier transform of $x(t)$ to find the Hilbert transform of $x(t) = e^{-t^2}$.

Because $\mathcal{F}(e^{-t^2}) = \sqrt{\pi}\, e^{-\omega^2/4}$,

$$\widehat{X}(\omega) = -i\sqrt{\pi}\, \text{sgn}(\omega) e^{-\omega^2/4}. \qquad (8.1.18)$$

Therefore,

$$\widehat{x}(t) = \frac{i}{2\sqrt{\pi}} \int_{-\infty}^{0} e^{it\omega - \omega^2/4} \, d\omega - \frac{i}{2\sqrt{\pi}} \int_{0}^{\infty} e^{it\omega - \omega^2/4} \, d\omega \qquad (8.1.19)$$

$$= \frac{i}{\sqrt{\pi}} \int_{-\infty}^{0} e^{2it\eta - \eta^2} \, d\eta - \frac{i}{\sqrt{\pi}} \int_{0}^{\infty} e^{2it\eta - \eta^2} \, d\eta \qquad (8.1.20)$$

$$= \frac{e^{-t^2}}{\sqrt{\pi}} \int_{-i\infty}^{t} e^{-s^2} \, ds - \frac{e^{-t^2}}{\sqrt{\pi}} \int_{t}^{i\infty} e^{-s^2} \, ds \qquad (8.1.21)$$

$$= \frac{2e^{-t^2}}{\sqrt{\pi}} \int_{0}^{t} e^{-s^2} \, ds, \qquad (8.1.22)$$

where $s = t + \eta i$. The integral in (8.1.22) is the well known *Dawson's integral.*[3] See Gautschi and Waldvogel[4] for an alternative derivation.

• Example 8.1.4: Numerical computation of the Hilbert transform

Recently André Weideman[5] devised a particularly efficient method for *numerically* computing the Hilbert transform when $x(t)$ is known exactly for any real t and enjoys the property that

$$\int_{-\infty}^{\infty} |x(t)|^2 \, dt < \infty. \qquad (8.1.23)$$

Given (8.1.23), the function $x(t)$ can be represented by the rational expansion

$$x(t) = \sum_{n=-\infty}^{\infty} a_n \rho_n(t), \qquad (8.1.24)$$

where $\rho_n(t)$ is the set of rational functions

$$\rho_n(t) = \frac{(1 + it)^n}{(1 - it)^{n+1}}, \qquad n = 0, \pm 1, \pm 2, \cdots, \qquad (8.1.25)$$

and

$$a_n = \frac{1}{\pi} \int_{-\infty}^{\infty} x(t) \rho_n^*(t) \, dt \qquad (8.1.26)$$

[3] Press, W. H., S. A. Teukolsky, W. T. Vetterling, and B. P. Flannery, 1992: *Numerical Recipes in Fortran: The Art of Scientific Computing.* Cambridge University Press, §6.10.

[4] Gautschi, W., and J. Waldvogel, 2000: Computing the Hilbert transform of the generalized Laguerre and Hermite weight functions. *BIT,* **41**, 490–503.

[5] Weideman, J. A. C., 1995: Computing the Hilbert transform on the real line. *Math. Comput.,* **64**, 745–762.

or

$$a_n = \frac{1}{2\pi} \int_{-\pi}^{\pi} \left[1 - i\tan\left(\tfrac{1}{2}\theta\right)\right] x\left[\tan\left(\tfrac{1}{2}\theta\right)\right] e^{-in\theta}\, d\theta, \tag{8.1.27}$$

if we introduce the substitution $t = \tan(\theta/2)$.

Why is (8.1.24) useful? Taking the Hilbert transform of both sides of (8.1.24),

$$\hat{x}(t) = \sum_{n=-\infty}^{\infty} a_n \hat{\rho}_n(t). \tag{8.1.28}$$

Using contour integration, we find that

$$\hat{\rho}_n(t) = \frac{1}{\pi} PV \int_{-\infty}^{\infty} \frac{(1+i\tau)^n}{(1-i\tau)^{n+1}(t-\tau)}\, d\tau = -i\, \mathrm{sgn}(n)\rho_n(t), \tag{8.1.29}$$

where $\mathrm{sgn}(t)$ is the signum function with $\mathrm{sgn}(0) = 1$. Therefore,

$$\hat{x}(t) = -i \sum_{n=-\infty}^{\infty} \mathrm{sgn}(n)\, a_n\, \rho_n(t). \tag{8.1.30}$$

We must now approximate (8.1.30) so that we can evaluate it numerically. We do this by introducing the following truncated version:

$$\hat{x}_N(t) = -i \sum_{n=-N}^{N-1} \mathrm{sgn}(n)\, A_n\, \rho_n(t). \tag{8.1.31}$$

This particular truncation was chosen because $\rho_n(t)$ and $\rho_{-n-1}(t)$ are a conjugate pair. The coefficient a_n has become A_n, which equals

$$A_n = \frac{1}{N} \sum_{j=-N+1}^{N-1} \left[1 - i\tan\left(\tfrac{1}{2}\theta_j\right)\right] x\left[\tan\left(\tfrac{1}{2}\theta_j\right)\right] e^{-in\theta_j}, \tag{8.1.32}$$

where $\theta_j = \pi j/N$. The terms corresponding to $j = \pm N$ have been set to zero because it is assumed that $x(t)$ vanishes rapidly with $t \to \pm\infty$. Finally, we substitute θ for t and transform (8.1.31) into

$$\hat{x}_N(t_j) = -\frac{i}{1 - i\tan(\theta_j)} \sum_{n=-N}^{N-1} \mathrm{sgn}(n)A_n e^{in\theta_j}. \tag{8.1.33}$$

The advantage of (8.1.32) and (8.1.33) is that they can be evaluated using fast Fourier transforms. For example, the following MATLAB script devised by Weideman illustrates his methods for $x(t) = 1/(1 + t^4)$:

```
% initialize parameters used in computation
b = 1; N = 8; n = [-N:N-1]';
```

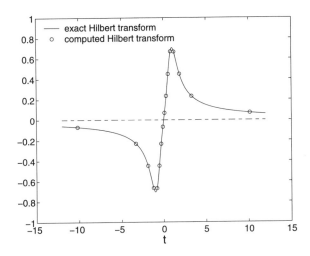

Figure 8.1.2: The Hilbert transform for $x(t) = 1/(1 + t^4)$ computed from Weideman's algorithm.

```
% set up collocation points and evaluate function there
t = b * tan(pi*(n+1/2)/(2*N)); F = 1./(1+t.^4);
% evaluate (8.1.32)
an = fftshift(fft(F.*(b-i*t)));
% compute hilbert transform via (8.1.33)
hilbert = ifft(fftshift(i*(sign(n+1/2).*an)))./(b-i*t);
hilbert = -real(hilbert);
% find points at which we will compute exact answer
tt = [-12:0.02:12];
% compute exact answer
answer = tt.*(1+tt.^2)./(1+tt.^4)./sqrt(2);
fzero = zeros(size(tt));
% plot both computed Hilbert transform and exact answer
plot(tt,answer,'-',t,hilbert,'o',tt,fzero,'--')
xlabel('t','Fontsize',20)
legend('exact Hilbert transform','computed Hilbert transform')
legend boxoff
```

Figure 8.1.2 illustrates Weideman's algorithm for numerically computing the Hilbert transform of $1/(1 + t^4)$.

There are two important points concerning Weideman's implementation of his algorithm. First, the collocation points originally given by $t_j = \tan[\pi j/(2N)]$, $j = -N, \ldots, N - 1$ have changed to $t_j = \tan[(j + \frac{1}{2})\pi/(2N)]$, $j = -N, \ldots, N - 1$. This change replaces the trapezoidal rule discretization for the Fourier coefficients with a midpoint rule. The advantages are twofold: First, it avoids the nuisance of dealing with a collocation point at infinity. Second, it actually yields more accurate results in many cases.

The discerning student will also notice that Weideman introduced a free parameter b which we set to one. This rescaling parameter can have a major influence on the accuracy. The interested student is referred to the bottom of page 756 in Weideman's paper for further details.

• Example 8.1.5: Discrete Hilbert transform

Quite often the function is given as discrete data points. How do we find the Hilbert transform in this case? We will now prove[6] that the equivalent *discrete* Hilbert transform is

$$
\mathcal{H}(f_n) = \widehat{f}_k = \begin{cases} \dfrac{2}{\pi} \displaystyle\sum_{n \text{ odd}} \dfrac{f_n}{k-n}, & k \text{ even}, \\[2mm] \dfrac{2}{\pi} \displaystyle\sum_{n \text{ even}} \dfrac{f_n}{k-n}, & k \text{ odd}, \end{cases} \tag{8.1.34}
$$

where f_n denotes a set of discrete data values that are sampled at $t = nT$ and both k and n run from $-\infty$ to ∞. The corresponding inverse is

$$
f_n = \begin{cases} \dfrac{2}{\pi} \displaystyle\sum_{k \text{ odd}} \dfrac{\widehat{f}_k}{k-n}, & n \text{ even}, \\[2mm] \dfrac{2}{\pi} \displaystyle\sum_{k \text{ even}} \dfrac{\widehat{f}_k}{k-n}, & n \text{ odd}. \end{cases} \tag{8.1.35}
$$

We begin our proof by inserting (8.1.34) into (8.1.35). For n even,

$$
f_n = \frac{2}{\pi} \sum_{k \text{ odd}} \frac{1}{k-n} \left(\frac{2}{\pi} \sum_{p \text{ even}} \frac{f_p}{k-p} \right) \tag{8.1.36}
$$

$$
= \frac{4}{\pi^2} \sum_{p \text{ even}} \sum_{k \text{ odd}} \frac{f_p}{(k-p)(k-n)} \tag{8.1.37}
$$

$$
= \frac{4}{\pi^2} \sum_{k \text{ odd}} \frac{f_n}{(k-n)^2} + \frac{4}{\pi^2} \sum_{p \text{ even}, p \neq n} \sum_{k \text{ odd}} (n-p) f_p \left\{ \frac{1}{k-n} - \frac{1}{k-p} \right\}. \tag{8.1.38}
$$

The term within the curly brackets equals zero as k runs through all of its values. Therefore, (8.1.38) reduces to

$$
f_n = \frac{8}{\pi^2} f_n \left(1 + \frac{1}{3^2} + \frac{1}{5^2} + + \frac{1}{7^2} + \cdots \right). \tag{8.1.39}
$$

[6] Kak, S. C., 1970: The discrete Hilbert transform. *Proc. IEEE*, **58**, 585–586. ©1970 IEEE. For an alternative derivation, see Kress, R., and E. Martensen, 1970: Anwendung der Rechteckregel auf die reelle Hilberttransformation mit unendlichem Intervall. *Zeit. Angew. Math. Mech.*, **50**, T61–T64.

However, the term in the brackets of (8.1.39) equals $\pi^2/8$. Therefore, (8.1.34)–(8.1.35) is proved for n even. An identical proof follows for n odd.

A popular alternative[7] to (8.1.34) involves the (fast) Fourier transform and the relationship that $\widehat{X}(\omega) = -i\,\text{sgn}(\omega)X(\omega)$, where $X(\omega)$ and $\widehat{X}(\omega)$ denote the Fourier transform of $x(t)$ and $\widehat{x}(t)$, respectively. In this technique, a fast Fourier transform is taken of the data. This transformed dataset is then multiplied by $-i\,\text{sgn}(\omega)$ and then back transformed to give the Hilbert transform.

Let $x(t)$ be a real, even function. Then $X(\omega)$, the Fourier transform of $x(t)$, is also an even function. Consequently,

$$\widehat{x}(t) = \frac{1}{2\pi} \int_{-\infty}^{\infty} \widehat{X}(\omega) e^{i\omega t}\, d\omega \tag{8.1.40}$$

$$= \frac{1}{2\pi} \int_{-\infty}^{\infty} -i\,\text{sgn}(\omega)X(\omega)\left[\cos(\omega t) + i\sin(\omega t)\right] d\omega \tag{8.1.41}$$

$$= -\frac{i}{2\pi} \int_{-\infty}^{\infty} \text{sgn}(\omega)X(\omega)\cos(\omega t)\, d\omega + \frac{1}{2\pi} \int_{-\infty}^{\infty} \text{sgn}(\omega)X(\omega)\sin(\omega t)\, d\omega \tag{8.1.42}$$

$$= \frac{1}{\pi} \int_{0}^{\infty} X(\omega)\sin(\omega t)\, d\omega. \tag{8.1.43}$$

Note that the Hilbert transform in this case is an odd function. Similarly, if $x(t)$ is a real, odd function,

$$\widehat{x}(t) = -\frac{i}{\pi} \int_{0}^{\infty} X(\omega)\cos(\omega t)\, d\omega, \tag{8.1.44}$$

and the Hilbert transform is an even function.

Problems

1. Show that the Hilbert transform of a constant function is zero.

2. Use (8.1.4) to compute the Hilbert transform of $\cos(\omega t)$, $\omega \neq 0$.

3. Use (8.1.4) to show that the Hilbert transform of the Dirac delta function $\delta(t)$ is $1/(\pi t)$.

4. Use (8.1.4) to show that the Hilbert transform of $1/(t^2 + 1)$ is $t/(t^2 + 1)$.

---

[7] Čížek, V., 1970: Discrete Hilbert transform. *IEEE Trans. Audio Electroacoust.*, **AU-18**, 340–343.

5. The output $y(t)$ from an ideal lowpass filter can be expressed by the convolution integral

$$y(t) = x(t) * \frac{\sin(2\pi\omega t)}{\pi t},$$

where $x(t)$ is the input signal. Show that this expression can also be expressed in terms of Hilbert transforms as

$$y(t) = \mathcal{H}[x(t)\cos(2\pi\omega t)]\sin(2\pi\omega t) - \mathcal{H}[x(t)\sin(2\pi\omega t)]\cos(2\pi\omega t).$$

Use (8.1.26) to find the Hilbert transforms of

6. $x(t) = \dfrac{1}{1+t^2}$

7. $x(t) = \begin{cases} 1, & -a < t < a \\ 0, & \text{otherwise} \end{cases}$

Using MATLAB, test Weideman's algorithm for the following cases. Why does the algorithm do well or not?

8. $\begin{cases} 1, & -1 < t < 1 \\ 0, & \text{otherwise} \end{cases}$

9. $\sin(t)$

10. $\dfrac{1}{t^2 + 1}$

11. $\dfrac{\sin(t)}{1 + t^4}$

For Problem 11, you will need

$$\mathcal{H}\left[\frac{\sin(t)}{t^4 + 1}\right] = \frac{e^{-1/\sqrt{2}}[\cos(1/\sqrt{2}) + \sin(1/\sqrt{2})t^2] - \cos(t)}{t^4 + 1}.$$

8.2 SOME USEFUL PROPERTIES

In principle we could construct any desired transform from the definition of the Hilbert transform. However, there are several general theorems that are much more effective in finding new transforms.

> Linearity

From the definition of the Hilbert transform, it immediately follows that if $z(t) = c_1 x(t) + c_2 y(t)$, where c_1 and c_2 are arbitrary constants, then $\hat{z}(t) = c_1 \hat{x}(t) + c_2 \hat{y}(t)$.

> The energy in a signal and its Hilbert transform are the same.

Consider the energy spectral densities at input and output of a quadrature phase shifter. The output equals

$$|\hat{X}(\omega)|^2 = \left|\mathcal{F}[\hat{x}(t)]\right|^2 = |-i\,\text{sgn}(\omega)|^2 |X(\omega)|^2 = |X(\omega)|^2. \qquad (8.2.1)$$

Because the energy spectral density at input and output are the same, so are the total energies.

> A signal and its Hilbert transform
> are orthogonal.

From Parseval's theorem,

$$\int_{-\infty}^{\infty} x(t)\widehat{x}(t)\, dt = \int_{-\infty}^{\infty} X(\omega)\widehat{X}^*(\omega)\, d\omega, \tag{8.2.2}$$

where $\widehat{X}(\omega) = \mathcal{F}[\widehat{x}(t)]$. Then,

$$\int_{-\infty}^{\infty} X(\omega)\widehat{X}^*(\omega)\, d\omega = \int_{-\infty}^{\infty} i\,\mathrm{sgn}(\omega)|X(\omega)|^2\, d\omega = 0, \tag{8.2.3}$$

because the integrand in the middle expression of (8.2.3) is odd. Thus,

$$\int_{-\infty}^{\infty} x(t)\widehat{x}(t)\, dt = 0. \tag{8.2.4}$$

The reason why a function and its Hilbert transform are orthogonal to each other follows from the fact that a Hilbert transformation of a function shifts the phase of each Fourier component of the function *forward* by $\pi/2$ for positive frequencies and *backward* for negative frequencies.

• Example 8.2.1

Let us verify the orthogonality condition for Hilbert transforms using $x(t) = 1/(1+t^2)$.
Because $\widehat{x}(t) = t/(1+t^2)$,

$$\int_{-\infty}^{\infty} x(t)\widehat{x}(t)\, dt = \int_{-\infty}^{\infty} \frac{t}{(1+t^2)^2}\, dt = 0, \tag{8.2.5}$$

since the integrand is an odd function.

> Shifting

Let us find the Hilbert transform of $x(t+a)$ if we know $\widehat{x}(t)$. From the definition of Hilbert transforms,

$$\mathcal{H}[x(t+a)] = \frac{1}{\pi}\int_{-\infty}^{\infty} \frac{x(\eta+a)}{t-\eta}\, d\eta = \frac{1}{\pi}\int_{-\infty}^{\infty} \frac{x(\tau)}{(t+a)-\tau}\, d\tau = \widehat{x}(t+a) \tag{8.2.6}$$

or $\mathcal{H}[x(t+a)] = \widehat{x}(t+a)$.

> ### Time scaling

Let $a > 0$. Then,

$$\mathcal{H}[x(at)] = \frac{1}{\pi} \int_{-\infty}^{\infty} \frac{x(a\eta)}{t-\eta} \, d\eta = \frac{1}{\pi} \int_{-\infty}^{\infty} \frac{x(\tau)}{at-\tau} \, d\tau = \widehat{x}(at). \qquad (8.2.7)$$

On the other hand, if $a < 0$,

$$\mathcal{H}[x(at)] = \frac{1}{\pi} \int_{-\infty}^{\infty} \frac{x(a\eta)}{t-\eta} \, d\eta = -\frac{1}{\pi} \int_{-\infty}^{\infty} \frac{x(\tau)}{at-\tau} \, d\tau = -\widehat{x}(at). \qquad (8.2.8)$$

Thus, we have that $\mathcal{H}[x(at)] = \operatorname{sgn}(a)\,\widehat{x}(at)$.

> ### Derivatives

Let us find the relationship between the nth derivative of $x(t)$ and its Hilbert transform. Using the derivative rule as it applies to Fourier transforms,

$$\mathcal{H}\left\{\mathcal{F}\left[\frac{d^n x}{dt^n}\right]\right\} = -i\operatorname{sgn}(\omega)(i\omega)^n X(\omega) = (i\omega)^n[-i\operatorname{sgn}(\omega)X(\omega)] \qquad (8.2.9)$$

$$= (i\omega)^n \widehat{X}(\omega) = \mathcal{F}\left[\frac{d^n \widehat{x}}{dt^n}\right]. \qquad (8.2.10)$$

Taking the inverse Fourier transforms, we have that

$$\mathcal{H}\left(\frac{d^n x}{dt^n}\right) = \frac{d^n \widehat{x}}{dt^n}. \qquad (8.2.11)$$

> ### Convolution

Hilbert transforms enjoy a similar, but not identical, property with Fourier transforms with respect to convolution. If

$$w(t) = u(t) * v(t) = \int_{-\infty}^{\infty} u(\tau)v(t-\tau)\,d\tau = \int_{-\infty}^{\infty} u(t-\tau)v(\tau)\,d\tau, \qquad (8.2.12)$$

then

$$\widehat{w}(t) = v(t) * \widehat{u}(t). \qquad (8.2.13)$$

Table 8.2.1: Some General Properties of Hilbert Transforms

	function, x(t)	Hilbert transform, $\widehat{x}(t)$
1.	$\widehat{x}(t)$	$-x(t)$
2.	$x(t) + y(t)$	$\widehat{x}(t) + \widehat{y}(t)$
3.	$x(t+a)$, $\quad a$ real	$\widehat{x}(t+a)$
4.	$\dfrac{d^n x(t)}{dt^n}$	$\dfrac{d^n \widehat{x}(t)}{dt^n}$
5.	$x(at)$	$\operatorname{sgn}(a)\,\widehat{x}(at)$
6.	$tx(t)$	$t\widehat{x}(t) + \frac{1}{\pi}\int_{-\infty}^{\infty} x(\tau)\,d\tau$
7.	$(t+a)x(t)$	$(t+a)\widehat{x}(t) + \frac{1}{\pi}\int_{-\infty}^{\infty} x(\tau)\,d\tau$

Proof: From the convolution theorem as it applies to Fourier transforms, $W(\omega) = V(\omega)U(\omega)$. Multiplying both sides of the equation by $-i\operatorname{sgn}(\omega)$,

$$\widehat{W}(\omega) = -i\operatorname{sgn}(\omega)W(\omega) = V(\omega)[-i\operatorname{sgn}(\omega)U(\omega)] = V(\omega)\widehat{U}(\omega). \quad (8.2.14)$$

Again, using the convolution theorem as it applies to Fourier transforms, we arrive at the final result. □

• **Example 8.2.2**

Given the functions $u(t) = \cos(t)$ and $v(t) = 1/(1+t^4)$, let us verify the convolution theorem as it applies to Hilbert transforms.

With $u(t) = \cos(t)$ and $v(t) = 1/(1+t^4)$,

$$w(t) = u(t) * v(t) = \int_{-\infty}^{\infty} \frac{\cos(t-x)}{1+x^4}\,dx \qquad (8.2.15)$$

$$= \int_{-\infty}^{\infty} \frac{\cos(t)\cos(x)}{1+x^4}\,dx + \int_{-\infty}^{\infty} \frac{\sin(t)\sin(x)}{1+x^4}\,dx \qquad (8.2.16)$$

$$= \frac{\pi}{\sqrt{2}}e^{-1/\sqrt{2}}\left[\cos\left(\frac{1}{\sqrt{2}}\right) + \sin\left(\frac{1}{\sqrt{2}}\right)\right]\cos(t) \qquad (8.2.17)$$

so that

$$\widehat{w}(t) = \frac{\pi}{\sqrt{2}}e^{-1/\sqrt{2}}\left[\cos\left(\frac{1}{\sqrt{2}}\right) + \sin\left(\frac{1}{\sqrt{2}}\right)\right]\sin(t). \qquad (8.2.18)$$

Because $\widehat{v}(t) = t(1 + t^2)/[\sqrt{2}\,(1 + t^4)]$,

$$u(t) * \widehat{v}(t) = \frac{1}{\sqrt{2}} \int_{-\infty}^{\infty} \cos(t - x) \frac{x(1 + x^2)}{1 + x^4}\, dx \qquad (8.2.19)$$

$$= \frac{1}{\sqrt{2}} \int_{-\infty}^{\infty} \frac{\cos(t)\cos(x)x(1 + x^2)}{1 + x^4}\, dx$$

$$+ \frac{1}{\sqrt{2}} \int_{-\infty}^{\infty} \frac{\sin(t)\sin(x)x(1 + x^2)}{1 + x^4}\, dx \qquad (8.2.20)$$

$$= \frac{1}{\sqrt{2}} \sin(t) \int_{-\infty}^{\infty} \frac{x(1 + x^2)\sin(x)}{1 + x^4}\, dx \qquad (8.2.21)$$

$$= \frac{\pi}{\sqrt{2}} e^{-1/\sqrt{2}} \left[\cos\left(\frac{1}{\sqrt{2}}\right) + \sin\left(\frac{1}{\sqrt{2}}\right) \right] \sin(t), \qquad (8.2.22)$$

and the convolution theorem for Hilbert transforms holds true in this case.

Product theorem

Let $f(t)$ and $g(t)$ denote complex functions with Fourier transforms $F(\omega)$ and $G(\omega)$, respectively. If

1) $F(\omega)$ vanishes for $|\omega| > a$, and $G(\omega)$ vanishes for $|\omega| < a$, where $a > 0$,

or

2) $f(t)$ and $g(t)$ are analytic functions (their real and imaginary parts are Hilbert pairs),

then the Hilbert transform of the product of $f(t)$ and $g(t)$ is

$$\mathcal{H}[f(t)g(t)] = f(t)\widehat{g}(t). \qquad (8.2.23)$$

Proof:[8] The product $f(t)g(t)$ can be expressed as

$$f(t)g(t) = \frac{1}{4\pi^2} \int_{-\infty}^{\infty} \int_{-\infty}^{\infty} F(u)G(v)e^{i(u+v)t}\, dv\, du. \qquad (8.2.24)$$

Because $\mathcal{H}(e^{ibt}) = i\,\mathrm{sgn}(b)e^{ibt}$,

$$\mathcal{H}[f(t)g(t)] = \frac{i}{4\pi^2} \int_{-\infty}^{\infty} \int_{-\infty}^{\infty} F(u)G(v)\,\mathrm{sgn}(u + v)e^{i(u+v)t}\, dv\, du. \qquad (8.2.25)$$

---

[8] Taken from Bedrosian, E., 1963: A product theorem for Hilbert transforms. *Proc. IEEE*, **51**, 868–869. ©1963 IEEE. This theorem has been extended to functions of n-dimensional real vectors by Stark, H., 1971: An extension of the Hilbert transform product theorem. *Proc. IEEE*, **59**, 1359–1360.

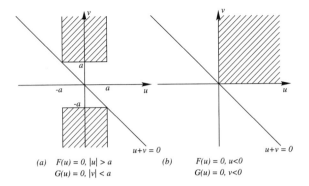

(a) $F(u) = 0, |u| > a$ (b) $F(u) = 0, u<0$
$G(u) = 0, |v| < a$ $G(u) = 0, v<0$

Figure 8.2.1: Region of integration in the proof of the product theorem.

The shaded regions of Figure 8.2.1 are those in which the product $F(u)G(v)$ is nonvanishing for the conditions of the theorem. In Figure 8.2.1(a) the nonoverlapping Fourier transforms yield two semi-infinite strips in which the product is nonvanishing. In Figure 8.2.1(b), for analytic functions, the Fourier transforms vanish for negative arguments[9] so that the product is nonvanishing only in the first quadrant. In both cases $\operatorname{sgn}(u + v) = \operatorname{sgn}(v)$ over the regions of integration in which the integrand is nonvanishing. Thus,

$$\mathcal{H}[f(t)g(t)] = \frac{i}{4\pi^2} \int_{-\infty}^{\infty} \int_{-\infty}^{\infty} F(u)G(v) \operatorname{sgn}(v)e^{i(u+v)t} \, dv \, du \qquad (8.2.26)$$

$$= f(t)\frac{i}{2\pi} \int_{-\infty}^{\infty} G(v) \operatorname{sgn}(v)e^{ivt} \, dv = f(t)\hat{g}(t). \qquad (8.2.27)$$

□

• Example 8.2.3: Hilbert Transforms of Band-Pass Functions

In communications, we have the double-sideband, amplitude modulated signal given by $a(t)\cos(\omega t + \varphi)$, where φ is constant. From the product theorem its Hilbert transform equals $a(t)\sin(\omega t+\varphi)$, $\omega > 0$, provided that the highest frequency component in $a(t)$ is less than ω. Paradoxically, the Hilbert transform of more general $a(t)\cos[\omega t + \varphi(t)]$, which equals $a(t)\sin[\omega t + \varphi(t)]$, has no such restriction.

Problems

Verify the orthogonality property of Hilbert transforms using

1. $x(t) = 1/(1 + t^4)$ 2. $x(t) = \sin(t)/(1 + t^2)$

---

[9] Titchmarsh, E. C., 1948: *Introduction to the Theory of Fourier Integrals*. Oxford University Press, p. 128.

3. $x(t) = \begin{cases} 1, & 0 < t < a \\ 0, & \text{otherwise} \end{cases}$

Verify the convolution theorem for Hilbert transforms using

4.

$$u(t) = \begin{cases} 1, & 0 < t < a, \\ 0, & \text{otherwise,} \end{cases} \qquad v(t) = \sin(t)$$

5.

$$u(t) = \cos(t), \qquad v(t) = \frac{1}{1 + t^2}$$

6. Use the product theorem to show that

$$\mathcal{H}[\sin(at)J_n(bt)] = -\cos(at)J_n(bt), \qquad 0 < b < a,$$

if $n = 0, 1, 2, 3, \ldots$.

Hint:

$$\mathcal{F}[J_n(bt)] = \frac{2(-1)^m}{\sqrt{b^2 - \omega^2}} T_n\left(\frac{|\omega|}{b}\right) H(b - |\omega|),$$

where $T_n(\ )$ is a Chebyshev polynomial of the first kind and $m = n/2$ or $(n-1)/2$, depending upon which definition gives an integer.

7. Given cosine and sine integrals:

$$Ci(x) = -\int_x^\infty \frac{\cos(t)}{t} \, dt, \qquad Si(x) = -\int_x^\infty \frac{\sin(t)}{t} \, dt,$$

and

$$\mathcal{H}[Ci(a|t|)] = -\text{sgn}(t)Si(a|t|), \qquad 0 < a,$$

use the product rule to show that

$$\mathcal{H}[\sin(bt)Ci(a|t|)] = -\text{sgn}(t)\sin(bt)Si(a|t|), \qquad 0 < b < a.$$

Hint:

$$\mathcal{F}[Ci(a|t|)] = \begin{cases} 0, & 0 < |\omega| < a, \\ -\pi/|\omega|, & a < |\omega| < \infty, \end{cases} \qquad 0 < a.$$

8. Prove that

$$\mathcal{H}[tx(t)] = t\hat{x}(t) - \frac{1}{\pi}\int_{-\infty}^\infty x(\tau) \, d\tau.$$

Hint:

$$\frac{\tau x(\tau)}{t - \tau} = \frac{tx(\tau)}{t - \tau} - x(\tau).$$

8.3 ANALYTIC SIGNALS

The monochromatic signal $A\cos(\omega_0 t + \varphi)$ appears in many physical and engineering applications. It is common to represent this signal by the complex representation $Ae^{i(\omega_0 t + \varphi)}$. These two representations are related to each other by

$$A\cos(\omega_0 t + \varphi) = \text{Re}\left[Ae^{i(\omega_0 t + \varphi)}\right] = \tfrac{1}{2}\left[Ae^{i(\omega_0 t + \varphi)} + Ae^{-i(\omega_0 t + \varphi)}\right]. \quad (\textbf{8.3.1})$$

Furthermore, the Fourier transform of $A\cos(\omega_0 t + \varphi)$ is

$$\mathcal{F}[A\cos(\omega_0 t + \varphi)] = \tfrac{1}{2}\left[Ae^{i\varphi}\delta(\omega - \omega_0) + Ae^{-i\varphi}\delta(\omega + \omega_0)\right], \quad (\textbf{8.3.2})$$

while the Fourier transform of $Ae^{i(\omega_0 t + \varphi)}$ is

$$\mathcal{F}\left[Ae^{i(\omega_0 t + \varphi)}\right] = Ae^{i\varphi}\delta(\omega - \omega_0). \quad (\textbf{8.3.3})$$

As (8.3.2) and (8.3.3) clearly show, in passing from the real signal to its complex representation, we double the strength of the positive frequencies and remove entirely the negative frequencies.

Let us generalize these concepts to nonmonochromatic signals. For the real signal $x(t)$ with Fourier transform $X(\omega)$ and the complex signal $z(t)$ with Fourier transform $Z(\omega)$, the previous paragraph shows that our generalization must have the property:

$$Z(\omega) = X(\omega) + \text{sgn}(\omega)X(\omega) \quad (\textbf{8.3.4})$$

or

$$Z(\omega) = \begin{cases} 2X(\omega), & \omega > 0, \\ X(\omega), & \omega = 0, \\ 0, & \omega < 0. \end{cases} \quad (\textbf{8.3.5})$$

Taking the inverse of (8.3.4), we have the definition of an *analytic signal* as

$$z(t) = x(t) + i\widehat{x}(t), \quad (\textbf{8.3.6})$$

where $x(t)$ is a real signal and $\widehat{x}(t)$ is its Hilbert transform.

● **Example 8.3.1**

In Figure 8.3.1 the amplitude spectrum of the analytic signal is graphed when $x(t)$ is the rectangular pulse (3.1.9). Note that the amplitude spectrum equals zero for $\omega < 0$ and twice the amplitude spectrum for $\omega > 0$.

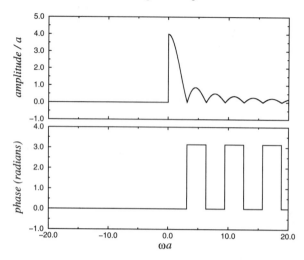

Figure 8.3.1: The spectrum of the analytic signal when $x(t)$ is the rectangular pulse given by (5.1.9).

• **Example 8.3.2**

Let us find the energy of an analytic signal.

The energy of an analytic signal is

$$\int_{-\infty}^{\infty} |z(t)|^2 \, dt = \int_{-\infty}^{\infty} x^2(t) \, dt + \int_{-\infty}^{\infty} \widehat{x}^2(t) \, dt \qquad (8.3.7)$$

$$= 2 \int_{-\infty}^{\infty} x^2(t) \, dt = 2 \int_{-\infty}^{\infty} |X(\omega)|^2 \, d\omega \qquad (8.3.8)$$

by Parseval's theorem. Thus, the analytic signal has twice the energy of the corresponding real signal.

Consider the function $x(t)$ whose amplitude spectrum $|X(\omega)|$ is shown in Figure 8.3.2(a). If we were to amplitude modulate $x(t)$ with $\cos(\omega_0 t)$, then the amplitude spectrum of this modulated signal would appear as pictured in Figure 8.3.2(b).

Consider now the signal

$$y(t) = x(t) \cos(\omega_0 t) - \widehat{x}(t) \sin(\omega_0 t) = \mathrm{Re}\left\{[x(t) + i\widehat{x}(t)]e^{i\omega_0 t}\right\} \quad (8.3.9)$$

$$= \mathrm{Re}\left\{z(t)e^{i\omega_0 t}\right\} = \tfrac{1}{2}\left[z(t)e^{i\omega_0 t} + z^*(t)e^{-i\omega_0 t}\right], \qquad (8.3.10)$$

where $z(t)$ is the analytic signal of $x(t)$. We have plotted the amplitude spectrum $|Z(\omega)|$ in Figure 8.3.2(c). If we compute the amplitude spectrum of $y(t)$, we would find that

$$Y(\omega) = \tfrac{1}{2}Z(\omega - \omega_0) + \tfrac{1}{2}Z(-\omega - \omega_0) \qquad (8.3.11)$$

$$= \begin{cases} X(\omega - \omega_0), & \omega_0 \le \omega \le \omega_0 + \omega_{\max}, \\ X^*(-\omega - \omega_0), & -\omega_0 - \omega_{\max} \le \omega \le \omega_0, \\ 0, & \text{otherwise.} \end{cases} \qquad (8.3.12)$$

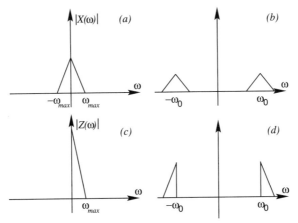

Figure 8.3.2: Given a function $x(t)$ with an amplitude spectrum shown in (a), frame (b) shows the amplitude spectrum of the amplitude modulated signal $x(t)\cos(\omega_0 t)$ while frames (c) and (d) give the amplitude spectrum of the analytic signal $z(t)$ and $x(t)\cos(\omega_0 t) - \widehat{x}(t)\sin(\omega_0 t)$, respectively.

We have sketched this amplitude spectrum $|Y(\omega)|$ in Figure 8.3.2(d). Each triangular part is called the *single sideband signal* because it contains the upper frequencies ($|\omega| > \omega_0$) of the modulated signal $x(t)\cos(\omega_0 t)$. Similarly, if we had used $x(t)\cos(\omega_0 t) + \widehat{x}(t)\sin(\omega_0 t)$, we would have only obtained the lower sidebands. Consequently, a communication system using $x(t)\cos(\omega_0 t) - \widehat{x}(t)\sin(\omega_0 t)$ or $x(t)\cos(\omega_0 t) + \widehat{x}(t)\sin(\omega_0 t)$ would realize a 50% savings in its frequency bandwidth over one transmitting $x(t)\cos(\omega_0 t)$.

Problems

1. Find the analytic signal corresponding to $x(t) = \cos(\omega t)$, $\omega > 0$.

2. Show that the polar form of an analytic signal can be written

$$z(t) = |z(t)|e^{i\varphi(t)},$$

where

$$|z(t)|^2 = x^2(t) + \widehat{x}^2(t), \quad \varphi(t) = \tan^{-1}\left[\frac{\widehat{x}(t)}{x(t)}\right].$$

3. Analytic signals are often used with narrow-band waveforms with carrier frequency ω_0. If $\varphi(t) = \omega_0 t + \varphi'(t)$, show that the analytic signal can be written $z(t) = r(t)e^{i\omega_0 t}$, where $r(t) = |z(t)|e^{i\varphi'(t)}$. The function $r(t)$ is called the *complex envelope* or the *phasor amplitude*; this is a generalization of the phasor idea beyond pure alternating currents.

8.4 CAUSALITY: THE KRAMERS-KRONIG RELATIONSHIP

Causality is the physical principle which states that an event cannot proceed its cause. In this section we explore what effect this principle has on Hilbert transforms.

We begin by introducing the concept of causal functions. A *causal function* is a function which equals zero for all $t < 0$. As with all functions we can write it in terms of an even $x_e(t)$ and an odd $x_o(t)$ part as $x(t) = x_e(t) + x_o(t)$. Because $x(t)$ is causal, $x_o(t) = \text{sgn}(t)x_e(t)$ and

$$x(t) = x_e(t) + \text{sgn}(t)x_e(t). \tag{8.4.1}$$

Taking the Fourier transform of (8.4.1), we find that the Fourier transform of *all* causal functions are of the form

$$X(\omega) = X_e(\omega) - i\widehat{X}_e(\omega), \tag{8.4.2}$$

where

$$\widehat{X}_e(\omega) = \frac{1}{\pi} \int_{-\infty}^{\infty} \frac{X_e(\tau)}{\omega - \tau} \, d\tau, \tag{8.4.3}$$

and

$$X_e(\omega) = -\frac{1}{\pi} \int_{-\infty}^{\infty} \frac{\widehat{X}_e(\tau)}{\omega - \tau} \, d\tau, \tag{8.4.4}$$

because

$$2\pi \mathcal{F}[x_e(t)\text{sgn}(t)] = \frac{2}{i\omega} * X_e(\omega) = \frac{2}{i} \int_{-\infty}^{\infty} \frac{X_e(\tau)}{\omega - \tau} d\tau. \tag{8.4.5}$$

Equations (8.4.3)–(8.4.4) first arose in dielectric theory and, taken together, are called the *Kramers*[10] *and Kronig*[11] *relation* after their discoverers who derived these relationships during their work on the dispersion of light by gaseous atoms or molecules.

• Example 8.4.1

Let us verify the Kramers-Kronig relation using the causal time function $x(t) = H(t)$.

Because $x_e(t) = \frac{1}{2}$ and $X_e(\omega) = \pi\delta(\omega)$ by (5.2.3),

$$\widehat{X}_e(\omega) = \frac{1}{\pi} \int_{-\infty}^{\infty} \frac{\pi\,\delta(\tau)}{\omega - \tau} \, d\tau = -\frac{1}{\omega}. \tag{8.4.6}$$

---

[10] Kramers, H. A., 1929: Die Dispersion und Absorption von Röntgenstrahlen. *Phys. Zeit.*, **30**, 522-523.

[11] Kronig, R. de L., 1926: On the theory of dispersion of x-rays. *J. Opt. Soc. Am.*, **12**, 547-551.

Consequently, by the Kramers-Kronig relation

$$\mathcal{F}[H(t)] = X_e(\omega) - i\widehat{X}_e(\omega) = \pi\delta(\omega) + \frac{i}{\omega}. \tag{8.4.7}$$

This agrees with the result given in Example 5.2.2.

- **Example 8.4.2**

A simple example of a causal function is the impulse response or Green's function introduced in earlier chapters. From (8.4.2) we have the result that the transfer function $G(\omega)$, the Fourier transform of the impulse response, must yield the Hilbert transform pair $G_e(\omega) - i\widehat{G}_e(\omega)$.

For example, if $g(t) = e^{-t}H(t)$, then $G(\omega) = 1/(1+i\omega)$. Because

$$\frac{1}{1+i\omega} = \frac{1}{\omega^2+1} - i\frac{\omega}{\omega^2+1}, \tag{8.4.8}$$

we have the Hilbert transform pair of

$$x(t) = \frac{1}{t^2+1} \quad \text{and} \quad \widehat{x}(t) = \frac{t}{t^2+1}. \tag{8.4.9}$$

- **Example 8.4.3**

Let us verify the Kramers-Kronig relation for the Hilbert transform pair

$$x(t) = \frac{1}{t^4+1} \quad \text{and} \quad \widehat{x}(t) = \frac{t(t^2+1)}{\sqrt{2}(t^4+1)} \tag{8.4.10}$$

by direct integration.

From (8.4.3), we have that

$$\frac{\omega(\omega^2+1)}{\sqrt{2}(\omega^4+1)} = \frac{1}{\pi} \int_{-\infty}^{\infty} \frac{d\tau}{(\tau^4+1)(\omega-\tau)}. \tag{8.4.11}$$

Applying the residue theorem to the right side of (8.4.11), we obtain

$$\frac{\omega(\omega^2+1)}{\sqrt{2}(\omega^4+1)} = i\,\mathrm{Res}\left[\frac{1}{(z^4+1)(\omega-z)};\omega\right] + 2i\,\mathrm{Res}\left[\frac{1}{(z^4+1)(\omega-z)};e^{\pi i/4}\right]$$
$$+ 2i\,\mathrm{Res}\left[\frac{1}{(z^4+1)(\omega-z)};e^{3\pi i/4}\right]. \tag{8.4.12}$$

We only include one half of the value of the residue at $\tau = \omega$ because the singularity lies on the path of integration and we must treat this integration along the lines of a Cauchy principal value. Evaluating the residues, we find

$$\mathrm{Res}\left[\frac{1}{(z^4+1)(\omega-z)};\omega\right] = -\frac{1}{\omega^4+1}, \tag{8.4.13}$$

$$\text{Res}\left[\frac{1}{(z^4+1)(\omega-z)};e^{\pi i/4}\right] = \frac{\sqrt{2}-(1+i)\omega}{4\sqrt{2}\left[\left(\omega-\frac{1}{\sqrt{2}}\right)^2+\frac{1}{2}\right]}, \qquad (8.4.14)$$

and

$$\text{Res}\left[\frac{1}{(z^4+1)(\omega-z)};e^{3\pi i/4}\right] = \frac{\sqrt{2}+(1-i)\omega}{4\sqrt{2}\left[\left(\omega+\frac{1}{\sqrt{2}}\right)^2+\frac{1}{2}\right]}. \qquad (8.4.15)$$

Substituting (8.4.13)–(8.4.15) into the right side of (8.4.12), we obtain the left side.

Problems

1. For a causal function $x(t)$, prove that $x_o(t) = \text{sgn}(t)x_e(t)$ and $x_e(t) = \text{sgn}(t)x_o(t)$.

2. Redo our analysis if $x(t)$ is a negative time function, i.e., $x(t) = 0$ if $t > 0$. Verify your result using $x(t) = e^t H(-t)$.

3. Using $g(t) = te^{-t}H(t)$, find the corresponding Hilbert transform pairs.

4. Using $g(t) = e^{-t}\cos(\omega t)H(t)$, find the corresponding Hilbert transform pairs.

5. Verify the Kramers-Kronig relation for the Hilbert transform pair

$$x(t) = \frac{1}{t^2+1} \qquad \text{and} \qquad \hat{x}(t) = \frac{t}{t^2+1}$$

by direct integration.

Chapter 9
The Sturm-Liouville Problem

In the next three chapters we will be solving partial differential equations using the technique of separation of variables. This technique requires that we expand a piece-wise continuous function $f(x)$ as a linear sum of *eigenfunctions*, much as we used sines and cosines to re-express $f(x)$ in a Fourier series. The purpose of this chapter is to explain and illustrate these eigenfunction expansions.

9.1 EIGENVALUES AND EIGENFUNCTIONS

Repeatedly, in the next three chapters on partial differential equations, we will solve the following second-order linear differential equation:

$$\frac{d}{dx}\left[p(x)\frac{dy}{dx}\right] + [q(x) + \lambda r(x)]y = 0, \quad a \leq x \leq b, \qquad \textbf{(9.1.1)}$$

together with the boundary conditions:

$$\alpha y(a) + \beta y'(a) = 0 \quad \text{and} \quad \gamma y(b) + \delta y'(b) = 0. \qquad \textbf{(9.1.2)}$$

423

Figure 9.1.1: By the time that Charles-François Sturm (1803–1855) met Joseph Liouville in the early 1830s, he had already gained fame for his work on the compression of fluids and his celebrated theorem on the number of real roots of a polynomial. An eminent teacher, Sturm spent most of his career teaching at various Parisian colleges. (Portrait courtesy of the Archives de l'Académie des sciences, Paris.)

In (9.1.1), $p(x)$, $q(x)$, and $r(x)$ are real functions of x; λ is a parameter; and $p(x)$ and $r(x)$ are functions that are continuous and positive on the interval $a \leq x \leq b$. Taken together, (9.1.1) and (9.1.2) constitute a regular *Sturm-Liouville problem*, named after the French mathematicians Sturm and Liouville[1] who first studied these equations in the 1830s. In the case when $p(x)$ or $r(x)$ vanishes at one of the endpoints of the interval $[a, b]$ or when the interval is of infinite length, the problem becomes a *singular Sturm-Liouville problem*.

Consider now the solutions to the regular Sturm-Liouville problem. Clearly there is the trivial solution $y = 0$ for all λ. However, nontrivial solutions exist only if λ takes on specific values; these values are called *characteristic values* or *eigenvalues*. The corresponding nontrivial solutions are

---

[1] For the complete history as well as the relevant papers, see Lützen, J., 1984: Sturm and Liouville's work on ordinary linear differential equations. The emergence of Sturm-Liouville theory. *Arch. Hist. Exact Sci.*, **29**, 309–376.

called the *characteristic functions* or *eigenfunctions*. In particular, we have the following theorems.

Theorem: *For a regular Sturm-Liouville problem with $p(x) > 0$, all of the eigenvalues are real if $p(x)$, $q(x)$, and $r(x)$ are real functions and the eigenfunctions are differentiable and continuous.*

Proof: Let $y(x) = u(x) + iv(x)$ be an eigenfunction corresponding to an eigenvalue $\lambda = \lambda_r + i\lambda_i$, where λ_r, λ_i are real numbers and $u(x), v(x)$ are real functions of x. Substituting into the Sturm-Liouville equation yields

$$\{p(x)[u'(x) + iv'(x)]\}' + [q(x) + (\lambda_r + i\lambda_i)r(x)][u(x) + iv(x)] = 0. \quad \textbf{(9.1.3)}$$

Separating the real and imaginary parts gives

$$[p(x)u'(x)]' + [q(x) + \lambda_r]u(x) - \lambda_i r(x)v(x) = 0, \quad \textbf{(9.1.4)}$$

and

$$[p(x)v'(x)]' + [q(x) + \lambda_r]v(x) + \lambda_i r(x)u(x) = 0. \quad \textbf{(9.1.5)}$$

If we multiply (9.1.4) by v and (9.1.5) by u and subtract the results, we find that

$$u(x)[p(x)v'(x)]' - v(x)[p(x)u'(x)]' + \lambda_i r(x)[u^2(x) + v^2(x)] = 0. \quad \textbf{(9.1.6)}$$

The derivative terms in (9.1.6) can be rewritten so that (9.1.6) becomes

$$\frac{d}{dx}\{[p(x)v'(x)]u(x) - [p(x)u'(x)]v(x)\} + \lambda_i r(x)[u^2(x) + v^2(x)] = 0. \quad \textbf{(9.1.7)}$$

Integrating from a to b, we find that

$$-\lambda_i \int_a^b r(x)[u^2(x) + v^2(x)]\,dx = \{p(x)[u(x)v'(x) - v(x)u'(x)]\}\big|_a^b. \quad \textbf{(9.1.8)}$$

From the boundary conditions (9.1.2),

$$\alpha[u(a) + iv(a)] + \beta[u'(a) + iv'(a)] = 0, \quad \textbf{(9.1.9)}$$

and

$$\gamma[u(b) + iv(b)] + \delta[u'(b) + iv'(b)] = 0. \quad \textbf{(9.1.10)}$$

Separating the real and imaginary parts yields

$$\alpha u(a) + \beta u'(a) = 0, \quad \text{and} \quad \alpha v(a) + \beta v'(a) = 0, \quad \textbf{(9.1.11)}$$

and

$$\gamma u(b) + \delta u'(b) = 0, \quad \text{and} \quad \gamma v(b) + \delta v'(b) = 0. \quad \textbf{(9.1.12)}$$

Figure 9.1.2: Although educated as an engineer, Joseph Liouville (1809–1882) would devote his life to teaching pure and applied mathematics in the leading Parisian institutions of higher education. Today he is most famous for founding and editing for almost 40 years the *Journal de Liouville*. (Portrait courtesy of the Archives de l'Académie des sciences, Paris.)

Both α and β cannot be zero; otherwise, there would be no boundary condition at $x = a$. Similar considerations hold for γ and δ. Therefore,

$$u(a)v'(a) - u'(a)v(a) = 0, \quad \text{and} \quad u(b)v'(b) - u'(b)v(b) = 0, \qquad (9.1.13)$$

if we treat α, β, γ, and δ as unknowns in a system of homogeneous equations (9.1.11)–(9.1.12) and require that the corresponding determinants equal zero. Applying (9.1.13) to the right side of (9.1.8), we obtain

$$\lambda_i \int_a^b r(x)[u^2(x) + v^2(x)]\,dx = 0. \qquad (9.1.14)$$

Because $r(x) > 0$, the integral is positive and $\lambda_i = 0$. Since $\lambda_i = 0$, λ is purely real. This implies that the eigenvalues are real. $\qquad \square$

If there is only one independent eigenfunction for each eigenvalue, that eigenvalue is *simple*. When more than one eigenfunction belongs to a single eigenvalue, the problem is *degenerate*.

Theorem: *The regular Sturm-Liouville problem has infinitely many real and simple eigenvalues* λ_n, $n = 0, 1, 2, \ldots$, *which can be arranged in a monotonically increasing sequence* $\lambda_0 < \lambda_1 < \lambda_2 < \cdots$ *such that* $\lim_{n \to \infty} \lambda_n = \infty$. *Every eigenfunction* $y_n(x)$ *associated with the corresponding eigenvalue* λ_n *has exactly* n *zeros in the interval* (a, b). *For each eigenvalue there exists only one eigenfunction (up to a multiplicative constant).*

The proof is beyond the scope of this book but may be found in more advanced treatises.[2]

In the following examples we illustrate how to find these real eigenvalues and their corresponding eigenfunctions.

• **Example 9.1.1**

Let us find the eigenvalues and eigenfunctions of

$$y'' + \lambda y = 0, \qquad (9.1.15)$$

subject to the boundary conditions

$$y(0) = 0, \qquad \text{and} \qquad y(\pi) - y'(\pi) = 0. \qquad (9.1.16)$$

Our first task is to check to see whether the problem is indeed a regular Sturm-Liouville problem. A comparison between (9.1.1) and (9.1.15) shows that they are the same if $p(x) = 1$, $q(x) = 0$, and $r(x) = 1$. Similarly, the boundary conditions (9.1.16) are identical to (9.1.2) if $\alpha = \gamma = 1$, $\delta = -1$, $\beta = 0$, $a = 0$, and $b = \pi$.

Because the form of the solution to (9.1.15) depends on λ, we consider three cases: λ negative, positive, or equal to zero. The general solution[3] of the differential equation is

$$y(x) = A \cosh(mx) + B \sinh(mx), \quad \text{if} \quad \lambda < 0, \qquad (9.1.17)$$

$$y(x) = C + Dx, \quad \text{if} \quad \lambda = 0, \qquad (9.1.18)$$

and

$$y(x) = E \cos(kx) + F \sin(kx), \quad \text{if} \quad \lambda > 0, \qquad (9.1.19)$$

---

[2] See, for example, Birkhoff, G., and G.-C. Rota, 1989: *Ordinary Differential Equations.* John Wiley & Sons, Chapters 10 and 11; Sagan, H., 1961: *Boundary and Eigenvalue Problems in Mathematical Physics.* John Wiley & Sons, Chapter 5.

[3] In many differential equations courses, the solution to $y'' - m^2 y = 0$, $m > 0$, is written $y(x) = c_1 e^{mx} + c_2 e^{-mx}$. However, we can rewrite this solution as $y(x) = (c_1 + c_2)\frac{1}{2}(e^{mx} + e^{-mx}) + (c_1 - c_2)\frac{1}{2}(e^{mx} - e^{-mx}) = A \cosh(mx) + B \sinh(mx)$, where $\cosh(mx) = (e^{mx} + e^{-mx})/2$ and $\sinh(mx) = (e^{mx} - e^{-mx})/2$. The advantage of using these hyperbolic functions over exponentials is the simplification that occurs when we substitute the hyperbolic functions into the boundary conditions.

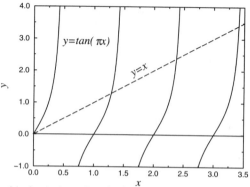

Figure 9.1.3: Graphical solution of $\tan(\pi x) = x$.

where for convenience $\lambda = -m^2 < 0$ in (9.1.17) and $\lambda = k^2 > 0$ in (9.1.19). Both k and m are real and positive by these definitions.

Turning to the condition that $y(0) = 0$, we find that $A = C = E = 0$. The other boundary condition $y(\pi) - y(\pi) = 0$ gives

$$B[\sinh(m\pi) - m\cosh(m\pi)] = 0, \tag{9.1.20}$$

$$D = 0, \tag{9.1.21}$$

and

$$F[\sin(k\pi) - k\cos(k\pi)] = 0. \tag{9.1.22}$$

If we graph $\sinh(m\pi) - m\cosh(m\pi)$ for all positive m, this quantity is always negative. Consequently, $B = 0$. However, in (9.1.22), a nontrivial solution (i.e., $F \neq 0$) occurs if

$$F\cos(k\pi)[\tan(k\pi) - k] = 0, \quad \text{or} \quad \tan(k\pi) = k. \tag{9.1.23}$$

In summary, we found nontrivial solutions only when $\lambda_n = k_n^2 > 0$, where k_n is the nth root of the transcendental equation (9.1.23). We can find the roots either graphically or through the use of a numerical algorithm. Figure 9.1.3 illustrates the graphical solution to the problem. We exclude the root $k = 0$ because λ must be greater than zero.

Let us now find the corresponding eigenfunctions. Because $A = B = C = D = E = 0$, we are left with $y(x) = F\sin(kx)$. Consequently, the eigenfunction, traditionally written without the arbitrary amplitude constant, is

$$y_n(x) = \sin(k_n x), \tag{9.1.24}$$

because k must equal k_n. Figure 9.1.4 shows the first four eigenfunctions.

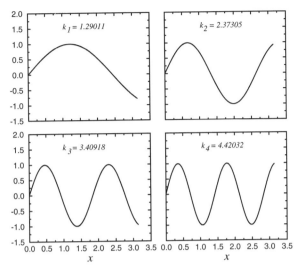

Figure 9.1.4: The first four eigenfunctions $\sin(k_n x)$ corresponding to the eigenvalue problem $\tan(k\pi) = k$.

• **Example 9.1.2**

For our second example let us solve the Sturm-Liouville problem,[4]

$$y'' + \lambda y = 0, \tag{9.1.25}$$

with the boundary conditions

$$y(0) - y'(0) = 0, \quad \text{and} \quad y(\pi) - y'(\pi) = 0. \tag{9.1.26}$$

Once again the three possible solutions to (9.1.25) are

$$y(x) = A\cosh(mx) + B\sinh(mx), \quad \text{if} \quad \lambda = -m^2 < 0, \tag{9.1.27}$$

$$y(x) = C + Dx, \quad \text{if} \quad \lambda = 0, \tag{9.1.28}$$

and

$$y(x) = E\cos(kx) + F\sin(kx), \quad \text{if} \quad \lambda = k^2 > 0. \tag{9.1.29}$$

Let us first check and see if there are any nontrivial solutions for $\lambda < 0$. Two simultaneous equations result from the substitution of (9.1.27) into (9.1.26):

$$A - mB = 0, \tag{9.1.30}$$

---

[4] Sosov and Theodosiou [Sosov, Y., and C. E. Theodosiou, 2002: On the complete solution of the Sturm-Liouville problem $(d^2 X/dx^2) + \lambda^2 X = 0$ over a closed interval. *J. Math. Phys.*, **43**, 2831–2843] have analyzed this problem with the general boundary conditions (9.1.2).

and

$$[\cosh(m\pi) - m\sinh(m\pi)]A + [\sinh(m\pi) - m\cosh(m\pi)]B = 0. \qquad \textbf{(9.1.31)}$$

The elimination of A between the two equations yields

$$\sinh(m\pi)(1 - m^2)B = 0. \qquad \textbf{(9.1.32)}$$

If (9.1.27) is a nontrivial solution, then $B \neq 0$, and

$$\sinh(m\pi) = 0, \qquad \textbf{(9.1.33)}$$

or

$$m^2 = 1. \qquad \textbf{(9.1.34)}$$

Equation (9.1.33) cannot hold because it implies $m = \lambda = 0$ which contradicts the assumption used in deriving (9.1.27) that $\lambda < 0$. On the other hand, (9.1.34) is quite acceptable. It corresponds to the eigenvalue $\lambda = -1$ and the eigenfunction is

$$y_0 = \cosh(x) + \sinh(x) = e^x, \qquad \textbf{(9.1.35)}$$

because it satisfies the differential equation

$$y_0'' - y_0 = 0, \qquad \textbf{(9.1.36)}$$

and the boundary conditions

$$y_0(0) - y_0'(0) = 0, \qquad \textbf{(9.1.37)}$$

and

$$y_0(\pi) - y_0'(\pi) = 0. \qquad \textbf{(9.1.38)}$$

An alternative method of finding m, which is quite popular because of its use in more difficult problems, follows from viewing (9.1.30) and (9.1.31) as a system of homogeneous linear equations, where A and B are the unknowns. It is well known[5] that for (9.1.30)–(9.1.31) to have a nontrivial solution (i.e., $A \neq 0$ and/or $B \neq 0$) the determinant of the coefficients must vanish:

$$\begin{vmatrix} 1 & -m \\ \cosh(m\pi) - m\sinh(m\pi) & \sinh(m\pi) - m\cosh(m\pi) \end{vmatrix} = 0. \qquad \textbf{(9.1.39)}$$

Expanding the determinant,

$$\sinh(m\pi)(1 - m^2) = 0, \qquad \textbf{(9.1.40)}$$

---

[5] See Chapter 14.

which leads directly to (9.1.33) and (9.1.34).

We consider next the case of $\lambda = 0$. Substituting (9.1.28) into (9.1.26), we find that

$$C - D = 0, \tag{9.1.41}$$

and

$$C + D\pi - D = 0. \tag{9.1.42}$$

This set of simultaneous equations yields $C = D = 0$ and we have only trivial solutions for $\lambda = 0$.

Finally, we examine the case when $\lambda > 0$. Substituting (9.1.29) into (9.1.26), we obtain

$$E - kF = 0, \tag{9.1.43}$$

and

$$[\cos(k\pi) + k\sin(k\pi)]E + [\sin(k\pi) - k\cos(k\pi)]F = 0. \tag{9.1.44}$$

The elimination of E from (9.1.43) and (9.1.44) gives

$$F(1 + k^2)\sin(k\pi) = 0. \tag{9.1.45}$$

If (9.1.29) is nontrivial, $F \neq 0$, and

$$k^2 = -1, \tag{9.1.46}$$

or

$$\sin(k\pi) = 0. \tag{9.1.47}$$

Condition (9.1.46) violates the assumption that k is real, which follows from the fact that $\lambda = k^2 > 0$. On the other hand, we can satisfy (9.1.47) if $k = 1, 2, 3, \ldots$; a negative k yields the same λ. Consequently we have the additional eigenvalues $\lambda_n = n^2$.

Let us now find the corresponding eigenfunctions. Because $E = kF$, $y(x) = F\sin(kx) + Fk\cos(kx)$ from (9.1.29). Thus, the eigenfunctions for $\lambda > 0$ are

$$y_n(x) = \sin(nx) + n\cos(nx). \tag{9.1.48}$$

Figure 9.1.5 illustrates some of the eigenfunctions given by (9.1.35) and (9.1.48).

• **Example 9.1.3**

Consider now the Sturm-Liouville problem

$$y'' + \lambda y = 0, \tag{9.1.49}$$

with

$$y(\pi) = y(-\pi), \quad \text{and} \quad y'(\pi) = y'(-\pi). \tag{9.1.50}$$

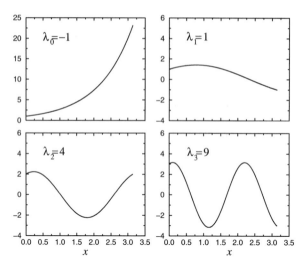

Figure 9.1.5: The first four eigenfunctions for the Sturm-Liouville problem (9.1.25)–(9.1.26).

This is *not* a regular Sturm-Liouville problem because the boundary conditions are periodic and do not conform to the canonical boundary condition (9.1.2).

The general solution to (9.1.49) is

$$y(x) = A\cosh(mx) + B\sinh(mx), \quad \text{if} \quad \lambda = -m^2 < 0, \qquad \textbf{(9.1.51)}$$

$$y(x) = C + Dx, \quad \text{if} \quad \lambda = 0, \qquad \textbf{(9.1.52)}$$

and

$$y(x) = E\cos(kx) + F\sin(kx), \quad \text{if} \quad \lambda = k^2 > 0. \qquad \textbf{(9.1.53)}$$

Substituting these solutions into the boundary condition (9.1.50),

$$A\cosh(m\pi) + B\sinh(m\pi) = A\cosh(-m\pi) + B\sinh(-m\pi), \qquad \textbf{(9.1.54)}$$

$$C + D\pi = C - D\pi, \qquad \textbf{(9.1.55)}$$

and

$$E\cos(k\pi) + F\sin(k\pi) = E\cos(-k\pi) + F\sin(-k\pi), \qquad \textbf{(9.1.56)}$$

or

$$B\sinh(m\pi) = 0, \quad D = 0, \quad \text{and} \quad F\sin(k\pi) = 0, \qquad \textbf{(9.1.57)}$$

because $\cosh(-m\pi) = \cosh(m\pi)$, $\sinh(-m\pi) = -\sinh(m\pi)$, $\cos(-k\pi) = \cos(k\pi)$, and $\sin(-k\pi) = -\sin(k\pi)$. Because m must be positive, $\sinh(m\pi)$ cannot equal zero and $B = 0$. On the other hand, if $\sin(k\pi) = 0$ or $k = n$, $n = 1, 2, 3, \ldots$, we have a nontrivial solution for positive λ and $\lambda_n = n^2$. Note that we still have A, C, E, and F as free constants.

From the boundary condition (9.1.50),

$$A \sinh(m\pi) = A \sinh(-m\pi), \tag{9.1.58}$$

and

$$-E \sin(k\pi) + F \cos(k\pi) = -E \sin(-k\pi) + F \cos(-k\pi). \tag{9.1.59}$$

The solution $y_0(x) = C$ identically satisfies the boundary condition (9.1.50) for all C. Because m and $\sinh(m\pi)$ must be positive, $A = 0$. From (9.1.57), we once again have $\sin(k\pi) = 0$, and $k = n$. Consequently, the eigenfunction solutions to (9.1.49)–(9.1.50) are

$$\lambda_0 = 0, \qquad y_0(x) = 1, \tag{9.1.60}$$

and

$$\lambda_n = n^2, \qquad y_n(x) = \begin{cases} \sin(nx), \\ \cos(nx), \end{cases} \tag{9.1.61}$$

and we have a degenerate set of eigenfunctions to the Sturm-Liouville problem (9.1.49) with the periodic boundary condition (9.1.50).

Problems

Find the eigenvalues and eigenfunctions for each of the following:

1. $y'' + \lambda y = 0$, $\quad y'(0) = 0$, $\quad y(L) = 0$

2. $y'' + \lambda y = 0$, $\quad y'(0) = 0$, $\quad y'(\pi) = 0$

3. $y'' + \lambda y = 0$, $\quad y(0) + y'(0) = 0$, $\quad y(\pi) + y'(\pi) = 0$

4. $y'' + \lambda y = 0$, $\quad y'(0) = 0$, $\quad y(\pi) - y'(\pi) = 0$

5. $y^{(iv)} + \lambda y = 0$, $\quad y(0) = y''(0) = 0$, $\quad y(L) = y''(L) = 0$

Find an equation from which you could find λ and give the form of the eigenfunction for each of the following:

6. $y'' + \lambda y = 0$, $\quad y(0) + y'(0) = 0$, $\quad y(1) = 0$

7. $y'' + \lambda y = 0$, $\quad y(0) = 0$, $\quad y(\pi) + y'(\pi) = 0$

8. $y'' + \lambda y = 0$, $\quad y'(0) = 0$, $\quad y(1) - y'(1) = 0$

9. $y'' + \lambda y = 0$, $\quad y(0) + y'(0) = 0$, $\quad y'(\pi) = 0$

10. $y'' + \lambda y = 0$, $y(0) + y'(0) = 0$, $y(\pi) - y'(\pi) = 0$

11. Find the eigenvalues and eigenfunctions of the Sturm-Liouville problem

$$\frac{d}{dx}\left(x\frac{dy}{dx}\right) + \frac{\lambda}{x}y = 0, \quad 1 \le x \le e$$

for each of the following boundary conditions: (a) $u(1) = u(e) = 0$, (b) $u(1) = u'(e) = 0$, and (c) $u'(1) = u'(e) = 0$.

Find the eigenvalues and eigenfunctions of the following Sturm-Liouville problems:

12.
$$x^2 y'' + 2xy' + \lambda y = 0, \quad y(1) = y(e) = 0, \quad 1 \le x \le e$$

13.
$$\frac{d}{dx}(x^3 y') + \lambda x y = 0, \quad y(1) = y(e^\pi) = 0, \quad 1 \le x \le e^\pi$$

14.
$$\frac{d}{dx}\left(\frac{1}{x}y'\right) + \frac{\lambda}{x}y = 0, \quad y(1) = y(e) = 0, \quad 1 \le x \le e$$

15.
$$y'''' - \lambda^4 y = 0, \quad y'''(0) = y''(0) = y'''(1) = y'(1) = 0, \quad 0 < x < 1$$

9.2 ORTHOGONALITY OF EIGENFUNCTIONS

In the previous section we saw how nontrivial solutions to the regular Sturm-Liouville problem consist of eigenvalues and eigenfunctions. The most important property of eigenfunctions is orthogonality.

Theorem: *Let the functions $p(x)$, $q(x)$, and $r(x)$ of the regular Sturm-Liouville problem (9.1.1)–(9.1.2) be real and continuous on the interval $[a, b]$. If $y_n(x)$ and $y_m(x)$ are continuously differentiable eigenfunctions corresponding to the distinct eigenvalues λ_n and λ_m, respectively, then $y_n(x)$ and $y_m(x)$ satisfy the* orthogonality *condition:*

$$\int_a^b r(x)y_n(x)y_m(x)\,dx = 0, \qquad \text{(9.2.1)}$$

if $\lambda_n \ne \lambda_m$. When (9.2.1) is satisfied, the eigenfunctions $y_n(x)$ and $y_m(x)$ are said to be *orthogonal* to each other with respect to the *weight function $r(x)$.*

The term *orthogonality* appears to be borrowed from linear algebra where a similar relationship holds between two perpendicular or orthogonal vectors.

Proof: Let $y_n(x)$ and $y_m(x)$ denote the eigenfunctions associated with two different eigenvalues λ_n and λ_m. Then

$$\frac{d}{dx}\left[p(x)\frac{dy_n}{dx}\right] + [q(x) + \lambda_n r(x)]y_n(x) = 0, \tag{9.2.2}$$

$$\frac{d}{dx}\left[p(x)\frac{dy_m}{dx}\right] + [q(x) + \lambda_m r(x)]y_m(x) = 0, \tag{9.2.3}$$

and both solutions satisfy the boundary conditions. Let us multiply the first differential equation by y_m; the second by y_n. Next, we subtract these two equations and move the terms containing $y_n y_m$ to the right side. The resulting equation is

$$y_n\frac{d}{dx}\left[p(x)\frac{dy_m}{dx}\right] - y_m\frac{d}{dx}\left[p(x)\frac{dy_n}{dx}\right] = (\lambda_n - \lambda_m)r(x)y_n y_m. \tag{9.2.4}$$

Integrating (9.2.4) from a to b yields

$$\int_a^b\left\{y_n\frac{d}{dx}\left[p(x)\frac{dy_m}{dx}\right] - y_m\frac{d}{dx}\left[p(x)\frac{dy_n}{dx}\right]\right\} dx = (\lambda_n - \lambda_m)\int_a^b r(x)y_n y_m\,dx. \tag{9.2.5}$$

We can simplify the left side of (9.2.5) by integrating by parts to give

$$\int_a^b\left\{y_n\frac{d}{dx}\left[p(x)\frac{dy_m}{dx}\right] - y_m\frac{d}{dx}\left[p(x)\frac{dy_n}{dx}\right]\right\} dx$$

$$= [p(x)y_m'y_n - p(x)y_n'y_m]_a^b - \int_a^b p(x)[y_n'y_m' - y_n'y_m']\,dx. \tag{9.2.6}$$

The second integral equals zero since the integrand vanishes identically. Because $y_n(x)$ and $y_m(x)$ satisfy the boundary condition at $x = a$,

$$\alpha y_n(a) + \beta y_n'(a) = 0, \tag{9.2.7}$$

and

$$\alpha y_m(a) + \beta y_m'(a) = 0. \tag{9.2.8}$$

These two equations are simultaneous equations in α and β. Hence, the determinant of the equations must be zero:

$$y_n'(a)y_m(a) - y_m'(a)y_n(a) = 0. \tag{9.2.9}$$

Similarly, at the other end,

$$y_n'(b)y_m(b) - y_m'(b)y_n(b) = 0. \tag{9.2.10}$$

Consequently, the right side of (9.2.6) vanishes and (9.2.5) reduces to (9.2.1).
□

• **Example 9.2.1**

Let us verify the orthogonality condition for the eigenfunctions that we
found in Example 9.1.1.

Because $r(x) = 1$, $a = 0$, $b = \pi$, and $y_n(x) = \sin(k_n x)$, we find that

$$\int_a^b r(x) y_n y_m \, dx = \int_0^\pi \sin(k_n x) \sin(k_m x) \, dx \qquad (9.2.11)$$

$$= \tfrac{1}{2} \int_0^\pi \{ \cos[(k_n - k_m)x] - \cos[(k_n + k_m)x] \} \, dx \qquad (9.2.12)$$

$$= \frac{\sin[(k_n - k_m)x]}{2(k_n - k_m)} \bigg|_0^\pi - \frac{\sin[(k_n + k_m)x]}{2(k_n + k_m)} \bigg|_0^\pi \qquad (9.2.13)$$

$$= \frac{\sin[(k_n - k_m)\pi]}{2(k_n - k_m)} - \frac{\sin[(k_n + k_m)\pi]}{2(k_n + k_m)} \qquad (9.2.14)$$

$$= \frac{\sin(k_n \pi)\cos(k_m \pi) - \cos(k_n \pi)\sin(k_m \pi)}{2(k_n - k_m)}$$

$$- \frac{\sin(k_n \pi)\cos(k_m \pi) + \cos(k_n \pi)\sin(k_m \pi)}{2(k_n + k_m)} \qquad (9.2.15)$$

$$= \frac{k_n \cos(k_n \pi)\cos(k_m \pi) - k_m \cos(k_n \pi)\cos(k_m \pi)}{2(k_n - k_m)}$$

$$- \frac{k_n \cos(k_n \pi)\cos(k_m \pi) + k_m \cos(k_n \pi)\cos(k_m \pi)}{2(k_n + k_m)} \qquad (9.2.16)$$

$$= \frac{(k_n - k_m)\cos(k_n \pi)\cos(k_m \pi)}{2(k_n - k_m)}$$

$$- \frac{(k_n + k_m)\cos(k_n \pi)\cos(k_m \pi)}{2(k_n + k_m)} = 0. \qquad (9.2.17)$$

We used the relationships $k_n = \tan(k_n \pi)$, and $k_m = \tan(k_m \pi)$ to simplify
(9.2.15). Note, however, that if $n = m$,

$$\int_0^\pi \sin(k_n x) \sin(k_n x) \, dx = \tfrac{1}{2} \int_0^\pi [1 - \cos(2k_n x)] \, dx \qquad (9.2.18)$$

$$= \frac{\pi}{2} - \frac{\sin(2k_n \pi)}{4k_n} \qquad (9.2.19)$$

$$= \tfrac{1}{2}[\pi - \cos^2(k_n \pi)] > 0, \qquad (9.2.20)$$

because $\sin(2A) = 2\sin(A)\cos(A)$, and $k_n = \tan(k_n \pi)$. That is, any eigenfunction *cannot* be orthogonal to itself.

In closing, we note that had we defined the eigenfunction in our example as

$$y_n(x) = \frac{\sin(k_n x)}{\sqrt{[\pi - \cos^2(k_n \pi)]/2}} \tag{9.2.21}$$

rather than $y_n(x) = \sin(k_n x)$, the orthogonality condition would read

$$\int_0^\pi y_n(x) y_m(x) \, dx = \begin{cases} 0, & m \neq n, \\ 1, & m = n. \end{cases} \tag{9.2.22}$$

This process of *normalizing* an eigenfunction so that the orthogonality condition becomes

$$\int_a^b r(x) y_n(x) y_m(x) \, dx = \begin{cases} 0, & m \neq n, \\ 1, & m = n, \end{cases} \tag{9.2.23}$$

generates *orthonormal* eigenfunctions. We will see the convenience of doing this in the next section.

Problems

1. The Sturm-Liouville problem $y'' + \lambda y = 0$, $y(0) = y(L) = 0$, has the eigenfunction solution $y_n(x) = \sin(n\pi x/L)$. By direct integration verify the orthogonality condition (9.2.1).

2. The Sturm-Liouville problem $y'' + \lambda y = 0$, $y'(0) = y'(L) = 0$, has the eigenfunction solutions $y_0(x) = 1$ and $y_n(x) = \cos(n\pi x/L)$. By direct integration verify the orthogonality condition (9.2.1).

3. The Sturm-Liouville problem $y'' + \lambda y = 0$, $y(0) = y'(L) = 0$, has the eigenfunction solution $y_n(x) = \sin[(2n-1)\pi x/(2L)]$. By direct integration verify the orthogonality condition (9.2.1).

4. The Sturm-Liouville problem $y'' + \lambda y = 0$, $y'(0) = y(L) = 0$, has the eigenfunction solution $y_n(x) = \cos[(2n-1)\pi x/(2L)]$. By direct integration verify the orthogonality condition (9.2.1).

9.3 EXPANSION IN SERIES OF EIGENFUNCTIONS

In calculus we learned that under certain conditions we could represent a function $f(x)$ by a linear and infinite sum of polynomials $(x-x_0)^n$. In this section we show that an analogous procedure exists for representing a piece-wise continuous function by a linear sum of eigenfunctions. These *eigenfunction expansions* will be used in the next three chapters to solve partial differential equations.

Let the function $f(x)$ be defined in the interval $a < x < b$. We wish to re-express $f(x)$ in terms of the eigenfunctions $y_n(x)$ given by a regular Sturm-Liouville problem. Assuming that the function $f(x)$ can be represented by a uniformly convergent series,[6] we write

$$f(x) = \sum_{n=1}^{\infty} c_n y_n(x). \qquad (9.3.1)$$

The orthogonality relation (9.2.1) gives us the method for computing the coefficients c_n. First we multiply both sides of (9.3.1) by $r(x)y_m(x)$, where m is a fixed integer, and then integrate from a to b. Because this series is uniformly convergent and $y_n(x)$ is continuous, we can integrate the series term by term or

$$\int_a^b r(x)f(x)y_m(x)\,dx = \sum_{n=1}^{\infty} c_n \int_a^b r(x)y_n(x)y_m(x)\,dx. \qquad (9.3.2)$$

The orthogonality relationship states that all of the terms on the right side of (9.3.2) must disappear except the one for which $n = m$. Thus, we are left with

$$\int_a^b r(x)f(x)y_m(x)\,dx = c_m \int_a^b r(x)y_m(x)y_m(x)\,dx \qquad (9.3.3)$$

or

$$c_n = \frac{\int_a^b r(x)f(x)y_n(x)\,dx}{\int_a^b r(x)y_n^2(x)\,dx}, \qquad (9.3.4)$$

if we replace m by n in (9.3.3).

Usually, both integrals in (9.3.4) are evaluated by direct integration. In the case when the evaluation of the denominator is very difficult, Lockshin[7] has shown that the denominator of (9.3.4) always equals

$$\int_a^b r(x)y^2(x)\,dx = p(x)\left[\frac{\partial y}{\partial x}\frac{\partial y}{\partial \lambda} - y\frac{\partial^2 y}{\partial \lambda \partial x}\right]\Bigg|_a^b, \qquad (9.3.5)$$

---

[6] If $S_n(x) = \sum_{k=1}^{n} u_k(x)$, $S(x) = \lim_{n\to\infty} S_n(x)$, and $0 < |S_n(x) - S(x)| < \epsilon$ for all $n > M > 0$, the series $\sum_{k=1}^{\infty} u_k(x)$ is uniformly convergent if M is dependent on ϵ alone and not x.

[7] Lockshin, J. L, 2001: Explicit closed-form expression for eigenfunction norms. *Appl. Math. Lett.*, **14**, 553–555.

for a regular Sturm-Liouville problem with eigenfunction solution y where $p(x)$, $q(x)$, and $r(x)$ are continuously differentiable on the interval $[a, b]$.

The series (9.3.1) with the coefficients found by (9.3.4) is a *generalized Fourier series* of the function $f(x)$ with respect to the eigenfunction $y_n(x)$. It is called a generalized Fourier series because we generalized the procedure of re-expressing a function $f(x)$ by sines and cosines into one involving solutions to regular Sturm-Liouville problems. Note that if we had used an orthonormal set of eigenfunctions, then the denominator of (9.3.4) would equal one and we reduce our work by half. The coefficients c_n are the *Fourier coefficients*.

One of the most remarkable facts about generalized Fourier series is their applicability even when the function has a finite number of bounded discontinuities in the range $[a, b]$. We may formally express this fact by the following theorem:

Theorem: *If both $f(x)$ and $f'(x)$ are piece-wise continuous in $a \le x \le b$, then $f(x)$ can be expanded in a uniformly convergent Fourier series (9.3.1), whose coefficients c_n are given by (9.3.4). It converges to $[f(x^+) + f(x^-)]/2$ at any point x in the open interval $a < x < b$.*

The proof is beyond the scope of this book but can be found in more advanced treatises.[8] If we are willing to include stronger constraints, we can make even stronger statements about convergence. For example,[9] if we require that $f(x)$ be a continuous function with a piece-wise continuous first derivative, then the eigenfunction expansion (9.3.1) converges to $f(x)$ uniformly and absolutely in $[a, b]$ if $f(x)$ satisfies the same boundary conditions as does $y_n(x)$.

In the case when $f(x)$ is discontinuous, we are not merely rewriting $f(x)$ in a new form. We are actually choosing the coefficients c_n so that the eigenfunction expansion fits $f(x)$ in the "least squares" sense that

$$\int_a^b r(x) \left| f(x) - \sum_{n=1}^{\infty} c_n y_n(x) \right|^2 dx = 0. \tag{9.3.6}$$

Consequently we should expect peculiar things, such as spurious oscillations, to occur in the neighborhood of the discontinuity. These are *Gibbs phenomena*,[10] the same phenomena discovered with Fourier series. See §4.2.

---

[8] For example, Titchmarsh, E. C., 1962: *Eigenfunction Expansions Associated with Second-Order Differential Equations. Part 1*. Oxford University Press, pp. 12–16.

[9] Tolstov, G. P., 1962: *Fourier Series*. Dover Publishers, p. 255.

[10] Apparently first discussed by Weyl, H., 1910: Die Gibbs'sche Erscheinung in der Theorie der Sturm-Liouvilleschen Reihen. *Rend. Circ. Mat. Palermo*, **29**, 321–323.

• **Example 9.3.1**

To illustrate the concept of an eigenfunction expansion, let us find the expansion for $f(x) = x$ over the interval $0 < x < \pi$ using the solution to the regular Sturm-Liouville problem of

$$y'' + \lambda y = 0, \qquad y(0) = y(\pi) = 0. \tag{9.3.7}$$

This problem arises when we solve the wave or heat equation by separation of variables in the next two chapters.

Because the eigenfunctions are $y_n(x) = \sin(nx)$, $n = 1, 2, 3, \ldots$, $r(x) = 1$, $a = 0$, and $b = \pi$, (9.3.4) yields

$$c_n = \frac{\int_0^\pi x \sin(nx)\, dx}{\int_0^\pi \sin^2(nx)\, dx} = \frac{-x\cos(nx)/n + \sin(nx)/n^2 \big|_0^\pi}{x/2 - \sin(2nx)/(4n)\big|_0^\pi} \tag{9.3.8}$$

$$= -\frac{2}{n}\cos(n\pi) = -\frac{2}{n}(-1)^n. \tag{9.3.9}$$

Equation (9.3.1) then gives

$$f(x) = -2\sum_{n=1}^{\infty} \frac{(-1)^n}{n}\sin(nx). \tag{9.3.10}$$

This particular example is in fact an example of a half-range sine expansion.

Finally we must state the values of x for which (9.3.10) is valid. At $x = \pi$ the series converges to zero while $f(\pi) = \pi$. At $x = 0$ both the series and the function converge to zero. Hence the series expansion (9.3.10) is valid for $0 \le x < \pi$.

• **Example 9.3.2**

For our second example let us find the expansion for $f(x) = x$ over the interval $0 \le x < \pi$ using the solution to the regular Sturm-Liouville problem of

$$y'' + \lambda y = 0, \qquad y(0) = y(\pi) - y'(\pi) = 0. \tag{9.3.11}$$

We will encounter this problem when we solve the heat equation with radiative boundary conditions by separation of variables.

Because $r(x) = 1$, $a = 0$, $b = \pi$ and the eigenfunctions are $y_n(x) = \sin(k_n x)$, where $k_n = \tan(k_n\pi)$, (9.3.4) yields

$$c_n = \frac{\int_0^\pi x \sin(k_n x)\, dx}{\int_0^\pi \sin^2(k_n x)\, dx} = \frac{\int_0^\pi x \sin(k_n x)\, dx}{\frac{1}{2}\int_0^\pi [1 - \cos(2k_n x)]\, dx} \tag{9.3.12}$$

$$= \frac{2\sin(k_n x)/k_n^2 - 2x\cos(k_n x)/k_n \big|_0^\pi}{x - \sin(2k_n x)/(2k_n)\big|_0^\pi} \tag{9.3.13}$$

$$= \frac{2\sin(k_n\pi)/k_n^2 - 2\pi\cos(k_n\pi)/k_n}{\pi - \sin(2k_n\pi)/(2k_n)} \tag{9.3.14}$$

$$= \frac{2(1-\pi)\cos(k_n\pi)/k_n}{\pi - \cos^2(k_n\pi)}, \tag{9.3.15}$$

where we used the property that $\sin(k_n\pi) = k_n\cos(k_n\pi)$. Equation (9.3.1) then gives

$$f(x) = 2(1 - \pi)\sum_{n=1}^{\infty} \frac{\cos(k_n\pi)}{k_n[\pi - \cos^2(k_n\pi)]}\sin(k_nx). \qquad (9.3.16)$$

To illustrate the use of (9.3.5), we note that

$$y(x) = \sin(\sqrt{\lambda}\,x), \quad \frac{\partial y}{\partial x} = \sqrt{\lambda}\cos(\sqrt{\lambda}\,x), \quad \frac{\partial y}{\partial \lambda} = \frac{x}{2\sqrt{\lambda}}\cos(\sqrt{\lambda}\,x), \quad (9.3.17)$$

and

$$\frac{\partial^2 y}{\partial \lambda \partial x} = \frac{\partial^2 y}{\partial x \partial \lambda} = \frac{1}{2\sqrt{\lambda}}\cos(\sqrt{\lambda}\,x) - \frac{x}{2}\sin(\sqrt{\lambda}\,x). \qquad (9.3.18)$$

Therefore,

$$\int_0^{\pi} r(x)y_n^2(x)\,dx = \left\{\frac{x}{2}\cos^2(k_nx) - \sin(k_nx)\left[\frac{1}{2k_n}\cos(k_nx) - \frac{x}{2}\sin(k_nx)\right]\right\}\Big|_0^{\pi}$$
$$(9.3.19)$$

$$= \left[\frac{x}{2} - \frac{1}{2k_n}\sin(k_nx)\cos(k_nx)\right]\Big|_0^{\pi} \qquad (9.3.20)$$

$$= \frac{\pi}{2} - \frac{\cos^2(k_n\pi)}{2}. \qquad (9.3.21)$$

Note that we set $\lambda = \lambda_n = k_n^2$ after taking the derivatives with respect to λ.

Problems

1. The Sturm-Liouville problem $y'' + \lambda y = 0$, $y(0) = y(L) = 0$, has the eigenfunction solution $y_n(x) = \sin(n\pi x/L)$. Find the eigenfunction expansion for $f(x) = x$ using this eigenfunction.

2. The Sturm-Liouville problem $y'' + \lambda y = 0$, $y'(0) = y'(L) = 0$, has the eigenfunction solutions $y_0(x) = 1$, and $y_n(x) = \cos(n\pi x/L)$. Find the eigenfunction expansion for $f(x) = x$ using these eigenfunctions.

3. The Sturm-Liouville problem $y'' + \lambda y = 0$, $y(0) = y'(L) = 0$, has the eigenfunction solution $y_n(x) = \sin[(2n - 1)\pi x/(2L)]$. Find the eigenfunction expansion for $f(x) = x$ using this eigenfunction.

4. The Sturm-Liouville problem $y'' + \lambda y = 0$, $y'(0) = y(L) = 0$, has the eigenfunction solution $y_n(x) = \cos[(2n - 1)\pi x/(2L)]$. Find the eigenfunction expansion for $f(x) = x$ using this eigenfunction.

5. Consider the eigenvalue problem

$$y'' + (\lambda - a^2)y = 0, \qquad 0 < x < 1,$$

with the boundary conditions

$$y'(0) + ay(0) = 0 \qquad \text{and} \qquad y'(1) + ay(1) = 0.$$

Step 1: Show that this is a regular Sturm-Liouville problem.

Step 2: Show that the eigenvalues and eigenfunctions are

$$\lambda_0 = 0, \qquad\qquad y_0(x) = e^{-ax},$$
$$\lambda_n = a^2 + n^2\pi^2, \qquad y_n(x) = a\sin(n\pi x) - n\pi\cos(n\pi x),$$

where $n = 1, 2, 3, \dots$.

Step 3: Given a function $f(x)$, show that we can expand it as follows:

$$f(x) = C_0 e^{-ax} + \sum_{n=1}^{\infty} C_n \left[a\sin(n\pi x) - n\pi\cos(n\pi x) \right],$$

where

$$\left(1 - e^{-2a} \right) C_0 = 2a \int_0^1 f(x) e^{-ax}\, dx,$$

and

$$(a^2 + n^2\pi^2) C_n = 2 \int_0^1 f(x) \left[a\sin(n\pi x) - n\pi\cos(n\pi x) \right]\, dx.$$

6. Consider the eigenvalue problem

$$y'''' + \lambda y'' = 0, \qquad 0 < x < 1,$$

with the boundary conditions $y(0) = y'(0) = y(1) = y'(1) = 0$. Prove the following points:

Step 1: Show that the eigenfunctions are

$$y_n(x) = 1 - \cos(k_n x) + \frac{1 - \cos(k_n)}{k_n - \sin(k_n)} [\sin(k_n x) - k_n x],$$

where k_n denotes the nth root of

$$2 - 2\cos(k) - k\sin(k) = \sin(k/2)[\sin(k/2) - (k/2)\cos(k/2)] = 0.$$

Step 2: Show that there are two classes of eigenfunctions:

$$\kappa_n = 2n\pi, \qquad y_n(x) = 1 - \cos(2n\pi x),$$

and

$$\tan(\kappa_n/2) = \kappa_n/2, \qquad y_n(x) = 1 - \cos(\kappa_n x) + \frac{2}{\kappa_n}[\sin(\kappa_n x) - \kappa_n x].$$

Step 3: Show that the orthogonality condition for this problem is

$$\int_0^1 y_n'(x)y_m'(x)\, dx = 0, \qquad n \neq m,$$

where $y_n(x)$ and $y_m(x)$ are two distinct eigenfunction solutions of this problem. Hint: Follow the proof in §9.2 and integrate repeatedly by parts to eliminate higher derivative terms.

Step 4: Show that we can construct an eigenfunction expansion for an arbitrary function $f(x)$ via

$$f(x) = \sum_{n=1}^\infty C_n y_n(x), \qquad 0 < x < 1,$$

provided

$$C_n = \frac{\int_0^1 f'(x)y_n'(x)\, dx}{\int_0^1 [y_n'(x)]^2\, dx}.$$

What are the condition(s) on $f(x)$?

9.4 A SINGULAR STURM-LIOUVILLE PROBLEM: LEGENDRE'S EQUATION

In the previous sections we used solutions to a regular Sturm-Liouville problem in the eigenfunction expansion of the function $f(x)$. The fundamental reason why we could form such an expansion was the orthogonality condition (9.2.1). This crucial property allowed us to solve for the Fourier coefficient c_n given by (9.3.4).

In the next few chapters, when we solve partial differential equations in cylindrical and spherical coordinates, we will find that $f(x)$ must be expanded in terms of eigenfunctions from singular Sturm-Liouville problems. Is this permissible? How do we compute the Fourier coefficients in this case? The final two sections of this chapter deal with these questions by examining the two most frequently encountered singular Sturm-Liouville problems, those involving Legendre's and Bessel's equations.

We begin by determining the orthogonality condition for singular Sturm-Liouville problems. Returning to the beginning portions of §9.2, we combine (9.2.5) and (9.2.6) to obtain

$$(\lambda_n - \lambda_m)\int_a^b r(x)y_n y_m\, dx = [p(b)y_m'(b)y_n(b) - p(b)y_n'(b)y_m(b)$$
$$- p(a)y_m'(a)y_n(a) + p(a)y_n'(a)y_m(a)]. \qquad (9.4.1)$$

Figure 9.4.1: Born into an affluent family, Adrien-Marie Legendre's (1752–1833) modest family fortune was sufficient to allow him to devote his life to research in celestial mechanics, number theory, and the theory of elliptic functions. In July 1784 he read before the *Académie des sciences* his *Recherches sur la figure des planètes*. It is in this paper that Legendre polynomials first appeared. (Portrait courtesy of the Archives de l'Académie des sciences, Paris.)

From (9.4.1) the right side vanishes and we preserve orthogonality if $y_n(x)$ is finite and $p(x)y_n'(x)$ tends to zero at both endpoints. This is not the only choice but let us see where it leads.

Consider now Legendre's equation:

$$(1 - x^2)\frac{d^2y}{dx^2} - 2x\frac{dy}{dx} + n(n+1)y = 0, \qquad (9.4.2)$$

or

$$\frac{d}{dx}\left[(1 - x^2)\frac{dy}{dx}\right] + n(n+1)y = 0, \qquad (9.4.3)$$

where we set $a = -1$, $b = 1$, $\lambda = n(n+1)$, $p(x) = 1 - x^2$, $q(x) = 0$, and $r(x) = 1$. This equation arises in the solution of partial differential equations involving spherical geometry. Because $p(-1) = p(1) = 0$, we are faced with a singular Sturm-Liouville problem. Before we can determine if any of its solutions can be used in an eigenfunction expansion, we must find them.

Equation (9.4.2) does not have a simple general solution. [If $n = 0$, then $y(x) = 1$ is a solution.] Consequently we try to solve it with the power series:

$$y(x) = \sum_{k=0}^{\infty} A_k x^k, \qquad (9.4.4)$$

$$y'(x) = \sum_{k=0}^{\infty} k A_k x^{k-1}, \qquad (9.4.5)$$

and

$$y''(x) = \sum_{k=0}^{\infty} k(k-1) A_k x^{k-2}. \qquad (9.4.6)$$

Substituting into (9.4.2),

$$\sum_{k=0}^{\infty} k(k-1) A_k x^{k-2} + \sum_{k=0}^{\infty} \left[n(n+1) - 2k - k(k-1) \right] A_k x^k = 0, \qquad (9.4.7)$$

which equals

$$\sum_{m=2}^{\infty} m(m-1) A_m x^{m-2} + \sum_{k=0}^{\infty} \left[n(n+1) - k(k+1) \right] A_k x^k = 0. \qquad (9.4.8)$$

If we define $k = m + 2$ in the first summation, then

$$\sum_{k=0}^{\infty} (k+2)(k+1) A_{k+2} x^k + \sum_{k=0}^{\infty} \left[n(n+1) - k(k+1) \right] A_k x^k = 0. \qquad (9.4.9)$$

Because (9.4.9) must be true for any x, each power of x must vanish separately. It then follows that

$$(k+2)(k+1) A_{k+2} = [k(k+1) - n(n+1)] A_k, \qquad (9.4.10)$$

or

$$A_{k+2} = \frac{[k(k+1) - n(n+1)]}{(k+1)(k+2)} A_k, \qquad (9.4.11)$$

where $k = 0, 1, 2, \ldots$. Note that we still have the two arbitrary constants A_0 and A_1 that are necessary for the general solution of (9.4.2).

The first few terms of the solution associated with A_0 are

$$u_p(x) = 1 - \frac{n(n+1)}{2!} x^2 + \frac{n(n-2)(n+1)(n+3)}{4!} x^4$$
$$- \frac{n(n-2)(n-4)(n+1)(n+3)(n+5)}{6!} x^6 + \cdots, \qquad (9.4.12)$$

while the first few terms associated with the A_1 coefficient are

$$
v_p(x) = x - \frac{(n-1)(n+2)}{3!}x^3 + \frac{(n-1)(n-3)(n+2)(n+4)}{5!}x^5
$$
$$
- \frac{(n-1)(n-3)(n-5)(n+2)(n+4)(n+6)}{7!}x^7 + \cdots. \qquad (9.4.13)
$$

If n is an *even* positive integer (including $n = 0$), then the series (9.4.12) terminates with the term involving x^n: the solution is a polynomial of degree n. Similarly, if n is an *odd* integer, the series (9.4.13) terminates with the term involving x^n. Otherwise, for n noninteger the expressions are infinite series.

For reasons that will become apparent, we restrict ourselves to positive integers n. Actually, this includes all possible integers because the negative integer $-n-1$ has the same Legendre's equation and solution as the positive integer n. These polynomials are *Legendre polynomials*[11] and we may compute them by the power series:

$$
P_n(x) = \sum_{k=0}^{m}(-1)^k \frac{(2n-2k)!}{2^n k!(n-k)!(n-2k)!}x^{n-2k}, \qquad (9.4.14)
$$

where $m = n/2$, or $m = (n-1)/2$, depending upon which is an integer. We chose to use (9.4.14) over (9.4.12) or (9.4.13) because (9.4.14) has the advantage that $P_n(1) = 1$. Table 9.4.1 gives the first ten Legendre polynomials.

The other solution, the infinite series, is the Legendre function of the second kind, $Q_n(x)$. Figure 9.4.2 illustrates the first four Legendre polynomials $P_n(x)$ while Figure 9.4.3 gives the first four Legendre functions of the second kind $Q_n(x)$. From this figure we see that $Q_n(x)$ becomes infinite at the points $x = \pm 1$. As shown earlier, this is important because we are only interested in solutions to Legendre's equation that are finite over the interval $[-1, 1]$. On the other hand, in problems where we exclude the points $x = \pm 1$, Legendre functions of the second kind will appear in the general solution.[12]

In the case that n is not an integer, we can construct a solution[13] that remains finite at $x = 1$ but not at $x = -1$. Furthermore, we can construct a solution which is finite at $x = -1$ but not at $x = 1$. Because our solutions

---

[11] Legendre, A. M., 1785: Sur l'attraction des sphéroïdes homogénes. *Mém. math. phys. présentés à l'Acad. sci. pars divers savants*, **10**, 411–434. The best reference on Legendre polynomials is Hobson, E. W., 1965:*The Theory of Spherical and Ellipsoidal Harmonics.* Chelsea Publishing Co., 500 pp.

[12] See Smythe, W. R., 1950: *Static and Dynamic Electricity.* McGraw-Hill, §5.215 for an example.

[13] See Carrier, G. F., M. Krook, and C. E. Pearson, 1966: *Functions of the Complex Variable: Theory and Technique.* McGraw-Hill, pp. 212–213.

Table 9.4.1: The First Ten Legendre Polynomials

$$P_0(x) = 1$$

$$P_1(x) = x$$

$$P_2(x) = \tfrac{1}{2}(3x^2 - 1)$$

$$P_3(x) = \tfrac{1}{2}(5x^3 - 3x)$$

$$P_4(x) = \tfrac{1}{8}(35x^4 - 30x^2 + 3)$$

$$P_5(x) = \tfrac{1}{8}(63x^5 - 70x^3 + 15x)$$

$$P_6(x) = \tfrac{1}{16}(231x^6 - 315x^4 + 105x^2 - 5)$$

$$P_7(x) = \tfrac{1}{16}(429x^7 - 693x^5 + 315x^3 - 35x)$$

$$P_8(x) = \tfrac{1}{128}(6435x^8 - 12012x^6 + 6930x^4 - 1260x^2 + 35)$$

$$P_9(x) = \tfrac{1}{128}(12155x^9 - 25740x^7 + 18018x^5 - 4620x^3 + 315x)$$

$$P_{10}(x) = \tfrac{1}{256}(46189x^{10} - 109395x^8 + 90090x^6 - 30030x^4 + 3465x^2 - 63)$$

must be finite at both endpoints so that we can use them in an eigenfunction expansion, we must reject these solutions from further consideration and are left only with Legendre polynomials. From now on, we will only consider the properties and uses of these polynomials.

Although we have the series (9.4.14) to compute $P_n(x)$, there are several alternative methods. We obtain the first method, known as *Rodrigues' formula*,[14] by writing (9.4.14) in the form

$$P_n(x) = \frac{1}{2^n n!} \sum_{k=0}^{n} (-1)^k \frac{n!}{k!(n-k)!} \frac{(2n-2k)!}{(n-2k)!} x^{n-2k} \qquad (9.4.15)$$

$$P_n(x) = \frac{1}{2^n n!} \frac{d^n}{dx^n} \left[\sum_{k=0}^{n} (-1)^k \frac{n!}{k!(n-k)!} x^{2n-2k} \right]. \qquad (9.4.16)$$

The last summation is the binomial expansion of $(x^2 - 1)^n$ so that

$$P_n(x) = \frac{1}{2^n n!} \frac{d^n}{dx^n} (x^2 - 1)^n. \qquad (9.4.17)$$

---

[14] Rodrigues, O., 1816: Mémoire sur l'attraction des sphéroïdes. *Correspond. l'Ecole Polytech.*, **3**, 361–385.

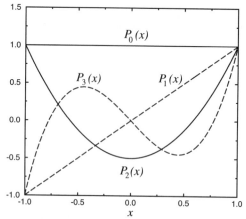

Figure 9.4.2: The first four Legendre functions of the first kind.

Another method for computing $P_n(x)$ involves the use of recurrence formulas. The first step in finding these formulas is to establish the fact that

$$(1 + h^2 - 2xh)^{-1/2} = P_0(x) + hP_1(x) + h^2 P_2(x) + \cdots. \qquad (9.4.18)$$

The function $(1 + h^2 - 2xh)^{-1/2}$ is the *generating function* for $P_n(x)$. We obtain the expansion via the formal binomial expansion

$$(1 + h^2 - 2xh)^{-1/2} = 1 + \tfrac{1}{2}(2xh - h^2) + \tfrac{1}{2}\tfrac{3}{2}\tfrac{1}{2!}(2xh - h^2)^2 + \cdots. \qquad (9.4.19)$$

Upon expanding the terms contained in $2x - h^2$ and grouping like powers of h,

$$(1 + h^2 - 2xh)^{-1/2} = 1 + xh + (\tfrac{3}{2}x^2 - \tfrac{1}{2})h^2 + \cdots. \qquad (9.4.20)$$

A direct comparison between the coefficients of each power of h and the Legendre polynomial $P_n(x)$ completes the demonstration. Note that these results hold only if $|x|$ and $|h| < 1$.

Next we define $W(x, h) = (1 + h^2 - 2xh)^{-1/2}$. A quick check shows that $W(x, h)$ satisfies the first-order partial differential equation

$$(1 - 2xh + h^2)\frac{\partial W}{\partial h} + (h - x)W = 0. \qquad (9.4.21)$$

The substitution of (9.4.18) into (9.4.21) yields

$$(1 - 2xh + h^2)\sum_{n=0}^{\infty} nP_n(x)h^{n-1} + (h - x)\sum_{n=0}^{\infty} P_n(x)h^n = 0. \qquad (9.4.22)$$

Setting the coefficients of h^n equal to zero, we find that

$$(n+1)P_{n+1}(x) - 2nxP_n(x) + (n-1)P_{n-1}(x) + P_{n-1}(x) - xP_n(x) = 0, \qquad (9.4.23)$$

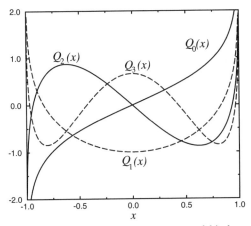

Figure 9.4.3: The first four Legendre functions of the second kind.

or

$$(n + 1)P_{n+1}(x) - (2n + 1)xP_n(x) + nP_{n-1}(x) = 0 \qquad \textbf{(9.4.24)}$$

with $n = 1, 2, 3, \ldots$.

Similarly, the first-order partial differential equation

$$(1 - 2xh + h^2)\frac{\partial W}{\partial x} - hW = 0 \qquad \textbf{(9.4.25)}$$

leads to

$$(1 - 2xh + h^2)\sum_{n=0}^{\infty} P'_n(x)h^n - \sum_{n=0}^{\infty} P_n(x)h^{n+1} = 0, \qquad \textbf{(9.4.26)}$$

which implies

$$P'_{n+1}(x) - 2xP'_n(x) + P'_{n-1}(x) - P_n(x) = 0. \qquad \textbf{(9.4.27)}$$

Differentiating (9.4.24), we first eliminate $P'_{n-1}(x)$ and then $P'_{n+1}(x)$ from the resulting equations and (9.4.27). This gives two further recurrence relationships:

$$P'_{n+1}(x) - xP'_n(x) - (n+1)P_n(x) = 0, \quad n = 0, 1, 2, \ldots, \qquad \textbf{(9.4.28)}$$

and

$$xP'_n(x) - P'_{n-1}(x) - nP_n(x) = 0, \quad n = 1, 2, 3, \ldots. \qquad \textbf{(9.4.29)}$$

Adding (9.4.28) and (9.4.29), we obtain the more symmetric formula

$$P'_{n+1}(x) - P'_{n-1}(x) = (2n+1)P_n(x), \quad n = 1, 2, 3, \ldots. \qquad (9.4.30)$$

Given any two of the polynomials $P_{n+1}(x)$, $P_n(x)$, and $P_{n-1}(x)$, (9.4.24) or (9.4.30) yields the third.

Having determined several methods for finding the Legendre polynomial $P_n(x)$, we now turn to the actual orthogonality condition.[15] Consider the integral

$$J = \int_{-1}^{1} \frac{dx}{\sqrt{1 + h^2 - 2xh} \sqrt{1 + t^2 - 2xt}}, \qquad |h|, |t| < 1 \qquad (9.4.31)$$

$$= \int_{-1}^{1} [P_0(x) + hP_1(x) + \cdots + h^n P_n(x) + \cdots]$$

$$\times [P_0(x) + tP_1(x) + \cdots + t^n P_n(x) + \cdots] \, dx \qquad (9.4.32)$$

$$= \sum_{n=0}^{\infty} \sum_{m=0}^{\infty} h^n t^m \int_{-1}^{1} P_n(x) P_m(x) \, dx. \qquad (9.4.33)$$

On the other hand, if $a = (1 + h^2)/2h$, and $b = (1 + t^2)/2t$, the integral J is

$$J = \int_{-1}^{1} \frac{dx}{\sqrt{1 + h^2 - 2xh} \sqrt{1 + t^2 - 2xt}} \qquad (9.4.34)$$

$$= \frac{1}{2\sqrt{ht}} \int_{-1}^{1} \frac{dx}{\sqrt{a-x} \sqrt{b-x}} = \frac{1}{\sqrt{ht}} \int_{-1}^{1} \frac{\frac{1}{2}\left(\frac{1}{\sqrt{a-x}} + \frac{1}{\sqrt{b-x}}\right)}{\sqrt{a-x} + \sqrt{b-x}} \, dx \qquad (9.4.35)$$

$$= -\frac{1}{\sqrt{ht}} \ln(\sqrt{a-x} + \sqrt{b-x})\Big|_{-1}^{1} = \frac{1}{\sqrt{ht}} \ln\left(\frac{\sqrt{a+1} + \sqrt{b+1}}{\sqrt{a-1} + \sqrt{b-1}}\right). \qquad (9.4.36)$$

But $a + 1 = (1 + h^2 + 2h)/2h = (1+h)^2/2h$, and $a - 1 = (1-h)^2/2h$. After a little algebra,

$$J = \frac{1}{\sqrt{ht}} \ln\left(\frac{1 + \sqrt{ht}}{1 - \sqrt{ht}}\right) = \frac{2}{\sqrt{ht}} \left(\sqrt{ht} + \frac{1}{3}\sqrt{(ht)^3} + \frac{1}{5}\sqrt{(ht)^5} + \cdots\right)$$

$$\qquad (9.4.37)$$

$$= 2\left(1 + \frac{ht}{3} + \frac{h^2 t^2}{5} + \cdots + \frac{h^n t^n}{2n+1} + \cdots\right). \qquad (9.4.38)$$

---

[15] From Symons, B., 1982: Legendre polynomials and their orthogonality. *Math. Gaz.*, **66**, 152–154 with permission.

As we noted earlier, the coefficients of $h^n t^m$ in this series is $\int_{-1}^{1} P_n(x) P_m(x)\, dx$. If we match the powers of $h^n t^m$, the orthogonality condition is

$$\int_{-1}^{1} P_n(x) P_m(x)\, dx = \begin{cases} 0, & m \neq n, \\ \frac{2}{2n+1}, & m = n. \end{cases} \qquad (9.4.39)$$

With the orthogonality condition (9.4.39) we are ready to show that we can represent a function $f(x)$, which is piece-wise differentiable in the interval $(-1, 1)$, by the series:

$$f(x) = \sum_{m=0}^{\infty} A_m P_m(x), \qquad -1 \leq x \leq 1. \qquad (9.4.40)$$

To find A_m we multiply both sides of (9.4.40) by $P_n(x)$ and integrate from -1 to 1:

$$\int_{-1}^{1} f(x) P_n(x)\, dx = \sum_{m=0}^{\infty} A_m \int_{-1}^{1} P_n(x) P_m(x)\, dx. \qquad (9.4.41)$$

All of the terms on the right side vanish except for $n = m$ because of the orthogonality condition (9.4.39). Consequently, the coefficient A_n is

$$A_n \int_{-1}^{1} P_n^2(x)\, dx = \int_{-1}^{1} f(x) P_n(x)\, dx, \qquad (9.4.42)$$

or

$$A_n = \frac{2n+1}{2} \int_{-1}^{1} f(x) P_n(x)\, dx. \qquad (9.4.43)$$

In the special case when $f(x)$ and its first n derivatives are continuous throughout the interval $(-1, 1)$, we may use Rodrigues' formula to evaluate

$$\int_{-1}^{1} f(x) P_n(x)\, dx = \frac{1}{2^n n!} \int_{-1}^{1} f(x) \frac{d^n (x^2 - 1)^n}{dx^n}\, dx \qquad (9.4.44)$$

$$= \frac{(-1)^n}{2^n n!} \int_{-1}^{1} (x^2 - 1)^n f^{(n)}(x)\, dx \qquad (9.4.45)$$

by integrating by parts n times. Consequently,

$$A_n = \frac{2n+1}{2^{n+1}n!} \int_{-1}^{1} (1-x^2)^n f^{(n)}(x)\, dx. \tag{9.4.46}$$

A particularly useful result follows from (9.4.46) if $f(x)$ is a polynomial of degree k. Because all derivatives of $f(x)$ of order n vanish identically when $n > k$, $A_n = 0$ if $n > k$. Consequently, any polynomial of degree k can be expressed as a linear combination of the first $k+1$ Legendre polynomials $[P_0(x), \ldots, P_k(x)]$. Another way of viewing this result is to recognize that any polynomial of degree k is an expansion in powers of x. When we expand in Legendre polynomials we are merely regrouping these powers of x into new groups that can be identified as $P_0(x), P_1(x), P_2(x), \ldots, P_k(x)$.

• **Example 9.4.1**

Let us use Rodrigues' formula to compute $P_2(x)$. From (9.4.17) with $n = 2$,

$$P_2(x) = \frac{1}{2^2 2!}\frac{d^2}{dx^2}[(x^2-1)^2] = \frac{1}{8}\frac{d^2}{dx^2}(x^4 - 2x^2 - 1) = \frac{1}{2}(3x^2 - 1). \tag{9.4.47}$$

• **Example 9.4.2**

Let us compute $P_3(x)$ from a recurrence relation. From (9.4.24) with $n = 2$,

$$3P_3(x) - 5xP_2(x) + 2P_1(x) = 0. \tag{9.4.48}$$

But $P_2(x) = (3x^2 - 1)/2$, and $P_1(x) = x$, so that

$$3P_3(x) = 5xP_2(x) - 2P_1(x) = 5x[(3x^2-1)/2] - 2x = \tfrac{15}{2}x^3 - \tfrac{9}{2}x, \tag{9.4.49}$$

or

$$P_3(x) = (5x^3 - 3x)/2. \tag{9.4.50}$$

• **Example 9.4.3**

We want to show that

$$\int_{-1}^{1} P_n(x)\, dx = 0, \qquad n > 0. \tag{9.4.51}$$

From (9.4.30),

$$(2n+1)\int_{-1}^{1} P_n(x)\, dx = \int_{-1}^{1} [P'_{n+1}(x) - P'_{n-1}(x)]\, dx \tag{9.4.52}$$

$$= P_{n+1}(x) - P_{n-1}(x)\big|_{-1}^{1} \tag{9.4.53}$$

$$= P_{n+1}(1) - P_{n-1}(1)$$

$$- P_{n+1}(-1) + P_{n-1}(-1) = 0, \tag{9.4.54}$$

because $P_n(1) = 1$ and $P_n(-1) = (-1)^n$.

• Example 9.4.4

Let us express $f(x) = x^2$ in terms of Legendre polynomials. The results from (9.4.46) mean that we need only worry about $P_0(x)$, $P_1(x)$, and $P_2(x)$:

$$x^2 = A_0 P_0(x) + A_1 P_1(x) + A_2 P_2(x). \tag{9.4.55}$$

Substituting for the Legendre polynomials,

$$x^2 = A_0 + A_1 x + \tfrac{1}{2} A_2 (3x^2 - 1), \tag{9.4.56}$$

and

$$A_0 = \tfrac{1}{3}, \quad A_1 = 0, \quad \text{and} \quad A_2 = \tfrac{2}{3}. \tag{9.4.57}$$

• Example 9.4.5

Let us find the expansion in Legendre polynomials of the function:

$$f(x) = \begin{cases} 0, & -1 < x < 0, \\ 1, & 0 < x < 1. \end{cases} \tag{9.4.58}$$

We could have done this expansion as a Fourier series but in the solution of partial differential equations on a sphere we must make the expansion in Legendre polynomials.

In this problem, we find that

$$A_n = \frac{2n+1}{2} \int_0^1 P_n(x) \, dx. \tag{9.4.59}$$

Therefore,

$$A_0 = \tfrac{1}{2} \int_0^1 1 \, dx = \tfrac{1}{2}, \qquad A_1 = \tfrac{3}{2} \int_0^1 x \, dx = \tfrac{3}{4}, \tag{9.4.60}$$

$$A_2 = \tfrac{5}{2} \int_0^1 \tfrac{1}{2}(3x^2 - 1) \, dx = 0, \quad \text{and} \quad A_3 = \tfrac{7}{2} \int_0^1 \tfrac{1}{2}(5x^3 - 3x) \, dx = -\tfrac{7}{16}, \tag{9.4.61}$$

so that

$$f(x) = \tfrac{1}{2} P_0(x) + \tfrac{3}{4} P_1(x) - \tfrac{7}{16} P_3(x) + \tfrac{11}{32} P_5(x) + \cdots. \tag{9.4.62}$$

Figure 9.4.4 illustrates the expansion (9.4.62) where we used only the first four terms. It was created using the MATLAB script

```
clear;
x = [-1:0.01:1]; % create x points in plot
f = zeros(size(x)); % initialize function f(x)
for k = 1:length(x) % construct function f(x)
    if x(k) < 0; f(k) = 0; else f(k) = 1; end;
end
% initialize Fourier-Legendre series with zeros
flegendre = zeros(size(x));
% read in Fourier coefficients
a(1) = 1/2; a(2) = 3/4; a(3) = 0;
a(4) = -7/16; a(5) = 0; a(6) = 11/32;
clf % clear any figures
for n = 1:6
% compute Legendre polynomial
N = n-1; P = legendre(N,x);
% compute Fourier-Legendre series
flegendre = flegendre + a(n) * P(1,:);
% create plot of truncated Fourier-Legendre series
%     with n terms
if n==1 subplot(2,2,1), plot(x,flegendre,x,f,'--');
    legend('one term','f(x)'); legend boxoff; end
if n==2 subplot(2,2,2), plot(x,flegendre,x,f,'--');
    legend('two terms','f(x)'); legend boxoff; end
if n==4 subplot(2,2,3), plot(x,flegendre,x,f,'--');
    legend('four terms','f(x)'); legend boxoff;
    xlabel('x','Fontsize',20); end
if n==6 subplot(2,2,4), plot(x,flegendre,x,f,'--');
    legend('six terms','f(x)'); legend boxoff;
    xlabel('x','Fontsize',20); end
axis([-1 1 -0.5 1.5])
end
```

As we add each additional term in the orthogonal expansion, the expansion fits $f(x)$ better in the "least squares" sense of (9.3.5). The spurious oscillations arise from trying to represent a discontinuous function by four continuous, oscillatory functions. Even if we add additional terms, the spurious oscillations persist, although located nearer to the discontinuity. This is another example of *Gibbs phenomena.*[16] See §4.2.

---

[16] Weyl, H., 1910: Die Gibbs'sche Erscheinung in der Theorie der Kugelfunktionen. *Rend. Circ. Mat. Palermo*, **29**, 308–321.

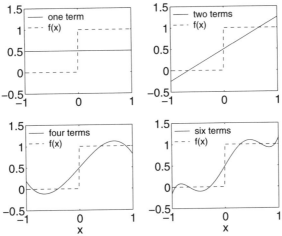

Figure 9.4.4: Representation of the function $f(x) = 1$ for $0 < x < 1$ and 0 for $-1 < x < 0$ by various partial summations of its Legendre polynomial expansion. The dashed lines denote the exact function.

• Example 9.4.6: Iterative solution of the radiative transfer equation

One of the fundamental equations of astrophysics is the integro-differential equation that describes radiative transfer (the propagation of energy by radiative, rather than conductive or convective, processes) in a gas.

Consider a gas that varies in only one spatial direction and that we divide into infinitesimally thin slabs. As radiation enters a slab, it is absorbed and scattered. If we assume that all of the radiation undergoes isotropic scattering, the radiative transfer equation is

$$\mu \frac{dI}{d\tau} = I - \tfrac{1}{2} \int_{-1}^{1} I \, d\mu, \qquad (9.4.63)$$

where I is the intensity of the radiation, τ is the optical depth (a measure of the absorbing power of the gas and related to the distance that you travel within the gas), $\mu = \cos(\theta)$, and θ is the angle at which radiation enters the slab. In this example, we show how the Fourier-Legendre expansion[17]

$$I(\tau, \mu) = \sum_{n=0}^{\infty} I_n(\tau) P_n(\mu) \qquad (9.4.64)$$

may be used to solve (9.4.63). Here $I_n(\tau)$ is the Fourier coefficient in the Fourier-Legendre expansion involving the Legendre polynomial $P_n(\mu)$.

---

[17] Chandrasekhar, S., 1944: On the radiative equilibrium of a stellar atmosphere. *Astrophys. J.*, **99**, 180–190. Published by University of Chicago Press, ©1944.

We begin by substituting (9.4.64) into (9.4.63),

$$\sum_{n=0}^{\infty} \frac{[(n+1)P_{n+1}(\mu) + nP_{n-1}(\mu)] \, dI_n}{2n+1} \frac{dI_n}{d\tau} = \sum_{n=0}^{\infty} I_n P_n(\mu) - I_0, \qquad (9.4.65)$$

where we used (9.4.24) to eliminate $\mu P_n(\mu)$. Note that only the $I_0(\tau)$ term remains after integrating because of the orthogonality condition:

$$\int_{-1}^{1} 1 \cdot P_n(\mu) \, d\mu = \int_{-1}^{1} P_0(\mu) P_n(\mu) \, d\mu = 0, \qquad (9.4.66)$$

if $n > 0$. Equating the coefficients of the various Legendre polynomials,

$$\frac{n}{2n-1} \frac{dI_{n-1}}{d\tau} + \frac{n+1}{2n+3} \frac{dI_{n+1}}{d\tau} = I_n, \qquad (9.4.67)$$

for $n = 1, 2, \ldots$ and

$$\frac{dI_1}{d\tau} = 0. \qquad (9.4.68)$$

Thus, the solution for I_1 is $I_1 = \text{constant} = 3F/4$, where F is the net integrated flux and an observable quantity.

For $n = 1$,

$$\frac{dI_0}{d\tau} + \frac{2}{5} \frac{dI_2}{d\tau} = I_1 = \frac{3F}{4}. \qquad (9.4.69)$$

Therefore,

$$I_0 + \tfrac{2}{5} I_2 = \tfrac{3}{4} F\tau + A. \qquad (9.4.70)$$

The next differential equation arises from $n = 2$ and equals

$$\frac{2}{3} \frac{dI_1}{d\tau} + \frac{3}{7} \frac{dI_3}{d\tau} = I_2. \qquad (9.4.71)$$

Because I_1 is a constant and we only retain I_0, I_1, and I_2 in the simplest approximation, we neglect $dI_3/d\tau$ and $I_2 = 0$. Thus, the simplest approximate solution is

$$I_0 = \tfrac{3}{4} F\tau + A, \quad I_1 = \tfrac{3}{4} F, \quad \text{and} \quad I_2 = 0. \qquad (9.4.72)$$

To complete our approximate solution, we must evaluate A. If we are dealing with a stellar atmosphere where we assume no external radiation incident on the star, $I(0, \mu) = 0$ for $-1 \le \mu < 0$. Therefore,

$$\int_{-1}^{1} I(\tau, \mu) P_n(\mu) \, d\mu = \sum_{m=0}^{\infty} I_m(\tau) \int_{-1}^{1} P_m(\mu) P_n(\mu) \, d\mu = \frac{2}{2n+1} I_n(\tau).$$

$$(9.4.73)$$

Taking the limit $\tau \to 0$ and using the boundary condition,

$$\frac{2}{2n+1} I_n(0) = \int_0^1 I(0,\mu) P_n(\mu)\, d\mu = \sum_{m=0}^{\infty} I_m(0) \int_0^1 P_n(\mu) P_m(\mu)\, d\mu.$$

(9.4.74)

Thus, we must satisfy, in principle, an infinite set of equations. For example, for $n = 0$, 1, and 2,

$$2I_0(0) = I_0(0) + \tfrac{1}{2} I_1(0) - \tfrac{1}{8} I_3(0) + \tfrac{1}{16} I_5(0) + \cdots,$$ (9.4.75)

$$\tfrac{2}{3} I_1(0) = \tfrac{1}{2} I_0(0) + \tfrac{1}{3} I_1(0) + \tfrac{1}{8} I_2(0) - \tfrac{1}{48} I_4(0) + \cdots,$$ (9.4.76)

and

$$\tfrac{2}{5} I_2(0) = \tfrac{1}{8} I_1(0) + \tfrac{1}{5} I_2(0) + \tfrac{1}{8} I_3(0) - \tfrac{5}{128} I_5(0) + \cdots.$$ (9.4.77)

Using $I_1(0) = 3F/4$,

$$\tfrac{1}{2} I_0(0) + \tfrac{1}{16} I_3(0) - \tfrac{1}{32} I_5(0) + \cdots = \tfrac{3}{16} F,$$ (9.4.78)

$$\tfrac{1}{2} I_0(0) + \tfrac{1}{8} I_2(0) - \tfrac{1}{48} I_4(0) + \cdots = \tfrac{1}{4} F,$$ (9.4.79)

and

$$\tfrac{2}{5} I_2(0) - \tfrac{1}{4} I_3(0) + \tfrac{5}{64} I_5(0) + \cdots = \tfrac{3}{16} F.$$ (9.4.80)

Of the two possible Equations (9.4.78)–(9.4.79), Chandrasekhar chose (9.4.79) from physical considerations. Thus, to first approximation, the solution is

$$I(\mu, \tau) = \tfrac{3}{4} F \left(\tau + \tfrac{2}{3}\right) + \tfrac{3}{4} F \mu + \cdots.$$ (9.4.81)

Better approximations can be obtained by including more terms; the interested reader is referred to the original article. In the early 1950s, Wang and Guth[18] improved the procedure for finding the successive approximations and formulating the approximate boundary conditions.

Problems

Find the first three nonvanishing coefficients in the Legendre polynomial expansion for the following functions:

1. $f(x) = \begin{cases} 0, & -1 < x < 0 \\ x, & 0 < x < 1 \end{cases}$

2. $f(x) = \begin{cases} 1/(2\epsilon), & |x| < \epsilon \\ 0, & \epsilon < |x| < 1 \\ x, & 0 < x < 1 \end{cases}$

3. $f(x) = |x|, \qquad |x| < 1$

4. $f(x) = x^3, \qquad |x| < 1$

5. $f(x) = \begin{cases} -1, & -1 < x < 0 \\ 1, & 0 < x < 1 \end{cases}$

6. $f(x) = \begin{cases} -1, & -1 < x < 0 \\ x, & 0 < x < 1 \end{cases}$

---

[18] Wang, M. C., and E. Guth, 1951: On the theory of multiple scattering, particularly of charged particles. *Phys. Rev.*, Ser. 2, **84**, 1092–1111.

Then use MATLAB to illustrate various partial sums of the Fourier-Legendre series.

7. Use Rodrigues' formula to show that $P_4(x) = \frac{1}{8}(35x^4 - 30x^2 + 3)$.

8. Given $P_5(x) = \frac{63}{8}x^5 - \frac{70}{8}x^3 + \frac{15}{8}x$ and $P_4(x)$ from Problem 7, use the recurrence formula for $P_{n+1}(x)$ to find $P_6(x)$.

9. Show that (a) $P_n(1) = 1$, (b) $P_n(-1) = (-1)^n$, (c) $P_{2n+1}(0) = 0$, and (d) $P_{2n}(0) = (-1)^n (2n)!/(2^{2n}n!n!)$.

10. Prove that

$$\int_x^1 P_n(t)\,dt = \frac{1}{2n+1}[P_{n-1}(x) - P_{n+1}(x)], \qquad n > 0.$$

11. Given[19]

$$P_n[\cos(\theta)] = \frac{2}{\pi} \int_0^\theta \frac{\cos[(n+\frac{1}{2})x]}{\sqrt{2[\cos(x) - \cos(\theta)]}}\,dx = \frac{2}{\pi} \int_\theta^\pi \frac{\sin[(n+\frac{1}{2})x]}{\sqrt{2[\cos(\theta) - \cos(x)]}}\,dx,$$

show that the following generalized Fourier series hold:

$$\frac{H(\theta - t)}{\sqrt{2\cos(t) - 2\cos(\theta)}} = \sum_{n=0}^\infty P_n[\cos(\theta)] \cos\left[\left(n + \tfrac{1}{2}\right)t\right], \quad 0 \le t < \theta \le \pi,$$

if we use the eigenfunction $y_n(x) = \cos\left[\left(n + \frac{1}{2}\right)x\right]$, $0 < x < \pi$, $r(x) = 1$ and $H(\ )$ is Heaviside's step function, and

$$\frac{H(t - \theta)}{\sqrt{2\cos(\theta) - 2\cos(t)}} = \sum_{n=0}^\infty P_n[\cos(\theta)] \sin\left[\left(n + \tfrac{1}{2}\right)t\right], \quad 0 \le \theta < t \le \pi,$$

if we use the eigenfunction $y_n(x) = \sin\left[\left(n + \frac{1}{2}\right)x\right]$, $0 < x < \pi$, $r(x) = 1$ and $H(\ )$ is Heaviside's step function.

12. The series given in Problem 11 are also expansions in Legendre polynomials. In that light, show that

$$\int_0^t \frac{P_n[\cos(\theta)] \sin(\theta)}{\sqrt{2\cos(\theta) - 2\cos(t)}}\,d\theta = \frac{\sin\left[\left(n + \frac{1}{2}\right)t\right]}{n + \frac{1}{2}},$$

---

[19] Hobson, E. W., 1965: *The Theory of Spherical and Ellipsoidal Harmonics.* Chelsea Publishing Co., pp. 26–27.

and
$$\int_t^\pi \frac{P_n[\cos(\theta)]\,\sin(\theta)}{\sqrt{2\cos(t) - 2\cos(\theta)}}\,d\theta = \frac{\cos\left[\left(n+\frac{1}{2}\right)t\right]}{n+\frac{1}{2}},$$

where $0 < t < \pi$.

13. (a) Use the generating function (9.4.18) to show that

$$\frac{1}{\sqrt{1 - 2tx + t^2}} = \sum_{n=0}^\infty t^{-n-1} P_n(x), \qquad |x| < 1,\ 1 < |t|.$$

(b) Use the results from part (a) to show that

$$\frac{1}{\sqrt{\cosh(\mu) - x}} = \sqrt{2}\sum_{n=0}^\infty e^{-(n+\frac{1}{2})|\mu|} P_n(x), \qquad |x| < 1.$$

Hint:

$$\frac{1}{\sqrt{\cosh(\mu) - x}} = \frac{\sqrt{2}}{\sqrt{e^{|\mu|} - 2x + e^{-|\mu|}}}.$$

14. The generating function (9.4.18) actually holds[20] for $|h| \le 1$ if $|x| < 1$. Using this relationship, show that

$$\sum_{n=0}^\infty P_n(x) = \frac{1}{\sqrt{2(1 - x)}}, \qquad |x| < 1,$$

and
$$\sum_{n=0}^\infty \frac{P_n(x)}{n + 1} = \ln\left[\frac{1 + \sqrt{(1-x)/2}}{\sqrt{(1-x)/2}}\right], \qquad |x| < 1.$$

Use these relationships to show that

$$\sum_{n=1}^\infty \frac{2n+1}{n+1} P_n(x) = 2\sum_{n=1}^\infty P_n(x) - \sum_{n=1}^\infty \frac{P_n(x)}{n+1}$$

$$= \frac{1}{\sqrt{(1-x)/2}} - \ln\left[\frac{1 + \sqrt{(1-x)/2}}{\sqrt{(1-x)/2}}\right] - 1,$$

if $|x| < 1$.

---

[20] *Ibid.*, p. 28.

Figure 9.5.1: It was Friedrich William Bessel's (1784–1846) apprenticeship to the famous mercantile firm of Kulenkamp that ignited his interest in mathematics and astronomy. As the founder of the German school of practical astronomy, Bessel discovered his functions while studying the problem of planetary motion. Bessel functions arose as coefficients in one of the series that described the gravitational interaction between the sun and two other planets in elliptic orbit. (Portrait courtesy of Photo AKG, London.)

9.5 ANOTHER SINGULAR STURM-LIOUVILLE PROBLEM: BESSEL'S EQUATION

In the previous section we discussed the solutions to Legendre's equation, especially with regard to their use in orthogonal expansions. In the section we consider another classic equation, Bessel's equation[21]

$$x^2 y'' + x y' + (\mu^2 x^2 - n^2) y = 0, \tag{9.5.1}$$

or

$$\frac{d}{dx}\left(x \frac{dy}{dx}\right) + \left(\mu^2 x - \frac{n^2}{x}\right) y = 0. \tag{9.5.2}$$

---

[21] Bessel, F. W., 1824: Untersuchung des Teils der planetarischen Störungen, welcher aus der Bewegung der Sonne entsteht. *Abh. d. K. Akad. Wiss. Berlin*, 1–52. See Dutka, J., 1995: On the early history of Bessel functions. *Arch. Hist. Exact Sci.*, **49**, 105–134. The classic reference on Bessel functions is Watson, G. N., 1966: *A Treatise on the Theory of Bessel Functions*. Cambridge University Press, 804 pp.

Once again, our ultimate goal is the use of its solutions in orthogonal expansions. These orthogonal expansions, in turn, are used in the solution of partial differential equations in cylindrical coordinates.

A quick check of Bessel's equation shows that it conforms to the canonical form of the Sturm-Liouville problem: $p(x) = x$, $q(x) = -n^2/x$, $r(x) = x$, and $\lambda = \mu^2$. Restricting our attention to the interval $[0, L]$, the Sturm-Liouville problem involving (9.5.2) is singular because $p(0) = 0$. From (9.4.1) in the previous section, the eigenfunctions to a singular Sturm-Liouville problem will still be orthogonal over the interval $[0, L]$ if (1) $y(x)$ is finite and $xy'(x)$ is zero at $x = 0$, and (2) $y(x)$ satisfies the homogeneous boundary condition (9.1.2) at $x = L$. Consequently, we only seek solutions that satisfy these conditions.

We cannot write down the solution to Bessel's equation in a simple closed form; as in the case with Legendre's equation, we must find the solution by power series. Because we intend to make the expansion about $x = 0$ and this point is a regular singular point, we must use the method of Frobenius, where n is an integer.[22] Moreover, because the quantity n^2 appears in (9.5.2), we may take n to be nonnegative without any loss of generality.

To simplify matters, we first find the solution when $\mu = 1$; the solution for $\mu \neq 1$ follows by substituting μx for x. Consequently, we seek solutions of the form

$$y(x) = \sum_{k=0}^{\infty} B_k x^{2k+s}, \tag{9.5.3}$$

$$y'(x) = \sum_{k=0}^{\infty} (2k+s) B_k x^{2k+s-1}, \tag{9.5.4}$$

and

$$y''(x) = \sum_{k=0}^{\infty} (2k+s)(2k+s-1) B_k x^{2k+s-2}, \tag{9.5.5}$$

where we formally assume that we can interchange the order of differentiation and summation. The substitution of (9.5.3)–(9.5.5) into (9.5.1) with $\mu = 1$ yields

$$\sum_{k=0}^{\infty} (2k+s)(2k+s-1) B_k x^{2k+s} + \sum_{k=0}^{\infty} (2k+s) B_k x^{2k+s}$$

$$+ \sum_{k=0}^{\infty} B_k x^{2k+s+2} - n^2 \sum_{k=0}^{\infty} B_k x^{2k+s} = 0, \tag{9.5.6}$$

or

$$\sum_{k=0}^{\infty} [(2k+s)^2 - n^2] B_k x^{2k} + \sum_{k=0}^{\infty} B_k x^{2k+2} = 0. \tag{9.5.7}$$

---

[22] This case is much simpler than for arbitrary n. See Hildebrand, F. B., 1962: *Advanced Calculus for Applications*. Prentice-Hall, §4.8.

If we explicitly separate the $k = 0$ term from the other terms in the first summation in (9.5.7),

$$(s^2 - n^2)B_0 + \sum_{m=1}^{\infty} [(2m+s)^2 - n^2]B_m x^{2m} + \sum_{k=0}^{\infty} B_k x^{2k+2} = 0. \qquad (9.5.8)$$

We now change the dummy integer in the first summation of (9.5.8) by letting $m = k + 1$ so that

$$(s^2 - n^2)B_0 + \sum_{k=0}^{\infty} \{[(2k+s+2)^2 - n^2]B_{k+1} + B_k\}x^{2k+2} = 0. \qquad (9.5.9)$$

Because (9.5.9) must be true for all x, each power of x must vanish identically. This yields $s = \pm n$, and

$$[(2k+s+2)^2 - n^2]B_{k+1} + B_k = 0. \qquad (9.5.10)$$

Since the difference of the larger indicial root from the lower root equals the integer $2n$, we are only guaranteed a power series solution of the form (9.5.3) for $s = n$. If we use this indicial root and the recurrence formula (9.5.10), this solution, known as the Bessel function of the first kind of order n and denoted by $J_n(x)$, is

$$J_n(x) = \sum_{k=0}^{\infty} \frac{(-1)^k (x/2)^{n+2k}}{k!(n+k)!}. \qquad (9.5.11)$$

To find the second general solution to Bessel's equation, the one corresponding to $s = -n$, the most economical method[23] is to express it in terms of partial derivatives of $J_n(x)$ with respect to its order n:

$$Y_n(x) = \left[\frac{\partial J_\nu(x)}{\partial \nu} - (-1)^n \frac{\partial J_{-\nu}(x)}{\partial \nu} \right]_{\nu=n}. \qquad (9.5.12)$$

Upon substituting the power series representation (9.5.11) into (9.5.12),

$$Y_n(x) = \frac{2}{\pi} J_n(x) \ln(x/2) - \frac{1}{\pi} \sum_{k=0}^{n-1} \frac{(n-k-1)!}{k!} \left(\frac{x}{2}\right)^{2k-n}$$

$$- \frac{1}{\pi} \sum_{k=0}^{\infty} \frac{(-1)^k (x/2)^{n+2k}}{k!(n+k)!} [\psi(k+1) + \psi(k+n+1)], \qquad (9.5.13)$$

[23] See Watson, G. N., 1966: *A Treatise on the Theory of Bessel Functions*. Cambridge University Press, §3.5 for the derivation.

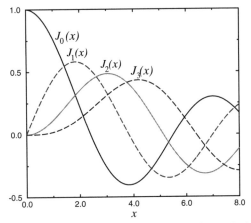

Figure 9.5.2: The first four Bessel functions of the first kind over $0 \leq x \leq 8$.

where

$$\psi(m+1) = -\gamma + 1 + \frac{1}{2} + \cdots + \frac{1}{m}, \qquad (9.5.14)$$

$\psi(1) = -\gamma$, and γ is Euler's constant (0.5772157). In the case $n = 0$, the first sum in $(9.5.13)$ disappears. This function $Y_n(x)$ is Neumann's Bessel function of the second kind of order n. Consequently, the general solution to $(9.5.1)$ is

$$y(x) = A J_n(\mu x) + B Y_n(\mu x). \qquad (9.5.15)$$

Figure 9.5.2 illustrates the functions $J_0(x)$, $J_1(x)$, $J_2(x)$, and $J_3(x)$ while Figure 9.5.3 gives $Y_0(x)$, $Y_1(x)$, $Y_2(x)$, and $Y_3(x)$.

An equation which is very similar to $(9.5.1)$ is

$$x^2 \frac{d^2 y}{dx^2} + x \frac{dy}{dx} - (n^2 + x^2) y = 0. \qquad (9.5.16)$$

It arises in the solution of partial differential equations in cylindrical coordinates. If we substitute $ix = t$ (where $i = \sqrt{-1}$) into $(9.5.16)$, it becomes Bessel's equation:

$$t^2 \frac{d^2 y}{dt^2} + t \frac{dy}{dt} + (t^2 - n^2) y = 0. \qquad (9.5.17)$$

Consequently, we may immediately write the solution to $(9.5.16)$ as

$$y(x) = c_1 J_n(ix) + c_2 Y_n(ix), \qquad (9.5.18)$$

if n is an integer. Traditionally the solution to $(9.5.16)$ has been written

$$y(x) = c_1 I_n(x) + c_2 K_n(x) \qquad (9.5.19)$$

rather than in terms of $J_n(ix)$ and $Y_n(ix)$, where

$$I_n(x) = \sum_{k=0}^{\infty} \frac{(x/2)^{2k+n}}{k!(k+n)!}, \qquad (9.5.20)$$

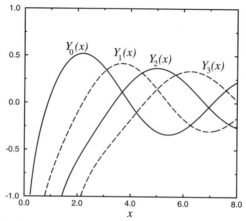

Figure 9.5.3: The first four Bessel functions of the second kind over $0 \le x \le 8$.

and

$$K_n(x) = \frac{\pi}{2} i^{n+1} \left[J_n(ix) + i Y_n(ix) \right]. \tag{9.5.21}$$

The function $I_n(x)$ is the modified Bessel function of the first kind, of order n, while $K_n(x)$ is the modified Bessel function of the second kind, of order n. Figure 9.5.4 illustrates $I_0(x)$, $I_1(x)$, $I_2(x)$, and $I_3(x)$ while in Figure 9.5.5 $K_0(x)$, $K_1(x)$, $K_2(x)$, and $K_3(x)$ are graphed. Note that $K_n(x)$ has no real zeros while $I_n(x)$ equals zero only at $x = 0$ for $n \ge 1$.

As our derivation suggests, modified Bessel functions are related to ordinary Bessel functions via complex variables. In particular, $J_n(iz) = i^n I_n(z)$, and $I_n(iz) = i^n J_n(z)$ for z complex.

Although we found solutions to Bessel's equation (9.5.1), as well as (9.5. 16), can we use any of them in an eigenfunction expansion? From Figures 9.5.2–9.5.5 we see that $J_n(x)$ and $I_n(x)$ remain finite at $x = 0$ while $Y_n(x)$ and $K_n(x)$ do not. Furthermore, the products $x J_n'(x)$ and $x I_n'(x)$ tend to zero at $x = 0$. Thus, both $J_n(x)$ and $I_n(x)$ satisfy the first requirement of an eigenfunction for a Fourier-Bessel expansion.

What about the second condition that the eigenfunction must satisfy the homogeneous boundary condition (9.1.2) at $x = L$? From Figure 9.5.4 we see that $I_n(x)$ can never satisfy this condition while from Figure 9.5.2 $J_n(x)$ can. For that reason, we discard $I_n(x)$ from further consideration and continue our analysis only with $J_n(x)$.

Before we can derive the expressions for a Fourier-Bessel expansion, we need to find how $J_n(x)$ is related to $J_{n+1}(x)$ and $J_{n-1}(x)$. Assuming that n is a positive integer, we multiply the series (9.5.11) by x^n and then differentiate with respect to x. This gives

$$\frac{d}{dx} \left[x^n J_n(x) \right] = \sum_{k=0}^{\infty} \frac{(-1)^k (2n + 2k) x^{2n+2k-1}}{2^{n+2k} k! (n + k)!} \tag{9.5.22}$$

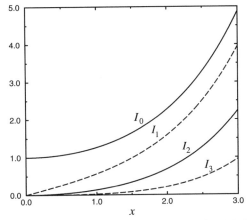

Figure 9.5.4: The first four modified Bessel functions of the first kind over $0 \leq x \leq 3$.

$$= x^n \sum_{k=0}^{\infty} \frac{(-1)^k (x/2)^{n-1+2k}}{k!(n-1+k)!} \tag{9.5.23}$$

$$= x^n J_{n-1}(x) \tag{9.5.24}$$

or

$$\frac{d}{dx}[x^n J_n(x)] = x^n J_{n-1}(x) \tag{9.5.25}$$

for $n = 1, 2, 3, \ldots$ Similarly, multiplying (9.5.11) by x^{-n}, we find that

$$\frac{d}{dx}[x^{-n} J_n(x)] = -x^{-n} J_{n+1}(x) \tag{9.5.26}$$

for $n = 0, 1, 2, 3, \ldots$ If we now carry out the differentiation on (9.5.25) and (9.5.26) and divide by the factors $x^{\pm n}$, we have that

$$J_n'(x) + \frac{n}{x} J_n(x) = J_{n-1}(x), \tag{9.5.27}$$

and

$$J_n'(x) - \frac{n}{x} J_n(x) = -J_{n+1}(x). \tag{9.5.28}$$

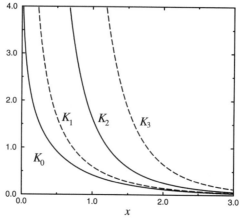

Figure 9.5.5: The first four modified Bessel functions of the second kind over $0 \le x \le 3$.

Equations (9.3.27)–(9.3.28) immediately yield the *recurrence relationships*

$$J_{n-1}(x) + J_{n+1}(x) = \frac{2n}{x} J_n(x) \qquad (9.5.29)$$

and

$$J_{n-1}(x) - J_{n+1}(x) = 2J_n'(x) \qquad (9.5.30)$$

for $n = 1, 2, 3, \ldots$. For $n = 0$, we replace (9.5.30) by $J_0'(x) = -J_1(x)$.

Let us now construct a Fourier-Bessel series. The exact form of the expansion depends upon the boundary condition at $x = L$. There are three possible cases. One of them is $y(L) = 0$ and results in the condition that $J_n(\mu_k L) = 0$. Another condition is $y'(L) = 0$ and gives $J_n'(\mu_k L) = 0$. Finally, if $hy(L) + y'(L) = 0$, then $hJ_n(\mu_k L) + \mu_k J_n'(\mu_k L) = 0$. In all of these cases, the eigenfunction expansion is the same, namely

$$f(x) = \sum_{k=1}^{\infty} A_k J_n(\mu_k x), \qquad (9.5.31)$$

where μ_k is the kth positive solution of either $J_n(\mu_k L) = 0$, $J_n'(\mu_k L) = 0$, or $hJ_n(\mu_k L) + \mu_k J_n'(\mu_k L) = 0$.

We now need a mechanism for computing A_k. We begin by multiplying (9.5.31) by $xJ_n(\mu_m x)\,dx$ and integrate from 0 to L. This yields

$$\sum_{k=1}^{\infty} A_k \int_0^L xJ_n(\mu_k x)J(\mu_m x)\,dx = \int_0^L xf(x)J_n(\mu_m x)\,dx. \qquad (9.5.32)$$

From the general orthogonality condition (9.2.1),

$$\int_0^L xJ_n(\mu_k x)J_n(\mu_m x)\,dx = 0, \qquad (9.5.33)$$

if $k \neq m$. Equation (9.5.32) then simplifies to

$$A_m \int_0^L xJ_n^2(\mu_m x)\,dx = \int_0^L xf(x)J_n(\mu_m x)\,dx, \qquad (9.5.34)$$

or

$$A_k = \frac{1}{C_k} \int_0^L xf(x)J_n(\mu_k x)\,dx, \qquad (9.5.35)$$

where

$$C_k = \int_0^L xJ_n^2(\mu_k x)\,dx, \qquad (9.5.36)$$

and k has replaced m in (9.5.34).

The factor C_k depends upon the nature of the boundary conditions at $x = L$. In all cases we start from Bessel's equation

$$[xJ_n'(\mu_k x)]' + \left(\mu_k^2 x - \frac{n^2}{x}\right)J_n(\mu_k x) = 0. \qquad (9.5.37)$$

If we multiply both sides of (9.5.37) by $2xJ_n'(\mu_k x)$, the resulting equation is

$$(\mu_k^2 x^2 - n^2)\left[J_n^2(\mu_k x)\right]' = -\frac{d}{dx}[xJ_n'(\mu_k x)]^2. \qquad (9.5.38)$$

An integration of (9.5.38) from 0 to L, followed by the subsequent use of integration by parts, results in

$$(\mu_k^2 x^2 - n^2)J_n^2(\mu_k x)\Big|_0^L - 2\mu_k^2 \int_0^L xJ_n^2(\mu_k x)\,dx = -\left[xJ_n'(\mu_k x)\right]^2\Big|_0^L. \qquad (9.5.39)$$

Because $J_n(0) = 0$ for $n > 0$, $J_0(0) = 1$ and $x J_n'(x) = 0$ at $x = 0$, the contribution from the lower limits vanishes. Thus,

$$C_k = \int_0^L x J_n^2(\mu_k x)\, dx \tag{9.5.40}$$

$$= \frac{1}{2\mu_k^2}\left[(\mu_k^2 L^2 - n^2) J_n^2(\mu_k L) + L^2 J_n'^2(\mu_k L)\right]. \tag{9.5.41}$$

Because

$$J_n'(\mu_k x) = \frac{n}{x} J_n(\mu_k x) - \mu_k J_{n+1}(\mu_k x) \tag{9.5.42}$$

from (9.5.28), C_k becomes

$$C_k = \tfrac{1}{2} L^2 J_{n+1}^2(\mu_k L), \tag{9.5.43}$$

if $J_n(\mu_k L) = 0$. Otherwise, if $J_n'(\mu_k L) = 0$, then

$$C_k = \frac{\mu_k^2 L^2 - n^2}{2\mu_k^2} J_n^2(\mu_k L). \tag{9.5.44}$$

Finally,

$$C_k = \frac{\mu_k^2 L^2 - n^2 + h^2 L^2}{2\mu_k^2} J_n^2(\mu_k L), \tag{9.5.45}$$

if $\mu_k J_n'(\mu_k L) = -h J_n(\mu_k L)$.

All of the preceding results must be slightly modified when $n = 0$ and the boundary condition is $J_0'(\mu_k L) = 0$ or $\mu_k J_1(\mu_k L) = 0$. This modification results from the additional eigenvalue $\mu_0 = 0$ being present and we must add the extra term A_0 to the expansion. For this case the series reads

$$f(x) = A_0 + \sum_{k=1}^{\infty} A_k J_0(\mu_k x), \tag{9.5.46}$$

where the equation for finding A_0 is

$$A_0 = \frac{2}{L^2} \int_0^L f(x)\, x\, dx, \tag{9.5.47}$$

and (9.5.35) and (9.5.44) with $n = 0$ give the remaining coefficients.

• **Example 9.5.1**

Starting with Bessel's equation, we show that the solution to

$$y'' + \frac{1 - 2a}{x}y' + \left(b^2 c^2 x^{2c-2} + \frac{a^2 - n^2 c^2}{x^2}\right)y = 0 \qquad (9.5.48)$$

is

$$y(x) = Ax^a J_n(bx^c) + Bx^a Y_n(bx^c), \qquad (9.5.49)$$

provided that $bx^c > 0$ so that $Y_n(bx^c)$ exists.

The general solution to

$$\xi^2 \frac{d^2\eta}{d\xi^2} + \xi \frac{d\eta}{d\xi} + (\xi^2 - n^2)\eta = 0 \qquad (9.5.50)$$

is

$$\eta = AJ_n(\xi) + BY_n(\xi). \qquad (9.5.51)$$

If we now let $\eta = y(x)/x^a$ and $\xi = bx^c$, then

$$\frac{d}{d\xi} = \frac{dx}{d\xi}\frac{d}{dx} = \frac{x^{1-c}}{bc}\frac{d}{dx}, \qquad (9.5.52)$$

$$\frac{d^2}{d\xi^2} = \frac{x^{2-2c}}{b^2 c^2}\frac{d^2}{dx^2} - \frac{(c-1)x^{1-2c}}{b^2 c^2}\frac{d}{dx}, \qquad (9.5.53)$$

$$\frac{d}{dx}\left(\frac{y}{x^a}\right) = \frac{1}{x^a}\frac{dy}{dx} - \frac{a}{x^{a+1}}y, \qquad (9.5.54)$$

and

$$\frac{d^2}{dx^2}\left(\frac{y}{x^a}\right) = \frac{1}{x^a}\frac{d^2y}{dx^2} - \frac{2a}{x^{a+1}}\frac{dy}{dx} + \frac{a(1+a)}{x^{a+2}}y. \qquad (9.5.55)$$

Substituting (9.5.52)–(9.5.55) into (9.5.50) and simplifying yields the desired result.

• **Example 9.5.2**

We show that

$$x^2 J_n''(x) = (n^2 - n - x^2)J_n(x) + xJ_{n+1}(x). \qquad (9.5.56)$$

From (9.5.28),

$$J_n'(x) = \frac{n}{x}J_n(x) - J_{n+1}(x), \qquad (9.5.57)$$

$$J_n''(x) = -\frac{n}{x^2}J_n(x) + \frac{n}{x}J_n'(x) - J_{n+1}'(x), \qquad (9.5.58)$$

and

$$J_n''(x) = -\frac{n}{x^2} J_n(x) + \frac{n}{x} \left[\frac{n}{x} J_n(x) - J_{n+1}(x) \right]$$
$$- \left[J_n(x) - \frac{n+1}{x} J_{n+1}(x) \right] \qquad (9.5.59)$$

after using (9.5.27) and (9.5.28). Simplifying,

$$J_n''(x) = \left(\frac{n^2 - n}{x^2} - 1 \right) J_n(x) + \frac{J_{n+1}(x)}{x}. \qquad (9.5.60)$$

After multiplying (9.5.60) by x^2, we obtain (9.5.56).

• **Example 9.5.3**

Show that

$$\int_0^a x^5 J_2(x)\, dx = a^5 J_3(a) - 2a^4 J_4(a). \qquad (9.5.61)$$

We begin by integrating (9.5.61) by parts. If $u = x^2$, and $dv = x^3 J_2(x)\, dx$, then

$$\int_0^a x^5 J_2(x)\, dx = x^5 J_3(x)\big|_0^a - 2 \int_0^a x^4 J_3(x)\, dx, \qquad (9.5.62)$$

because $d[x^3 J_3(x)]/dx = x^2 J_2(x)$ by (9.5.25). Finally,

$$\int_0^a x^5 J_2(x)\, dx = a^5 J_3(a) - 2x^4 J_4(x)\big|_0^a = a^5 J_3(a) - 2a^4 J_4(a), \qquad (9.5.63)$$

since $x^4 J_3(x) = d[x^4 J_4(x)]/dx$ by (9.5.25).

• **Example 9.5.4**

Let us expand $f(x) = x$, $0 < x < 1$, in the series

$$f(x) = \sum_{k=1}^{\infty} A_k J_1(\mu_k x), \qquad (9.5.64)$$

where μ_k denotes the kth zero of $J_1(\mu)$. From (9.5.35) and (9.5.43),

$$A_k = \frac{2}{J_2^2(\mu_k)} \int_0^1 x^2 J_1(\mu_k x)\, dx. \qquad (9.5.65)$$

However, from (9.5.25),

$$\frac{d}{dx}\left[x^2 J_2(x) \right] = x^2 J_1(x), \qquad (9.5.66)$$

if $n = 2$. Therefore, (9.5.65) becomes

$$A_k = \frac{2x^2 J_2(x)}{\mu_k^3 J_2^2(\mu_k)}\bigg|_0^{\mu_k} = \frac{2}{\mu_k J_2(\mu_k)}, \qquad (9.5.67)$$

and the resulting expansion is

$$x = 2 \sum_{k=1}^{\infty} \frac{J_1(\mu_k x)}{\mu_k J_2(\mu_k)}, \qquad 0 \leq x < 1. \qquad (9.5.68)$$

Figure 9.5.6 shows the Fourier-Bessel expansion of $f(x) = x$ in truncated form when we only include one, two, three, and four terms. It was created using the MATLAB script

```
clear;
x = [0:0.01:1]; % create x points in plot
f = x; % construct function f(x)
% initialize Fourier-Bessel series
fbessel = zeros(size(x));
% read in the first four zeros of J_1(mu) = 0
mu(1) =  3.83171; mu(2) =  7.01559;
mu(3) = 10.17347; mu(4) = 13.32369;
clf % clear any figures
for n = 1:4
% Fourier coefficient
factor = 2 / (mu(n) * besselj(2,mu(n)));
% compute Fourier-Bessel series
fbessel = fbessel + factor * besselj(1,mu(n)*x);
% create plot of truncated Fourier-Bessel series
%      with n terms
subplot(2,2,n), plot(x,fbessel,x,f,'--')
axis([0 1 -0.25 1.25])
if n == 1 legend('1 term','f(x)'); legend boxoff;
else legend([num2str(n) ' terms'],'f(x)'); legend boxoff;
end
if n > 2 xlabel('x','Fontsize',20); end
end
```

• Example 9.5.5

Let us expand the function $f(x) = x^2$, $0 < x < 1$, in the series

$$f(x) = \sum_{k=1}^{\infty} A_k J_0(\mu_k x), \qquad (9.5.69)$$

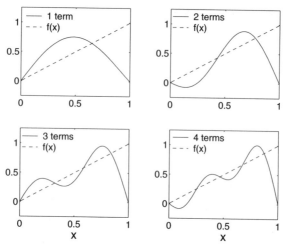

Figure 9.5.6: The Fourier-Bessel series representation (9.5.68) for $f(x) = x$, $0 < x < 1$, when we truncate the series so that it includes only the first, first two, first three, and first four terms.

where μ_k denotes the kth positive zero of $J_0(\mu)$. From (9.5.35) and (9.5.43),

$$A_k = \frac{2}{J_1^2(\mu_k)} \int_0^1 x^3 J_0(\mu_k x)\,dx. \qquad (9.5.70)$$

If we let $t = \mu_k x$, the integration (9.5.70) becomes

$$A_k = \frac{2}{\mu_k^4 J_1^2(\mu_k)} \int_0^{\mu_k} t^3 J_0(t)\,dt. \qquad (9.5.71)$$

We now let $u = t^2$ and $dv = t J_0(t)\,dt$ so that integration by parts results in

$$A_k = \frac{2}{\mu_k^4 J_1^2(\mu_k)} \left[t^3 J_1(t) \big|_0^{\mu_k} - 2 \int_0^{\mu_k} t^2 J_1(t)\,dt \right] \qquad (9.5.72)$$

$$= \frac{2}{\mu_k^4 J_1^2(\mu_k)} \left[\mu_k^3 J_1(\mu_k) - 2 \int_0^{\mu_k} t^2 J_1(t)\,dt \right], \qquad (9.5.73)$$

because $v = t J_1(t)$ from (9.5.25). If we integrate by parts once more, we find that

$$A_k = \frac{2}{\mu_k^4 J_1^2(\mu_k)} \left[\mu_k^3 J_1(\mu_k) - 2\mu_k^2 J_2(\mu_k) \right] \qquad (9.5.74)$$

$$= \frac{2}{J_1^2(\mu_k)} \left[\frac{J_1(\mu_k)}{\mu_k} - \frac{2 J_2(\mu_k)}{\mu_k^2} \right]. \qquad (9.5.75)$$

However, from (9.5.29) with $n = 1$,

$$J_1(\mu_k) = \tfrac{1}{2}\mu_k \left[J_2(\mu_k) + J_0(\mu_k) \right], \qquad (9.5.76)$$

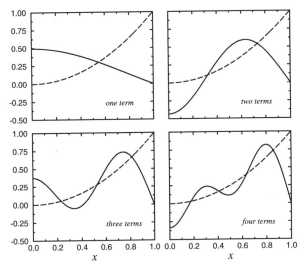

Figure 9.5.7: The Fourier-Bessel series representation (9.5.79) for $f(x) = x^2$, $0 < x < 1$, when we truncate the series so that it includes only the first, first two, first three, and first four terms.

or

$$J_2(\mu_k) = \frac{2J_1(\mu_k)}{\mu_k}, \tag{9.5.77}$$

because $J_0(\mu_k) = 0$. Therefore,

$$A_k = \frac{2(\mu_k^2 - 4)J_1(\mu_k)}{\mu_k^3 J_1^2(\mu_k)}, \tag{9.5.78}$$

and

$$x^2 = 2\sum_{k=1}^{\infty} \frac{(\mu_k^2 - 4)J_0(\mu_k x)}{\mu_k^3 J_1(\mu_k)}, \qquad 0 < x < 1. \tag{9.5.79}$$

Figure 9.5.7 shows the representation of x^2 by the Fourier-Bessel series (9.5.79) when we truncate it so that it includes only one, two, three, or four terms. As we add each additional term in the orthogonal expansion, the expansion fits $f(x)$ better in the "least squares" sense of (9.3.5).

Problems

1. Show from the series solution that

$$\frac{d}{dx}\left[J_0(kx)\right] = -kJ_1(kx).$$

From the recurrence formulas, show these following relations:

2. $2J_0'''(x) = J_2(x) - J_0(x)$

3. $J_2(x) = J_0''(x) - J_0'(x)/x$

4.
$$J_0'''(x) = \frac{J_0(x)}{x} + \left(\frac{2}{x^2} - 1\right) J_0'(x)$$

5.
$$\frac{J_2(x)}{J_1(x)} = \frac{1}{x} - \frac{J_0''(x)}{J_0'(x)} = \frac{2}{x} - \frac{J_0(x)}{J_1(x)} = \frac{2}{x} + \frac{J_0(x)}{J_0'(x)}$$

6.
$$J_4(x) = \left(\frac{48}{x^3} - \frac{8}{x}\right) J_1(x) - \left(\frac{24}{x^2} - 1\right) J_0(x)$$

7.
$$J_{n+2}(x) = \left[2n + 1 - \frac{2n(n^2 - 1)}{x^2}\right] J_n(x) + 2(n+1)J_n''(x)$$

8.
$$J_3(x) = \left(\frac{8}{x^2} - 1\right) J_1(x) - \frac{4}{x}J_0(x)$$

9.
$$4J_n''(x) = J_{n-2}(x) - 2J_n(x) + J_{n+2}(x)$$

10. Show that the maximum and minimum values of $J_n(x)$ occur when

$$x = \frac{nJ_n(x)}{J_{n+1}(x)}, \quad x = \frac{nJ_n(x)}{J_{n-1}(x)}, \quad \text{and} \quad J_{n-1}(x) = J_{n+1}(x).$$

Show that

11.
$$\frac{d}{dx}\left[x^2 J_3(2x)\right] = -xJ_3(2x) + 2x^2 J_2(2x)$$

12.
$$\frac{d}{dx}\left[xJ_0(x^2)\right] = J_0(x^2) - 2x^2 J_1(x^2)$$

13.
$$\int x^3 J_2(3x)\, dx = \tfrac{1}{3}x^3 J_3(3x) + C$$

14.
$$\int x^{-2} J_3(2x)\, dx = -\tfrac{1}{2}x^{-2} J_2(2x) + C$$

15.
$$\int x \ln(x) J_0(x)\, dx = J_0(x) + x \ln(x) J_1(x) + C$$

16.
$$\int_0^a x J_0(kx)\, dx = \frac{a^2 J_1(ka)}{ka}$$

17.
$$\int_0^1 x(1 - x^2) J_0(kx)\, dx = \frac{4}{k^3} J_1(k) - \frac{2}{k^2} J_0(k)$$

18.
$$\int_0^1 x^3 J_0(kx)\, dx = \frac{k^2 - 4}{k^3} J_1(k) + \frac{2}{k^2} J_0(k)$$

19. Show that
$$1 = 2 \sum_{k=1}^{\infty} \frac{J_0(\mu_k x)}{\mu_k J_1(\mu_k)}, \qquad 0 \le x < 1,$$

where μ_k is the kth positive root of $J_0(\mu) = 0$. Then use MATLAB to illustrate various partial sums of the Fourier-Bessel series.

20. Show that
$$\frac{1 - x^2}{8} = \sum_{k=1}^{\infty} \frac{J_0(\mu_k x)}{\mu_k^3 J_1(\mu_k)}, \qquad 0 \le x \le 1,$$

where μ_k is the kth positive root of $J_0(\mu) = 0$. Then use MATLAB to illustrate various partial sums of the Fourier-Bessel series.

21. Show that
$$4x - x^3 = -16 \sum_{k=1}^{\infty} \frac{J_1(\mu_k x)}{\mu_k^3 J_0(2\mu_k)}, \qquad 0 \le x \le 2,$$

where μ_k is the kth positive root of $J_1(2\mu) = 0$. Then use MATLAB to illustrate various partial sums of the Fourier-Bessel series.

22. Show that

$$x^3 = 2 \sum_{k=1}^{\infty} \frac{(\mu_k^2 - 8) J_1(\mu_k x)}{\mu_k^3 J_2(\mu_k)}, \qquad 0 \le x \le 1,$$

where μ_k is the kth positive root of $J_1(\mu) = 0$. Then use MATLAB to illustrate various partial sums of the Fourier-Bessel series.

23. Show that

$$x = 2 \sum_{k=1}^{\infty} \frac{\mu_k J_2(\mu_k) J_1(\mu_k x)}{(\mu_k^2 - 1) J_1^2(\mu_k)}, \qquad 0 \le x \le 1,$$

where μ_k is the kth positive root of $J_1'(\mu) = 0$. Then use MATLAB to illustrate various partial sums of the Fourier-Bessel series.

24. Show that

$$1 - x^4 = 32 \sum_{k=1}^{\infty} \frac{(\mu_k^2 - 4) J_0(\mu_k x)}{\mu_k^5 J_1(\mu_k)}, \qquad 0 \le x \le 1,$$

where μ_k is the kth positive root of $J_0(\mu) = 0$. Then use MATLAB to illustrate various partial sums of the Fourier-Bessel series.

25. Show that

$$1 = 2\alpha L \sum_{k=1}^{\infty} \frac{J_0(\mu_k x/L)}{(\mu_k^2 + \alpha^2 L^2) J_0(\mu_k)}, \qquad 0 \le x \le L,$$

where μ_k is the kth positive root of $\mu J_1(\mu) = \alpha L J_0(\mu)$. Then use MATLAB to illustrate various partial sums of the Fourier-Bessel series.

26. Using the relationship[24]

$$\int_0^a J_\nu(\alpha r) J_\nu(\beta r)\, r\, dr = \frac{a\beta J_\nu(\alpha a) J_\nu'(\beta a) - a\alpha J_\nu(\beta a) J_\nu'(\alpha a)}{\alpha^2 - \beta^2},$$

show that

$$-\frac{J_0(bx) - J_0(ba)}{J_0(ba)} = \frac{2b^2}{a} \sum_{k=1}^{\infty} \frac{J_0(\mu_k x)}{\mu_k(\mu_k^2 - b^2) J_1(\mu_k a)}, \qquad 0 \le x \le a,$$

where μ_k is the kth positive root of $J_0(\mu a) = 0$ and b is a constant.

---

[24] Watson, *op. cit.*, §5.11, Equation 8.

27. Given the definite integral[25]

$$\int_0^1 \frac{x\,J_0(bx)}{\sqrt{1-x^2}}\,dx = \frac{\sin(b)}{b}, \qquad 0 < b,$$

show that

$$\frac{H(t-x)}{\sqrt{t^2-x^2}} = 2\sum_{k=1}^{\infty} \frac{\sin(\mu_k t)\,J_0(\mu_k x)}{\mu_k\,J_1^2(\mu_k)}, \qquad 0 < x < 1, \quad 0 < t \le 1,$$

where μ_k is the kth positive root of $J_0(\mu) = 0$ and $H(\ )$ is Heaviside's step function.

28. Using the same definite integral from the previous problem, show[26] that

$$\frac{H(a-x)}{\sqrt{a^2-x^2}} = \frac{2}{b}\sum_{n=1}^{\infty} \frac{\sin(\mu_n a/b)\,J_0(\mu_n x/b)}{\mu_n\,J_0^2(\mu_n)}, \qquad 0 \le x < b,$$

where $a < b$, μ_n is the nth positive root of $J_0'(\mu) = -J_1(\mu) = 0$, and $H(\ )$ is Heaviside's step function.

29. Given the definite integral[27]

$$\int_0^a \cos(cx)\,J_0\left(b\sqrt{a^2-x^2}\right)\,dx = \frac{\sin\left(a\sqrt{b^2+c^2}\right)}{\sqrt{b^2+c^2}}, \qquad 0 < b,$$

show that

$$\frac{\cosh\left(b\sqrt{t^2-x^2}\right)}{\sqrt{t^2-x^2}}\,H(t-x) = \frac{2}{a^2}\sum_{k=1}^{\infty} \frac{\sin\left(t\sqrt{\mu_k^2-b^2}\right)\,J_0(\mu_k x)}{\sqrt{\mu_k^2-b^2}\,J_1^2(\mu_k a)},$$

where $0 < x < a$, μ_k is the kth positive root of $J_0(\mu a) = 0$, $H(\ )$ is Heaviside's step function, and b is a constant.

30. Using the integral definition of the Bessel function[28] for $J_1(z)$:

$$J_1(z) = \frac{2}{\pi}\int_0^1 \frac{t\,\sin(zt)}{\sqrt{1-t^2}}\,dt, \qquad 0 < z,$$

---

[25] Gradshteyn, I. S., and I. M. Ryzhik, 1965: *Table of Integrals, Series, and Products.* Academic Press, §6.567, Formula 1 with $\nu = 0$ and $\mu = -1/2$.

[26] Reprinted from *Int. J. Solids Struct.*, **37**, X. X. Wei and K. T. Chau, Finite solid circular cylinders subjected to arbitrary surface load. Part II–Application to double-punch test, 5733–5744, ©2000, with permission of Elsevier Science.

[27] Gradshteyn and Ryzhik, *op. cit.*, §6.677, Formula 6.

[28] Gradshteyn and Ryzhik, *op. cit.*, §3.753, Formula 5.

show that

$$\frac{x}{t\sqrt{t^2 - x^2}} H(t - x) = \frac{\pi}{L} \sum_{n=1}^{\infty} J_1\left(\frac{n\pi t}{L}\right) \sin\left(\frac{n\pi x}{L}\right), \qquad 0 \le x < L,$$

where $H(\ )$ is Heaviside's step function. [Hint: Treat this as a Fourier half-range sine expansion.]

31. Show that

$$\delta(x - b) = \frac{2b}{a^2} \sum_{k=1}^{\infty} \frac{J_0(\mu_k b/a) J_0(\mu_k x/a)}{J_1^2(\mu_k)}, \qquad 0 \le x, b < a,$$

where μ_k is the kth positive root of $J_0(\mu) = 0$ and $\delta(\ )$ is the Dirac delta function.

32. Show that

$$\frac{\delta(x)}{2\pi x} = \frac{1}{\pi a^2} \sum_{k=1}^{\infty} \frac{J_0(\mu_k x/a)}{J_1^2(\mu_k)}, \qquad 0 \le x < a,$$

where μ_k is the kth positive root of $J_0(\mu) = 0$ and $\delta(\ )$ is the Dirac delta function.

Chapter 10

The Wave Equation

In this chapter we will study problems associated with the equation

$$\frac{\partial^2 u}{\partial t^2} = c^2 \frac{\partial^2 u}{\partial x^2}, \qquad (10.0.1)$$

where $u = u(x,t)$, x and t are the two independent variables, and c is a constant. This equation, called the *wave equation*, serves as the prototype for a wider class of *hyperbolic equations*

$$a(x,t)\frac{\partial^2 u}{\partial x^2} + b(x,t)\frac{\partial^2 u}{\partial x \partial t} + c(x,t)\frac{\partial^2 u}{\partial t^2} = f\left(x, t, u, \frac{\partial u}{\partial x}, \frac{\partial u}{\partial t}\right), \qquad (10.0.2)$$

where $b^2 > 4ac$. It arises in the study of many important physical problems involving wave propagation, such as the transverse vibrations of an elastic string and the longitudinal vibrations or torsional oscillations of a rod.

10.1 THE VIBRATING STRING

The motion of a string of length L and constant density ρ (mass per unit *length*) is a simple example of a physical system described by the wave equation. See Figure 10.1.1. Assuming that the equilibrium position of the string and the interval $[0, L]$ along the x-axis coincide, the equation of motion which

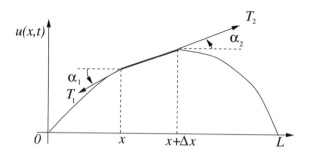

Figure 10.1.1: The vibrating string.

describes the vertical displacement $u(x,t)$ of the string follows by considering a short piece whose ends are at x and $x + \Delta x$ and applying Newton's second law.

If we assume that the string is perfectly flexible and offers no resistance to bending, Figure 10.1.1 shows the forces on an element of the string. Applying Newton's second law in the x-direction, the sum of forces equals

$$-T(x)\cos(\alpha_1) + T(x + \Delta x)\cos(\alpha_2), \qquad (10.1.1)$$

where $T(x)$ denotes the tensile force. If we assume that a point on the string moves only in the vertical direction, the sum of forces in (10.1.1) equals zero and the horizontal component of tension is constant:

$$-T(x)\cos(\alpha_1) + T(x + \Delta x)\cos(\alpha_2) = 0, \qquad (10.1.2)$$

and

$$T(x)\cos(\alpha_1) = T(x + \Delta x)\cos(\alpha_2) = T, \text{ a constant.} \qquad (10.1.3)$$

If gravity is the only external force, Newton's law in the vertical direction gives

$$-T(x)\sin(\alpha_1) + T(x + \Delta x)\sin(\alpha_2) - mg = m\frac{\partial^2 u}{\partial t^2}, \qquad (10.1.4)$$

where u_{tt} is the acceleration. Because

$$T(x) = \frac{T}{\cos(\alpha_1)}, \quad \text{and} \quad T(x + \Delta x) = \frac{T}{\cos(\alpha_2)}, \qquad (10.1.5)$$

then

$$-T\tan(\alpha_1) + T\tan(\alpha_2) - \rho g \Delta x = \rho \Delta x \frac{\partial^2 u}{\partial t^2}. \qquad (10.1.6)$$

The quantities $\tan(\alpha_1)$ and $\tan(\alpha_2)$ equal the slope of the string at x and $x + \Delta x$, respectively; that is,

$$\tan(\alpha_1) = \frac{\partial u(x,t)}{\partial x}, \quad \text{and} \quad \tan(\alpha_2) = \frac{\partial u(x + \Delta x, t)}{\partial x}. \qquad (10.1.7)$$

Substituting (10.1.7) into (10.1.6),

$$T\left[\frac{\partial u(x+\Delta x,t)}{\partial x} - \frac{\partial u(x,t)}{\partial x}\right] = \rho\Delta x\left(\frac{\partial^2 u}{\partial t^2} + g\right). \qquad (\mathbf{10.1.8})$$

After dividing through by Δx, we have a difference quotient on the left:

$$\frac{T}{\Delta x}\left[\frac{\partial u(x+\Delta x,t)}{\partial x} - \frac{\partial u(x,t)}{\partial x}\right] = \rho\left(\frac{\partial^2 u}{\partial t^2} + g\right). \qquad (\mathbf{10.1.9})$$

In the limit as $\Delta x \to 0$, this difference quotient becomes a partial derivative with respect to x, leaving Newton's second law in the form

$$T\frac{\partial^2 u}{\partial x^2} = \rho\frac{\partial^2 u}{\partial t^2} + \rho g, \qquad (\mathbf{10.1.10})$$

or

$$\frac{\partial^2 u}{\partial x^2} = \frac{1}{c^2}\frac{\partial^2 u}{\partial t^2} + \frac{g}{c^2}, \qquad (\mathbf{10.1.11})$$

where $c^2 = T/\rho$. Because u_{tt} is generally much larger than g, we can neglect the last term, giving the equation of the vibrating string as

$$\frac{\partial^2 u}{\partial x^2} = \frac{1}{c^2}\frac{\partial^2 u}{\partial t^2}. \qquad (\mathbf{10.1.12})$$

Equation (10.1.12) is the one-dimensional *wave equation*.

As a second example[1] we derive the threadline equation which describes how a thread composed of yard vibrates as we draw it between two eyelets spaced a distance L apart. We assume that the tension in the thread is constant, the vibrations are small, the thread is perfectly flexible, the effects of gravity and air drag are negligible, and the mass of the thread per unit length is constant. Unlike the vibrating string between two fixed ends, we draw the threadline through the eyelets at a speed V so that a segment of thread experiences motion in both the x and y directions as it vibrates about its equilibrium position. The eyelets may move in the vertical direction.

From Newton's second law,

$$\frac{d}{dt}\left(m\frac{dy}{dt}\right) = \sum \text{forces}, \qquad (\mathbf{10.1.13})$$

where m is the mass of the thread. But

$$\frac{dy}{dt} = \frac{\partial y}{\partial t} + \frac{dx}{dt}\frac{\partial y}{\partial x}. \qquad (\mathbf{10.1.14})$$

---

[1] Reprinted from *J. Franklin Inst.*, **275**, R. D. Swope and W. F. Ames, Vibrations of a moving threadline, 36–55, ©1963, with kind permission from Elsevier Science Ltd, The Boulevard, Langford Lane, Kidlington OX5 1GB, UK.

Because $dx/dt = V$,

$$\frac{dy}{dt} = \frac{\partial y}{\partial t} + V\frac{\partial y}{\partial x}, \tag{10.1.15}$$

and

$$\frac{d}{dt}\left(m\frac{dy}{dt}\right) = \frac{\partial}{\partial t}\left[m\left(\frac{\partial y}{\partial t} + V\frac{\partial y}{\partial x}\right)\right] + V\frac{\partial}{\partial x}\left[m\left(\frac{\partial y}{\partial t} + V\frac{\partial y}{\partial x}\right)\right]. \tag{10.1.16}$$

Because both m and V are constant, it follows that

$$\frac{d}{dt}\left(m\frac{dy}{dt}\right) = m\frac{\partial^2 y}{\partial t^2} + 2mV\frac{\partial^2 y}{\partial x\partial t} + mV^2\frac{\partial^2 y}{\partial x^2}. \tag{10.1.17}$$

The sum of the forces again equals

$$T\frac{\partial^2 y}{\partial x^2}\Delta x \tag{10.1.18}$$

so that the threadline equation is

$$T\frac{\partial^2 y}{\partial x^2}\Delta x = m\frac{\partial^2 y}{\partial t^2} + 2mV\frac{\partial^2 y}{\partial x\partial t} + mV^2\frac{\partial^2 y}{\partial x^2}, \tag{10.1.19}$$

or

$$\frac{\partial^2 y}{\partial t^2} + 2V\frac{\partial^2 y}{\partial x\partial t} + \left(V^2 - \frac{gT}{\rho}\right)\frac{\partial^2 y}{\partial x^2} = 0, \tag{10.1.20}$$

where ρ is the density of the thread. Although (10.1.20) is *not* the classic wave equation given in (10.1.12), it is an example of a hyperbolic equation. As we shall see, the solutions to hyperbolic equations share the same behavior, namely, wave-like motion.

10.2 INITIAL CONDITIONS: CAUCHY PROBLEM

Any mathematical model of a physical process must include not only the governing differential equation but also any conditions that are imposed on the solution. For example, in time-dependent problems the solution must conform to the initial condition of the modeled process. Finding those solutions that satisfy the initial conditions (initial data) is called the *Cauchy problem*.

In the case of partial differential equations with second-order derivatives in time, such as the wave equation, we correctly pose the *Cauchy boundary condition* if we specify the value of the solution $u(x, t_0) = f(t)$ *and* its time derivative $u_t(x, t_0) = g(t)$ at some initial time t_0, usually taken to be $t_0 = 0$. The functions $f(t)$ and $g(t)$ are called the *Cauchy data*. We require two conditions involving time because the differential equation has two time derivatives.

In addition to the initial conditions, we must specify boundary conditions in the spatial direction. For example, we may require that the end of the string be fixed. In the next chapter, we discuss boundary conditions in greater depth. However, one boundary condition that is uniquely associated with the wave equation on an open domain is the *radiation condition*. It requires that the waves radiate off to infinity and remain finite as they propagate there.

In summary, Cauchy boundary conditions, along with the appropriate spatial boundary conditions, uniquely determine the solution to the wave equation; any additional information is extraneous. Having developed the differential equation and initial conditions necessary to solve the wave equation, let us now turn to the actual methods used to solve this equation.

10.3 SEPARATION OF VARIABLES

Separation of variables is the most popular method for solving the wave equation. Despite its current widespread use, its initial application to the vibrating string problem was controversial because of the use of a half-range Fourier sine series to represent the initial conditions. On one side, Daniel Bernoulli claimed (in 1775) that he could represent any general initial condition with this technique. To d'Alembert and Euler, however, the half-range Fourier sine series, with its period of $2L$, could not possibly represent any arbitrary function.[2] However, by 1807 Bernoulli was proven correct by the use of separation of variables in the heat conduction problem and it rapidly grew in acceptance.[3] In the following examples we show how to apply this method.

Separation of variables consists of four distinct steps which convert a second-order partial differential equation into two ordinary differential equations. First, we *assume* that the solution equals the product $X(x)T(t)$. Direct substitution into the partial differential equation and boundary conditions yields two ordinary differential equations and the corresponding boundary conditions. Step two involves solving a boundary-value problem of the Sturm-Liouville type. In step three we find the corresponding time dependence. Finally we construct the complete solution as a sum of all product solutions. Upon applying the initial conditions, we have an eigenfunction expansion and must compute the Fourier coefficients. The substitution of these coefficients into the summation yields the complete solution.

• **Example 10.3.1**

Let us solve the wave equation for the special case when we clamp the string at $x = 0$ and $x = L$. Mathematically, we find the solution to the wave

---

[2] See Hobson, E. W., 1957: *The Theory of Functions of a Real Variable and the Theory of Fourier's Series*, Vol. 2. Dover Publishers, §§312–314.

[3] Lützen, J., 1984: Sturm and Liouville's work on ordinary linear differential equations. The emergence of Sturm-Liouville theory. *Arch. Hist. Exact Sci.*, **29**, 317.

equation

$$\frac{\partial^2 u}{\partial t^2} = c^2 \frac{\partial^2 u}{\partial x^2}, \quad 0 < x < L, \quad 0 < t, \tag{10.3.1}$$

which satisfies the initial conditions

$$u(x,0) = f(x), \quad \frac{\partial u(x,0)}{\partial t} = g(x), \quad 0 < x < L, \tag{10.3.2}$$

and the boundary conditions

$$u(0,t) = u(L,t) = 0, \quad 0 < t. \tag{10.3.3}$$

For the present, we leave the Cauchy data quite arbitrary.

We begin by assuming that the solution $u(x,t)$ equals the product $X(x)T(t)$. (Here T no longer denotes tension.) Because

$$\frac{\partial^2 u}{\partial t^2} = X(x)T''(t), \tag{10.3.4}$$

and

$$\frac{\partial^2 u}{\partial x^2} = X''(x)T(t), \tag{10.3.5}$$

the wave equation becomes

$$c^2 X''T = T''X, \tag{10.3.6}$$

or

$$\frac{X''}{X} = \frac{T''}{c^2 T}, \tag{10.3.7}$$

after dividing through by $c^2 X(x)T(t)$. Because the left side of (10.3.7) depends only on x and the right side depends only on t, both sides must equal a constant. We write this separation constant $-\lambda$ and separate (10.3.7) into two ordinary differential equations:

$$T'' + c^2 \lambda T = 0, \quad 0 < t, \tag{10.3.8}$$

and

$$X'' + \lambda X = 0, \quad 0 < x < L. \tag{10.3.9}$$

We now rewrite the boundary conditions in terms of $X(x)$ by noting that the boundary conditions become

$$u(0,t) = X(0)T(t) = 0, \tag{10.3.10}$$

and

$$u(L,t) = X(L)T(t) = 0 \tag{10.3.11}$$

for $0 < t$. If we were to choose $T(t) = 0$, then we would have a trivial solution for $u(x, t)$. Consequently,

$$X(0) = X(L) = 0. \tag{10.3.12}$$

This concludes the first step.

In the second step we consider three possible values for λ: $\lambda < 0$, $\lambda = 0$, and $\lambda > 0$. Turning first to $\lambda < 0$, we set $\lambda = -m^2$ so that square roots of λ will not appear later on and m is real. The general solution of (10.3.9) is

$$X(x) = A \cosh(mx) + B \sinh(mx). \tag{10.3.13}$$

Because $X(0) = 0$, $A = 0$. On the other hand, $X(L) = B \sinh(mL) = 0$. The function $\sinh(mL)$ does not equal to zero since $mL \neq 0$ (recall $m > 0$). Thus, $B = 0$ and we have trivial solutions for a positive separation constant.

If $\lambda = 0$, the general solution now becomes

$$X(x) = C + Dx. \tag{10.3.14}$$

The condition $X(0) = 0$ yields $C = 0$ while $X(L) = 0$ yields $DL = 0$ or $D = 0$. Hence, we have a trivial solution for the $\lambda = 0$ separation constant.

If $\lambda = k^2 > 0$, the general solution to (10.3.9) is

$$X(x) = E \cos(kx) + F \sin(kx). \tag{10.3.15}$$

The condition $X(0) = 0$ results in $E = 0$. On the other hand, $X(L) = F \sin(kL) = 0$. If we wish to avoid a trivial solution in this case ($F \neq 0$), $\sin(kL) = 0$, or $k_n = n\pi/L$, and $\lambda_n = n^2\pi^2/L^2$. The x-dependence equals $X_n(x) = F_n \sin(n\pi x/L)$. We added the n subscript to k and λ to indicate that these quantities depend on n. This concludes the second step.

Turning to (10.3.8) for the third step, the solution to the $T(t)$ equation is

$$T_n(t) = G_n \cos(k_n ct) + H_n \sin(k_n ct), \tag{10.3.16}$$

where G_n and H_n are arbitrary constants. For each $n = 1, 2, 3, \ldots$, a particular solution that satisfies the wave equation and prescribed boundary conditions is

$$u_n(x, t) = F_n \sin\left(\frac{n\pi x}{L}\right) \left[G_n \cos\left(\frac{n\pi ct}{L}\right) + H_n \sin\left(\frac{n\pi ct}{L}\right) \right], \tag{10.3.17}$$

or

$$u_n(x, t) = \sin\left(\frac{n\pi x}{L}\right) \left[A_n \cos\left(\frac{n\pi ct}{L}\right) + B_n \sin\left(\frac{n\pi ct}{L}\right) \right], \tag{10.3.18}$$

where $A_n = F_n G_n$ and $B_n = F_n H_n$. This concludes the third step.

An equivalent method of finding the product solution is to treat (10.3.9) along with $X(0) = X(L) = 0$ as a Sturm-Liouville problem. In this method we obtain the spatial dependence by solving the Sturm-Liouville problem and finding the corresponding eigenvalues λ_n and eigenfunctions. Next we solve for $T_n(t)$. Finally we form the product solution $u_n(x,t)$ by multiplying the eigenfunction times the temporal dependence.

For any choice of A_n and B_n, (10.3.18) is a solution of the partial differential equation (10.3.1) also satisfying the boundary conditions (10.3.3). Therefore, any linear combination of $u_n(x,t)$ also satisfies the partial differential equation and the boundary conditions. In making this linear combination we need no new constants because A_n and B_n are still arbitrary. We have, then,

$$u(x,t) = \sum_{n=1}^{\infty} \sin\left(\frac{n\pi x}{L}\right) \left[A_n \cos\left(\frac{n\pi ct}{L}\right) + B_n \sin\left(\frac{n\pi ct}{L}\right) \right]. \qquad (\mathbf{10.3.19})$$

Our method of using particular solutions to build up the general solution illustrates the powerful *principle of linear superposition*, which is applicable to any *linear* system. This principle states that if u_1 and u_2 are any solutions of a linear homogeneous partial differential equation in any region, then $u = c_1 u_1 + c_2 u_2$ is also a solution of that equation in that region, where c_1 and c_2 are any constants. We can generalize this to an infinite sum. It is extremely important because it allows us to construct general solutions to partial differential equations from particular solutions to the same problem.

Our fourth and final task remains to determine A_n and B_n. At $t = 0$,

$$u(x,0) = \sum_{n=1}^{\infty} A_n \sin\left(\frac{n\pi x}{L}\right) = f(x), \qquad (\mathbf{10.3.20})$$

and

$$u_t(x,0) = \sum_{n=1}^{\infty} \frac{n\pi c}{L} B_n \sin\left(\frac{n\pi x}{L}\right) = g(x). \qquad (\mathbf{10.3.21})$$

Both of these series are Fourier half-range sine expansions over the interval $(0, L)$. Applying the results from §4.3,

$$A_n = \frac{2}{L} \int_0^L f(x) \sin\left(\frac{n\pi x}{L}\right) dx, \qquad (\mathbf{10.3.22})$$

and

$$\frac{n\pi c}{L} B_n = \frac{2}{L} \int_0^L g(x) \sin\left(\frac{n\pi x}{L}\right) dx, \qquad (\mathbf{10.3.23})$$

or

$$B_n = \frac{2}{n\pi c} \int_0^L g(x) \sin\left(\frac{n\pi x}{L}\right) dx. \qquad (\mathbf{10.3.24})$$

At this point we might ask ourselves whether the Fourier series solution to the wave equation always converges. For the case $g(x) = 0$, Carslaw[4] showed that if the initial position of the string forms a curve so that $f(x)$ or the slope $f'(x)$ is continuous between $x = 0$ and $x = L$, then the series converges uniformly.

As an example, let us take the initial conditions

$$f(x) = \begin{cases} 0, & 0 < x \le L/4, \\ 4h\left(\frac{x}{L} - \frac{1}{4}\right), & L/4 \le x \le L/2, \\ 4h\left(\frac{3}{4} - \frac{x}{L}\right), & L/2 \le x \le 3L/4, \\ 0, & 3L/4 \le x < L, \end{cases} \qquad (10.3.25)$$

and

$$g(x) = 0, \qquad 0 < x < L. \qquad (10.3.26)$$

In this particular example, $B_n = 0$ for all n because $g(x) = 0$. On the other hand,

$$A_n = \frac{8h}{L} \int_{L/4}^{L/2} \left(\frac{x}{L} - \frac{1}{4}\right) \sin\left(\frac{n\pi x}{L}\right) dx$$

$$+ \frac{8h}{L} \int_{L/2}^{3L/4} \left(\frac{3}{4} - \frac{x}{L}\right) \sin\left(\frac{n\pi x}{L}\right) dx \qquad (10.3.27)$$

$$= \frac{8h}{n^2\pi^2} \left[2\sin\left(\frac{n\pi}{2}\right) - \sin\left(\frac{3n\pi}{4}\right) - \sin\left(\frac{n\pi}{4}\right)\right] \qquad (10.3.28)$$

$$= \frac{8h}{n^2\pi^2} \left[2\sin\left(\frac{n\pi}{2}\right) - 2\sin\left(\frac{n\pi}{2}\right)\cos\left(\frac{n\pi}{4}\right)\right] \qquad (10.3.29)$$

$$= \frac{16h}{n^2\pi^2} \sin\left(\frac{n\pi}{2}\right) \left[1 - \cos\left(\frac{n\pi}{4}\right)\right] \qquad (10.3.30)$$

$$= \frac{32h}{n^2\pi^2} \sin\left(\frac{n\pi}{2}\right) \sin^2\left(\frac{n\pi}{8}\right), \qquad (10.3.31)$$

because $\sin(A) + \sin(B) = 2\sin[\frac{1}{2}(A + B)]\cos[\frac{1}{2}(A - B)]$, and $1 - \cos(2A) = 2\sin^2(A)$. Therefore,

$$u(x, t) = \frac{32h}{\pi^2} \sum_{n=1}^{\infty} \sin\left(\frac{n\pi}{2}\right) \sin^2\left(\frac{n\pi}{8}\right) \frac{1}{n^2} \sin\left(\frac{n\pi x}{L}\right) \cos\left(\frac{n\pi ct}{L}\right). \qquad (10.3.32)$$

Because $\sin(n\pi/2)$ vanishes for n even, so does A_n. If (10.3.32) were evaluated on a computer, considerable time and effort would be wasted. Consequently it is preferable to rewrite (10.3.32) so that we eliminate these vanishing terms. The most convenient method introduces the general expression $n = 2m - 1$ for any odd integer, where $m = 1, 2, 3, \ldots$, and notes that

[4] Carslaw, H. S., 1902: Note on the use of Fourier's series in the problem of the transverse vibrations of strings. *Proc. Edinburgh Math. Soc., Ser. 1*, **20**, 23–28.

$\sin[(2m - 1)\pi/2] = (-1)^{m+1}$. Therefore, (10.3.32) becomes

$$u(x,t) = \frac{32h}{\pi^2} \sum_{m=1}^{\infty} \frac{(-1)^{m+1}}{(2m - 1)^2} \sin^2\left[\frac{(2m - 1)\pi}{8}\right] \sin\left[\frac{(2m - 1)\pi x}{L}\right]$$

$$\times \cos\left[\frac{(2m - 1)\pi ct}{L}\right]. \quad (10.3.33)$$

Although we completely solved the problem, it is useful to rewrite (10.3.33) as

$$u(x,t) = \frac{1}{2} \sum_{n=1}^{\infty} A_n \left\{ \sin\left[\frac{n\pi}{L}(x - ct)\right] + \sin\left[\frac{n\pi}{L}(x + ct)\right] \right\} \quad (\mathbf{10.3.34})$$

through the application of the trigonometric identity $\sin(A)\cos(B) = \frac{1}{2}\sin(A - B) + \frac{1}{2}\sin(A + B)$. From general physics we find expressions like $\sin[k_n(x - ct)]$ or $\sin(kx - \omega t)$ arising in studies of simple wave motions. The quantity $\sin(kx - \omega t)$ is the mathematical description of a propagating wave in the sense that we must move to the right at the speed c if we wish to keep in the same position relative to the nearest crest and trough. The quantities k, ω, and c are the wavenumber, frequency, and phase speed or wave-velocity, respectively. The relationship $\omega = kc$ holds between the frequency and phase speed.

It may seem paradoxical that we are talking about traveling waves in a problem dealing with waves confined on a string of length L. Actually we are dealing with standing waves because at the same time that a wave is propagating to the right its mirror image is running to the left so that there is no resultant progressive wave motion. Figures 10.3.1 and 10.3.2 illustrate our solution. Figure 10.3.1 gives various cross sections. The single large peak at $t = 0$ breaks into two smaller peaks which race towards the two ends. At each end, they reflect and turn upside down as they propagate back towards $x = L/2$ at $ct/L = 1$. This large, negative peak at $x = L/2$ again breaks apart, with the two smaller peaks propagating towards the endpoints. They reflect and again become positive peaks as they propagate back to $x = L/2$ at $ct/L = 2$. After that time, the whole process repeats itself.

MATLAB can used to examine the solution in its totality. The script

```
% set parameters for the calculation
clear; M = 50; dx = 0.02; dt = 0.02;
% compute Fourier coefficients
sign = 32;
for m = 1:M
temp1 = (2*m-1)*pi; temp2 = sin(temp1/8);
a(m) = sign * temp2 * temp2 / (temp1 * temp1);
sign = -sign;
end
```

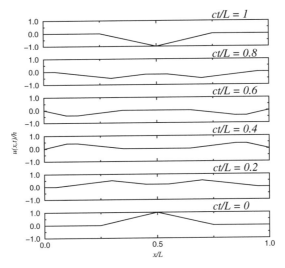

Figure 10.3.1: The vibration of a string $u(x,t)/h$ at various positions x/L at the times $ct/L = 0, 0.2, 0.4, 0.6, 0.8$, and 1. For times $1 < ct/L < 2$ the pictures appear in reverse time order.

```
% compute grid and initialize solution
X = [0:dx:1]; T = [0:dt:2];
u = zeros(length(T),length(X));
XX = repmat(X,[length(T) 1]);
TT = repmat(T',[1 length(X)]);
% compute solution from (10.3.33)
for m = 1:M
temp1 = (2*m-1)*pi;
u = u + a(m) .* sin(temp1*XX) .* cos(temp1*TT);
end
% plot space/time picture of the solution
surf(XX,TT,u)
xlabel('DISTANCE','Fontsize',20); ylabel('TIME','Fontsize',20)
zlabel('SOLUTION','Fontsize',20)
```

gives a three-dimensional view of (10.3.33). The solution can be viewed in many different prospects using the interactive capacity of MATLAB.

An important dimension to the vibrating string problem is the fact that the wavenumber k_n is not a free parameter but has been restricted to the values of $n\pi/L$. This restriction on wavenumber is common in wave problems dealing with limited domains (for example, a building, ship, lake, or planet) and these oscillations are given the special name of *normal modes* or *natural vibrations.*

In our problem of the vibrating string, all of the components propagate with the same phase speed. That is, all of the waves, regardless of wavenumber

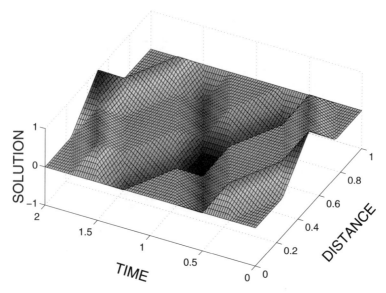

Figure 10.3.2: Two-dimensional plot of the vibration of a string $u(x,t)/h$ at various times ct/L and positions x/L.

k_n, move the characteristic distance $c\Delta t$ or $-c\Delta t$ after the time interval Δt elapsed. In the next example we will see that this is not always true.

• Example 10.3.2: Dispersion

In the preceding example, the solution to the vibrating string problem consisted of two simple waves, each propagating with a phase speed c to the right and left. In many problems where the equations of motion are a little more complicated than (10.3.1), all of the harmonics no longer propagate with the same phase speed but at a speed that depends upon the wavenumber. In such systems the phase relation varies between the harmonics and these systems are referred to as *dispersive*.

A modification of the vibrating string problem provides a simple illustration. We now subject each element of the string to an additional applied force which is proportional to its displacement:

$$\frac{\partial^2 u}{\partial t^2} = c^2 \frac{\partial^2 u}{\partial x^2} - hu, \quad 0 < x < L, \quad 0 < t, \tag{10.3.35}$$

where $h > 0$ is constant. For example, if we embed the string in a thin sheet of rubber, then in addition to the restoring force due to tension, there is a restoring force due to the rubber on each portion of the string. From its use in the quantum mechanics of "scalar" mesons, (10.3.35) is often referred to as the *Klein-Gordon* equation.

We shall again look for particular solutions of the form $u(x,t) = X(x)T(t)$. This time, however,

$$XT'' - c^2 X''T + hXT = 0, \tag{10.3.36}$$

or

$$\frac{T''}{c^2 T} + \frac{h}{c^2} = \frac{X''}{X} = -\lambda, \tag{10.3.37}$$

which leads to two ordinary differential equations

$$X'' + \lambda X = 0, \tag{10.3.38}$$

and

$$T'' + (\lambda c^2 + h)T = 0. \tag{10.3.39}$$

If we attach the string at $x = 0$ and $x = L$, the $X(x)$ solution is

$$X_n(x) = \sin\left(\frac{n\pi x}{L}\right) \tag{10.3.40}$$

with $k_n = n\pi/L$, and $\lambda_n = n^2\pi^2/L^2$. On the other hand, the $T(t)$ solution becomes

$$T_n(t) = A_n \cos\left(\sqrt{k_n^2 c^2 + h}\, t\right) + B_n \sin\left(\sqrt{k_n^2 c^2 + h}\, t\right), \tag{10.3.41}$$

so that the product solution is

$$u_n(x,t) = \sin\left(\frac{n\pi x}{L}\right)\left[A_n \cos\left(\sqrt{k_n^2 c^2 + h}\, t\right) + B_n \sin\left(\sqrt{k_n^2 c^2 + h}\, t\right)\right]. \tag{10.3.42}$$

Finally, the general solution becomes

$$u(x,t) = \sum_{n=1}^{\infty} \sin\left(\frac{n\pi x}{L}\right)\left[A_n \cos\left(\sqrt{k_n^2 c^2 + h}\, t\right) + B_n \sin\left(\sqrt{k_n^2 c^2 + h}\, t\right)\right] \tag{10.3.43}$$

from the principle of linear superposition. Let us consider the case when $B_n = 0$. Then we can write (10.3.43) as

$$u(x,t) = \sum_{n=1}^{\infty} \frac{A_n}{2}\left[\sin\left(k_n x + \sqrt{k_n^2 c^2 + h}\, t\right) + \sin\left(k_n x - \sqrt{k_n^2 c^2 + h}\, t\right)\right]. \tag{10.3.44}$$

Comparing our results with (10.3.34), the distance that a particular mode k_n moves during the time interval Δt depends not only upon external parameters such as h, the tension and density of the string, but also upon its wavenumber (or equivalently, wavelength). Furthermore, the frequency of a particular harmonic is larger than that when $h = 0$. This result is not surprising, because the added stiffness of the medium should increase the natural frequencies.

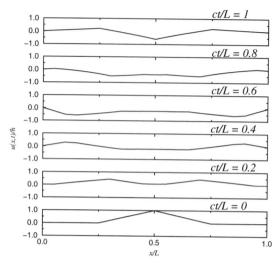

Figure 10.3.3: The vibration of a string $u(x,t)/h$ embedded in a thin sheet of rubber at various positions x/L at the times $ct/L = 0$, 0.2, 0.4, 0.6, 0.8, and 1 for $hL^2/c^2 = 10$. The same parameters were used as in Figure 10.3.1.

The importance of dispersion lies in the fact that if the solution $u(x,t)$ is a superposition of progressive waves in the same direction, then the phase relationship between the different harmonics changes with time. Because most signals consist of an infinite series of these progressive waves, dispersion causes the signal to become garbled. We show this by comparing the solution (10.3.43) given in Figures 10.3.3 and 10.3.4 for the initial conditions (10.3.25) and (10.3.26) with $hL^2/c^2 = 10$ to the results given in Figures 10.3.1 and 10.3.2. In the case of Figure 10.3.4, the MATLAB script line
```
u = u + a(m) .* sin(temp1*XX) .* cos(temp1*TT);
```
has been replaced with
```
temp2 = temp1 * sqrt(1 + H/(temp1*temp1));
u = u + a(m) .* sin(temp1*XX) .* cos(temp2*TT);
```
where H = 10 is defined earlier in the script. Note how garbled the picture becomes at $ct/L = 2$ in Figure 10.3.4 compared to the nondispersive solution at the same time in Figure 10.3.2.

• Example 10.3.3: Damped wave equation

In the previous example a slight modification of the wave equation resulted in a wave solution where each Fourier harmonic propagates with its own particular phase speed. In this example we introduce a modification of the wave equation that results not only in dispersive waves but also in the exponential decay of the amplitude as the wave propagates.

So far we neglected the reaction of the surrounding medium (air or water, for example) on the motion of the string. For small-amplitude motions this

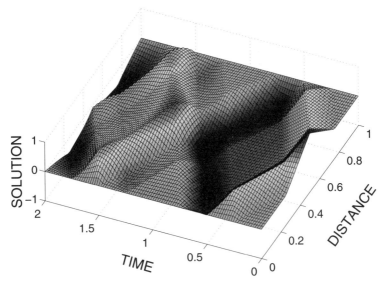

Figure 10.3.4: The two-dimensional plot of the vibration of a string $u(x,t)/h$ embedded in a thin sheet of rubber at various times ct/L and positions x/L for $hL^2/c^2 = 10$.

reaction opposes the motion of each element of the string and is proportional to the element's velocity. The equation of motion, when we account for the tension and friction in the medium but not its stiffness or internal friction, is

$$\frac{\partial^2 u}{\partial t^2} + 2h\frac{\partial u}{\partial t} = c^2\frac{\partial^2 u}{\partial x^2}, \quad 0 < x < L, \quad 0 < t. \tag{10.3.45}$$

Because (10.3.45) first arose in the mathematical description of the telegraph,[5] it is generally known as the *equation of telegraphy*. The effect of friction is, of course, to damp out the free vibration.

Let us assume a solution of the form $u(x,t) = X(x)T(t)$ and separate the variables to obtain the two ordinary differential equations:

$$X'' + \lambda X = 0, \tag{10.3.46}$$

and

$$T'' + 2hT' + \lambda c^2 T = 0 \tag{10.3.47}$$

with $X(0) = X(L) = 0$. Friction does not affect the shape of the normal modes; they are still

$$X_n(x) = \sin\left(\frac{n\pi x}{L}\right) \tag{10.3.48}$$

---

[5] The first published solution was by Kirchhoff, G., 1857: Über die Bewegung der Electrität in Drähten. *Ann. Phys. Chem.*, **100**, 193–217. English translation: Kirchhoff, G., 1857: On the motion of electricity in wires. *Philos. Mag.*, *Ser. 4*, **13**, 393–412.

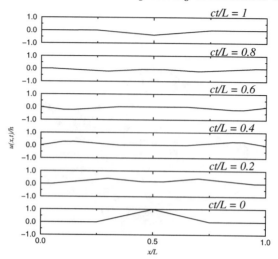

Figure 10.3.5: The vibration of a string $u(x,t)/h$ with frictional dissipation at various positions x/L at the times $ct/L = 0$, 0.2, 0.4, 0.6, 0.8, and 1 for $hL/c = 1$. The same parameters were used as in Figure 10.3.1.

with $k_n = n\pi/L$ and $\lambda_n = n^2\pi^2/L^2$.

The solution for the $T(t)$ equation is

$$T_n(t) = e^{-ht}\left[A_n \cos\left(\sqrt{k_n^2 c^2 - h^2}\, t\right) + B_n \sin\left(\sqrt{k_n^2 c^2 - h^2}\, t\right)\right] \quad (\mathbf{10.3.49})$$

with the condition that $k_n c > h$. If we violate this condition, the solutions are two exponentially decaying functions in time. Because most physical problems usually fulfill this condition, we concentrate on this solution.

From the principle of linear superposition, the general solution is

$$u(x,t) = e^{-ht}\sum_{n=1}^{\infty}\sin\left(\frac{n\pi x}{L}\right)\left[A_n \cos\left(\sqrt{k_n^2 c^2 - h^2}\, t\right) + B_n \sin\left(\sqrt{k_n^2 c^2 - h^2}\, t\right)\right],$$

$$(\mathbf{10.3.50})$$

where $\pi c > hL$. From (10.3.50) we see two important effects. First, the presence of friction slows all of the harmonics. Furthermore, friction dampens all of the harmonics. Figures 10.3.5 and 10.3.6 illustrate the solution using the initial conditions given by (10.3.25) and (10.3.26) with $hL/c = 1$. In the case of Figure 10.3.6, the script line that produced Figure 10.3.2:

```
u = u + a(m) .* sin(temp1*XX) .* cos(temp1*TT);
```
has been replaced with
```
temp2 = temp1 * sqrt(1 - (H*H)/(temp1*temp1));
u = u + a(m) .* exp(-H*TT) .* sin(temp1*XX) .* cos(temp2*TT);
```
where H = 1 is defined earlier in the script. Because this is a rather large coefficient of friction, Figures 10.3.5 and 10.3.6 exhibit rapid damping as well as dispersion.

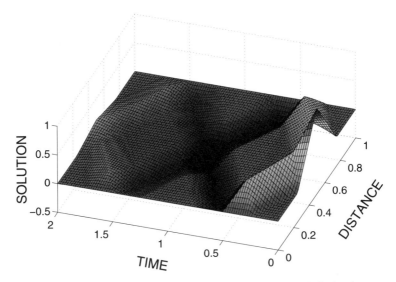

Figure 10.3.6: The vibration of a string $u(x,t)/h$ with frictional dissipation at various times ct/L and positions x/L for $hL/c = 1$.

This damping and dispersion of waves also occurs in solutions of the equation of telegraphy where the solutions are progressive waves. Because early telegraph lines were short, time delay effects were negligible. However, when engineers laid the first transoceanic cables in the 1850s, the time delay became seconds and differences in the velocity of propagation of different frequencies, as predicted by (10.3.50), became noticeable to the operators. Table 10.3.1 gives the transmission rate for various transatlantic submarine telegraph lines. As it shows, increases in the transmission rates during the nineteenth century were due primarily to improvements in terminal technology.

When they instituted long-distance telephony just before the turn of the twentieth century, this difference in velocity between frequencies should have limited the circuits to a few tens of miles.[6] However, in 1899, Prof. Michael Pupin, at Columbia University, showed that by adding inductors ("loading coils") to the line at regular intervals the velocities at the different frequencies could be equalized.[7] Heaviside[8] and the French engineer Vaschy[9] made

---

[6] Rayleigh, J. W., 1884: On telephoning through a cable. *Br. Assoc. Rep.*, 632–633; Jordan, D. W., 1982: The adoption of self-induction by telephony, 1886–1889. *Ann. Sci.*, **39**, 433–461.

[7] There is considerable controversy concerning who is exactly the inventor. See Brittain, J. E., 1970: The introduction of the loading coil: George A. Campbell and Michael I. Pupin. *Tech. Culture*, **11**, 36–57.

[8] First published 3 June 1887. Reprinted in Heaviside, O., 1970: *Electrical Papers, Vol. 2.* Chelsea Publishing, pp. 119–124.

[9] See Devaux-Charbonnel, X. G. F., 1917: La contribution des ingénieurs français à

Table 10.3.1: Technological Innovation on Transatlantic Telegraph Cables

Year	Technological Innovation	Performance (words/min)
1857–58	Mirror galvanometer	3–7
1870	Condensers	12
1872	Siphon recorder	17
1879	Duplex	24
1894	Larger diameter cable	72–90
1915–20	Brown drum repeater and Heurtley magnifier	100
1923–28	Magnetically loaded lines	300–320
1928–32	Electronic signal shaping amplifiers and time division multiplexing	480
1950	Repeaters on the continental shelf	100–300
1956	Repeater telephone cables	21600

From Coates, V. T., and B. Finn, 1979: *A Retrospective Technology Assessment: Submarine Telegraphy. The Transatlantic Cable of 1866.* San Francisco Press, Inc., 268 pp.

similar suggestions in the nineteenth century. Thus, adding resistance and inductance, which would seem to make things worse, actually made possible long-distance telephony. Today you can see these loading coils as you drive along the street; they are the black cylinders, approximately one between each pair of telephone poles, spliced into the telephone cable. The loading of long submarine telegraph cables had to wait for the development of permalloy and mu-metal materials of high magnetic induction.

• **Example 10.3.4: Axisymmetric vibrations of a circular membrane**

The wave equation

$$\frac{\partial^2 u}{\partial r^2} + \frac{1}{r}\frac{\partial u}{\partial r} = \frac{1}{c^2}\frac{\partial^2 u}{\partial t^2}, \quad 0 \leq r < a, \quad 0 < t \qquad (10.3.51)$$

governs axisymmetric vibrations of a circular membrane, where $u(r,t)$ is the vertical displacement of the membrane, r is the radial distance, t is time, c is the square root of the ratio of the tension of the membrane to its density, and a is the radius of the membrane. We will solve (10.3.51) when the membrane is initially at rest, $u(r,0) = 0$, and struck so that its initial velocity is

$$\frac{\partial u(r,0)}{\partial t} = \begin{cases} P/(\pi\epsilon^2\rho), & 0 \leq r < \epsilon, \\ 0, & \epsilon < r < a. \end{cases} \qquad (10.3.52)$$

la téléphonie à grande distance par câbles souterrains: Vaschy et Barbarat. *Rev. Gén. Électr.*, **2**, 288–295.

If this problem can be solved by separation of variables, then $u(r,t) = R(r)T(t)$. Following the substitution of this $u(r,t)$ into (10.3.51), separation of variables leads to

$$\frac{1}{rR}\frac{d}{dr}\left(r\frac{dR}{dr}\right) = \frac{1}{c^2 T}\frac{d^2 T}{dt^2} = -k^2, \qquad (10.3.53)$$

or

$$\frac{1}{r}\frac{d}{dr}\left(r\frac{dR}{dr}\right) + k^2 R = 0, \qquad (10.3.54)$$

and

$$\frac{d^2 T}{dt^2} + k^2 c^2 T = 0. \qquad (10.3.55)$$

The separation constant $-k^2$ must be negative so that we obtain solutions that remain bounded in the region $0 \le r < a$ and can satisfy the boundary condition. This boundary condition is $u(a,t) = R(a)T(t) = 0$, or $R(a) = 0$.

The solutions of (10.3.54)–(10.3.55), subject to the boundary condition, are

$$R_n(r) = J_0\left(\frac{\lambda_n r}{a}\right), \qquad (10.3.56)$$

and

$$T_n(t) = A_n \sin\left(\frac{\lambda_n ct}{a}\right) + B_n \cos\left(\frac{\lambda_n ct}{a}\right), \qquad (10.3.57)$$

where λ_n satisfies the equation $J_0(\lambda) = 0$. Because $u(r,0) = 0$, and $T_n(0) = 0$, $B_n = 0$. Consequently, the product solution is

$$u(r,t) = \sum_{n=1}^{\infty} A_n J_0\left(\frac{\lambda_n r}{a}\right) \sin\left(\frac{\lambda_n ct}{a}\right). \qquad (10.3.58)$$

To determine A_n, we use the condition

$$\frac{\partial u(r,0)}{\partial t} = \sum_{n=1}^{\infty} \frac{\lambda_n c}{a} A_n J_0\left(\frac{\lambda_n r}{a}\right) = \begin{cases} P/(\pi\epsilon^2 \rho), & 0 \le r < \epsilon, \\ 0, & \epsilon < r < a. \end{cases} \qquad (10.3.59)$$

Equation (10.3.59) is a Fourier-Bessel expansion in the orthogonal function $J_0(\lambda_n r/a)$, where

$$\frac{\lambda_n c}{a} A_n = \frac{2}{a^2 J_1^2(\lambda_n)} \int_0^\epsilon \frac{P}{\pi\epsilon^2 \rho} J_0\left(\frac{\lambda_n r}{a}\right) r\, dr \qquad (10.3.60)$$

from (9.5.35) and (9.5.43) in §9.5. Carrying out the integration,

$$A_n = \frac{2P J_1(\lambda_n \epsilon/a)}{c\pi\epsilon\rho\lambda_n^2 J_1^2(\lambda_n)}, \qquad (10.3.61)$$

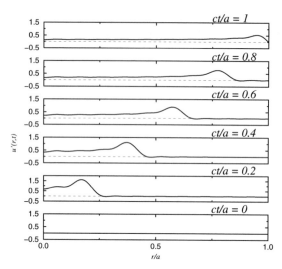

Figure 10.3.7: The axisymmetric vibrations $u'(r,t) = ca\rho u(r,t)/P$ of a circular membrane at various positions r/a at the times $ct/a = 0$, 0.2, 0.4, 0.6, 0.8, and 1 for $\epsilon = a/4$. Initially the membrane is struck by a hammer.

or

$$u(r,t) = \frac{2P}{c\pi\epsilon\rho} \sum_{n=1}^{\infty} \frac{J_1(\lambda_n\epsilon/a)}{\lambda_n^2 J_1^2(\lambda_n)} J_0\left(\frac{\lambda_n r}{a}\right) \sin\left(\frac{\lambda_n ct}{a}\right). \qquad (10.3.62)$$

Figures 10.3.7, 10.3.8, and 10.3.9 illustrate the solution (10.3.62) for various times and positions when $\epsilon = a/4$, and $\epsilon = a/20$. They were generated using the MATLAB script

```
% initialize parameters
clear; eps_over_a = 0.25; M = 20; dr = 0.02; dt = 0.02;
% load in zeros of J_0
zero( 1) =  2.40483; zero( 2) =  5.52008; zero( 3) =  8.65373;
zero( 4) = 11.79153; zero( 5) = 14.93092; zero( 6) = 18.07106;
zero( 7) = 21.21164; zero( 8) = 24.35247; zero( 9) = 27.49347;
zero(10) = 30.63461; zero(11) = 33.77582; zero(12) = 36.91710;
zero(13) = 40.05843; zero(14) = 43.19979; zero(15) = 46.34119;
zero(16) = 49.48261; zero(17) = 52.62405; zero(18) = 55.76551;
zero(19) = 58.90698; zero(20) = 62.04847;
% compute Fourier-Bessel coefficients
for m = 1:M
a(m) = 2 * besselj(1,eps_over_a*zero(m)) ...
     / (eps_over_a*pi*zero(m)*zero(m)*besselj(1,zero(m))^2);
end
R = [0:dr:1]; T = [0:dt:4];
u = zeros(length(T),length(R));
RR = repmat(R,[length(T) 1]);
```

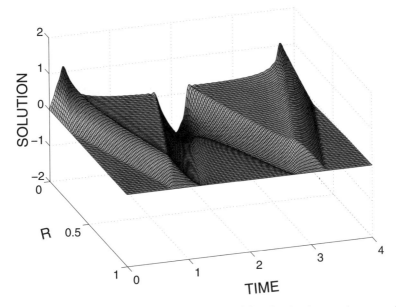

Figure 10.3.8: The axisymmetric vibrations $c a \rho u(r,t)/P$ of a circular membrane resulting from an initial hammer blow with $\epsilon = a/4$. The solution is plotted at various times ct/a and positions r/a.

```
TT = repmat(T',[1 length(R)]);
% compute solution from series solution
for m = 1:M
u = u + a(m) .* besselj(0,zero(m)*RR) .* sin(zero(m)*TT);
end
% plot results
surf(RR,TT,u)
xlabel('R','Fontsize',20); ylabel('TIME','Fontsize',20)
zlabel('SOLUTION','Fontsize',20)
```

Figures 10.3.8 and 10.3.9 show that striking the membrane with a hammer generates a pulse that propagates out to the rim, reflects, inverts, and propagates back to the center. This process then repeats forever.

Problems

Solve the wave equation $u_{tt} = c^2 u_{xx}$, $0 < x < L$, $0 < t$, subject to the boundary conditions that $u(0,t) = u(L,t) = 0, 0 < t$, and the following initial conditions for $0 < x < L$. Use MATLAB to illustrate your solution.

1. $u(x,0) = 0$, $u_t(x,0) = 1$

2. $u(x,0) = 1$, $u_t(x,0) = 0$

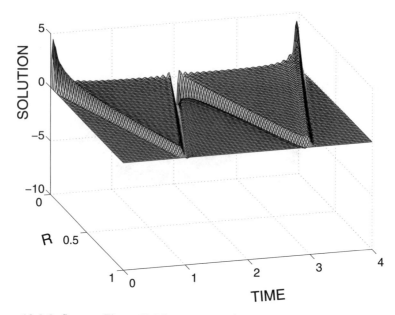

Figure 10.3.9: Same as Figure 10.3.8 except $\epsilon = a/20$.

3. $u(x,0) = \begin{cases} 3hx/2L, & 0 < x < 2L/3, \\ 3h(L-x)/L, & 2L/3 < x < L, \end{cases} \quad u_t(x,0) = 0$

4. $u(x,0) = [3\sin(\pi x/L) - \sin(3\pi x/L)]/4, \quad u_t(x,0) = 0,$

5. $u(x,0) = \sin(\pi x/L), \quad u_t(x,0) = \begin{cases} 0, & 0 < x < L/4 \\ a, & L/4 < x < 3L/4 \\ 0, & 3L/4 < x < L \end{cases}$

6. $u(x,0) = 0, \quad u_t(x,0) = \begin{cases} ax/L, & 0 < x < L/2 \\ a(L-x)/L, & L/2 < x < L \end{cases}$

7. $u(x,0) = \begin{cases} x, & 0 < x < L/2, \\ L-x, & L/2 < x < L, \end{cases} \quad u_t(x,0) = 0$

8. Solve the wave equation

$$\frac{\partial^2 u}{\partial t^2} = c^2 \frac{\partial^2 u}{\partial x^2}, \qquad 0 < x < \pi, \qquad 0 < t,$$

subject to the boundary conditions

$$\frac{\partial u(0,t)}{\partial x} = \frac{\partial u(\pi,t)}{\partial x} = 0, \qquad 0 < t,$$

and the initial conditions

$$u(x,0) = 0, \qquad \frac{\partial u(x,0)}{\partial t} = 1 + \cos^3(x), \qquad 0 < x < \pi.$$

[Hint: You must include the separation constant of zero.]

9. Solve[10] the wave equation

$$\frac{\partial^2 u}{\partial t^2} = \frac{\partial}{\partial x}\left(x\frac{\partial u}{\partial x}\right), \qquad 0 \leq x < 1, \qquad 0 < t,$$

subject to the boundary conditions

$$\lim_{x \to 0} |u(x,t)| < \infty, \qquad u(1,t) = 0, \qquad 0 < t,$$

and the initial conditions

$$u(x,0) = 0, \quad 0 \leq x \leq 1, \qquad \frac{\partial u(x,0)}{\partial t} = \begin{cases} 1, & 0 \leq x < a, \\ 0, & a < x \leq 1, \end{cases}$$

where $a < 1$. Hint: Use the substitution $4x = r^2$.

10. The differential equation for the longitudinal vibrations of a rod within a viscous fluid is

$$\frac{\partial^2 u}{\partial t^2} + 2h\frac{\partial u}{\partial t} = c^2\frac{\partial^2 u}{\partial x^2}, \qquad 0 < x < L, \qquad 0 < t,$$

where c is the velocity of sound in the rod and h is the damping coefficient. If the rod is fixed at $x = 0$ so that $u(0,t) = 0$, and allowed to freely oscillate at the other end $x = L$, so that $u_x(L,t) = 0$, find the vibrations for any location x and subsequent time t if the rod has the initial displacement of $u(x,0) = x$ and the initial velocity $u_t(x,0) = 0$ for $0 < x < L$. Assume that $h < c\pi/(2L)$. Why?

11. A closed pipe of length L contains air whose density is slightly greater than that of the outside air in the ratio of $1 + s_0$ to 1. Everything being at rest, we suddenly draw aside the disk closing one end of the pipe. We want to determine what happens *inside* the pipe after we remove the disk.

As the air rushes outside, it generates sound waves within the pipe. The wave equation

$$\frac{\partial^2 u}{\partial t^2} = c^2\frac{\partial^2 u}{\partial x^2}$$

---

[10] Solved in a slightly different manner by Bailey, H., 2000: Motions of a hanging chain after the free end is given an initial velocity. *Am. J. Phys.*, **68**, 764–767.

governs these waves, where c is the speed of sound and $u(x,t)$ is the velocity potential. Without going into the fluid mechanics of the problem, the boundary conditions are

a. No flow through the closed end: $u_x(0,t) = 0$.

b. No infinite acceleration at the open end: $u_{xx}(L,t) = 0$.

c. Air is initially at rest: $u_x(x,0) = 0$.

d. Air initially has a density greater than the surrounding air by the amount s_0: $u_t(x,0) = -c^2 s_0$.

Find the velocity potential at all positions within the pipe and all subsequent times.

12. One of the classic applications of the wave equation has been the explanation of the acoustic properties of string instruments. Usually we excite a string in one of three ways: by plucking (as in the harp, zither, etc.), by striking with a hammer (piano), or by bowing (violin, violoncello, etc.). In all of these cases, the governing partial differential equation is

$$\frac{\partial^2 u}{\partial t^2} = c^2 \frac{\partial^2 u}{\partial x^2}$$

with the boundary conditions $u(0,t) = u(L,t) = 0$, $0 < t$. For each of the following methods of exciting a string instrument, find the complete solution to the problem:

(a) Plucked string

For the initial conditions:

$$u(x,0) = \begin{cases} \beta x/a, & 0 < x < a, \\ \beta(L-x)/(L-a), & a < x < L, \end{cases}$$

and

$$u_t(x,0) = 0, \qquad 0 < x < L,$$

show that

$$u(x,t) = \frac{2\beta L^2}{\pi^2 a(L-a)} \sum_{n=1}^{\infty} \frac{1}{n^2} \sin\left(\frac{n\pi a}{L}\right) \sin\left(\frac{n\pi x}{L}\right) \cos\left(\frac{n\pi ct}{L}\right).$$

We note that the harmonics are absent where $\sin(n\pi a/L) = 0$. Thus, if we pluck the string at the center, all of the harmonics of even order are absent. Furthermore, the intensity of the successive harmonics varies as n^{-2}. The higher harmonics (overtones) are therefore relatively feeble compared to the $n = 1$ term (the fundamental).

(b) String excited by impact

The effect of the impact of a hammer depends upon the manner and duration of the contact, and is more difficult to estimate. However, as a first estimate, let

$$u(x,0) = 0, \qquad 0 < x < L,$$

and

$$u_t(x,0) = \begin{cases} \mu, & a - \epsilon < x < a + \epsilon, \\ 0, & \text{otherwise}, \end{cases}$$

where $\epsilon \ll 1$. Show that the solution in this case is

$$u(x,t) = \frac{4\mu L}{\pi^2 c} \sum_{n=1}^{\infty} \frac{1}{n^2} \sin\left(\frac{n\pi\epsilon}{L}\right) \sin\left(\frac{n\pi a}{L}\right) \sin\left(\frac{n\pi x}{L}\right) \sin\left(\frac{n\pi ct}{L}\right).$$

As in part (a), the nth mode is absent if the origin is at a node. The intensity of the overtones are now of the same order of magnitude; higher harmonics (overtones) are relatively more in evidence than in part (a).

(c) Bowed violin string

The theory of the vibration of a string when excited by bowing is poorly understood. The bow drags the string for a time until the string springs back. After awhile the process repeats. It can be shown[11] that the proper initial conditions are

$$u(x,0) = 0, \qquad 0 < x < L,$$

and

$$u_t(x,0) = 4\beta c(L - x)/L^2, \qquad 0 < x < L,$$

where β is the maximum displacement. Show that the solution is now

$$u(x,t) = \frac{8\beta}{\pi^2} \sum_{n=1}^{\infty} \frac{1}{n^2} \sin\left(\frac{n\pi x}{L}\right) \sin\left(\frac{n\pi ct}{L}\right).$$

10.4 D'ALEMBERT'S FORMULA

In the previous section we sought solutions to the homogeneous wave equation in the form of a product $X(x)T(t)$. For the one-dimensional wave

---

[11] See Lamb, H., 1960: *The Dynamical Theory of Sound*. Dover Publishers, §27.

Figure 10.4.1: Although largely self-educated in mathematics, Jean Le Rond d'Alembert (1717–1783) gained equal fame as a mathematician and *philosophe* of the continental Enlightenment. By the middle of the eighteenth century, he stood with such leading European mathematicians and mathematical physicists as Clairaut, D. Bernoulli, and Euler. Today we best remember him for his work in fluid dynamics and applying partial differential equations to problems in physics. (Portrait courtesy of the Archives de l'Académie des sciences, Paris.)

equation there is a more general method for constructing the solution, published by D'Alembert[12] in 1747.

Let us determine a solution to the homogeneous wave equation

$$\frac{\partial^2 u}{\partial t^2} = c^2 \frac{\partial^2 u}{\partial x^2}, \quad -\infty < x < \infty, \quad 0 < t, \tag{10.4.1}$$

which satisfies the initial conditions

$$u(x,0) = f(x), \quad \frac{\partial u(x,0)}{\partial t} = g(x), \quad -\infty < x < \infty. \tag{10.4.2}$$

We begin by introducing two new variables ξ, η defined by $\xi = x + ct$, and $\eta = x - ct$, and set $u(x,t) = w(\xi, \eta)$. The variables ξ and η are called the *characteristics* of the wave equation. Using the chain rule,

$$\frac{\partial}{\partial x} = \frac{\partial \xi}{\partial x} \frac{\partial}{\partial \xi} + \frac{\partial \eta}{\partial x} \frac{\partial}{\partial \eta} = \frac{\partial}{\partial \xi} + \frac{\partial}{\partial \eta} \tag{10.4.3}$$

---

[12] D'Alembert, J., 1747: Recherches sur la courbe que forme une corde tenduë mise en vibration. *Hist. Acad. R. Sci. Belles Lett.*, Berlin, 214–219.

$$\frac{\partial}{\partial t} = \frac{\partial \xi}{\partial t}\frac{\partial}{\partial \xi} + \frac{\partial \eta}{\partial t}\frac{\partial}{\partial \eta} = c\frac{\partial}{\partial \xi} - c\frac{\partial}{\partial \eta} \tag{10.4.4}$$

$$\frac{\partial^2}{\partial x^2} = \frac{\partial \xi}{\partial x}\frac{\partial}{\partial \xi}\left(\frac{\partial}{\partial \xi} + \frac{\partial}{\partial \eta}\right) + \frac{\partial \eta}{\partial x}\frac{\partial}{\partial \eta}\left(\frac{\partial}{\partial \xi} + \frac{\partial}{\partial \eta}\right) \tag{10.4.5}$$

$$= \frac{\partial^2}{\partial \xi^2} + 2\frac{\partial^2}{\partial \xi \partial \eta} + \frac{\partial^2}{\partial \eta^2}, \tag{10.4.6}$$

and similarly

$$\frac{\partial^2}{\partial t^2} = c^2\left(\frac{\partial^2}{\partial \xi^2} - 2\frac{\partial^2}{\partial \xi \partial \eta} + \frac{\partial^2}{\partial \eta^2}\right), \tag{10.4.7}$$

so that the wave equation becomes

$$\frac{\partial^2 w}{\partial \xi \partial \eta} = 0. \tag{10.4.8}$$

The general solution of (10.4.8) is

$$w(\xi, \eta) = F(\xi) + G(\eta). \tag{10.4.9}$$

Thus, the general solution of (10.4.1) is of the form

$$u(x,t) = F(x + ct) + G(x - ct), \tag{10.4.10}$$

where F and G are arbitrary functions of one variable and are assumed to be twice differentiable. Setting $t = 0$ in (10.4.10) and using the initial condition that $u(x,0) = f(x)$,

$$F(x) + G(x) = f(x). \tag{10.4.11}$$

The partial derivative of (10.4.10) with respect to t yields

$$\frac{\partial u(x,t)}{\partial t} = cF'(x + ct) - cG'(x - ct). \tag{10.4.12}$$

Here primes denote differentiation with respect to the argument of the function. If we set $t = 0$ in (10.4.12) and apply the initial condition that $u_t(x,0) = g(x)$,

$$cF'(x) - cG'(x) = g(x). \tag{10.4.13}$$

Integrating (10.4.13) from 0 to any point x gives

$$F(x) - G(x) = \frac{1}{c}\int_0^x g(\tau)\,d\tau + C, \tag{10.4.14}$$

where C is the constant of integration. Combining this result with (10.4.11),

$$F(x) = \frac{f(x)}{2} + \frac{1}{2c}\int_0^x g(\tau)\,d\tau + \frac{C}{2}, \tag{10.4.15}$$

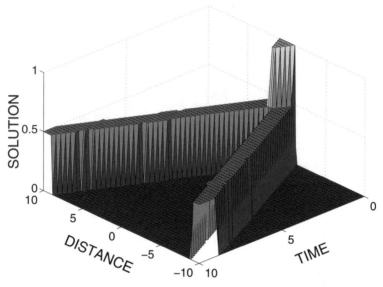

Figure 10.4.2: D'Alembert's solution (10.4.18) to the wave equation.

and

$$G(x) = \frac{g(x)}{2} - \frac{1}{2c} \int_0^x g(\tau)\,d\tau - \frac{C}{2}. \qquad \textbf{(10.4.16)}$$

If we replace the variable x in the expression for F and G by $x+ct$ and $x-ct$, respectively, and substitute the results into (10.4.10), we finally arrive at the formula

$$u(x,t) = \frac{f(x+ct) + f(x-ct)}{2} + \frac{1}{2c} \int_{x-ct}^{x+ct} g(\tau)\,d\tau. \qquad \textbf{(10.4.17)}$$

This is known as *d'Alembert's formula* for the solution of the wave equation (10.4.1) subject to the initial conditions (10.4.2). It gives a *representation* of the solution in terms of *known* initial conditions.

• **Example 10.4.1**

To illustrate d'Alembert's formula, let us find the solution to the wave equation (10.4.1) satisfying the initial conditions $u(x,0) = H(x+1) - H(x-1)$ and $u_t(x,0) = 0$, $-\infty < x < \infty$. By d'Alembert's formula (10.4.17),

$$u(x,t) = \tfrac{1}{2}\left[H(x + ct + 1) + H(x - ct + 1) - H(x + ct - 1) - H(x - ct - 1)\right]. \qquad \textbf{(10.4.18)}$$

We illustrate this solution in Figure 10.4.2 generated by the MATLAB script

```
% set mesh size for solution
clear; dx = 0.1; dt = 0.1;
```

```
% compute grid
X=[-10:dx:10]; T = [0:dt:10];
for j=1:length(T); t = T(j);
for i=1:length(X); x = X(i);
% compute characteristics
characteristic_1 = x + t; characteristic_2 = x - t;
% compute solution
XX(i,j) = x; TT(i,j) = t;
u(i,j ) = 0.5 * (stepfun(characteristic_1,-1) ...
          + stepfun(characteristic_2,-1) ...
          - stepfun(characteristic_1, 1) ...
          - stepfun(characteristic_2, 1));
end; end
surf(XX,TT,u); colormap autumn;
xlabel('DISTANCE','Fontsize',20); ylabel('TIME','Fontsize',20)
zlabel('SOLUTION','Fontsize',20)
```

In this figure, you can clearly see the characteristics as they emanate from the discontinuities at $x = \pm 1$.

- **Example 10.4.2**

Let us find the solution to the wave equation (10.4.1) when $u(x,0) = 0$, and $u_t(x,0) = \sin(2x)$, $-\infty < x < \infty$. By d'Alembert's formula, the solution is

$$u(x,t) = \frac{1}{2c} \int_{x-ct}^{x+ct} \sin(2\tau)\, d\tau = \frac{\sin(2x)\sin(2ct)}{2}. \qquad (\mathbf{10.4.19})$$

In addition to providing a method of solving the wave equation, d'Alembert's solution can also provide physical insight into the vibration of a string. Consider the case when we release a string with zero velocity after giving it an initial displacement of $f(x)$. According to (10.4.17), the displacement at a point x at any time t is

$$u(x,t) = \frac{f(x+ct) + f(x-ct)}{2}. \qquad (\mathbf{10.4.20})$$

Because the function $f(x - ct)$ is the same as the function of $f(x)$ translated to the right by a distance equal to ct, $f(x - ct)$ represents a wave of form $f(x)$ traveling to the right with the velocity c, a forward wave. Similarly, we can interpret the function $f(x + ct)$ as representing a wave with the shape $f(x)$ traveling to the left with the velocity c, a backward wave. Thus, the solution (10.4.17) is a superposition of forward and backward waves traveling with the same velocity c and having the shape of the initial profile $f(x)$ with half of the

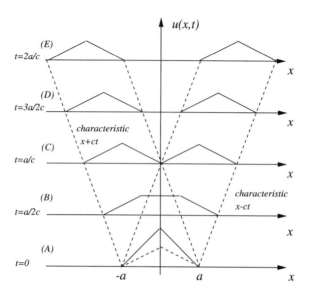

Figure 10.4.3: The propagation of waves due to an initial displacement according to d'Alembert's formula.

amplitude. Clearly the characteristics $x + ct$ and $x - ct$ give the propagation paths along which the waveform $f(x)$ propagates.

● **Example 10.4.3**

To illustrate our physical interpretation of d'Alembert's solution, suppose that the string has an initial displacement defined by

$$f(x) = \begin{cases} a - |x|, & -a \leq x \leq a, \\ 0, & \text{otherwise.} \end{cases} \qquad (10.4.21)$$

In Figure 10.4.3(A) the forward and backward waves, indicated by the dashed line, coincide at $t = 0$. As time advances, both waves move in opposite directions. In particular, at $t = a/(2c)$, they moved through a distance $a/2$, resulting in the displacement of the string shown in Figure 10.4.3(B). Eventually, at $t = a/c$, the forward and backward waves completely separate. Finally, Figures 10.4.3(D) and 10.4.3(E) show how the waves radiate off to infinity at the speed of c. Note that at each point the string returns to its original position of rest after the passage of each wave.

Consider now the opposite situation when $u(x,0) = 0$, and $u_t(x,0) = g(x)$. The displacement is

$$u(x, t) = \frac{1}{2c} \int_{x-ct}^{x+ct} g(\tau) \, d\tau. \qquad (10.4.22)$$

If we introduce the function

$$\varphi(x) = \frac{1}{2c} \int_0^x g(\tau)\, d\tau, \qquad (\mathbf{10.4.23})$$

then we can write (10.4.22) as

$$u(x,t) = \varphi(x + ct) - \varphi(x - ct), \qquad (\mathbf{10.4.24})$$

which again shows that the solution is a superposition of a forward wave $-\varphi(x - ct)$ and a backward wave $\varphi(x + ct)$ traveling with the same velocity c. The function φ, which we compute from (10.4.23) and the initial velocity $g(x)$, determines the exact form of these waves.

• Example 10.4.4: Vibration of a moving threadline

The characterization and analysis of the oscillations of a string or yarn have an important application in the textile industry because they describe the way that yarn winds on a bobbin.[13] As we showed in §10.4.1, the governing equation, the "threadline equation," is

$$\frac{\partial^2 u}{\partial t^2} + \alpha \frac{\partial^2 u}{\partial x \partial t} + \beta \frac{\partial^2 u}{\partial x^2} = 0, \qquad (\mathbf{10.4.25})$$

where $\alpha = 2V$, $\beta = V^2 - gT/\rho$, V is the windup velocity, g is the gravitational attraction, T is the tension in the yarn, and ρ is the density of the yarn. We now introduce the characteristics $\xi = x + \lambda_1 t$, and $\eta = x + \lambda_2 t$, where λ_1 and λ_2 are yet undetermined. Upon substituting ξ and η into (10.4.25),

$$(\lambda_1^2 + 2V\lambda_1 + V^2 - gT/\rho)u_{\xi\xi} + (\lambda_2^2 + 2V\lambda_2 + V^2 - gT/\rho)u_{\eta\eta}$$
$$+ [2V^2 - 2gT/\rho + 2V(\lambda_1 + \lambda_2) + 2\lambda_1\lambda_2]u_{\xi\eta} = 0. \qquad (\mathbf{10.4.26})$$

If we choose λ_1 and λ_2 to be roots of the equation

$$\lambda^2 + 2V\lambda + V^2 - gT/\rho = 0, \qquad (\mathbf{10.4.27})$$

(10.4.26) reduces to the simple form

$$u_{\xi\eta} = 0, \qquad (\mathbf{10.4.28})$$

which has the general solution

$$u(x,t) = F(\xi) + G(\eta) = F(x + \lambda_1 t) + G(x + \lambda_2 t). \qquad (\mathbf{10.4.29})$$

---

[13] Reprinted from *J. Franklin Inst.*, **275**, R. D. Swope and W. F. Ames, Vibrations of a moving threadline, 36–55, ©1963, with kind permission from Elsevier Science Ltd, The Boulevard, Langford Lane, Kidlington OX5 1GB, UK.

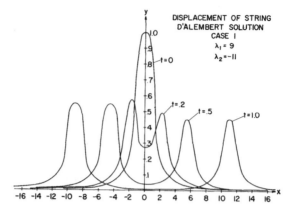

Figure 10.4.4: Displacement of an infinite, moving threadline when $c = 10$, and $V = 1$.

Solving (10.4.27) yields

$$\lambda_1 = c - V, \qquad \text{and} \qquad \lambda_2 = -c - V, \qquad (10.4.30)$$

where $c = \sqrt{gT/\rho}$. If the initial conditions are

$$u(x,0) = f(x), \qquad \text{and} \qquad u_t(x,0) = g(x), \qquad (10.4.31)$$

then

$$u(x,t) = \frac{1}{2c}\left[\lambda_1 f(x + \lambda_2 t) - \lambda_2 f(x + \lambda_1 t) + \int_{x+\lambda_2 t}^{x+\lambda_1 t} g(\tau)\,d\tau\right]. \qquad (10.4.32)$$

Because λ_1 does not generally equal to λ_2, the two waves that constitute the motion of the string move with different speeds and have different shapes and forms. For example, if

$$f(x) = \frac{1}{x^2 + 1}, \qquad \text{and} \quad g(x) = 0, \qquad (10.4.33)$$

$$u(x,t) = \frac{1}{2c}\left\{\frac{c - V}{1 + [x - (c + V)t]^2} + \frac{c + V}{1 + [x - (c - V)t]^2}\right\}. \qquad (10.4.34)$$

Figures 10.4.4 and 10.4.5 illustrate this solution for several different parameters.

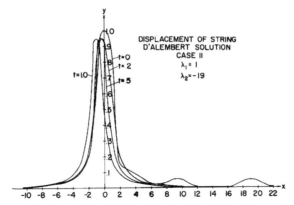

Figure 10.4.5: Displacement of an infinite, moving threadline when $c = 11$, and $V = 10$.

Problems

Use d'Alembert's formula to solve the wave equation (10.4.1) for the following initial conditions defined for $|x| < \infty$. Then illustrate your solution using MATLAB.

1. $u(x,0) = 2\sin(x)\cos(x)$ $u_t(x,0) = \cos(x)$
2. $u(x,0) = x\sin(x)$ $u_t(x,0) = \cos(2x)$
3. $u(x,0) = 1/(x^2 + 1)$ $u_t(x,0) = e^x$
4. $u(x,0) = e^{-x}$ $u_t(x,0) = 1/(x^2 + 1)$
5. $u(x,0) = \cos(\pi x/2)$ $u_t(x,0) = \sinh(ax)$
6. $u(x,0) = \sin(3x)$ $u_t(x,0) = \sin(2x) - \sin(x)$

7. Assuming that the functions F and G are differentiable, show by direct substitution that

$$u(x,t) = EF(x+ct) - EG(x-ct) - \tfrac{1}{8}kc^2t^2 + \tfrac{3}{8}kx^2,$$

and

$$v(x,t) = cF(x+ct) + cG(x-ct) - \frac{kc^2xt}{4E}$$

are the D'Alembert solutions to the hyperbolic system

$$\frac{\partial u}{\partial t} = E\frac{\partial v}{\partial x}, \qquad \frac{\partial u}{\partial x} = \rho\frac{\partial v}{\partial t} + kx, \qquad -\infty < x < \infty, \quad 0 < t,$$

where $c^2 = E/\rho$ and E, k, and ρ are constants.

8. D'Alembert's solution can also be used in problems over the limited domain $0 < x < L$. To illustrate this, let us solve the wave equation (10.4.1) with the initial conditions $u(x,0) = 0$, $u_t(x,0) = V_{max}(1 - x/L)$, $0 < x < L$, and the boundary conditions $u(0,t) = u(L,t) = 0$, $0 < t$.

Step 1: Show that the solution to this problem is

$$u(x, t) = \tfrac{1}{2}[V_0(x + ct) - V_0(x - ct)],$$

where

$$V_0(\chi) = \frac{1}{c} \int_0^\chi u_t(\xi, 0) \, d\xi = \frac{V_{max}\chi}{c} \left(1 - \frac{\chi}{2L}\right), \qquad 0 < \chi < L,$$

along with the periodicity conditions $V_0(\chi) = V_0(-\chi)$, and $V_0(L + \chi) = V_0(L - \chi)$ to take care of those cases when the argument of $V_0(\ )$ is outside of $(0, L)$. Hint: Substitute the solution into the boundary conditions.

Step 2: Show that at any point x within the interval $(0, L)$, the solution repeats with a period of $2L/c$ if $ct > 2L$. Therefore, if we know the behavior of the solution for the time interval $0 < ct < 2L$, we know the behavior for any other time.

Step 3: Show that the solution at any point x within the interval $(0, L)$ and time $t + L/c$, where $0 < ct < L$, is the mirror image (about $u = 0$) of the solution at the point $L - x$ and time t, where $0 < ct < L$.

Step 4: Show that the maximum value of $u(x, t)$ occurs at $x = ct$, where $0 < x < L$ and when $0 < ct < L$. At that point,

$$u_{max} = \frac{V_{max}x}{c} \left(1 - \frac{x}{L}\right),$$

where u_{max} equals the largest magnitude of $u(x, t)$ for any time t. Plot u_{max} as a function x and show that it a parabola. Hint: Find the maximum value of $u(x, t)$ when $0 < x \leq ct$ and $ct \leq x < L$ with $0 < x + ct < L$ or $L < x + ct < 2L$.

10.5 THE LAPLACE TRANSFORM METHOD

The solution of linear partial differential equations by Laplace transforms is the most commonly employed analytic technique after separation of variables. Because the transform consists solely of an integration with respect to time, the transform $U(x, s)$ of the solution of the wave equation $u(x, t)$ is

$$U(x, s) = \int_0^\infty u(x, t)e^{-st} \, dt, \tag{10.5.1}$$

assuming that the wave equation only varies in a single spatial variable x and time t.

Partial derivatives involving time have transforms similar to those that we encountered in the case of functions of a single variable. They include

$$\mathcal{L}[u_t(x, t)] = sU(x, s) - u(x, 0), \tag{10.5.2}$$

and

$$\mathcal{L}[u_{tt}(x,t)] = s^2 U(x,s) - su(x,0) - u_t(x,0). \qquad (10.5.3)$$

These transforms introduce the initial conditions via $u(x,0)$ and $u_t(x,0)$. On the other hand, derivatives involving x become

$$\mathcal{L}[u_x(x,t)] = \frac{d}{dx}\{\mathcal{L}[u(x,t)]\} = \frac{dU(x,s)}{dx}, \qquad (10.5.4)$$

and

$$\mathcal{L}[u_{xx}(x,t)] = \frac{d^2}{dx^2}\{\mathcal{L}[u(x,t)]\} = \frac{d^2U(x,s)}{dx^2}. \qquad (10.5.5)$$

Because the transformation eliminates the time variable, only $U(x,s)$ and its derivatives remain in the equation. Consequently, we transform the partial differential equation into a boundary-value problem involving an ordinary differential equation. Because this equation is often easier to solve than a partial differential equation, the use of Laplace transforms considerably simplifies the original problem. Of course, the Laplace transforms must exist for this technique to work.

The following schematic summarizes the Laplace transform method:

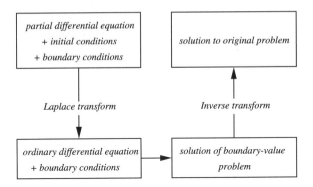

In the following examples, we illustrate transform methods by solving the classic equation of telegraphy as it applies to a uniform transmission line. The line has a resistance R, an inductance L, a capacitance C, and a leakage conductance G per unit length. We denote the current in the direction of positive x by I; V is the voltage drop across the transmission line at the point x. The dependent variables I and V are functions of both distance x along the line and time t.

To derive the differential equations that govern the current and voltage in the line, consider the points A at x and B at $x + \Delta x$ in Figure 10.5.1. The current and voltage at A are $I(x,t)$ and $V(x,t)$; at B, $I + \frac{\partial I}{\partial x}\Delta x$ and $V + \frac{\partial V}{\partial x}\Delta x$. Therefore, the voltage drop from A to B is $-\frac{\partial V}{\partial x}\Delta x$ and the

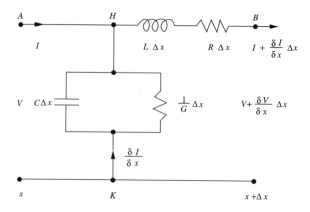

Figure 10.5.1: Schematic of an uniform transmission line.

current in the line is $I + \frac{\partial I}{\partial x}\Delta x$. Neglecting terms that are proportional to $(\Delta x)^2$,

$$\left(L\frac{\partial I}{\partial t} + RI\right)\Delta x = -\frac{\partial V}{\partial x}\Delta x. \tag{10.5.6}$$

The voltage drop over the parallel portion HK of the line is V while the current in this portion of the line is $-\frac{\partial I}{\partial x}\Delta x$. Thus,

$$\left(C\frac{\partial V}{\partial t} + GV\right)\Delta x = -\frac{\partial I}{\partial x}\Delta x. \tag{10.5.7}$$

Therefore, the differential equations for I and V are

$$L\frac{\partial I}{\partial t} + RI = -\frac{\partial V}{\partial x}, \tag{10.5.8}$$

and

$$C\frac{\partial V}{\partial t} + GV = -\frac{\partial I}{\partial x}. \tag{10.5.9}$$

Turning to the initial conditions, we solve these simultaneous partial differential equations with the initial conditions

$$I(x,0) = I_0(x), \tag{10.5.10}$$

and

$$V(x,0) = V_0(x) \tag{10.5.11}$$

for $0 < t$. There are also boundary conditions at the ends of the line; we will introduce them for each specific problem. For example, if the line is short-circuited at $x = a$, $V = 0$ at $x = a$; if there is an open circuit at $x = a$, $I = 0$ at $x = a$.

To solve (10.5.8)–(10.5.9) by Laplace transforms, we take the Laplace transform of both sides of these equations, which yields

$$(Ls + R)\overline{I}(x, s) = -\frac{d\overline{V}(x, s)}{dx} + LI_0(x), \tag{10.5.12}$$

and

$$(Cs + G)\overline{V}(x, s) = -\frac{d\overline{I}(x, s)}{dx} + CV_0(x). \tag{10.5.13}$$

Eliminating $\overline{I}$ gives an ordinary differential equation in $\overline{V}$

$$\frac{d^2\overline{V}}{dx^2} - q^2\overline{V} = L\frac{dI_0(x)}{dx} - C(Ls + R)V_0(x), \tag{10.5.14}$$

where $q^2 = (Ls + R)(Cs + G)$. After finding $\overline{V}$, we may compute $\overline{I}$ from

$$\overline{I} = -\frac{1}{Ls + R}\frac{d\overline{V}}{dx} + \frac{LI_0(x)}{Ls + R}. \tag{10.5.15}$$

At this point we treat several classic cases.

• Example 10.5.1: The semi-infinite transmission line

We consider the problem of a semi-infinite line $0 < x$ with no initial current and charge. The end $x = 0$ has a constant voltage E for $0 < t$.

In this case,

$$\frac{d^2\overline{V}}{dx^2} - q^2\overline{V} = 0, \qquad 0 < x. \tag{10.5.16}$$

The boundary conditions at the ends of the line are

$$V(0, t) = E, \qquad 0 < t, \tag{10.5.17}$$

and $V(x, t)$ is finite as $x \to \infty$. The transform of these boundary conditions is

$$\overline{V}(0, s) = E/s, \quad \text{and} \quad \lim_{x \to \infty} |\overline{V}(x, s)| < \infty. \tag{10.5.18}$$

The general solution of (10.5.16) is

$$\overline{V}(x, s) = Ae^{-qx} + Be^{qx}. \tag{10.5.19}$$

The requirement that $\overline{V}$ remains finite as $x \to \infty$ forces $B = 0$. The boundary condition at $x = 0$ gives $A = E/s$. Thus,

$$\overline{V}(x, s) = \frac{E}{s}\exp[-\sqrt{(Ls + R)(Cs + G)}\,x]. \tag{10.5.20}$$

We discuss the general case later. However, for the so-called "lossless" line, where $R = G = 0$,

$$\overline{V}(x, s) = \frac{E}{s} \exp(-sx/c), \tag{10.5.21}$$

where $c = 1/\sqrt{LC}$. Consequently,

$$V(x, t) = EH\left(t - \frac{x}{c}\right), \tag{10.5.22}$$

where $H(t)$ is Heaviside's step function. The physical interpretation of this solution is as follows: $V(x, t)$ is zero up to the time x/c at which time a wave traveling with speed c from $x = 0$ would arrive at the point x. $V(x, t)$ has the constant value E afterwards.

For the so-called "distortionless" line,[14] $R/L = G/C = \rho$,

$$V(x, t) = Ee^{-\rho x/c}H\left(t - \frac{x}{c}\right). \tag{10.5.23}$$

In this case, the disturbance not only propagates with velocity c but also attenuates as we move along the line.

Suppose now, that instead of applying a constant voltage E at $x = 0$, we apply a time-dependent voltage, $f(t)$. The only modification is that in place of (10.5.20),

$$\overline{V}(x, s) = F(s)e^{-qx}. \tag{10.5.24}$$

In the case of the distortionless line, $q = (s + \rho)/c$, this becomes

$$\overline{V}(x, s) = F(s)e^{-(s+\rho)x/c} \tag{10.5.25}$$

and

$$V(x, t) = e^{-\rho x/c}f\left(t - \frac{x}{c}\right)H\left(t - \frac{x}{c}\right). \tag{10.5.26}$$

Thus, our solution shows that the voltage at x is zero up to the time x/c. Afterwards $V(x, t)$ follows the voltage at $x = 0$ with a time lag of x/c and decreases in magnitude by $e^{-\rho x/c}$.

● **Example 10.5.2: The finite transmission line**

We now discuss the problem of a finite transmission line $0 < x < l$ with zero initial current and charge. We ground the end $x = 0$ and maintain the end $x = l$ at constant voltage E for $0 < t$.

The transformed partial differential equation becomes

$$\frac{d^2\overline{V}}{dx^2} - q^2\overline{V} = 0, \qquad 0 < x < l. \tag{10.5.27}$$

---

[14] Prechtl and Schürhuber [Prechtl, A., and R. Schürhuber, 2000: Nonuniform distortionless transmission lines. *Electr. Engng*, **82**, 127–134] have generalized this problem to nonuniform transmission lines.

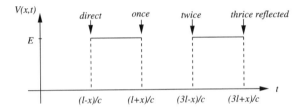

Figure 10.5.2: The voltage within a lossless, finite transmission line of length l as a function of time t.

The boundary conditions are

$$V(0,t) = 0, \quad \text{and} \quad V(l,t) = E, \qquad 0 < t. \tag{10.5.28}$$

The Laplace transform of these boundary conditions is

$$\overline{V}(0,s) = 0, \quad \text{and} \quad \overline{V}(l,s) = E/s. \tag{10.5.29}$$

The solution of (10.5.27) which satisfies the boundary conditions is

$$\overline{V}(x,s) = \frac{E \, \sinh(qx)}{s \, \sinh(ql)}. \tag{10.5.30}$$

Let us rewrite (10.5.30) in a form involving negative exponentials and expand the denominator by the binomial theorem,

$$\overline{V}(x,s) = \frac{E}{s} e^{-q(l-x)} \frac{1 - e^{-2qx}}{1 - e^{-2ql}} \tag{10.5.31}$$

$$= \frac{E}{s} e^{-q(l-x)} \left(1 - e^{-2qx}\right) \left(1 + e^{-2ql} + e^{-4ql} + \cdots\right) \tag{10.5.32}$$

$$= \frac{E}{s} \left[e^{-q(l-x)} - e^{-q(l+x)} + e^{-q(3l-x)} - e^{-q(3l+x)} + \cdots\right]. \tag{10.5.33}$$

In the special case of the lossless line where $q = s/c$,

$$\overline{V}(x,s) = \frac{E}{s} \left[e^{-s(l-x)/c} - e^{-s(l+x)/c} + e^{-s(3l-x)/c} - e^{-s(3l+x)/c} + \cdots\right], \tag{10.5.34}$$

or

$$V(x,t) = E\left[H\left(t - \frac{l-x}{c}\right) - H\left(t - \frac{l+x}{c}\right)\right.$$
$$\left. + H\left(t - \frac{3l-x}{c}\right) - H\left(t - \frac{3l+x}{c}\right) + \cdots\right]. \tag{10.5.35}$$

We illustrate (10.5.35) in Figure 10.5.2. The voltage at x is zero up to the time $(l - x)/c$, at which time a wave traveling directly from the end $x = l$ would reach the point x. The voltage then has the constant value E up to the time $(l + x)/c$, at which time a wave traveling from the end $x = l$ and reflected back from the end $x = 0$ would arrive. From this time up to the time of arrival of a twice-reflected wave, it has the value zero, and so on.

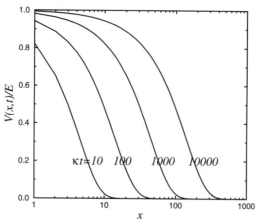

Figure 10.5.3: The voltage within a submarine cable as a function of distance for various values of κt.

• Example 10.5.3: The semi-infinite transmission line reconsidered

In the first example, we showed that the transform of the solution for the semi-infinite line is

$$\overline{V}(x, s) = \frac{E}{s} e^{-qx}, \tag{10.5.36}$$

where $q^2 = (Ls + R)(Cs + G)$. In the case of a lossless line $(R = G = 0)$, we found traveling wave solutions.

In this example, we shall examine the case of a submarine cable,[15] where $L = G = 0$. In this special case,

$$\overline{V}(x, s) = \frac{E}{s} e^{-x\sqrt{s/\kappa}}, \tag{10.5.37}$$

where $\kappa = 1/(RC)$. From a table of Laplace transforms,[16] we can immediately invert (10.5.37) and find that

$$V(x, t) = E \operatorname{erfc}\left(\frac{x}{2\sqrt{\kappa t}}\right), \tag{10.5.38}$$

where erfc is the complementary error function. Unlike the traveling wave solution, the voltage diffuses into the cable as time increases. We illustrate (10.5.38) in Figure 10.5.3.

---

[15] First solved by Thomson, W., 1855: On the theory of the electric telegraph. *Proc. R. Soc. London, Ser. A*, **7**, 382–399.

[16] See Churchill, R. V., 1972: *Operational Mathematics*. McGraw-Hill Book, §27.

• Example 10.5.4: A short-circuited, finite transmission line

Let us find the voltage of a lossless transmission line of length l that initially has the constant voltage E. At $t = 0$, we ground the line at $x = 0$ while we leave the end $x = l$ insulated.

The transformed partial differential equation now becomes

$$\frac{d^2 \overline{V}}{dx^2} - \frac{s^2}{c^2} \overline{V} = -\frac{sE}{c^2}, \tag{10.5.39}$$

where $c = 1/\sqrt{LC}$. The boundary conditions are

$$\overline{V}(0, s) = 0, \tag{10.5.40}$$

and

$$\overline{I}(l, s) = -\frac{1}{Ls} \frac{d\overline{V}(l, s)}{dx} = 0 \tag{10.5.41}$$

from (10.5.15).

The solution to this boundary-value problem is

$$\overline{V}(x, s) = \frac{E}{s} - \frac{E \cosh[s(l - x)/c]}{s \cosh(sl/c)}. \tag{10.5.42}$$

The first term on the right side of (10.5.42) is easy to invert and the inversion equals E. The second term is much more difficult to handle. We will use Bromwich's integral.

In §6.10 we showed that

$$\mathcal{L}^{-1}\left\{ \frac{\cosh[s(l - x)/c]}{s \cosh(sl/c)} \right\} = \frac{1}{2\pi i} \int_{c-\infty i}^{c+\infty i} \frac{\cosh[z(l - x)/c]e^{tz}}{z \cosh(zl/c)} \, dz. \tag{10.5.43}$$

To evaluate this integral we must first locate and then classify the singularities. Using the product formula for the hyperbolic cosine,

$$\frac{\cosh[z(l - x)/c]}{z \cosh(zl/c)} = \frac{[1 + \frac{4z^2(l-x)^2}{c^2\pi^2}][1 + \frac{4z^2(l-x)^2}{9c^2\pi^2}]\cdots}{z[1 + \frac{4z^2l^2}{c^2\pi^2}][1 + \frac{4z^2l^2}{9c^2\pi^2}]\cdots}. \tag{10.5.44}$$

This shows that we have an infinite number of simple poles located at $z = 0$, and $z_n = \pm(2n - 1)\pi ci/(2l)$, where $n = 1, 2, 3, \ldots$. Therefore, Bromwich's contour can lie along, and just to the right of, the imaginary axis. By Jordan's lemma we close the contour with a semicircle of infinite radius in the left half of the complex plane. Computing the residues,

$$\text{Res}\left\{ \frac{\cosh[z(l - x)/c]e^{tz}}{z \cosh(zl/c)}; 0 \right\} = \lim_{z \to 0} \frac{\cosh[z(l - x)/c]e^{tz}}{\cosh(zl/c)} = 1, \tag{10.5.45}$$

and

$$\mathrm{Res}\left\{\frac{\cosh[z(l-x)/c]e^{tz}}{z\cosh(zl/c)};z_n\right\}$$

$$=\lim_{z\to z_n}\frac{(z-z_n)\cosh[z(l-x)/c]e^{tz}}{z\cosh(zl/c)} \qquad (10.5.46)$$

$$=\frac{\cosh[(2n-1)\pi(l-x)i/(2l)]\exp[\pm(2n-1)\pi cti/(2l)]}{[(2n-1)\pi i/2]\sinh[(2n-1)\pi i/2]} \qquad (10.5.47)$$

$$=\frac{2(-1)^n}{(2n-1)\pi}\cos\left[\frac{(2n-1)\pi(l-x)}{2l}\right]\exp\left[\pm\frac{(2n-1)\pi cti}{2l}\right]. \qquad (10.5.48)$$

Summing the residues and using the relationship that $\cos(t)=(e^{ti}+e^{-ti})/2$,

$$V(x,t)=E-E\left\{1-\frac{4}{\pi}\sum_{n=1}^{\infty}\frac{(-1)^{n+1}}{2n-1}\cos\left[\frac{(2n-1)\pi(l-x)}{2l}\right]\right.$$

$$\left.\times\cos\left[\frac{(2n-1)\pi ct}{2l}\right]\right\} \qquad (10.5.49)$$

$$=\frac{4E}{\pi}\sum_{n=1}^{\infty}\frac{(-1)^{n+1}}{2n-1}\cos\left[\frac{(2n-1)\pi(l-x)}{2l}\right]\cos\left[\frac{(2n-1)\pi ct}{2l}\right]. \qquad (10.5.50)$$

An alternative to contour integration is to rewrite (10.5.42) as

$$\overline{V}(x,s)=\frac{E}{s}\left\{1-\frac{e^{-sx/c}\left[1+e^{-2s(l-x)/c}\right]}{1+e^{-2sl/c}}\right\} \qquad (10.5.51)$$

$$=\frac{E}{s}\left[1-e^{-sx/c}-e^{-s(2l-x)/c}+e^{-s(2l+x)/c}+\cdots\right] \qquad (10.5.52)$$

so that

$$V(x,t)=E\left[1-H\left(t-\frac{x}{c}\right)-H\left(t-\frac{2l-x}{c}\right)+H\left(t-\frac{2l+x}{c}\right)+\cdots\right]. \qquad (10.5.53)$$

● **Example 10.5.5: The general solution of the equation of telegraphy**

In this example we solve the equation of telegraphy without any restrictions on R, C, G, or L. We begin by eliminating the dependent variable $I(x,t)$ from the set of equations (10.5.8)–(10.5.9). This yields

$$CL\frac{\partial^2 V}{\partial t^2}+(GL+RC)\frac{\partial V}{\partial t}+RGV=\frac{\partial^2 V}{\partial x^2}. \qquad (10.5.54)$$

We next take the Laplace transform of (10.5.54) assuming that $V(x,0) = f(x)$, and $V_t(x,0) = g(x)$. The transformed version of (10.5.54) is

$$\frac{d^2\overline{V}}{dx^2} - [CLs^2 + (GL + RC)s + RG]\overline{V} = -CLg(x) - (CLs + GL + RC)f(x),$$
(10.5.55)

or

$$\frac{d^2\overline{V}}{dx^2} - \frac{(s+\rho)^2 - \sigma^2}{c^2}\overline{V} = -\frac{g(x)}{c^2} - \left(\frac{s}{c^2} + \frac{2\rho}{c^2}\right)f(x),$$
(10.5.56)

where $c^2 = 1/LC$, $\rho = c^2(RC + GL)/2$, and $\sigma = c^2(RC - GL)/2$.

We solve (10.5.56) by Fourier transforms (see §5.6) with the requirement that the solution dies away as $|x| \to \infty$. The most convenient way of expressing this solution is the convolution product (see §5.5)

$$\overline{V}(x,s) = \left[\frac{g(x)}{c} + \left(\frac{s}{c} + \frac{2\rho}{c}\right)f(x)\right] * \frac{\exp[-|x|\sqrt{(s+\rho)^2 - \sigma^2}/c]}{2\sqrt{(s+\rho)^2 - \sigma^2}}. \quad (10.5.57)$$

From a table of Laplace transforms,

$$\mathcal{L}^{-1}\left[\frac{\exp\left(-b\sqrt{s^2 - a^2}\right)}{\sqrt{s^2 - a^2}}\right] = I_0\left(a\sqrt{t^2 - b^2}\right)H(t - b), \quad (10.5.58)$$

where $b > 0$ and $I_0(\ )$ is the zeroth order modified Bessel function of the first kind. Therefore, by the first shifting theorem,

$$\mathcal{L}^{-1}\left\{\frac{\exp\left[-|x|\sqrt{(s+\rho)^2 - \sigma^2}/c\right]}{\sqrt{(s+\rho)^2 - \sigma^2}}\right\} = e^{-\rho t}I_0\left[\sigma\sqrt{t^2 - (x/c)^2}\right]H\left(t - \frac{|x|}{c}\right).$$
(10.5.59)

Using (10.5.59) to invert (10.5.57), we have that

$$\begin{aligned}
V(x,t) = &\tfrac{1}{2c}e^{-\rho t}g(x) * I_0\left[\sigma\sqrt{t^2 - (x/c)^2}\right]H(t - |x|/c) \\
&+ \tfrac{1}{2c}e^{-\rho t}f(x) * \frac{\partial}{\partial t}\left\{I_0[\sigma\sqrt{t^2 - (x/c)^2}]\right\}H(t - |x|/c) \\
&+ \tfrac{\rho}{c}e^{-\rho t}f(x) * I_0\left[\sigma\sqrt{t^2 - (x/c)^2}\right]H(t - |x|/c) \\
&+ \tfrac{1}{2}e^{-\rho t}[f(x + ct) + f(x - ct)].
\end{aligned}$$
(10.5.60)

The last term in (10.5.60) arises from noting that $sF(s) = \mathcal{L}[f(t)] + f(0)$. If we explicitly write out the convolution, the final form of the solution is

$$\begin{aligned}
V(x,t) = &\tfrac{1}{2}e^{-\rho t}[f(x + ct) + f(x - ct)] \\
&+ \tfrac{1}{2c}e^{-\rho t}\int_{x-ct}^{x+ct}[g(\eta) + 2\rho f(\eta)]I_0\left[\sigma\sqrt{c^2t^2 - (x - \eta)^2}\Big/c\right]d\eta \\
&+ \tfrac{1}{2c}e^{-\rho t}\int_{x-ct}^{x+ct}f(\eta)\frac{\partial}{\partial t}\left\{I_0\left[\sigma\sqrt{c^2t^2 - (x - \eta)^2}\Big/c\right]\right\}d\eta. \quad (10.5.61)
\end{aligned}$$

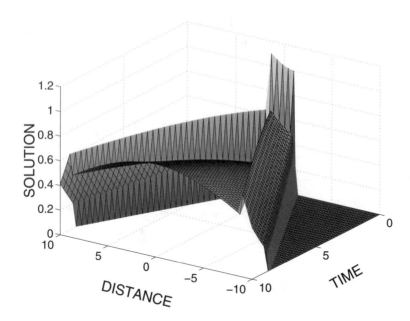

Figure 10.5.4: The evolution of the voltage with time given by the general equation of telegraphy for initial conditions and parameters stated in the text.

The physical interpretation of the first line of (10.5.61) is straightforward. It represents damped progressive waves; one is propagating to the right and the other to the left. In addition to these progressive waves, there is a contribution from the integrals, even after the waves pass. These integrals include all of the points where $f(x)$ and $g(x)$ are nonzero within a distance ct from the point in question. This effect persists through all time, although dying away, and constitutes a residue or tail. Figure 10.5.4 illustrates this for $\rho = 0.1$, $\sigma = 0.2$, and $c = 1$. This figure was obtained using the MATLAB script:

```
% initialize parameters in calculation
clear; dx = 0.1; dt = 0.5; rho_over_c = 0.1; sigma_over_c = 0.2;
%
X=[-10:dx:10]; T = [0:dt:10]; % compute locations of x and t
%
for j=1:length(T); t = T(j);
for i=1:length(X); x = X(i);
XX(i,j) = x; TT(i,j) = t; deta_i = 0.05 % set up grid
%
% compute characteristics x+ct and x-ct
%
characteristic_1 = x - t; characteristic_2 = x + t;
%
% compute first term in (10.5.61)
```

```
%
F = inline('stepfun(x,-1.0001)-stepfun(x,1.0001)');
u(i,j ) = F(characteristic_1) + F(characteristic_2);
%
% find the upper and lower limits of the integration
%
upper = characteristic_2; lower = characteristic_1;
%
if t > 0 & upper > -1 & lower < 1
if upper > 1 upper = 1; end
if lower < -1 lower = -1; end
%
% set up parameters needed for integration
%
interval = upper-lower;
NN = interval / deta_i;
if mod(NN,2) > 0 NN = NN + 1; end;
deta = interval / NN;
%
% compute integrals in (10.5.61) by Simpson's rule
% sum1 deals with the first integral while sum2 is the second
%
sum1 = 0; sum2 = 0; eta = lower;
for k = 0:2:NN-2
arg = sigma_over_c * sqrt(t*t-(x-eta)*(x-eta));
sum1 = sum1 + besseli(0,arg);
if (arg == 0) sum2 = sum2 + 0.5 * sigma_over_c * t;
else sum2 = sum2 + t * besseli(1,arg) / arg; end
eta = eta + deta;
arg = sigma_over_c * sqrt(t*t-(x-eta)*(x-eta));
sum1 = sum1 + 4*besseli(0,arg);
if (arg == 0) sum2 = sum2 + 4 * 0.5 * sigma_over_c * t;
else sum2 = sum2 + 4 * t * besseli(1,arg) / arg; end
eta = eta + deta;
arg = sigma_over_c * sqrt(t*t-(x-eta)*(x-eta));
sum1 = sum1 + besseli(0,arg);
if (arg == 0) sum2 = sum2 + 0.5 * sigma_over_c * t;
else sum2 = sum2 + t * besseli(1,arg) / arg; end
end
u(i,j) = u(i,j) + 2 * rho_over_c * deta * sum1 / 3 ...
        + sigma_over_c * deta * sum2 / 3;
end
%
% multiply final answer by damping coefficient
%
```

```
u(i,j) = 0.5 * exp(-rho_over_c * t) * u(i,j);
%
end;end;
%
% plot results
%
mesh(XX,TT,real(u)); colormap spring;
xlabel('DISTANCE','Fontsize',20); ylabel('TIME','Fontsize',20)
zlabel('SOLUTION','Fontsize',20)
```

We evaluated the integrals by Simpson's rule for the initial conditions $f(x) = H(x+1) - H(x-1)$, and $g(x) = 0$. If there was no loss, then two pulses would propagate to the left and right. However, with resistance and leakage the waves leave a residue after their leading edge has passed.

• Example 10.5.6: Cutoff frequency

A powerful method for solving certain partial differential equations is the joint application of Laplace and Fourier transforms. To illustrate this *joint transform method*, let us find the *Green's function* for the Klein-Gordon equation

$$\frac{\partial^2 u}{\partial x^2} - \frac{1}{c^2}\left(\frac{\partial^2 u}{\partial t^2} + a^2 u\right) = -\delta(x)\delta(t), \qquad -\infty < x < \infty, \quad 0 < t, \quad (\mathbf{10.5.62})$$

subject to the boundary condition $\lim_{x\to\pm\infty}|u(x,t)| < \infty$, $0 < t$, and the initial conditions $u(x,0) = u_t(x,0) = 0$, $-\infty < x < \infty$.

We begin by taking the Laplace transform of (10.5.62) and find that

$$\frac{d^2 U(x,s)}{dx^2} - \frac{s^2 + a^2}{c^2}U(x,s) = -\delta(x), \qquad -\infty < x < \infty, \qquad (\mathbf{10.5.63})$$

with the boundary condition $\lim_{x\to\pm\infty}|U(x,s)| < \infty$. Assuming that the Fourier transform of $U(x,s)$, $\overline{U}(k,s)$, exists, we take the Fouier transform of (10.5.63) and obtain

$$-k^2\overline{U}(k,s) - \frac{s^2 + a^2}{c^2}\overline{U}(x,s) = -1, \qquad (\mathbf{10.5.64})$$

or

$$\overline{U}(k,s) = \frac{c^2}{s^2 + k^2 c^2 + a^2}, \qquad (\mathbf{10.5.65})$$

where $\overline{U}(k,s) = \int_{-\infty}^{\infty} U(x,s)e^{-ikx}\,dx$.

Inverting the Laplace transform first, we have that

$$\overline{u}(k,t) = c^2\frac{\sin\left(t\sqrt{k^2 c^2 + a^2}\right)}{\sqrt{k^2 c^2 + a^2}}. \qquad (\mathbf{10.5.66})$$

Consequently,

$$u(x,t) = \frac{c^2}{2\pi} \int_{-\infty}^{\infty} \frac{\sin\left(t\sqrt{k^2c^2 + a^2}\right)}{\sqrt{k^2c^2 + a^2}} e^{ikx}\, dk \qquad (\mathbf{10.5.67})$$

$$= \frac{c^2}{\pi} \int_{0}^{\infty} \frac{\sin\left(t\sqrt{k^2c^2 + a^2}\right)}{\sqrt{k^2c^2 + a^2}} \cos(kx)\, dk \qquad (\mathbf{10.5.68})$$

$$= \frac{c}{2} J_0\left(a\sqrt{t^2 - x^2/c^2}\right) H(ct - |x|), \qquad (\mathbf{10.5.69})$$

where $J_0(\ )$ is the zeroth order Bessel function of the first kind. Thus, forcing the Klein-Gorden equation by an impulse forcing yields waves that propagate to the right and left from $x = 0$ with the wave front located at $x = \pm ct$. At a given point, after the passage of the wave front, the solution vibrates with an ever decreasing amplitude and at a frequency that approaches a — the so-called *cutoff frequency* — at $t \to \infty$.

Why is a called a cutoff frequency? From (10.5.67), we see that, although the spectral representation includes all of the wavenumbers k running from $-\infty$ to ∞, the frequency $\omega = \sqrt{c^2k^2 + a^2}$ is restrictred to the range $\omega \geq a$. Thus, a is the lowest possible frequency that a wave solution to the Klein-Gorden equation may have for a real value of k.

Problems

1. Use transform methods to solve the wave equation

$$\frac{\partial^2 u}{\partial t^2} = \frac{\partial^2 u}{\partial x^2}, \quad 0 < x < 1, \quad 0 < t,$$

with the boundary conditions $u(0,t) = u(1,t) = 0$, $0 < t$, and the initial conditions $u(x,0) = 0$, $u_t(0,t) = 1$, $0 < x < 1$.

2. Use transform methods to solve the wave equation

$$\frac{\partial^2 u}{\partial t^2} = \frac{\partial^2 u}{\partial x^2}, \quad 0 < x < 1, \quad 0 < t,$$

with the boundary conditions $u(0,t) = u_x(1,t) = 0$, $0 < t$, and the initial conditions $u(x,0) = 0$, $u_t(0,t) = x$, $0 < x < 1$.

3. Use transform methods to solve the wave equation

$$\frac{\partial^2 u}{\partial t^2} = \frac{\partial^2 u}{\partial x^2}, \quad 0 < x < 1, \quad 0 < t,$$

with the boundary conditions $u(0,t) = u(1,t) = 0$, $0 < t$, and the initial conditions $u(x,0) = \sin(\pi x)$, $u_t(x,0) = -\sin(\pi x)$, $0 < x < 1$.

4. Use transform methods to solve the wave equation

$$\frac{\partial^2 u}{\partial t^2} = c^2 \frac{\partial^2 u}{\partial x^2}, \quad 0 < x < a, \quad 0 < t,$$

with the boundary conditions $u(0, t) = \sin(\omega t)$, $u(a, t) = 0$, $0 < t$, and the initial conditions $u(x, 0) = u_t(x, 0) = 0$, $0 < x < a$. Assume that $\omega a/c$ is *not* an integer multiple of π. Why?

5. Use transform methods to solve the wave equation

$$\frac{\partial^2 u}{\partial t^2} = c^2 \frac{\partial^2 u}{\partial x^2}, \quad 0 < x < L, \quad 0 < t,$$

with the boundary conditions $u_x(0, t) = -f(t)$, $u_x(L, t) = 0$, $0 < t$, and the initial conditions $u(x, 0) = u_t(x, 0) = 0$, $0 < x < L$. Hint: Invert the Laplace transform following the procedure used in Example 10.5.2.

6. Use transform methods to solve the wave equation

$$\frac{\partial^2 u}{\partial t^2} = c^2 \frac{\partial^2 u}{\partial x^2} - q'(t), \quad a < x < b, \quad 0 < t,$$

with the boundary conditions $u(a, t) = 0$, $u_x(b, t) = 0$, $0 < t$, and the initial conditions $u(x, 0) = 0$, $u_t(x, 0) = -q(0)$, $a < x < b$. Hint: To find $U(x, s)$, express both $U(x, s)$ and the right side of the ordinary differential equation governing $U(x, s)$ in an eigenfunction expansion using $\sin\{(2n+1)\pi(x-a)/[2(b-a)]\}$. These eigenfunctions satisfy the boundary conditions.

7. Use transform methods to solve the wave equation

$$\frac{\partial^2 u}{\partial t^2} - \frac{\partial^2 u}{\partial x^2} = te^{-x}, \quad 0 < x < \infty, \quad 0 < t,$$

with the boundary conditions

$$u(0, t) = 1 - e^{-t}, \quad \lim_{x \to \infty} |u(x, t)| \sim x^n, \; n \text{ finite}, \quad 0 < t,$$

and the initial conditions $u(x, 0) = 0$, $u_t(x, 0) = x$, $0 < x < \infty$.

8. Use transform methods to solve the wave equation

$$\frac{\partial^2 u}{\partial t^2} - \frac{\partial^2 u}{\partial x^2} = xe^{-t}, \quad 0 < x < \infty, \quad 0 < t,$$

with the boundary conditions

$$u(0, t) = \cos(t), \quad \lim_{x \to \infty} |u(x, t)| \sim x^n, \; n \text{ finite}, \quad 0 < t,$$

and the initial conditions $u(x,0) = 1$, $u_t(x,0) = 0$, $0 < x < \infty$.

9. Use transform methods to solve the wave equation

$$\frac{\partial^2 u}{\partial t^2} = \frac{\partial^2 u}{\partial x^2}, \quad 0 < x < L, \quad 0 < t,$$

with the boundary conditions

$$u(0,t) = 0, \quad \frac{\partial^2 u(L,t)}{\partial t^2} + \frac{k}{m}\frac{\partial u(L,t)}{\partial x} = g, \quad 0 < t,$$

and the initial conditions $u(x,0) = u_t(x,0) = 0$, $0 < x < L$, where k, m, and g are constants.

10. Use transform methods[17] to solve the wave equation

$$\frac{\partial^2 u}{\partial t^2} = c^2 \frac{\partial}{\partial x}\left(x\frac{\partial u}{\partial x}\right), \quad 0 < x < 1, \quad 0 < t,$$

with the boundary conditions

$$\lim_{x \to 0} |u(x,t)| < \infty, \quad u(1,t) = A\sin(\omega t), \quad 0 < t,$$

and the initial conditions $u(x,0) = u_t(x,0) = 0$, $0 < x < 1$. Assume that $2\omega \neq c\beta_n$, where $J_0(\beta_n) = 0$. [Hint: The ordinary differential equation

$$\frac{d}{dx}\left(x\frac{dU}{dx}\right) - \frac{s^2}{c^2}U = 0$$

has the solution

$$U(x,s) = c_1 I_0\left(\frac{s}{c}\sqrt{x}\right) + c_2 K_0\left(\frac{s}{c}\sqrt{x}\right),$$

where $I_0(x)$ and $K_0(x)$ are modified Bessel functions of the first and second kind, respectively. Note that $J_n(iz) = i^n I_n(z)$ and $I_n(iz) = i^n J_n(z)$ for complex z.]

11. A lossless transmission line of length ℓ has a constant voltage E applied to the end $x = 0$ while we insulate the other end $[V_x(\ell,t) = 0]$. Find the voltage at any point on the line if the initial current and charge are zero.

---

[17] Suggested by a problem solved by Brown, J., 1975: Stresses in towed cables during re-entry. *J. Spacecr. Rockets*, **12**, 524–527.

12. Solve the equation of telegraphy without leakage

$$\frac{\partial^2 u}{\partial x^2} = CR\frac{\partial u}{\partial t} + CL\frac{\partial^2 u}{\partial t^2}, \qquad 0 < x < \ell, \qquad 0 < t,$$

subject to the boundary conditions $u(0,t) = 0$, $u(\ell,t) = E$, $0 < t$, and the initial conditions $u(x,0) = u_t(x,0) = 0$, $0 < x < \ell$. Assume that $4\pi^2 L/CR^2\ell^2 > 1$. Why?

13. The pressure and velocity oscillations from water hammer in a pipe without friction[18] are given by the equations

$$\frac{\partial p}{\partial t} = -\rho c^2 \frac{\partial u}{\partial x}, \qquad \text{and} \qquad \frac{\partial u}{\partial t} = -\frac{1}{\rho}\frac{\partial p}{\partial x},$$

where $p(x,t)$ denotes the pressure perturbation, $u(x,t)$ is the velocity perturbation, c is the speed of sound in water, and ρ is the density of water. These two first-order partial differential equations can be combined to yield

$$\frac{\partial^2 p}{\partial t^2} = c^2 \frac{\partial^2 p}{\partial x^2}.$$

Find the solution to this partial differential equation if $p(0,t) = p_0$, and $u(L,t) = 0$, and the initial conditions are $p(x,0) = p_0$, $p_t(x,0) = 0$, and $u(x,0) = u_0$.

14. Use Laplace transforms to solve the wave equation[19]

$$\frac{\partial^2 u}{\partial t^2} = c^2 \left(\frac{\partial^2 u}{\partial r^2} + \frac{2}{r}\frac{\partial u}{\partial r} - \frac{2u}{r^2} \right), \qquad a < r < \infty, \quad 0 < t,$$

subject to the boundary conditions that

$$u(a,t) = A\left(1 - e^{-ct/a}\right) H(t), \qquad \lim_{r\to\infty} |u(r,t)| < \infty, \quad 0 < t,$$

and the initial conditions that $u(r,0) = u_t(r,0) = 0$, $a < r < \infty$. Hint: The homogeneous solution to the ordinary differential equation

$$\frac{d^2 y}{dr^2} + \frac{2}{r}\frac{dy}{dr} - \frac{2y}{r^2} - b^2 y = 0$$

[18] See Rich, G. R., 1945: Water-hammer analysis by the Laplace-Mellin transformation. *Trans. ASME*, **67**, 361–376.

[19] Reprinted from *Soil Dynam. Earthq. Engng.*, **5**, J. P. Wolf and G. R. Darbre, Time-domain boundary element method in visco-elasticity with application to a spherical cavity, 138–148, ©1986, with permission from Elsevier Science.

is

$$y(r) = C_1 \left[\frac{\cosh(br)}{br} - \frac{\sinh(br)}{b^2 r^2} \right] + C_2 \left(\frac{1}{br} + \frac{1}{b^2 r^2} \right) e^{-br}.$$

15. Use Laplace transforms to solve the wave equation[20]

$$\frac{\partial^2 (ru)}{\partial t^2} = c^2 \frac{\partial^2 (ru)}{\partial r^2}, \qquad a < r < \infty, \qquad 0 < t,$$

subject to the boundary conditions that

$$-\rho c^2 \left(\frac{\partial^2 u}{\partial r^2} + \frac{2}{3r} \frac{\partial u}{\partial r} \right) \Big|_{r=a} = p_0 e^{-\alpha t} H(t), \qquad \lim_{r \to \infty} |u(r,t)| < \infty, \quad 0 < t,$$

where $\alpha > 0$, and the initial conditions that $u(r,0) = u_t(r,0) = 0$, $a < r < \infty$.

16. Consider a vertical rod or column of length L that is supported at both ends. The elastic waves that arise when the support at the bottom is suddenly removed are governed by the wave equation[21]

$$\frac{\partial^2 u}{\partial t^2} = c^2 \frac{\partial^2 u}{\partial x^2} + g, \qquad 0 < x < L, \qquad 0 < t,$$

where g denotes the gravitational acceleration, $c^2 = E/\rho$, E is Young's modulus, and ρ is the mass density. Find the wave solution if the boundary conditions are $u_x(0,t) = u_x(L,t) = 0$, $0 < t$, and the initial conditions are

$$u(x,0) = -\frac{gx^2}{2c^2}, \qquad \frac{\partial u(x,0)}{\partial t} = 0, \qquad 0 < x < L.$$

17. Use Laplace transforms to solve the hyperbolic equation

$$\frac{\partial^2 u}{\partial t^2} - \frac{\partial^2 u}{\partial x^2} + 1 = 0, \qquad 0 < x < 1, \qquad 0 < t,$$

subject to the boundary conditions that $u_x(0,t) = 0$, $u_x(1,t) = 1$, $0 < t$, and the initial conditions that $u(x,0) = u_t(x,0) = 0$, $0 < x < 1$.

---

[20] Originally solved using Fourier transforms by Sharpe, J. A., 1942: The production of elastic waves by explosion pressures. I. Theory and empirical field observations. *Geophysics*, **7**, 144–154.

[21] Abstracted with permission from Hall, L. H., 1953: Longitudinal vibrations of a vertical column by the method of Laplace transform. *Am. J. Phys.*, **21**, 287–292. ©1953 American Association of Physics Teachers.

18. Solve the telegraph-like equation[22]

$$\frac{\partial^2 u}{\partial t^2} + k\frac{\partial u}{\partial t} = c^2\left(\frac{\partial^2 u}{\partial x^2} + a\frac{\partial u}{\partial x}\right), \quad 0 < x < \infty, \quad 0 \le t$$

subject to the boundary conditions

$$\frac{\partial u(0,t)}{\partial x} = -u_0\delta(t), \quad \lim_{x\to\infty}|u(x,t)| < \infty, \quad 0 \le t,$$

and the initial conditions $u(x,0) = u_0$, $u_t(x,0) = 0$, $0 < x < \infty$, with $ac > k$.

Step 1: Take the Laplace transform of the partial differential equation and boundary conditions and show that

$$\frac{d^2U(x,s)}{dx^2} + a\frac{dU(x,s)}{dx} - \left(\frac{s^2 + ks}{c^2}\right)U(x,s) = -\left(\frac{s+k}{c^2}\right)u_0,$$

with $U'(0,s) = -u_0$, and $\lim_{x\to\infty}|U(x,s)| < \infty$.

Step 2: Show that the solution to the previous step is

$$U(x,s) = \frac{u_0}{s} + u_0 e^{-ax/2}\frac{\exp\left[-x\sqrt{\left(s+\frac{k}{2}\right)^2 + a^2}/c\right]}{\frac{\alpha}{2} + \sqrt{(s+\frac{k}{2})^2 + a^2}/c},$$

where $4a^2 = \alpha^2 c^2 - k^2 > 0$.

Step 3: Using the first and second shifting theorems and the property that

$$F\left(\sqrt{s^2 + a^2}\right) = \mathcal{L}\left[f(t) - a\int_0^t \frac{J_1\left(a\sqrt{t^2 - \tau^2}\right)}{\sqrt{t^2 - \tau^2}}\tau f(\tau)\,d\tau\right],$$

show that

$$u(x,t) = u_0 + u_0 c e^{-kt/2}H(t - x/c)$$
$$\times\left[e^{-\alpha ct/2} - a\int_{x/c}^t \frac{J_1\left(a\sqrt{t^2 - \tau^2}\right)}{\sqrt{t^2 - \tau^2}}\tau e^{-\alpha c\tau/2}\,d\tau\right].$$

19. As an electric locomotive travels down a track at the speed V, the pantograph (the metallic framework that connects the overhead power lines to the

---

[22] From Abbott, M. R., 1959: The downstream effect of closing a barrier across an estuary with particular reference to the Thames. *Proc. R. Soc. London, Ser. A*, **251**, 426–439 with permission.

locomotive) pushes up the line with a force P. Let us find the behavior[23] of the overhead wire as a pantograph passes between two supports of the electrical cable that are located a distance L apart. We model this system as a vibrating string with a point load:

$$\frac{\partial^2 u}{\partial t^2} = c^2 \frac{\partial^2 u}{\partial x^2} + \frac{P}{\rho V} \delta\left(t - \frac{x}{V}\right), \quad 0 < x < L, \quad 0 < t.$$

Let us assume that the wire is initially at rest $[u(x,0) = u_t(x,0) = 0$ for $0 < x < L]$ and fixed at both ends $[u(0,t) = u(L,t) = 0$ for $0 < t]$.

Step 1: Take the Laplace transform of the partial differential equation and show that

$$s^2 U(x,s) = c^2 \frac{d^2 U(x,s)}{dx^2} + \frac{P}{\rho V} e^{-xs/V}.$$

Step 2: Solve the ordinary differential equation in Step 1 as a Fourier half-range sine series

$$U(x,s) = \sum_{n=1}^{\infty} B_n(s) \sin\left(\frac{n\pi x}{L}\right),$$

where

$$B_n(s) = \frac{2P\beta_n}{\rho L(\beta_n^2 - \alpha_n^2)} \left[\frac{1}{s^2 + \alpha_n^2} - \frac{1}{s^2 + \beta_n^2}\right] \left[1 - (-1)^n e^{-Ls/V}\right],$$

$\alpha_n = n\pi c/L$ and $\beta_n = n\pi V/L$. This solution satisfies the boundary conditions.

Step 3: By inverting the solution in Step 2, show that

$$u(x,t) = \frac{2P}{\rho L} \sum_{n=1}^{\infty} \left[\frac{\sin(\beta_n t)}{\alpha_n^2 - \beta_n^2} - \frac{V}{c} \frac{\sin(\alpha_n t)}{\alpha_n^2 - \beta_n^2}\right] \sin\left(\frac{n\pi x}{L}\right)$$

$$- \frac{2P}{\rho L} H\left(t - \frac{L}{V}\right) \sum_{n=1}^{\infty} (-1)^n \sin\left(\frac{n\pi x}{L}\right)$$

$$\times \left\{\frac{\sin[\beta_n(t - L/V)]}{\alpha_n^2 - \beta_n^2} - \frac{V}{c} \frac{\sin[\alpha_n(t - L/V)]}{\alpha_n^2 - \beta_n^2}\right\}$$

or

$$u(x,t) = \frac{2P}{\rho L} \sum_{n=1}^{\infty} \left[\frac{\sin(\beta_n t)}{\alpha_n^2 - \beta_n^2} - \frac{V}{c} \frac{\sin(\alpha_n t)}{\alpha_n^2 - \beta_n^2}\right] \sin\left(\frac{n\pi x}{L}\right)$$

$$- \frac{2P}{\rho L} H\left(t - \frac{L}{V}\right) \sum_{n=1}^{\infty} \sin\left(\frac{n\pi x}{L}\right)$$

$$\times \left\{\frac{\sin(\beta_n t)}{\alpha_n^2 - \beta_n^2} - \frac{V}{c}(-1)^n \frac{\sin[\alpha_n(t - L/V)]}{\alpha_n^2 - \beta_n^2}\right\}.$$

---

[23] From Oda, O., and Y. Ooura, 1976: Vibrations of catenary overhead wire. *Q. Rep., (Tokyo) Railway Tech. Res. Inst.*, **17**, 134–135 with permission.

The first term in both summations represents the static uplift on the line; this term disappears after the pantograph passes. The second term in both summations represents the vibrations excited by the traveling force. Even after the pantograph passes, they continue to exist.

20. Solve the wave equation

$$\frac{1}{c^2}\frac{\partial^2 u}{\partial t^2} - \frac{\partial^2 u}{\partial r^2} - \frac{1}{r}\frac{\partial u}{\partial r} + \frac{u}{r^2} = \frac{\delta(r-\alpha)}{\alpha^2}, \qquad 0 \le r < a, \quad 0 < t,$$

where $0 < \alpha < a$, subject to the boundary conditions

$$\lim_{r \to 0} |u(r,t)| < \infty, \qquad \frac{\partial u(a,t)}{\partial r} + \frac{h}{a}u(a,t) = 0, \quad 0 < t,$$

and the initial conditions $u(r,0) = u_t(r,0) = 0$, $0 \le r < a$.

Step 1: Take the Laplace transform of the partial differential equation and show that

$$\frac{d^2 U(r,s)}{dr^2} + \frac{1}{r}\frac{dU(r,s)}{dr} - \left(\frac{s^2}{c^2} + \frac{1}{r^2}\right)U(r,s) = -\frac{\delta(r-\alpha)}{s\alpha^2}, \qquad 0 \le r < a,$$

with

$$\lim_{r \to 0} |U(r,s)| < \infty, \qquad \frac{dU(a,s)}{dr} + \frac{h}{a}U(a,s) = 0.$$

Step 2: Show that the Dirac delta function can be reexpressed as the Fourier-Bessel series

$$\delta(r-\alpha) = \frac{2\alpha}{a^2} \sum_{n=1}^{\infty} \frac{\beta_n^2 J_1(\beta_n\alpha/a)}{(\beta_n^2 + h^2 - 1)J_1^2(\beta_n)}J_1(\beta_n r/a), \qquad 0 \le r < a,$$

where β_n is the nth root of $\beta J_1'(\beta) + h\,J_1(\beta) = \beta J_0(\beta) + (h-1)J_1(\beta) = 0$ and $J_0(\ )$, $J_1(\ )$ are the zeroth and first-order Bessel functions of the first kind, respectively.

Step 3: Show that solution to the ordinary differential equation in Step 1 is

$$U(r,s) = \frac{2}{\alpha} \sum_{n=1}^{\infty} \frac{J_1(\beta_n\alpha/a)J_1(\beta_n r/a)}{(\beta_n^2 + h^2 - 1)\,J_1^2(\beta_n)}\left[\frac{1}{s} - \frac{s}{s^2 + c^2\beta_n^2/a^2}\right].$$

Note that this solution satisfies the boundary conditions.

Step 4: Taking the inverse of the Laplace transform in Step 3, show that the solution to the partial differential equation is

$$u(r,t) = \frac{2}{\alpha} \sum_{n=1}^{\infty} \frac{J_1(\beta_n\alpha/a)J_1(\beta_n r/a)}{(\beta_n^2 + h^2 - 1)\,J_1^2(\beta_n)}\left[1 - \cos\left(\frac{c\beta_n t}{a}\right)\right].$$

21. Solve the hyperbolic equation

$$\frac{\partial^2 u}{\partial x \partial t} + u = 0, \qquad 0 < x, t,$$

subject to the boundary conditions $u(0, t) = e^{-t}$, $\lim_{x \to \infty} |u(x, t)| < \infty$, $0 < t$, and $u(x, 0) = 1$, $\lim_{t \to \infty} |u(x, t)| < Me^{kt}$, $0 < k, M, x, t$.

Step 1: Take the Laplace transform of the partial differential equation and show that

$$s\frac{dU(x, s)}{dx} + U = 0, \qquad U(0, s) = \frac{1}{s + 1}, \qquad \lim_{x \to \infty} |U(x, s)| < \infty.$$

Step 2: Show that

$$U(x, s) = \frac{e^{-x/s}}{s + 1} = \frac{e^{-x/s}}{s} - \frac{e^{-x/s}}{s(s + 1)}.$$

Step 3: Using tables and the convolution theorem, show that the solution is

$$u(x, t) = J_0(2\sqrt{xt}) - e^{-t} \int_0^t e^\tau J_0(2\sqrt{x\tau}) \, d\tau,$$

where $J_0(\ )$ is the Bessel function of the first kind and order zero.

22. Solve the hyperbolic equation

$$\frac{\partial^2 u}{\partial x \partial t} + a\frac{\partial u}{\partial t} + b\frac{\partial u}{\partial x} = 0, \qquad 0 < a, b, x, t,$$

subject to the boundary conditions $u(0, t) = e^{ct}$, $\lim_{x \to \infty} |u(x, t)| < \infty$, $0 < t$, and $u(x, 0) = 1$, $\lim_{t \to \infty} |u(x, t)| < Me^{kt}$, $0 < k, M, t, x$.

Step 1: Take the Laplace transform of the partial differential equation and show that

$$(s + b)\frac{dU(x, s)}{dx} + asU = a, \qquad U(0, s) = \frac{1}{s - c}, \qquad \lim_{x \to \infty} |U(x, s)| < \infty.$$

Step 2: Show that

$$U(x, s) = \frac{1}{s} + \frac{ce^{-ax}}{s(s - c)} \exp\left(\frac{bx}{s + b}\right).$$

Step 3: Using tables, the first shifting theorem, and the convolution theorem, show that the solution is

$$u(x,t) = 1 + c\, e^{ct-ax} \int_0^t e^{-(b+c)\tau} I_0(2\sqrt{bx\tau}\,)\, d\tau,$$

where $I_0(\ )$ is the modified Bessel function of the first kind and order zero.

10.6 NUMERICAL SOLUTION OF THE WAVE EQUATION

Despite the powerful techniques shown in the previous sections for solving the wave equation, often these analytic techniques fail and we must resort to numerical techniques. In contrast to the continuous solutions, finite difference methods, a type of numerical solution technique, give discrete numerical values at a specific location (x_m, t_n), called a *grid point*. These numerical values represent a numerical approximation of the continuous solution over the region $(x_m - \Delta x/2, x_m + \Delta x/2)$ and $(t_n - \Delta t/2, t_n + \Delta t/2)$, where Δx and Δt are the distance and time intervals between grid points, respectively. Clearly, in the limit of $\Delta x, \Delta t \to 0$, we recover the continuous solution. However, practical considerations such as computer memory or execution time often require that Δx and Δt, although small, are not negligibly small.

The first task in the numerical solution of a partial differential equation is the replacement of its continuous derivatives with finite differences. The most popular approach employs Taylor expansions. If we focus on the x-derivative, then the value of the solution at $u[(m+1)\Delta x, n\Delta t]$ in terms of the solution at $(m\Delta x, n\Delta t)$ is

$$u[(m+1)\Delta x, n\Delta t] = u(x_m, t_n) + \frac{\Delta x}{1!}\frac{\partial u(x_m, t_n)}{\partial x} + \frac{(\Delta x)^2}{2!}\frac{\partial^2 u(x_m, t_n)}{\partial x^2}$$
$$+ \frac{(\Delta x)^3}{3!}\frac{\partial^3 u(x_m, t_n)}{\partial x^3} + \frac{(\Delta x)^4}{4!}\frac{\partial^4 u(x_m, t_n)}{\partial x^4} + \cdots \quad (10.6.1)$$
$$= u(x_m, t_n) + \Delta x\frac{\partial u(x_m, t_n)}{\partial x} + O[(\Delta x)^2], \quad (10.6.2)$$

where $O[(\Delta x)^2]$ gives a measure of the magnitude of neglected terms.[24]

From (10.6.2), one possible approximation for u_x is

$$\frac{\partial u(x_m, t_n)}{\partial x} = \frac{u_{m+1}^n - u_m^n}{\Delta x} + O(\Delta x), \quad (10.6.3)$$

where we use the standard notation that $u_m^n = u(x_m, t_n)$. This is an example of a *one-sided finite difference* approximation of the partial derivative u_x. The error in using this approximation grows as Δx.

---

[24] The symbol O is a mathematical notation indicating relative magnitude of terms, namely that $f(\epsilon) = O(\epsilon^n)$ provided $\lim_{\epsilon\to 0}|f(\epsilon)/\epsilon^n| < \infty$. For example, as $\epsilon \to 0$, $\sin(\epsilon) = O(\epsilon)$, $\sin(\epsilon^2) = O(\epsilon^2)$, and $\cos(\epsilon) = O(1)$.

Another possible approximation for the derivative arises from using $u(m\Delta x, n\Delta t)$ and $u[(m-1)\Delta x, n\Delta t]$. From the Taylor expansion:

$$u[(m-1)\Delta x, n\Delta t] = u(x_m, t_n) - \frac{\Delta x}{1!}\frac{\partial u(x_m, t_n)}{\partial x} + \frac{(\Delta x)^2}{2!}\frac{\partial^2 u(x_m, t_n)}{\partial x^2}$$
$$- \frac{(\Delta x)^3}{3!}\frac{\partial^3 u(x_m, t_n)}{\partial x^3} + \frac{(\Delta x)^4}{4!}\frac{\partial^4 u(x_m, t_n)}{\partial x^4} - \cdots,$$

$$(10.6.4)$$

we can also obtain the one-sided difference formula

$$\frac{u(x_m, t_n)}{\partial x} = \frac{u_m^n - u_{m-1}^n}{\Delta x} + O(\Delta x). \qquad (10.6.5)$$

A third possibility arises from subtracting (10.6.4) from (10.6.1):

$$u_{m+1}^n - u_{m-1}^n = 2\Delta x\frac{\partial u(x_m, t_n)}{\partial x} + O[(\Delta x)^3], \qquad (10.6.6)$$

or

$$\frac{\partial u(x_m, t_n)}{\partial x} = \frac{u_{m+1}^n - u_{m-1}^n}{2\Delta x} + O[(\Delta x)^2]. \qquad (10.6.7)$$

Thus, the choice of the finite differencing scheme can produce profound differences in the accuracy of the results. In the present case, *centered finite differences* can yield results that are markedly better than using one-sided differences.

To solve the wave equation, we need to approximate u_{xx}. If we add (10.6.1) and (10.6.4),

$$u_{m+1}^n + u_{m-1}^n = 2u_m^n + \frac{\partial^2 u(x_m, t_n)}{\partial x^2}(\Delta x)^2 + O[(\Delta x)^4], \qquad (10.6.8)$$

or

$$\frac{\partial^2 u(x_m, t_n)}{\partial x^2} = \frac{u_{m+1}^n - 2u_m^n + u_{m-1}^n}{(\Delta x)^2} + O[(\Delta x)^2]. \qquad (10.6.9)$$

Similar considerations hold for the time derivative. Thus, by neglecting errors of $O[(\Delta x)^2]$ and $O[(\Delta t)^2]$, we may approximate the wave equation by

$$\frac{u_m^{n+1} - 2u_m^n + u_m^{n-1}}{(\Delta t)^2} = c^2\frac{u_{m+1}^n - 2u_m^n + u_{m-1}^n}{(\Delta x)^2}. \qquad (10.6.10)$$

Because the wave equation represents evolutionary change of some quantity, (10.6.10) is generally used as a predictive equation where we forecast u_m^{n+1} by

$$u_m^{n+1} = 2u_m^n - u_m^{n-1} + \left(\frac{c\Delta t}{\Delta x}\right)^2\left(u_{m+1}^n - 2u_m^n + u_{m-1}^n\right). \qquad (10.6.11)$$

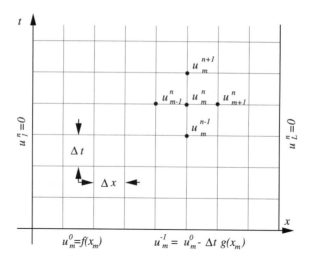

Figure 10.6.1: Schematic of the numerical solution of the wave equation with fixed end points.

Figure 10.6.1 illustrates this numerical scheme.

The greatest challenge in using (10.6.11) occurs with the very first prediction. When $n = 0$, clearly u_{m+1}^0, u_m^0, and u_{m-1}^0 are specified from the initial condition $u(m\Delta x, 0) = f(x_m)$. But what about u_m^{-1}? Recall that we still have $u_t(x, 0) = g(x)$. If we use the backward difference formula (10.6.5),

$$\frac{u_m^0 - u_m^{-1}}{\Delta t} = g(x_m). \tag{10.6.12}$$

Solving for u_m^{-1},

$$u_m^{-1} = u_m^0 - \Delta t g(x_m). \tag{10.6.13}$$

One disadvantage of using the backward finite-difference formula is the larger error associated with this term compared to those associated with the finite-differenced form of the wave equation. In the case of the barotropic vorticity equation, a partial differential equation with wave-like solutions, this inconsistency eventually leads to a separation of solution between adjacent time levels.[25] This difficulty is avoided by stopping after a certain number of time steps, averaging the solution, and starting again.

A better solution for computing that first time step employs the centered difference form

$$\frac{u_m^1 - u_m^{-1}}{2\Delta t} = g(x_m), \tag{10.6.14}$$

---

[25] Gates, W. L., 1959: On the truncation error, stability, and convergence of difference solutions of the barotropic vorticity equation. *J. Meteorol.*, **16**, 556–568. See §4.

along with the wave equation

$$\frac{u_m^1 - 2u_m^0 + u_m^{-1}}{(\Delta t)^2} = c^2 \frac{u_{m+1}^0 - 2u_m^0 + u_{m-1}^0}{(\Delta x)^2}, \tag{10.6.15}$$

so that

$$u_m^1 = \left(\frac{c\Delta t}{\Delta x}\right)^2 \frac{f(x_{m+1}) + f(x_{m-1})}{2} + \left[1 - \left(\frac{c\Delta t}{\Delta x}\right)^2\right] f(x_m) + \Delta t g(x_m). \tag{10.6.16}$$

Although it appears that we are ready to start calculating, we need to check whether our numerical scheme possesses three properties: convergence, stability, and consistency. By *consistency* we mean that the difference equations approach the differential equation as $\Delta x, \Delta t \to 0$. To prove consistency, we first write u_{m+1}^n, u_{m-1}^n, u_m^{n-1}, and u_m^{n+1} in terms of $u(x,t)$ and its derivatives evaluated at (x_m, t_n). From Taylor expansions,

$$u_{m+1}^n = u_m^n + \Delta x \frac{\partial u}{\partial x}\Big|_n^m + \frac{1}{2}(\Delta x)^2 \frac{\partial^2 u}{\partial x^2}\Big|_n^m + \frac{1}{6}(\Delta x)^3 \frac{\partial^3 u}{\partial x^3}\Big|_n^m + \cdots, \tag{10.6.17}$$

$$u_{m-1}^n = u_m^n - \Delta x \frac{\partial u}{\partial x}\Big|_n^m + \frac{1}{2}(\Delta x)^2 \frac{\partial^2 u}{\partial x^2}\Big|_n^m - \frac{1}{6}(\Delta x)^3 \frac{\partial^3 u}{\partial x^3}\Big|_n^m + \cdots, \tag{10.6.18}$$

$$u_m^{n+1} = u_m^n + \Delta t \frac{\partial u}{\partial t}\Big|_n^m + \frac{1}{2}(\Delta t)^2 \frac{\partial^2 u}{\partial t^2}\Big|_n^m + \frac{1}{6}(\Delta t)^3 \frac{\partial^3 u}{\partial t^3}\Big|_n^m + \cdots, \tag{10.6.19}$$

and

$$u_m^{n-1} = u_m^n - \Delta t \frac{\partial u}{\partial t}\Big|_n^m + \frac{1}{2}(\Delta t)^2 \frac{\partial^2 u}{\partial t^2}\Big|_n^m - \frac{1}{6}(\Delta t)^3 \frac{\partial^3 u}{\partial t^3}\Big|_n^m + \cdots. \tag{10.6.20}$$

Substituting (10.6.17)–(10.6.20) into (10.6.10), we obtain

$$\frac{u_m^{n+1} - 2u_m^n + u_m^{n-1}}{(\Delta t)^2} - c^2 \frac{u_{m+1}^n - 2u_m^n + u_{m-1}^n}{(\Delta x)^2}$$
$$= \left(\frac{\partial^2 u}{\partial t^2} - c^2 \frac{\partial^2 u}{\partial x^2}\right)\Big|_n^m + \frac{1}{12}(\Delta t)^2 \frac{\partial^4 u}{\partial t^4}\Big|_n^m - \frac{1}{12}(c\Delta x)^2 \frac{\partial^4 u}{\partial x^4}\Big|_n^m + \cdots. \tag{10.6.21}$$

The first term on the right side of (10.6.21) vanishes because $u(x,t)$ satisfies the wave equation. As $\Delta x \to 0$, $\Delta t \to 0$, the remaining terms on the right side of (10.6.21) tend to zero and (10.6.10) is a consistent finite difference approximation of the wave equation.

Stability is another question. Under certain conditions the small errors inherent in fixed precision arithmetic (round off) can grow for certain choices of Δx and Δt. During the 1920s the mathematicians Courant, Friedrichs, and

Lewy[26] found that if $c\Delta t/\Delta x > 1$, then our scheme is unstable. This CFL criterion has its origin in the fact that if $c\Delta t > \Delta x$, then we are asking signals in the numerical scheme to travel faster than their real-world counterparts and this unrealistic expectation leads to instability!

One method of determining *stability*, commonly called the von Neumann method,[27] involves examining solutions to (10.6.11) that have the form

$$u_m^n = e^{im\theta}e^{in\lambda}, \tag{10.6.22}$$

where θ is an arbitrary real number and λ is a yet undetermined complex number. Our choice of (10.6.22) is motivated by the fact that the initial condition u_m^0 can be represented by a Fourier series where a typical term behaves as $e^{im\theta}$.

If we substitute (10.6.22) into (10.6.10) and divide out the common factor $e^{im\theta}e^{in\lambda}$, we have that

$$\frac{e^{i\lambda} - 2 + e^{-i\lambda}}{(\Delta t)^2} = c^2 \frac{e^{i\theta} - 2 + e^{-i\theta}}{(\Delta x)^2}, \tag{10.6.23}$$

or

$$\sin^2\left(\frac{\lambda}{2}\right) = \left(\frac{c\Delta t}{\Delta x}\right)^2 \sin^2\left(\frac{\theta}{2}\right). \tag{10.6.24}$$

The behavior of u_m^n is determined by the values of λ given by (10.6.24). If $c\Delta t/\Delta x \le 1$, then λ is real and u_m^n is bounded for all θ as $n \to \infty$. If $c\Delta t/\Delta x > 1$, then it is possible to find a value of θ such that the right side of (10.6.24) exceeds unity and the corresponding values of λ occur as complex conjugate pairs. The λ with the negative imaginary part produces a solution with exponential growth because $n = t_n/\Delta t \to \infty$ as $\Delta t \to 0$ for a fixed t_n and $c\Delta t/\Delta x$. Thus, the value of u_m^n becomes infinitely large, even though the initial data may be arbitrarily small.

Finally, we must check for convergence. A numerical scheme is *convergent* if the numerical solution approaches the continuous solution as $\Delta x, \Delta t \to 0$. The general procedure for proving convergence involves the evolution of the error term e_m^n which gives the difference between the true solution $u(x_m, t_n)$ and the finite difference solution u_m^n. From (10.6.21),

$$e_m^{n+1} = \left(\frac{c\Delta t}{\Delta x}\right)^2 (e_{m+1}^n + e_{m-1}^n) + 2\left[1 - \left(\frac{c\Delta t}{\Delta x}\right)^2\right]e_m^n - e_m^{n-1}$$
$$+ O[(\Delta t)^4] + O[(\Delta x)^2(\Delta t)^2]. \tag{10.6.25}$$

---

[26] Courant, R., K. O. Friedrichs, and H. Lewy, 1928: Über die partiellen Differenzengleichungen der mathematischen Physik. *Math. Annalen*, **100**, 32–74. Translated into English in *IBM J. Res. Dev.*, **11**, 215–234.

[27] After its inventor, J. von Neumann. See O'Brien, G. G., M. A. Hyman, and S. Kaplan, 1950: A study of the numerical solution of partial differential equations. *J. Math. Phys.* (Cambridge, MA), **29**, 223–251.

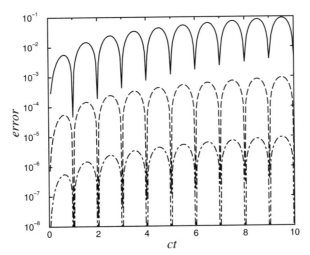

Figure 10.6.2: The growth of error $||e_n||$ as a function of ct for various resolutions. For the top line, $\Delta x = 0.1$; for the middle line, $\Delta x = 0.01$; and for the bottom line, $\Delta x = 0.001$.

Let us apply (10.6.25) to work backwards from the point (x_m, t_n) by changing n to $n-1$. The nonvanishing terms in e_m^n reduce to a sum of $n+1$ values on the line $n = 1$ plus $\frac{1}{2}(n+1)n$ terms of the form $A(\Delta x)^4$. If we define the max norm $||e_n|| = \max_m |e_m^n|$, then

$$||e_n|| \leq nB(\Delta x)^3 + \tfrac{1}{2}(n+1)nA(\Delta x)^4. \tag{10.6.26}$$

Because $n\Delta x \leq ct_n$, (10.6.26) simplifies to

$$||e_n|| \leq ct_n B(\Delta x)^2 + \tfrac{1}{2}c^2 t_n^2 A(\Delta x)^2. \tag{10.6.27}$$

Thus, the error tends to zero as $\Delta x \to 0$, verifying convergence. We illustrate (10.6.27) by using the finite difference equation (10.6.11) to compute $||e_n||$ during a numerical experiment that used $c\Delta t/\Delta x = 0.5$, $f(x) = \sin(\pi x)$, and $g(x) = 0$; $||e_n||$ is plotted in Figure 10.6.2. Note how each increase of resolution by 10 results in a drop in the error by 100.

In the following examples we apply our scheme to solve a few simple initial and boundary conditions:

• Example 10.6.1

For our first example, we resolve (10.3.1)–(10.3.3) and (10.3.25)–(10.3.26) numerically using MATLAB. The MATLAB code is

```
clear
coeff = 0.5; coeffsq = coeff * coeff % coeff = cΔt/Δx
dx = 0.04; dt = coeff * dx; N = 100; x = 0:dx:1;
M = 1/dx + 1; % M = number of spatial grid points
```

```
% introduce the initial conditions via F and G
F = zeros(M,1); G = zeros(M,1);
for m = 1:M
if x(m) >= 0.25 & x(m) <= 0.5
    F(m) = 4 * x(m) - 1; end
if x(m) >= 0.5 & x(m) <= 0.75
    F(m) = 3 - 4 * x(m); end; end
% at t = 0, the solution is:
tplot(1) = 0; u = zeros(M,N+1); u(1:M,1) = F(1:M);
% at t = Δt, the solution is given by (10.6.16)
tplot(2) = dt;
for m = 2:M-1
u(m,2) = 0.5 * coeffsq * (F(m+1) + F(m-1)) ...
        + (1 - coeffsq) * F(m) + dt * G(m);
end
% in general, the solution is given by (10.6.11)
for n = 2:N
tplot(n+1) = dt * n;
for m = 2:M-1
u(m,n+1) = 2 * u(m,n) - u(m,n-1) ...
        + coeffsq * (u(m+1,n) - 2 * u(m,n) + u(m-1,n));
end; end
X = x' * ones(1,length(tplot)); T = ones(M,1) * tplot;
surf(X,T,u)
xlabel('DISTANCE','Fontsize',20); ylabel('TIME','Fontsize',20)
zlabel('SOLUTION','Fontsize',20)
```

Overall, the numerical solution shown in Figure 10.6.3 approximates the exact or analytic solution well. However, we note small-scale noise in the numerical solution at later times. Why does this occur? Recall that the exact solution could be written as an infinite sum of sines in the x dimension. Each successive harmonic adds a contribution from waves of shorter and shorter wavelength. In the case of the numerical solution, the longer-wavelength harmonics are well represented by the numerical scheme because there are many grid points available to resolve a given wavelength. As the wavelengths become shorter, the higher harmonics are poorly resolved by the numerical scheme, move at incorrect phase speeds, and their misplacement (dispersion) creates the small-scale noise that you observe rather than giving the sharp angular features of the exact solution. The only method for avoiding this problem is to devise schemes that minimize dispersion.

• Example 10.6.2

Let us redo Example 10.6.1 except that we introduce the boundary condition that $u_x(L,t) = 0$. This corresponds to a string where we fix the left end and allow the right end to freely move up and down. This requires a

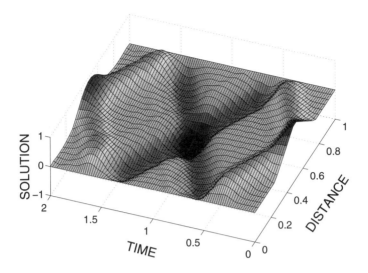

Figure 10.6.3: The numerical solution $u(x,t)/h$ of the wave equation with $c\Delta t/\Delta x = \frac{1}{2}$ using (10.6.11) at various positions $x' = x/L$ and times $t' = ct/L$. The exact solution is plotted in Figure 10.3.2.

new difference condition along the right boundary. If we employ centered differencing,

$$\frac{u_{L+1}^n - u_{L-1}^n}{2\Delta x} = 0, \tag{10.6.28}$$

and

$$u_L^{n+1} = 2u_L^n - u_L^{n-1} + \left(\frac{c\Delta t}{\Delta x}\right)^2 \left(u_{L+1}^n - 2u_L^n + u_{L-1}^n\right). \tag{10.6.29}$$

Eliminating u_{L+1}^n between (10.6.28)–(10.6.29),

$$u_L^{n+1} = 2u_L^n - u_L^{n-1} + \left(\frac{c\Delta t}{\Delta x}\right)^2 \left(2u_{L-1}^n - 2u_L^n\right). \tag{10.6.30}$$

For the special case of $n = 1$, (10.6.30) becomes

$$u_L^1 = f(x_L) + \left(\frac{c\Delta t}{\Delta x}\right)^2 [f(x_{L-1}) - f(x_L)] + \Delta t f(x_L). \tag{10.6.31}$$

The MATLAB code used to numerically solve the wave equation with a Neumann boundary condition is very similar to the one used in the previous example that we must add the line

```
u(M,2) = coeffsq * F(M-1) + (1-coeffsq) * F(M) + dt*G(M);
```

after

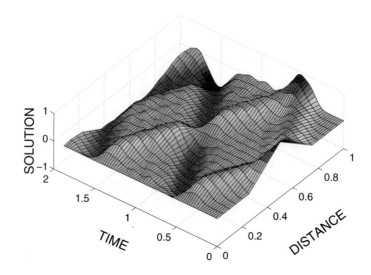

Figure 10.6.4: The numerical solution $u(x,t)/h$ of the wave equation when the right end moves freely with $c\Delta t/\Delta x = \frac{1}{2}$ using (10.6.11) and (10.6.30) at various positions $x' = x/L$ and times $t' = ct/L$.

```
for m = 2:M-1
u(m,2) = 0.5 * coeffsq * (F(m+1) + F(m-1)) ...
         + (1 - coeffsq) * F(m) + dt * G(m);
end
```

and

```
u(M,n+1) = 2 * u(M,n) - u(M,n-1) ...
           + 2 * coeffsq * (u(M-1,n)-u(M,n));
```

after

```
for m = 2:M-1
u(m,n+1) = 2 * u(m,n) - u(m,n-1) ...
           + coeffsq * (u(m+1,n) - 2 * u(m,n) + u(m-1,n));
end
```

Figure 10.6.4 shows the results. The numerical solution agrees well with the exact solution

$$u(x,t) = \frac{32h}{\pi^2} \sum_{n=1}^{\infty} \frac{1}{(2n-1)^2} \sin\left[\frac{(2n-1)\pi x}{2L}\right] \cos\left[\frac{(2n-1)\pi ct}{2L}\right]$$

$$\times \left\{ 2\sin\left[\frac{(2n-1)\pi}{4}\right] - \sin\left[\frac{3(2n-1)\pi}{8}\right] - \sin\left[\frac{(2n-1)\pi}{8}\right] \right\}. \textbf{(10.6.32)}$$

The results are also consistent with those presented in Example 10.6.1, especially with regard to small-scale noise due to dispersion.

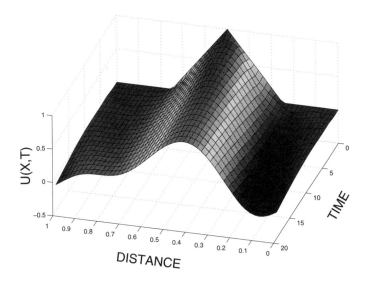

Figure 10.6.5: The numerical solution $u(x,t)$ of the first-order hyperbolic partial differential equation $u_t + u_x = 0$ using the Lax-Wendroff formula as observed at $t = 0, 1, 2, \ldots, 20$. The initial conditions are given by (10.3.25) with $h = 1$, $\Delta t / \Delta x = \frac{2}{3}$, and $\Delta x = 0.02$.

Project: Numerical Solution of First-Order Hyperbolic Equations

The equation $u_t + u_x = 0$ is the simplest possible hyperbolic partial differential equation. Indeed the classic wave equation consists of a system of these equations: $u_t + cv_x = 0$, and $v_t + cu_x = 0$. In this project you will examine several numerical schemes for solving such a partial differential equation using MATLAB.

Step 1: One of the simplest numerical schemes is the forward-in-time, centered-in-space of

$$\frac{u_m^{n+1} - u_m^n}{\Delta t} + \frac{u_{m+1}^n - u_{m-1}^n}{2\Delta x} = 0.$$

Use von Neumann's stability analysis to show that this scheme is *always* unstable.

Step 2: The most widely used method for numerically integrating first-order hyperbolic equations is the *Lax-Wendroff* method:[28]

$$u_m^{n+1} = u_m^n - \frac{\Delta t}{2\Delta x}\left(u_{m+1}^n - u_{m-1}^n\right) + \frac{(\Delta t)^2}{2(\Delta x)^2}\left(u_{m+1}^n - 2u_m^n + u_{m-1}^n\right).$$

---

[28] Lax, P. D., and B. Wendroff, 1960: Systems of conservative laws. *Comm. Pure Appl. Math.*, **13**, 217–237.

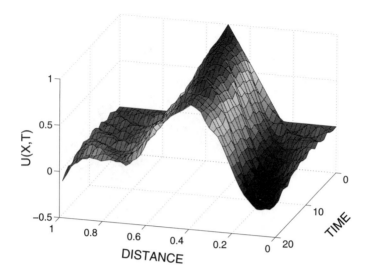

Figure 10.6.6: Same as Figure 10.6.5 except that the centered-in-time, centered-in-space scheme was used.

This method introduces errors of $O[(\Delta t)^2]$ and $O[(\Delta x)^2]$. Show that this scheme is stable if it satisfies the CFL criteria of $\Delta t/\Delta x \leq 1$.

Using the initial condition given by (10.3.25), write a MATLAB code that uses this scheme to numerically integrate $u_t + u_x = 0$. Plot the results for various $\Delta t/\Delta x$ over the interval $0 \leq x \leq 1$ given the *periodic* boundary conditions of $u(0, t) = u(1, t)$ for the temporal interval $0 \leq t \leq 20$. Discuss the strengths and weaknesses of the scheme with respect to dissipation or damping of the numerical solution and preserving the phase of the solution. Most numerical methods books discuss this.[29]

Step 3: Another simple scheme is the centered-in-time, centered-in-space of

$$\frac{u_m^{n+1} - u_m^{n-1}}{2\Delta t} + \frac{u_{m+1}^n - u_{m-1}^n}{2\Delta x} = 0.$$

This method introduces errors of $O[(\Delta t)^2]$ and $O[(\Delta x)^2]$.

Repeat the analysis from Step 1 for this scheme. One of the difficulties is taking the first time step. Use the scheme in Step 1 to take this first time step.

---

[29] For example, Lapidus, L., and G. F. Pinder, 1982: *Numerical Solution of Partial Differential Equations in Science and Engineering*. John Wiley & Sons, 677 pp.

Chapter 11

The Heat Equation

In this chapter we deal with the linear parabolic differential equation

$$\frac{\partial u}{\partial t} = a^2 \frac{\partial^2 u}{\partial x^2} \tag{11.0.1}$$

in the two independent variables x and t. This equation, known as the one-dimensional heat equation, serves as the prototype for a wider class of *parabolic equations*

$$a(x,t)\frac{\partial^2 u}{\partial x^2} + b(x,t)\frac{\partial^2 u}{\partial x \partial t} + c(x,t)\frac{\partial^2 u}{\partial t^2} = f\left(x,t,u,\frac{\partial u}{\partial x}, \frac{\partial u}{\partial t}\right), \tag{11.0.2}$$

where $b^2 = 4ac$. It arises in the study of heat conduction in solids as well as in a variety of diffusive phenomena. The heat equation is similar to the wave equation in that it is also an equation of evolution. However, the heat equation is not "conservative" because if we reverse the sign of t, we obtain a different solution. This reflects the presence of entropy which must always increase during heat conduction.

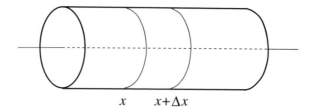

$$x \qquad x+\Delta x$$

Figure 11.1.1: Heat conduction in a thin bar.

11.1 DERIVATION OF THE HEAT EQUATION

To derive the heat equation, consider a heat-conducting homogeneous rod, extending from $x = 0$ to $x = L$ along the x-axis (see Figure 11.1.1). The rod has uniform cross section A and constant density ρ, is insulated laterally so that heat flows only in the x-direction, and is sufficiently thin so that the temperature at all points on a cross section is constant. Let $u(x, t)$ denote the temperature of the cross section at the point x at any instant of time t, and let c denote the specific heat of the rod (the amount of heat required to raise the temperature of a unit mass of the rod by a degree). In the segment of the rod between the cross section at x and the cross section at $x + \Delta x$, the amount of heat is

$$Q(t) = \int_{x}^{x+\Delta x} c\rho A u(s, t)\, ds. \tag{11.1.1}$$

On the other hand, the rate at which heat flows into the segment across the cross section at x is proportional to the cross section and the gradient of the temperature at the cross section (Fourier's law of heat conduction):

$$-\kappa A \frac{\partial u(x, t)}{\partial x}, \tag{11.1.2}$$

where κ denotes the thermal conductivity of the rod. The sign in (11.1.2) indicates that heat flows in the direction of decreasing temperature. Similarly, the rate at which heat flows out of the segment through the cross section at $x + \Delta x$ equals

$$-\kappa A \frac{\partial u(x + \Delta x, t)}{\partial x}. \tag{11.1.3}$$

The difference between the amount of heat that flows in through the cross section at x and the amount of heat that flows out through the cross section at $x + \Delta x$ must equal the change in the heat content of the segment $x \leq s \leq x + \Delta x$. Hence, by subtracting (11.1.3) from (11.1.2) and equating the result to the time derivative of (11.1.1),

$$\frac{\partial Q}{\partial t} = \int_{x}^{x+\Delta x} c\rho A \frac{\partial u(s, t)}{\partial t}\, ds = \kappa A \left[\frac{\partial u(x + \Delta x, t)}{\partial x} - \frac{\partial u(x, t)}{\partial x} \right]. \tag{11.1.4}$$

Assuming that the integrand in (11.1.4) is a continuous function of s, then by the mean value theorem for integrals,

$$\int_x^{x+\Delta x} \frac{\partial u(s,t)}{\partial t}\, ds = \frac{\partial u(\xi,t)}{\partial t}\Delta x, \qquad x < \xi < x + \Delta x, \qquad (11.1.5)$$

so that (11.1.4) becomes

$$c\rho\Delta x \frac{\partial u(\xi,t)}{\partial t} = \kappa\left[\frac{\partial u(x+\Delta x,t)}{\partial x} - \frac{\partial u(x,t)}{\partial x}\right]. \qquad (11.1.6)$$

Dividing both sides of (11.1.6) by $c\rho\Delta x$ and taking the limit as $\Delta x \to 0$,

$$\frac{\partial u(x,t)}{\partial t} = a^2 \frac{\partial^2 u(x,t)}{\partial x^2} \qquad (11.1.7)$$

with $a^2 = \kappa/(c\rho)$. Equation (11.1.7) is called the one-dimensional *heat equation*. The constant a^2 is called the *diffusivity* within the solid.

If an external source supplies heat to the rod at a rate $f(x,t)$ per unit volume per unit time, we must add the term $\int_x^{x+\Delta x} f(s,t)\, ds$ to the time derivative term of (11.1.4). Thus, in the limit $\Delta x \to 0$,

$$\frac{\partial u(x,t)}{\partial t} - a^2 \frac{\partial^2 u(x,t)}{\partial x^2} = F(x,t), \qquad (11.1.8)$$

where $F(x,t) = f(x,t)/(c\rho)$ is the source density. This equation is called the *nonhomogeneous heat equation*.

11.2 INITIAL AND BOUNDARY CONDITIONS

In the case of heat conduction in a thin rod, the temperature function $u(x,t)$ must satisfy not only the heat equation (11.1.7) but also how the two ends of the rod exchange heat energy with the surrounding medium. If (1) there is no heat source, (2) the function $f(x)$, $0 < x < L$, describes the temperature in the rod at $t = 0$, and (3) we maintain both ends at zero temperature for all time, then the partial differential equation

$$\frac{\partial u}{\partial t} = a^2 \frac{\partial^2 u}{\partial x^2}, \qquad 0 < x < L, \quad 0 < t, \qquad (11.2.1)$$

describes the temperature distribution $u(x,t)$ in the rod at any later time $0 < t$ subject to the conditions

$$u(x,0) = f(x), \qquad 0 < x < L, \qquad (11.2.2)$$

and

$$u(0,t) = u(L,t) = 0, \qquad 0 < t. \qquad (11.2.3)$$

Equations (11.2.1)–(11.2.3) describe the *initial-boundary-value problem* for this particular heat conduction problem; (11.2.3) is the boundary condition while (11.2.2) gives the initial condition. Note that in the case of the heat equation, the problem only demands the initial value of $u(x,t)$ and not $u_t(x,0)$, as with the wave equation.

Historically most linear boundary conditions have been classified in one of three ways. The condition (11.2.3) is an example of a *Dirichlet problem*[1] or *condition of the first kind*. This type of boundary condition gives the value of the solution (which is not necessarily equal to zero) along a boundary.

The next simplest condition involves derivatives. If we insulate both ends of the rod so that no heat flows from the ends, then according to (11.1.2) the boundary condition assumes the form

$$\frac{\partial u(0,t)}{\partial x} = \frac{\partial u(L,t)}{\partial x} = 0, \qquad 0 < t. \tag{11.2.4}$$

This is an example of a *Neumann problem*[2] or *condition of the second kind*. This type of boundary condition specifies the value of the normal derivative (which may not be equal to zero) of the solution along the boundary.

Finally, if there is radiation of heat from the ends of the rod into the surrounding medium, we shall show that the boundary condition is of the form

$$\frac{\partial u(0,t)}{\partial x} - hu(0,t) = \text{a constant}, \tag{11.2.5}$$

and

$$\frac{\partial u(L,t)}{\partial x} + hu(L,t) = \text{another constant} \tag{11.2.6}$$

for $0 < t$, where h is a positive constant. This is an example of a *condition of the third kind* or *Robin problem*[3] and is a linear combination of Dirichlet and Neumann conditions.

11.3 SEPARATION OF VARIABLES

As with the wave equation, the most popular and widely used technique for solving the heat equation is separation of variables. Its success depends on our ability to express the solution $u(x,t)$ as the product $X(x)T(t)$. If we cannot achieve this separation, then the technique must be abandoned for others. In the following examples we show how to apply this technique.

---

[1] Dirichlet, P. G. L., 1850: Über einen neuen Ausdruck zur Bestimmung der Dichtigkeit einer unendlich dünnen Kugelschale, wenn der Werth des Potentials derselben in jedem Punkte ihrer Oberfläche gegeben ist. *Abh. Königlich. Preuss. Akad. Wiss.*, 99–116.

[2] Neumann, C. G., 1877: *Untersuchungen über das Logarithmische und Newton'sche Potential.*

[3] Robin, G., 1886: Sur la distribution de l'électricité à la surface des conducteurs fermés et des conducteurs ouverts. *Ann. Sci. l'Ecole Norm. Sup.*, Ser. 3, **3**, S1–S58.

● **Example 11.3.1**

Let us find the solution to the homogeneous heat equation

$$\frac{\partial u}{\partial t} = a^2 \frac{\partial^2 u}{\partial x^2}, \qquad 0 < x < L, \quad 0 < t, \tag{11.3.1}$$

which satisfies the initial condition

$$u(x,0) = f(x), \quad 0 < x < L, \tag{11.3.2}$$

and the boundary conditions

$$u(0,t) = u(L,t) = 0, \quad 0 < t. \tag{11.3.3}$$

This system of equations models heat conduction in a thin metallic bar where both ends are held at the constant temperature of zero and the bar initially has the temperature $f(x)$.

We shall solve this problem by the method of separation of variables. Accordingly, we seek particular solutions of (11.3.1) of the form

$$u(x,t) = X(x)T(t), \tag{11.3.4}$$

which satisfy the boundary conditions (11.3.3). Because

$$\frac{\partial u}{\partial t} = X(x)T'(t), \tag{11.3.5}$$

and

$$\frac{\partial^2 u}{\partial x^2} = X''(x)T(t), \tag{11.3.6}$$

(11.3.1) becomes

$$T'(t)X(x) = a^2 X''(x)T(t). \tag{11.3.7}$$

Dividing both sides of (11.3.7) by $a^2 X(x)T(t)$ gives

$$\frac{T'}{a^2 T} = \frac{X''}{X} = -\lambda, \tag{11.3.8}$$

where $-\lambda$ is the separation constant. Equation (11.3.8) immediately yields two ordinary differential equations:

$$X'' + \lambda X = 0, \tag{11.3.9}$$

and

$$T' + a^2 \lambda T = 0 \tag{11.3.10}$$

for the functions $X(x)$ and $T(t)$, respectively.

We now rewrite the boundary conditions in terms of $X(x)$ by noting that the boundary conditions are $u(0,t) = X(0)T(t) = 0$, and $u(L,t) = X(L)T(t) = 0$ for $0 < t$. If we were to choose $T(t) = 0$, then we would have a trivial solution for $u(x,t)$. Consequently, $X(0) = X(L) = 0$.

We now solve (11.3.9). There are three possible cases: $\lambda = -m^2$, $\lambda = 0$, and $\lambda = k^2$. If $\lambda = -m^2 < 0$, then we must solve the boundary-value problem

$$X'' - m^2 X = 0, \qquad X(0) = X(L) = 0. \tag{11.3.11}$$

The general solution to (11.3.11) is

$$X(x) = A\cosh(mx) + B\sinh(mx). \tag{11.3.12}$$

Because $X(0) = 0$, it follows that $A = 0$. The condition $X(L) = 0$ yields $B\sinh(mL) = 0$. Since $\sinh(mL) \neq 0$, $B = 0$, and we have a trivial solution for $\lambda < 0$.

If $\lambda = 0$, the corresponding boundary-value problem is

$$X''(x) = 0, \qquad X(0) = X(L) = 0. \tag{11.3.13}$$

The general solution is

$$X(x) = C + Dx. \tag{11.3.14}$$

From $X(0) = 0$, we have that $C = 0$. From $X(L) = 0$, $DL = 0$, or $D = 0$. Again, we obtain a trivial solution.

Finally, we assume that $\lambda = k^2 > 0$. The corresponding boundary-value problem is

$$X'' + k^2 X = 0, \qquad X(0) = X(L) = 0. \tag{11.3.15}$$

The general solution to (11.3.15) is

$$X(x) = E\cos(kx) + F\sin(kx). \tag{11.3.16}$$

Because $X(0) = 0$, it follows that $E = 0$; from $X(L) = 0$, we obtain $F\sin(kL) = 0$. For a nontrivial solution, $F \neq 0$ and $\sin(kL) = 0$. This implies that $k_n L = n\pi$, where $n = 1, 2, 3, \dots$. In summary, the x-dependence of the solution is

$$X_n(x) = F_n \sin\left(\frac{n\pi x}{L}\right), \tag{11.3.17}$$

where $\lambda_n = n^2\pi^2/L^2$.

Turning to the time dependence, we use $\lambda_n = n^2\pi^2/L^2$ in (11.3.10)

$$T_n' + \frac{a^2 n^2 \pi^2}{L^2} T_n = 0. \tag{11.3.18}$$

The corresponding general solution is

$$T_n(t) = G_n \exp\left(-\frac{a^2 n^2 \pi^2}{L^2} t\right). \tag{11.3.19}$$

Thus, the functions

$$u_n(x, t) = B_n \sin\left(\frac{n\pi x}{L}\right) \exp\left(-\frac{a^2 n^2 \pi^2}{L^2} t\right), n = 1, 2, 3, \ldots, \tag{11.3.20}$$

where $B_n = F_n G_n$, are particular solutions of (11.3.1) and satisfy the homogeneous boundary conditions (11.3.3).

As we noted in the case of the wave equation, we can solve the x-dependence equation as a regular Sturm-Liouville problem. After finding the eigenvalue λ_n and eigenfunction, we solve for $T_n(t)$. The product solution $u_n(x, t)$ equals the product of the eigenfunction and $T_n(t)$.

Having found particular solutions to our problem, the most general solution equals a linear sum of these particular solutions:

$$u(x, t) = \sum_{n=1}^{\infty} B_n \sin\left(\frac{n\pi x}{L}\right) \exp\left(-\frac{a^2 n^2 \pi^2}{L^2} t\right). \tag{11.3.21}$$

The coefficient B_n is chosen so that (11.3.21) yields the initial condition (11.3.2) if $t = 0$. Thus, setting $t = 0$ in (11.3.21), we see from (11.3.2) that the coefficients B_n must satisfy the relationship

$$f(x) = \sum_{n=1}^{\infty} B_n \sin\left(\frac{n\pi x}{L}\right), \qquad 0 < x < L. \tag{11.3.22}$$

This is precisely a Fourier half-range sine series for $f(x)$ on the interval $(0, L)$. Therefore, the formula

$$B_n = \frac{2}{L} \int_0^L f(x) \sin\left(\frac{n\pi x}{L}\right) dx, \qquad n = 1, 2, 3, \ldots \tag{11.3.23}$$

gives the coefficients B_n. For example, if $L = \pi$ and $u(x, 0) = x(\pi - x)$, then

$$B_n = \frac{2}{\pi} \int_0^\pi x(\pi - x) \sin(nx) \, dx \tag{11.3.24}$$

$$= 2 \int_0^\pi x \sin(nx) \, dx - \frac{2}{\pi} \int_0^\pi x^2 \sin(nx) \, dx \tag{11.3.25}$$

$$= 4 \frac{1 - (-1)^n}{n^3 \pi}. \tag{11.3.26}$$

Hence,

$$u(x, t) = \frac{8}{\pi} \sum_{n=1}^{\infty} \frac{\sin[(2n - 1)x]}{(2n - 1)^3} e^{-(2n-1)^2 a^2 t}. \tag{11.3.27}$$

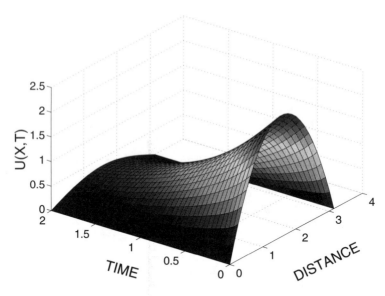

Figure 11.3.1: The temperature $u(x,t)$ within a thin bar as a function of position x and time $a^2 t$ when we maintain both ends at zero and the initial temperature equals $x(\pi - x)$.

Figure 11.3.1 illustrates (11.3.27) for various times. It was created using the MATLAB script

```
clear
M = 20; dx = pi/25; dt = 0.05;
% compute grid and initialize solution
X = [0:dx:pi]; T = [0:dt:2];
u = zeros(length(T),length(X));
XX = repmat(X,[length(T) 1]); TT = repmat(T',[1 length(X)]);
% compute solution from (11.3.27)
for m = 1:M
temp1 = 2*m-1; coeff = 8 / (pi * temp1 * temp1 * temp1);
u = u + coeff * sin(temp1*XX) .* exp(-temp1 * temp1 * TT);
end
surf(XX,TT,u)
xlabel('DISTANCE','Fontsize',20); ylabel('TIME','Fontsize',20)
zlabel('U(X,T)','Fontsize',20)
```

Note that both ends of the bar satisfy the boundary conditions, namely that the temperature equals zero. As time increases, heat flows out from the center of the bar to both ends where it is removed. This process is reflected in the collapse of the original parabolic shape of the temperature profile toward zero as time increases.

• Example 11.3.2

As a second example, let us solve the heat equation

$$\frac{\partial u}{\partial t} = a^2 \frac{\partial^2 u}{\partial x^2}, \qquad 0 < x < L, \quad 0 < t, \qquad (11.3.28)$$

which satisfies the initial condition

$$u(x, 0) = x, \qquad 0 < x < L, \qquad (11.3.29)$$

and the boundary conditions

$$\frac{\partial u(0, t)}{\partial x} = u(L, t) = 0, \qquad 0 < t. \qquad (11.3.30)$$

The condition $u_x(0, t) = 0$ expresses mathematically the constraint that no heat flows through the left boundary (insulated end condition).

Once again, we employ separation of variables; as in the previous example, the positive and zero separation constants yield trivial solutions. For a negative separation constant, however,

$$X'' + k^2 X = 0, \qquad (11.3.31)$$

with

$$X'(0) = X(L) = 0, \qquad (11.3.32)$$

because $u_x(0, t) = X'(0)T(t) = 0$, and $u(L, t) = X(L)T(t) = 0$. This regular Sturm-Liouville problem has the solution

$$X_n(x) = \cos\left[\frac{(2n - 1)\pi x}{2L}\right], \qquad n = 1, 2, 3, \ldots. \qquad (11.3.33)$$

The temporal solution then becomes

$$T_n(t) = B_n \exp\left[-\frac{a^2(2n - 1)^2 \pi^2 t}{4L^2}\right]. \qquad (11.3.34)$$

Consequently, a linear superposition of the particular solutions gives the total solution which equals

$$u(x, t) = \sum_{n=1}^{\infty} B_n \cos\left[\frac{(2n - 1)\pi x}{2L}\right] \exp\left[-\frac{a^2(2n - 1)^2 \pi^2}{4L^2} t\right]. \qquad (11.3.35)$$

Our final task remains to find the coefficients B_n. Evaluating (11.3.35) at $t = 0$,

$$u(x, 0) = x = \sum_{n=1}^{\infty} B_n \cos\left[\frac{(2n - 1)\pi x}{2L}\right], \qquad 0 < x < L. \qquad (11.3.36)$$

Equation (11.3.36) is not a half-range cosine expansion; it is an expansion in the orthogonal functions $\cos[(2n-1)\pi x/(2L)]$ corresponding to the regular Sturm-Liouville problem (11.3.31)–(11.3.32). Consequently, B_n is given by (9.3.4) with $r(x) = 1$ as

$$B_n = \frac{\int_0^L x \cos[(2n-1)\pi x/(2L)]\, dx}{\int_0^L \cos^2[(2n-1)\pi x/(2L)]\, dx} \tag{11.3.37}$$

$$= \frac{\frac{4L^2}{(2n-1)^2\pi^2}\cos\left[\frac{(2n-1)\pi x}{2L}\right]\Big|_0^L + \frac{2Lx}{(2n-1)\pi}\sin\left[\frac{(2n-1)\pi x}{2L}\right]\Big|_0^L}{\frac{x}{2}\Big|_0^L + \frac{L}{2(2n-1)\pi}\sin\left[\frac{(2n-1)\pi x}{L}\right]\Big|_0^L} \tag{11.3.38}$$

$$= \frac{8L}{(2n-1)^2\pi^2}\left\{\cos\left[\frac{(2n-1)\pi}{2}\right]-1\right\} + \frac{4L}{(2n-1)\pi}\sin\left[\frac{(2n-1)\pi}{2}\right] \tag{11.3.39}$$

$$= -\frac{8L}{(2n-1)^2\pi^2} - \frac{4L(-1)^n}{(2n-1)\pi}, \tag{11.3.40}$$

as $\cos[(2n-1)\pi/2] = 0$, and $\sin[(2n-1)\pi/2] = (-1)^{n+1}$. Consequently, the complete solution is

$$u(x,t) = -\frac{4L}{\pi}\sum_{n=1}^{\infty}\left[\frac{2}{(2n-1)^2\pi} + \frac{(-1)^n}{2n-1}\right]\cos\left[\frac{(2n-1)\pi x}{2L}\right]$$

$$\times \exp\left[-\frac{(2n-1)^2\pi^2 a^2 t}{4L^2}\right]. \tag{11.3.41}$$

Figure 11.3.2 illustrates the evolution of the temperature field with time. It was generated using the MATLAB script

```
clear
M = 200; dx = 0.02; dt = 0.05;
% compute fourier coefficients
sign = -1;
for m = 1:M
temp1 = 2*m-1;
a(m) = 2/(pi*temp1*temp1) + sign/temp1;
sign = - sign;
end
% compute grid and initialize solution
X = [0:dx:1]; T = [0:dt:1];
u = zeros(length(T),length(X));
XX = repmat(X,[length(T) 1]);
TT = repmat(T',[1 length(X)]);
% compute solution from (11.3.41)
for m = 1:M
```

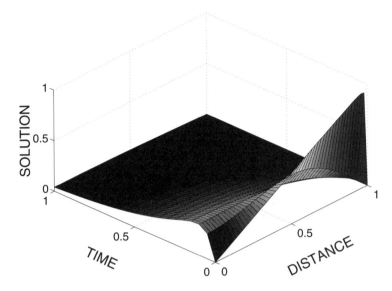

Figure 11.3.2: The temperature $u(x,t)/L$ within a thin bar as a function of position x/L and time a^2t/L^2 when we insulate the left end and hold the right end at the temperature of zero. The initial temperature equals x.

```
temp1 = (2*m-1)*pi/2;
u = u + a(m) * cos(temp1*XX) .* exp(-temp1 * temp1 * TT);
end
u = - (4/pi) * u;
surf(XX,TT,u); axis([0 1 0 1 0 1]);
xlabel('DISTANCE','Fontsize',20); ylabel('TIME','Fontsize',20)
zlabel('SOLUTION','Fontsize',20)
```

Initially, heat near the center of the bar flows toward the cooler, insulated end, resulting in an increase of temperature there. On the right side, heat flows out of the bar because the temperature is maintained at zero at $x = L$. Eventually the heat that has accumulated at the left end flows rightward because of the continual heat loss on the right end. In the limit of $t \to \infty$, all of the heat has left the bar.

• **Example 11.3.3**

A slight variation on Example 11.3.1 is

$$\frac{\partial u}{\partial t} = a^2 \frac{\partial^2 u}{\partial x^2}, \qquad 0 < x < L, \quad 0 < t, \tag{11.3.42}$$

where

$$u(x,0) = u(0,t) = 0, \quad \text{and} \quad u(L,t) = \theta. \tag{11.3.43}$$

We begin by blindly employing the technique of separation of variables. Once again, we obtain the ordinary differential equation (11.3.9) and (11.3.10). The initial and boundary conditions become, however,

$$X(0) = T(0) = 0, \tag{11.3.44}$$

and

$$X(L)T(t) = \theta. \tag{11.3.45}$$

Although (11.3.44) is acceptable, (11.3.45) gives us an impossible condition because $T(t)$ cannot be constant. If it were, it would have to equal to zero by (11.3.44).

To find a way around this difficulty, suppose that we want the solution to our problem at a time long after $t = 0$. From experience we know that heat conduction with time-independent boundary conditions eventually results in an evolution from the initial condition to some time-independent (steady-state) equilibrium. If we denote this steady-state solution by $w(x)$, it must satisfy the heat equation

$$a^2 w''(x) = 0, \tag{11.3.46}$$

and the boundary conditions

$$w(0) = 0, \quad \text{and} \quad w(L) = \theta. \tag{11.3.47}$$

We can integrate (11.3.46) immediately to give

$$w(x) = A + Bx, \tag{11.3.48}$$

and the boundary condition (11.3.47) results in

$$w(x) = \frac{\theta x}{L}. \tag{11.3.49}$$

Clearly (11.3.49) cannot hope to satisfy the initial conditions; that was never expected of it. However, if we add a time-varying (transient) solution $v(x,t)$ to $w(x)$ so that

$$u(x,t) = w(x) + v(x,t), \tag{11.3.50}$$

we could satisfy the initial condition if

$$v(x,0) = u(x,0) - w(x), \tag{11.3.51}$$

and $v(x,t)$ tends to zero as $t \to \infty$. Furthermore, because $w''(x) = w(0) = 0$, and $w(L) = \theta$,

$$\frac{\partial v}{\partial t} = a^2 \frac{\partial^2 v}{\partial x^2}, \quad 0 < x < L, \quad 0 < t, \tag{11.3.52}$$

with the boundary conditions

$$v(0, t) = v(L, t) = 0, \quad 0 < t. \tag{11.3.53}$$

We can solve (11.3.51), (11.3.52), and (11.3.53) by separation of variables; we did it in Example 11.3.1. However, in place of $f(x)$ we now have $u(x, 0) - w(x)$, or $-w(x)$ because $u(x, 0) = 0$. Therefore, the solution $v(x, t)$ is

$$v(x, t) = \sum_{n=1}^{\infty} B_n \sin\left(\frac{n\pi x}{L}\right) \exp\left(-\frac{a^2 n^2 \pi^2}{L^2} t\right) \tag{11.3.54}$$

with

$$B_n = \frac{2}{L} \int_0^L -w(x) \sin\left(\frac{n\pi x}{L}\right) dx \tag{11.3.55}$$

$$= \frac{2}{L} \int_0^L -\frac{\theta x}{L} \sin\left(\frac{n\pi x}{L}\right) dx \tag{11.3.56}$$

$$= -\frac{2\theta}{L^2} \left[\frac{L^2}{n^2 \pi^2} \sin\left(\frac{n\pi x}{L}\right) - \frac{xL}{n\pi} \cos\left(\frac{n\pi x}{L}\right) \right]_0^L \tag{11.3.57}$$

$$= (-1)^n \frac{2\theta}{n\pi}. \tag{11.3.58}$$

Thus, the entire solution is

$$u(x, t) = \frac{\theta x}{L} + \frac{2\theta}{\pi} \sum_{n=1}^{\infty} \frac{(-1)^n}{n} \sin\left(\frac{n\pi x}{L}\right) \exp\left(-\frac{a^2 n^2 \pi^2}{L^2} t\right). \tag{11.3.59}$$

The quantity $a^2 t / L^2$ is the *Fourier number*.

Figure 11.3.3 illustrates our solution and was created with the MATLAB script

```
clear
M = 1000; dx = 0.01; dt = 0.01;
% compute grid and initialize solution
X = [0:dx:1]; T = [0:dt:0.2];
XX = repmat(X,[length(T) 1]); TT = repmat(T',[1 length(X)]);
u = XX;
% compute solution from (11.3.59)
sign = -2/pi;
for m = 1:M
coeff = sign/m;
u = u + coeff * sin((m*pi)*XX) .* exp(-(m*m*pi*pi) * TT);
sign = -sign;
end
surf(XX,TT,u); axis([0 1 0 0.2 0 1]);
```

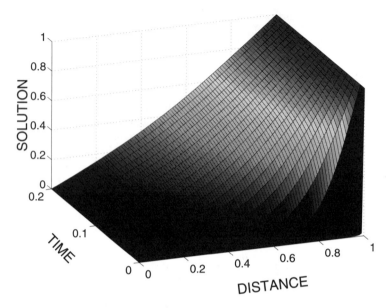

Figure 11.3.3: The temperature $u(x,t)/\theta$ within a thin bar as a function of position x/L and time a^2t/L^2 with the left end held at a temperature of zero and right end held at a temperature θ while the initial temperature of the bar is zero.

```
xlabel('DISTANCE','Fontsize',20); ylabel('TIME','Fontsize',20)
zlabel('SOLUTION','Fontsize',20)
```

Clearly it satisfies the boundary conditions. Initially, heat flows rapidly from right to left. As time increases, the rate of heat transfer decreases until the final equilibrium (steady-state) is established and no more heat flows.

● **Example 11.3.4**

Let us find the solution to the heat equation

$$\frac{\partial u}{\partial t} = a^2 \frac{\partial^2 u}{\partial x^2}, \qquad 0 < x < L, \quad 0 < t, \tag{11.3.60}$$

subject to the Neumann boundary conditions

$$\frac{\partial u(0,t)}{\partial x} = \frac{\partial u(L,t)}{\partial x} = 0, \qquad 0 < t, \tag{11.3.61}$$

and the initial condition that

$$u(x,0) = x, \qquad 0 < x < L. \tag{11.3.62}$$

We have now insulated *both* ends of the bar.

Assuming that $u(x,t) = X(x)T(t)$,

$$\frac{T'}{a^2 T} = \frac{X''}{X} = -k^2, \tag{11.3.63}$$

where we have presently assumed that the separation constant is negative. The Neumann conditions give $u_x(0,t) = X'(0)T(t) = 0$, and $u_x(L,t) = X'(L)T(t) = 0$ so that $X'(0) = X'(L) = 0$.

The Sturm-Liouville problem

$$X'' + k^2 X = 0, \tag{11.3.64}$$

and

$$X'(0) = X'(L) = 0 \tag{11.3.65}$$

gives the x-dependence. The eigenfunction solution is

$$X_n(x) = \cos\left(\frac{n\pi x}{L}\right), \tag{11.3.66}$$

where $k_n = n\pi/L$ and $n = 1, 2, 3, \dots$.

The corresponding temporal part equals the solution of

$$T_n' + a^2 k_n^2 T_n = T_n' + \frac{a^2 n^2 \pi^2}{L^2} T_n = 0, \tag{11.3.67}$$

which is

$$T_n(t) = A_n \exp\left(-\frac{a^2 n^2 \pi^2}{L^2} t\right). \tag{11.3.68}$$

Thus, the product solution given by a negative separation constant is

$$u_n(x,t) = X_n(x)T_n(t) = A_n \cos\left(\frac{n\pi x}{L}\right) \exp\left(-\frac{a^2 n^2 \pi^2}{L^2} t\right). \tag{11.3.69}$$

Unlike our previous problems, there is a nontrivial solution for a separation constant that equals zero. In this instance, the x-dependence equals

$$X(x) = Ax + B. \tag{11.3.70}$$

The boundary conditions $X'(0) = X'(L) = 0$ force A to be zero but B is completely free. Consequently, the eigenfunction in this particular case is

$$X_0(x) = 1. \tag{11.3.71}$$

Because $T_0'(t) = 0$ in this case, the temporal part equals a constant which we shall take to be $A_0/2$. Therefore, the product solution corresponding to the zero separation constant is

$$u_0(x,t) = X_0(x)T_0(t) = A_0/2. \tag{11.3.72}$$

The most general solution to our problem equals the sum of all of the possible solutions:

$$u(x,t) = \frac{A_0}{2} + \sum_{n=1}^{\infty} A_n \cos\left(\frac{n\pi x}{L}\right) \exp\left(-\frac{a^2 n^2 \pi^2}{L^2}t\right). \qquad (11.3.73)$$

Upon substituting $t = 0$ into (11.3.73), we can determine A_n because

$$u(x,0) = x = \frac{A_0}{2} + \sum_{n=1}^{\infty} A_n \cos\left(\frac{n\pi x}{L}\right) \qquad (11.3.74)$$

is merely a half-range Fourier cosine expansion of the function x over the interval $(0, L)$. From (2.1.23)–(2.1.24),

$$A_0 = \frac{2}{L} \int_0^L x\,dx = L, \qquad (11.3.75)$$

and

$$A_n = \frac{2}{L} \int_0^L x \cos\left(\frac{n\pi x}{L}\right) dx \qquad (11.3.76)$$

$$= \frac{2}{L} \left[\frac{L^2}{n^2\pi^2} \cos\left(\frac{n\pi x}{L}\right) + \frac{xL}{n\pi} \sin\left(\frac{n\pi x}{L}\right) \right]_0^L \qquad (11.3.77)$$

$$= \frac{2L}{n^2\pi^2} [(-1)^n - 1]. \qquad (11.3.78)$$

The complete solution is

$$u(x,t) = \frac{L}{2} - \frac{4L}{\pi^2} \sum_{m=1}^{\infty} \frac{1}{(2m-1)^2} \cos\left[\frac{(2m-1)\pi x}{L}\right] \exp\left[-\frac{a^2(2m-1)^2\pi^2}{L^2}t\right],$$

$$(11.3.79)$$

because all of the even harmonics vanish and we may rewrite the odd harmonics using $n = 2m - 1$, where $m = 1, 2, 3, 4, \ldots$.

Figure 11.3.4 illustrates (11.3.79) for various positions and times. It was generated using the MATLAB script

```
clear
M = 100; dx = 0.01; dt = 0.01;
% compute grid and initialize solution
X = [0:dx:1]; T = [0:dt:0.3];
u = zeros(length(T),length(X)); u = 0.5;
XX = repmat(X,[length(T) 1]); TT = repmat(T',[1 length(X)]);
% compute solution from (11.3.79)
for m = 1:M
temp1 = (2*m-1) * pi;
```

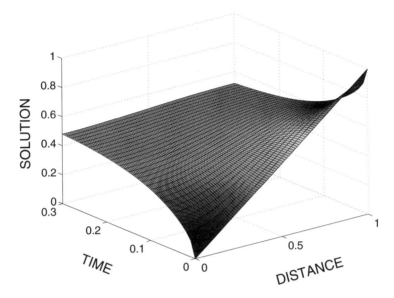

Figure 11.3.4: The temperature $u(x,t)/L$ within a thin bar as a function of position x/L and time a^2t/L^2 when we insulate both ends. The initial temperature of the bar is x.

```
coeff = 4 / (temp1*temp1);
u = u - coeff * cos(temp1*XX) .* exp(-temp1 * temp1 * TT);
end
surf(XX,TT,u); axis([0 1 0 0.3 0 1]);
xlabel('DISTANCE','Fontsize',20); ylabel('TIME','Fontsize',20)
zlabel('SOLUTION','Fontsize',20)
```

The physical interpretation is quite simple. Since heat cannot flow in or out of the rod because of the insulation, it can only redistribute itself. Thus, heat flows from the warm right end to the cooler left end. Eventually the temperature achieves steady-state when the temperature is uniform throughout the bar.

• Example 11.3.5

So far we have dealt with problems where the temperature or flux of heat has been specified at the ends of the rod. In many physical applications, one or both of the ends may radiate to free space at temperature u_0. According to Stefan's law, the amount of heat radiated from a given area dA in a given time interval dt is

$$\sigma(u^4 - u_0^4)\, dA\, dt, \qquad (11.3.80)$$

where σ is called the Stefan-Boltzmann constant. On the other hand, the amount of heat that reaches the surface from the interior of the body, assuming that we are at the right end of the bar, equals

$$-\kappa \frac{\partial u}{\partial x}\, dA\, dt, \tag{11.3.81}$$

where κ is the thermal conductivity. Because these quantities must be equal,

$$-\kappa \frac{\partial u}{\partial x} = \sigma(u^4 - u_0^4) = \sigma(u - u_0)(u^3 + u^2 u_0 + u u_0^2 + u_0^3). \tag{11.3.82}$$

If u and u_0 are nearly equal, we may approximate the second bracketed term on the right side of (11.3.82) as $4u_0^3$. We write this approximate form of (11.3.82) as

$$-\frac{\partial u}{\partial x} = h(u - u_0), \tag{11.3.83}$$

where h, the *surface conductance* or the *coefficient of surface heat transfer*, equals $4\sigma u_0^3/\kappa$. Equation (11.3.83) is a "radiation" boundary condition. Sometimes someone will refer to it as "Newton's law" because (11.3.83) is mathematically identical to Newton's law of cooling of a body by forced convection.

Let us now solve the problem of a rod that we initially heat to the uniform temperature of 100. We then allow it to cool by maintaining the temperature at zero at $x = 0$ and radiatively cooling to the surrounding air at the temperature of zero[4] at $x = L$. We may restate the problem as

$$\frac{\partial u}{\partial t} = a^2 \frac{\partial^2 u}{\partial x^2}, \qquad 0 < x < L, \quad 0 < t, \tag{11.3.84}$$

with

$$u(x,0) = 100, \quad 0 < x < L, \tag{11.3.85}$$

$$u(0,t) = 0, \quad 0 < t, \tag{11.3.86}$$

and

$$\frac{\partial u(L,t)}{\partial x} + h u(L,t) = 0, \quad 0 < t. \tag{11.3.87}$$

Once again, we assume a product solution $u(x,t) = X(x)T(t)$ with a negative separation constant so that

$$\frac{X''}{X} = \frac{T'}{a^2 T} = -k^2. \tag{11.3.88}$$

We obtain for the x-dependence that

$$X'' + k^2 X = 0, \tag{11.3.89}$$

---

[4] Although this would appear to make $h = 0$, we have merely chosen a temperature scale so that the air temperature is zero and the absolute temperature used in Stefan's law is nonzero.

The Heat Equation

Table 11.3.1: The First Ten Roots of (11.3.93) and C_n for $hL = 1$

n	α_n	Approximate α_n	C_n
1	2.0288	2.2074	118.9221
2	4.9132	4.9246	31.3414
3	7.9787	7.9813	27.7549
4	11.0855	11.0865	16.2891
5	14.2074	14.2079	14.9916
6	17.3364	17.3366	10.8362
7	20.4692	20.4693	10.2232
8	23.6043	23.6044	8.0999
9	26.7409	26.7410	7.7479
10	29.8786	29.8786	6.4626

but the boundary conditions are now

$$X(0) = 0, \quad \text{and} \quad X'(L) + hX(L) = 0. \tag{11.3.90}$$

The most general solution of (11.3.89) is

$$X(x) = A\cos(kx) + B\sin(kx). \tag{11.3.91}$$

However, $A = 0$, because $X(0) = 0$. On the other hand,

$$k\cos(kL) + h\sin(kL) = kL\cos(kL) + hL\sin(kL) = 0, \tag{11.3.92}$$

if $B \neq 0$. The nondimensional number hL is the *Biot number* and depends completely upon the physical characteristics of the rod.

In Chapter 9 we saw how to find the roots of the transcendental equation

$$\alpha + hL\tan(\alpha) = 0, \tag{11.3.93}$$

where $\alpha = kL$. Consequently, if α_n is the nth root of (11.3.93), then the eigenfunction is

$$X_n(x) = \sin(\alpha_n x/L). \tag{11.3.94}$$

In Table 11.3.1, we list the first ten roots of (11.3.93) for $hL = 1$.

In general, we must solve (11.3.93) either numerically or graphically. If

α is large, however, we can find approximate values[5] by noting that

$$\cot(\alpha) = -hL/\alpha \approx 0, \qquad (11.3.95)$$

or

$$\alpha_n = (2n-1)\pi/2, \qquad (11.3.96)$$

where $n = 1, 2, 3, \ldots$. We can obtain a better approximation by setting

$$\alpha_n = (2n-1)\pi/2 - \epsilon_n, \qquad (11.3.97)$$

where $\epsilon_n \ll 1$. Substituting into (11.3.95),

$$[(2n-1)\pi/2 - \epsilon_n] \cot[(2n-1)\pi/2 - \epsilon_n] + hL = 0. \qquad (11.3.98)$$

We can simplify (11.3.98) to

$$\epsilon_n^2 + (2n-1)\pi\epsilon_n/2 + hL = 0, \qquad (11.3.99)$$

because $\cot[(2n-1)\pi/2 - \theta] = \tan(\theta)$, and $\tan(\theta) \approx \theta$ for $\theta \ll 1$. Solving for ϵ_n,

$$\epsilon_n \approx -\frac{2hL}{(2n-1)\pi}, \qquad (11.3.100)$$

and

$$\alpha_n \approx \frac{(2n-1)\pi}{2} + \frac{2hL}{(2n-1)\pi}. \qquad (11.3.101)$$

In Table 11.3.1 we compare the approximate roots given by (11.3.101) with the actual roots.

The temporal part equals

$$T_n(t) = C_n \exp(-k_n^2 a^2 t) = C_n \exp\left(-\frac{\alpha_n^2 a^2 t}{L^2}\right). \qquad (11.3.102)$$

---

[5] Using the same technique, Stevens and Luck [Stevens, J. W., and R. Luck, 1999: Explicit approximations for all eigenvalues of the 1-D transient heat conduction equations. *Heat Transfer Engng.*, **20(2)**, 35–41] have found approximate solutions to $\zeta_n \tan(\zeta_n) = Bi$. They showed that

$$\zeta_n \approx z_n + \frac{-B + \sqrt{B^2 - 4C}}{2},$$

where

$$B = z_n + (1 + Bi)\tan(z_n), \qquad C = Bi - z_n \tan(z_n),$$

$$z_n = c_n + \frac{\pi}{4}\left(\frac{Bi - c_n}{Bi + c_n}\right), \qquad c_n = \left(n - \frac{3}{4}\right)\pi.$$

Consequently, the general solution is

$$u(x,t) = \sum_{n=1}^{\infty} C_n \sin\left(\frac{\alpha_n x}{L}\right) \exp\left(-\frac{\alpha_n^2 a^2 t}{L^2}\right), \qquad (11.3.103)$$

where α_n is the nth root of (11.3.93).

To determine C_n, we use the initial condition (11.3.85) and find that

$$100 = \sum_{n=1}^{\infty} C_n \sin\left(\frac{\alpha_n x}{L}\right). \qquad (11.3.104)$$

Equation (11.3.104) is an eigenfunction expansion of 100 employing the eigenfunctions from the Sturm-Liouville problem

$$X'' + k^2 X = 0, \qquad (11.3.105)$$

and

$$X(0) = X'(L) + hX(L) = 0. \qquad (11.3.106)$$

Thus, the coefficient C_n is given by (9.3.4) or

$$C_n = \frac{\int_0^L 100 \sin(\alpha_n x/L)\, dx}{\int_0^L \sin^2(\alpha_n x/L)\, dx}, \qquad (11.3.107)$$

as $r(x) = 1$. Performing the integrations,

$$C_n = \frac{100L[1 - \cos(\alpha_n)]/\alpha_n}{\frac{1}{2}[L - L\sin(2\alpha_n)/(2\alpha_n)]} = \frac{200[1 - \cos(\alpha_n)]}{\alpha_n[1 + \cos^2(\alpha_n)/(hL)]}, \qquad (11.3.108)$$

because $\sin(2\alpha_n) = 2\cos(\alpha_n)\sin(\alpha_n)$, and $\alpha_n = -hL\tan(\alpha_n)$. The complete solution is

$$u(x,t) = \sum_{n=1}^{\infty} \frac{200[1 - \cos(\alpha_n)]}{\alpha_n[1 + \cos^2(\alpha_n)/(hL)]} \sin\left(\frac{\alpha_n x}{L}\right) \exp\left(-\frac{\alpha_n^2 a^2 t}{L^2}\right). \qquad (11.3.109)$$

Figure 11.3.5 illustrates this solution for $hL = 1$ at various times and positions. It was generated using the MATLAB script

```
clear
hL = 1; M = 200; dx = 0.02; dt = 0.02;
% create initial guess at alpha_n
zero = zeros(M,1);
for n = 1:M
temp = (2*n-1)*pi; zero(n) = 0.5*temp + 2*hL/temp;
end;
% use Newton-Raphson method to improve values of alpha_n
```

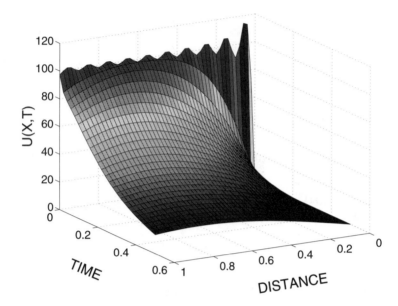

Figure 11.3.5: The temperature $u(x,t)$ within a thin bar as a function of position x/L and time a^2t/L^2 when we allow the bar to radiatively cool at $x = L$ while the temperature is zero at $x = 0$. Initially the temperature was 100.

```
for n = 1:M; for k = 1:10
f = zero(n) + hL * tan(zero(n)); fp =1 + hL * sec(zero(n))^2;
zero(n) = zero(n) - f / fp;
end; end;
% compute Fourier coefficients
for m = 1:M
a(m) = 200*(1-cos(zero(m)))/(zero(m)*(1+cos(zero(m))^2/hL));
end
% compute grid and initialize solution
X = [0:dx:1]; T = [0:dt:0.5];
u = zeros(length(T),length(X));
XX = repmat(X,[length(T) 1]);
TT = repmat(T',[1 length(X)]);
% compute solution from (11.3.109)
for m = 1:M
u = u + a(m) * sin(zero(m)*XX) .* exp(-zero(m)*zero(m)*TT);
end
surf(XX,TT,u)
xlabel('DISTANCE','Fontsize',20); ylabel('TIME','Fontsize',20)
zlabel('U(X,T)','Fontsize',20)
```

It is similar to Example 11.3.1 in that the heat lost to the environment occurs either because the temperature at an end is zero or because it radiates heat to space which has the temperature of zero.

• Example 11.3.6: Refrigeration of apples

Some decades ago, shiploads of apples, going from Australia to England, deteriorated from a disease called "brown heart," which occurred under insufficient cooling conditions. Apples, when placed on shipboard, are usually warm and must be cooled to be carried in cold storage. They also generate heat by their respiration. It was suspected that this heat generation effectively counteracted the refrigeration of the apples, resulting in the "brown heart."

This was the problem which induced Awberry[6] to study the heat distribution within a sphere in which heat is being generated. Awberry first assumed that the apples are initially at a uniform temperature. We can take this temperature to be zero by the appropriate choice of temperature scale. At time $t = 0$, the skins of the apples assume the temperature θ immediately when we introduce them into the hold.

Because of the spherical geometry, the nonhomogeneous heat equation becomes

$$\frac{1}{a^2} \frac{\partial u}{\partial t} = \frac{1}{r^2} \frac{\partial}{\partial r} \left(r^2 \frac{\partial u}{\partial r} \right) + \frac{G}{\kappa}, \qquad 0 \le r < b, \quad 0 < t, \qquad \textbf{(11.3.110)}$$

where a^2 is the thermal diffusivity, b is the radius of the apple, κ is the thermal conductivity, and G is the heating rate (per unit time per unit volume).

If we try to use separation of variables on (11.3.110), we find that it does not work because of the G/κ term. To circumvent this difficulty, we ask the simpler question of what happens after a very long time. We anticipate that a balance will eventually be established where conduction transports the heat produced within the apple to the surface of the apple where the surroundings absorb it. Consequently, just as we introduced a steady-state solution in Example 11.3.3, we again anticipate a steady-state solution $w(r)$ where the heat conduction removes the heat generated within the apples. The ordinary differential equation

$$\frac{1}{r^2} \frac{d}{dr} \left(r^2 \frac{dw}{dr} \right) = -\frac{G}{\kappa} \qquad \textbf{(11.3.111)}$$

gives the steady-state. Furthermore, just as we introduced a transient solution which allowed our solution to satisfy the initial condition, we must also have one here and the governing equation is

$$\frac{\partial v}{\partial t} = \frac{a^2}{r^2} \frac{\partial}{\partial r} \left(r^2 \frac{\partial v}{\partial r} \right). \qquad \textbf{(11.3.112)}$$

---

[6] Awberry, J. H., 1927: The flow of heat in a body generating heat. *Philos. Mag., Ser.* 7, **4**, 629–638.

Solving (11.3.111) first,

$$w(r) = C + \frac{D}{r} - \frac{Gr^2}{6\kappa}. \tag{11.3.113}$$

The constant D equals zero because the solution must be finite at $r = 0$. Since the steady-state solution must satisfy the boundary condition $w(b) = \theta$,

$$C = \theta + \frac{Gb^2}{6\kappa}. \tag{11.3.114}$$

Turning to the transient problem, we introduce a new dependent variable $y(r, t) = rv(r, t)$. This new dependent variable allows us to replace (11.3.112) with

$$\frac{\partial y}{\partial t} = a^2 \frac{\partial^2 y}{\partial r^2}, \tag{11.3.115}$$

which we can solve. If we assume that $y(r, t) = R(r)T(t)$ and we only have a negative separation constant, the $R(r)$ equation becomes

$$\frac{d^2 R}{dr^2} + k^2 R = 0, \tag{11.3.116}$$

which has the solution

$$R(r) = A \cos(kr) + B \sin(kr). \tag{11.3.117}$$

The constant A equals zero because the solution (11.3.117) must vanish at $r = 0$ so that $v(0, t)$ remains finite. However, because $\theta = w(b) + v(b, t)$ for all time and $v(b, t) = R(b)T(t)/b = 0$, then $R(b) = 0$. Consequently, $k_n = n\pi/b$, and

$$v_n(r, t) = \frac{B_n}{r} \sin\left(\frac{n\pi r}{b}\right) \exp\left(-\frac{n^2 \pi^2 a^2 t}{b^2}\right). \tag{11.3.118}$$

Superposition gives the total solution which equals

$$u(r, t) = \theta + \frac{G}{6\kappa}(b^2 - r^2) + \sum_{n=1}^{\infty} \frac{B_n}{r} \sin\left(\frac{n\pi r}{b}\right) \exp\left(-\frac{n^2 \pi^2 a^2 t}{b^2}\right). \tag{11.3.119}$$

Finally, we determine the coefficients B_n by the initial condition that $u(r, 0) = 0$. Therefore,

$$B_n = -\frac{2}{b} \int_0^b r \left[\theta + \frac{G}{6\kappa}(b^2 - r^2)\right] \sin\left(\frac{n\pi r}{b}\right) dr \tag{11.3.120}$$

$$= \frac{2\theta b}{n\pi}(-1)^n + \frac{2G}{\kappa}\left(\frac{b}{n\pi}\right)^3 (-1)^n. \tag{11.3.121}$$

The complete solution is

$$u(r,t) = \theta + \frac{2\theta b}{r\pi} \sum_{n=1}^{\infty} \frac{(-1)^n}{n} \sin\left(\frac{n\pi r}{b}\right) \exp\left(-\frac{n^2\pi^2 a^2 t}{b^2}\right) \tag{11.3.122}$$

$$+ \frac{G}{6\kappa}(b^2 - r^2) + \frac{2Gb^3}{r\kappa\pi^3} \sum_{n=1}^{\infty} \frac{(-1)^n}{n^3} \sin\left(\frac{n\pi r}{b}\right) \exp\left(-\frac{n^2\pi^2 a^2 t}{b^2}\right).$$

The first line of (11.3.122) gives the temperature distribution due to the imposition of the temperature θ on the surface of the apple while the second line gives the rise in the temperature due to the interior heating.

Returning to our original problem of whether the interior heating is strong enough to counteract the cooling by refrigeration, we merely use the second line of (11.3.122) to find how much the temperature deviates from what we normally expect. Because the highest temperature exists at the center of each apple, its value there is the only one of interest in this problem. Assuming $b = 4$ cm as the radius of the apple, $a^2 G/\kappa = 1.33 \times 10^{-5}$ °C/s, and $a^2 = 1.55 \times 10^{-3}$ cm^2/s, the temperature effect of the heat generation is very small, only 0.0232 °C when, after about 2 hours, the temperatures within the apples reach equilibrium. Thus, we must conclude that heat generation within the apples is not the cause of brown heart.

We now know that brown heart results from an excessive concentration of carbon dioxide and an insufficient amount of oxygen in the storage hold.[7] Presumably this atmosphere affects the metabolic activities that are occurring in the apple[8] and leads to low-temperature breakdown.

• Example 11.3.7

In this example we illustrate how separation of variables can be employed in solving the axisymmetric heat equation in an infinitely long cylinder. In circular coordinates the heat equation is

$$\frac{\partial u}{\partial t} = a^2 \left(\frac{\partial^2 u}{\partial r^2} + \frac{1}{r}\frac{\partial u}{\partial r}\right), \qquad 0 \leq r < b, \quad 0 < t, \tag{11.3.123}$$

where r denotes the radial distance and a^2 denotes the thermal diffusivity. Let us assume that we heated this cylinder of radius b to the uniform temperature T_0 and then allowed it to cool by having its surface held at the temperature of zero starting from the time $t = 0$.

---

[7] Thornton, N. C., 1931: The effect of carbon dioxide on fruits and vegetables in storage. *Contrib. Boyce Thompson Inst.*, **3**, 219–244.

[8] Fidler, J. C., and C. J. North, 1968: The effect of conditions of storage on the respiration of apples. IV. Changes in concentration of possible substrates of respiration, as related to production of carbon dioxide and uptake of oxygen by apples at low temperatures. *J. Hortic. Sci.*, **43**, 429–439.

We begin by assuming that the solution is of the form $u(r,t) = R(r)T(t)$ so that

$$\frac{1}{R}\left(\frac{d^2R}{dr^2} + \frac{1}{r}\frac{dR}{dr}\right) = \frac{1}{a^2T}\frac{dT}{dt} = -\frac{k^2}{b^2}. \tag{11.3.124}$$

The only values of the separation constant that yield nontrivial solutions are negative. The nontrivial solutions are $R(r) = J_0(kr/b)$, where J_0 is the Bessel function of the first kind and zeroth order. A separation constant of zero gives $R(r) = \ln(r)$ which becomes infinite at the origin. Positive separation constants yield the modified Bessel function $I_0(kr/b)$. Although this function is finite at the origin, it cannot satisfy the boundary condition that $u(b,t) = R(b)T(t) = 0$, or $R(b) = 0$.

The boundary condition that $R(b) = 0$ requires that $J_0(k) = 0$. This transcendental equation yields an infinite number of constants k_n. For each k_n, the temporal part of the solution satisfies the differential equation

$$\frac{dT_n}{dt} + \frac{k_n^2 a^2}{b^2}T_n = 0, \tag{11.3.125}$$

which has the solution

$$T_n(t) = A_n \exp\left(-\frac{k_n^2 a^2}{b^2}t\right). \tag{11.3.126}$$

Consequently, the product solutions are

$$u_n(r,t) = A_n J_0\left(k_n\frac{r}{b}\right)\exp\left(-\frac{k_n^2 a^2}{b^2}t\right). \tag{11.3.127}$$

The total solution is a linear superposition of all of the particular solutions or

$$u(r,t) = \sum_{n=1}^{\infty} A_n J_0\left(k_n\frac{r}{b}\right)\exp\left(-\frac{k_n^2 a^2}{b^2}t\right). \tag{11.3.128}$$

Our final task remains to determine A_n. From the initial condition that $u(r,0) = T_0$,

$$u(r,0) = T_0 = \sum_{n=1}^{\infty} A_n J_0\left(k_n\frac{r}{b}\right). \tag{11.3.129}$$

From (9.5.35) and (9.5.43),

$$A_n = \frac{2T_0}{J_1^2(k_n)b^2}\int_0^b rJ_0\left(k_n\frac{r}{b}\right)dr \tag{11.3.130}$$

$$= \frac{2T_0}{k_n^2 J_1^2(k_n)}\left.\left(\frac{k_n r}{b}\right)J_1\left(k_n\frac{r}{b}\right)\right|_0^b = \frac{2T_0}{k_n J_1(k_n)} \tag{11.3.131}$$

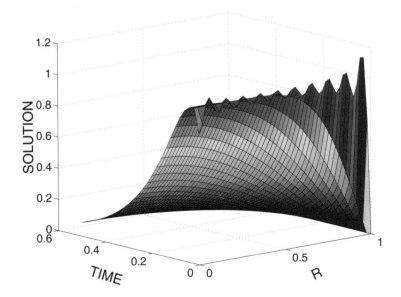

Figure 11.3.6: The temperature $u(r,t)/T_0$ within an infinitely long cylinder at various positions r/b and times a^2t/b^2 that we initially heated to the uniform temperature T_0 and then allowed to cool by forcing its surface to equal zero.

from (9.5.25). Thus, the complete solution is

$$u(r,t) = 2T_0 \sum_{n=1}^{\infty} \frac{1}{k_n J_1(k_n)} J_0\left(k_n \frac{r}{b}\right) \exp\left(-\frac{k_n^2 a^2}{b^2}t\right). \qquad \textbf{(11.3.132)}$$

Figure 11.3.6 illustrates the solution (11.3.132) for various Fourier numbers a^2t/b^2. It was generated using the MATLAB script

```
clear
M = 20; dr = 0.02; dt = 0.02;
% load in zeros of J_0
zero( 1) =  2.40482; zero( 2) =  5.52007; zero( 3) =  8.65372;
zero( 4) = 11.79153; zero( 5) = 14.93091; zero( 6) = 18.07106;
zero( 7) = 21.21164; zero( 8) = 24.35247; zero( 9) = 27.49347;
zero(10) = 30.63461; zero(11) = 33.77582; zero(12) = 36.91710;
zero(13) = 40.05843; zero(14) = 43.19979; zero(15) = 46.34119;
zero(16) = 49.48261; zero(17) = 52.62405; zero(18) = 55.76551;
zero(19) = 58.90698; zero(20) = 62.04847;
% compute Fourier coefficients
for m = 1:M
a(m) = 2 / (zero(m)*besselj(1,zero(m)));
end
% compute grid and initialize solution
R = [0:dr:1]; T = [0:dt:0.5];
```

```
u = zeros(length(T),length(R));
RR = repmat(R,[length(T) 1]);
TT = repmat(T',[1 length(R)]);
% compute solution from (11.3.132)
for m = 1:M
u = u + a(m)*besselj(0,zero(m)*RR).*exp(-zero(m)*zero(m)*TT);
end
surf(RR,TT,u)
xlabel('R','Fontsize',20); ylabel('TIME','Fontsize',20)
zlabel('SOLUTION','Fontsize',20)
```

It is similar to Example 11.3.1 except that we are in cylindrical coordinates. Heat flows from the interior and is removed at the cylinder's surface where the temperature equals zero. The initial oscillations of the solution result from Gibbs phenomena because we have a jump in the temperature field at $r = b$.

• **Example 11.3.8**

In this example[9] we find the evolution of the temperature field within a cylinder of radius b as it radiatively cools from an initial uniform temperature T_0. The heat equation is

$$\frac{\partial u}{\partial t} = a^2 \left(\frac{\partial^2 u}{\partial r^2} + \frac{1}{r} \frac{\partial u}{\partial r} \right), \qquad 0 \le r < b, \quad 0 < t, \qquad (11.3.133)$$

which we will solve by separation of variables $u(r,t) = R(r)T(t)$. Therefore,

$$\frac{1}{R} \left(\frac{d^2 R}{dr^2} + \frac{1}{r} \frac{dR}{dr} \right) = \frac{1}{a^2 T} \frac{dT}{dt} = -\frac{k^2}{b^2}, \qquad (11.3.134)$$

because only a negative separation constant yields a $R(r)$ which is finite at the origin and satisfies the boundary condition. This solution is $R(r) = J_0(kr/b)$, where J_0 is the Bessel function of the first kind and zeroth order.

The radiative boundary condition can be expressed as

$$\frac{\partial u(b,t)}{\partial r} + hu(b,t) = T(t) \left[\frac{dR(b)}{dr} + hR(b) \right] = 0. \qquad (11.3.135)$$

Because $T(t) \ne 0$,

$$k J_0'(k) + hb J_0(k) = -k J_1(k) + hb J_0(k) = 0, \qquad (11.3.136)$$

---

[9] For another example of solving the heat equation with Robin boundary conditions, see §3.2 in Balakotaiah, V., N. Gupta, and D. H. West, 2000: A simplified model for analyzing catalytic reactions in short monoliths. *Chem. Engng. Sci.*, **55**, 5367–5383.

where the product hb is the Biot number. The solution of the transcendental equation (11.3.136) yields an infinite number of distinct constants k_n. For each k_n, the temporal part equals the solution of

$$\frac{dT_n}{dt} + \frac{k_n^2 a^2}{b^2} T_n = 0, \tag{11.3.137}$$

or

$$T_n(t) = A_n \exp\left(-\frac{k_n^2 a^2}{b^2} t\right). \tag{11.3.138}$$

The product solution is, therefore,

$$u_n(r,t) = A_n J_0\left(k_n \frac{r}{b}\right) \exp\left(-\frac{k_n^2 a^2}{b^2} t\right) \tag{11.3.139}$$

and the most general solution is a sum of these product solutions

$$u(r,t) = \sum_{n=1}^{\infty} A_n J_0\left(k_n \frac{r}{b}\right) \exp\left(-\frac{k_n^2 a^2}{b^2} t\right). \tag{11.3.140}$$

Finally, we must determine A_n. From the initial condition that $u(r,0) = T_0$,

$$u(r,0) = T_0 = \sum_{n=1}^{\infty} A_n J_0\left(k_n \frac{r}{b}\right), \tag{11.3.141}$$

where

$$A_n = \frac{2k_n^2 T_0}{b^2[k_n^2 + b^2 h^2]J_0^2(k_n)} \int_0^b r J_0\left(k_n \frac{r}{b}\right) dr \tag{11.3.142}$$

$$= \frac{2T_0}{[k_n^2 + b^2 h^2]J_0^2(k_n)} \left(\frac{k_n r}{b}\right) J_1\left(k_n \frac{r}{b}\right)\Big|_0^b \tag{11.3.143}$$

$$= \frac{2k_n T_0 J_1(k_n)}{[k_n^2 + b^2 h^2]J_0^2(k_n)} = \frac{2k_n T_0 J_1(k_n)}{k_n^2 J_0^2(k_n) + b^2 h^2 J_0^2(k_n)} \tag{11.3.144}$$

$$= \frac{2k_n T_0 J_1(k_n)}{k_n^2 J_0^2(k_n) + k_n^2 J_1^2(k_n)} = \frac{2T_0 J_1(k_n)}{k_n[J_0^2(k_n) + J_1^2(k_n)]}, \tag{11.3.145}$$

which follows from (9.5.25), (9.5.35), (9.5.45), and (11.3.136). Consequently, the complete solution is

$$u(r,t) = 2T_0 \sum_{n=1}^{\infty} \frac{J_1(k_n)}{k_n[J_0^2(k_n) + J_1^2(k_n)]} J_0\left(k_n \frac{r}{b}\right) \exp\left(-\frac{k_n^2 a^2}{b^2} t\right). \tag{11.3.146}$$

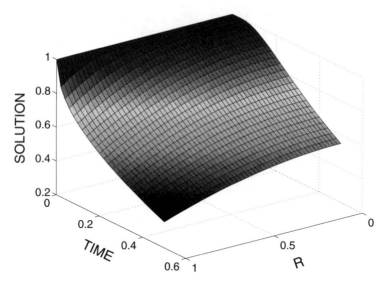

Figure 11.3.7: The temperature $u(r,t)/T_0$ within an infinitely long cylinder at various positions r/b and times $a^2 t/b^2$ that we initially heated to the temperature T_0 and then allowed to radiatively cool with $hb = 1$.

Figure 11.3.7 illustrates the solution (11.3.146) for various Fourier numbers $a^2 t/b^2$ with $hb = 1$. It was created using the MATLAB script

```
clear
hb = 1; m=0; M = 100; dr = 0.02; dt = 0.02;
% find k_n which satisfies hb J_0(k) = k J_1(k)
for n = 1:10000
k1 = 0.05*n; k2 = 0.05*(n+1);
y1 = hb * besselj(0,k1) - k1 * besselj(1,k1);
y2 = hb * besselj(0,k2) - k2 * besselj(1,k2);
if y1*y2 <= 0; m = m+1; zero(m) = k1; end;
end;
%
% use Newton-Raphson method to improve values of k_n
%
for n = 1:M; for k = 1:5
term0 = besselj(0,zero(n));
term1 = besselj(1,zero(n));
term2 = besselj(2,zero(n));
f = hb * term0 - zero(n) * term1;
fp = 0.5*zero(n)*(term2-term0) - (1+hb)*term1;
zero(n) = zero(n) - f / fp;
end; end;
% compute Fourier coefficients
for m = 1:M
```

```
denom = zero(m)*(besselj(0,zero(m))^2+besselj(1,zero(m))^2);
a(m) = 2 * besselj(1,zero(m)) / denom;
end
% compute grid and initialize solution
R = [0:dr:1]; T = [0:dt:0.5];
u = zeros(length(T),length(R));
RR = repmat(R,[length(T) 1]);
TT = repmat(T',[1 length(R)]);
% compute solution from (11.3.146)
for m = 1:M
u = u + a(m)*besselj(0,zero(m)*RR).*exp(-zero(m)*zero(m)*TT);
end
surf(RR,TT,u)
xlabel('R','Fontsize',20); ylabel('TIME','Fontsize',20)
zlabel('SOLUTION','Fontsize',20)
```

These results are similar to Example 11.3.5 except that we are in cylindrical coordinates. Heat flows from the interior and is removed at the cylinder's surface where it radiates to space at the temperature zero. Note that we do *not* suffer from Gibbs phenomena in this case because there is no initial jump in the temperature distribution.

• Example 11.3.9: Temperature within an electrical cable

In the design of cable installations we need the temperature reached within an electrical cable as a function of current and other parameters. To this end,[10] let us solve the nonhomogeneous heat equation in cylindrical coordinates with a radiation boundary condition.

The derivation of the heat equation follows from the conservation of energy:

$$\text{heat generated} = \text{heat dissipated} + \text{heat stored},$$

or

$$I^2 R N \, dt = -\kappa \left[2\pi r \left. \frac{\partial u}{\partial r} \right|_r - 2\pi(r + \Delta r) \left. \frac{\partial u}{\partial r} \right|_{r+\Delta r} \right] dt + 2\pi r \Delta r c \rho \, du,$$

$$(11.3.147)$$

where I is the current through each wire, R is the resistance of each conductor, N is the number of conductors in the shell between radii r and $r + \Delta r = 2\pi m r \Delta r / (\pi b^2)$, b is the radius of the cable, m is the total number of conductors in the cable, κ is the thermal conductivity, ρ is the density, c is

---

[10] Iskenderian, H. P., and W. J. Horvath, 1946: Determination of the temperature rise and the maximum safe current through multiconductor electric cables. *J. Appl. Phys.*, **17**, 255–262.

the average specific heat, and u is the temperature. In the limit of $\Delta r \to 0$, (11.3.147) becomes

$$\frac{\partial u}{\partial t} = A + \frac{a^2}{r} \frac{\partial}{\partial r} \left(r \frac{\partial u}{\partial r} \right), \qquad 0 \le r < b, \quad 0 < t, \qquad (11.3.148)$$

where $A = I^2 Rm/(\pi b^2 c\rho)$, and $a^2 = \kappa/(\rho c)$.

Equation (11.3.148) is the nonhomogeneous heat equation for an infinitely long, axisymmetric cylinder. From Example 11.3.3, we know that we must write the temperature as the sum of a steady-state and transient solution: $u(r,t) = w(r) + v(r,t)$. The steady-state solution $w(r)$ satisfies

$$\frac{1}{r} \frac{d}{dr} \left(r \frac{dw}{dr} \right) = -\frac{A}{a^2}, \qquad (11.3.149)$$

or

$$w(r) = T_c - \frac{Ar^2}{4a^2}, \qquad (11.3.150)$$

where T_c is the (yet unknown) temperature in the center of the cable.

The transient solution $v(r,t)$ is governed by

$$\frac{\partial v}{\partial t} = a^2 \frac{1}{r} \frac{\partial}{\partial r} \left(r \frac{\partial v}{\partial r} \right), \qquad 0 \le r < b, \quad 0 < t, \qquad (11.3.151)$$

with the initial condition that $u(r,0) = T_c - Ar^2/(4a^2) + v(0,t) = 0$. At the surface $r = b$, heat radiates to free space so that the boundary condition is $u_r = -hu$, where h is the surface conductance. Because the temperature equals the steady-state solution when all transient effects die away, $w(r)$ must satisfy this radiation boundary condition regardless of the transient solution. This requires that

$$T_c = \frac{A}{a^2} \left(\frac{b^2}{4} + \frac{b}{2h} \right). \qquad (11.3.152)$$

Therefore, $v(r,t)$ must satisfy $v_r(b,t) = -hv(b,t)$ at $r = b$.

We find the transient solution $v(r,t)$ by separation of variables $v(r,t) = R(r)T(t)$. Substituting into (11.3.151),

$$\frac{1}{rR} \frac{d}{dr} \left(r \frac{dR}{dr} \right) = \frac{1}{a^2 T} \frac{dT}{dt} = -k^2, \qquad (11.3.153)$$

or

$$\frac{d}{dr} \left(r \frac{dR}{dr} \right) + k^2 rR = 0, \qquad (11.3.154)$$

and

$$\frac{dT}{dt} + k^2 a^2 T = 0, \qquad (11.3.155)$$

with $R'(b) = -hR(b)$. The only solution of (11.3.154) which remains finite at $r = 0$ and satisfies the boundary condition is $R(r) = J_0(kr)$, where J_0 is the zero-order Bessel function of the first kind. Substituting $J_0(kr)$ into the boundary condition, the transcendental equation is

$$kbJ_1(kb) - hbJ_0(kb) = 0. \qquad (11.3.156)$$

For a given value of h and b, (11.3.156) yields an infinite number of unique zeros k_n.

The corresponding temporal solution to the problem is

$$T_n(t) = A_n \exp(-a^2 k_n^2 t), \qquad (11.3.157)$$

so that the sum of the product solutions is

$$v(r, t) = \sum_{n=1}^{\infty} A_n J_0(k_n r) \exp(-a^2 k_n^2 t). \qquad (11.3.158)$$

Our final task remains to compute A_n. By evaluating (11.3.158) at $t = 0$,

$$v(r, 0) = \frac{Ar^2}{4a^2} - T_c = \sum_{n=1}^{\infty} A_n J_0(k_n r), \qquad (11.3.159)$$

which is a Fourier-Bessel series in $J_0(k_n r)$. In §9.5 we showed that the coefficient of a Fourier-Bessel series with the orthogonal function $J_0(k_n r)$ and the boundary condition (11.3.156) equals

$$A_n = \frac{2k_n^2}{(k_n^2 b^2 + h^2 b^2) J_0^2(k_n b)} \int_0^b r \left(\frac{Ar^2}{4a^2} - T_c \right) J_0(k_n r) \, dr \qquad (11.3.160)$$

from (9.5.35) and (9.5.45). Carrying out the indicated integrations,

$$A_n = \frac{2}{(k_n^2 + h^2) J_0^2(k_n b)} \left[\left(\frac{Ak_n b}{4a^2} - \frac{A}{k_n ba^2} - \frac{T_c k_n}{b} \right) J_1(k_n b) + \frac{A}{2a^2} J_0(k_n b) \right]. \qquad (11.3.161)$$

We obtained (11.3.161) by using (9.5.25) and integrating by parts as shown in Example 9.5.5.

To illustrate this solution, let us compute it for the typical parameters $b = 4$ cm, $hb = 1$, $a^2 = 1.14$ cm^2/s, $A = 2.2747$ °C/s, and $T_c = 23.94$°C. The value of A corresponds to 37 wires of #6 AWG copper wire within a cable carrying a current of 22 amp.

Figure 11.3.8 illustrates the solution as a function of radius at various times. It was created using the MATLAB script

```
clear
asq = 1.14; A = 2.2747; b = 4; dr = 0.02; dt = 0.02;
```

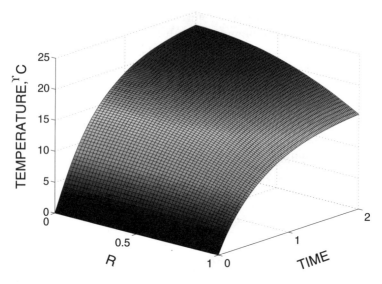

Figure 11.3.8: The temperature field (in degrees Celsius) within an electric copper cable containing 37 wires and a current of 22 amperes at various positions r/b and times a^2t/b^2. Initially the temperature was zero and then we allow the cable to cool radiatively as it is heated. The parameters are $hb = 1$ and the radius of the cable $b = 4$ cm.

```
hb = 1; m=0; M = 10; T_c = 23.94;
const1 = A * b * b / (4 * asq); const2 = A * b * b / asq;
const3 = A * b * b / (2 * asq);
% find k_nb which satisfies hb J_0(kb) = kb J_1(kb)
for n = 1:10000
k1 = 0.05*n; k2 = 0.05*(n+1);
y1 = hb * besselj(0,k1) - k1 * besselj(1,k1);
y2 = hb * besselj(0,k2) - k2 * besselj(1,k2);
if y1*y2 <= 0; m = m+1; zero(m) = k1; end;
end;
%
% use Newton-Raphson method to improve values of k_n
%
for n = 1:M; for k = 1:5
term0 = besselj(0,zero(n));
term1 = besselj(1,zero(n));
term2 = besselj(2,zero(n));
f = hb * term0 - zero(n) * term1;
fp = 0.5*zero(n)*(term2-term0) - (1+hb)*term1;
zero(n) = zero(n) - f / fp;
end; end;
for m = 1:M
denom = (zero(m)*zero(m)+hb*hb)*besselj(0,zero(m))^2;
```

```
a(m) = ((const1-T_c)*zero(m) ...
      - const2/zero(m))*besselj(1,zero(m)) ...
      + const3 * besselj(0,zero(m)));
a(m) = 2 * a(m) / denom;
end
% compute grid and initialize solution
R = [0:dr:1]; T = [0:dt:2];
u = T_c * ones(length(T),length(R));
RR = repmat(R,[length(T) 1]);
TT = repmat(T',[1 length(R)]);
% compute solution from (11.3.132)
u = u - const1 * RR .* RR;
for m = 1:M
u = u + a(m)*besselj(0,zero(m)*RR).*exp(-zero(m)*zero(m)*TT);
end
surf(RR,TT,u); axis([0 1 0 2 0 25]);
xlabel('R','Fontsize',20); ylabel('TIME','Fontsize',20)
zlabel('TEMPERATURE,^\circ C','Fontsize',20)
```

From an initial temperature of zero, the temperature rises due to the constant electrical heating. After a short period of time, it reaches its steady-state distribution given by (11.3.150). The cable is coolest at the surface where heat is radiating away. Heat flows from the interior to replace the heat lost by radiation.

Problems

For Problems 1–5, solve the heat equation $u_t = a^2 u_{xx}$, $0 < x < \pi$, $0 < t$, subject to the boundary conditions that $u(0,t) = u(\pi,t) = 0$, $0 < t$, and the following initial conditions for $0 < x < \pi$. Then plot your results using MATLAB.

1. $u(x,0) = A$, a constant

2. $u(x,0) = \sin^3(x) = [3\sin(x) - \sin(3x)]/4$

3. $u(x,0) = x$

4. $u(x,0) = \pi - x$

5. $u(x,0) = \begin{cases} x, & 0 < x < \pi/2 \\ \pi - x, & \pi/2 < x < \pi \end{cases}$

For Problems 6–10, solve the heat equation $u_t = a^2 u_{xx}$, $0 < x < \pi$, $0 < t$, subject to the boundary conditions that $u_x(0,t) = u_x(\pi,t) = 0$, $0 < t$, and the following initial conditions for $0 < x < \pi$. Then plot your results using MATLAB.

6. $u(x,0) = 1$

7. $u(x,0) = x$

8. $u(x,0) = \cos^2(x) = [1 + \cos(2x)]/2$

9. $u(x,0) = \pi - x$

10. $u(x,0) = \begin{cases} T_0, & 0 < x < \pi/2 \\ T_1, & \pi/2 < x < \pi \end{cases}$

For Problems 11–17, solve the heat equation $u_t = a^2 u_{xx}$, $0 < x < \pi$, $0 < t$, subject to the following boundary conditions and initial condition. Then plot your results using MATLAB.

11. $u_x(0,t) = u(\pi,t) = 0$, $0 < t$; $u(x,0) = x^2 - \pi^2$, $0 < x < \pi$

12. $u(0,t) = u(\pi,t) = T_0$, $0 < t$; $u(x,0) = T_1 \neq T_0$, $0 < x < \pi$

13. $u(0,t) = 0$, $u_x(\pi,t) = 0$, $0 < t$; $u(x,0) = 1$, $0 < x < \pi$

14. $u(0,t) = 0$, $u_x(\pi,t) = 0$, $0 < t$; $u(x,0) = x$, $0 < x < \pi$

15. $u(0,t) = 0$, $u_x(\pi,t) = 0$, $0 < t$; $u(x,0) = \pi - x$, $0 < x < \pi$

16. $u(0,t) = T_0$, $u_x(\pi,t) = 0$, $0 < t$; $u(x,0) = T_1 \neq T_0$, $0 < x < \pi$

17. $u(0,t) = 0$, $u(\pi,t) = T_0$, $0 < t$; $u(x,0) = T_0$, $0 < x < \pi$

18. It is well known that a room with masonry walls is often very difficult to heat. Consider a wall of thickness L, conductivity κ, and diffusivity a^2 which we heat by a surface heat flux at a constant rate H. The temperature of the outside (out-of-doors) face of the wall remains constant at T_0 and the entire wall initially has the uniform temperature T_0. Let us find the temperature of the inside face as a function of time.[11]

We begin by solving the heat conduction problem

$$\frac{\partial u}{\partial t} = a^2 \frac{\partial^2 u}{\partial x^2}, \qquad 0 < x < L, \quad 0 < t,$$

---

[11] Reproduced with acknowledgment to Taylor and Francis, Publishers, from Dufton, A. F., 1927: The warming of walls. *Philos. Mag.*, Ser. 7, **4**, 888–889.

subject to the boundary conditions that

$$\frac{\partial u(0,t)}{\partial x} = -\frac{H}{\kappa}, \qquad \text{and} \qquad u(L,t) = T_0,$$

and the initial condition that $u(x,0) = T_0$. Show that the temperature field equals

$$u(x,t) = T_0 + \frac{HL}{\kappa}\left\{1 - \frac{x}{L} - \frac{8}{\pi^2}\sum_{n=1}^{\infty}\frac{1}{(2n-1)^2}\cos\left[\frac{(2n-1)\pi x}{2L}\right]\right.$$
$$\left. \times \exp\left[-\frac{(2n-1)^2\pi^2 a^2 t}{4L^2}\right]\right\}.$$

Therefore, the rise of temperature at the interior wall $x = 0$ is

$$\frac{HL}{\kappa}\left\{1 - \frac{8}{\pi^2}\sum_{n=1}^{\infty}\frac{1}{(2n-1)^2}\exp\left[-\frac{(2n-1)^2\pi^2 a^2 t}{4L^2}\right]\right\},$$

or

$$\frac{8HL}{\kappa\pi^2}\sum_{n=1}^{\infty}\frac{1}{(2n-1)^2}\left\{1 - \exp\left[-\frac{(2n-1)^2\pi^2 a^2 t}{4L^2}\right]\right\}.$$

For $a^2 t/L^2 \le 1$, this last expression can be approximated[12] by $2Hat^{1/2}/\pi^{1/2}\kappa$. We thus see that the temperature will initially rise as the square root of time and diffusivity and inversely with conductivity. For an average

---

[12] Let us define the function:

$$f(t) = \sum_{n=1}^{\infty}\frac{1 - \exp[-(2n-1)^2\pi^2 a^2 t/L^2]}{(2n-1)^2}.$$

Then

$$f'(t) = \frac{a^2\pi^2}{L^2}\sum_{n=1}^{\infty}\exp[-(2n-1)^2\pi^2 a^2 t/L^2].$$

Consider now the integral

$$\int_0^{\infty}\exp\left(-\frac{a^2\pi^2 t}{L^2}x^2\right)dx = \frac{L}{2a\sqrt{\pi t}}.$$

If we approximate this integral by using the trapezoidal rule with $\Delta x = 2$, then

$$\int_0^{\infty}\exp\left(-\frac{a^2\pi^2 t}{L^2}x^2\right)dx \approx 2\sum_{n=1}^{\infty}\exp[-(2n-1)^2\pi^2 a^2 t/L^2],$$

and $f'(t) \approx a\pi^{3/2}/(4Lt^{1/2})$. Integrating and using $f(0) = 0$, we finally have $f(t) \approx a\pi^{3/2}t^{1/2}/(2L)$. The smaller $a^2 t/L^2$ is, the smaller the error will be. For example, if $t = L^2/a^2$, then the error is 2.4% .

rock $\kappa = 0.0042$ g/cm-s, and $a^2 = 0.0118$ cm^2/s, while for wood (Spruce) $\kappa = 0.0003$ g/cm-s, and $a^2 = 0.0024$ cm^2/s.

The same set of equations applies to heat transfer within a transistor operating at low frequencies.[13] At the junction $(x = 0)$ heat is produced at the rate of H and flows to the transistor's supports $(x = \pm L)$ where it is removed. The supports are maintained at the temperature T_0 which is also the initial temperature of the transistor.

19. The linearized Boussinesq equation[14]

$$\frac{\partial u}{\partial t} = \frac{\partial^2 u}{\partial x^2}, \qquad 0 < x < L, \quad 0 < t,$$

governs the height of the water table $u(x,t)$ above some reference point, where a^2 is the product of the storage coefficient times the hydraulic coefficient divided by the aquifer thickness. A typical value of a^2 is 10 m^2/min. Consider the problem of a strip of land of width L that separates two reservoirs of depth h_1. Initially the height of the water table would be h_1. Suddenly we lower the reservoir on the right $x = L$ to a depth h_2 $[u(0,t) = h_1,\ u(L,t) = h_2$, and $u(x,0) = h_1]$. Find the height of the water table at any position x within the aquifer and any time $t > 0$.

20. The equation (see Problem 19)

$$\frac{\partial u}{\partial t} = \frac{\partial^2 u}{\partial x^2}, \qquad 0 < x < L, \quad 0 < t,$$

governs the height of the water table $u(x,t)$. Consider the problem[15] of a piece of land that suddenly has two drains placed at the points $x = 0$ and $x = L$ so that $u(0,t) = u(L,t) = 0$. If the water table initially has the profile $u(x,0) = 8H(L^3 x - 3L^2 x^2 + 4Lx^3 - 2x^4)/L^4$, find the height of the water table at any point within the aquifer and any time $t > 0$.

21. We want to find the rise of the water table of an aquifer which we sandwich between a canal and impervious rocks if we suddenly raise the water level in the canal h_0 units above its initial elevation and then maintain the canal at this level. The linearized Boussinesq equation (see Problem 19)

$$\frac{\partial u}{\partial t} = \frac{\partial^2 u}{\partial x^2}, \qquad 0 < x < L, \quad 0 < t,$$

---

[13] Mortenson, K. E., 1957: Transistor junction temperature as a function of time. *Proc. IRE*, **45**, 504–513. Equation 2a should read $T_x = -F/k$.

[14] See, for example, Van Schilfgaarde, J., 1970: Theory of flow to drains. *Advances in Hydroscience*, No. 6, Academic Press, 81–85.

[15] For a similar problem, see Dumm, L. D., 1954: New formula for determining depth and spacing of subsurface drains in irrigated lands. *Agric. Eng.*, **35**, 726–730.

governs the level of the water table with the boundary conditions $u(0, t) = h_0$, and $u_x(L, t) = 0$, and the initial condition $u(x, 0) = 0$. Find the height of the water table at any point in the aquifer and any time $t > 0$.

22. Solve the nonhomogeneous heat equation

$$\frac{\partial u}{\partial t} - a^2 \frac{\partial^2 u}{\partial x^2} = e^{-x}, \qquad 0 < x < \pi, \quad 0 < t,$$

subject to the boundary conditions $u(0, t) = u_x(\pi, t) = 0$, $0 < t$, and the initial condition $u(x, 0) = f(x)$, $0 < x < \pi$.

23. Solve the nonhomogeneous heat equation

$$\frac{\partial u}{\partial t} - \frac{\partial^2 u}{\partial x^2} = -1, \qquad 0 < x < 1, \quad 0 < t,$$

subject to the boundary conditions $u_x(0, t) = u_x(1, t) = 0$, $0 < t$, and the initial condition $u(x, 0) = \frac{1}{2}(1 - x^2)$, $0 < x < 1$. [Hint: Note that any function of time satisfies the boundary conditions.]

24. Solve the nonhomogeneous heat equation

$$\frac{\partial u}{\partial t} - a^2 \frac{\partial^2 u}{\partial x^2} = A \cos(\omega t), \qquad 0 < x < \pi, \quad 0 < t,$$

subject to the boundary conditions $u_x(0, t) = u_x(\pi, t) = 0$, $0 < t$, and the initial condition $u(x, 0) = f(x)$, $0 < x < \pi$. [Hint: Note that any function of time satisfies the boundary conditions.]

25. Solve the nonhomogeneous heat equation

$$\frac{\partial u}{\partial t} - \frac{\partial^2 u}{\partial x^2} = \begin{cases} x, & 0 < x \leq \pi/2, \\ \pi - x, & \pi/2 \leq x < \pi, \end{cases} \qquad 0 < x < \pi, \quad 0 < t,$$

subject to the boundary conditions $u(0, t) = u(\pi, t) = 0$, $0 < t$, and the initial condition $u(x, 0) = 0$, $0 < x < \pi$. [Hint: Represent the forcing function as a half-range Fourier sine expansion over the interval $(0, \pi)$.]

26. A uniform, conducting rod of length L and thermometric diffusivity a^2 is initially at temperature zero. We supply heat uniformly throughout the rod so that the heat conduction equation is

$$a^2 \frac{\partial^2 u}{\partial x^2} = \frac{\partial u}{\partial t} - P, \qquad 0 < x < L, \quad 0 < t,$$

where P is the rate at which the temperature would rise if there was no conduction. If we maintain the ends of the rod at the temperature of zero, find the temperature at any position and subsequent time. How would the solution change if the boundary conditions became $u(0,t) = u(L,t) = A \neq 0$, $0 < t$, and the initial conditions read $u(x,0) = A$, $0 < x < L$?

27. Solve the nonhomogeneous heat equation

$$\frac{\partial u}{\partial t} = a^2 \frac{\partial^2 u}{\partial x^2} + \frac{A_0}{c\rho}, \qquad 0 < x < L, \quad 0 < t,$$

where $a^2 = \kappa/c\rho$, with the boundary conditions that

$$\frac{\partial u(0,t)}{\partial x} = 0, \qquad \kappa \frac{\partial u(L,t)}{\partial x} + hu(L,t) = 0, \quad 0 < t,$$

and the initial condition that $u(x,0) = 0$, $0 < x < L$.

28. Find the solution of

$$\frac{\partial u}{\partial t} = \frac{\partial^2 u}{\partial x^2} - u, \qquad 0 < x < L, \quad 0 < t,$$

with the boundary conditions $u(0,t) = 1$, and $u(L,t) = 0$, $0 < t$, and the initial condition $u(x,0) = 0$, $0 < x < L$.

29. Solve[16]

$$\frac{\partial u}{\partial t} + k_1 u = \frac{\partial^2 u}{\partial x^2}, \qquad 0 < x < L, \quad 0 < t,$$

with the boundary conditions

$$\frac{\partial u(0,t)}{\partial x} = 0, \qquad a^2 \frac{\partial u(L,t)}{\partial x} + k_2 u(L,t) = 0, \quad 0 < t,$$

and the initial condition $u(x,0) = u_0$, $0 < x < L$.

30. Solve

$$\frac{\partial u}{\partial t} + \frac{\partial u}{\partial x} = a^2 \frac{\partial^2 u}{\partial x^2}, \qquad 0 < x < 1, \quad 0 < t,$$

with the boundary conditions

$$a^2 \frac{\partial u(0,t)}{\partial x} = u(0,t), \qquad \frac{\partial u(1,t)}{\partial x} = 0, \qquad 0 < t,$$

and the initial condition $u(x,0) = 1$, $0 < x < 1$. Hint: Let $u(x,t) = v(x,t) \exp[(2x - t)/(4a^2)]$ so that the problem becomes

$$\frac{\partial v}{\partial t} = a^2 \frac{\partial^2 v}{\partial x^2}, \qquad 0 < x < 1, \quad 0 < t,$$

[16] Motivated by problems solved in Gomer, R., 1951: Wall reactions and diffusion in static and flow systems. *J. Chem. Phys.*, **19**, 284–289.

with the boundary conditions

$$2a^2 \frac{\partial v(0,t)}{\partial x} = v(0,t), \quad 2a^2 \frac{\partial v(1,t)}{\partial x} = -v(1,t), \qquad 0 < t,$$

and the initial condition $v(x,0) = \exp[-x/(2a^2)]$, $0 < x < 1$.

31. Solve the heat equation in spherical coordinates

$$\frac{\partial u}{\partial t} = \frac{a^2}{r^2} \frac{\partial}{\partial r} \left(r^2 \frac{\partial u}{\partial r} \right) = \frac{a^2}{r} \frac{\partial^2 (ru)}{\partial r^2}, \qquad 0 \le r < 1, \quad 0 < t,$$

subject to the boundary conditions $\lim_{r \to 0} |u(r,t)| < \infty$, and $u(1,t) = 0$, $0 < t$, and the initial condition $u(r,0) = 1$, $0 \le r < 1$.

32. Solve the heat equation in spherical coordinates

$$\frac{\partial u}{\partial t} = \frac{a^2}{r^2} \frac{\partial}{\partial r} \left(r^2 \frac{\partial u}{\partial r} \right) = \frac{a^2}{r} \frac{\partial^2 (ru)}{\partial r^2}, \qquad \alpha < r < \beta, \quad 0 < t,$$

subject to the boundary conditions $u(\alpha,t) = u_r(\beta,t) = 0$, $0 < t$, and the initial condition $u(r,0) = u_0$, $\alpha < r < \beta$.

33. Solve[17] the heat equation in spherical coordinates

$$\frac{\partial u}{\partial t} = a^2 \left(\frac{\partial^2 u}{\partial r^2} + \frac{2}{r} \frac{\partial u}{\partial r} \right) = \frac{a^2}{r} \frac{\partial^2 (ru)}{\partial r^2}, \qquad 0 \le r < b, \quad 0 < t,$$

subject to the boundary conditions

$$\lim_{r \to 0} \frac{\partial u(r,t)}{\partial r} \to 0, \quad \text{and} \quad \frac{\partial u(b,t)}{\partial r} = -\frac{A}{b} u(b,t), \qquad 0 < t,$$

and the initial condition $u(r,0) = u_0$, $0 \le r < b$.

34. Solve[18] the heat equation in cylindrical coordinates

$$\frac{\partial u}{\partial t} = a^2 \left(\frac{\partial^2 u}{\partial r^2} + \frac{1}{r} \frac{\partial u}{\partial r} \right), \qquad 0 \le r < b, \quad 0 < t,$$

---

[17] Reprinted from *Chem. Engng. Sci.*, **57**, H. Zhou, S. Abanades, G. Flamant, D. Gauthier, and J. Lu, Simulation of heavy metal vaporization dynamics of a fluidized bed, 2603–2614, ©2002, with permission from Elsevier Science. See also Mantell, C., M. Rodriguez, and E. Martinez de la Ossa, 2002: Semi-batch extraction of anthocyanins from red grape pomace in packed beds: Experimental results and process modelling. *Chem. Engng. Sci.*, **57**, 3831–3838.

[18] Taken from Destriau, G., 1946: Propagation des charges électriques sur les pellicules faiblement conductrices "problèm plan." *J. Phys. Radium*, **7**, 43–48.

subject to the boundary conditions $\lim_{r \to 0} |u(r,t)| < \infty$, and $u(b,t) = u_0$, $0 < t$, and the initial condition $u(r,0) = 0$, $0 \le r < b$.

35. Solve the heat equation in cylindrical coordinates

$$\frac{\partial u}{\partial t} = \frac{a^2}{r} \frac{\partial}{\partial r} \left(r \frac{\partial u}{\partial r} \right), \qquad 0 \le r < b, \quad 0 < t,$$

subject to the boundary conditions $\lim_{r \to 0} |u(r,t)| < \infty$, and $u(b,t) = \theta$, $0 < t$, and the initial condition $u(r,0) = 1$, $0 \le r < b$.

36. Solve the heat equation in cylindrical coordinates

$$\frac{\partial u}{\partial t} = \frac{a^2}{r} \frac{\partial}{\partial r} \left(r \frac{\partial u}{\partial r} \right), \qquad 0 \le r < 1, \quad 0 < t,$$

subject to the boundary conditions $\lim_{r \to 0} |u(r,t)| < \infty$, and $u(1,t) = 0$, $0 < t$, and the initial condition

$$u(x,0) = \begin{cases} A, & 0 \le r < b, \\ B, & b < r < 1. \end{cases}$$

37. The equation[19]

$$\frac{\partial u}{\partial t} = \frac{G}{\rho} + \nu \left(\frac{\partial^2 u}{\partial r^2} + \frac{1}{r} \frac{\partial u}{\partial r} \right), \qquad 0 \le r < b, \quad 0 < t,$$

governs the velocity $u(r,t)$ of an incompressible fluid of density ρ and kinematic viscosity ν flowing in a long circular pipe of radius b with an imposed, constant pressure gradient $-G$. If the fluid is initially at rest $u(r,0) = 0$, $0 \le r < b$, and there is no slip at the wall $u(b,t) = 0$, $0 < t$, find the velocity at any subsequent time and position.

38. Solve the heat equation in cylindrical coordinates

$$\frac{\partial u}{\partial t} = \frac{a^2}{r} \frac{\partial}{\partial r} \left(r \frac{\partial u}{\partial r} \right), \qquad 0 \le r < b, \quad 0 < t,$$

subject to the boundary conditions $\lim_{r \to 0} |u(r,t)| < \infty$, and $u_r(b,t) = -h\, u(b,t)$, $0 < t$, and the initial condition $u(r,0) = b^2 - r^2$, $0 \le r < b$.

---

[19] Reprinted from *J. Math. Pures Appl.*, *Ser.* 9, **11**, P. Szymanski, Quelques solutions exactes des équations de l'hydrodynamique du fluide visqueux dans le cas d'un tube cylindrique, 67–107, ©1932, with permission from Elsevier Science.

39. Solve[20] the heat equation in cylindrical coordinates

$$\frac{\partial u}{\partial t} = a^2 \left(\frac{\partial^2 u}{\partial r^2} + \frac{1}{r} \frac{\partial u}{\partial r} \right) - \kappa u, \qquad 0 \le r < L, \quad 0 < t,$$

subject to the boundary conditions $\lim_{r \to 0} |u(r,t)| < \infty$, and $u_r(L,t) = -hu(L,t)$, $0 < t$, and the initial condition

$$u(r,0) = \begin{cases} 0, & 0 \le r < b, \\ T_0, & b < r \le L, \end{cases}$$

where $b < L$, and $0 < h, \kappa$.

40. In their study of heat conduction within a thermocouple through which a steady current flows, Reich and Madigan[21] solved the following nonhomogeneous heat conduction problem:

$$\frac{\partial u}{\partial t} - a^2 \frac{\partial^2 u}{\partial x^2} = J - P\,\delta(x - b), \qquad 0 < x < L, \quad 0 < t, \quad 0 < b < L,$$

where J represents the Joule heating generated by the steady current and the P term represents the heat loss from Peltier cooling.[22] Find $u(x,t)$ if both ends are kept at zero $[u(0,t) = u(L,t) = 0]$ and initially the temperature is zero $[u(x,0) = 0]$. The interesting aspect of this problem is the presence of the delta function.

Step 1: Assuming that $u(x,t)$ equals the sum of a steady-state solution $w(x)$ and a transient solution $v(x,t)$, show that the steady-state solution is governed by

$$a^2 \frac{d^2 w}{dx^2} = P\,\delta(x - b) - J, \qquad w(0) = w(L) = 0.$$

Step 2: Show that the steady-state solution is

$$w(x) = \begin{cases} Jx(L - x)/2a^2 + Ax, & 0 < x < b, \\ Jx(L - x)/2a^2 + B(L - x), & b < x < L. \end{cases}$$

Step 3: The temperature must be continuous at $x = b$; otherwise, we would have infinite heat conduction there. Use this condition to show that $Ab = B(L - b)$.

---

[20] Mack, W., M. Plöchl, and U. Gamer, 2000: Effects of a temperature cycle on an elastic-plastic shrink fit with solid inclusion. *Chinese J. Mech.*, **16**, 23–30.

[21] Reich, A. D., and J. R. Madigan, 1961: Transient response of a thermocouple circuit under steady currents. *J. Appl. Phys.*, **32**, 294–301.

[22] In 1834 Jean Charles Athanase Peltier (1785–1845) discovered that there is a heating or cooling effect, quite apart from ordinary resistance heating, whenever an electric current flows through the junction between two different metals.

Step 4: To find a second relationship between A and B, integrate the steady-state differential equation across the interface at $x = b$ and show that

$$\lim_{\epsilon \to 0} a^2 \left. \frac{dw}{dx} \right|_{b-\epsilon}^{b+\epsilon} = P.$$

Step 5: Using the result from Step 4, show that $A + B = -P/a^2$, and

$$w(x) = \begin{cases} Jx(L-x)/2a^2 - Px(L-b)/a^2 L, & 0 < x < b, \\ Jx(L-x)/2a^2 - Pb(L-x)/a^2 L, & b < x < L. \end{cases}$$

Step 6: Re-express $w(x)$ as a half-range Fourier sine expansion and show that

$$w(x) = \frac{4JL^2}{a^2 \pi^3} \sum_{m=1}^{\infty} \frac{\sin[(2m-1)\pi x/L]}{(2m-1)^3} - \frac{2LP}{a^2 \pi^2} \sum_{n=1}^{\infty} \frac{\sin(n\pi b/L)\sin(n\pi x/L)}{n^2}.$$

Step 7: Use separation of variables to find the transient solution by solving

$$\frac{\partial v}{\partial t} = a^2 \frac{\partial^2 v}{\partial x^2}, \qquad 0 < x < L, \quad 0 < t,$$

subject to the boundary conditions $v(0, t) = v(L, t) = 0$, $0 < t$, and the initial condition $v(x, 0) = -w(x), 0 < x < L$.

Step 8: Add the steady-state and transient solutions together and show that

$$u(x, t) = \frac{4JL^2}{a^2 \pi^3} \sum_{m=1}^{\infty} \frac{\sin[(2m-1)\pi x/L]}{(2m-1)^3} \left[1 - e^{-a^2(2m-1)^2\pi^2 t/L^2}\right]$$

$$- \frac{2LP}{a^2 \pi^2} \sum_{n=1}^{\infty} \frac{\sin(n\pi b/L)\sin(n\pi x/L)}{n^2} \left[1 - e^{-a^2 n^2 \pi^2 t/L^2}\right].$$

41. Use separation of variables to solve[23] the partial differential equation

$$\frac{\partial u}{\partial t} = \frac{\partial^2 u}{\partial x^2} + 2a \frac{\partial u}{\partial x}, \qquad 0 < x < 1, \quad 0 < t,$$

subject to the boundary conditions that $u_x(0, t) + 2au(0, t) = 0$, $u_x(1, t) + 2au(1, t) = 0$, $0 < t$, and the initial condition that $u(x, 0) = 1$, $0 < x < 1$.

Step 1: Introducing $u(x, t) = e^{-ax}v(x, t)$, show that the problem becomes

$$\frac{\partial v}{\partial t} = \frac{\partial^2 v}{\partial x^2} - a^2 v, \qquad 0 < x < 1, \quad 0 < t,$$

---

[23] Reprinted from *Physica*, **9**, S. R. DeGroot, Théorie phénoménologique de l'effet Soret, 699–707, ©1942, with permission from Elsevier Science.

subject to the boundary conditions that $v_x(0,t) + av(0,t) = 0$, $v_x(1,t) + av(1,t) = 0$, $0 < t$, and the initial condition that $u(x,0) = e^{ax}$, $0 < x < 1$.

Step 2: Assuming that $v(x,t) = X(x)T(t)$, show that the problem reduces to the ordinary differential equations

$$X'' + (\lambda - a^2)X = 0, \qquad X'(0) + aX(0) = 0, \quad X'(1) + aX(1) = 0,$$

and $T' + \lambda T = 0$, where λ is the separation constant.

Step 3: Solve the eigenvalue problem and show that $\lambda_0 = 0$, $X_0(x) = e^{-ax}$, $T_0(t) = A_0$, and $\lambda_n = a^2 + n^2\pi^2$, $X_n(x) = a\sin(n\pi x) - n\pi\cos(n\pi x)$, and $T_n(t) = A_n e^{-(a^2+n^2\pi^2)t}$, where $n = 1, 2, 3, \ldots$, so that

$$v(x,t) = A_0 e^{-ax} + \sum_{n=1}^{\infty} A_n \left[a\sin(n\pi x) - n\pi\cos(n\pi x)\right] e^{-(a^2+n^2\pi^2)t}.$$

Step 4: Evaluate A_0 and A_n and show that

$$u(x,t) = \frac{2ae^{-2ax}}{1 - e^{-2a}}$$
$$+ 4a\pi \sum_{n=1}^{\infty} \frac{n\left[1 - (-1)^n e^a\right]}{(a^2 + n^2\pi^2)^2} \left[a\sin(n\pi x) - n\pi\cos(n\pi x)\right] e^{-ax-(a^2+n^2\pi^2)t}.$$

42. Use separation of variables to solve[24] the partial differential equation

$$\frac{\partial^3 u}{\partial x^2 \partial t} = \frac{\partial^4 u}{\partial x^4}, \qquad 0 < x < 1, \quad 0 < t,$$

subject to the boundary conditions that $u(0,t) = u_x(0,t) = u(1,t) = u_x(1,t) = 0$, $0 < t$, and the initial condition that $u(x,0) = Ax/2 - (1-A)x^2\left(\frac{3}{2} - x\right)$, $0 \le x \le 1$.

Step 1: Assuming that $u(x,t) = X(x)T(t)$, show that the problem reduces to the ordinary differential equations $X'''' + k^2 X'' = 0$, $X(0) = X'(0) = X(1) = X'(1) = 0$, and $T' + k^2 T = 0$, where k^2 is the separation constant.

Step 2: Solving the eigenvalue problem first, show that

$$X_n(x) = 1 - \cos(k_n x) + \frac{1 - \cos(k_n)}{k_n - \sin(k_n)}[\sin(k_n x) - k_n x],$$

where k_n denotes the nth root of

$$2 - 2\cos(k) - k\sin(k) = \sin(k/2)[\sin(k/2) - (k/2)\cos(k/2)] = 0.$$

---

[24] Taken from Hamza, E. A., 1999: Impulsive squeezing with suction and injection. *J. Appl. Mech.*, **66**, 945–951.

Step 3: Using the results from Step 2, show that there are two classes of eigenfunctions: $\kappa_n = 2n\pi$, $X_n(x) = 1 - \cos(2n\pi x)$, and

$$\tan(\kappa_n/2) = \kappa_n/2, \qquad X_n(x) = 1 - \cos(\kappa_n x) + \frac{2}{\kappa_n}[\sin(\kappa_n x) - \kappa_n x].$$

Step 4: Consider the eigenvalue problem

$$X'''' + \lambda X'' = 0, \qquad 0 < x < 1,$$

with the boundary conditions $X(0) = X'(0) = X(1) = X'(1) = 0$. Show that the orthogonality condition for this problem is

$$\int_0^1 X_n'(x) X_m'(x)\, dx = 0, \qquad n \neq m,$$

where $X_n(x)$ and $X_m(x)$ are two distinct eigenfunctions of this problem. Then show that we can construct an eigenfunction expansion for an arbitrary function $f(x)$ via

$$f(x) = \sum_{n=1}^{\infty} C_n X_n(x), \qquad \text{provided} \qquad C_n = \frac{\int_0^1 f'(x) X_n'(x)\, dx}{\int_0^1 [X_n'(x)]^2\, dx}$$

and $f'(x)$ exists over the interval $(0, 1)$. Hint: Follow the proof in §9.2 and integrate repeatedly by parts to eliminate the higher derivative terms.

Step 5: Show that

$$\int_0^1 [X_n'(x)]^2\, dx = 2n^2\pi^2,$$

if $X_n(x) = 1 - \cos(2n\pi x)$, and

$$\int_0^1 [X_n'(x)]^2\, dx = \kappa_n^2/2,$$

if $X_n(x) = 1 - \cos(\kappa_n x) + 2[\sin(\kappa_n x) - \kappa_n x]/\kappa_n$. Hint: $\sin(\kappa_n) = \kappa_n[1 + \cos(\kappa_n)]/2$.

Step 6: Use the above results to show that

$$u(x, t) = \sum_{n=1}^{\infty} A_n[1 - \cos(2n\pi x)]e^{-4n^2\pi^2 t}$$

$$+ \sum_{n=1}^{\infty} B_n \left\{ 1 - \cos(\kappa_n x) - \frac{2}{\kappa_n}[\sin(\kappa_n x) - \kappa_n x] \right\} e^{-\kappa_n^2 t},$$

where A_n is the Fourier coefficient corresponding to the eigenfunction $1 - \cos(2n\pi x)$ while B_n is the Fourier coefficient corresponding to the eigenfunction $1 - \cos(\kappa_n x) - 2[\sin(\kappa_n x) - \kappa_n x]/\kappa_n$.

Step 7: Show that $A_n = 0$ and $B_n = 2(1 - A)/\kappa_n^2$, so that

$$u(x,t) = 2(1 - A) \sum_{n=1}^{\infty} \left\{ 1 - \cos(\kappa_n x) - \frac{2}{\kappa_n}[\sin(\kappa_n x) - \kappa_n x] \right\} \frac{e^{-\kappa_n^2 t}}{\kappa_n^2}.$$

Hint: $\sin(\kappa_n) = \kappa_n[1+\cos(\kappa_n)]/2$, $\sin(\kappa_n) = 2[1-\cos(\kappa_n)]/\kappa_n$, and $\cos(\kappa_n) = (4 - \kappa_n^2)/(4 + \kappa_n^2)$.

11.4 THE LAPLACE TRANSFORM METHOD

In the previous chapter we showed that we can solve the wave equation by the method of Laplace transforms. This is also true for the heat equation. Once again, we take the Laplace transform with respect to time. From the definition of Laplace transforms,

$$\mathcal{L}[u(x,t)] = U(x,s), \tag{11.4.1}$$

$$\mathcal{L}[u_t(x,t)] = sU(x,s) - u(x,0), \tag{11.4.2}$$

and

$$\mathcal{L}[u_{xx}(x,t)] = \frac{d^2U(x,s)}{dx^2}. \tag{11.4.3}$$

We next solve the resulting ordinary differential equation, known as the *auxiliary equation*, along with the corresponding Laplace transformed boundary conditions. The initial condition gives us the value of $u(x,0)$. The final step is the inversion of the Laplace transform $U(x,s)$. We typically use the inversion integral.

• Example 11.4.1

To illustrate these concepts, we solve a heat conduction problem[25] in a plane slab of thickness $2L$. Initially the slab has a constant temperature of unity. For $0 < t$, we allow both faces of the slab to radiatively cool in a medium which has a temperature of zero.

If $u(x,t)$ denotes the temperature, a^2 is the thermal diffusivity, h is the relative emissivity, t is the time, and x is the distance perpendicular to the face of the slab and measured from the middle of the slab, then the governing equation is

$$\frac{\partial u}{\partial t} = a^2 \frac{\partial^2 u}{\partial x^2}, \qquad -L < x < L, \quad 0 < t, \tag{11.4.4}$$

with the initial condition

$$u(x,0) = 1, \qquad -L < x < L, \tag{11.4.5}$$

---

[25] Goldstein, S., 1932: The application of Heaviside's operational method to the solution of a problem in heat conduction. *Zeit. Angew. Math. Mech.*, **12**, 234–243.

and boundary conditions

$$\frac{\partial u(L, t)}{\partial x} + hu(L, t) = 0, \quad \frac{\partial u(-L, t)}{\partial x} + hu(-L, t) = 0, \quad 0 < t. \quad \textbf{(11.4.6)}$$

Taking the Laplace transform of (11.4.4) and substituting the initial condition,

$$a^2 \frac{d^2 U(x, s)}{dx^2} - sU(x, s) = -1. \quad \textbf{(11.4.7)}$$

If we write $s = a^2 q^2$, (11.4.7) becomes

$$\frac{d^2 U(x, s)}{dx^2} - q^2 U(x, s) = -\frac{1}{a^2}. \quad \textbf{(11.4.8)}$$

From the boundary conditions $U(x, s)$ is an even function in x and we may conveniently write the solution as

$$U(x, s) = \frac{1}{s} + A\cosh(qx). \quad \textbf{(11.4.9)}$$

From (11.4.6),

$$qA\sinh(qL) + \frac{h}{s} + hA\cosh(qL) = 0, \quad \textbf{(11.4.10)}$$

and

$$U(x, s) = \frac{1}{s} - \frac{h\cosh(qx)}{s[q\sinh(qL) + h\cosh(qL)]}. \quad \textbf{(11.4.11)}$$

The inverse of $U(x, s)$ consists of two terms. The first term is simply unity. We will invert the second term by contour integration.

We begin by examining the nature and location of the singularities in the second term. Using the product formulas for the hyperbolic cosine and sine functions, the second term equals

$$\frac{h\left(1 + \frac{4q^2 x^2}{\pi^2}\right)\left(1 + \frac{4q^2 x^2}{9\pi^2}\right)\cdots}{s\left[q^2 L\left(1 + \frac{q^2 L^2}{\pi^2}\right)\left(1 + \frac{q^2 L^2}{4\pi^2}\right)\cdots + h\left(1 + \frac{4q^2 L^2}{\pi^2}\right)\left(1 + \frac{4q^2 L^2}{9\pi^2}\right)\cdots\right]}. \quad \textbf{(11.4.12)}$$

Because $q^2 = s/a^2$, (11.4.12) shows that we do not have any $\sqrt{s}$ in the transform and we need not concern ourselves with branch points and cuts. Furthermore, we have only simple poles: one located at $s = 0$ and the others where

$$q\sinh(qL) + h\cosh(qL) = 0. \quad \textbf{(11.4.13)}$$

If we set $q = i\lambda$, (11.4.13) becomes

$$h\cos(\lambda L) - \lambda\sin(\lambda L) = 0, \quad \textbf{(11.4.14)}$$

or

$$\lambda L \tan(\lambda L) = hL. \tag{11.4.15}$$

From Bromwich's integral,

$$\mathcal{L}^{-1}\left\{\frac{h\cosh(qx)}{s[q\sinh(qL) + h\cosh(qL)]}\right\} = \frac{1}{2\pi i}\oint_C \frac{h\cosh(qx)e^{tz}}{z[q\sinh(qL) + h\cosh(qL)]}\,dz, \tag{11.4.16}$$

where $q = z^{1/2}/a$ and the closed contour C consists of Bromwich's contour plus a semicircle of infinite radius in the left half of the z-plane. The residue at $z = 0$ is 1 while at $z_n = -a^2\lambda_n^2$,

$$\mathrm{Res}\left\{\frac{h\cosh(qx)e^{tz}}{z[q\sinh(qL) + h\cosh(qL)]}; z_n\right\}$$

$$= \lim_{z\to z_n}\frac{h(z + a^2\lambda_n^2)\cosh(qx)e^{tz}}{z[q\sinh(qL) + h\cosh(qL)]} \tag{11.4.17}$$

$$= \lim_{z\to z_n}\frac{h\cosh(qx)e^{tz}}{z[(1 + hL)\sinh(qL) + qL\cosh(qL)]/(2a^2q)} \tag{11.4.18}$$

$$= \frac{2ha^2\lambda_n i\cosh(i\lambda_n x)\exp(-\lambda_n^2 a^2 t)}{(-a^2\lambda_n^2)[(1 + hL)i\sin(\lambda_n L) + i\lambda_n L\cos(\lambda_n L)]} \tag{11.4.19}$$

$$= -\frac{2h\cos(\lambda_n x)\exp(-a^2\lambda_n^2 t)}{\lambda_n[(1 + hL)\sin(\lambda_n L) + \lambda_n L\cos(\lambda_n L)]}. \tag{11.4.20}$$

Therefore, the inversion of $U(x, s)$ is

$$u(x, t) = 1 - \left\{1 - 2h\sum_{n=1}^{\infty}\frac{\cos(\lambda_n x)\exp(-a^2\lambda_n^2 t)}{\lambda_n[(1 + hL)\sin(\lambda_n L) + \lambda_n L\cos(\lambda_n L)]}\right\}, \tag{11.4.21}$$

or

$$u(x, t) = 2h\sum_{n=1}^{\infty}\frac{\cos(\lambda_n x)\exp(-a^2\lambda_n^2 t)}{\lambda_n[(1 + hL)\sin(\lambda_n L) + \lambda_n L\cos(\lambda_n L)]}. \tag{11.4.22}$$

We can further simplify (11.4.22) by using $h/\lambda_n = \tan(\lambda_n L)$, and $hL = \lambda_n L\tan(\lambda_n L)$. Substituting these relationships into (11.4.22) and simplifying,

$$u(x, t) = 2\sum_{n=1}^{\infty}\frac{\sin(\lambda_n L)\cos(\lambda_n x)\exp(-a^2\lambda_n^2 t)}{\lambda_n L + \sin(\lambda_n L)\cos(\lambda_n L)}. \tag{11.4.23}$$

Figure 11.4.1 illustrates (11.4.23). It was created using the MATLAB script

```
clear
hL = 1; m = 0; M = 100; dx = 0.05; dt = 0.05;
%
```

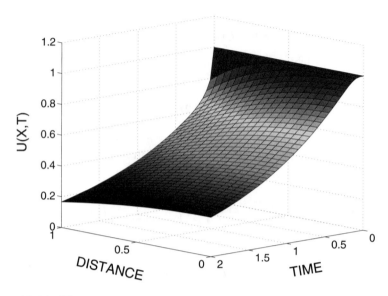

Figure 11.4.1: The temperature within the portion of a slab $0 < x/L < 1$ at various times $a^2 t/L^2$ if the faces of the slab radiate to free space at temperature zero and the slab initially has the temperature 1. The parameter $hL = 1$.

```
% create initial guess at zero_n
%
zero = zeros(length(M));
for n = 1:10000
k1 = 0.1*n; k2 = 0.1*(n+1);
prod = k1 * tan(k1); y1 = hL - prod; y2 = hL - k2 * tan(k2);
if (y1*y2 <= 0 & prod < 2 & m < M) m = m+1; zero(m) = k1; end;
end;
%
% use Newton-Raphson method to improve values of zero_n
%
for n = 1:M; for k = 1:10
f = hL - zero(n) * tan(zero(n));
fp = - tan(zero(n)) - zero(n) * sec(zero(n))^2;
zero(n) = zero(n) - f / fp;
end; end;
% compute Fourier coefficients
for m = 1:M
a(m) = 2 * sin(zero(m)) / (zero(m) + sin(zero(m))*cos(zero(m)));
end
% compute grid and initialize solution
X = [0:dx:1]; T = [0:dt:2];
u = zeros(length(T),length(X));
```

```
XX = repmat(X,[length(T) 1]); TT = repmat(T',[1 length(X)]);
% compute solution from (11.4.23)
for m = 1:M
u = u + a(m) * cos(zero(m)*XX) .* exp(-zero(m)*zero(m)*TT);
end
surf(XX,TT,u)
xlabel('distance','Fontsize',20); ylabel('time','Fontsize',20)
zlabel('U(X,T)','Fontsize',20)
```

• Example 11.4.2: Heat dissipation in disc brakes

Disc brakes consist of two blocks of frictional material known as pads which press against each side of a rotating annulus, usually made of a ferrous material. In this problem we determine the transient temperatures reached in a disc brake during a single brake application.[26] If we ignore the errors introduced by replacing the cylindrical portion of the drum by a rectangular plate, we can model our disc brakes as a one-dimensional solid which friction heats at both ends. Assuming symmetry about $x = 0$, the boundary condition there is $u_x(0, t) = 0$. To model the heat flux from the pads, we assume a uniform disc deceleration that generates heat from the frictional surfaces at the rate $N(1 - Mt)$, where M and N are experimentally determined constants.

If $u(x, t)$, κ, and a^2 denote the temperature, thermal conductivity, and diffusivity of the rotating annulus, respectively, then the heat equation is

$$\frac{\partial u}{\partial t} = a^2 \frac{\partial^2 u}{\partial x^2}, \qquad 0 < x < L, \quad 0 < t, \qquad (11.4.24)$$

with the boundary conditions

$$\frac{\partial u(0, t)}{\partial x} = 0, \quad \kappa \frac{\partial u(L, t)}{\partial x} = N(1 - Mt), \qquad 0 < t. \qquad (11.4.25)$$

The boundary condition at $x = L$ gives the frictional heating of the disc pads. Introducing the Laplace transform of $u(x, t)$, defined as

$$U(x, s) = \int_0^\infty u(x, t)e^{-st} \, dt, \qquad (11.4.26)$$

the equation to be solved becomes

$$\frac{d^2 U}{dx^2} - \frac{s}{a^2} U = 0, \qquad (11.4.27)$$

---

[26] From Newcomb, T. P., 1958: The flow of heat in a parallel-faced infinite solid. *Brit. J. Appl. Phys.*, **9**, 370–372. See also Newcomb, T. P., 1958/59: Transient temperatures in brake drums and linings. *Proc. Inst. Mech. Eng., Auto. Div.*, 227–237; Newcomb, T. P., 1959: Transient temperatures attained in disk brakes. *Brit. J. Appl. Phys.*, **10**, 339–340.

subject to the boundary conditions that

$$\frac{dU(0,s)}{dx} = 0, \quad \text{and} \quad \frac{dU(L,s)}{dx} = \frac{N}{\kappa}\left(\frac{1}{s} - \frac{M}{s^2}\right). \tag{11.4.28}$$

The solution of (11.4.27) is

$$U(x,s) = A\cosh(qx) + B\sinh(qx), \tag{11.4.29}$$

where $q = s^{1/2}/a$. Using the boundary conditions, the solution becomes

$$U(x,s) = \frac{N}{\kappa}\left(\frac{1}{s} - \frac{M}{s^2}\right)\frac{\cosh(qx)}{q\sinh(qL)}. \tag{11.4.30}$$

It now remains to invert the transform (11.4.30). We will invert $\cosh(qx)/[sq\sinh(qL)]$; the inversion of the second term follows by analog.

Our first concern is the presence of $s^{1/2}$ because this is a multivalued function. However, when we replace the hyperbolic cosine and sine functions with their Taylor expansions, $\cosh(qx)/[sq\sinh(qL)]$ contains only powers of s and is, in fact, a single-valued function.

From Bromwich's integral,

$$\mathcal{L}^{-1}\left[\frac{\cosh(qx)}{sq\sinh(qL)}\right] = \frac{1}{2\pi i}\int_{c-\infty i}^{c+\infty i}\frac{\cosh(qx)e^{tz}}{zq\sinh(qL)}\,dz, \tag{11.4.31}$$

where $q = z^{1/2}/a$. Just as in the previous example, we replace the hyperbolic cosine and sine with their product expansion to determine the nature of the singularities. The point $z = 0$ is a second-order pole. The remaining poles are located where $z_n^{1/2}L/a = n\pi i$, or $z_n = -n^2\pi^2 a^2/L^2$, where $n = 1,2,3,\dots$. We have chosen the positive sign because $z^{1/2}$ must be single-valued; if we had chosen the negative sign the answer would have been the same. Our expansion also shows that the poles are simple.

Having classified the poles, we now close Bromwich's contour, which lies slightly to the right of the imaginary axis, with an infinite semicircle in the left half-plane, and use the residue theorem. The values of the residues are

$$\text{Res}\left[\frac{\cosh(qx)e^{tz}}{zq\sinh(qL)};0\right] = \frac{1}{1!}\lim_{z\to 0}\frac{d}{dz}\left\{\frac{(z-0)^2\cosh(qx)e^{tz}}{zq\sinh(qL)}\right\} \tag{11.4.32}$$

$$= \lim_{z\to 0}\frac{d}{dz}\left\{\frac{z\cosh(qx)e^{tz}}{q\sinh(qL)}\right\} \tag{11.4.33}$$

$$= \frac{a^2}{L}\lim_{z\to 0}\frac{d}{dz}\left\{\frac{z\left[1 + \frac{zx^2}{2!a^2} + \cdots\right]\left[1 + tz + \frac{t^2z^2}{2!} + \cdots\right]}{z + \frac{L^2z^2}{3!a^2} + \cdots}\right\} \tag{11.4.34}$$

$$= \frac{a^2}{L}\lim_{z\to 0}\frac{d}{dz}\left\{1 + tz + \frac{zx^2}{2a^2} - \frac{zL^2}{3!a^2} + \cdots\right\} \tag{11.4.35}$$

$$= \frac{a^2}{L}\left\{t + \frac{x^2}{2a^2} - \frac{L^2}{6a^2}\right\}, \tag{11.4.36}$$

and

$$\text{Res}\left[\frac{\cosh(qx)e^{tz}}{zq\sinh(qL)}; z_n\right] = \left[\lim_{z\to z_n}\frac{\cosh(qx)}{zq}e^{tz}\right]\left[\lim_{z\to z_n}\frac{z-z_n}{\sinh(qL)}\right] \qquad (11.4.37)$$

$$= \lim_{z\to z_n}\frac{\cosh(qx)e^{tz}}{zq\cosh(qL)L/(2a^2q)} \qquad (11.4.38)$$

$$= \frac{\cosh(n\pi xi/L)\exp(-n^2\pi^2a^2t/L^2)}{(-n^2\pi^2a^2/L^2)\cosh(n\pi i)L/(2a^2)} \qquad (11.4.39)$$

$$= -\frac{2L(-1)^n}{n^2\pi^2}\cos(n\pi x/L)e^{-n^2\pi^2a^2t/L^2}. \qquad (11.4.40)$$

When we sum all of the residues from both inversions, the solution is

$$u(x,t) = \frac{a^2N}{\kappa L}\left\{t + \frac{x^2}{2a^2} - \frac{L^2}{6a^2}\right\} - \frac{2LN}{\kappa\pi^2}\sum_{n=1}^{\infty}\frac{(-1)^n}{n^2}\cos(n\pi x/L)e^{-n^2\pi^2a^2t/L^2}$$

$$- \frac{a^2NM}{\kappa L}\left\{\frac{t^2}{2} + \frac{tx^2}{2a^2} - \frac{tL^2}{6a^2} + \frac{x^4}{24a^4} - \frac{x^2L^2}{12a^4} + \frac{7L^4}{360a^4}\right\}$$

$$- \frac{2L^3NM}{a^2\kappa\pi^4}\sum_{n=1}^{\infty}\frac{(-1)^n}{n^4}\cos(n\pi x/L)e^{-n^2\pi^2a^2t/L^2}. \qquad (11.4.41)$$

Figure 11.4.2 shows the temperature in the brake lining at various places within the lining $[x' = x/L]$ if $a^2 = 3.3 \times 10^{-3}$ cm^2/sec, $\kappa = 1.8 \times 10^{-3}$ cal/(cm sec°C), $L = 0.48$ cm, and $N = 1.96$ cal/(cm^2 sec). Initially the frictional heating results in an increase in the disc brake's temperature. As time increases, the heating rate decreases and radiative cooling becomes sufficiently large that the temperature begins to fall.

• **Example 11.4.3**

In the previous example we showed that Laplace transforms are particularly useful when the boundary conditions are time dependent. Consider now the case when one of the boundaries is moving.

We wish to solve the heat equation

$$\frac{\partial u}{\partial t} = a^2\frac{\partial^2 u}{\partial x^2}, \qquad \beta t < x < \infty, \quad 0 < t, \qquad (11.4.42)$$

subject to the boundary conditions

$$u(x,t)\big|_{x=\beta t} = f(t), \quad \text{and} \quad \lim_{x\to\infty}|u(x,t)| < \infty, \quad 0 < t, \qquad (11.4.43)$$

and the initial condition

$$u(x,0) = 0, \qquad 0 < x < \infty. \qquad (11.4.44)$$

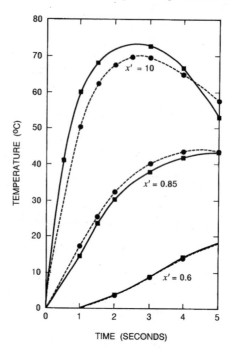

Figure 11.4.2: Typical curves of transient temperature at different locations in a brake lining. Circles denote computed values while squares are experimental measurements. (From Newcomb, T. P., 1958: The flow of heat in a parallel-faced infinite solid. *Brit. J. Appl. Phys.*, **9**, 372 with permission.)

This type of problems arises in combustion problems where the boundary moves due to the burning of the fuel.

We begin by introducing the coordinate $\eta = x - \beta t$. Then the problem can be reformulated as

$$\frac{\partial u}{\partial t} - \beta \frac{\partial u}{\partial \eta} = a^2 \frac{\partial^2 u}{\partial \eta^2}, \quad 0 < \eta < \infty, \quad 0 < t, \tag{11.4.45}$$

subject to the boundary conditions

$$u(0,t) = f(t), \quad \lim_{\eta \to \infty} |u(\eta,t)| < \infty, \quad 0 < t, \tag{11.4.46}$$

and the initial condition

$$u(\eta, 0) = 0, \quad 0 < \eta < \infty. \tag{11.4.47}$$

Taking the Laplace transform of (11.4.45), we have that

$$\frac{d^2 U(\eta, s)}{d\eta^2} + \frac{\beta}{a^2} \frac{dU(\eta, s)}{d\eta} - \frac{s}{a^2} U(\eta, s) = 0, \tag{11.4.48}$$

with

$$U(0, s) = F(s), \quad \text{and} \quad \lim_{\eta \to \infty} |U(\eta, s)| < \infty. \tag{11.4.49}$$

The solution to (11.4.48)–(11.4.49) is

$$U(\eta, s) = F(s) \exp\left(-\frac{\beta\eta}{2a^2} - \frac{\eta}{a}\sqrt{s + \frac{\beta^2}{4a^2}}\right). \tag{11.4.50}$$

Because

$$\mathcal{L}\left[\Phi(\eta, t)\right] = \exp\left(-\frac{\eta}{a}\sqrt{s + \frac{\beta^2}{4a^2}}\right), \tag{11.4.51}$$

where

$$\Phi(\eta, t) = \frac{1}{2}\left[e^{-\beta\eta/2a^2}\operatorname{erfc}\left(\frac{\eta}{2a\sqrt{t}} - \frac{\beta\sqrt{t}}{2a}\right) + e^{\beta\eta/2a^2}\operatorname{erfc}\left(\frac{\eta}{2a\sqrt{t}} + \frac{\beta\sqrt{t}}{2a}\right)\right], \tag{11.4.52}$$

and

$$\operatorname{erfc}(x) = 1 - \frac{2}{\sqrt{\pi}}\int_0^x e^{-\eta^2}\, d\eta, \tag{11.4.53}$$

we have by the convolution theorem that

$$u(\eta, t) = e^{-\beta\eta/2a^2}\int_0^t f(t - \tau)\Phi(\eta, \tau)\, d\tau, \tag{11.4.54}$$

or

$$u(x, t) = e^{-\beta(x - \beta t)/2a^2}\int_0^t f(t - \tau)\Phi(x - \beta\tau, \tau)\, d\tau. \tag{11.4.55}$$

Problems

1. Solve
$$\frac{\partial u}{\partial t} = \frac{\partial^2 u}{\partial x^2} - a^2(u - T_0), \qquad 0 < x < 1, \quad 0 < t,$$
subject to the boundary conditions $u_x(0, t) = u_x(1, t) = 0$, $0 < t$, and the initial condition $u(x, 0) = 0$, $0 < x < 1$.

2. Solve
$$\frac{\partial u}{\partial t} = \frac{\partial^2 u}{\partial x^2}, \qquad 0 < x < 1, \quad 0 < t,$$
subject to the boundary conditions $u_x(0, t) = 0$, $u(1, t) = t$, $0 < t$, and the initial condition $u(x, 0) = 0$, $0 < x < 1$.

3. Solve
$$\frac{\partial u}{\partial t} = \frac{\partial^2 u}{\partial x^2}, \qquad 0 < x < 1, \quad 0 < t,$$

subject to the boundary conditions $u(0,t) = 0$, $u(1,t) = 1$, $0 < t$, and the initial condition $u(x,0) = 0$, $0 < x < 1$.

4. Solve

$$\frac{\partial u}{\partial t} = \frac{\partial^2 u}{\partial x^2}, \qquad -\tfrac{1}{2} < x < \tfrac{1}{2}, \quad 0 \le t,$$

subject to the boundary conditions $u_x\left(-\tfrac{1}{2},t\right) = 0$, $u_x\left(\tfrac{1}{2},t\right) = \delta(t)$, $0 \le t$, and the initial condition $u(x,0) = 0$, $-\tfrac{1}{2} < x < \tfrac{1}{2}$.

5. Solve

$$\frac{\partial u}{\partial t} - \frac{\partial^2 u}{\partial x^2} = 1, \qquad 0 < x < 1, \quad 0 < t,$$

subject to the boundary conditions $u(0,t) = u(1,t) = 0$, $0 < t$, and the initial condition $u(x,0) = 0$, $0 < x < 1$.

6. Solve[27]

$$\frac{\partial u}{\partial t} = a^2 \frac{\partial^2 u}{\partial x^2}, \qquad 0 < x < \infty, \quad 0 < t,$$

subject to the boundary conditions

$$u(0,t) = 1, \qquad \lim_{x \to \infty} |u(x,t)| < \infty, \quad 0 < t,$$

and the initial condition $u(x,0) = 0$, $0 < x < \infty$. [Hint: Use tables to invert the Laplace transform.]

7. Solve

$$\frac{\partial u}{\partial t} = \frac{\partial^2 u}{\partial x^2}, \qquad 0 < x < \infty, \quad 0 < t,$$

subject to the boundary conditions

$$\frac{\partial u(0,t)}{\partial x} = 1, \qquad \lim_{x \to \infty} |u(x,t)| < \infty, \quad 0 < t,$$

and the initial condition $u(x,0) = 0$, $0 < x < \infty$. [Hint: Use tables to invert the Laplace transform.]

8. Solve

$$\frac{\partial u}{\partial t} = \frac{\partial^2 u}{\partial x^2}, \qquad 0 < x < \infty, \quad 0 < t,$$

---

[27] If $u(x,t)$ denotes the Eulerian velocity of a viscous fluid in the half space $x > 0$ and parallel to the wall located at $x = 0$, then this problem was first solved by Stokes, G. G., 1850: On the effect of the internal friction of fluids on the motions of pendulums. *Proc. Cambridge Philos. Soc.*, **9**, **Part II**, [8]–[106].

subject to the boundary conditions

$$u(0,t) = 1, \quad \lim_{x \to \infty} |u(x,t)| < \infty, \quad 0 < t,$$

and the initial condition $u(x,0) = e^{-x}$, $0 < x < \infty$. [Hint: Use tables to invert the Laplace transform.]

9. Solve

$$\frac{\partial u}{\partial t} = a^2 \left[\frac{\partial^2 u}{\partial x^2} + (1 + \delta)\frac{\partial u}{\partial x} + \delta u \right], \quad 0 < x < \infty, \quad 0 < t,$$

where δ is a constant, subject to the boundary conditions

$$u(0,t) = u_0, \quad \lim_{x \to \infty} |u(x,t)| < \infty, \quad 0 < t,$$

and the initial condition $u(x,0) = 0$, $0 < x < \infty$. Note that

$$\mathcal{L}^{-1}\left[\frac{1}{s} \exp\left(-2\alpha\sqrt{s + \beta^2} \right) \right] = \tfrac{1}{2}e^{2\alpha\beta}\mathrm{erfc}\left(\frac{\alpha}{\sqrt{t}} + \beta\sqrt{t} \right)$$
$$+ \tfrac{1}{2}e^{-2\alpha\beta}\mathrm{erfc}\left(\frac{\alpha}{\sqrt{t}} - \beta\sqrt{t} \right),$$

where erfc is the complementary error function.

10. During their modeling of a chemical reaction with a back reaction, Agmon et al.[28] solved

$$\frac{\partial u}{\partial t} = a^2 \frac{\partial^2 u}{\partial x^2}, \quad 0 < x < \infty, \quad 0 < t,$$

subject to the boundary conditions

$$\kappa_d + a^2 u_x(0,t) + a^2 \kappa_d \int_0^t u_x(0,\tau)\, d\tau = \kappa_r u(0,t),$$

$$\lim_{x \to \infty} |u(x,t)| < \infty, \quad 0 < t,$$

and the initial condition $u(x,0) = 0$, $0 < x < \infty$, where κ_d and κ_r denote the intrinsic dissociation and recombination rate coefficients, respectively. What should they have found?

---

[28] Reprinted with permission from Agmon, N., E. Pines, and D. Huppert, 1988: Geminate recombination in proton-transfer reactions. II. Comparison of diffusional and kinetic schemes. *J. Chem. Phys.*, **88**, 5631–5638. ©1988, American Institute of Physics.

11. Solve[29]

$$\frac{\partial u}{\partial t} = \frac{\partial^2 u}{\partial x^2} - \beta u, \qquad 0 < x < \infty, \quad 0 < t,$$

subject to the boundary conditions

$$\rho u(0,t) - u_x(0,t) = e^{(\sigma^2 - \beta)t}, \qquad \lim_{x \to \infty} |u(x,t)| < \infty, \quad 0 < t,$$

and the initial condition $u(x,0) = 0$, $0 < x < \infty$, where β, ρ, and σ are constants and $\sigma \neq \rho$.

12. Solve

$$\frac{\partial u}{\partial t} = a^2 \frac{\partial^2 u}{\partial x^2} + Ae^{-kx}, \qquad 0 < x < \infty, \quad 0 < t,$$

subject to the boundary conditions

$$\frac{\partial u(0,t)}{\partial x} = 0, \qquad \lim_{x \to \infty} u(x,t) = u_0, \quad 0 < t,$$

and the initial condition $u(x,0) = u_0$, $0 < x < \infty$.

13. Solve

$$\frac{\partial u}{\partial t} = \frac{\partial^2 u}{\partial x^2} - P, \qquad 0 < x < L, \quad 0 < t,$$

subject to the boundary conditions $u(0,t) = t$, $u(L,t) = 0$, $0 < t$, and the initial condition $u(x,0) = 0$, $0 < x < L$.

14. Solve

$$\frac{\partial u}{\partial t} = a^2 \frac{\partial^2 u}{\partial x^2} + ku, \qquad 0 < x < L, \quad 0 < t, \quad 0 < k,$$

subject to the boundary conditions $u(0,t) = u(L,t) = T_0$, $0 < t$, and the initial condition $u(x,0) = T_0$, $0 < x < L$.

15. An electric fuse protects electrical devices by using resistance heating to melt an enclosed wire when excessive current passes through it. A knowledge of the distribution of temperature along the wire is important in the design of the fuse. If the temperature rises to the melting point only over a small interval of the element, the melt will produce a small gap, resulting in an unnecessary prolongation of the fault and a considerable release of energy.

[29] Reprinted from *Bull. Math. Biol.*, **49**, G. M. Saidel, E. D. Morris, and G. M. Chisolm, Transport of macromolecules in arterial wall *in vivo*: A mathematical model and analytic solutions, 153–169. ©1987, with permission from Elsevier Science.

Therefore, the desirable temperature distribution should melt most of the wire. For this reason, Guile and Carne[30] solved the heat conduction equation

$$\frac{\partial u}{\partial t} = a^2 \frac{\partial^2 u}{\partial x^2} + q(1 + \alpha u), \qquad -L < x < L, \quad 0 < t,$$

to understand the temperature structure within the fuse just before meltdown. The second term on the right side of the heat conduction equation gives the resistance heating which is assumed to vary linearly with temperature. If the terminals at $x = \pm L$ remain at a constant temperature, which we can take to be zero, the boundary conditions are $u(-L, t) = u(L, t) = 0$, $0 < t$. The initial condition is $u(x, 0) = 0$, $-L < x < L$. Find the temperature field as a function of the parameters a, q, and α.

16. Solve[31]

$$\frac{\partial u}{\partial t} = \frac{\partial^2 u}{\partial r^2} + \frac{2}{r} \frac{\partial u}{\partial r}, \qquad 0 \le r < 1, \quad 0 < t,$$

subject to the boundary conditions

$$\lim_{r \to 0} |u(r, t)| < \infty, \qquad \frac{\partial u(1, t)}{\partial r} = 1, \quad 0 < t,$$

and the initial condition $u(r, 0) = 0$, $0 \le r < 1$. [Hint: Use the new dependent variable $v(r, t) = ru(r, t)$.]

17. Solve[32]

$$\frac{\partial u}{\partial t} = a^2 \left(\frac{\partial^2 u}{\partial r^2} + \frac{2}{r} \frac{\partial u}{\partial r} \right) + q(t) = \frac{a^2}{r} \frac{\partial^2 (ru)}{\partial r^2} + q(t), \qquad b < r < \infty, \quad 0 < t,$$

subject to the boundary conditions

$$\frac{\partial u(b, t)}{\partial r} = u(b, t), \qquad \lim_{r \to \infty} u(r, t) = u_0 + \int_0^t q(\tau) \, d\tau, \quad 0 < t,$$

and the initial condition $u(r, 0) = u_0$, $b < r < \infty$.

---

30 From Guile, A. E., and E. B. Carne, 1954: An analysis of an analogue solution applied to the heat conduction problem in a cartridge fuse. *AIEE Trans., Part 1*, **72**, 861–868. ©AIEE (now IEEE).

31 From Reismann, H., 1962: Temperature distribution in a spinning sphere during atmospheric entry. *J. Aerosp. Sci.*, **29**, 151–159 with permission.

32 Reprinted with permission from Frisch, H. L, and F. C. Collins, 1952: Diffusional processes in the growth of aerosol particles. *J. Chem. Phys.*, **20**, 1797–1803. ©1952, American Institute of Physics.

18. Consider[33] a viscous fluid located between two fixed walls $x = \pm L$. At $x = 0$ we introduce a thin, infinitely long rigid barrier of mass m per unit area and let it fall under the force of gravity which points in the direction of positive x. We wish to find the velocity of the fluid $u(x,t)$. The fluid is governed by the partial differential equation

$$\frac{\partial u}{\partial t} = \nu \frac{\partial^2 u}{\partial x^2}, \qquad 0 < x < L, \quad 0 < t,$$

subject to the boundary conditions

$$u(L,t) = 0, \qquad \frac{\partial u(0,t)}{\partial t} - \frac{2\mu}{m}\frac{\partial u(0,t)}{\partial x} = g, \qquad 0 < t,$$

and the initial condition $u(x,0) = 0$, $0 < x < L$.

19. Consider[34] a viscous fluid located between two fixed walls $x = \pm L$. At $x = 0$ we introduce a thin, infinitely long rigid barrier of mass m per unit area. The barrier is acted upon by an elastic force in such a manner that it would vibrate with a frequency ω if the liquid were absent. We wish to find the barrier's deviation from equilibrium, $y(t)$. The fluid is governed by the partial differential equation

$$\frac{\partial u}{\partial t} = \nu \frac{\partial^2 u}{\partial x^2}, \qquad 0 < x < L, \quad 0 < t.$$

The boundary conditions are

$$u(L,t) = m\frac{d^2y}{dt^2} - 2\mu\frac{\partial u(0,t)}{\partial x} + m\omega^2 y = 0, \qquad \frac{dy}{dt} = u(0,t), \quad 0 < t,$$

and the initial conditions are $u(x,0) = 0$, $0 < x < L$, and $y(0) = A$, $y'(0) = 0$.

20. Solve[35]

$$\frac{\partial u}{\partial t} = a^2 \frac{\partial^2 u}{\partial x^2}, \qquad 0 < x < L, \quad 0 < t,$$

subject to the boundary conditions $u_x(0,t) = 0$, $a^2 u_x(L,t) + \alpha u(L,t) = F$, $0 < t$, and the initial condition $u(x,0) = 0$, $0 < x < L$.

---

[33] Reproduced with acknowledgment to Taylor and Francis, Publishers, from Havelock, T. H., 1921: The solution of an integral equation occurring in certain problems of viscous fluid motion. *Philos. Mag., Ser. 6*, **42**, 620–628.

[34] Reproduced with acknowledgment to Taylor and Francis, Publishers, from Havelock, T. H., 1921: On the decay of oscillation of a solid body in a viscous fluid. *Philos. Mag., Ser. 6*, **42**, 628–634.

[35] Taken from McCarthy, T. A., and H. J. Goldsmid, 1970: Electro-deposited copper in bismuth telluride. *J. Phys. D*, **3**, 697–706. Reprinted with the permission of IOP Publishing Limited.

21. Solve

$$\frac{\partial u}{\partial t} = \frac{\partial^2 u}{\partial x^2}, \qquad 0 \le x < 1, \quad 0 \le t,$$

subject to the boundary conditions

$$u(0,t) = 0, \quad 3a \left[\frac{\partial u(1,t)}{\partial x} - u(1,t) \right] + \frac{\partial u(1,t)}{\partial t} = \delta(t), \quad 0 \le t,$$

and the initial condition $u(x,0) = 0$, $0 \le x < 1$.

22. Solve[36] the partial differential equation

$$\frac{\partial u}{\partial t} + V \frac{\partial u}{\partial x} = \frac{\partial^2 u}{\partial x^2}, \qquad 0 < x < 1, \quad 0 < t,$$

where V is a constant, subject to the boundary conditions

$$u(0,t) = 1, \quad u_x(1,t) = 0, \qquad 0 < t,$$

and the initial condition $u(x,0) = 0$, $0 < x < 1$.

23. Solve

$$\frac{1}{r} \frac{\partial}{\partial r} \left(r \frac{\partial u}{\partial r} \right) - \frac{\partial u}{\partial t} = \delta(t), \qquad 0 \le r < a, \quad 0 \le t,$$

subject to the boundary conditions

$$\lim_{r \to 0} |u(r,t)| < \infty, \quad u(a,t) = 0, \quad 0 \le t,$$

and the initial condition $u(r,0) = 0$, $0 \le r < a$, where $\delta(t)$ is the Dirac delta function. Note that $J_n(iz) = i^n I_n(z)$ and $I_n(iz) = i^n J_n(z)$ for all complex z.

24. Solve

$$\frac{\partial u}{\partial t} = \frac{1}{r} \frac{\partial}{\partial r} \left(r \frac{\partial u}{\partial r} \right) + H(t), \qquad 0 \le r < a, \quad 0 < t,$$

subject to the boundary conditions

$$\lim_{r \to 0} |u(r,t)| < \infty, \quad u(a,t) = 0, \quad 0 < t,$$

and the initial condition $u(r,0) = 0$, $0 \le r < a$. Note that $J_n(iz) = i^n I_n(z)$ and $I_n(iz) = i^n J_n(z)$ for all complex z.

---

[36] Reprinted from *Solar Energy*, **56**, H. Yoo and E.-T. Pak, Analytical solutions to a one-dimensional finite-domain model for stratified thermal storage tanks, 315–322, ©1996, with kind permission from Elsevier Science Ltd, The Boulevard, Langford Lane, Kidlington OX5 1GB, UK.

25. Solve

$$\frac{\partial u}{\partial t} = \frac{1}{r}\frac{\partial}{\partial r}\left(r\frac{\partial u}{\partial r}\right), \qquad 0 \le r < a, \quad 0 < t,$$

subject to the boundary conditions

$$\lim_{r\to 0}|u(r,t)| < \infty, \quad u(a,t) = e^{-t/\tau_0}, \quad 0 < t,$$

and the initial condition $u(r,0) = 1$, $0 \le r < a$. Note that $J_n(iz) = i^n I_n(z)$ and $I_n(iz) = i^n J_n(z)$ for all complex z.

26. Solve

$$\frac{\partial u}{\partial t} = a^2\left(\frac{\partial^2 u}{\partial r^2} + \frac{1}{r}\frac{\partial u}{\partial r}\right), \qquad 0 \le r < b, \quad 0 < t,$$

subject to the boundary conditions

$$\lim_{r\to 0}|u(r,t)| < \infty, \quad u(b,t) = kt, \quad 0 < t,$$

and the initial condition $u(r,0) = 0$, $0 \le r < b$. Note that $J_n(iz) = i^n I_n(z)$ and $I_n(iz) = i^n J_n(z)$ for all complex z.

27. Solve the nonhomogeneous heat equation for the spherical shell[37]

$$\frac{\partial u}{\partial t} = a^2\left(\frac{\partial^2 u}{\partial r^2} + \frac{2}{r}\frac{\partial u}{\partial r} + \frac{A}{r^4}\right), \qquad \alpha < r < \beta, \quad 0 < t,$$

subject to the boundary conditions

$$\frac{\partial u(\alpha,t)}{\partial r} = u(\beta,t) = 0, \qquad 0 < t,$$

and the initial condition $u(r,0) = 0$, $\alpha < r < \beta$.
Step 1: By introducing $v(r,t) = r\,u(r,t)$, show that the problem simplifies to

$$\frac{\partial v}{\partial t} = a^2\left(\frac{\partial^2 v}{\partial r^2} + \frac{A}{r^3}\right), \qquad \alpha < r < \beta, \quad 0 < t,$$

subject to the boundary conditions

$$\frac{\partial v(\alpha,t)}{\partial r} - \frac{v(\alpha,t)}{\alpha} = v(\beta,t) = 0, \qquad 0 < t,$$

---

[37] Abstracted with permission from Malkovich, R. Sh., 1977: Heating of a spherical shell by a radial current. *Sov. Phys. Tech. Phys.*, **22**, 636. ©1977 American Institute of Physics.

and the initial condition

$$v(r,0) = 0, \qquad \alpha < r < \beta.$$

Step 2: Using Laplace transforms and variation of parameters, show that the Laplace transform of $u(r,t)$ is

$$U(r,s) = \frac{A}{srq}\left\{ \frac{\sinh[q(\beta - r)]}{\alpha q \cosh(q\ell) + \sinh(q\ell)} \int_0^\ell \frac{\alpha q \cosh(q\eta) + \sinh(q\eta)}{(\alpha + \eta)^3}\, d\eta \right.$$
$$\left. - \int_0^{\beta-r} \frac{\sinh(q\eta)}{(r + \eta)^3}\, d\eta \right\},$$

where $q = \sqrt{s}/a$, and $\ell = \beta - \alpha$.

Step 3: Take the inverse of $U(r,s)$ and show that

$$u(r,t) = A\left\{ \left(\frac{1}{r} - \frac{1}{\beta}\right)\left[\frac{1}{\alpha} - \frac{1}{2}\left(\frac{1}{r} + \frac{1}{\beta}\right)\right] \right.$$
$$\left. - \frac{2\alpha^2}{r\ell^2} \sum_{n=0}^\infty \frac{\sin[\gamma_n(\beta - r)]\exp(-a^2\gamma_n^2 t)}{\sin^2(\gamma_n\ell)(\beta + \alpha^2\ell\gamma_n^2)} \int_0^1 \frac{\sin(\gamma_n\ell\eta)}{(\delta - \eta)^3}\, d\eta \right\},$$

where γ_n is the nth root of $\alpha\gamma + \tan(\ell\gamma) = 0$, and $\delta = 1 + \alpha/\ell$.

11.5 THE FOURIER TRANSFORM METHOD

We now consider the problem of one-dimensional heat flow in a rod of infinite length with insulated sides. Although there are no boundary conditions because the slab is of infinite extent, we do require that the solution remains bounded as we go to either positive or negative infinity. The initial temperature within the rod is $u(x,0) = f(x)$.

Employing the product solution technique of §11.3, $u(x,t) = X(x)T(t)$ with

$$T' + a^2\lambda T = 0, \tag{11.5.1}$$

and

$$X'' + \lambda X = 0. \tag{11.5.2}$$

Solutions to (11.5.1)–(11.5.2) which remain finite over the entire x-domain are

$$X(x) = E\cos(kx) + F\sin(kx), \tag{11.5.3}$$

and

$$T(t) = C\exp(-k^2 a^2 t). \tag{11.5.4}$$

Because we do not have any boundary conditions, we must include *all* possible values of k. Thus, when we sum all of the product solutions according to the principle of linear superposition, we obtain the integral

$$u(x,t) = \int_0^\infty [A(k)\cos(kx) + B(k)\sin(kx)]e^{-k^2a^2t}\, dk. \qquad (11.5.5)$$

We can satisfy the initial condition by choosing

$$A(k) = \frac{1}{\pi} \int_{-\infty}^\infty f(x)\cos(kx)\, dx, \qquad (11.5.6)$$

and

$$B(k) = \frac{1}{\pi} \int_{-\infty}^\infty f(x)\sin(kx)\, dx, \qquad (11.5.7)$$

because the initial condition has the form of a Fourier integral

$$f(x) = \int_0^\infty [A(k)\cos(kx) + B(k)\sin(kx)]\, dk, \qquad (11.5.8)$$

when $t = 0$.

Several important results follow by rewriting (11.5.8) as

$$u(x,t) = \frac{1}{\pi} \int_0^\infty \left[\int_{-\infty}^\infty f(\xi)\cos(k\xi)\cos(kx)\, d\xi \right.$$
$$\left. + \int_{-\infty}^\infty f(\xi)\sin(k\xi)\sin(kx)\, d\xi \right] e^{-k^2a^2t}\, dk. \qquad (11.5.9)$$

Combining terms,

$$u(x,t) = \frac{1}{\pi} \int_0^\infty \left\{ \int_{-\infty}^\infty f(\xi)[\cos(k\xi)\cos(kx) \right.$$
$$\left. + \sin(k\xi)\sin(kx)]\, d\xi \right\} e^{-k^2a^2t}\, dk \qquad (11.5.10)$$
$$= \frac{1}{\pi} \int_0^\infty \left[\int_{-\infty}^\infty f(\xi)\cos[k(\xi-x)]\, d\xi \right] e^{-k^2a^2t}\, dk. \qquad (11.5.11)$$

Reversing the order of integration,

$$u(x,t) = \frac{1}{\pi} \int_{-\infty}^\infty f(\xi) \left[\int_0^\infty \cos[k(\xi-x)]e^{-k^2a^2t}\, dk \right] d\xi. \qquad (11.5.12)$$

The inner integral is called the *source function*. We may compute its value through an integration on the complex plane; it equals

$$\int_0^\infty \cos[k(\xi-x)]\exp(-k^2a^2t)\, dk = \left(\frac{\pi}{4a^2t}\right)^{1/2} \exp\left[-\frac{(\xi-x)^2}{4a^2t}\right], \qquad (11.5.13)$$

if $0 < t$. This gives the final form for the temperature distribution:

$$u(x,t) = \frac{1}{\sqrt{4a^2\pi t}} \int_{-\infty}^{\infty} f(\xi) \exp\left[-\frac{(\xi - x)^2}{4a^2t}\right] d\xi. \qquad (11.5.14)$$

- **Example 11.5.1**

    Let us find the temperature field if the initial distribution is

    $$u(x,0) = \begin{cases} T_0, & x > 0, \\ -T_0, & x < 0. \end{cases} \qquad (11.5.15)$$

Then

$$u(x,t) = \frac{T_0}{\sqrt{4a^2\pi t}} \left\{ \int_0^{\infty} \exp\left[-\frac{(\xi - x)^2}{4a^2t}\right] d\xi - \int_{-\infty}^0 \exp\left[-\frac{(\xi - x)^2}{4a^2t}\right] d\xi \right\} \qquad (11.5.16)$$

$$= \frac{T_0}{\sqrt{\pi}} \left[\int_{-x/2a\sqrt{t}}^{\infty} e^{-\tau^2} d\tau - \int_{x/2a\sqrt{t}}^{\infty} e^{-\tau^2} d\tau \right] \qquad (11.5.17)$$

$$= \frac{T_0}{\sqrt{\pi}} \int_{-x/2a\sqrt{t}}^{x/2a\sqrt{t}} e^{-\tau^2} d\tau = \frac{2T_0}{\sqrt{\pi}} \int_0^{x/2a\sqrt{t}} e^{-\tau^2} d\tau \qquad (11.5.18)$$

$$= T_0 \operatorname{erf}\left(\frac{x}{2a\sqrt{t}}\right), \qquad (11.5.19)$$

where erf is the error function.

- **Example 11.5.2: Kelvin's estimate of the age of the earth**

    In the middle of the nineteenth century Lord Kelvin[38] estimated the age of the earth using the observed vertical temperature gradient at the earth's surface. He hypothesized that the earth was initially formed at a uniform high temperature T_0 and that its surface was subsequently maintained at the lower temperature of T_S. Assuming that most of the heat conduction occurred near the earth's surface, he reasoned that he could neglect the curvature of the earth, consider the earth's surface planar, and employ our one-dimensional heat conduction model in the vertical direction to compute the observed heat flux.

    Following Kelvin, we model the earth's surface as a flat plane with an infinitely deep earth below ($z > 0$). Initially the earth has the temperature T_0. Suddenly we drop the temperature at the surface to T_S. We wish to find

---

[38] Thomson, W., 1863: On the secular cooling of the earth. *Philos. Mag.*, *Ser. 4*, **25**, 157–170.

the heat flux across the boundary at $z = 0$ from the earth into an infinitely deep atmosphere.

The first step is to redefine our temperature scale $v(z, t) = u(z, t) + T_S$, where $v(z, t)$ is the observed temperature so that $u(0, t) = 0$ at the surface. Next, in order to use (11.5.14), we must define our initial state for $z < 0$. To maintain the temperature $u(0, t) = 0$, the initial temperature field $f(z)$ must be an odd function or

$$f(z) = \begin{cases} T_0 - T_S, & z > 0, \\ T_S - T_0, & z < 0. \end{cases} \tag{11.5.20}$$

From (11.5.14)

$$u(z, t) = \frac{T_0 - T_S}{\sqrt{4a^2 \pi t}} \left\{ \int_0^\infty \exp\left[-\frac{(\xi - z)^2}{4a^2 t} \right] d\xi - \int_{-\infty}^0 \exp\left[-\frac{(\xi - z)^2}{4a^2 t} \right] d\xi \right\} \tag{11.5.21}$$

$$= (T_0 - T_S) \, \text{erf}\left(\frac{z}{2a\sqrt{t}} \right), \tag{11.5.22}$$

following the work in the previous example.

The heat flux q at the surface $z = 0$ is obtained by differentiating (11.5.22) according to Fourier's law and evaluating the result at $z = 0$:

$$q = -\kappa \frac{\partial v}{\partial z} \bigg|_{z=0} = \frac{\kappa (T_S - T_0)}{a\sqrt{\pi t}}. \tag{11.5.23}$$

The surface heat flux is infinite at $t = 0$ because of the sudden application of the temperature T_S at $t = 0$. After that time, the heat flux decreases with time. Consequently, the time t at which we have the temperature gradient $\partial v(0, t)/\partial z$ is

$$t = \frac{(T_0 - T_S)^2}{\pi a^2 [\partial v(0, t)/\partial z]^2}. \tag{11.5.24}$$

For the present near-surface thermal gradient of 25 K/km, $T_0 - T_S = 2000$ K, and $a^2 = 1$ mm^2/s, the age of the earth from (11.5.24) is 65 million years.

Although Kelvin realized that this was a very rough estimate, his calculation showed that the earth had a finite age. This was in direct contradiction to the contemporary geological principle of *uniformitarianism* that the earth's surface and upper crust had remained unchanged in temperature and other physical quantities for millions and millions of years. The resulting debate would rage throughout the latter half of the nineteenth century and feature such luminaries as Kelvin, Charles Darwin, Thomas Huxley, and Oliver Heaviside.[39] Eventually Kelvin's arguments would prevail and uniformitarianism would fade into history.

---

[39] See Burchfield, J. D., 1975: *Lord Kelvin and the Age of the Earth*. Science History Publ., 260 pp.

Today, Kelvin's estimate is of academic interest because of the discovery of radioactivity at the turn of the twentieth century. During the first half of the twentieth century, geologists assumed that the radioactivity was uniformly distributed around the globe and restricted to the upper few tens of kilometers of the crust. Using this model they would then use observed heat fluxes to compute the distribution of radioactivity within the solid earth.[40] Today we know that the interior of the earth is quite dynamic; the oceans and continents are mobile and interconnected according to the theory of plate tectonics. However, geophysicists still use measured surface heat fluxes to infer the interior[41] of the earth.

• Example 11.5.3

So far we have shown how a simple application of separation of variables and the Fourier transform yields solutions to the heat equation over the semi-infinite interval $(0, \infty)$ via (11.5.5). Can we still use this technique for more complicated versions of the heat equation? The answer is yes but the procedure is more complicated. We illustrate it by solving[42]

$$\frac{\partial u}{\partial t} = \alpha \frac{\partial^3 u}{\partial t \partial x^2} + a^2 \frac{\partial^2 u}{\partial x^2}, \qquad 0 < x < \infty, \quad 0 < t, \qquad (11.5.25)$$

subject to the boundary conditions

$$u(0,t) = f(t), \quad \lim_{x \to 0} |u(x,t)| < \infty, \qquad 0 < t, \qquad (11.5.26)$$

$$\lim_{x \to \infty} u(x,t) \to 0, \ \lim_{x \to \infty} u_x(x,t) \to 0, \qquad 0 < t, \qquad (11.5.27)$$

and the initial condition

$$u(x,0) = 0, \qquad 0 < x < \infty. \qquad (11.5.28)$$

We begin by multiplying (11.5.25) by $\sin(kx)$ and integrating over x from 0 to ∞:

$$\alpha \int_0^\infty u_{txx} \sin(kx)\, dx + a^2 \int_0^\infty u_{xx} \sin(kx)\, dx = \int_0^\infty u_t \sin(kx)\, dx. \quad (11.5.29)$$

---

[40] See Slichter, L. B., 1941: Cooling of the earth. *Bull. Geol. Soc. Am.*, **52**, 561–600.

[41] Sclater, J. G., C. Jaupart, and D. Galson, 1980: The heat flow through oceanic and continental crust and the heat loss of the earth. *Rev. Geophys. Space Phys.*, **18**, 269–311.

[42] Taken from Fetecău, C., and J. Zierep, 2001: On a class of exact solutions of the equations of motion of a second grade fluid. *Acta Mech.*, **150**, 135–138.

Next, we integrate by parts. For example,

$$\int_0^\infty u_{xx} \sin(kx)\, dx = u_x \sin(kx)\Big|_0^\infty - k \int_0^\infty u_x \cos(kx)\, dx \qquad (11.5.30)$$

$$= -k \int_0^\infty u_x \cos(kx)\, dx \qquad (11.5.31)$$

$$= -ku(x,t)\cos(kx)\Big|_0^\infty - k^2 \int_0^\infty u(x,t)\sin(kx)\, dx$$
$$(11.5.32)$$

$$= kf(t) - k^2 U(k,t), \qquad (11.5.33)$$

where

$$U(k,t) = \int_0^\infty u(x,t)\sin(kx)\, dx, \qquad (11.5.34)$$

and the boundary conditions have been used to simplify (11.5.30) and (11.5. 32). Equation (11.5.34) is the definition of the *Fourier sine transform*. It and its mathematical cousin, the *Fourier cosine transform* $\int_0^\infty u(x,t)\cos(kx)\, dx$, are analogous to the half-range sine and cosine expansions that appear in solving the heat equation over the finite interval $(0, L)$. The difference here is that our range runs from 0 to ∞.

Applying the same technique to the other terms, we obtain

$$\alpha[kf'(t) - k^2 U'(k,t)] + a^2[kf(t) - k^2 U(k,t)] = U'(k,t) \qquad (11.5.35)$$

with $U(k,0) = 0$, where the primes denote differentiation with respect to time. Solving (11.5.35),

$$e^{a^2 k^2 t/(1+\alpha k^2)} U(k,t) = \frac{\alpha k}{1+\alpha k^2} \int_0^t f'(\tau) e^{a^2 k^2 \tau/(1+\alpha k^2)}\, d\tau$$
$$+ \frac{a^2 k}{1+\alpha k^2} \int_0^t f(\tau) e^{a^2 k^2 \tau/(1+\alpha k^2)}\, d\tau. \qquad (11.5.36)$$

Using integration by parts on the second integral in (11.5.36), we find that

$$U(k,t) = \frac{1}{k}\left[f(t) - f(0)e^{-a^2 k^2 t/(1+\alpha k^2)} \right.$$
$$\left. - \frac{1}{1+\alpha k^2} \int_0^t f'(\tau) e^{-a^2 k^2 (t-\tau)/(1+\alpha k^2)}\, d\tau \right]. \qquad (11.5.37)$$

Because

$$u(x,t) = \frac{2}{\pi} \int_0^\infty U(k,t)\sin(kx)\, dk, \qquad (11.5.38)$$

$$u(x,t) = \frac{2}{\pi} f(t) \int_0^\infty \frac{\sin(kx)}{k} \, dk$$

$$- \frac{2}{\pi} \int_0^\infty \frac{\sin(kx)}{k} e^{-a^2 k^2 t/(1+\alpha k^2)} \, dk$$

$$\times \left[f(0) + \frac{1}{1+\alpha k^2} \int_0^t f'(\tau) e^{a^2 k^2 \tau/(1+\alpha k^2)} \, d\tau \right] \qquad \textbf{(11.5.39)}$$

$$u(x,t) = f(t) - \frac{2}{\pi} \int_0^\infty \frac{\sin(kx)}{k} e^{-a^2 k^2 t/(1+\alpha k^2)} \, dk$$

$$\times \left[f(0) + \frac{1}{1+\alpha k^2} \int_0^t f'(\tau) e^{a^2 k^2 \tau/(1+\alpha k^2)} \, d\tau \right]. \qquad \textbf{(11.5.40)}$$

Problems

For Problems 1–4, find the solution of the heat equation

$$\frac{\partial u}{\partial t} = a^2 \frac{\partial^2 u}{\partial x^2}, \qquad -\infty < x < \infty, \quad 0 < t,$$

subject to the stated initial conditions.

1. $u(x,0) = \begin{cases} 1, & |x| < b \\ 0, & |x| > b \end{cases}$

2. $u(x,0) = e^{-b|x|}$

3. $u(x,0) = \begin{cases} 0, & -\infty < x < 0 \\ T_0, & 0 < x < b \\ 0, & b < x < \infty \end{cases}$

4. $u(x,0) = \delta(x)$

Lovering[43] has applied the solution to Problem 1 to cases involving the cooling of lava.

5. Solve the spherically symmetric equation of diffusion,[44]

$$\frac{\partial u}{\partial t} = a^2 \left(\frac{\partial^2 u}{\partial r^2} + \frac{2}{r} \frac{\partial u}{\partial r} \right), \qquad 0 \le r < \infty, \quad 0 < t,$$

with $u(r,0) = u_0(r)$.

---

[43] Lovering, T. S., 1935: Theory of heat conduction applied to geological problems. *Bull. Geol. Soc. Am.*, **46**, 69–94.

[44] From Shklovskii, I. S., and V. G. Kurt, 1960: Determination of atmospheric density at a height of 430 km by means of the diffusion of sodium vapors. *ARS J.*, **30**, 662–667 with permission.

Step 1: Assuming $v(r, t) = r\, u(r, t)$, show that the problem can be recast as

$$\frac{\partial v}{\partial t} = a^2 \frac{\partial^2 v}{\partial r^2}, \qquad 0 \leq r < \infty, \quad 0 < t,$$

with $v(r, 0) = r\, u_0(r)$.

Step 2: Using (11.5.14), show that the general solution is

$$u(r, t) = \frac{1}{2ar\sqrt{\pi t}} \int_0^\infty u_0(\rho) \left\{ \exp\left[-\frac{(r-\rho)^2}{4a^2 t}\right] - \exp\left[-\frac{(r+\rho)^2}{4a^2 t}\right] \right\} \rho\, d\rho.$$

Hint: What is the constraint on (11.5.14) so that the solution remains radially symmetric?

Step 3: For the initial concentration of

$$u_0(r) = \begin{cases} N_0, & 0 \leq r < r_0, \\ 0, & r_0 < r, \end{cases}$$

show that

$$\begin{aligned} u(r, t) = \tfrac{1}{2} N_0 \Bigg[&\operatorname{erf}\left(\frac{r_0 - r}{2a\sqrt{t}}\right) + \operatorname{erf}\left(\frac{r_0 + r}{2a\sqrt{t}}\right) \\ &+ \frac{2a\sqrt{t}}{r\sqrt{\pi}} \left\{ \exp\left[-\frac{(r_0 + r)^2}{4a^2 t}\right] - \exp\left[-\frac{(r_0 - r)^2}{4a^2 t}\right] \right\} \Bigg], \end{aligned}$$

where erf is the error function.

11.6 THE SUPERPOSITION INTEGRAL

In our study of Laplace transforms, we showed that we can construct solutions to ordinary differential equations with a general forcing $f(t)$ by first finding the solution to a similar problem where the forcing equals Heaviside's step function. Then we can write the general solution in terms of a superposition integral according to Duhamel's theorem. In this section we show that similar considerations hold in solving the heat equation with time-dependent boundary conditions or forcings.

Let us solve the heat condition problem

$$\frac{\partial u}{\partial t} = a^2 \frac{\partial^2 u}{\partial x^2}, \qquad 0 < x < L, \quad 0 < t, \tag{11.6.1}$$

with the boundary conditions

$$u(0, t) = 0, \quad u(L, t) = f(t), \quad 0 < t, \tag{11.6.2}$$

and the initial condition

$$u(x,0) = 0, \quad 0 < x < L. \tag{11.6.3}$$

The solution of (11.6.1)–(11.6.3) is difficult because of the time-dependent boundary condition. Instead of solving this system directly, let us solve the easier problem

$$\frac{\partial A}{\partial t} = a^2 \frac{\partial^2 A}{\partial x^2}, \quad 0 < x < L, \quad 0 < t, \tag{11.6.4}$$

with the boundary conditions

$$A(0,t) = 0, \quad A(L,t) = 1, \quad 0 < t, \tag{11.6.5}$$

and the initial condition

$$A(x,0) = 0, \quad 0 < x < L. \tag{11.6.6}$$

Separation of variables yields the solution

$$A(x,t) = \frac{x}{L} + \frac{2}{\pi} \sum_{n=1}^{\infty} \frac{(-1)^n}{n} \sin\left(\frac{n\pi x}{L}\right) \exp\left(-\frac{a^2 n^2 \pi^2 t}{L^2}\right). \tag{11.6.7}$$

Consider the following case. Suppose that we maintain the temperature at zero at the end $x = L$ until $t = \tau_1$ and then raise it to the value of unity. The resulting temperature distribution equals zero everywhere when $t < \tau_1$ and equals $A(x, t - \tau_1)$ for $t > \tau_1$. We have merely shifted our time axis so that the initial condition occurs at $t = \tau_1$.

Consider an analogous, but more complicated, situation of the temperature at the end position $x = L$ held at $f(0)$ from $t = 0$ to $t = \tau_1$ at which time we abruptly change it by the amount $f(\tau_1) - f(0)$ to the value $f(\tau_1)$. This temperature remains until $t = \tau_2$ when we again abruptly change it by an amount $f(\tau_2) - f(\tau_1)$. We can imagine this process continuing up to the instant $t = \tau_n$. Because of linear superposition, the temperature distribution at any given time equals the sum of these temperature increments:

$$u(x,t) = f(0)A(x,t) + [f(\tau_1) - f(0)]A(x, t - \tau_1) + [f(\tau_2) - f(\tau_1)]A(x, t - \tau_2)$$
$$+ \cdots + [f(\tau_n) - f(\tau_{n-1})]A(x, t - \tau_n), \tag{11.6.8}$$

where τ_n is the time of the most recent temperature change. If we write

$$\Delta f_k = f(\tau_k) - f(\tau_{k-1}), \quad \text{and} \quad \Delta \tau_k = \tau_k - \tau_{k-1}, \tag{11.6.9}$$

(11.6.8) becomes

$$u(x,t) = f(0)A(x,t) + \sum_{k=1}^{n} A(x, t - \tau_k) \frac{\Delta f_k}{\Delta \tau_k} \Delta \tau_k. \tag{11.6.10}$$

Consequently, in the limit of $\Delta\tau_k \to 0$, (11.6.10) becomes

$$u(x, t) = f(0)A(x, t) + \int_0^t A(x, t - \tau)f'(\tau)\, d\tau, \qquad (11.6.11)$$

assuming that $f(t)$ is differentiable. Equation (11.6.11) is the *superposition integral.* We can obtain an alternative form by integration by parts:

$$u(x, t) = f(t)A(x, 0) - \int_0^t f(\tau)\frac{\partial A(x, t - \tau)}{\partial \tau}\, d\tau, \qquad (11.6.12)$$

or

$$u(x, t) = f(t)A(x, 0) + \int_0^t f(\tau)\frac{\partial A(x, t - \tau)}{\partial t}\, d\tau, \qquad (11.6.13)$$

because

$$\frac{\partial A(x, t - \tau)}{\partial \tau} = -\frac{\partial A(x, t - \tau)}{\partial t}. \qquad (11.6.14)$$

To illustrate the superposition integral, suppose $f(t) = t$. Then, by (11.6.11),

$$u(x, t) = \int_0^t \left\{ \frac{x}{L} + \frac{2}{\pi}\sum_{n=1}^\infty \frac{(-1)^n}{n}\sin\left(\frac{n\pi x}{L}\right)\exp\left[-\frac{a^2n^2\pi^2}{L^2}(t - \tau)\right] \right\} d\tau \qquad (11.6.15)$$

$$= \frac{xt}{L} - \frac{2L^2}{a^2\pi^3}\sum_{n=1}^\infty \frac{(-1)^n}{n^3}\sin\left(\frac{n\pi x}{L}\right)\left[1 - \exp\left(-\frac{a^2n^2\pi^2 t}{L^2}\right)\right]. \qquad (11.6.16)$$

• Example 11.6.1: Temperature oscillations in a wall heated by an alternating current

In addition to finding solutions to heat conduction problems with time-dependent boundary conditions, we can also apply the superposition integral to the nonhomogeneous heat equation when the source depends on time. Jeglic[45] used this technique in obtaining the temperature distribution within a slab heated by alternating electric current. If we assume that the flat plate has a surface area A and depth L, then the heat equation for the plate when electrically heated by an alternating current of frequency ω is

$$\frac{\partial u}{\partial t} - a^2\frac{\partial^2 u}{\partial x^2} = \frac{2q}{\rho C_p A L}\sin^2(\omega t), \quad 0 < x < L, \quad 0 < t, \qquad (11.6.17)$$

---

[45] Jeglic, F. A., 1962: An analytical determination of temperature oscillations in a wall heated by alternating current. *NASA Tech. Note No. D-1286.* In a similar vein, Al-Nimr and Abdallah [Al-Nimr, M. A., and M. R. Abdallah, 1999: Thermal behavior of insulated electric wires producing pulsating signals. *Heat Transfer Engng.*, **20(4)**, 62–74] have found the heat transfer with an insulated wire that carries an alternating current.

where q is the average heat rate caused by the current, ρ is the density, C_p is the specific heat at constant pressure, and a^2 is the diffusivity of the slab. We will assume that we insulated the inner wall so that

$$\frac{\partial u(0,t)}{\partial x} = 0, \qquad 0 < t, \tag{11.6.18}$$

bigskip while we allow the outer wall to radiatively cool to free space at the temperature of zero or

$$\kappa \frac{\partial u(L,t)}{\partial x} + hu(L,t) = 0, \qquad 0 < t, \tag{11.6.19}$$

where κ is the thermal conductivity and h is the heat transfer coefficient. The slab is initially at the temperature of zero or

$$u(x,0) = 0, \qquad 0 < x < L. \tag{11.6.20}$$

To solve the heat equation, we first solve the simpler problem of

$$\frac{\partial A}{\partial t} - a^2 \frac{\partial^2 A}{\partial x^2} = 1, \quad 0 < x < L, \quad 0 < t, \tag{11.6.21}$$

with the boundary conditions

$$\frac{\partial A(0,t)}{\partial x} = 0, \quad \kappa \frac{\partial A(L,t)}{\partial x} + hA(L,t) = 0, \quad 0 < t, \tag{11.6.22}$$

and the initial condition

$$A(x,0) = 0, \qquad 0 < x < L. \tag{11.6.23}$$

The solution $A(x,t)$ is the *indicial admittance* because it is the response of a system to forcing by the step function $H(t)$.

We solve (11.6.21)–(11.6.23) by separation of variables. We begin by assuming that $A(x,t)$ consists of a steady-state solution $w(x)$ plus a transient solution $v(x,t)$, where

$$a^2 w''(x) = -1, \quad w'(0) = 0, \quad \kappa w'(L) + hw(L) = 0, \tag{11.6.24}$$

$$\frac{\partial v}{\partial t} = a^2 \frac{\partial^2 v}{\partial x^2}, \quad \frac{\partial v(0,t)}{\partial x} = 0, \quad \kappa \frac{\partial v(L,t)}{\partial x} + hv(L,t) = 0, \tag{11.6.25}$$

and

$$v(x,0) = -w(x). \tag{11.6.26}$$

Solving (11.6.24),

$$w(x) = \frac{L^2 - x^2}{2a^2} + \frac{\kappa L}{ha^2}. \tag{11.6.27}$$

Table 11.6.1: The First Six Roots of the Equation $k_n \tan(k_n) = h^*$

h^*	k_1	k_2	k_3	k_4	k_5	k_6
0.001	0.03162	3.14191	6.28334	9.42488	12.56645	15.70803
0.002	0.04471	3.14223	6.28350	9.42499	12.56653	15.70809
0.005	0.07065	3.14318	6.28398	9.42531	12.56677	15.70828
0.010	0.09830	3.14477	6.28478	9.42584	12.56717	15.70860
0.020	0.14095	3.14795	6.28637	9.42690	12.56796	15.70924
0.050	0.22176	3.15743	6.29113	9.43008	12.57035	15.71115
0.100	0.31105	3.17310	6.29906	9.43538	12.57432	15.71433
0.200	0.43284	3.20393	6.31485	9.44595	12.58226	15.72068
0.500	0.65327	3.29231	6.36162	9.47748	12.60601	15.73972
1.000	0.86033	3.42562	6.43730	9.52933	12.64529	15.77128
2.000	1.07687	3.64360	6.57833	9.62956	12.72230	15.83361
5.000	1.31384	4.03357	6.90960	9.89275	12.93522	16.01066
10.000	1.42887	4.30580	7.22811	10.20026	13.21418	16.25336
20.000	1.49613	4.49148	7.49541	10.51167	13.54198	16.58640
∞	1.57080	4.71239	7.85399	10.99557	14.13717	17.27876

Turning to the transient solution $v(x,t)$, we use separation of variables and find that

$$v(x,t) = \sum_{n=1}^{\infty} C_n \cos\left(\frac{k_n x}{L}\right) \exp\left(-\frac{a^2 k_n^2 t}{L^2}\right), \qquad (11.6.28)$$

where k_n is the nth root of the transcendental equation: $k_n \tan(k_n) = hL/\kappa = h^*$. Table 11.6.1 gives the first six roots for various values of hL/κ.

Our final task is to compute C_n. After substituting $t = 0$ into (11.6.28), we are left with a orthogonal expansion of $-w(x)$ using the eigenfunctions $\cos(k_n x/L)$. From (9.3.4),

$$C_n = \frac{\int_0^L -w(x) \cos(k_n x/L)\, dx}{\int_0^L \cos^2(k_n x/L)\, dx} = \frac{-L^3 \sin(k_n)/(a^2 k_n^3)}{L[k_n + \sin(2k_n)/2]/(2k_n)} \qquad (11.6.29)$$

$$= -\frac{2L^2 \sin(k_n)}{a^2 k_n^2 [k_n + \sin(2k_n)/2]}. \qquad (11.6.30)$$

Combining (11.6.28) and (11.6.30),

$$v(x,t) = -\frac{2L^2}{a^2} \sum_{n=1}^{\infty} \frac{\sin(k_n) \cos(k_n x/L)}{k_n^2 [k_n + \sin(2k_n)/2]} \exp\left(-\frac{a^2 k_n^2 t}{L^2}\right). \qquad (11.6.31)$$

Consequently, $A(x,t)$ equals

$$A(x,t) = \frac{L^2 - x^2}{2a^2} + \frac{\kappa L}{ha^2} - \frac{2L^2}{a^2} \sum_{n=1}^{\infty} \frac{\sin(k_n) \cos(k_n x/L)}{k_n^2 [k_n + \sin(2k_n)/2]} \exp\left(-\frac{a^2 k_n^2 t}{L^2}\right).$$
$$\qquad (11.6.32)$$

We now wish to use the solution (11.6.32) to find the temperature distribution within the slab when it is heated by a time-dependent source $f(t)$. As in the case of time-dependent boundary conditions, we imagine that we can break the process into an infinite number of small changes to the heating which occur at the times $t = \tau_1$, $t = \tau_2$, etc. Consequently, the temperature distribution at the time t following the change at $t = \tau_n$ and before the change at $t = \tau_{n+1}$ is

$$u(x,t) = f(0)A(x,t) + \sum_{k=1}^{n} A(x, t - \tau_k) \frac{\Delta f_k}{\Delta \tau_k} \Delta \tau_k, \tag{11.6.33}$$

where

$$\Delta f_k = f(\tau_k) - f(\tau_{k-1}), \quad \text{and} \quad \Delta \tau_k = \tau_k - \tau_{k-1}. \tag{11.6.34}$$

In the limit of $\Delta \tau_k \to 0$,

$$u(x,t) = f(0)A(x,t) + \int_0^t A(x, t - \tau) f'(\tau) \, d\tau \tag{11.6.35}$$

$$= f(t)A(x,0) + \int_0^t f(\tau) \frac{\partial A(x, t - \tau)}{\partial \tau} \, d\tau. \tag{11.6.36}$$

In our present problem,

$$f(t) = \frac{2q}{\rho C_p AL} \sin^2(\omega t), \qquad f'(t) = \frac{2q\omega}{\rho C_p AL} \sin(2\omega t). \tag{11.6.37}$$

Therefore,

$$u(x,t) = \frac{2q\omega}{\rho C_p AL} \int_0^t \sin(2\omega \tau) \left\{ \frac{L^2 - x^2}{2a^2} + \frac{\kappa L}{ha^2} \right.$$
$$- \frac{2L^2}{a^2} \sum_{n=1}^{\infty} \frac{\sin(k_n)}{k_n^2 [k_n + \sin(2k_n)/2]} \cos\left(\frac{k_n x}{L}\right)$$
$$\left. \times \exp\left[-\frac{a^2 k_n^2 (t - \tau)}{L^2} \right] \right\} d\tau \tag{11.6.38}$$

$$= -\frac{q}{\rho C_p AL} \left(\frac{L^2 - x^2}{2a^2} + \frac{\kappa L}{ha^2} \right) \cos(2\omega \tau)|_0^t$$
$$- \frac{4L^2 q\omega}{a^2 \rho C_p AL} \sum_{n=1}^{\infty} \frac{\sin(k_n) \exp(-a^2 k_n^2 t/L^2)}{k_n^2 [k_n + \sin(2k_n)/2]} \cos\left(\frac{k_n x}{L}\right)$$
$$\times \int_0^t \sin(2\omega \tau) \exp\left(\frac{a^2 k_n^2 \tau}{L^2} \right) d\tau \tag{11.6.39}$$

$$= \frac{qL}{a^2 A\rho C_p} \left\{ \left[\frac{L^2 - x^2}{2L^2} + \frac{\kappa}{hL} \right] [1 - \cos(2\omega t)] \right.$$
$$- \sum_{n=1}^{\infty} \frac{4 \sin(k_n) \cos(k_n x/L)}{k_n^2 [k_n + \sin(2k_n)/2][4 + a^4 k_n^4/(L^4 \omega^2)]}$$
$$\left. \times \left[\frac{a^2 k_n^2}{\omega L^2} \sin(2\omega t) - 2\cos(2\omega t) + 2\exp\left(-\frac{a^2 k_n^2 t}{L^2} \right) \right] \right\}. \tag{11.6.40}$$

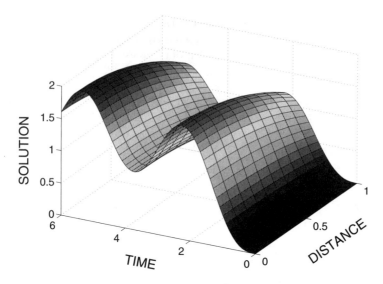

Figure 11.6.1: The nondimensional temperature $a^2 A\rho C_p u(x,t)/qL$ within a slab that we heat by alternating electric current as a function of position x/L and time $a^2 t/L^2$ when we insulate the $x = 0$ end and let the $x = L$ end radiate to free space at temperature zero. The initial temperature is zero, $hL/\kappa = 1$, and $a^2/(L^2\omega) = 1$.

Figure 11.6.1 illustrates (11.6.40) for $hL/\kappa = 1$, and $a^2/(L^2\omega) = 1$. This figure was created using the MATLAB script

```
clear
asq_over_omegaL2 = 1; h_star = 1; m = 0; M = 10;
dx = 0.1; dt = 0.1;
%
% create initial guess at k_n
%
zero = zeros(length(M));
for n = 1:10000
k1 = 0.1*n; k2 = 0.1*(n+1);
prod = k1 * tan(k1);
y1 = h_star - prod; y2 = h_star - k2 * tan(k2);
if (y1*y2 <= 0 & prod < 2 & m < M) m = m+1; zero(m) = k1; end;
end;
%
% use Newton-Raphson method to improve values of k_n
%
for n = 1:M; for k = 1:10
f = h_star - zero(n) * tan(zero(n));
fp = - tan(zero(n)) - zero(n) * sec(zero(n))^2;
zero(n) = zero(n) - f / fp;
end; end;
```

```
% compute grid and initialize solution
X = [0:dx:1]; T = [0:dt:6];
temp1 = (0.5 + 1/h_star)*ones(1,length(X)) - 0.5*X.*X;
temp2 = ones(1,length(T)) - cos(2*T);
u = temp1' * temp2;
XX = X' * ones(1,length(T));
TT = ones(1,length(X))' * T;
% compute solution from (11.6.40)
for m = 1:M
xtemp1 = zero(m) * zero(m);
xtemp2 = 4 + asq_over_omegaL2*asq_over_omegaL2*xtemp1*xtemp1;
xtemp3 = asq_over_omegaL2 * xtemp1;
xtemp4 = zero(m) + sin(2*zero(m))/2;
xtemp5 = asq_over_omegaL2 * xtemp1;
aaaaa = 4 * sin(zero(m)) / (xtemp1 * xtemp2 * xtemp4);
u = u - aaaaa * cos(zero(m)*X)' ...
    * (xtemp5 * sin(2*T) - 2 * cos(2*T) + 2 * exp(-xtemp5 * T));
end
surf(XX,TT,u)
xlabel('DISTANCE','Fontsize',20); ylabel('TIME','Fontsize',20)
zlabel('SOLUTION','Fontsize',20)
```

The oscillating solution, reflecting the periodic heating by the alternating current, rapidly reaches equilibrium. Because heat is radiated to space at $x = L$, the temperature is maximum at $x = 0$ at any given instant as heat flows from $x = 0$ to $x = L$.

- **Example 11.6.2**

Consider the following heat conduction problem with time-dependent forcing and/or boundary conditions:

$$\frac{\partial u}{\partial t} = a^2 L(u) + f(P,t), \qquad 0 < t, \qquad (11.6.41)$$

$$B(u) = g(Q,t), \qquad 0 < t, \qquad (11.6.42)$$

and

$$u(P,0) = h(P), \qquad (11.6.43)$$

where

$$L(u) = C_0 + C_1 \frac{\partial}{\partial x_1}\left(K_1 \frac{\partial u}{\partial x_1}\right) + C_2 \frac{\partial}{\partial x_2}\left(K_2 \frac{\partial u}{\partial x_2}\right) + C_3 \frac{\partial}{\partial x_3}\left(K_3 \frac{\partial u}{\partial x_3}\right),$$
$$(11.6.44)$$

$$B(u) = c_0 + c_1 \frac{\partial u}{\partial x_1} + c_2 \frac{\partial u}{\partial x_2} + c_3 \frac{\partial u}{\partial x_3}, \qquad (11.6.45)$$

P denotes an arbitrary interior point at (x_1, x_2, x_3) of a region R, and Q is any point on the boundary of R. Here c_i, C_i, and K_i are functions of x_1, x_2, and x_3 only.

Many years ago, Bartels and Churchill[46] extended Duhumel's theorem to solve this heat conduction problem. They did this by first introducing the simpler initial-boundary-value problem:

$$\frac{\partial v}{\partial t} = a^2 L(v) + f(P, t_1), \qquad 0 < t, \tag{11.6.46}$$

$$B(v) = g(Q, t_1), \qquad 0 < t, \tag{11.6.47}$$

and

$$v(P, 0) = h(P), \tag{11.6.48}$$

which has a constant forcing and boundary conditions in place of the time-dependent ones. Here t_1 denotes an arbitrary but *fixed* instant of time. Then Bartels and Churchill proved that the solution to the original problem is given by the convolution integral

$$u(P, t) = \frac{\partial}{\partial t}\left[\int_0^t v(P, t - \tau, \tau)\, d\tau\right]. \tag{11.6.49}$$

To illustrate[47] this technique, let us solve

$$\frac{\partial u}{\partial t} = a^2\left(\frac{\partial^2 u}{\partial r^2} + \frac{2}{r}\frac{\partial u}{\partial r}\right) = \frac{a^2}{r}\frac{\partial^2(ru)}{\partial r^2}, \qquad \alpha < r < \beta, \quad 0 < t, \tag{11.6.50}$$

subject to the boundary conditions

$$u(\alpha, t) = u_0 e^{-ct}, \qquad \frac{\partial u(\beta, t)}{\partial r} = 0, \qquad 0 < t, \tag{11.6.51}$$

and the initial condition $u(r, 0) = u_0$, $\alpha < r < \beta$.

We begin by solving the alternative problem

$$\frac{\partial v}{\partial t} = a^2\left(\frac{\partial^2 v}{\partial r^2} + \frac{2}{r}\frac{\partial v}{\partial r}\right) = \frac{a^2}{r}\frac{\partial^2(rv)}{\partial r^2}, \qquad \alpha < r < \beta, \quad 0 < t, \tag{11.6.52}$$

subject to the boundary conditions

$$v(\alpha, t, t') = u_0 e^{-ct'}, \qquad \frac{\partial v(\beta, t, t')}{\partial r} = 0, \qquad 0 < t, \tag{11.6.53}$$

---

[46] Bartels, R. C. F., and R. V. Churchill, 1942: Resolution of boundary problems by the use of a generalized convolution. *Am. Math. Soc. Bull.*, **48**, 276–282.

[47] Reprinted with permission from Reiss, H., and V. K. LaMer, 1950: Diffusional boundary value problems involving moving boundaries, connected with the growth of colloidal particles. *J. Chem. Phys.*, **18**, 1–12. ©1950, American Institute of Physics.

and the initial condition $v(r, 0, t') = u_0$, $\alpha < r < \beta$, or equivalently

$$\frac{\partial w}{\partial t} = a^2 \left(\frac{\partial^2 w}{\partial r^2} + \frac{2}{r} \frac{\partial w}{\partial r} \right) = \frac{a^2}{r} \frac{\partial^2 (rw)}{\partial r^2}, \qquad \alpha < r < \beta, \quad 0 < t, \quad \textbf{(11.6.54)}$$

subject to the boundary conditions

$$w(\alpha, t, t') = 0, \qquad \frac{\partial w(\beta, t, t')}{\partial r} = 0, \qquad 0 < t, \qquad \textbf{(11.6.55)}$$

and the initial condition $w(r, 0, t') = u_0(1 - e^{-ct'})$, $\alpha < r < \beta$, where $v(r, t, t') = u_0 e^{-ct'} + w(r, t, t')$.

The heat condition problem (11.6.54)–(11.6.55) can be solved using separation of variables. Following example 11.3.6, we find that

$$w(r, t, t') = \frac{\alpha u_0 (1 - e^{-ct'})}{r} \sum_{n=1}^{\infty} \frac{\sin[k_n(r - \alpha)]}{k_n c_n} e^{-a^2 k_n^2 t}, \qquad \textbf{(11.6.56)}$$

where k_n is the nth root of $\beta k = \tan[k(\beta - \alpha)]$, and $2c_n = \{\beta \sin^2[k_n(\beta - \alpha)] - \alpha\}$. Therefore,

$$u(x, t) = \frac{\alpha u_0}{r} \frac{\partial}{\partial t} \left\{ \int_0^t (1 - e^{-c\tau}) \sum_{n=1}^{\infty} \frac{\sin[k_n(r - \alpha)]}{k_n c_n} e^{-a^2 k_n^2 (t - \tau)} \, d\tau \right\} \textbf{(11.6.57)}$$

$$= \frac{\alpha u_0}{r} \sum_{n=1}^{\infty} \frac{\sin[k_n(r - \alpha)]}{k_n c_n} \frac{\partial}{\partial t} \left\{ \int_0^t e^{-a^2 k_n^2 (t - \tau)} - e^{-a^2 k_n^2 (t - \tau) - c\tau} \, d\tau \right\}$$

$$\textbf{(11.6.58)}$$

$$= \frac{\alpha u_0}{r} \sum_{n=1}^{\infty} \frac{\sin[k_n(r - \alpha)]}{k_n c_n} \frac{\partial}{\partial t} \left\{ \frac{1 - e^{-a^2 k_n^2 t}}{a^2 k_n^2} - \frac{e^{-ct} - e^{-a^2 k_n^2 t}}{a^2 k_n^2 - c} \right\}$$

$$\textbf{(11.6.59)}$$

$$= \frac{\alpha c u_0}{r} \sum_{n=1}^{\infty} \frac{\sin[k_n(r - \alpha)]}{k_n c_n} \frac{e^{-a^2 k_n^2 t} - e^{-ct}}{c - a^2 k_n^2}, \qquad \textbf{(11.6.60)}$$

and the final answer is

$$u(x, t) = u_0 e^{-ct} + \frac{\alpha c u_0}{r} \sum_{n=1}^{\infty} \frac{e^{-a^2 k_n^2 t} - e^{-ct}}{(c - a^2 k_n^2) k_n c_n} \sin[k_n(r - \alpha)]. \qquad \textbf{(11.6.61)}$$

Problems

1. Solve the heat equation[48]

$$\frac{\partial u}{\partial t} = a^2 \frac{\partial^2 u}{\partial x^2}, \qquad 0 < x < L, \quad 0 < t,$$

---

[48] From Tao, L. N., 1960: Magnetohydrodynamic effects on the formation of Couette flow. *J. Aerosp. Sci.*, **27**, 334–338 with permission.

subject to the boundary conditions $u(0,t) = u(L,t) = f(t)$, $0 < t$, and the initial condition $u(x,0) = 0$, $0 < x < L$.

Step 1: First solve the heat conduction problem

$$\frac{\partial A}{\partial t} = a^2 \frac{\partial^2 A}{\partial x^2}, \qquad 0 < x < L, \quad 0 < t,$$

subject to the boundary conditions $A(0,t) = A(L,t) = 1$, $0 < t$, and the initial condition $A(x,0) = 0$, $0 < x < L$. Show that

$$A(x,t) = 1 - \frac{4}{\pi} \sum_{n=1}^{\infty} \frac{\sin[(2n-1)\pi x/L]}{2n-1} e^{-a^2(2n-1)^2\pi^2 t/L^2}.$$

Step 2: Use Duhamel's theorem and show that

$$u(x,t) = \frac{4\pi a^2}{L^2} \sum_{n=1}^{\infty} (2n-1) \sin\left[\frac{(2n-1)\pi x}{L}\right] e^{-a^2(2n-1)^2\pi^2 t/L^2}$$
$$\times \int_0^t f(\tau) e^{a^2(2n-1)^2\pi^2 \tau/L^2} \, d\tau.$$

2. A thermometer measures temperature by the thermal expansion of a liquid (usually mercury or alcohol) stored in a bulb into a glass stem containing an empty cylindrical channel. Under normal conditions, temperature changes occur sufficiently slow so that the temperature within the liquid is uniform. However, for rapid temperature changes (such as those that would occur during the rapid ascension of an airplane or meteorological balloon), significant errors could occur. In such situations the recorded temperature would lag behind the actual temperature because of the time needed for the heat to conduct in or out of the bulb. During his investigation of this question, McLeod[49] solved

$$\frac{\partial u}{\partial t} = a^2 \frac{1}{r} \frac{\partial}{\partial r}\left(r \frac{\partial u}{\partial r}\right), \qquad 0 \le r < b, \quad 0 < t,$$

subject to the boundary conditions $\lim_{r \to 0} |u(r,t)| < \infty$, and $u(b,t) = \varphi(t)$, $0 < t$, and the initial condition $u(r,0) = 0$, $0 \le r < b$. The analysis was as follows:

---

[49] Reproduced with acknowledgment to Taylor and Francis, Publishers, from McLeod, A. R., 1919: On the lags of thermometers with spherical and cylindrical bulbs in a medium whose temperature is changing at a constant rate. *Philos. Mag.*, Ser. *6*, **37**, 134–144. See also Bromwich, T. J. I'A., 1919: Examples of operational methods in mathematical physics. *Philos. Mag.*, Ser. *6*, **37**, 407–419; McLeod, A. R., 1922: On the lags of thermometers. *Philos. Mag.*, Ser. *6*, **43**, 49–70.

Step 1: First solve the heat conduction problem

$$\frac{\partial A}{\partial t} = \frac{a^2}{r} \frac{\partial}{\partial r} \left(r \frac{\partial A}{\partial r} \right), \qquad 0 \le r < b, \quad 0 < t,$$

subject to the boundary conditions $\lim_{r \to 0} |A(r,t)| < \infty$, and $A(b,t) = 1$, $0 < t$, and the initial condition $A(r,0) = 0$, $0 \le r < b$. Show that

$$A(r,t) = 1 - 2 \sum_{n=1}^{\infty} \frac{J_0(k_n r/b)}{k_n J_1(k_n)} e^{-a^2 k_n^2 t/b^2},$$

where $J_0(k_n) = 0$.

Step 2: Use Duhamel's theorem and show that

$$u(r,t) = \frac{2a^2}{b^2} \sum_{n=1}^{\infty} \frac{k_n J_0(k_n r/b)}{J_1(k_n)} \int_0^t \varphi(\tau) e^{-a^2 k_n^2 (t-\tau)/b^2} \, d\tau.$$

Step 3: If $\varphi(t) = Gt$, show that

$$u(r,t) = 2G \sum_{n=1}^{\infty} \frac{J_0(k_n r/b)}{k_n J_1(k_n)} \left[t + \frac{b^2}{a^2 k_n^2} \left(e^{-a^2 k_n^2 t/b^2} - 1 \right) \right].$$

McLeod found that for a mercury thermometer of 10-cm length a lag of 0.01 °C would occur for a warming rate of 0.032°C s^{-1} (a warming gradient of 1.9°C per thousand feet and a descent of one thousand feet per minute). Although this is a very small number, when he included the surface conductance of the glass tube, the lag increased to 0.85°C. Similar problems plague bimetal thermometers[50] and thermistors[51] used in radiosondes (meteorological sounding balloons).

3. A classic problem[52] in fluid mechanics is the motion of a semi-infinite viscous fluid that results from the sudden movement of the adjacent wall starting at $t = 0$. Initially the fluid is at rest. If we denote the velocity of the fluid parallel to the wall by $u(x,t)$, the governing equation is

$$\frac{\partial u}{\partial t} = \nu \frac{\partial^2 u}{\partial x^2}, \qquad 0 < x < \infty, \quad 0 < t,$$

---

[50] Mitra, H., and M. B. Datta, 1954: Lag coefficient of bimetal thermometer of chronometric radiosonde. *Indian J. Meteorol. Geophys.*, **5**, 257–261.

[51] Badgley, F. I., 1957: Response of radiosonde thermistors. *Rev. Sci. Instrum.*, **28**, 1079–1084.

[52] This problem was first posed and partially solved by Stokes, G. G., 1850: On the effect of the internal friction of fluids on the motions of pendulums. *Proc. Cambridge Philos. Soc.*, **9, Part II**, [8]–[106].

with the boundary conditions $u(0,t) = V(t)$, $\lim_{x\to\infty} u(x,t) \to 0$, $0 < t$, and the initial condition $u(x,0) = 0$, $0 < x < \infty$.

Step 1: Find the step response by solving

$$\frac{\partial A}{\partial t} = \nu \frac{\partial^2 A}{\partial x^2}, \qquad 0 < x < \infty, \quad 0 < t,$$

subject to the boundary conditions

$$A(0,t) = 1, \qquad \lim_{x\to\infty} A(x,t) \to 0, \qquad 0 < t,$$

and the initial condition $A(x,0) = 0$, $0 < x < \infty$. Show that

$$A(x,t) = \operatorname{erfc}\left(\frac{x}{2\sqrt{\nu t}}\right) = \frac{2}{\sqrt{\pi}} \int_{x/2\sqrt{\nu t}}^{\infty} e^{-\eta^2} \, d\eta,$$

where erfc is the complementary error function. Hint: Use Laplace transforms.

Step 2: Use Duhamel's theorem and show that the solution is

$$u(x,t) = \int_0^t V(t-\tau) \frac{x \exp(-x^2/4\nu\tau)}{2\sqrt{\pi\nu\tau^3}} \, d\tau = \frac{2}{\pi} \int_{x/\sqrt{4\nu t}}^{\infty} V\left(t - \frac{x^2}{4\nu\eta^2}\right) e^{-\eta^2} \, d\eta.$$

4. During their study of the propagation of a temperature step in a nearly supercritical, van der Waals gas, Zappoli and Durand-Daubin[53] solved

$$\frac{\partial u}{\partial t} = a^2 \frac{\partial^2 u}{\partial x^2}, \qquad 0 < x < \infty, \quad 0 < t,$$

with the boundary conditions $u(0,t) = u_0 - \frac{2}{3} f(t)$, $\lim_{x\to\infty} u(x,t) \to 0$, $0 < t$, and the initial condition $u(x,0) = 0$, $0 < x < \infty$, where u_0 is a constant.

Step 1: Find the step response by solving

$$\frac{\partial A}{\partial t} = a^2 \frac{\partial^2 A}{\partial x^2}, \qquad 0 < x < \infty, \quad 0 < t,$$

subject to the boundary conditions $A(0,t) = 1$, $\lim_{x\to\infty} A(x,t) \to 0$, $0 < t$, and the initial condition $A(x,0) = 0$, $0 < x < \infty$. Show that

$$A(x,t) = \operatorname{erfc}\left(\frac{x}{2a\sqrt{t}}\right) = \frac{2}{\sqrt{\pi}} \int_{x/(2a\sqrt{t})}^{\infty} e^{-\eta^2} \, d\eta,$$

[53] Reprinted with permission from Zappoli, B., and A. Durand-Daubin, 1994: Heat- and mass transport in a near supercritical fluid. *Phys. Fluids*, **6**, 1929–1936. ©1994, American Institute of Physics.

where erfc is the complementary error function. Hint: Use Laplace transforms.

Step 2: Use Duhamel's theorem and show that the solution is

$$u(x,t) = u_0 \, \text{erfc}\left(\frac{x}{2a\sqrt{t}}\right) - \frac{4}{3\sqrt{\pi}} \int_{x/(2a\sqrt{t})}^{\infty} f\left(t - \frac{x^2}{4a^2\eta^2}\right) e^{-\eta^2} \, d\eta.$$

5. Solve the heat equation

$$\frac{\partial u}{\partial t} = \frac{\partial^2 u}{\partial x^2}, \qquad 0 < x < 1, \quad 0 < t,$$

subject to the boundary conditions $u(0,t) = f(t)$, $u_x(1,t) = -hu(1,t)$, $0 < t$, and the initial condition $u(x,0) = 0$, $0 < x < 1$.

Step 1: First solve the heat conduction problem

$$\frac{\partial A}{\partial t} = \frac{\partial^2 A}{\partial x^2}, \qquad 0 < x < 1, \quad 0 < t,$$

subject to the boundary conditions $A(0,t) = 1$, $A_x(1,t) = -hA(1,t)$, $0 < t$, and the initial condition $A(x,0) = 0$, $0 < x < 1$. Show that

$$A(x,t) = 1 - \frac{hx}{1+h} - 2\sum_{n=1}^{\infty} \frac{k_n^2 + h^2}{k_n \left(k_n^2 + h^2 + h\right)} \sin(k_n x) e^{-k_n^2 t},$$

where k_n is the nth root of $k \cot(k) = -h$.

Step 2: Use Duhamel's theorem and show that

$$u(x,t) = 2\sum_{n=1}^{\infty} \frac{k_n(k_n^2 + h^2)}{k_n^2 + h^2 + h} \sin(k_n x) e^{-k_n^2 t} \int_0^t f(\tau) e^{k_n^2 \tau} \, d\tau.$$

11.7 NUMERICAL SOLUTION OF THE HEAT EQUATION

In the previous chapter we showed how we may use finite difference techniques to solve the wave equation. In this section we show that similar considerations hold for the heat equation.

Starting with the heat equation

$$\frac{\partial u}{\partial t} = a^2 \frac{\partial^2 u}{\partial x^2}, \tag{11.7.1}$$

we must first replace the exact derivatives with finite differences. Drawing upon our work in §10.6,

$$\frac{\partial u(x_m, t_n)}{\partial t} = \frac{u_m^{n+1} - u_m^n}{\Delta t} + O(\Delta t), \tag{11.7.2}$$

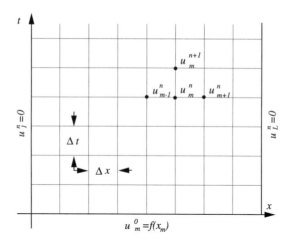

Figure 11.7.1: Schematic of the numerical solution of the heat equation when we hold both ends at a temperature of zero.

and

$$\frac{\partial^2 u(x_m, t_n)}{\partial x^2} = \frac{u^n_{m+1} - 2u^n_m + u^n_{m-1}}{(\Delta x)^2} + O[(\Delta x)^2], \tag{11.7.3}$$

where the notation u^n_m denotes $u(x_m, t_n)$. Figure 11.7.1 illustrates our numerical scheme when we hold both ends at the temperature of zero. Substituting (11.7.2)–(11.7.3) into (11.7.1) and rearranging,

$$u^{n+1}_m = u^n_m + \frac{a^2 \Delta t}{(\Delta x)^2} \left(u^n_{m+1} - 2u^n_m + u^n_{m-1} \right). \tag{11.7.4}$$

The numerical integration begins with $n = 0$ and the value of u^0_{m+1}, u^0_m, and u^0_{m-1} are given by $f(m\Delta x)$.

   Once again we must check the *convergence*, *stability*, and *consistency* of our scheme. We begin by writing u^n_{m+1}, u^n_{m-1}, and u^{n+1}_m in terms of the exact solution u and its derivatives evaluated at the point $x_m = m\Delta x$ and $t_n = n\Delta t$. By Taylor's expansion,

$$u^n_{m+1} = u^n_m + \Delta x \left.\frac{\partial u}{\partial x}\right|^m_n + \tfrac{1}{2}(\Delta x)^2 \left.\frac{\partial^2 u}{\partial x^2}\right|^m_n + \tfrac{1}{6}(\Delta x)^3 \left.\frac{\partial^3 u}{\partial x^3}\right|^m_n + \cdots, \tag{11.7.5}$$

$$u^n_{m-1} = u^n_m - \Delta x \left.\frac{\partial u}{\partial x}\right|^m_n + \tfrac{1}{2}(\Delta x)^2 \left.\frac{\partial^2 u}{\partial x^2}\right|^m_n - \tfrac{1}{6}(\Delta x)^3 \left.\frac{\partial^3 u}{\partial x^3}\right|^m_n + \cdots, \tag{11.7.6}$$

and

$$u^{n+1}_m = u^n_m + \Delta t \left.\frac{\partial u}{\partial t}\right|^m_n + \tfrac{1}{2}(\Delta t)^2 \left.\frac{\partial^2 u}{\partial t^2}\right|^m_n + \tfrac{1}{6}(\Delta t)^3 \left.\frac{\partial^3 u}{\partial t^3}\right|^m_n + \cdots. \tag{11.7.7}$$

Substituting into (11.7.4), we obtain

$$\frac{u_m^{n+1} - u_m^n}{\Delta t} - a^2 \frac{u_{m+1}^n - 2u_m^n + u_{m-1}^n}{(\Delta x)^2}$$

$$= \left(\frac{\partial u}{\partial t} - a^2 \frac{\partial^2 u}{\partial x^2} \right)\Big|_n^m + \tfrac{1}{2}\Delta t \frac{\partial^2 u}{\partial t^2}\Big|_n^m - \tfrac{1}{12}(a\Delta x)^2 \frac{\partial^4 u}{\partial x^4}\Big|_n^m + \cdots .$$

$$(11.7.8)$$

The first term on the right side of (11.7.8) vanishes because $u(x,t)$ satisfies the heat equation. Thus, in the limit of $\Delta x \to 0$, $\Delta t \to 0$, the right side of (11.7.8) vanishes and the scheme is *consistent*.

To determine the *stability* of the explicit scheme, we again use the Fourier method. Assuming a solution of the form:

$$u_n^m = e^{im\theta} e^{in\lambda}, \tag{11.7.9}$$

we substitute (11.7.9) into (11.7.4) and find that

$$\frac{e^{i\lambda} - 1}{\Delta t} = a^2 \frac{e^{i\theta} - 2 + e^{-i\theta}}{(\Delta x)^2}, \tag{11.7.10}$$

or

$$e^{i\lambda} = 1 - 4 \frac{a^2 \Delta t}{(\Delta x)^2} \sin^2\left(\frac{\theta}{2}\right). \tag{11.7.11}$$

The quantity $e^{i\lambda}$ will grow exponentially unless

$$-1 \le 1 - 4 \frac{a^2 \Delta t}{(\Delta x)^2} \sin^2\left(\frac{\theta}{2}\right) < 1. \tag{11.7.12}$$

The right inequality is trivially satisfied if $a^2 \Delta t/(\Delta x)^2 > 0$, while the left inequality yields

$$\frac{a^2 \Delta t}{(\Delta x)^2} \le \frac{1}{2\sin^2(\theta/2)}, \tag{11.7.13}$$

leading to the stability condition $0 < a^2\Delta t/(\Delta x)^2 \le \tfrac{1}{2}$. This is a rather restrictive condition because doubling the resolution (halving Δx) requires that we reduce the time step by a quarter. Thus, for many calculations the required time step may be unacceptably small. For this reason, many use an implicit form of the finite differencing (Crank-Nicholson implicit method[54]):

$$\frac{u_m^{n+1} - u_m^n}{\Delta t} = \frac{a^2}{2}\left[\frac{u_{m+1}^n - 2u_m^n + u_{m-1}^n}{(\Delta x)^2} + \frac{u_{m+1}^{n+1} - 2u_m^{n+1} + u_{m-1}^{n+1}}{(\Delta x)^2}\right],$$

$$(11.7.14)$$

---

[54] Crank, J., and P. Nicholson, 1947: A practical method for numerical evaluation of solutions of partial differential equations of the heat-conduction type. *Proc. Cambridge. Philos. Soc.*, **43**, 50–67.

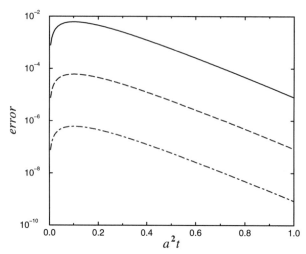

Figure 11.7.2: The growth of error $||e_n||$ as a function of $a^2 t$ for various resolutions. For the top line, $\Delta x = 0.1$; for the middle line, $\Delta x = 0.01$; and for the bottom line, $\Delta x = 0.001$.

although it requires the solution of a simultaneous set of linear equations. However, there are several efficient methods for their solution.

Finally we must check and see if our explicit scheme *converges* to the true solution. If we let e_m^n denote the difference between the exact and our finite differenced solution to the heat equation, we can use (11.7.8) to derive the equation governing e_m^n and find that

$$e_m^{n+1} = e_m^n + \frac{a^2 \Delta t}{(\Delta x)^2} \left(e_{m+1}^n - 2e_m^n + e_{m-1}^n \right) + O[(\Delta t)^2 + \Delta t (\Delta x)^2], \quad (11.7.15)$$

for $m = 1, 2, \ldots, M$. Assuming that $a^2 \Delta t / (\Delta x)^2 \leq \frac{1}{2}$, then

$$|e_m^{n+1}| \leq \frac{a^2 \Delta t}{(\Delta x)^2} |e_{m-1}^n| + \left[1 - 2\frac{a^2 \Delta t}{(\Delta x)^2} \right] |e_m^n| + \frac{a^2 \Delta t}{(\Delta x)^2} |e_{m+1}^n|$$

$$+ A[(\Delta t)^2 + \Delta t (\Delta x)^2] \qquad\qquad (11.7.16)$$

$$\leq ||e_n|| + A[(\Delta t)^2 + \Delta t (\Delta x)^2], \qquad\qquad (11.7.17)$$

where $||e_n|| = \max_{m=0,1,\ldots,M} |e_m^n|$. Consequently,

$$||e_{n+1}|| \leq ||e_n|| + A[(\Delta t)^2 + \Delta t (\Delta x)^2]. \qquad\qquad (11.7.18)$$

Because $||e_0|| = 0$ and $n\Delta t \leq t_n$, we find that

$$||e_{n+1}|| \leq An[(\Delta t)^2 + \Delta t (\Delta x)^2] \leq At_n[\Delta t + (\Delta x)^2]. \qquad\qquad (11.7.19)$$

As $\Delta x \to 0$, $\Delta t \to 0$, the errors tend to zero and we have convergence. We have illustrated (11.7.19) in Figure 11.7.2 by using the finite difference equation (11.7.4) to compute $||e_n||$ during a numerical experiment that used $a^2 \Delta t / (\Delta x)^2 = 0.5$, and $f(x) = \sin(\pi x)$. Note how each increase of resolution by 10 results in a drop in the error by 100.

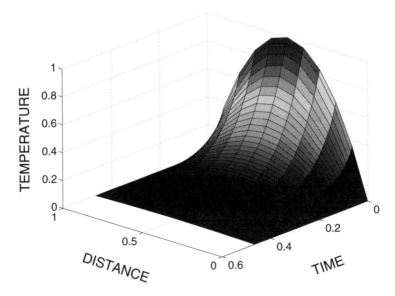

Figure 11.7.3: The numerical solution $u(x,t)$ of the heat equation with $a^2\Delta t/(\Delta x)^2 = 0.47$ at various positions $x' = x/L$ and times $t' = a^2 t/L^2$ using (11.7.4). The initial temperature $u(x,0)$ equals $4x'(1-x')$ and we hold both ends at a temperature of zero.

The following examples illustrate the use of numerical methods.

● **Example 11.7.1**

For our first example, we redo Example 11.3.1 with $a^2\Delta t/(\Delta x)^2 = 0.47$ and 0.53. Our numerical solution was computed using the MATLAB script

```
clear
coeff = 0.47; % coeff = a²Δt/(Δx)²
ncount = 1; dx = 0.1; dt = coeff * dx * dx;
N = 99; x = 0:dx:1;
M = 1/dx + 1; % M = number of spatial grid points
tplot(1) = 0; u = zeros(M,N+1);
for m = 1:M; u(m,1)=4*x(m)*(1-x(m)); temp(m,1)=u(m,1); end
% integrate forward in time
for n = 1:N
t = dt * n;
for m = 2:M-1
u(m,n+1) = u(m,n) + coeff * (u(m+1,n) - 2 * u(m,n) + u(m-1,n));
end
if mod(n+1,2) == 0
ncount = ncount + 1; tplot(ncount) = t;
for m = 1:M; temp(m,ncount) = u(m,n+1); end
end; end
```

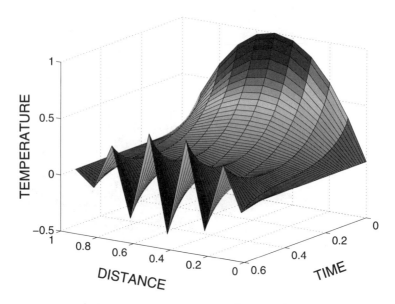

Figure 11.7.4: Same as Figure 11.7.3 except that $a^2 \Delta t / (\Delta x)^2 = 0.53$.

```
% plot the numerical solution
X = x' * ones(1,length(tplot)); T = ones(M,1) * tplot;
surf(X,T,temp)
xlabel('DISTANCE','Fontsize',20); ylabel('TIME','Fontsize',20)
zlabel('TEMPERATURE','Fontsize',20)
```

As Figure 11.7.3 shows, the solution with $a^2 \Delta t / (\Delta x)^2 < 1/2$ performs well. On the other hand, Figure 11.7.4 shows small-scale, growing disturbances when $a^2 \Delta t / (\Delta x)^2 > 1/2$. It should be noted that for the reasonable $\Delta x = L/100$, it takes approximately *20,000* time steps before we reach $a^2 t / L^2 = 1$.

• **Example 11.7.2**

In this example, we redo the previous example with an insulated end at $x = L$. Using the centered differencing formula,

$$u_{M+1}^n - u_{M-1}^n = 0, \qquad (11.7.20)$$

because $u_x(L, t) = 0$. Also, at $i = M$,

$$u_M^{n+1} = u_M^n + \frac{a^2 \Delta t}{(\Delta x)^2} \left(u_{M+1}^n - 2u_M^n + u_{M-1}^n \right). \qquad (11.7.21)$$

Eliminating u_{M+1}^n between the two equations,

$$u_M^{n+1} = u_M^n + \frac{a^2 \Delta t}{(\Delta x)^2} \left(2u_{M-1}^n - 2u_M^n \right). \qquad (11.7.22)$$

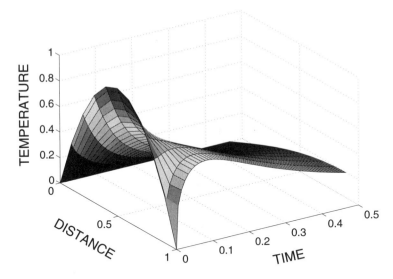

Figure 11.7.5: Same as Figure 11.7.4 except that we now have an insulated boundary condition $u_x(L,t) = 0$.

To implement this new boundary condition in our MATLAB script, we add the line

```
u(M,n+1) = u(M,n) + 2 * coeff * (u(M-1,n) - u(M,n));
```

after the lines

```
for m = 2:M-1
u(m,n+1) = u(m,n) + coeff * (u(m+1,n) - 2 * u(m,n) + u(m-1,n));
end
```

Figure 11.7.5 illustrates our numerical solution at various positions and times.

Project: Implicit Numerical Integration of the Heat Equation

The difficulty in using explicit time differencing to solve the heat equation is the very small time step that must be taken at moderate spatial resolutions to ensure stability. This small time step translates into an unacceptably long execution time. In this project you will investigate the Crank-Nicholson implicit scheme which allows for a much more reasonable time step.

Step 1: Develop a MATLAB script that uses the Crank-Nicholson equation (11.7.14) to numerically integrate the heat equation. To do this, you will need a tridiagonal solver to find u_m^{n+1}. This is explained at the end of §14.1. However, many numerical methods books[55] actually have code already developed for your use. You might as well use this code.

---

[55] For example, Press, W. H., B. P. Flannery, S. A. Teukolsky, and W. T. Vetterling, 1986: *Numerical Recipes: The Art of Scientific Computing*. Cambridge University Press, §2.6.

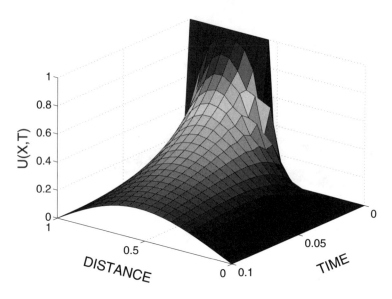

Figure 11.7.6: The numerical solution $u(x,t)$ of the heat equation $u_t = a^2 u_{xx}$ using the Crank-Nicholson method. The parameters used in the numerical solution are $a^2 \Delta t = 0.005$ and $\Delta x = 0.05$. Both ends are held at zero with an initial condition of $u(x,0) = 0$ for $0 \le x < \frac{1}{2}$, and $u(x,0) = 1$ for $\frac{1}{2} < x \le 1$.

Step 2: Test your code by solving the heat equation given the initial condition $u(x,0) = \sin(\pi x)$, and the boundary conditions $u(0,t) = u(1,t) = 0$. Find the solution for various values of Δt with $\Delta x = 0.01$. Compare this numerical solution against the exact solution which you can find. How does the error (between the numerical and exact solutions) change with Δt? For small Δt, the errors should be small. If not, then you have a mistake in your code.

Step 3: Once you have confidence in your code, discuss the behavior of the scheme for various values of Δx and Δt for the initial condition $u(x,0) = 0$ for $0 \le x < \frac{1}{2}$, and $u(x,0) = 1$ for $\frac{1}{2} < x \le 1$ with the boundary conditions $u(0,t) = u(1,t) = 0$. Although you can take quite a large Δt, what happens? Did a similar problem arise in Step 2? Explain your results. Zvan *et al.*[56] have reported a similar problem in the numerical integration of the Black-Scholes equation from mathematical finance.

---

[56] Zvan, R., K. Vetzal, and P. Forsyth, 1998: Swing low, swing high. *Risk*, **11(3)**, 71–75.

Chapter 12

Laplace's Equation

In the previous chapter we solved the one-dimensional heat equation. Quite often we found that the transient solution died away, leaving a steady state. The partial differential equation that describes the steady state for two-dimensional heat conduction is Laplace's equation

$$\frac{\partial^2 u}{\partial x^2} + \frac{\partial^2 u}{\partial y^2} = 0. \tag{12.0.1}$$

In general, this equation governs physical processes where *equilibrium* has been reached. It also serves as the prototype for a wider class of *elliptic equations*

$$a(x,t)\frac{\partial^2 u}{\partial x^2} + b(x,t)\frac{\partial^2 u}{\partial x \partial t} + c(x,t)\frac{\partial^2 u}{\partial t^2} = f\left(x,t,u,\frac{\partial u}{\partial x},\frac{\partial u}{\partial t}\right), \tag{12.0.2}$$

where $b^2 < 4ac$. Unlike the heat and wave equations, there are no initial conditions and the boundary conditions completely specify the solution. In this chapter we present some of the common techniques for solving this equation.

12.1 DERIVATION OF LAPLACE'S EQUATION

Imagine a thin, flat plate of heat-conducting material between two sheets of insulation. Sufficient time has passed so that the temperature depends only on the spatial coordinates x and y. Let us now apply the law of conservation of energy (in rate form) to a small rectangle with sides Δx and Δy.

If $q_x(x, y)$ and $q_y(x, y)$ denote the heat flow rates in the x- and y-direction, respectively, conservation of energy requires that the heat flow into the slab equals the heat flow out of the slab if there is no storage or generation of heat. Now

$$\text{rate in} = q_x(x, y + \Delta y/2)\Delta y + q_y(x + \Delta x/2, y)\Delta x, \qquad (12.1.1)$$

and

$$\text{rate out} = q_x(x + \Delta x, y + \Delta y/2)\Delta y + q_y(x + \Delta x/2, y + \Delta y)\Delta x. \quad (12.1.2)$$

If the plate has unit thickness,

$$[q_x(x, y + \Delta y/2) - q_x(x + \Delta x, y + \Delta y/2)]\Delta y$$
$$+ [q_y(x + \Delta x/2, y) - q_y(x + \Delta x/2, y + \Delta y)]\Delta x = 0. \qquad (12.1.3)$$

Upon dividing through by $\Delta x \Delta y$, we obtain two differences quotients on the left side of (12.1.3). In the limit as $\Delta x, \Delta y \to 0$, they become partial derivatives, giving

$$\frac{\partial q_x}{\partial x} + \frac{\partial q_y}{\partial y} = 0 \qquad (12.1.4)$$

for any point (x, y).

We now employ Fourier's law to eliminate the rates q_x and q_y, yielding

$$\frac{\partial}{\partial x}\left(a^2 \frac{\partial u}{\partial x}\right) + \frac{\partial}{\partial y}\left(a^2 \frac{\partial u}{\partial y}\right) = 0, \qquad (12.1.5)$$

if we have an isotropic (same in all directions) material. Finally, if a^2 is constant, (12.1.5) reduces to

$$\frac{\partial^2 u}{\partial x^2} + \frac{\partial^2 u}{\partial y^2} = 0, \qquad (12.1.6)$$

which is the two-dimensional, steady-state heat equation (i.e., $u_t \approx 0$ as $t \to \infty$).

Solutions of Laplace's equation (called *harmonic functions*) differ fundamentally from those encountered with the heat and wave equations. These latter two equations describe the evolution of some phenomena. Laplace's equation, on the other hand, describes things at equilibrium. Consequently, any change in the boundary conditions affects to some degree the *entire* domain because a change to any one point causes its neighbors to change in order to reestablish the equilibrium. Those points will, in turn, affect others.

Figure 12.1.1: Today we best remember Pierre-Simon Laplace (1749–1827) for his work in celestial mechanics and probability. In his five volumes *Traité de Mécanique céleste* (1799–1825), he accounted for the theoretical orbits of the planets and their satellites. Laplace's equation arose during this study of gravitational attraction. (Portrait courtesy of the Archives de l'Académie des sciences, Paris.)

Because all of these points are in equilibrium, this modification must occur instantaneously.

Further insight follows from the *maximum principle*. If Laplace's equation governs a region, then its solution cannot have a relative maximum or minimum *inside* the region unless the solution is constant.[1] If we think of the solution as a steady-state temperature distribution, this principle is clearly true because at any one point the temperature cannot be greater than at all other nearby points. If that were so, heat would flow away from the hot point to cooler points nearby, thus eliminating the hot spot when equilibrium was once again restored.

It is often useful to consider the two-dimensional Laplace's equation in other coordinate systems. In polar coordinates, where $x = r\cos(\theta)$, $y =$

---

[1] For the proof, see Courant, R., and D. Hilbert, 1962: *Methods of Mathematical Physics, Vol. 2: Partial Differential Equations.* Interscience, pp. 326–331.

$r\sin(\theta)$, and $z = z$, Laplace's equation becomes

$$\frac{\partial^2 u}{\partial r^2} + \frac{1}{r}\frac{\partial u}{\partial r} + \frac{\partial^2 u}{\partial z^2} = 0, \qquad (12.1.7)$$

if the problem possesses axisymmetry. On the other hand, if the solution is independent of z, Laplace's equation becomes

$$\frac{\partial^2 u}{\partial r^2} + \frac{1}{r}\frac{\partial u}{\partial r} + \frac{1}{r^2}\frac{\partial^2 u}{\partial \theta^2} = 0. \qquad (12.1.8)$$

In spherical coordinates, $x = r\cos(\varphi)\sin(\theta)$, $y = r\sin(\varphi)\sin(\theta)$, and $z = r\cos(\theta)$, where $r^2 = x^2 + y^2 + z^2$, θ is the angle measured *down* to the point from the z-axis (colatitude) and φ is the angle made between the x-axis and the projection of the point on the xy plane. In the case of axisymmetry (no φ dependence), Laplace's equation becomes

$$\frac{\partial}{\partial r}\left(r^2\frac{\partial u}{\partial r}\right) + \frac{1}{\sin(\theta)}\frac{\partial}{\partial \theta}\left[\sin(\theta)\frac{\partial u}{\partial \theta}\right] = 0. \qquad (12.1.9)$$

12.2 BOUNDARY CONDITIONS

Because Laplace's equation involves time-independent phenomena, we must only specify boundary conditions. As we discussed in §11.2, we can classify these boundary conditions as follows:

1. Dirichlet condition: u given

2. Neumann condition: $\dfrac{\partial u}{\partial n}$ given, where n is the unit normal direction

3. Robin condition: $u + \alpha\dfrac{\partial u}{\partial n}$ given

along any section of the boundary. In the case of Laplace's equation, if all of the boundaries have Neumann conditions, then the solution is not unique. This follows from the fact that if $u(x,y)$ is a solution, so is $u(x,y) + c$, where c is any constant.

Finally we note that we must specify the boundary conditions along each side of the boundary. These sides may be at infinity as in problems with semi-infinite domains. We must specify values along the entire boundary because we could not have an equilibrium solution if any portion of the domain was undetermined.

12.3 SEPARATION OF VARIABLES

As in the case of the heat and wave equations, separation of variables is the most popular technique for solving Laplace's equation. Although the same general procedure carries over from the previous two chapters, the following examples fill out the details.

• Example 12.3.1: Groundwater flow in a valley

Over a century ago, a French hydraulic engineer named Henri-Philibert-Gaspard Darcy (1803–1858) published the results of a laboratory experiment on the flow of water through sand. He showed that the *apparent* fluid velocity **q** relative to the sand grains is directly proportional to the gradient of the hydraulic potential $-k\nabla\varphi$, where the hydraulic potential φ equals the sum of the elevation of the point of measurement plus the pressure potential $(p/\rho g)$. In the case of steady flow, the combination of Darcy's law with conservation of mass $\nabla \cdot \mathbf{q} = 0$ yields Laplace's equation $\nabla^2\varphi = 0$ if the aquifer is isotropic (same in all directions) and homogeneous.

To illustrate how separation of variables can be used to solve Laplace's equation, we will determine the hydraulic potential within a small drainage basin that lies in a shallow valley. See Figure 12.3.1. Following Tóth,[2] the governing equation is the two-dimensional Laplace equation

$$\frac{\partial^2 u}{\partial x^2} + \frac{\partial^2 u}{\partial y^2} = 0, \quad 0 < x < L, \quad 0 < y < z_0, \tag{12.3.1}$$

along with the boundary conditions

$$u(x, z_0) = gz_0 + gcx, \tag{12.3.2}$$

$$u_x(0, y) = u_x(L, y) = 0, \quad \text{and} \quad u_y(x, 0) = 0, \tag{12.3.3}$$

where $u(x, y)$ is the hydraulic potential, g is the acceleration due to gravity, and c gives the slope of the topography. The conditions $u_x(L, y) = 0$, and $u_y(x, 0) = 0$ specify a no-flow condition through the bottom and sides of the aquifer. The condition $u_x(0, y) = 0$ ensures symmetry about the $x = 0$ line. Equation (12.3.2) gives the fluid potential at the water table, where z_0 is the elevation of the water table above the standard datum. The term gcx in (12.3.2) expresses the increase of the potential from the valley bottom toward the water divide. On average it closely follows the topography.

Following the pattern set in the previous two chapters, we assume that $u(x, y) = X(x)Y(y)$. Then (12.3.1) becomes

$$X''Y + XY'' = 0. \tag{12.3.4}$$

---

[2] Tóth, J., *J. Geophys. Res.*, **67**, 4375–4387, 1962, copyright by the American Geophysical Union.

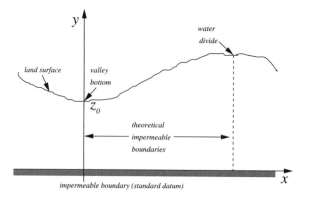

Figure 12.3.1: Cross section of a valley.

Separating the variables yields

$$\frac{X''}{X} = -\frac{Y''}{Y}. \tag{12.3.5}$$

Both sides of (12.3.5) must be constant, but the sign of that constant is not obvious. From previous experience we anticipate that the ordinary differential equation in the x-direction leads to a Sturm-Liouville problem because it possesses homogeneous boundary conditions. Proceeding along this line of reasoning, we consider three separation constants.

Trying a positive constant (say, m^2), (12.3.5) separates into the two ordinary differential equations

$$X'' - m^2 X = 0, \quad \text{and} \quad Y'' + m^2 Y = 0, \tag{12.3.6}$$

which have the solutions

$$X(x) = A \cosh(mx) + B \sinh(mx), \tag{12.3.7}$$

and

$$Y(y) = C \cos(my) + D \sin(my). \tag{12.3.8}$$

Because the boundary conditions (12.3.3) imply $X'(0) = X'(L) = 0$, both A and B must be zero, leading to the trivial solution $u(x, y) = 0$.

When the separation constant equals zero, we find a nontrivial solution given by the eigenfunction $X_0(x) = 1$, and $Y_0(y) = \frac{1}{2}A_0 + B_0 y$. However, because $Y_0'(0) = 0$ from (12.3.3), $B_0 = 0$. Thus, the particular solution for a zero separation constant is $u_0(x, y) = A_0/2$.

Finally, taking both sides of (12.3.5) equal to $-k^2$,

$$X'' + k^2 X = 0, \quad \text{and} \quad Y'' - k^2 Y = 0. \tag{12.3.9}$$

The first of these equations, along with the boundary conditions $X'(0) = X'(L) = 0$, gives the eigenfunction $X_n(x) = \cos(k_n x)$, with $k_n = n\pi/L$, $n = 1, 2, 3, \ldots$. The function $Y_n(y)$ for the same separation constant is

$$Y_n(y) = A_n \cosh(k_n y) + B_n \sinh(k_n y). \qquad (12.3.10)$$

We must take $B_n = 0$ because $Y_n'(0) = 0$.

We now have the product solution $X_n(x)Y_n(y)$, which satisfies Laplace's equation and all of the boundary conditions except (12.3.2). By the principle of superposition, the general solution is

$$u(x, y) = \frac{A_0}{2} + \sum_{n=1}^{\infty} A_n \cos\left(\frac{n\pi x}{L}\right) \cosh\left(\frac{n\pi y}{L}\right). \qquad (12.3.11)$$

Applying (12.3.2), we find that

$$u(x, z_0) = g z_0 + g c x = \frac{A_0}{2} + \sum_{n=1}^{\infty} A_n \cos\left(\frac{n\pi x}{L}\right) \cosh\left(\frac{n\pi z_0}{L}\right), \qquad (12.3.12)$$

which we recognize as a Fourier half-range cosine series such that

$$A_0 = \frac{2}{L} \int_0^L (g z_0 + g c x) \, dx, \qquad (12.3.13)$$

and

$$\cosh\left(\frac{n\pi z_0}{L}\right) A_n = \frac{2}{L} \int_0^L (g z_0 + g c x) \cos\left(\frac{n\pi x}{L}\right) \, dx. \qquad (12.3.14)$$

Performing the integrations,

$$A_0 = 2 g z_0 + g c L, \qquad (12.3.15)$$

and

$$A_n = -\frac{2 g c L [1 - (-1)^n]}{n^2 \pi^2 \cosh(n\pi z_0/L)}. \qquad (12.3.16)$$

Finally, the complete solution is

$$u(x, y) = g z_0 + \frac{g c L}{2} - \frac{4 g c L}{\pi^2} \sum_{m=1}^{\infty} \frac{\cos[(2m-1)\pi x/L] \cosh[(2m-1)\pi y/L]}{(2m-1)^2 \cosh[(2m-1)\pi z_0/L]}. \qquad (12.3.17)$$

Figure 12.3.2 presents two graphs by Tóth for two different aquifers. We see that the solution satisfies the boundary condition at the bottom and side

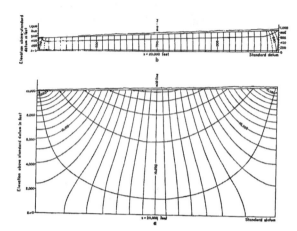

Figure 12.3.2: Two-dimensional potential distribution and flow patterns for different depths of the horizontally impermeable boundary.

boundaries. Water flows from the elevated land (on the right) into the valley (on the left), from regions of high to low hydraulic potential.

• Example 12.3.2

In the previous example, we had the advantage of homogeneous boundary conditions along $x = 0$ and $x = L$. In a different hydraulic problem, Kirkham[3] solved the more difficult problem of

$$\frac{\partial^2 u}{\partial x^2} + \frac{\partial^2 u}{\partial y^2} = 0, \quad 0 < x < L, \quad 0 < y < h, \tag{12.3.18}$$

subject to the Dirichlet boundary conditions

$$u(x, 0) = Rx, \quad u(x, h) = RL, \quad u(L, y) = RL, \tag{12.3.19}$$

and

$$u(0, y) = \begin{cases} 0, & 0 < y < a, \\ \frac{RL}{b-a}(y - a), & a < y < b, \\ RL, & b < y < h. \end{cases} \tag{12.3.20}$$

This problem arises in finding the steady flow within an aquifer resulting from the introduction of water at the top due to a steady rainfall and its removal along the sides by drains. The parameter L equals half of the distance between the drains, h is the depth of the aquifer, and R is the rate of rainfall.

---

[3] Kirkham, D., *Trans. Am. Geophys. Union*, **39**, 892–908, 1958, copyright by the American Geophysical Union.

The point of this example is: *We need homogeneous boundary conditions along either the x or y boundaries for separation of variables to work.* We achieve this by breaking the original problem into two parts, namely

$$u(x,y) = v(x,y) + w(x,y) + RL, \qquad (12.3.21)$$

where

$$\frac{\partial^2 v}{\partial x^2} + \frac{\partial^2 v}{\partial y^2} = 0, \quad 0 < x < L, \quad 0 < y < h, \qquad (12.3.22)$$

with

$$v(0,y) = v(L,y) = 0, \qquad v(x,h) = 0, \qquad (12.3.23)$$

and

$$v(x,0) = R(x - L); \qquad (12.3.24)$$

$$\frac{\partial^2 w}{\partial x^2} + \frac{\partial^2 w}{\partial y^2} = 0, \quad 0 < x < L, \quad 0 < y < h, \qquad (12.3.25)$$

with

$$w(x,0) = w(x,h) = 0, \qquad w(L,y) = 0, \qquad (12.3.26)$$

and

$$w(0,y) = \begin{cases} -RL, & 0 < y < a, \\ \frac{RL}{b-a}(y-a) - RL, & a < y < b, \\ 0, & b < y < h. \end{cases} \qquad (12.3.27)$$

Employing the same technique as in Example 12.3.1, we find that

$$v(x,y) = \sum_{n=1}^{\infty} A_n \sin\left(\frac{n\pi x}{L}\right) \frac{\sinh[n\pi(h-y)/L]}{\sinh(n\pi h/L)}, \qquad (12.3.28)$$

where

$$A_n = \frac{2}{L} \int_0^L R(x-L) \sin\left(\frac{n\pi x}{L}\right) dx = -\frac{2RL}{n\pi}. \qquad (12.3.29)$$

Similarly, the solution to $w(x,y)$ is found to be

$$w(x,y) = \sum_{n=1}^{\infty} B_n \sin\left(\frac{n\pi y}{h}\right) \frac{\sinh[n\pi(L-x)/h]}{\sinh(n\pi L/h)}, \qquad (12.3.30)$$

where

$$B_n = \frac{2}{h}\left[-RL \int_0^a \sin\left(\frac{n\pi y}{h}\right) dy + RL \int_a^b \left(\frac{y-a}{b-a} - 1\right) \sin\left(\frac{n\pi y}{h}\right) dy\right] \qquad (12.3.31)$$

$$= \frac{2RL}{\pi}\left\{\frac{h}{(b-a)n^2\pi}\left[\sin\left(\frac{n\pi b}{h}\right) - \sin\left(\frac{n\pi a}{h}\right)\right] - \frac{1}{n}\right\}. \qquad (12.3.32)$$

The complete solution consists of substituting (12.3.28) and (12.3.30) into (12.3.21).

• Example 12.3.3

The *electrostatic potential* is defined as the amount of work which must be done against electric forces to bring a unit charge from a reference point to a given point. It is readily shown[4] that the electrostatic potential is described by Laplace's equation if there is no charge within the domain. Let us find the electrostatic potential $u(r, z)$ inside a closed cylinder of length L and radius a. The base and lateral surfaces have the potential 0 while the upper surface has the potential V.

Because the potential varies in only r and z, Laplace's equation in cylindrical coordinates reduces to

$$\frac{1}{r}\frac{\partial}{\partial r}\left(r\frac{\partial u}{\partial r}\right) + \frac{\partial^2 u}{\partial z^2} = 0, \quad 0 \leq r < a, \quad 0 < z < L, \qquad (12.3.33)$$

subject to the boundary conditions

$$u(a, z) = u(r, 0) = 0, \quad \text{and} \quad u(r, L) = V. \qquad (12.3.34)$$

To solve this problem by separation of variables,[5] let $u(r, z) = R(r)Z(z)$ and

$$\frac{1}{rR}\frac{d}{dr}\left(r\frac{dR}{dr}\right) = -\frac{1}{Z}\frac{d^2 Z}{dz^2} = -\frac{k^2}{a^2}. \qquad (12.3.35)$$

Only a negative separation constant yields nontrivial solutions in the radial direction. In that case, we have that

$$\frac{1}{r}\frac{d}{dr}\left(r\frac{dR}{dr}\right) + \frac{k^2}{a^2}R = 0. \qquad (12.3.36)$$

The solutions of (12.3.36) are the Bessel functions $J_0(kr/a)$ and $Y_0(kr/a)$. Because $Y_0(kr/a)$ becomes infinite at $r = 0$, the only permissible solution is $J_0(kr/a)$. The condition that $u(a, z) = R(a)Z(z) = 0$ forces us to choose values of k such that $J_0(k) = 0$. Therefore, the solution in the radial direction is $J_0(k_n r/a)$, where k_n is the nth root of $J_0(k) = 0$.

---

[4] For static fields, $\nabla \times \mathbf{E} = \mathbf{0}$, where $\mathbf{E}$ is the electric force. From §13.4, we can introduce a potential φ such that $\mathbf{E} = \nabla\varphi$. From Gauss' law, $\nabla \cdot \mathbf{E} = \nabla^2\varphi = 0$.

[5] Wang and Liu [Wang, M.-L., and B.-L. Liu, 1995: Solution of Laplace equation by the method of separation of variables. *J. Chinese Inst. Eng.*, **18**, 731–739] have written a review article on the solutions to (12.3.33) based upon which order the boundary conditions are satisfied.

In the z direction,

$$\frac{d^2 Z_n}{dz^2} + \frac{k_n^2}{a^2} Z_n = 0. \qquad (12.3.37)$$

The general solution to (12.3.37) is

$$Z_n(z) = A_n \sinh\left(\frac{k_n z}{a}\right) + B_n \cosh\left(\frac{k_n z}{a}\right). \qquad (12.3.38)$$

Because $u(r,0) = R(r)Z(0) = 0$ and $\cosh(0) = 1$, B_n must equal zero. Therefore, the general product solution is

$$u(r,z) = \sum_{n=1}^{\infty} A_n J_0\left(\frac{k_n r}{a}\right) \sinh\left(\frac{k_n z}{a}\right). \qquad (12.3.39)$$

The condition that $u(r,L) = V$ determines the arbitrary constant A_n. Along $z = L$,

$$u(r,L) = V = \sum_{n=1}^{\infty} A_n J_0\left(\frac{k_n r}{a}\right) \sinh\left(\frac{k_n L}{a}\right), \qquad (12.3.40)$$

where

$$\sinh\left(\frac{k_n L}{a}\right) A_n = \frac{2V}{a^2 J_1^2(k_n)} \int_0^L r\, J_0\left(\frac{k_n r}{a}\right) dr \qquad (12.3.41)$$

from (9.5.35) and (9.5.43). Thus,

$$\sinh\left(\frac{k_n L}{a}\right) A_n = \frac{2V}{k_n^2 J_1^2(k_n)} \left(\frac{k_n r}{a}\right) J_1\left(\frac{k_n r}{a}\right)\Big|_0^a = \frac{2V}{k_n J_1(k_n)}. \qquad (12.3.42)$$

The solution is then

$$u(r,z) = 2V \sum_{n=1}^{\infty} \frac{J_0(k_n r/a)}{k_n J_1(k_n)} \frac{\sinh(k_n z/a)}{\sinh(k_n L/a)}. \qquad (12.3.43)$$

Figure 12.3.3 illustrates (12.3.43) for the case when $L = a$ where we included the first 20 terms of the series. It was created using the MATLAB script

```
clear
L_over_a = 1; M = 20; dr = 0.02; dz = 0.02;
% load in zeros of J_0
zero( 1) =  2.40482; zero( 2) =  5.52007; zero( 3) =  8.65372;
zero( 4) = 11.79153; zero( 5) = 14.93091; zero( 6) = 18.07106;
zero( 7) = 21.21164; zero( 8) = 24.35247; zero( 9) = 27.49347;
zero(10) = 30.63461; zero(11) = 33.77582; zero(12) = 36.91710;
```

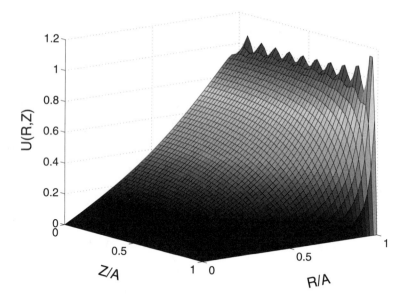

Figure 12.3.3: The steady-state potential (divided by V) within a cylinder of equal radius and height a when the top has the potential V while the lateral side and bottom are at potential 0.

```
zero(13) = 40.05843; zero(14) = 43.19979; zero(15) = 46.34119;
zero(16) = 49.48261; zero(17) = 52.62405; zero(18) = 55.76551;
zero(19) = 58.90698; zero(20) = 62.04847;
% compute Fourier coefficients
for m = 1:M
a(m) = 2/(zero(m)*besselj(1,zero(m))*sinh(L_over_a * zero(m)));
end
% compute grid and initialize solution
R_over_a = [0:dr:1]; Z_over_a = [0:dz:1];
u = zeros(length(Z_over_a),length(R_over_a));
RR_over_a = repmat(R_over_a,[length(Z_over_a) 1]);
ZZ_over_a = repmat(Z_over_a',[1 length(R_over_a)]);
% compute solution from (12.3.43)
for m = 1:M
u=u+a(m).*besselj(0,zero(m)*RR_over_a).*sinh(zero(m)*ZZ_over_a);
end
surf(RR_over_a,ZZ_over_a,u)
xlabel('R/A','Fontsize',20); ylabel('Z/A','Fontsize',20)
zlabel('U(R,Z)','Fontsize',20)
```

Of particular interest are the ripples along the line $z = L$. Along that line, the solution must jump from V to 0 at $r = a$. For that reason our solution suffers from Gibbs phenomena along this boundary. As we move away from that region the electrostatic potential varies smoothly.

• Example 12.3.4

Let us now consider a similar, but slightly different, version of example 12.3.3, where the ends are held at zero potential while the lateral side has the value V. Once again, the governing equation is (12.3.33) with the boundary conditions

$$u(r,0) = u(r,L) = 0, \quad \text{and} \quad u(a,z) = V. \tag{12.3.44}$$

Separation of variables yields

$$\frac{1}{rR}\frac{d}{dr}\left(r\frac{dR}{dr}\right) = -\frac{1}{Z}\frac{d^2Z}{dz^2} = \frac{k^2}{L^2} \tag{12.3.45}$$

with $Z(0) = Z(L) = 0$. We chose a positive separation constant because a negative constant would give hyperbolic functions in z which cannot satisfy the boundary conditions. A separation constant of zero would give a straight line for $Z(z)$. Applying the boundary conditions gives a trivial solution. Consequently, the only solution in the z direction which satisfies the boundary conditions is $Z_n(z) = \sin(n\pi z/L)$.

In the radial direction, the differential equation is

$$\frac{1}{r}\frac{d}{dr}\left(r\frac{dR_n}{dr}\right) - \frac{n^2\pi^2}{L^2}R_n = 0. \tag{12.3.46}$$

As we showed in §9.5, the general solution is

$$R_n(r) = A_n I_0\left(\frac{n\pi r}{L}\right) + B_n K_0\left(\frac{n\pi r}{L}\right), \tag{12.3.47}$$

where I_0 and K_0 are modified Bessel functions of the first and second kind, respectively, of order zero. Because $K_0(x)$ behaves as $-\ln(x)$ as $x \to 0$, we must discard it and our solution in the radial direction becomes $R_n(r) = A_n I_0(n\pi r/L)$. Hence, the product solution is

$$u_n(r,z) = A_n I_0\left(\frac{n\pi r}{L}\right)\sin\left(\frac{n\pi z}{L}\right), \tag{12.3.48}$$

and the general solution is a sum of these particular solutions, namely

$$u(r,z) = \sum_{n=1}^{\infty} A_n I_0\left(\frac{n\pi r}{L}\right)\sin\left(\frac{n\pi z}{L}\right). \tag{12.3.49}$$

Finally, we use the boundary conditions that $u(a, z) = V$ to compute A_n. This condition gives

$$u(a, z) = V = \sum_{n=1}^{\infty} A_n I_0 \left(\frac{n\pi a}{L} \right) \sin \left(\frac{n\pi z}{L} \right), \tag{12.3.50}$$

so that

$$I_0 \left(\frac{n\pi a}{L} \right) A_n = \frac{2}{L} \int_0^L V \sin \left(\frac{n\pi z}{L} \right) dz = \frac{2V[1 - (-1)^n]}{n\pi}. \tag{12.3.51}$$

Therefore, the final answer is

$$u(r, z) = \frac{4V}{\pi} \sum_{m=1}^{\infty} \frac{I_0[(2m-1)\pi r/L] \sin[(2m-1)\pi z/L]}{(2m-1)I_0[(2m-1)\pi a/L]}. \tag{12.3.52}$$

Figure 12.3.4 illustrates the solution (12.3.52) for the case when $L = a$. It was created using the MATLAB script

```
clear
a_over_L = 1; M = 200; dr = 0.02; dz = 0.02;
%
% compute grid and initialize solution
%
R_over_L = [0:dr:1]; Z_over_L = [0:dz:1];
u = zeros(length(Z_over_L),length(R_over_L));
RR_over_L = repmat(R_over_L,[length(Z_over_L) 1]);
ZZ_over_L = repmat(Z_over_L',[1 length(R_over_L)]);
%
for m = 1:M
temp = (2*m-1)*pi; prod1 = temp*a_over_L;
%
% compute modified bessel functions in (12.3.52)
%
for j = 1:length(Z_over_L); for i = 1:length(R_over_L);
prod2 = temp*RR_over_L(i,j);
%
if prod2 - prod1 > -10
%
if prod2 < 20
ratio(i,j) = besseli(0,prod2) / besseli(0,prod1);
else
% for large values of prod, use asymptotic expansion
%          for modified bessel function
ratio(i,j) = sqrt(prod1/prod2) * exp(prod2-prod1); end;
%
```

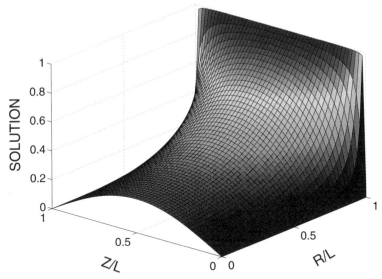

Figure 12.3.4: Potential (divided by V) within a conducting cylinder when the top and bottom have a potential 0 while the lateral side has a potential V.

```
else
ratio(i,j) = 0; end
%
end; end;
%
% compute solution from (12.3.52)
%
u = u + (4/temp) * ratio .* sin(temp*ZZ_over_L);
end
surf(RR_over_L,ZZ_over_L,u)
xlabel('R/L','Fontsize',20); ylabel('Z/L','Fontsize',20)
zlabel('SOLUTION','Fontsize',20)
```

Once again, there is a convergence of equipotentials at the corners along the right side. If we had plotted more contours, we would have observed Gibbs phenomena in the solution along the top and bottom of the cylinder.

• **Example 12.3.5**

In the previous examples, the domain was always of finite extent. Assuming axial symmetry, let us now solve Laplace's equation

$$\frac{1}{r}\frac{\partial}{\partial r}\left(r\frac{\partial u}{\partial r}\right) + \frac{\partial^2 u}{\partial z^2} = 0, \quad 0 \le r < \infty, \quad 0 < z < \infty, \qquad (\mathbf{12.3.53})$$

in the half-plane $z > 0$ subject to the boundary conditions

$$\lim_{z \to \infty} |u(r, z)| < \infty, \qquad u(r, 0) = \begin{cases} u_0, & r < a, \\ 0, & r > a, \end{cases} \tag{12.3.54}$$

$$\lim_{r \to 0} |u(r, z)| < \infty, \qquad \text{and} \qquad \lim_{r \to \infty} |u(r, z)| < \infty. \tag{12.3.55}$$

This problem gives the steady-state temperature distribution in the half-space $z > 0$ where the temperature on the bounding plane $z = 0$ equals u_0 within a circle of radius a and equals 0 outside of the circle.

As before we begin by assuming the product solution $u(r, z) = R(r)Z(z)$ and separate the variables. Again, the separation constant may be positive, negative, or zero. Turning to the positive separation constant first, we have that

$$\frac{R''}{R} + \frac{1}{r}\frac{R'}{R} = -\frac{Z''}{Z} = m^2. \tag{12.3.56}$$

Focusing on the R equation,

$$\frac{R''}{R} + \frac{1}{r}\frac{R'}{R} - m^2 = 0, \quad \text{or} \quad r^2 R'' + rR - m^2 r^2 R = 0. \tag{12.3.57}$$

The solution to (12.3.57) is

$$R(r) = A_1 I_0(mr) + A_2 K_0(mr), \tag{12.3.58}$$

where $I_0(\ )$ and $K_0(\ )$ denote modified Bessel functions of order zero and the first and second kind, respectively. Because $u(r, z)$, and hence $R(r)$, must be bounded as $r \to 0$, $A_2 = 0$. Similarly, since $u(r, z)$ must also be bounded as $r \to \infty$, $A_1 = 0$ because $\lim_{r \to \infty} I_0(mr) \to \infty$. Thus, there is only a trivial solution for a positive separation constant.

We next try the case when the separation constant equals 0. This yields

$$\frac{R''}{R} + \frac{1}{r}\frac{R'}{R} = 0, \quad \text{or} \quad r^2 R'' + rR = 0. \tag{12.3.59}$$

The solution here is

$$R(r) = A_1 + A_2 \ln(r). \tag{12.3.60}$$

Again, boundedness as $r \to 0$ requires that $A_2 = 0$. What about A_1? Clearly, for any arbitrary value of z, the amount of internal energy must be finite. This corresponds to

$$\int_0^\infty |u(r, z)| \, dr < \infty \quad \text{or} \quad \int_0^\infty |R(r)| \, dr < \infty \tag{12.3.61}$$

and $A_1 = 0$. The choice of the zero separation constant yields a trivial solution.

Finally, when the separation constant equals $-k^2$, the equations for $R(r)$ and $Z(z)$ are

$$r^2 R'' + rR + k^2 r^2 R = 0, \qquad \text{and} \qquad Z'' - k^2 Z = 0, \qquad (\mathbf{12.3.62})$$

respectively. Solving for $R(r)$ first, we have that

$$R(r) = A_1 J_0(kr) + A_2 Y_0(kr), \qquad (\mathbf{12.3.63})$$

where $J_0(\ )$ and $Y_0(\ )$ denote Bessel functions of order zero and the first and second kind, respectively. The requirement that $u(r,z)$, and hence $R(r)$, is bounded as $r \to 0$ forces us to take $A_2 = 0$, leaving $R(r) = A_1 J_0(kr)$. From the equation for $Z(z)$, we conclude that

$$Z(z) = B_1 e^{kz} + B_2 e^{-kz}. \qquad (\mathbf{12.3.64})$$

Since $u(r,z)$, and hence $Z(z)$, must be bounded as $z \to \infty$, it follows that $B_1 = 0$, leaving $Z(z) = B_2 e^{-kz}$.

Presently our analysis has followed closely those for a finite domain. However, we have satisfied all of the boundary conditions and yet there is still *no* restriction on k. Consequently, we conclude that k is completely arbitrary and any product solution

$$u_k(r,z) = A_1 B_2 J_0(kr) e^{-kz} \qquad (\mathbf{12.3.65})$$

is a solution to our partial differential equation and satisfies the boundary conditions. From the principle of linear superposition, the most general solution equals the sum of *all* of the possible solutions or

$$u(r,z) = \int_0^\infty A(k) \, k \, J_0(kr) \, e^{-kz} \, dk, \qquad (\mathbf{12.3.66})$$

where we have written the arbitrary constant $A_1 B_2$ as $A(k)k$. Our final task remains to compute $A(k)$.

Before we can find $A(k)$, we must derive an intermediate result. If we define our Fourier transform in an appropriate manner, we can write the two-dimensional Fourier transform pair as

$$f(x,y) = \frac{1}{2\pi} \int_{-\infty}^{\infty} \int_{-\infty}^{\infty} F(k,\ell) \, e^{i(kx+\ell y)} \, dk \, d\ell, \qquad (\mathbf{12.3.67})$$

where

$$F(k,\ell) = \frac{1}{2\pi} \int_{-\infty}^{\infty} \int_{-\infty}^{\infty} f(x,y) \, e^{-i(kx+\ell y)} \, dx \, dy. \qquad (\mathbf{12.3.68})$$

Consider now the special case where $f(x,y)$ is only a function of $r = \sqrt{x^2 + y^2}$, so that $f(x,y) = g(r)$. Then, changing to polar coordinates through the substitution $x = r\cos(\theta)$, $y = r\sin(\theta)$, $k = \rho\cos(\varphi)$, and $\ell = \rho\sin(\varphi)$, we have that

$$kx + \ell y = r\rho[\cos(\theta)\cos(\varphi) + \sin(\theta)\sin(\varphi)] = r\rho\cos(\theta - \varphi), \qquad (\mathbf{12.3.69})$$

and

$$dA = dx\, dy = r\, dr\, d\theta. \qquad (\mathbf{12.3.70})$$

Therefore, the integral in (12.3.68) becomes

$$F(k, \ell) = \frac{1}{2\pi} \int_0^\infty \int_0^{2\pi} g(r)\, e^{-ir\rho\cos(\theta-\varphi)} r\, dr\, d\theta \qquad (\mathbf{12.3.71})$$

$$= \frac{1}{2\pi} \int_0^\infty r\, g(r) \left[\int_0^{2\pi} e^{-ir\rho\cos(\theta-\varphi)}\, d\theta \right] dr. \qquad (\mathbf{12.3.72})$$

If we introduce $\lambda = \theta - \varphi$, the integral

$$\int_0^{2\pi} e^{-ir\rho\cos(\theta-\varphi)}\, d\theta = \int_{-\varphi}^{2\pi-\varphi} e^{-ir\rho\cos(\lambda)}\, d\lambda \qquad (\mathbf{12.3.73})$$

$$= \int_0^{2\pi} e^{-ir\rho\cos(\lambda)}\, d\lambda \qquad (\mathbf{12.3.74})$$

$$= 2\pi J_0(\rho r). \qquad (\mathbf{12.3.75})$$

Integral (12.3.74) is equivalent to (12.3.73) because the integral of a periodic function over one full period is the same regardless of where the integration begins. Equation (12.3.75) follows from the integral definition of the Bessel function.[6] Therefore,

$$F(k, \ell) = \int_0^\infty r\, g(r)\, J_0(\rho r)\, dr. \qquad (\mathbf{12.3.76})$$

Finally, because (12.3.76) is clearly a function of $\rho = \sqrt{k^2 + \ell^2}$, $F(k, \ell) = G(\rho)$ and

$$G(\rho) = \int_0^\infty r\, g(r)\, J_0(\rho r)\, dr. \qquad (\mathbf{12.3.77})$$

Conversely, if we begin with (12.3.67), make the same substitution, and integrate over the $k\ell$ plane, we have that

$$f(x, y) = g(r) = \frac{1}{2\pi} \int_0^\infty \int_0^{2\pi} F(k, \ell)\, e^{ir\rho\cos(\theta-\varphi)} \rho\, d\rho\, d\varphi \qquad (\mathbf{12.3.78})$$

$$= \frac{1}{2\pi} \int_0^\infty \rho\, G(\rho) \left[\int_0^{2\pi} e^{ir\rho\cos(\theta-\varphi)}\, d\varphi \right] d\rho \qquad (\mathbf{12.3.79})$$

$$= \int_0^\infty \rho\, G(\rho)\, J_0(\rho r)\, d\rho. \qquad (\mathbf{12.3.80})$$

---

[6] Watson, G. N., 1966: *A Treatise on the Theory of Bessel Functions.* Cambridge University Press, §2.2, Equation 5.

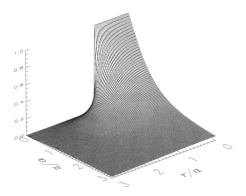

Figure 12.3.5: The axisymmetric potential $u(r,z)/u_0$ in the half-space $z > 0$ when $u(r,0) = u_0$ if $r < a$ and $u(r,0) = 0$ if $r > a$.

Thus, we obtain the result that if $\int_0^\infty |F(r)|\, dr$ exists, then

$$g(r) = \int_0^\infty \rho\, G(\rho)\, J_0(\rho r)\, d\rho, \qquad (12.3.81)$$

where

$$G(\rho) = \int_0^\infty r\, g(r)\, J_0(\rho r)\, dr. \qquad (12.3.82)$$

Taken together, (12.3.81) and (12.3.82) constitute the *Hankel transform pair for Bessel function of order 0*. The function $G(\rho)$ is called the Hankel transform of $g(r)$.

   Why did we introduce Hankel transforms? First, setting $z = 0$ in (12.3.66), we find that

$$u(r,0) = \int_0^\infty A(k)\, k\, J_0(kr)\, dk. \qquad (12.3.83)$$

If we now compare (12.3.83) with (12.3.81), we recognize that $A(k)$ is the Hankel transform of $u(r,0)$. Therefore,

$$A(k) = \int_0^\infty r\, u(r,0)\, J_0(kr)\, dr = u_0 \int_0^a r\, J_0(kr)\, dr \qquad (12.3.84)$$

$$= \frac{u_0}{k}\, r\, J_1(kr)\big|_0^a = \frac{au_0}{k}\, J_1(ka). \qquad (12.3.85)$$

Thus, the complete solution is

$$u(r,z) = au_0 \int_0^\infty J_1(ka)\, J_0(kr)\, e^{-kz}\, dk. \qquad (12.3.86)$$

Equation (12.3.86) is illustrated in Figure 12.3.5.

• Example 12.3.6: Mixed boundary-value problem

In all of our previous examples, the boundary condition along any specific boundary remained the same. In this example, we relax this condition and consider a *mixed boundary-value problem*.

Consider[7] the axisymmetric Laplace equation

$$\frac{1}{r}\frac{\partial}{\partial r}\left(r\frac{\partial u}{\partial r}\right) + \frac{\partial^2 u}{\partial z^2} = 0, \quad 0 \le r < \infty, \quad 0 < z < 1, \tag{12.3.87}$$

subject to the boundary conditions

$$\lim_{r\to 0}|u(r,z)| < \infty, \qquad \lim_{r\to\infty}|u(r,z)| < \infty, \qquad u(r,0) = 0, \tag{12.3.88}$$

and

$$\begin{cases} u(r,1) = 1, & 0 < r \le a, \\[2mm] u(r,1) + \dfrac{u_z(r,1)}{\sigma} = 1, & a < r < \infty. \end{cases} \tag{12.3.89}$$

The interesting aspect of this example is the mixture of boundary conditions along the boundary $z = 1$. For $r \le a$, we have a Dirichlet boundary condition which becomes a Robin boundary condition when $r > a$.

Our analysis begins as it did in the previous examples with separation of variables and a superposition of solutions. In the present case the solution is

$$u(r,z) = \frac{\sigma z}{1+\sigma} + \frac{a}{1+\sigma}\int_0^\infty A(k,a)\sinh(kz)J_0(kr)\,dk. \tag{12.3.90}$$

The first term on the right side of (12.3.90) arises from a separation constant that equals zero while the second term is the contribution from a negative separation constant. Note that (12.3.90) satisfies all of the boundary conditions given in (12.3.88). Substitution of (12.3.90) into (12.3.89) leads to the *dual integral equations*:

$$a\int_0^\infty A(k,a)\sinh(k)J_0(kr)\,dk = 1, \tag{12.3.91}$$

if $0 < r \le a$, and

$$\int_0^\infty A(k,a)\left[\sinh(k) + \frac{k\cosh(k)}{\sigma}\right]J_0(kr)\,dk = 0, \tag{12.3.92}$$

if $a < r < \infty$.

---

[7] Reprinted from *J. Theor. Biol.*, **81**, A. Nir and R. Pfeffer, Transport of macro-molecules across arterial wall in the presence of local endothial injury, 685–711, ©1979, with permission from Elsevier Science.

What sets this problem from the routine separation of variables is the solution of dual integral equations;[8] in general, they are very difficult to solve. The process usually begins with finding a solution that satisfies (12.3.92) via the orthogonality condition involving Bessel functions. This is the technique employed by Tranter[9] who proved that the dual integral equations:

$$\int_0^\infty G(\lambda)f(\lambda)J_0(\lambda a)\, d\lambda = g(a), \tag{12.3.93}$$

and

$$\int_0^\infty f(\lambda)J_0(\lambda a)\, d\lambda = 0 \tag{12.3.94}$$

have the solution

$$f(\lambda) = \lambda^{1-\kappa}\sum_{n=0}^\infty A_n J_{2m+\kappa}(\lambda), \tag{12.3.95}$$

if $G(\lambda)$ and $g(a)$ are known. The value of κ is chosen so that the difference $G(\lambda) - \lambda^{2\kappa-2}$ is fairly small. In the present case, $f(\lambda) = \sinh(\lambda)A(\lambda, a)$, $g(a) = 1$, and $G(\lambda) = 1 + \lambda\coth(\lambda)/\sigma$.

What is the value of κ here? Clearly we would like our solution to be valid for a wide range of σ. Because $G(\lambda) \to 1$ as $\sigma \to \infty$, a reasonable choice is $\kappa = 1$. Therefore, we take

$$\sinh(k)A(k, a) = \sum_{n=1}^\infty \frac{A_n}{1 + k\coth(k)/\sigma} J_{2n-1}(ka). \tag{12.3.96}$$

Our final task remains to find A_n.

We begin by writing

$$\frac{A_n}{1 + k\coth(k)/\sigma} J_{2n-1}(ka) = \sum_{m=1}^\infty B_{mn} J_{2m-1}(ka), \tag{12.3.97}$$

where B_{mn} depends only on a and σ. Multiplying (12.3.97) by $J_{2p-1}(ka)\, dk/k$ and integrating,

$$\int_0^\infty \frac{A_n}{1 + k\coth(k)/\sigma} J_{2n-1}(ka)\, J_{2p-1}(ka)\, \frac{dk}{k}$$
$$= \int_0^\infty \sum_{m=1}^\infty B_{nm} J_{2m-1}(ka)\, J_{2p-1}(ka)\, \frac{dk}{k}. \tag{12.3.98}$$

---

[8] The standard references is Sneddon, I. N., 1966: *Mixed Boundary Value Problems in Potential Theory*. Wiley, 283 pp.

[9] Tranter, C. J., 1950: On some dual integral equations occurring in potential problems with axial symmetry. *Quart. J. Mech. Appl. Math.*, **3**, 411–419.

Because[10]

$$\int_0^\infty J_{2n-1}(ka)\, J_{2p-1}(ka)\, \frac{dk}{k} = \frac{\delta_{mp}}{2(2m-1)}, \qquad (12.3.99)$$

where δ_{mp} is the Kronecker delta:

$$\delta_{mp} = \begin{cases} 1, & m = p, \\ 0, & m \neq p, \end{cases} \qquad (12.3.100)$$

(12.3.98) reduces to

$$A_n \int_0^\infty \frac{J_{2n-1}(ka) J_{2p-1}(ka)}{1 + k\coth(k)/\sigma}\, \frac{dk}{k} = \frac{B_{mn}}{2(2m-1)}. \qquad (12.3.101)$$

If we define

$$\int_0^\infty \frac{J_{2n-1}(ka)\, J_{2m-1}(ka)}{1 + k\coth(k)/\sigma}\, \frac{dk}{k} = S_{mn}, \qquad (12.3.102)$$

then we can rewrite (12.3.101) as

$$A_n S_{mn} = \frac{B_{mn}}{2(2m-1)}. \qquad (12.3.103)$$

Because[11]

$$a \int_0^\infty J_0(kr)\, J_{2m-1}(ka)\, dk = P_{m-1}\left(1 - \frac{2r^2}{a^2}\right), \qquad (12.3.104)$$

if $r < a$, where $P_m(\ )$ is the Legendre polynomial of order m, (12.3.91) can be rewritten

$$\sum_{n=1}^\infty \sum_{m=1}^\infty B_{mn} P_{m-1}\left(1 - \frac{2r^2}{a^2}\right) = 1. \qquad (12.3.105)$$

Equation (12.3.105) follows from the substitution of (12.3.96) into (12.3.91) and then using (12.3.104). Multiplying (12.3.105) by $P_{m-1}(\xi)\, d\xi$, integrating between -1 and 1, and using the orthogonality properties of the Legendre polynomial, we have that

$$\sum_{n=1}^\infty B_{mn} \int_{-1}^1 [P_{m-1}(\xi)]^2\, d\xi = \int_{-1}^1 P_{m-1}(\xi)\, d\xi \qquad (12.3.106)$$

$$= \int_{-1}^1 P_0(\xi) P_{m-1}(\xi)\, d\xi, \qquad (12.3.107)$$

---

[10] Gradshteyn, I. S., and I. M. Ryzhik, 1965: *Table of Integrals, Series, and Products.* Academic Press, §6.538, Formula 2.

[11] *Ibid.*, §6.512, Formula 4.

Table 12.3.1: The Convergence of the Coefficients A_n Given by (12.3.110) Where S_{mn} Has Nonzero Values for $1 \leq m, n \leq N$

N	A_1	A_2	A_3	A_4	A_5	A_6	A_7	A_8
1	2.9980							
2	3.1573	−1.7181						
3	3.2084	−2.0329	1.5978					
4	3.2300	−2.1562	1.9813	−1.4517				
5	3.2411	−2.2174	2.1548	−1.8631	1.3347			
6	3.2475	−2.2521	2.2495	−2.0670	1.7549	−1.2399		
7	3.2515	−2.2738	2.3073	−2.1862	1.9770	−1.6597	1.1620	
8	3.2542	−2.2882	2.3452	−2.2626	2.1133	−1.8925	1.5772	−1.0972

which shows that only $m = 1$ yields a nontrivial sum. Thus,

$$\sum_{n=1}^{\infty} B_{mn} = 2(2m - 1) \sum_{n=1}^{\infty} A_n S_{mn} = 0, \quad m \geq 2, \qquad (12.3.108)$$

and

$$\sum_{n=1}^{\infty} B_{1n} = 2 \sum_{n=1}^{\infty} A_n S_{1n} = 1, \qquad (12.3.109)$$

or

$$\sum_{n=1}^{\infty} S_{mn} A_n = \tfrac{1}{2} \delta_{m1}. \qquad (12.3.110)$$

Thus, we have reduced the problem to the solution of an infinite number of linear equations which yield A_n — a common occurrence in the solution of dual integral equations. Selecting some maximum value for n and m, say N, each term in the matrix S_{mn}, $1 \leq m, n \leq N$, is evaluated numerically for a given value of a and σ. By inverting (12.3.110), we obtain the coefficients A_n for $n = 1, \ldots, N$. Because we solved a truncated version (12.3.110), they will only be approximate. To find more accurate values, we can increase N by 1 and again invert (12.3.110). In addition to the new A_{N+1}, the previous coefficients will become more accurate. We can repeat this process of increasing N until the coefficients converge to their correct value. This is illustrated in Table 12.3.1 when $\sigma = a = 1$.

Once we have computed the coefficients A_n necessary for the desired accuracy, we use (12.3.96) to find $A(k, a)$ and then obtain $u(r, z)$ from (12.3.90) via numerical integration. Figure 12.3.6 illustrates the solution when $\sigma = 1$ and $a = 2$.

Mixed boundary-value problems over a finite domain can be solved in a

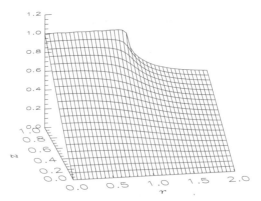

Figure 12.3.6: The solution of the axisymmetric Laplace's equation (12.3.87) with $u(r,0)$ $= 0$ and the mixed boundary condition (12.3.89). Here we have chosen $\sigma = 1$ and $a = 2$.

similar manner. Consider the partial differential equation[12]

$$\frac{\partial^2 u}{\partial r^2} + \frac{1}{r}\frac{\partial u}{\partial r} - \frac{u}{r^2} + \frac{\partial^2 u}{\partial z^2} = 0, \quad 0 \le r < a, \quad 0 < z < 1, \qquad (12.3.111)$$

subject to the boundary conditions

$$\lim_{r \to 0} |u(r,z)| < \infty, \quad u(a,z) = 0, \quad 0 \le z \le 1, \qquad (12.3.112)$$

and

$$u(r,0) = 0, \quad 0 \le r \le a, \quad \begin{cases} u(r,1) = r, & 0 \le r < 1, \\ u_z(r,1) = 0, & 1 < r \le a. \end{cases} \qquad (12.3.113)$$

We begin by solving (12.3.111) via separation of variables. This yields

$$u(r,z) = \sum_{n=1}^{\infty} A_n \sinh(k_n z)\, J_1(k_n r), \qquad (12.3.114)$$

where k_n is the nth root of $J_1(ka) = 0$. Note that (12.3.114) satisfies all of the boundary conditions except those along $z = 1$. Substituting (12.3.114) into (12.3.113), we find that

$$\sum_{n=1}^{\infty} A_n \sinh(k_n)\, J_1(k_n r) = r, \quad 0 \le r < 1, \qquad (12.3.115)$$

---

[12] Reprinted from *Chem. Engng. Sci.*, **46**, J. S. Vrentas, D. C. Venerus, and C. M. Vrentes, An exact analysis of reservoir effects for rotational viscometers, 33–37, ©1991, with permission from Elsevier Science.

Other examples include:

Sherwood, J. D., and H. A. Stone, 1997: Added mass of a disc accelerating within a pipe. *Phys. Fluids*, **9**, 3141–3148.

Galceran, J., J. Cecília, E. Companys, J. Salvador, and J. Puy, 2000: Analytical expressions for feedback currents at the scanning electrochemical microscope. *J. Phys. Chem. B*, **104**, 7993–8000.

and

$$\sum_{n=1}^{\infty} k_n A_n \cosh(k_n) J_1(k_n r) = 0, \quad 1 < r \le a. \tag{12.3.116}$$

Equations (12.3.115) and (12.3.116) show that in place of dual integral equations, we now have *dual Fourier-Bessel series*. Cooke and Tranter[13] have shown that the dual Fourier-Bessel series

$$\sum_{n=1}^{\infty} a_n J_\nu(k_n r) = 0, \quad 1 < r < a, \quad -1 < \nu, \tag{12.3.117}$$

where $J_\nu(k_n a) = 0$, will be automatically satisfied if

$$k_n^{1+p/2} J_{\nu+1}^2(k_n a) a_n = \sum_{m=0}^{\infty} b_m J_{\nu+2m+1+p/2}(k_m), \tag{12.3.118}$$

where $|p| \le 1$. Because $a_n = k_n A_n \cosh(k_n)$ and $\nu = 1$ here, A_n is given by

$$k_n^2 \cosh(k_n) J_2^2(k_n a) A_n = \sum_{m=1}^{\infty} B_m J_{2m}(k_n), \tag{12.3.119}$$

if we take $p = 0$.

Substitution of (12.3.119) into (12.3.115) gives

$$\sum_{m=1}^{\infty} B_m \sum_{n=1}^{\infty} \frac{\sinh(k_n) J_{2m}(k_n) J_1(k_n r)}{k_n^2 \cosh(k_n) J_2^2(k_n a)} = r. \tag{12.3.120}$$

Multiplying both sides of (12.3.120) by $r J_1(k_p r)\, dr$, $p = 1, 2, 3, \ldots$, and integrating from 0 to 1, we find that

$$\sum_{m=1}^{\infty} B_m \sum_{n=1}^{\infty} \frac{\sinh(k_n) J_{2m}(k_n) Q_{pn}}{k_n^2 \cosh(k_n) J_2^2(k_n a)} = \int_0^1 r^2 J_p(k_p r)\, dr, \tag{12.3.121}$$

where

$$Q_{pn} = \int_0^1 J_1(k_p r) J_1(k_n r)\, r\, dr \tag{12.3.122}$$

$$= \begin{cases} \frac{k_p J_1(k_n) J_0(k_p) - k_n J_1(k_p) J_0(k_n)}{k_n^2 - k_p^2}, & n \ne p, \\[2mm] \frac{J_1^2(k_p) - J_0(k_p) J_2(k_p)}{2}, & n = p. \end{cases} \tag{12.3.123}$$

---

[13] Cooke, J. C., and C. J. Tranter, 1959: Dual Fourier-Bessel series. *Quart. J. Mech. Appl. Math.*, **12**, 379–386.

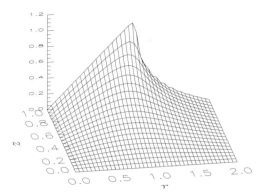

Figure 12.3.7: The solution of (12.3.111) which satisfies the boundary condition (12.3.112) and the mixed boundary condition (12.3.113). Here we have chosen $a = 2$.

Carrying out the integration, (12.3.120) yields the infinite set of equations

$$\sum_{m=1}^{\infty} M_{pm} B_m = \frac{J_2(k_p)}{k_p}, \qquad (12.3.124)$$

where

$$M_{pm} = \sum_{n=1}^{\infty} \frac{\sinh(k_n) J_{2m}(k_n) Q_{pn}}{k_n^2 \cosh(k_n) J_2^2(k_n a)}. \qquad (12.3.125)$$

Once again, we compute B_m by truncating (12.3.124) to M terms and inverting the systems of equations. Increasing the value of M yields more accurate results. Once we have B_m, we use (12.3.119) to find A_n. Finally, $u(r, z)$ follows from (12.3.114). Figure 12.3.7 illustrates $u(r, z)$ when $a = 2$.

• **Example 12.3.7**

Let us find the potential at any point P within a conducting sphere of radius a. At the surface, the potential is held at V_0 in the hemisphere $0 < \theta < \pi/2$, and $-V_0$ for $\pi/2 < \theta < \pi$.

Laplace's equation in spherical coordinates is

$$\frac{\partial}{\partial r}\left(r^2 \frac{\partial u}{\partial r}\right) + \frac{1}{\sin(\theta)} \frac{\partial}{\partial \theta}\left[\sin(\theta) \frac{\partial u}{\partial \theta}\right] = 0, \quad 0 \le r < a, \quad 0 \le \theta \le \pi.$$
$$(12.3.126)$$

To solve (12.3.126) we use the separation of variables $u(r, \theta) = R(r)\Theta(\theta)$. Substituting into (12.3.126), we have that

$$\frac{1}{R}\frac{d}{dr}\left(r^2 \frac{dR}{dr}\right) = -\frac{1}{\sin(\theta)\Theta}\frac{d}{d\theta}\left[\sin(\theta)\frac{d\Theta}{d\theta}\right] = k^2, \qquad (12.3.127)$$

or

$$r^2 R'' + 2r R' - k^2 R = 0, \qquad (12.3.128)$$

and

$$\frac{1}{\sin(\theta)} \frac{d}{d\theta} \left[\sin(\theta) \frac{d\Theta}{d\theta} \right] + k^2 \Theta = 0. \tag{12.3.129}$$

A common substitution replaces θ with $\mu = \cos(\theta)$. Then, as θ varies from 0 to π, μ varies from 1 to -1. With this substitution (12.3.129) becomes

$$\frac{d}{d\mu} \left[(1 - \mu^2) \frac{d\Theta}{d\mu} \right] + k^2 \Theta = 0. \tag{12.3.130}$$

This is Legendre's equation which we examined in §9.4. Consequently, because the solution must remain finite at the poles, $k^2 = n(n+1)$, and

$$\Theta_n(\theta) = P_n(\mu) = P_n[\cos(\theta)], \tag{12.3.131}$$

where $n = 0, 1, 2, 3, \ldots$.

Turning to (12.3.128), this equation is the equidimensional or Euler-Cauchy linear differential equation. One method of solving this equation consists of introducing a new independent variable s so that $r = e^s$, or $s = \ln(r)$. Because

$$\frac{d}{dr} = \frac{ds}{dr} \frac{d}{ds} = e^{-s} \frac{d}{ds}, \tag{12.3.132}$$

it follows that

$$\frac{d^2}{dr^2} = \frac{d}{dr} \left(e^{-s} \frac{d}{ds} \right) = e^{-s} \frac{d}{ds} \left(e^{-s} \frac{d}{ds} \right) = e^{-2s} \left(\frac{d^2}{ds^2} - \frac{d}{ds} \right). \tag{12.3.133}$$

Substituting into (12.3.128),

$$\frac{d^2 R_n}{ds^2} + \frac{dR_n}{ds} - n(n+1)R_n = 0. \tag{12.3.134}$$

Equation (12.3.134) is a second-order, constant coefficient ordinary differential equation which has the solution

$$R_n(s) = C_n e^{ns} + D_n e^{-(n+1)s} \tag{12.3.135}$$

$$= C_n \exp[n \ln(r)] + D_n \exp[-(n+1)\ln(r)] \tag{12.3.136}$$

$$= C_n \exp[\ln(r^n)] + D_n \exp[\ln(r^{-1-n})] \tag{12.3.137}$$

$$= C_n r^n + D_n r^{-1-n}. \tag{12.3.138}$$

A more convenient form of the solution is

$$R_n(r) = A_n \left(\frac{r}{a} \right)^n + B_n \left(\frac{r}{a} \right)^{-1-n}, \tag{12.3.139}$$

where $A_n = a^n C_n$ and $B_n = D_n/a^{n+1}$. We introduced the constant a, the radius of the sphere, to simplify future calculations.

Using the results from (12.3.131) and (12.3.139), the solution to Laplace's equation in axisymmetric problems is

$$u(r, \theta) = \sum_{n=0}^{\infty} \left[A_n \left(\frac{r}{a} \right)^n + B_n \left(\frac{r}{a} \right)^{-1-n} \right] P_n[\cos(\theta)]. \qquad (\mathbf{12.3.140})$$

In our particular problem we must take $B_n = 0$ because the solution becomes infinite at $r = 0$ otherwise. If the problem had involved the domain $a < r < \infty$, then $A_n = 0$ because the potential must remain finite as $r \to \infty$.

Finally, we must evaluate A_n. Finding the potential at the surface,

$$u(a, \mu) = \sum_{n=0}^{\infty} A_n P_n(\mu) = \begin{cases} V_0, & 0 < \mu \leq 1, \\ -V_0, & -1 \leq \mu < 0. \end{cases} \qquad (\mathbf{12.3.141})$$

Upon examining (12.3.141), it is merely an expansion in Legendre polynomials of the function

$$f(\mu) = \begin{cases} V_0, & 0 < \mu \leq 1, \\ -V_0, & -1 \leq \mu < 0. \end{cases} \qquad (\mathbf{12.3.142})$$

Consequently, from (12.3.142),

$$A_n = \frac{2n + 1}{2} \int_{-1}^{1} f(\mu) P_n(\mu) \, d\mu. \qquad (\mathbf{12.3.143})$$

Because $f(\mu)$ is an odd function, $A_n = 0$ if n is even. When n is odd, however,

$$A_n = (2n + 1) \int_{0}^{1} V_0 P_n(\mu) \, d\mu. \qquad (\mathbf{12.3.144})$$

We can further simplify (12.3.144) by using the relationship that

$$\int_{x}^{1} P_n(t) \, dt = \frac{1}{2n + 1} \left[P_{n-1}(x) - P_{n+1}(x) \right], \qquad (\mathbf{12.3.145})$$

where $n \geq 1$. In our problem, then,

$$A_n = \begin{cases} V_0[P_{n-1}(0) - P_{n+1}(0)], & n \text{ odd}, \\ 0, & n \text{ even}. \end{cases} \qquad (\mathbf{12.3.146})$$

The first few terms are $A_1 = 3V_0/2$, $A_3 = -7V_0/8$, and $A_5 = 11V_0/16$.

Figure 12.3.8 illustrates our solution. It was created using the MATLAB script

```
clear
N = 51; dr = 0.05; dtheta = pi / 15;
% compute grid and set solution equal to zero
r = [0:dr:1]; theta = [0:dtheta:2*pi];
```

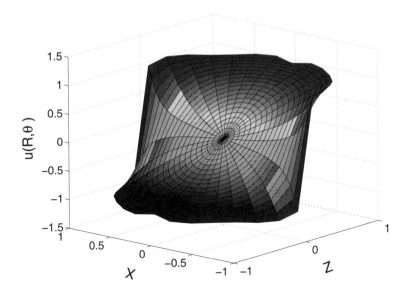

Figure 12.3.8: Electrostatic potential within a conducting sphere when the upper hemispheric surface has the potential 1 and the lower surface has the potential −1.

```
mu = cos(theta); Z = r' * mu;
for L = 1:2
if L == 1 X = r' * sin(theta);
    else X = -r' * sin(theta); end
u = zeros(size(X));
% compute solution from (12.3.140)
rfactor = r;
for n = 1:2:N
A = legendre(n-1,0); B = legendre(n+1,0); coeff = A(1) - B(1);
C = legendre(n,mu); Theta = C(1,:);
u = u + coeff * rfactor' * Theta;
rfactor = rfactor .* r .* r;
end
surf(Z,X,u); hold on; end
xlabel('Z','Fontsize',20); ylabel('X','Fontsize',20)
zlabel('u(R,\theta )','Fontsize',20);
```

Here we have the convergence of the equipotentials along the equator and at the surface. The slow rate at which the coefficients are approaching zero suggests that the solution suffers from Gibbs phenomena along the surface.

● **Example 12.3.8**

We now find the steady-state temperature field within a metallic sphere of radius a, which we place in direct sunlight and allow to radiatively cool.

This classic problem, first solved by Rayleigh,[14] requires the use of spherical coordinates with its origin at the center of sphere and its z-axis pointing toward the sun. With this choice for the coordinate system, the incident sunlight is

$$D(\theta) = \begin{cases} D(0)\cos(\theta), & 0 \le \theta \le \pi/2, \\ 0, & \pi/2 \le \theta \le \pi. \end{cases} \tag{12.3.147}$$

If heat dissipation takes place at the surface $r = a$ according to Newton's law of cooling and the temperature of the surrounding medium is zero, the solar heat absorbed by the surface dA must balance the Newtonian cooling at the surface plus the energy absorbed into the sphere's interior. This physical relationship is

$$(1 - \rho)D(\theta)\, dA = \epsilon u(a, \theta)\, dA + \kappa \frac{\partial u(a, \theta)}{\partial r}\, dA, \tag{12.3.148}$$

where ρ is the reflectance of the surface (the albedo), ϵ is the surface conductance or coefficient of surface heat transfer, and κ is the thermal conductivity. Simplifying (12.3.148), we have that

$$\frac{\partial u(a, \theta)}{\partial r} = \frac{1 - \rho}{\kappa} D(\theta) - \frac{\epsilon}{\kappa} u(a, \theta) \tag{12.3.149}$$

for $r = a$.

If the sphere has reached thermal equilibrium, Laplace's equation describes the temperature field within the sphere. In the previous example, we showed that the solution to Laplace's equation in axisymmetric problems is

$$u(r, \theta) = \sum_{n=0}^{\infty} \left[A_n \left(\frac{r}{a} \right)^n + B_n \left(\frac{r}{a} \right)^{-1-n} \right] P_n[\cos(\theta)]. \tag{12.3.150}$$

In this problem, $B_n = 0$ because the solution would become infinite at $r = 0$ otherwise. Therefore,

$$u(r, \theta) = \sum_{n=0}^{\infty} A_n \left(\frac{r}{a} \right)^n P_n[\cos(\theta)]. \tag{12.3.151}$$

Differentiation gives

$$\frac{\partial u}{\partial r} = \sum_{n=0}^{\infty} A_n \frac{n r^{n-1}}{a^n} P_n[\cos(\theta)]. \tag{12.3.152}$$

---

[14] Rayleigh, J. W., 1870: On the values of the integral $\int_0^1 Q_n Q_{n'}\, d\mu$, Q_n, $Q_{n'}$ being Laplace's coefficients of the orders n, n', with application to the theory of radiation. *Philos. Trans. R. Soc. London, Ser. A*, **160**, 579–590.

Substituting into the boundary condition leads to

$$\sum_{n=0}^{\infty} A_n \left(\frac{n}{a} + \frac{\epsilon}{\kappa}\right) P_n[\cos(\theta)] = \left(\frac{1-\rho}{\kappa}\right) D(\theta), \qquad (12.3.153)$$

or

$$D(\mu) = \sum_{n=0}^{\infty} \left[\frac{n\kappa + \epsilon a}{a(1-\rho)}\right] A_n P_n(\mu) = \sum_{n=0}^{\infty} C_n P_n(\mu), \qquad (12.3.154)$$

where

$$C_n = \left[\frac{n\kappa + \epsilon a}{a(1-\rho)}\right] A_n, \quad \text{and} \quad \mu = \cos(\theta). \qquad (12.3.155)$$

We determine the coefficients by

$$C_n = \frac{2n+1}{2} \int_{-1}^{1} D(\mu) P_n(\mu)\, d\mu = \frac{2n+1}{2} D(0) \int_{0}^{1} \mu P_n(\mu)\, d\mu. \qquad (12.3.156)$$

Evaluation of the first few coefficients gives

$$A_0 = \frac{(1-\rho)D(0)}{4\epsilon}, \quad A_1 = \frac{a(1-\rho)D(0)}{2(\kappa + \epsilon a)}, \quad A_2 = \frac{5a(1-\rho)D(0)}{16(2\kappa + \epsilon a)}, \qquad (12.3.157)$$

$$A_3 = 0, \quad A_4 = -\frac{3a(1-\rho)D(0)}{32(4\kappa + \epsilon a)}, \quad A_5 = 0 \qquad (12.3.158)$$

$$A_6 = \frac{13a(1-\rho)D(0)}{256(6\kappa + \epsilon a)}, \quad A_7 = 0, \quad A_8 = -\frac{17a(1-\rho)D(0)}{512(8\kappa + \epsilon a)}, \qquad (12.3.159)$$

$$A_9 = 0, \quad \text{and} \quad A_{10} = \frac{49a(1-\rho)D(0)}{2048(10\kappa + \epsilon a)}. \qquad (12.3.160)$$

Figure 12.3.9 illustrates the temperature field within the sphere with $D(0) = 1200$ W/m², $\kappa = 45$ W/m K, $\epsilon = 5$ W/m² K, $\rho = 0$, and $a = 0.1$ m. This corresponds to a cast iron sphere with blackened surface in sunlight. This figure was created by the MATLAB script

```
clear
dr = 0.05; dtheta = pi / 15;
D_0 = 1200; kappa = 45; epsilon = 5; rho = 0; a = 0.1;
% compute grid and set solution equal to zero
r = [0:dr:1]; theta = [0:dtheta:pi];
mu = cos(theta); Z = r' * mu;
aaaa = (1-rho) * D_0 / ( 4 * epsilon);
aa(1) = a * (1-rho) * D_0 / ( 2 * ( kappa+epsilon*a));
aa(2) = 5 * a * (1-rho) * D_0 / ( 16 * (2*kappa+epsilon*a));
aa(3) = 0;
aa(4) = - 3 * a * (1-rho) * D_0 / ( 32 * (4*kappa+epsilon*a));
```

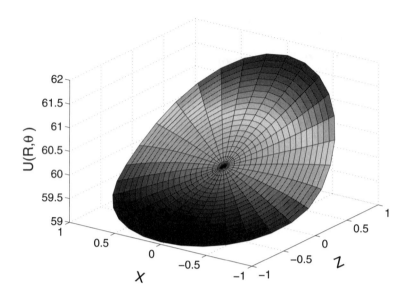

Figure 12.3.9: The difference (in °C) between the temperature field within a blackened iron surface of radius 0.1 m and the surrounding medium when we heat the surface by sunlight and allow it to radiatively cool.

```
aa(5) = 0;
aa(6) = 13 * a * (1-rho) * D_0 / ( 256 * (6*kappa+epsilon*a));
aa(7) = 0;
aa(8) = -17 * a * (1-rho) * D_0 / ( 512 * (8*kappa+epsilon*a));
aa(9) = 0;
aa(10) = 49 * a * (1-rho) * D_0 / (2048 * (10*kappa+epsilon*a));
for L = 1:2
if L == 1 X = r' * sin(theta);
   else X = -r' * sin(theta); end
u = aaaa * ones(size(X));
rfactor = r;
for n = 1:10
A = legendre(n,mu); Theta = A(1,:);
u = u + aa(n) * rfactor' * Theta;
rfactor = rfactor .* r;
end
surf(Z,X,u); hold on; end
xlabel('Z','Fontsize',20); ylabel('X','Fontsize',20);
zlabel('U(R,\theta )','Fontsize',20);
```

The temperature is quite warm with the highest temperature located at the position where the solar radiation is largest; the coolest temperatures are located in the shadow region.

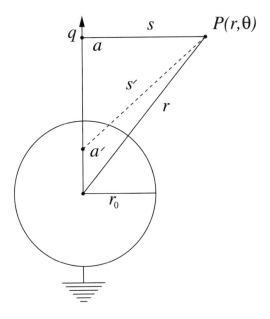

Figure 12.3.10: Point charge $+q$ in the presence of a grounded conducting sphere.

- **Example 12.3.9**

In this example we find the potential at any point P exterior to a conducting, grounded sphere centered at $z = 0$ after we place a point charge $+q$ at $z = a$ on the z-axis. See Figure 12.3.10. From the principle of linear superposition, the total potential $u(r, \theta)$ equals the sum of the potential from the point charge and the potential $v(r, \theta)$ due to the induced charge on the sphere

$$u(r, \theta) = \frac{q}{s} + v(r, \theta). \tag{12.3.161}$$

In common with the first term q/s, $v(r, \theta)$ must be a solution of Laplace's equation. In Example 12.3.7 we showed that the general solution to Laplace's equation in axisymmetric problems is

$$v(r, \theta) = \sum_{n=0}^{\infty} \left[A_n \left(\frac{r}{r_0} \right)^n + B_n \left(\frac{r}{r_0} \right)^{-1-n} \right] P_n[\cos(\theta)]. \tag{12.3.162}$$

Because the solutions must be valid *anywhere* outside of the sphere, $A_n = 0$; otherwise, the solution would not remain finite as $r \to \infty$. Hence,

$$v(r, \theta) = \sum_{n=0}^{\infty} B_n \left(\frac{r}{r_0} \right)^{-1-n} P_n[\cos(\theta)]. \tag{12.3.163}$$

We determine the coefficient B_n by the condition that $u(r_0, \theta) = 0$, or

$$\frac{q}{s}\bigg|_{\text{on sphere}} + \sum_{n=0}^{\infty} B_n P_n[\cos(\theta)] = 0. \tag{12.3.164}$$

We need to expand the first term on the left side of (12.3.164) in terms of Legendre polynomials. From the law of cosines,

$$s = \sqrt{r^2 + a^2 - 2ar\cos(\theta)}. \tag{12.3.165}$$

Consequently, if $a > r$, then

$$\frac{1}{s} = \frac{1}{a}\left[1 - 2\cos(\theta)\frac{r}{a} + \left(\frac{r}{a}\right)^2\right]^{-1/2}. \tag{12.3.166}$$

In §9.4, we showed that

$$(1 - 2xz + z^2)^{-1/2} = \sum_{n=0}^{\infty} P_n(x)z^n. \tag{12.3.167}$$

Therefore,

$$\frac{1}{s} = \frac{1}{a}\sum_{n=0}^{\infty} P_n[\cos(\theta)]\left(\frac{r}{a}\right)^n. \tag{12.3.168}$$

From (12.3.164),

$$\sum_{n=0}^{\infty}\left[\frac{q}{a}\left(\frac{r_0}{a}\right)^n + B_n\right]P_n[\cos(\theta)] = 0. \tag{12.3.169}$$

We can only satisfy (12.3.169) if the square-bracketed term vanishes identically so that

$$B_n = -\frac{q}{a}\left(\frac{r_0}{a}\right)^n. \tag{12.3.170}$$

On substituting (12.3.170) back into (12.3.163),

$$v(r, \theta) = -\frac{qr_0}{ra}\sum_{n=0}^{\infty}\left(\frac{r_0^2}{ar}\right)^n P_n[\cos(\theta)]. \tag{12.3.171}$$

The physical interpretation of (12.3.171) is as follows: Consider a point, such as a' (see Figure 12.3.10) on the z-axis. If $r > a'$, the Legendre expansion of $1/s'$ is

$$\frac{1}{s'} = \frac{1}{r}\sum_{n=0}^{\infty} P_n[\cos(\theta)]\left(\frac{a'}{r}\right)^n, \quad r > a'. \tag{12.3.172}$$

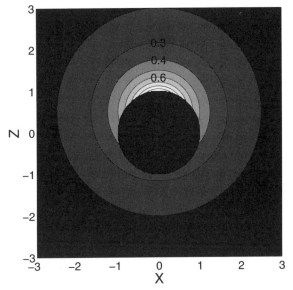

Figure 12.3.11: Electrostatic potential outside of a grounded conducting sphere in the presence of a point charge located at $a/r_0 = 2$. Contours are in units of $-q/r_0$.

Using (12.3.172), we can rewrite (12.3.171) as

$$v(r, \theta) = -\frac{qr_0}{as'}, \qquad (\mathbf{12.3.173})$$

if we set $a' = r_0^2/a$. Our final result is then

$$u(r, \theta) = \frac{q}{s} - \frac{q'}{s'}, \qquad (\mathbf{12.3.174})$$

provided that q' equals r_0q/a. In other words, when we place a grounded conducting sphere near a point charge $+q$, it changes the potential in the same manner as would a point charge of the opposite sign and magnitude $q' = r_0q/a$, placed at the point $a' = r_0^2/a$. The charge q' is the *image* of q.

Figure 12.3.11 illustrates the solution (12.3.171) and was created using the MATLAB script

```
clear
a_over_r0 = 2;
% set up x-z array
dx = 0.02; x = -3:dx:3; dz = 0.02; z = -3:dz:3;
u = 1000 * zeros(length(x),length(z));
X = x' * ones(1,length(z)); Z = ones(length(x),1) * z;
% compute r and theta
rr = sqrt(X .* X + Z .* Z);
theta = atan2(X,Z);
```

```
% find the potential
r_over_aprime = a_over_r0 * rr;
s = 1 + r_over_aprime .* r_over_aprime ...
    - 2 * r_over_aprime .* cos(theta);
for j = 1:length(z); for i = 1:length(x);
if rr(i,j) >= 1; u(i,j) = 1 ./ sqrt(s(i,j)); end;
end; end
% plot the solution
[cs,h] = contourf(X,Z,u); colormap(hot); brighten(hot,0.5);
axis square; clabel(cs,h,'manual','Fontsize',16);
xlabel('X','Fontsize',20); ylabel('Z','Fontsize',20);
```

Because the charge is located directly above the sphere, the electrostatic potential for any fixed r is largest at the point $\theta = 0$ and weakest at $\theta = \pi$.

• Example 12.3.10: Poisson's integral formula

In this example we find the solution to Laplace's equation within a unit disc. The problem can be posed as

$$\frac{\partial^2 u}{\partial r^2} + \frac{1}{r}\frac{\partial u}{\partial r} + \frac{1}{r^2}\frac{\partial^2 u}{\partial \varphi^2} = 0, \quad 0 \le r < 1, \quad 0 \le \varphi \le 2\pi, \qquad (12.3.175)$$

with the boundary condition $u(1,\varphi) = f(\varphi)$.

We begin by assuming the separable solution $u(r,\varphi) = R(r)\Phi(\varphi)$ so that

$$\frac{r^2 R'' + rR'}{R} = -\frac{\Phi''}{\Phi} = k^2. \qquad (12.3.176)$$

The solution to $\Phi'' + k^2\Phi = 0$ is

$$\Phi(\varphi) = A\cos(k\varphi) + B\sin(k\varphi). \qquad (12.3.177)$$

The solution to $R(r)$ is

$$R(r) = Cr^k + Dr^{-k}. \qquad (12.3.178)$$

Because the solution must be bounded for all r and periodic in φ, we must take $D = 0$ and $k = n$, where $n = 0, 1, 2, 3, \ldots$. Then, the most general solution is

$$u(r,\varphi) = \tfrac{1}{2}a_0 + \sum_{n=1}^{\infty} [a_n \cos(n\varphi) + b_n \sin(n\varphi)]\, r^n, \qquad (12.3.179)$$

where a_n and b_n are chosen to satisfy

$$u(1,\varphi) = f(\varphi) = \tfrac{1}{2}a_0 + \sum_{n=1}^{\infty} a_n \cos(n\varphi) + b_n \sin(n\varphi). \qquad (12.3.180)$$

Because

$$a_n = \frac{1}{\pi} \int_{-\pi}^{\pi} f(\theta) \cos(n\theta)\, d\theta, \quad b_n = \frac{1}{\pi} \int_{-\pi}^{\pi} f(\theta) \sin(n\theta)\, d\theta, \qquad \textbf{(12.3.181)}$$

we may write $u(r, \varphi)$ as

$$u(r, \varphi) = \frac{1}{\pi} \int_{-\pi}^{\pi} f(\theta) \left\{ \tfrac{1}{2} + \sum_{n=1}^{\infty} r^n \cos[n(\theta - \varphi)] \right\} d\theta. \qquad \textbf{(12.3.182)}$$

If we let $\alpha = \theta - \varphi$, and $z = r[\cos(\alpha) + i\sin(\alpha)]$, then

$$\sum_{n=0}^{\infty} r^n \cos(n\alpha) = \mathrm{Re}\left(\sum_{n=0}^{\infty} z^n \right) = \mathrm{Re}\left(\frac{1}{1-z} \right) \qquad \textbf{(12.3.183)}$$

$$= \mathrm{Re}\left[\frac{1}{1 - r\cos(\alpha) - ir\sin(\alpha)} \right] \qquad \textbf{(12.3.184)}$$

$$= \mathrm{Re}\left[\frac{1 - r\cos(\alpha) + ir\sin(\alpha)}{1 - 2r\cos(\alpha) + r^2} \right] \qquad \textbf{(12.3.185)}$$

for all r such that $|r| < 1$. Consequently,

$$\sum_{n=0}^{\infty} r^n \cos(n\alpha) = \frac{1 - r\cos(\alpha)}{1 - 2r\cos(\alpha) + r^2} \qquad \textbf{(12.3.186)}$$

$$\frac{1}{2} + \sum_{n=1}^{\infty} r^n \cos(n\alpha) = \frac{1 - r\cos(\alpha)}{1 - 2r\cos(\alpha) + r^2} - \frac{1}{2} \qquad \textbf{(12.3.187)}$$

$$= \frac{1}{2} \frac{1 - r^2}{1 - 2r\cos(\alpha) + r^2}. \qquad \textbf{(12.3.188)}$$

Substituting (12.3.188) into (12.3.182), we finally have that

$$u(r, \varphi) = \frac{1}{2\pi} \int_{-\pi}^{\pi} f(\theta) \frac{1 - r^2}{1 - 2r\cos(\theta - \varphi) + r^2}\, d\theta. \qquad \textbf{(12.3.189)}$$

This solution to Laplace's equation within the unit circle is referred to as *Poisson's integral formula*.[15]

---

[15] Poisson, S. D., 1820: Mémoire sur la manière d'exprimer les fonctions par des séries de quantités périodiques, et sur l'usage de cette transformation dans la résolution de différens problèmes. *J. École Polytech.*, **18**, 417–489.

Problems

Solve Laplace's equation over the rectangular region $0 < x < a, 0 < y < b$ with the following boundary conditions. Illustrate your solution using MATLAB.

1. $u(x,0) = u(x,b) = u(a,y) = 0, u(0,y) = 1$

2. $u(x,0) = u(0,y) = u(a,y) = 0, u(x,b) = x$

3. $u(x,0) = u(0,y) = u(a,y) = 0, u(x,b) = x - a$

4. $u(x,0) = u(0,y) = u(a,y) = 0,$
$$u(x,b) = \begin{cases} 2x/a, & 0 < x < a/2 \\ 2(a-x)/a, & a/2 < x < a \end{cases}$$

5. $u_x(0,y) = u(a,y) = u(x,0) = 0, u(x,b) = 1$

6. $u_y(x,0) = u(x,b) = u(a,y) = 0, u(0,y) = 1$

7. $u_y(x,0) = u_y(x,b) = 0, u(0,y) = u(a,y) = 1$

8. $u_x(a,y) = u_y(x,b) = 0, u(0,y) = u(x,0) = 1$

9. $u_y(x,0) = u(x,b) = 0, u(0,y) = u(a,y) = 1$

10. $u(a,y) = u(x,b) = 0, u(0,y) = u(x,0) = 1$

11. $u_x(0,y) = 0, u(a,y) = u(x,0) = u(x,b) = 1$

12. $u_x(0,y) = u_x(a,y) = 0, u(x,b) = u_1,$
$$u(x,0) = \begin{cases} f(x), & 0 < x < \alpha \\ 0, & \alpha < x < a \end{cases}$$

13. Variations in the earth's surface temperature can arise as a result of topographic undulations and the altitude dependence of the atmospheric temperature. These variations, in turn, affect the temperature within the solid earth. To show this, solve Laplace's equation with the surface boundary condition that
$$u(x,0) = T_0 + \Delta T \cos(2\pi x/\lambda),$$

where λ is the wavelength of the spatial temperature variation. What must be the condition on $u(x,y)$ as we go towards the center of the earth (i.e., $y \to \infty$)?

14. Tóth[16] generalized his earlier analysis of groundwater in an aquifer when the water table follows the topography. Find the groundwater potential if it varies as $u(x, z_0) = g[z_0 + cx + a\sin(bx)]$ at the surface $y = z_0$, while $u_x(0, y) = u_x(L, y) = u_y(x, 0) = 0$, where g is the acceleration due to gravity. Assume that $bL \neq n\pi$, where $n = 1, 2, 3, \ldots$.

15. Solve

$$\frac{\partial^2 u}{\partial r^2} + \frac{1}{r}\frac{\partial u}{\partial r} + \frac{\partial^2 u}{\partial z^2} = 0, \quad 0 \leq r < a, \quad -L < z < L,$$

with

$$u(a, z) = 0, \quad \text{and} \quad \frac{\partial u(r, -L)}{\partial z} = \frac{\partial u(r, L)}{\partial z} = 1.$$

16. During their study of the role that diffusion plays in equalizing gas concentrations within that portion of the lung that is connnected to terminal bronchioles, Chang *et al.*[17] solved Laplace's equation in cylindrical coordinates

$$\frac{\partial^2 u}{\partial r^2} + \frac{1}{r}\frac{\partial u}{\partial r} + \frac{\partial^2 u}{\partial z^2} = 0, \quad 0 \leq r < b, \quad -L < z < L,$$

subject to the boundary conditions that

$$\lim_{r \to 0} |u(r, z)| < \infty, \quad \frac{\partial u(b, z)}{\partial r} = 0, \quad -L < z < L,$$

and

$$\frac{\partial u(r, -L)}{\partial z} = \frac{\partial u(r, L)}{\partial z} = \begin{cases} A, & 0 \leq r < a, \\ 0, & a < r < b. \end{cases}$$

What should they have found?

17. Solve[18]

$$\frac{\partial^2 u}{\partial r^2} + \frac{1}{r}\frac{\partial u}{\partial r} + \frac{\partial^2 u}{\partial z^2} = 0, \quad 0 \leq r < b, \quad 0 < z < L,$$

with the boundary conditions

$$\lim_{r \to 0} |u(r, z)| < \infty, \quad \frac{\partial u(b, z)}{\partial r} = 0, \quad 0 \leq z \leq L,$$

---

[16] Tóth, J., *J. Geophys. Res.*, **68**, 4795–4812, 1963, copyright by the American Geophysical Union.

[17] Reprinted from *Math. Biosci.*, **29**, D. B. Chang, S. M. Lewis, and A. C. Young, A theoretical discussion of diffusion and convection in the lung, 331–349, ©1976, with permission from Elsevier Science.

[18] Reprinted from *Math. Biosci.*, **1**, K. H. Keller and T. R. Stein, A two-dimensional analysis of porous membrane transfer, 421–437, ©1967, with permission from Elsevier Science.

$$u(r, L) = A, \qquad 0 \le r \le b,$$

and

$$\frac{\partial u(r, 0)}{\partial z} = \begin{cases} B, & 0 \le r < a, \\ 0, & a < r < b. \end{cases}$$

18. Solve

$$\frac{\partial^2 u}{\partial r^2} + \frac{1}{r} \frac{\partial u}{\partial r} + \frac{\partial^2 u}{\partial z^2} = 0, \quad 0 \le r < a, \quad 0 < z < h,$$

with

$$\frac{\partial u(a, z)}{\partial r} = u(r, h) = 0$$

and

$$\frac{\partial u(r, 0)}{\partial z} = \begin{cases} 1, & 0 \le r < r_0, \\ 0, & r_0 < r < a. \end{cases}$$

19. Solve

$$\frac{\partial^2 u}{\partial r^2} + \frac{1}{r} \frac{\partial u}{\partial r} + \frac{\partial^2 u}{\partial z^2} = 0, \quad 0 \le r < 1, \quad 0 < z < d,$$

with

$$\frac{\partial u(1, z)}{\partial r} = \frac{\partial u(r, 0)}{\partial z} = 0,$$

and

$$u(r, d) = \begin{cases} -1, & 0 \le r < a, \quad b < r < 1, \\ 1/(b^2 - a^2) - 1, & a < r < b. \end{cases}$$

20. Solve

$$\frac{\partial^2 u}{\partial r^2} + \frac{1}{r} \frac{\partial u}{\partial r} - \frac{u}{r^2} + \frac{\partial^2 u}{\partial z^2} = 0, \quad 0 \le r < a, \quad 0 < z < h,$$

with

$$\lim_{r \to 0} |u(r, z)| < \infty, \quad u(r, 0) = u(a, z) = 0, \quad \text{and} \quad \frac{\partial u(r, h)}{\partial z} = Ar.$$

21. Solve

$$\frac{\partial^2 u}{\partial r^2} + \frac{1}{r} \frac{\partial u}{\partial r} - \frac{u}{r^2} + \frac{\partial^2 u}{\partial z^2} = 0, \quad 0 \le r < a, \quad 0 < z < 1,$$

with

$$\lim_{r \to 0} |u(r, z)| < \infty, \quad u(r, 0) = u(r, 1) = 0, \quad \text{and} \quad u(a, z) = z.$$

22. Solve

$$\frac{\partial^2 u}{\partial r^2} + \frac{1}{r}\frac{\partial u}{\partial r} - \frac{u}{r^2} + \frac{\partial^2 u}{\partial z^2} = 0, \quad 0 \le r < a, \quad 0 < z < h,$$

with

$$\lim_{r \to 0} |u(r,z)| < \infty, \quad \frac{\partial u(a,z)}{\partial r} = u(r,0) = 0, \quad \text{and} \quad \frac{\partial u(r,h)}{\partial z} = r.$$

23. Solve

$$\frac{\partial^2 u}{\partial r^2} + \frac{1}{r}\frac{\partial u}{\partial r} - \frac{u}{r^2} + \frac{\partial^2 u}{\partial z^2} = 0, \quad 0 \le r < 1, \quad -a < z < a,$$

with the boundary conditions

$$\lim_{r \to 0} |u(r,z)| < \infty, \quad \frac{\partial u(1,z)}{\partial r} = u(1,z), \quad -a < z < a,$$

and

$$-\frac{\partial u(r,-a)}{\partial z} = \frac{\partial u(r,a)}{\partial z} = r, \quad 0 \le r < 1.$$

24. Solve Laplace's equation in cylindrical coordinates

$$\frac{\partial^2 u}{\partial r^2} + \frac{1}{r}\frac{\partial u}{\partial r} + \frac{\partial^2 u}{\partial z^2} = 0, \quad 0 \le r < b, \quad 0 < z < \infty,$$

subject to the boundary conditions that

$$\lim_{r \to 0} |u(r,z)| < \infty, \quad \frac{\partial u(b,z)}{\partial r} = 0, \quad 0 < z < \infty,$$

and

$$\lim_{z \to \infty} |u(r,z)| < \infty, \quad u(r,0) = \begin{cases} A, & 0 \le r < a, \\ 0, & a < r < b. \end{cases}$$

25. Solve[19]

$$\frac{\partial^2 u}{\partial r^2} + \frac{1}{r}\frac{\partial u}{\partial r} + \frac{\partial^2 u}{\partial z^2} - \frac{\partial u}{\partial z} = 0, \quad 0 \le r < 1, \quad 0 < z < \infty,$$

---

[19] Reprinted from *Int. J. Heat Mass Transfer*, **19**, J. Kern and J. O. Hansen, Transient heat conduction in cylindrical systems with an axially moving boundary, 707–714, ©1976, with kind permission from Elsevier Science Ltd., The Boulevard, Langford Lane, Kidlington OX5 1GB, UK.

with the boundary conditions

$$\lim_{r \to 0} |u(r, z)| < \infty, \qquad \frac{\partial u(1, z)}{\partial r} = -Bu(1, z), \qquad 0 < z,$$

and

$$u(r, 0) = 1, \qquad \lim_{z \to \infty} |u(r, z)| < \infty, \qquad 0 \leq r < 1,$$

where B is a constant.

26. Solve[20]

$$\frac{\partial^2 u}{\partial r^2} + \frac{1}{r} \frac{\partial u}{\partial r} + \frac{\partial^2 u}{\partial z^2} - \frac{1}{H} \frac{\partial u}{\partial z} = 0, \qquad 0 \leq r < 1, \quad 0 < z < \infty,$$

with the boundary conditions

$$\lim_{r \to 0} |u(r, z)| < \infty, \qquad \frac{\partial u(b, z)}{\partial r} = -hu(b, z), \qquad 0 < z,$$

$$\lim_{z \to \infty} |u(r, z)| < \infty, \qquad 0 \leq r < b,$$

and

$$\frac{u(r, 0)}{H} - u_z(r, 0) = \begin{cases} Q, & 0 \leq r < a, \\ 0, & a \leq r < b, \end{cases}$$

where $b > a$.

27. Solve[21]

$$\frac{\partial^2 u}{\partial r^2} + \frac{1}{r} \frac{\partial u}{\partial r} + \frac{\partial^2 u}{\partial z^2} = 0, \qquad 0 \leq r < \infty, \quad 0 < z < \infty,$$

subject to the boundary conditions

$$\lim_{r \to 0} |u(r, z)| < \infty, \qquad \lim_{r \to \infty} |u(r, z)| < \infty, \qquad 0 < z < \infty,$$

$$\lim_{z \to \infty} u(r, z) \to u_\infty, \qquad 0 \leq r < \infty,$$

and

$$\begin{cases} u(r, 0) = u_\infty - \Delta u, & 0 \leq r < a, \\ u_z(r, 0) = 0, & a \leq r < \infty, \end{cases}$$

where u_∞ and Δu are constants.

---

[20] Taken from Smirnova, E. V., and I. A. Krinberg, 1970: Spatial distribution of the atoms of an impurity element in an arc discharge. I. *J. Appl. Spectroscopy*, **13**, 859–864.

[21] Reprinted from *J. Electroanal. Chem.*, **222**, M. Fleischmann and S. Pons, The behavior of microdisk and microring electrodes, 107–115, ©1987, with permission from Elsevier Science.

Step 1: Show that

$$u(r,z) = u_\infty - \int_0^\infty A(k)e^{-kz} J_0(kr)\, dk$$

satisfies the partial differential equation and the boundary conditions as $r \to 0$, $r \to \infty$, and $z \to \infty$.

Step 2: Show that

$$\int_0^\infty k A(k) J_0(kr)\, dk = 0, \qquad a < r < \infty.$$

Step 3: Using the relationship[22]

$$\int_0^\infty \sin(ka) J_0(kr)\, dk = \begin{cases} (a^2 - r^2)^{-\frac{1}{2}}, & r < a, \\ 0, & r > a, \end{cases}$$

show that $kA(k) = C\sin(ka)$.

Step 4: Using the relationship[23]

$$\int_0^\infty \sin(ka) J_0(kr)\, \frac{dk}{k} = \begin{cases} \pi/2, & r \le a, \\ \sin^{-1}(a/r), & r \ge a, \end{cases}$$

show that

$$u(r,z) = u_\infty - \frac{2\Delta u}{\pi} \int_0^\infty e^{-kz} \sin(ka) J_0(kr)\, \frac{dk}{k}.$$

28. Find the steady-state temperature within a sphere of radius a if the temperature along its surface is maintained at the temperature $u(a,\theta) = 100[\cos(\theta) - \cos^5(\theta)]$.

29. Find the steady-state temperature within a sphere if the upper half of the exterior surface at radius a is maintained at the temperature 100 while the lower half is maintained at the temperature 0.

30. The surface of a sphere of radius a has a temperature of zero everywhere except in a spherical cap at the north pole (defined by the cone $\theta = \alpha$), where it equals T_0. Find the steady-state temperature within the sphere.

---

[22] Gradshteyn and Ryzhik, *op. cit.*, §6.671, Formula 7.

[23] *Ibid.*, §6.693, Formula 1 with $\nu = 0$.

31. Using the relationship

$$\int_0^{2\pi} \frac{d\varphi}{1 - b\cos(\varphi)} = \frac{2\pi}{\sqrt{1 - b^2}}, \qquad |b| < 1$$

and Poisson's integral formula, find the solution to Laplace's equation within a unit disc if $u(1, \varphi) = f(\varphi) = T_0$, a constant.

12.4 THE SOLUTION OF LAPLACE'S EQUATION ON THE UPPER HALF-PLANE

In this section we shall use Fourier integrals and convolution to find the solution of Laplace's equation on the upper half-plane $y > 0$. We require that the solution remains bounded over the entire domain and specify it along the x-axis, $u(x, 0) = f(x)$. Under these conditions, we can take the Fourier transform of Laplace's equation and find that

$$\int_{-\infty}^{\infty} \frac{\partial^2 u}{\partial x^2} e^{-i\omega x} \, dx + \int_{-\infty}^{\infty} \frac{\partial^2 u}{\partial y^2} e^{-i\omega x} \, dx = 0. \qquad (12.4.1)$$

If everything is sufficiently differentiable, we may successively integrate by parts the first integral in (12.4.1) which yields

$$\int_{-\infty}^{\infty} \frac{\partial^2 u}{\partial x^2} e^{-i\omega x} \, dx = \frac{\partial u}{\partial x} e^{-i\omega x} \Big|_{-\infty}^{\infty} + i\omega \int_{-\infty}^{\infty} \frac{\partial u}{\partial x} e^{-i\omega x} \, dx \qquad (12.4.2)$$

$$= i\omega \, u(x, y) e^{-i\omega x} \Big|_{-\infty}^{\infty} - \omega^2 \int_{-\infty}^{\infty} u(x, y) e^{-i\omega x} \, dx$$

$$\qquad\qquad (12.4.3)$$

$$= -\omega^2 U(\omega, y), \qquad (12.4.4)$$

where

$$U(\omega, y) = \int_{-\infty}^{\infty} u(x, y) e^{-i\omega x} \, dx. \qquad (12.4.5)$$

The second integral becomes

$$\int_{-\infty}^{\infty} \frac{\partial^2 u}{\partial y^2} e^{-i\omega x} \, dx = \frac{d^2}{dy^2} \left[\int_{-\infty}^{\infty} u(x, y) e^{-i\omega x} \, dx \right] = \frac{d^2 U(\omega, y)}{dy^2}, \qquad (12.4.6)$$

along with the boundary condition that

$$F(\omega) = U(\omega, 0) = \int_{-\infty}^{\infty} f(x) e^{-i\omega x} \, dx. \qquad (12.4.7)$$

Consequently we reduced Laplace's equation, a partial differential equation, to an ordinary differential equation in y, where ω is merely a parameter:

$$\frac{d^2 U(\omega, y)}{dy^2} - \omega^2 U(\omega, y) = 0, \qquad (12.4.8)$$

with the boundary condition $U(\omega, 0) = F(\omega)$. The solution to (12.4.8) is

$$U(\omega, y) = A(\omega)e^{|\omega|y} + B(\omega)e^{-|\omega|y}, \quad 0 \le y. \qquad \textbf{(12.4.9)}$$

We must discard the $e^{|\omega|y}$ term because it becomes unbounded as we go to infinity along the y-axis. The boundary condition results in $B(\omega) = F(\omega)$. Consequently,

$$U(\omega, y) = F(\omega)e^{-|\omega|y}. \qquad \textbf{(12.4.10)}$$

The inverse of the Fourier transform $e^{-|\omega|y}$ equals

$$\frac{1}{2\pi} \int_{-\infty}^{\infty} e^{-|\omega|y} e^{i\omega x} \, d\omega = \frac{1}{2\pi} \int_{-\infty}^{0} e^{\omega y} e^{i\omega x} \, d\omega + \frac{1}{2\pi} \int_{0}^{\infty} e^{-\omega y} e^{i\omega x} \, d\omega$$

$$\textbf{(12.4.11)}$$

$$= \frac{1}{2\pi} \int_{0}^{\infty} e^{-\omega y} e^{-i\omega x} \, d\omega + \frac{1}{2\pi} \int_{0}^{\infty} e^{-\omega y} e^{i\omega x} \, d\omega$$

$$\textbf{(12.4.12)}$$

$$= \frac{1}{\pi} \int_{0}^{\infty} e^{-\omega y} \cos(\omega x) \, d\omega \qquad \textbf{(12.4.13)}$$

$$= \frac{1}{\pi} \left\{ \frac{\exp(-\omega y)}{x^2 + y^2} \left[-y\cos(\omega x) + x\sin(\omega x) \right] \right\} \Bigg|_{0}^{\infty}$$

$$\textbf{(12.4.14)}$$

$$= \frac{1}{\pi} \frac{y}{x^2 + y^2}. \qquad \textbf{(12.4.15)}$$

Furthermore, because (12.4.10) is a convolution of two Fourier transforms, its inverse is

$$u(x, y) = \frac{1}{\pi} \int_{-\infty}^{\infty} \frac{y f(t)}{(x - t)^2 + y^2} \, dt. \qquad \textbf{(12.4.16)}$$

Equation (12.4.16) is *Poisson's integral formula*[24] for the half-plane $y > 0$ or *Schwarz' integral formula*.[25]

• Example 12.4.1

As an example, let $u(x, 0) = 1$ if $|x| < 1$ and $u(x, 0) = 0$ otherwise. Then,

$$u(x, y) = \frac{1}{\pi} \int_{-1}^{1} \frac{y}{(x - t)^2 + y^2} \, dt \qquad \textbf{(12.4.17)}$$

$$= \frac{1}{\pi} \left[\tan^{-1}\left(\frac{1 - x}{y}\right) + \tan^{-1}\left(\frac{1 + x}{y}\right) \right]. \qquad \textbf{(12.4.18)}$$

---

[24] Poisson, S. D., 1823: Suite du mémoire sur les intégrales définies et sur la sommation des séries. *J. École Polytech.*, **19**, 404–509. See pg. 462.

[25] Schwarz, H. A., 1870: Über die Integration der partiellen Differentialgleichung $\partial^2 u/\partial x^2 + \partial^2 u/\partial y^2 = 0$ für die Fläche eines Kreises. *Vierteljahrsschr. Naturforsch. Ges. Zürich*, **15**, 113–128.

Problems

Find the solution to Laplace's equation in the upper half-plane for the following boundary conditions:

1. $u(x,0) = \begin{cases} 1, & 0 < x < 1 \\ 0, & \text{otherwise} \end{cases}$

2. $u(x,0) = \begin{cases} 1, & x > 0 \\ -1, & x < 0 \end{cases}$

3. $u(x,0) = \begin{cases} T_0, & x < 0 \\ 0, & x > 0 \end{cases}$

4. $u(x,0) = \begin{cases} 2T_0, & x < -1 \\ T_0, & -1 < x < 1 \\ 0, & 1 < x \end{cases}$

5.
$$u(x,0) = \begin{cases} T_0, & -1 < x < 0 \\ T_0 + (T_1 - T_0)x, & 0 < x < 1 \\ 0, & \text{otherwise} \end{cases}$$

6.
$$u(x,0) = \begin{cases} T_0, & x < a_1 \\ T_1, & a_1 < x < a_2 \\ T_2, & a_2 < x < a_3 \\ \vdots & \vdots \\ T_n, & a_n < x \end{cases}$$

12.5 POISSON'S EQUATION ON A RECTANGLE

Poisson's equation[26] is Laplace's equation with a source term:

$$\frac{\partial^2 u}{\partial x^2} + \frac{\partial^2 u}{\partial y^2} = f(x,y). \qquad (12.5.1)$$

It arises in such diverse areas as groundwater flow, electromagnetism, and potential theory. Let us solve it if $u(0,y) = u(a,y) = u(x,0) = u(x,b) = 0$.

We begin by solving a similar partial differential equation:

$$\frac{\partial^2 u}{\partial x^2} + \frac{\partial^2 u}{\partial y^2} = \lambda u, \quad 0 < x < a, \quad 0 < y < b, \qquad (12.5.2)$$

by separation of variables. If $u(x,y) = X(x)Y(y)$, then

$$\frac{X''}{X} + \frac{Y''}{Y} = \lambda. \qquad (12.5.3)$$

---

[26] Poisson, S. D., 1813: Remarques sur une équation qui se présente dans la théorie des attractions des sphéroïdes. *Nouv. Bull. Soc. Philomath. Paris*, **3**, 388–392.

Figure 12.5.1: Siméon-Denis Poisson (1781–1840) was a product as well as a member of the French scientific establishment of his day. Educated at the École Polytechnique, he devoted his life to teaching, both in the classroom and with administrative duties, and to scientific research. Poisson's equation dates from 1813 when Poisson sought to extend Laplace's work on gravitational attraction. (Portrait courtesy of the Archives de l'Académie des sciences, Paris.)

Because we must satisfy the boundary conditions that $X(0) = X(a) = Y(0) = Y(b) = 0$, we have the following eigenfunction solutions:

$$X_n(x) = \sin\left(\frac{n\pi x}{a}\right), \qquad Y_m(x) = \sin\left(\frac{m\pi y}{b}\right) \qquad (\mathbf{12.5.4})$$

with $\lambda_{nm} = -n^2\pi^2/a^2 - m^2\pi^2/b^2$; otherwise, we would only have trivial solutions. The corresponding particular solutions are

$$u_{nm} = A_{nm} \sin\left(\frac{n\pi x}{a}\right) \sin\left(\frac{m\pi y}{b}\right), \qquad (\mathbf{12.5.5})$$

where $n = 1, 2, 3, \ldots$, and $m = 1, 2, 3, \ldots$.

For a fixed y, we can expand $f(x, y)$ in the half-range Fourier sine series

$$f(x, y) = \sum_{n=1}^{\infty} A_n(y) \sin\left(\frac{n\pi x}{a}\right), \qquad (\mathbf{12.5.6})$$

where

$$A_n(y) = \frac{2}{a} \int_0^a f(x, y) \sin\left(\frac{n\pi x}{a}\right) dx. \qquad (12.5.7)$$

However, we can also expand $A_n(y)$ in a half-range Fourier sine series

$$A_n(y) = \sum_{m=1}^{\infty} a_{nm} \sin\left(\frac{m\pi y}{b}\right), \qquad (12.5.8)$$

where

$$a_{nm} = \frac{2}{b} \int_0^b A_n(y) \sin\left(\frac{m\pi y}{b}\right) dy \qquad (12.5.9)$$

$$= \frac{4}{ab} \int_0^b \int_0^a f(x, y) \sin\left(\frac{n\pi x}{a}\right) \sin\left(\frac{m\pi y}{b}\right) dx\, dy, \qquad (12.5.10)$$

and

$$f(x, y) = \sum_{n=1}^{\infty} \sum_{m=1}^{\infty} a_{nm} \sin\left(\frac{n\pi x}{a}\right) \sin\left(\frac{m\pi y}{b}\right). \qquad (12.5.11)$$

In other words, we re-expressed $f(x, y)$ in terms of a *double Fourier series*.

Because (12.5.2) must hold for each particular solution,

$$\frac{\partial^2 u_{nm}}{\partial x^2} + \frac{\partial^2 u_{nm}}{\partial y^2} = \lambda_{nm} u_{nm} = a_{nm} \sin\left(\frac{n\pi x}{a}\right) \sin\left(\frac{m\pi y}{b}\right), \qquad (12.5.12)$$

if we now associate (12.5.1) with (12.5.2). Therefore, the solution to Poisson's equation on a rectangle where the boundaries are held at zero is the double Fourier series

$$u(x, y) = -\sum_{n=1}^{\infty} \sum_{m=1}^{\infty} \frac{a_{nm}}{n^2\pi^2/a^2 + m^2\pi^2/b^2} \sin\left(\frac{n\pi x}{a}\right) \sin\left(\frac{m\pi y}{b}\right). \qquad (12.5.13)$$

Problems

1. The equation

$$\frac{\partial^2 u}{\partial x^2} + \frac{\partial^2 u}{\partial y^2} = -\frac{R}{T}, \qquad |x| < a, \quad |y| < b,$$

describes the hydraulic potential (elevation of the water table) $u(x, y)$ within a rectangular island on which a recharging well is located at $(0, 0)$. Here R is the rate of recharging and T is the product of the hydraulic conductivity and aquifer thickness. If the water table is at sea level around the island so that $u(-a, y) = u(a, y) = u(x, -b) = u(x, b) = 0$, find $u(x, y)$ everywhere in the island. [Hint: Use symmetry and redo the above analysis with the boundary conditions: $u_x(0, y) = u(a, y) = u_y(x, 0) = u(x, b) = 0$.]

2. Let us apply the same approach that we used to find the solution of Poisson's equation on a rectangle to solve the axisymmetric Poisson equation inside a circular cylinder

$$\frac{1}{r} \frac{\partial}{\partial r} \left(r \frac{\partial u}{\partial r} \right) + \frac{\partial^2 u}{\partial z^2} = f(r, z), \quad 0 \le r < a, \quad |z| < b,$$

subject to the boundary conditions

$$\lim_{r \to 0} |u(r, z)| < \infty, \qquad u(a, z) = 0, \qquad |z| < b,$$

and

$$u(r, -b) = u(r, b) = 0, \qquad 0 \le r < a.$$

Step 1: Replace the original problem with

$$\frac{1}{r} \frac{\partial}{\partial r} \left(r \frac{\partial u}{\partial r} \right) + \frac{\partial^2 u}{\partial z^2} = \lambda u, \quad 0 \le r < a, \quad |z| < b,$$

subject to the same boundary conditions. Use separation of variables to show that the solution to this new problem is

$$u_{nm}(r, z) = A_{nm} J_0 \left(k_n \frac{r}{a} \right) \cos \left[\frac{\left(m + \frac{1}{2} \right) \pi z}{b} \right],$$

where k_n is the nth zero of $J_0(k) = 0$, $n = 1, 2, 3, \ldots$, and $m = 0, 1, 2, \ldots$.

Step 2: Show that $f(r, z)$ can be expressed as

$$f(r, z) = \sum_{n=1}^{\infty} \sum_{m=0}^{\infty} a_{nm} J_0 \left(k_n \frac{r}{a} \right) \cos \left[\frac{\left(m + \frac{1}{2} \right) \pi z}{b} \right],$$

where

$$a_{nm} = \frac{2}{a^2 b J_1^2(k_n)} \int_{-b}^{b} \int_0^a f(r, z) J_0 \left(k_n \frac{r}{a} \right) \cos \left[\frac{\left(m + \frac{1}{2} \right) \pi z}{b} \right] r \, dr \, dz.$$

Step 3: Show that the general solution is

$$u(r, z) = -\sum_{n=1}^{\infty} \sum_{m=0}^{\infty} a_{nm} \frac{J_0(k_n r/a) \cos \left[\left(m + \frac{1}{2} \right) \pi z/b \right]}{(k_n/a)^2 + \left[\left(m + \frac{1}{2} \right) \pi/b \right]^2}.$$

12.6 THE LAPLACE TRANSFORM METHOD

Laplace transforms are useful in solving Laplace's or Poisson's equation over a semi-infinite strip. The following problem illustrates this technique.

Let us solve Poisson's equation within a semi-infinite circular cylinder

$$\frac{1}{r}\frac{\partial}{\partial r}\left(r\frac{\partial u}{\partial r}\right) + \frac{\partial^2 u}{\partial z^2} = \frac{2}{b}n(z)\delta(r-b), \quad 0 \le r < a, \quad 0 < z < \infty, \quad (\mathbf{12.6.1})$$

subject to the boundary conditions

$$u(r,0) = 0, \quad \lim_{z\to\infty} |u(r,z)| < \infty, \quad 0 \le r < a, \tag{12.6.2}$$

and

$$u(a,z) = 0, \qquad 0 < z < \infty, \tag{12.6.3}$$

where $0 < b < a$. This problem gives the electrostatic potential within a semi-infinite cylinder of radius a that is grounded and has the charge density of $n(z)$ within an infinitesimally thin shell located at $r = b$.

Because the domain is semi-infinite in the z direction, we introduce the Laplace transform

$$U(r,s) = \int_0^\infty u(r,z)\, e^{-sz}\, dz. \tag{12.6.4}$$

Thus, taking the Laplace transform of (12.6.1), we have that

$$\frac{1}{r}\frac{d}{dr}\left[r\frac{dU(r,s)}{dr}\right] + s^2 U(r,s) - su(r,0) - u_z(r,0) = \frac{2}{b}N(s)\delta(r-b). \tag{12.6.5}$$

Although $u(r,0) = 0$, $u_z(r,0)$ is unknown and we denote its value by $f(r)$. Therefore, (12.6.5) becomes

$$\frac{1}{r}\frac{d}{dr}\left[r\frac{dU(r,s)}{dr}\right] + s^2 U(r,s) = f(r) + \frac{2}{b}N(s)\delta(r-b), \quad 0 \le r < a, \tag{12.6.6}$$

with $\lim_{r\to 0} |U(r,s)| < \infty$, and $U(a,s) = 0$.

To solve (12.6.6) we first assume that we can rewrite $f(r)$ as the Fourier-Bessel series

$$f(r) = \sum_{n=1}^\infty A_n J_0(k_n r/a), \tag{12.6.7}$$

where k_n is the nth root of the $J_0(k) = 0$, and

$$A_n = \frac{2}{a^2 J_1^2(k_n)}\int_0^a f(r)\, J_0(k_n r/a)\, r\, dr. \tag{12.6.8}$$

Similarly, the expansion for the delta function is

$$\delta(r-b) = \frac{2b}{a^2} \sum_{n=1}^{\infty} \frac{J_0(k_n b/a) J_0(k_n r/a)}{J_1^2(k_n)}, \qquad (12.6.9)$$

because

$$\int_0^a \delta(r-b) J_0(k_n r/a)\, r\, dr = b\, J_0(k_n b/a). \qquad (12.6.10)$$

Why we chose this particular expansion will become apparent shortly.

Thus, (12.6.6) may be rewritten as

$$\frac{1}{r}\frac{d}{dr}\left[r\frac{dU(r,s)}{dr}\right] + s^2 U(r,s) = \frac{2}{a^2} \sum_{n=1}^{\infty} \frac{2N(s)J_0(k_n b/a) + a_k}{J_1^2(k_n)} J_0(k_n r/a),$$
$$(12.6.11)$$

where $a_k = \int_0^a f(r)\, J_0(k_n r/a)\, r\, dr$.

The form of the right side of (12.6.11) suggests that we seek solutions of the form

$$U(r,s) = \sum_{n=1}^{\infty} B_n J_0(k_n r/a), \qquad 0 \le r < a. \qquad (12.6.12)$$

We now understand why we rewrote the right side of (12.6.6) as a Fourier-Bessel series; the solution $U(r,s)$ automatically satisfies the boundary condition $U(a,s) = 0$. Substituting (12.6.12) into (12.6.11), we find that

$$U(r,s) = \frac{2}{a^2} \sum_{n=1}^{\infty} \frac{2N(s)J_0(k_n b/a) + a_k}{(s^2 - k_n^2/a^2)J_1^2(k_n)} J_0(k_n r/a), \qquad 0 \le r < a. \quad (12.6.13)$$

We have not yet determined a_k. Note, however, that in order for the inverse of (12.6.13) *not* to grow as $e^{k_n z/a}$, the numerator must vanish when $s = k_n/a$ and $s = k_n/a$ is a removable pole. Thus, $a_k = -2N(k_n/a)J_0(k_n b/a)$, and

$$U(r,s) = \frac{4}{a^2} \sum_{n=1}^{\infty} \frac{[N(s) - N(k_n/a)]J_0(k_n b/a)}{(s^2 - k_n^2/a^2)J_1^2(k_n)} J_0(k_n r/a), \qquad 0 \le r < a.$$
$$(12.6.14)$$

The inverse of $U(r,s)$ then follows directly from simple inversions, the convolution theorem, and the definition of the Laplace transform. The complete solution is

$$u(r,z) = \frac{2}{a} \sum_{n=1}^{\infty} \frac{J_0(k_n b/a) J_0(k_n r/a)}{k_n J_1^2(k_n)}$$
$$\times \left[\int_0^z n(\tau)e^{k_n(z-\tau)/a}\, d\tau - \int_0^z n(\tau)e^{-k_n(z-\tau)/a}\, d\tau \right]$$

$$- \int_0^\infty n(\tau) e^{-k_n \tau/a} e^{k_n z/a} \, d\tau + \int_0^\infty n(\tau) e^{-k_n \tau/a} e^{-k_n z/a} \, d\tau \Bigg]$$

$$\tag{12.6.15}$$

$$u(r,z) = \frac{2}{a} \sum_{n=1}^\infty \frac{J_0(k_n b/a) J_0(k_n r/a)}{k_n J_1^2(k_n)}$$

$$\times \left[\int_0^\infty n(\tau) e^{-k_n(z+\tau)/a} \, d\tau - \int_0^z n(\tau) e^{-k_n(z-\tau)/a} \, d\tau \right.$$

$$\left. - \int_z^\infty n(\tau) e^{-k_n(\tau-z)/a} \, d\tau \right].$$

$$\tag{12.6.16}$$

Problems

1. Use Laplace transforms to solve

$$\frac{\partial^2 u}{\partial x^2} + \frac{\partial^2 u}{\partial y^2} = 0, \quad 0 < x < \infty, \quad 0 < y < a,$$

subject to the boundary conditions

$$u(0,y) = 1, \quad \lim_{x \to \infty} |u(x,y)| < \infty, \quad 0 < y < a,$$

and

$$u(x,0) = u(x,a) = 0, \quad 0 < x < \infty.$$

2. Use Laplace transforms to solve

$$\frac{1}{r} \frac{\partial}{\partial r} \left(r \frac{\partial u}{\partial r} \right) + \frac{\partial^2 u}{\partial z^2} = 0, \quad 0 \le r < a, \quad 0 < z < \infty,$$

subject to the boundary conditions

$$u(r,0) = 1, \quad \lim_{z \to \infty} |u(r,z)| < \infty, \quad 0 \le r < a,$$

and

$$\lim_{r \to 0} |u(r,z)| < \infty, \quad u(a,z) = 0, \quad 0 < z < \infty.$$

12.7 NUMERICAL SOLUTION OF LAPLACE'S EQUATION

As in the case of the heat and wave equations, numerical methods can be used to solve elliptic partial differential equations when analytic techniques fail or are too cumbersome. They are also employed when the domain differs from simple geometries.

The numerical analysis of an elliptic partial differential equation begins by replacing the continuous partial derivatives by finite-difference formulas. Employing centered differencing,

$$\frac{\partial^2 u}{\partial x^2} = \frac{u_{m+1,n} - 2u_{m,n} + u_{m-1,n}}{(\Delta x)^2} + O[(\Delta x)^2], \qquad (\mathbf{12.7.1})$$

and

$$\frac{\partial^2 u}{\partial y^2} = \frac{u_{m,n+1} - 2u_{m,n} + u_{m,n-1}}{(\Delta y)^2} + O[(\Delta y)^2], \qquad (\mathbf{12.7.2})$$

where $u_{m,n}$ denotes the solution value at the grid point m, n. If $\Delta x = \Delta y$, Laplace's equation becomes the difference equation

$$u_{m+1,n} + u_{m-1,n} + u_{m,n+1} + u_{m,n-1} - 4u_{m,n} = 0. \qquad (\mathbf{12.7.3})$$

Thus, we must now solve a set of simultaneous linear equations that yield the value of the solution at each grid point.

The solution of (12.7.3) is best done using techniques developed by algebraists. Later on, in Chapter 14, we will show that a very popular method for directly solving systems of linear equations is Gaussian elimination. However, for many grids at a reasonable resolution, the number of equations are generally in the tens of thousands. Because most of the coefficients in the equations are zero, Gaussian elimination is unsuitable, both from the point of view of computational expense and accuracy. For this reason alternative methods have been developed that generally use successive corrections or iterations. The most common of these point iterative methods are the Jacobi method, unextrapolated Liebmann or Gauss-Seidel method, and extrapolated Liebmann or successive over-relaxation (SOR). None of these approaches is completely satisfactory because of questions involving convergence and efficiency. Because of its simplicity we will focus on the Gauss-Seidel method.

We may illustrate the Gauss-Seidel method by considering the system:

$$10x + y + z = 39, \qquad (\mathbf{12.7.4})$$

$$2x + 10y + z = 51, \qquad (\mathbf{12.7.5})$$

and

$$2x + 2y + 10z = 64. \qquad (\mathbf{12.7.6})$$

An important aspect of this system is the dominance of the coefficient of x in the first equation of the set and that the coefficients of y and z are dominant in the second and third equations, respectively.

The Gauss-Seidel method may be outlined as follows:

• Assign an initial value for each unknown variable. If possible, make a good first guess. If not, any arbitrarily selected values may be chosen. The initial value will not affect the convergence but will affect the number of iterations until convergence.

• Starting with (12.7.4), solve that equation for a new value of the unknown which has the largest coefficient in that equation, using the assumed values for the other unknowns.

• Go to (12.7.5) and employ the same technique used in the previous step to compute the unknown that has the largest coefficient in that equation. Where possible, use the latest values.

• Proceed to the remaining equations, always solving for the unknown having the largest coefficient in the particular equation and always using the *most recently* calculated values for the other unknowns in the equation. When the last equation (12.7.6) has been solved, you have completed a single iteration.

• Iterate until the value of each unknown does not change within a predetermined value.

Usually a compromise must be struck between the accuracy of the solution and the desired rate of convergence. The more accurate the solution is, the longer it will take for the solution to converge.

To illustrate this method, let us solve our system (12.7.4)–(12.7.6) with the initial guess $x = y = z = 0$. The first iteration yields $x = 3.9$, $y = 4.32$, and $z = 4.756$. The second iteration yields $x = 2.9924$, $y = 4.02592$, and $z = 4.996336$. As can be readily seen, the solution is converging to the correct solution of $x = 3$, $y = 4$, and $z = 5$.

Applying these techniques to (12.7.3),

$$u_{m,n}^{k+1} = \tfrac{1}{4} \left(u_{m+1,n}^{k} + u_{m-1,n}^{k+1} + u_{m,n+1}^{k} + u_{m,n-1}^{k+1} \right), \qquad (\mathbf{12.7.7})$$

where we assume that the calculations occur in order of increasing m and n.

• **Example 12.7.1**

To illustrate the numerical solution of Laplace's equation, let us redo Example 12.3.1 with the boundary condition along $y = H$ simplified to $u(x, H) = 1 + x/L$.

We begin by finite-differencing the boundary conditions. The condition $u_x(0, y) = u_x(L, y) = 0$ leads to $u_{1,n} = u_{-1,n}$ and $u_{M+1,n} = u_{M-1,n}$ if we employ centered differences at $m = 0$ and $m = M$. Substituting these values in (12.7.7), we have the following equations for the left and right boundaries:

$$u_{0,n}^{k+1} = \tfrac{1}{4} \left(2u_{1,n}^{k} + u_{0,n+1}^{k} + u_{0,n-1}^{k+1} \right) \qquad (\mathbf{12.7.8})$$

and

$$u_{M,n}^{k+1} = \tfrac{1}{4}\left(2u_{M-1,n}^{k+1} + u_{M,n+1}^{k} + u_{M,n-1}^{k+1}\right). \qquad (12.7.9)$$

On the other hand, $u_y(x,0) = 0$ yields $u_{m,1} = u_{m,-1}$, and

$$u_{m,0}^{k+1} = \tfrac{1}{4}\left(u_{m+1,0}^{k} + u_{m-1,0}^{k+1} + 2u_{m,1}^{k}\right). \qquad (12.7.10)$$

At the bottom corners, (12.7.8)–(12.7.10) simplify to

$$u_{0,0}^{k+1} = \tfrac{1}{2}\left(u_{1,0}^{k} + u_{0,1}^{k}\right) \qquad (12.7.11)$$

and

$$u_{L,0}^{k+1} = \tfrac{1}{2}\left(u_{L-1,0}^{k+1} + u_{L,1}^{k}\right). \qquad (12.7.12)$$

These equations along with (12.7.7) were solved using the Gauss-Seidel method using the MATLAB script

```
clear
dx = 0.1; x = 0:dx:1; M = 1/dx+1; % M = number of x grid points
dy = 0.1; y = 0:dy:1; N = 1/dy+1; % N = number of y grid points
X = x' * ones(1,N); Y = ones(M,1) * y;
u = zeros(M,N); % create initial guess for the solution
% introduce boundary condition along y = H
for m = 1:M; u(m,N) = 1 + x(m); end
%
% start Gauss-Seidel method for Laplace's equation
%
for iter = 1:256
%
% do the interior first
%
for n = 2:N-1; for m = 2:M-1;
u(m,n) = (u(m+1,n)+u(m-1,n)+u(m,n+1)+u(m,n-1)) / 4;
end; end
%
% now do the x = 0 and x = L sides
%
for n = 2:N-1
u(1,n) = (2*u( 2 ,n)+u(1,n+1)+u(1,n-1)) / 4;
u(M,n) = (2*u(M-1,n)+u(M,n+1)+u(M,n-1)) / 4;
end
%
% now do the y = 0 side
%
for m = 2:M-1
u(m,1) = (u(m+1,1)+u(m-1,1)+2*u(m,2)) / 4;
```

```
end
%
% finally do the corners
%
u(1,1) = (u(2,1)+u(1,2))/2; u(M,1) = (u(M-1,1)+u(M,2))/2;
%
% plot the solution
%
if (iter == 4) subplot(2,2,1), [cs,h] = contourf(X,Y,u);
    clabel(cs,h,[0.2 0.6 1 1.4],'Fontsize',16)
    axis tight; title('after 4 iterations','Fontsize',20);
    ylabel('Y/H','Fontsize',20); end
if (iter == 16) subplot(2,2,2), [cs,h] = contourf(X,Y,u);
    clabel(cs,h,'Fontsize',16)
    axis tight; title('after 16 iterations','Fontsize',20);
    ylabel('Y/H','Fontsize',20); end
if (iter == 64) subplot(2,2,3), [cs,h] = contourf(X,Y,u);
    clabel(cs,h,'Fontsize',16)
    axis tight; title('after 64 iterations','Fontsize',20);
    xlabel('X/L','Fontsize',20); ylabel('Y/H','Fontsize',20);
end
if (iter == 256) subplot(2,2,4), [cs,h] = contourf(X,Y,u);
    clabel(cs,h,'Fontsize',16)
    axis tight; title('after 256 iterations','Fontsize',20);
    xlabel('X/L','Fontsize',20); ylabel('Y/H','Fontsize',20);
end
end
```

The initial guess everywhere except along the top boundary was zero. In Figure 12.7.1 we illustrate the numerical solution after 4, 16, 64, and 256 iterations where we have taken 11 grid points in the x and y directions.

Project: Successive Over-Relaxation

The fundamental difficulty with relaxation methods used in solving Laplace's equation is the rate of convergence. Assuming $\Delta x = \Delta y$, the most popular method for accelerating convergence of these techniques is *successive over-relaxation (SOR)*:

$$u_{m,n}^{k+1} = (1 - \omega)u_{m,n}^k + \omega R_{m,n},$$

where

$$R_{m,n} = \tfrac{1}{4}\left(u_{m+1,n}^k + u_{m-1,n}^{k+1} + u_{m,n+1}^k + u_{m,n-1}^{k+1}\right).$$

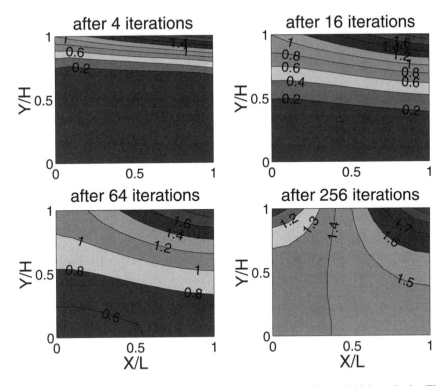

Figure 12.7.1: The solution to Laplace's equation by the Gauss-Seidel method. The boundary conditions are $u_x(0,y) = u_x(L,y) = u_y(x,0) = 0$, and $u(x,H) = 1 + x/L$.

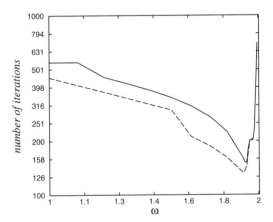

Figure 12.7.2: The number of iterations required so that $|R_{m,n}| \leq 10^{-3}$ as a function of ω during the iterative solution of the problem posed in the project. We used $\Delta x = \Delta y = 0.01$, and $L = z_0 = 1$. The iteration count for the boundary conditions stated in Step 1 is given by the solid line while the iteration count for the boundary conditions given in Step 2 is shown by the dotted line. The initial guess equaled zero.

Most numerical methods books dealing with partial differential equations discuss the theoretical reasons behind this technique;[27] the optimum value always
lies between one and two.

Step 1: Write a MATLAB script that uses the Gauss-Seidel method to numerically solve Laplace's equation for $0 \leq x \leq L$, $0 \leq y \leq z_0$ with the following
boundary conditions: $u(x,0) = 0$, $u(x,z_0) = 1 + x/L$, $u(0,y) = y/z_0$, and
$u(L,y) = 2y/z_0$. Because this solution will act as "truth" in this project, you
should iterate until the solution does not change.

Step 2: Now redo the calculation using successive over-relaxation. Count the
number of iterations until $|R_{m,n}| \leq 10^{-3}$ for *all* m and n. Plot the number of
iterations as a function of ω. How does the curve change with resolution Δx?

Step 3: Redo Steps 1 and 2 with the exception of $u(0,y) = u(L,y) = 0$. How
has the convergence rate changed? Can you explain why? How sensitive are
your results to the first guess?

---

[27] For example, Young, D. M., 1971: *Iterative Solution of Large Linear Systems*. Academic Press, 570 pp.

Chapter 13

Vector Calculus

Physicists invented vectors and vector operations to facilitate their mathematical expression of such diverse topics as mechanics and electromagnetism. In this chapter we focus on multivariable differentiations and integrations of vector fields, such as the velocity of a fluid, where the vector field is solely a function of its position.

13.1 REVIEW

The physical sciences and engineering abound with vectors and scalars. *Scalars* are physical quantities which only possess magnitude. Examples include mass, temperature, density, and pressure. *Vectors* are physical quantities that possess both magnitude and direction. Examples include velocity, acceleration, and force. We shall denote vectors by boldfaced letters.

Two vectors are equal if they have the same magnitude and direction. From the limitless number of possible vectors, two special cases are the *zero vector* $\mathbf{0}$ which has no magnitude and unspecified direction and the *unit vector* which has unit magnitude.

The most convenient method for expressing a vector analytically is in terms of its components. A vector $\mathbf{a}$ in three-dimensional real space is any order triplet of real numbers (*components*) a_1, a_2, and a_3 such that $\mathbf{a} = a_1\mathbf{i} + a_2\mathbf{j} + a_3\mathbf{k}$, where $a_1\mathbf{i}$, $a_2\mathbf{j}$, and $a_3\mathbf{k}$ are vectors which lie along the coordinate axes and have their origin at a common initial point. The *magnitude, length,* or *norm* of a vector $\mathbf{a}$, $|\mathbf{a}|$, equals $\sqrt{a_1^2 + a_2^2 + a_3^2}$. A particularly important vector is the *position vector*, defined by $\mathbf{r} = x\mathbf{i} + y\mathbf{j} + z\mathbf{k}$.

As in the case of scalars, certain arithmetic rules hold. Addition and subtraction are very similar to their scalar counterparts:

$$\mathbf{a} + \mathbf{b} = (a_1 + b_1)\mathbf{i} + (a_2 + b_2)\mathbf{j} + (a_3 + b_3)\mathbf{k}, \tag{13.1.1}$$

and

$$\mathbf{a} - \mathbf{b} = (a_1 - b_1)\mathbf{i} + (a_2 - b_2)\mathbf{j} + (a_3 - b_3)\mathbf{k}. \tag{13.1.2}$$

In contrast to its scalar counterpart, there are two types of multiplication. The *dot product* is defined as

$$\mathbf{a} \cdot \mathbf{b} = |\mathbf{a}||\mathbf{b}| \cos(\theta) = a_1 b_1 + a_2 b_2 + a_3 b_3, \tag{13.1.3}$$

where θ is the angle between the vector such that $0 \le \theta \le \pi$. The dot product yields a scalar answer. A particularly important case is $\mathbf{a} \cdot \mathbf{b} = 0$ with $|\mathbf{a}| \ne 0$, and $|\mathbf{b}| \ne 0$. In this case the vectors are orthogonal (perpendicular) to each other.

The other form of multiplication is the *cross product* which is defined by $\mathbf{a} \times \mathbf{b} = |\mathbf{a}||\mathbf{b}| \sin(\theta)\mathbf{n}$, where θ is the angle between the vectors such that $0 \le \theta \le \pi$ and $\mathbf{n}$ is a unit vector perpendicular to the plane of $\mathbf{a}$ and $\mathbf{b}$ with the direction given by the right-hand rule. A convenient method for computing the cross product from the scalar components of $\mathbf{a}$ and $\mathbf{b}$ is

$$\mathbf{a} \times \mathbf{b} = \begin{vmatrix} \mathbf{i} & \mathbf{j} & \mathbf{k} \\ a_1 & a_2 & a_3 \\ b_1 & b_2 & b_3 \end{vmatrix} = \begin{vmatrix} a_2 & a_3 \\ b_2 & b_3 \end{vmatrix} \mathbf{i} - \begin{vmatrix} a_1 & a_3 \\ b_1 & b_3 \end{vmatrix} \mathbf{j} + \begin{vmatrix} a_1 & a_2 \\ b_1 & b_2 \end{vmatrix} \mathbf{k}. \tag{13.1.4}$$

Two nonzero vectors $\mathbf{a}$ and $\mathbf{b}$ are *parallel* if and only if $\mathbf{a} \times \mathbf{b} = \mathbf{0}$.

Most of the vectors that we will use are vector-valued functions. These functions are vectors that vary either with a single parametric variable t or multiple variables, say x, y, and z.

The most commonly encountered example of a vector-valued function which varies with a single independent variable involves the trajectory of particles. If a *space curve* is parameterized by the equations $x = f(t)$, $y = g(t)$, and $z = h(t)$ with $a \le t \le b$, the position vector $\mathbf{r}(t) = f(t)\mathbf{i} + g(t)\mathbf{j} + h(t)\mathbf{k}$ gives the location of a point P as it moves from its initial position to its final position. Furthermore, because the increment quotient $\Delta\mathbf{r}/\Delta t$ is in the direction of a secant line, then the limit of this quotient as $\Delta t \to 0$, $\mathbf{r}'(t)$, gives the tangent to the curve at P.

• **Example 13.1.1: Foucault pendulum**

One of the great experiments of mid-nineteenth century physics was the demonstration by J. B. L. Foucault (1819–1868) in 1851 of the earth's rotation by designing a (spherical) pendulum, supported by a long wire, that essentially swings in an nonaccelerating coordinate system. This problem demonstrates many of the fundamental concepts of vector calculus.

The total force[1] acting on the bob of the pendulum is $\mathbf{F} = \mathbf{T} + m\mathbf{G}$, where $\mathbf{T}$ is the tension in the pendulum and $\mathbf{G}$ is the gravitational attraction per unit mass. Using Newton's second law,

$$\left.\frac{d^2\mathbf{r}}{dt^2}\right|_{\text{inertial}} = \frac{\mathbf{T}}{m} + \mathbf{G}, \tag{13.1.5}$$

where $\mathbf{r}$ is the position vector from a fixed point in an inertial coordinate system to the bob. This system is inconvenient because we live on a rotating coordinate system. Employing the conventional geographic coordinate system,[2] (13.1.5) becomes

$$\frac{d^2\mathbf{r}}{dt^2} + 2\boldsymbol{\Omega} \times \frac{d\mathbf{r}}{dt} + \boldsymbol{\Omega} \times (\boldsymbol{\Omega} \times \mathbf{r}) = \frac{\mathbf{T}}{m} + \mathbf{G}, \tag{13.1.6}$$

where $\boldsymbol{\Omega}$ is the angular rotation vector of the earth and $\mathbf{r}$ now denotes a position vector in the rotating reference system with its origin at the center of the earth and terminal point at the bob. If we define the gravity vector $\mathbf{g} = \mathbf{G} - \boldsymbol{\Omega} \times (\boldsymbol{\Omega} \times \mathbf{r})$, then the dynamical equation is

$$\frac{d^2\mathbf{r}}{dt^2} + 2\boldsymbol{\Omega} \times \frac{d\mathbf{r}}{dt} = \frac{\mathbf{T}}{m} + \mathbf{g}, \tag{13.1.7}$$

where the second term on the left side of (13.1.7) is called the *Coriolis force*.

Because the equation is *linear*, let us break the position vector $\mathbf{r}$ into two separate vectors: $\mathbf{r}_0$ and $\mathbf{r}_1$, where $\mathbf{r} = \mathbf{r}_0 + \mathbf{r}_1$. The vector $\mathbf{r}_0$ extends from the center of the earth to the pendulum's point of support and $\mathbf{r}_1$ extends from the support point to the bob. Because $\mathbf{r}_0$ is a constant in the geographic system,

$$\frac{d^2\mathbf{r}_1}{dt^2} + 2\boldsymbol{\Omega} \times \frac{d\mathbf{r}_1}{dt} = \frac{\mathbf{T}}{m} + \mathbf{g}. \tag{13.1.8}$$

If the length of the pendulum is L, then for small oscillations $\mathbf{r}_1 \approx x\mathbf{i} + y\mathbf{j} + L\mathbf{k}$ and the equations of motion are

$$\frac{d^2x}{dt^2} + 2\Omega \sin(\lambda)\frac{dy}{dt} = \frac{T_x}{m}, \tag{13.1.9}$$

$$\frac{d^2y}{dt^2} - 2\Omega \sin(\lambda)\frac{dx}{dt} = \frac{T_y}{m}, \tag{13.1.10}$$

---

[1] From Broxmeyer, C., 1960: Foucault pendulum effect in a Schuler-tuned system. *J. Aerosp. Sci.*, **27**, 343–347 with permission.

[2] For the derivation, see Marion, J. B., 1965: *Classical Dynamics of Particles and Systems*. Academic Press, §§12.2–12.3.

and

$$2\Omega \cos(\lambda) \frac{dy}{dt} - g = \frac{T_z}{m}, \tag{13.1.11}$$

where λ denotes the latitude of the point and Ω is the rotation rate of the earth. The relationships between the components of tension are $T_x = xT_z/L$, and $T_y = yT_z/L$. From (13.1.11),

$$\frac{T_z}{m} + g = 2\Omega \cos(\lambda) \frac{dy}{dt} \approx 0. \tag{13.1.12}$$

Substituting the definitions of T_x, T_y, and (13.1.12) into (13.1.9) and (13.1.10),

$$\frac{d^2 x}{dt^2} + \frac{g}{L} x + 2\Omega \sin(\lambda) \frac{dy}{dt} = 0, \tag{13.1.13}$$

and

$$\frac{d^2 y}{dt^2} + \frac{g}{L} y - 2\Omega \sin(\lambda) \frac{dx}{dt} = 0. \tag{13.1.14}$$

The approximate solution to these coupled differential equations is

$$x(t) = A_0 \cos[\Omega \sin(\lambda)t] \sin\left(\sqrt{g/L}\, t\right), \tag{13.1.15}$$

and

$$y(t) = A_0 \sin[\Omega \sin(\lambda)t] \sin\left(\sqrt{g/L}\, t\right), \tag{13.1.16}$$

if $\Omega^2 \ll g/L$. Thus, we have a pendulum that swings with an angular frequency $\sqrt{g/L}$. However, depending upon the *latitude* λ, the direction in which the pendulum swings changes counterclockwise with time, completing a full cycle in $2\pi/[\Omega \sin(\lambda)]$. This result is most clearly seen when $\lambda = \pi/2$ and we are at the North Pole. There the earth is turning underneath the pendulum. If initially we set the pendulum swinging along the 0° longitude, the pendulum will shift with time to longitudes east of the Greenwich median. Eventually, after 24 hours, the process repeats itself.

Consider now vector-valued functions that vary with several variables. A *vector function of position* assigns a vector value for every value of x, y, and z within some domain. Examples include the velocity field of a fluid at a given instant:

$$\mathbf{v} = u(x, y, z)\mathbf{i} + v(x, y, z)\mathbf{j} + w(x, y, z)\mathbf{k}. \tag{13.1.17}$$

Another example arises in electromagnetism where electric and magnetic fields often vary as a function of the space coordinates. For us, however, probably the most useful example involves the vector differential operator, *del* or *nabla*,

$$\nabla = \frac{\partial}{\partial x}\mathbf{i} + \frac{\partial}{\partial y}\mathbf{j} + \frac{\partial}{\partial z}\mathbf{k} \tag{13.1.18}$$

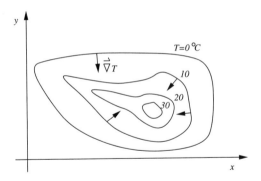

Figure 13.1.1: For a two-dimensional field $T(x, y)$, the gradient is a vector that is perpendicular to the isotherms $T(x, y) =$ constant and points in the direction of most rapidly increasing temperatures.

which we apply to the multivariable differentiable scalar function $F(x, y, z)$ to give the *gradient* ∇F.

An important geometric interpretation of the gradient—one which we shall use frequently—is the fact that ∇f is perpendicular (normal) to the level surface at a given point P. To prove this, let the equation $F(x, y, z) = c$ describe a three-dimensional surface. If the differentiable functions $x = f(t)$, $y = g(t)$, and $z = h(t)$ are the parametric equations of a curve on the surface, then the derivative of $F[f(t), g(t), h(t)] = c$ is

$$\frac{\partial F}{\partial x}\frac{dx}{dt} + \frac{\partial F}{\partial y}\frac{dy}{dt} + \frac{\partial F}{\partial z}\frac{dz}{dt} = 0, \qquad (13.1.19)$$

or

$$\nabla F \cdot \mathbf{r}' = 0. \qquad (13.1.20)$$

When $\mathbf{r}' \neq \mathbf{0}$, the vector ∇F is orthogonal to the tangent vector. Because our argument holds for any differentiable curve that passes through the arbitrary point (x, y, z), then ∇F is normal to the level surface at that point.

Figure 13.1.1 gives a common application of the gradient. Consider a two-dimensional temperature field $T(x, y)$. The level curves $T(x, y) =$ constant are lines that connect points where the temperature is the same (isotherms). The gradient in this case ∇T is a vector that is perpendicular or normal to these isotherms and points in the direction of most rapidly increasing temperature.

- **Example 13.1.2**

Let us find the gradient of the function $f(x, y, z) = x^2 z^2 \sin(4y)$. Using the definition of gradient,

$$\nabla f = \frac{\partial [x^2 z^2 \sin(4y)]}{\partial x}\mathbf{i} + \frac{\partial [x^2 z^2 \sin(4y)]}{\partial y}\mathbf{j} + \frac{\partial [x^2 z^2 \sin(4y)]}{\partial z}\mathbf{k} \qquad (13.1.21)$$

$$= 2xz^2 \sin(4y)\mathbf{i} + 4x^2 z^2 \cos(4y)\mathbf{j} + 2x^2 z \sin(4y)\mathbf{k}. \qquad (13.1.22)$$

• **Example 13.1.3**

    Let us find the unit normal to the unit sphere at any arbitrary point (x, y, z).

    The surface of a unit sphere is defined by the equation $f(x, y, z) = x^2 + y^2 + z^2 = 1$. Therefore, the normal is given by the gradient

$$\mathbf{N} = \nabla f = 2x\mathbf{i} + 2y\mathbf{j} + 2z\mathbf{k}, \qquad (\mathbf{13.1.23})$$

and the unit normal

$$\mathbf{n} = \frac{\nabla f}{|\nabla f|} = \frac{2x\mathbf{i} + 2y\mathbf{j} + 2z\mathbf{k}}{\sqrt{4x^2 + 4y^2 + 4z^2}} = x\mathbf{i} + y\mathbf{j} + z\mathbf{k}, \qquad (\mathbf{13.1.24})$$

because $x^2 + y^2 + z^2 = 1$.

• **Example 13.1.4**

    In Figure 13.1.2 MATLAB has been used to illustrate the unit normal of the surface $z = 4 - x^2 - y^2$. Here $f(x, y, z) = z + x^2 + y^2 = 4$ so that $\nabla f = 2x\mathbf{i} + 2y\mathbf{j} + \mathbf{k}$. The corresponding script is:

```
clear % clear variables
clf % clear figures
[x,y] = meshgrid(-2:0.5:2); % create the grid
z = 4 - x.^2 - y.^2; % compute surface within domain
% compute the gradient of f(x,y,z) = z + x^2 + y^2 = 4
% the x, y, and z components are u, v, and w
u = 2*x; v = 2*y; w = 1;
% find magnitude of gradient at each point
magnitude = sqrt(u.*u + v.*v + w.*w);
% compute unit gradient vector
u = u./magnitude; v = v./magnitude; w = w./magnitude;
mesh(x,y,z) % plot the surface
axis square
xlabel('x'); ylabel('y')
hold on
% plot the unit gradient vector
quiver3(x,y,z,u,v,w,0)
```

This figure clearly shows that gradient gives a vector which is perpendicular to the surface.

    A popular method for visualizing a vector field $\mathbf{F}$ is to draw space curves which are tangent to the vector field at each x, y, z. In fluid mechanics these lines are called *streamlines* while in physics they are generally called *lines of force* or *flux lines* for an electric, magnetic, or gravitational field. For a fluid

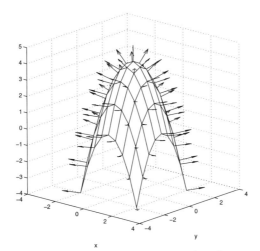

Figure 13.1.2: MATLAB plot of the function $z = 4 - x^2 - y^2$. The arrows give the unit normal to this surface.

with a velocity field that does not vary with time, the streamlines give the paths along which small parcels of the fluid move.

To find the streamlines of a given vector field $\mathbf{F}$ with components $P(x, y, z)$, $Q(x, y, z)$, and $R(x, y, z)$, we assume that we can parameterize the streamlines in the form $\mathbf{r}(t) = x(t)\mathbf{i} + y(t)\mathbf{j} + z(t)\mathbf{k}$. Then the tangent line is $\mathbf{r}'(t) = x'(t)\mathbf{i} + y'(t)\mathbf{j} + z'(t)\mathbf{k}$. Because the streamline must be parallel to the vector field at any t, $\mathbf{r}'(t) = \lambda \mathbf{F}$, or

$$\frac{dx}{dt} = \lambda P(x, y, z), \quad \frac{dy}{dt} = \lambda Q(x, y, z), \quad \text{and} \quad \frac{dz}{dt} = \lambda R(x, y, z), \quad (13.1.25)$$

or

$$\frac{dx}{P(x, y, z)} = \frac{dy}{Q(x, y, z)} = \frac{dz}{R(x, y, z)}. \quad (13.1.26)$$

The solution of this system of differential equations yields the streamlines.

• Example 13.1.5

Let us find the streamlines for the vector field $\mathbf{F} = \sec(x)\mathbf{i} - \cot(y)\mathbf{j} + \mathbf{k}$ that passes through the point $(\pi/4, \pi, 1)$. In this particular example, $\mathbf{F}$ represents a measured or computed fluid's velocity at a particular instant.

From (13.1.26),

$$\frac{dx}{\sec(x)} = -\frac{dy}{\cot(y)} = \frac{dz}{1}. \quad (13.1.27)$$

This yields two differential equations:

$$\cos(x)\, dx = -\frac{\sin(y)}{\cos(y)}\, dy, \quad \text{and} \quad dz = -\frac{\sin(y)}{\cos(y)}\, dy. \quad (13.1.28)$$

Integrating these equations gives

$$\sin(x) = \ln|\cos(y)| + c_1, \quad \text{and} \quad z = \ln|\cos(y)| + c_2. \tag{13.1.29}$$

Substituting for the given point, we finally have that

$$\sin(x) = \ln|\cos(y)| + \sqrt{2}/2, \quad \text{and} \quad z = \ln|\cos(y)| + 1. \tag{13.1.30}$$

• **Example 13.1.6**

Let us find the streamlines for the vector field $\mathbf{F} = \sin(z)\mathbf{j} + e^y\mathbf{k}$ that passes through the point $(2,0,0)$.
From (13.1.26),

$$\frac{dx}{0} = \frac{dy}{\sin(z)} = \frac{dz}{e^y}. \tag{13.1.31}$$

This yields two differential equations:

$$dx = 0, \quad \text{and} \quad \sin(z)\,dz = e^y\,dy. \tag{13.1.32}$$

Integrating these equations gives

$$x = c_1, \quad \text{and} \quad e^y = -\cos(z) + c_2. \tag{13.1.33}$$

Substituting for the given point, we finally have that

$$x = 2, \quad \text{and} \quad e^y = 2 - \cos(z). \tag{13.1.34}$$

Note that (13.1.34) only applies for a certain strip in the yz-plane.

Problems

Given the following vectors $\mathbf{a}$ and $\mathbf{b}$, verify that $\mathbf{a}\cdot(\mathbf{a}\times\mathbf{b}) = 0$, and $\mathbf{b}\cdot(\mathbf{a}\times\mathbf{b}) = 0$:

1. $\mathbf{a} = 4\mathbf{i} - 2\mathbf{j} + 5\mathbf{k}, \quad \mathbf{b} = 3\mathbf{i} + \mathbf{j} - \mathbf{k}$

2. $\mathbf{a} = \mathbf{i} - 3\mathbf{j} + \mathbf{k}, \quad \mathbf{b} = 2\mathbf{i} + 4\mathbf{k}$

3. $\mathbf{a} = \mathbf{i} + \mathbf{j} + \mathbf{k}, \quad \mathbf{b} = -5\mathbf{i} + 2\mathbf{j} + 3\mathbf{k}$

4. $\mathbf{a} = 8\mathbf{i} + \mathbf{j} - 6\mathbf{k}, \quad \mathbf{b} = \mathbf{i} - 2\mathbf{j} + 10\mathbf{k}$

5. $\mathbf{a} = 2\mathbf{i} + 7\mathbf{j} - 4\mathbf{k}, \quad \mathbf{b} = \mathbf{i} + \mathbf{j} - \mathbf{k}.$

6. Prove $\mathbf{a} \times (\mathbf{b} \times \mathbf{c}) = (\mathbf{a} \cdot \mathbf{c})\mathbf{b} - (\mathbf{a} \cdot \mathbf{b})\mathbf{c}.$

7. Prove $\mathbf{a} \times (\mathbf{b} \times \mathbf{c}) + \mathbf{b} \times (\mathbf{c} \times \mathbf{a}) + \mathbf{c} \times (\mathbf{a} \times \mathbf{b}) = \mathbf{0}$.

Find the gradient of the following functions:

8. $f(x, y, z) = xy^2/z^3$

9. $f(x, y, z) = xy \cos(yz)$

10. $f(x, y, z) = \ln(x^2 + y^2 + z^2)$

11. $f(x, y, z) = x^2 y^2 (2z + 1)^2$

12. $f(x, y, z) = 2x - y^2 + z^2$.

Use MATLAB to illustrate the following surfaces as well as the the unit normal.

13. $z = 3$

14. $x^2 + y^2 = 4$

15. $z = x^2 + y^2$

16. $z = \sqrt{x^2 + y^2}$

17. $z = y$

18. $x + y + z = 1$

19. $z = x^2$.

Find the streamlines for the following vector fields that pass through the specified point:

20. $\mathbf{F} = \mathbf{i} + \mathbf{j} + \mathbf{k}$; $(0, 1, 1)$

21. $\mathbf{F} = 2\mathbf{i} - y^2\mathbf{j} + z\mathbf{k}$; $(1, 1, 1)$

22. $\mathbf{F} = 3x^2\mathbf{i} - y^2\mathbf{j} + z^2\mathbf{k}$; $(2, 1, 3)$

23. $\mathbf{F} = x^2\mathbf{i} + y^2\mathbf{j} - z^3\mathbf{k}$; $(1, 1, 1)$

24. $\mathbf{F} = (1/x)\mathbf{i} + e^y\mathbf{j} - \mathbf{k}$; $(2, 0, 4)$

25. Solve the differential equations (13.1.13)–(13.1.14) with the initial conditions $x(0) = y(0) = y'(0) = 0$, and $x'(0) = A_0\sqrt{g/L}$ assuming that $\Omega^2 \ll g/L$.

26. If a fluid is bounded by a fixed surface $f(x, y, z) = c$, show that the fluid must satisfy the boundary condition $\mathbf{v} \cdot \nabla f = 0$, where $\mathbf{v}$ is the velocity of the fluid.

27. A sphere of radius a is moving in a fluid with the constant velocity $\mathbf{u}$. Show that the fluid satisfies the boundary condition $(\mathbf{v} - \mathbf{u}) \cdot (\mathbf{r} - \mathbf{u}t) = 0$ at the surface of the sphere, if the center of the sphere coincides with the origin at $t = 0$ and $\mathbf{v}$ denotes the velocity of the fluid.

13.2 DIVERGENCE AND CURL

Consider a vector field $\mathbf{v}$ defined in some region of three-dimensional space. The function $\mathbf{v}(\mathbf{r})$ can be resolved into components along the $\mathbf{i}$, $\mathbf{j}$, and $\mathbf{k}$ directions or

$$\mathbf{v}(\mathbf{r}) = u(x, y, z)\mathbf{i} + v(x, y, z)\mathbf{j} + w(x, y, z)\mathbf{k}. \tag{13.2.1}$$

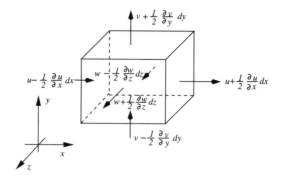

Figure 13.2.1: Divergence of a vector function $\mathbf{v}(x, y, z)$.

If $\mathbf{v}$ is a fluid's velocity field, then we can compute the flow rate through a small (differential) rectangular box defined by increments $(\Delta x, \Delta y, \Delta z)$ centered at the point (x, y, z). See Figure 13.2.1. The flow out from the box through the face with the outwardly pointing normal $\mathbf{n} = -\mathbf{j}$ is

$$\mathbf{v} \cdot (-\mathbf{j}) = -v(x, y - \Delta y/2, z)\Delta x\Delta z, \qquad (13.2.2)$$

and the flow through the face with the outwardly pointing normal $\mathbf{n} = \mathbf{j}$ is

$$\mathbf{v} \cdot \mathbf{j} = v(x, y + \Delta y/2, z)\Delta x\Delta z. \qquad (13.2.3)$$

The net flow through the two faces is

$$[v(x, y + \Delta y/2, z) - v(x, y - \Delta y/2, z)]\Delta x\Delta z \approx v_y(x, y, z)\Delta x\Delta y\Delta z. \quad (13.2.4)$$

A similar analysis of the other faces and combination of the results give the approximate total flow from the box as

$$[u_x(x, y, z) + v_y(x, y, z) + w_z(x, y, z)]\Delta x\Delta y\Delta z. \qquad (13.2.5)$$

Dividing by the volume $\Delta x\Delta y\Delta z$ and taking the limit as the dimensions of the box tend to zero yield $u_x + v_y + w_z$ as the flow out from (x, y, z) per unit volume per unit time. This scalar quantity is called the *divergence* of the vector $\mathbf{v}$:

$$\operatorname{div}(\mathbf{v}) = \nabla \cdot \mathbf{v} = \left(\frac{\partial}{\partial x}\mathbf{i} + \frac{\partial}{\partial y}\mathbf{j} + \frac{\partial}{\partial z}\mathbf{k}\right) \cdot (u\mathbf{i} + v\mathbf{j} + w\mathbf{k}) = u_x + v_y + w_z. \quad (13.2.6)$$

Thus, if the divergence is positive, either the fluid is expanding and its density at the point is falling with time, or the point is a *source* at which fluid is entering the field. When the divergence is negative, either the fluid is contracting and its density is rising at the point, or the point is a negative source or *sink* at which fluid is leaving the field.

If the divergence of a vector field is zero everywhere within a domain, then the flux entering any element of space exactly balances that leaving it and the vector field is called *nondivergent* or *solenoidal* (from a Greek word meaning a tube). For a fluid, if there are no sources or sinks, then its density cannot change.

Some useful properties of the divergence operator are

$$\nabla \cdot (\mathbf{F} + \mathbf{G}) = \nabla \cdot \mathbf{F} + \nabla \cdot \mathbf{G}, \tag{13.2.7}$$

$$\nabla \cdot (\varphi \mathbf{F}) = \varphi \nabla \cdot \mathbf{F} + \mathbf{F} \cdot \nabla \varphi \tag{13.2.8}$$

and

$$\nabla^2 \varphi = \nabla \cdot \nabla \varphi = \varphi_{xx} + \varphi_{yy} + \varphi_{zz}. \tag{13.2.9}$$

The expression (13.2.9) is very important in physics and is given the special name of the *Laplacian*.[3]

- **Example 13.2.1**

If $\mathbf{F} = x^2 z \mathbf{i} - 2y^3 z^2 \mathbf{j} + xy^2 z \mathbf{k}$, compute the divergence of $\mathbf{F}$.

$$\nabla \cdot \mathbf{F} = \frac{\partial}{\partial x}\left(x^2 z\right) + \frac{\partial}{\partial y}\left(-2y^3 z^2\right) + \frac{\partial}{\partial z}\left(xy^2 z\right) \tag{13.2.10}$$

$$= 2xz - 6y^2 z^2 + xy^2. \tag{13.2.11}$$

- **Example 13.2.2**

If $\mathbf{r} = x\mathbf{i} + y\mathbf{j} + z\mathbf{k}$, show that $\mathbf{r}/|\mathbf{r}|^3$ is nondivergent.

$$\nabla \cdot \left(\frac{\mathbf{r}}{|\mathbf{r}|^3}\right) = \frac{\partial}{\partial x}\left[\frac{x}{(x^2 + y^2 + z^2)^{3/2}}\right] + \frac{\partial}{\partial y}\left[\frac{y}{(x^2 + y^2 + z^2)^{3/2}}\right]$$

$$+ \frac{\partial}{\partial z}\left[\frac{z}{(x^2 + y^2 + z^2)^{3/2}}\right] \tag{13.2.12}$$

$$= \frac{3}{(x^2 + y^2 + z^2)^{3/2}} - \frac{3x^2 + 3y^2 + 3z^2}{(x^2 + y^2 + z^2)^{5/2}} = 0. \tag{13.2.13}$$

Another important vector function involving the vector field $\mathbf{v}$ is the curl of $\mathbf{v}$, written curl$(\mathbf{v})$ or rot$(\mathbf{v})$ in some older textbooks. In fluid flow problems

---

[3] Some mathematicians write Δ instead of ∇^2.

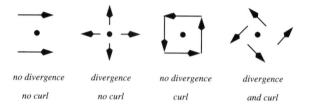

no divergence   divergence   no divergence   divergence

no curl   no curl   curl   and curl

Figure 13.2.2: Examples of vector fields with and without divergence and curl.

it is proportional to the instantaneous angular velocity of a fluid element. In rectangular coordinates,

$$\text{curl}(\mathbf{v}) = \nabla \times \mathbf{v} = (w_y - v_z)\mathbf{i} + (u_z - w_x)\mathbf{j} + (v_x - u_y)\mathbf{k}, \qquad (13.2.14)$$

where $\mathbf{v} = u\mathbf{i} + v\mathbf{j} + w\mathbf{k}$ as before. However, it is best remembered in the mnemonic form:

$$\nabla \times \mathbf{F} = \begin{vmatrix} \mathbf{i} & \mathbf{j} & \mathbf{k} \\ \frac{\partial}{\partial x} & \frac{\partial}{\partial y} & \frac{\partial}{\partial z} \\ u & v & w \end{vmatrix} = (w_y - v_z)\mathbf{i} + (u_z - w_x)\mathbf{j} + (v_x - u_y)\mathbf{k}. \quad (13.2.15)$$

If the curl of a vector field is zero everywhere within a region, then the field is *irrotational*.

    Figure 13.2.2 illustrates graphically some vector fields that do and do not possess divergence and curl. Let the vectors that are illustrated represent the motion of fluid particles. In the case of divergence only, fluid is streaming from the point, at which the density is falling. Alternatively the point could be a source. In the case where there is only curl, the fluid rotates about the point and the fluid is incompressible. Finally, the point that possesses both divergence and curl is a compressible fluid with rotation.

    Some useful computational formulas exist for both the divergence and curl operations:

$$\nabla \times (\mathbf{F} + \mathbf{G}) = \nabla \times \mathbf{F} + \nabla \times \mathbf{G}, \qquad (13.2.16)$$

$$\nabla \times \nabla \varphi = \mathbf{0}, \qquad (13.2.17)$$

$$\nabla \cdot \nabla \times \mathbf{F} = 0, \qquad (13.2.18)$$

$$\nabla \times (\varphi \mathbf{F}) = \varphi \nabla \times \mathbf{F} + \nabla \varphi \times \mathbf{F}, \qquad (13.2.19)$$

$$\nabla(\mathbf{F} \cdot \mathbf{G}) = (\mathbf{F} \cdot \nabla)\mathbf{G} + (\mathbf{G} \cdot \nabla)\mathbf{F} + \mathbf{F} \times (\nabla \times \mathbf{G}) + \mathbf{G} \times (\nabla \times \mathbf{F}), \quad (13.2.20)$$

$$\nabla \times (\mathbf{F} \times \mathbf{G}) = (\mathbf{G} \cdot \nabla)\mathbf{F} - (\mathbf{F} \cdot \nabla)\mathbf{G} + \mathbf{F}(\nabla \cdot \mathbf{G}) - \mathbf{G}(\nabla \cdot \mathbf{F}), \quad (13.2.21)$$

$$\nabla \times (\nabla \times \mathbf{F}) = \nabla(\nabla \cdot \mathbf{F}) - (\nabla \cdot \nabla)\mathbf{F}, \qquad (13.2.22)$$

and

$$\nabla \cdot (\mathbf{F} \times \mathbf{G}) = \mathbf{G} \cdot \nabla \times \mathbf{F} - \mathbf{F} \cdot \nabla \times \mathbf{G}. \qquad (13.2.23)$$

In this book the operation $\nabla \mathbf{F}$ is undefined.

• **Example 13.2.3**

If $\mathbf{F} = xz^3\mathbf{i} - 2x^2yz\mathbf{j} + 2yz^4\mathbf{k}$, compute the curl of $\mathbf{F}$ and verify that $\nabla \cdot \nabla \times \mathbf{F} = 0$.

From the definition of curl,

$$\nabla \times \mathbf{F} = \begin{vmatrix} \mathbf{i} & \mathbf{j} & \mathbf{k} \\ \frac{\partial}{\partial x} & \frac{\partial}{\partial y} & \frac{\partial}{\partial z} \\ xz^3 & -2x^2yz & 2yz^4 \end{vmatrix} \qquad (13.2.24)$$

$$= \left[\frac{\partial}{\partial y}\left(2yz^4\right) - \frac{\partial}{\partial z}\left(-2x^2yz\right) \right]\mathbf{i} - \left[\frac{\partial}{\partial x}\left(2yz^4\right) - \frac{\partial}{\partial z}\left(xz^3\right) \right]\mathbf{j}$$

$$+ \left[\frac{\partial}{\partial x}\left(-2x^2yz\right) - \frac{\partial}{\partial y}\left(xz^3\right) \right]\mathbf{k} \qquad (13.2.25)$$

$$= (2z^4 + 2x^2y)\mathbf{i} - (0 - 3xz^2)\mathbf{j} + (-4xyz - 0)\mathbf{k} \qquad (13.2.26)$$

$$= (2z^4 + 2x^2y)\mathbf{i} + 3xz^2\mathbf{j} - 4xyz\mathbf{k}. \qquad (13.2.27)$$

From the definition of divergence and (13.2.27),

$$\nabla \cdot \nabla \times \mathbf{F} = \frac{\partial}{\partial x}(2z^4 + 2x^2y) + \frac{\partial}{\partial y}(3xz^2) + \frac{\partial}{\partial z}(-4xyz) = 4xy + 0 - 4xy = 0.$$
$$(13.2.28)$$

• **Example 13.2.4: Potential flow theory**

One of the topics in most elementary fluid mechanics courses is the study of irrotational and nondivergent fluid flows. Because the fluid is irrotational, the velocity vector field $\mathbf{v}$ satisfies $\nabla \times \mathbf{v} = \mathbf{0}$. From (13.2.17) we can introduce a potential φ such that $\mathbf{v} = \nabla\varphi$. Because the flow field is nondivergent, $\nabla \cdot \mathbf{v} = \nabla^2\varphi = 0$. Thus, the fluid flow can be completely described in terms of solutions to Laplace's equation. This area of fluid mechanics is called *potential flow theory*.

Problems

Compute $\nabla \cdot \mathbf{F}, \nabla \times \mathbf{F}, \nabla \cdot (\nabla \times \mathbf{F})$, and $\nabla(\nabla \cdot \mathbf{F})$, for the following vector fields:

1. $\mathbf{F} = x^2z\mathbf{i} + yz^2\mathbf{j} + xy^2\mathbf{k}$

2. $\mathbf{F} = 4x^2y^2\mathbf{i} + (2x + 2yz)\mathbf{j} + (3z + y^2)\mathbf{k}$

3. $\mathbf{F} = (x - y)^2\mathbf{i} + e^{-xy}\mathbf{j} + xze^{2y}\mathbf{k}$

4. $\mathbf{F} = 3xy\mathbf{i} + 2xz^2\mathbf{j} + y^3\mathbf{k}$

5. $\mathbf{F} = 5yz\mathbf{i} + x^2 z\mathbf{j} + 3x^3\mathbf{k}$

6. $\mathbf{F} = y^3\mathbf{i} + (x^3 y^2 - xy)\mathbf{j} - (x^3 yz - xz)\mathbf{k}$

7. $\mathbf{F} = xe^{-y}\mathbf{i} + yz^2\mathbf{j} + 3e^{-z}\mathbf{k}$

8. $\mathbf{F} = y\ln(x)\mathbf{i} + (2 - 3yz)\mathbf{j} + xyz^3\mathbf{k}$

9. $\mathbf{F} = xyz\mathbf{i} + x^3 yze^z\mathbf{j} + xye^z\mathbf{k}$

10. $\mathbf{F} = (xy^3 - z^4)\mathbf{i} + 4x^4 y^2 z\mathbf{j} - y^4 z^5\mathbf{k}$

11. $\mathbf{F} = xy^2\mathbf{i} + xyz^2\mathbf{j} + xy\cos(z)\mathbf{k}$

12. $\mathbf{F} = xy^2\mathbf{i} + xyz^2\mathbf{j} + xy\sin(z)\mathbf{k}$

13. $\mathbf{F} = xy^2\mathbf{i} + xyz\mathbf{j} + xy\cos(z)\mathbf{k}$

14. (a) Assuming continuity of all partial derivatives, show that

$$\nabla \times (\nabla \times \mathbf{F}) = \nabla(\nabla \cdot \mathbf{F}) - \nabla^2 \mathbf{F}.$$

(b) Using $\mathbf{F} = 3xy\mathbf{i} + 4yz\mathbf{j} + 2xz\mathbf{k}$, verify the results in part (a).

15. If $\mathbf{E} = \mathbf{E}(x, y, z, t)$ and $\mathbf{B} = \mathbf{B}(x, y, z, t)$ represent the electric and magnetic fields in a vacuum, Maxwell's field equations are:

$$\nabla \cdot \mathbf{E} = 0, \qquad \nabla \times \mathbf{E} = -\frac{1}{c}\frac{\partial \mathbf{B}}{\partial t},$$

$$\nabla \cdot \mathbf{B} = 0, \qquad \nabla \times \mathbf{B} = \frac{1}{c}\frac{\partial \mathbf{E}}{\partial t},$$

where c is the speed of light. Using the results from Problem 14, show that $\mathbf{E}$ and $\mathbf{B}$ satisfy

$$\nabla^2 \mathbf{E} = \frac{1}{c^2}\frac{\partial^2 \mathbf{E}}{\partial t^2}, \quad \text{and} \quad \nabla^2 \mathbf{B} = \frac{1}{c^2}\frac{\partial^2 \mathbf{B}}{\partial t^2}.$$

16. If f and g are continuously differentiable scalar fields, show that $\nabla f \times \nabla g$ is solenoidal. Hint: Show that $\nabla f \times \nabla g = \nabla \times (f\nabla g)$.

17. An inviscid (frictionless) fluid in equilibrium obeys the relationship $\nabla p = \rho \mathbf{F}$, where ρ denotes the density of the fluid, p denotes the pressure, and $\mathbf{F}$ denotes the body forces (such as gravity). Show that $\mathbf{F} \cdot \nabla \times \mathbf{F} = 0$.

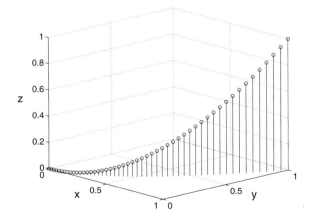

Figure 13.3.1: Diagram for the line integration in Example 13.3.1.

13.3 LINE INTEGRALS

Line integrals are ubiquitous in physics. In mechanics they are used to compute work. In electricity and magnetism, they provide simple methods for computing the electric and magnetic fields for simple geometries.

The line integral most frequently encountered is an *oriented* one in which the path C is directed and the integrand is the dot product between the vector function $\mathbf{F}(\mathbf{r})$ and the tangent of the path $d\mathbf{r}$. It is usually written in the economical form

$$\int_C \mathbf{F} \cdot d\mathbf{r} = \int_C P(x,y,z)\,dx + Q(x,y,z)\,dy + R(x,y,z)\,dz, \qquad (\mathbf{13.3.1})$$

where $\mathbf{F} = P(x,y,z)\mathbf{i} + Q(x,y,z)\mathbf{j} + R(x,y,z)\mathbf{k}$. If the starting and terminal points are the same so that the contour is closed, then this *closed contour integral* will be denoted by $\oint_C$. In the following examples we show how to evaluate the line integrals along various types of curves.

• **Example 13.3.1**

If $\mathbf{F} = (3x^2 + 6y)\mathbf{i} - 14yz\mathbf{j} + 20xz^2\mathbf{k}$, let us evaluate the line integral $\int_C \mathbf{F} \cdot d\mathbf{r}$ along the parametric curves $x(t) = t$, $y(t) = t^2$, and $z(t) = t^3$ from the point $(0,0,0)$ to $(1,1,1)$. Using the MATLAB commands

```
>> clear
>> t = 0:0.02:1
>> stem3(t,t.^2,t.^3); xlabel('x','Fontsize',20); ...
   ylabel('y','Fontsize',20); zlabel('z','Fontsize',20);
```

we illustrate these parametric curves in Figure 13.3.1.

We begin by finding the values of t which give the corresponding end points. A quick check shows that $t = 0$ gives $(0,0,0)$ while $t = 1$ yields $(1,1,1)$.

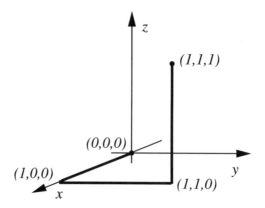

Figure 13.3.2: Diagram for the line integration in Example 13.3.2.

It should be noted that the same value of t must give the correct coordinates in each direction. Failure to do so suggests an error in the parameterization. Therefore,

$$\int_C \mathbf{F} \cdot d\mathbf{r} = \int_0^1 (3t^2 + 6t^2)\, dt - 14t^2(t^3)\, d(t^2) + 20t(t^3)^2 d(t^3) \qquad (13.3.2)$$

$$= \int_0^1 9t^2\, dt - 28t^6\, dt + 60t^9\, dt \qquad (13.3.3)$$

$$= \left(3t^3 - 4t^7 + 6t^{10}\right)\big|_0^1 = 5. \qquad (13.3.4)$$

• **Example 13.3.2**

Let us redo the previous example with a contour that consists of three "dog legs," namely straight lines from $(0,0,0)$ to $(1,0,0)$, from $(1,0,0)$ to $(1,1,0)$, and from $(1,1,0)$ to $(1,1,1)$. See Figure 13.3.2.

In this particular problem we break the integration down into integrals along each of the legs:

$$\int_C \mathbf{F} \cdot d\mathbf{r} = \int_{C_1} \mathbf{F} \cdot d\mathbf{r} + \int_{C_2} \mathbf{F} \cdot d\mathbf{r} + \int_{C_3} \mathbf{F} \cdot d\mathbf{r}. \qquad (13.3.5)$$

For C_1, $y = z = dy = dz = 0$, and

$$\int_{C_1} \mathbf{F} \cdot d\mathbf{r} = \int_0^1 (3x^2 + 6\cdot 0)\, dx - 14\cdot 0\cdot 0\cdot 0 + 20x\cdot 0^2\cdot 0 = \int_0^1 3x^2\, dx = 1. \ (13.3.6)$$

For C_2, $x = 1$ and $z = dx = dz = 0$, so that

$$\int_{C_2} \mathbf{F} \cdot d\mathbf{r} = \int_0^1 (3\cdot 1^2 + 6y)\cdot 0 - 14y\cdot 0\cdot dy + 20\cdot 1\cdot 0^2\cdot 0 = 0. \quad (13.3.7)$$

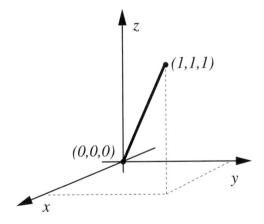

Figure 13.3.3: Diagram for the line integration in Example 13.3.3.

For C_3, $x = y = 1$ and $dx = dy = 0$, so that

$$\int_{C_3} \mathbf{F} \cdot d\mathbf{r} = \int_0^1 (3 \cdot 1^2 + 6 \cdot 1) \cdot 0 - 14 \cdot 1 \cdot z \cdot 0 + 20 \cdot 1 \cdot z^2 \, dz = \int_0^1 20z^2 \, dz = \tfrac{20}{3}.$$
$$(13.3.8)$$

Therefore,

$$\int_C \mathbf{F} \cdot d\mathbf{r} = \tfrac{23}{3}. \qquad (13.3.9)$$

• **Example 13.3.3**

For our third calculation, we redo the first example where the contour is a straight line. The parameterization in this case is $x = y = z = t$ with $0 \le t \le 1$. See Figure 13.3.3. Then,

$$\int_C \mathbf{F} \cdot d\mathbf{r} = \int_0^1 (3t^2 + 6t) \, dt - 14(t)(t) \, dt + 20t(t)^2 \, dt \qquad (13.3.10)$$

$$= \int_0^1 (3t^2 + 6t - 14t^2 + 20t^3) \, dt = \tfrac{13}{3}. \qquad (13.3.11)$$

An interesting aspect of these three examples is that, although we used a common vector field and moved from $(0, 0, 0)$ to $(1, 1, 1)$ in each case, we obtained a different answer in each case. Thus, for this vector field, the line integral is *path dependent*. This is generally true. In the next section we will meet *conservative vector fields* where the results will be path independent.

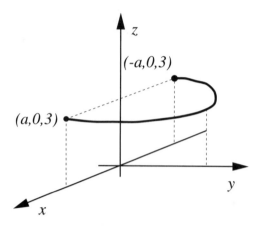

Figure 13.3.4: Diagram for the line integration in Example 13.3.4.

• **Example 13.3.4**

If $\mathbf{F} = (x^2 + y^2)\mathbf{i} - 2xy\mathbf{j} + x\mathbf{k}$, let us evaluate $\int_C \mathbf{F} \cdot d\mathbf{r}$ if the contour is that portion of the circle $x^2 + y^2 = a^2$ from the point $(a, 0, 3)$ to $(-a, 0, 3)$. See Figure 13.3.4.

The parametric equations for this example are $x = a\cos(\theta)$, $dx = -a\sin(\theta)\,d\theta$, $y = a\sin(\theta)$, $dy = a\cos(\theta)\,d\theta$, $z = 3$, and $dz = 0$ with $0 \le \theta \le \pi$. Therefore,

$$
\int_C \mathbf{F} \cdot d\mathbf{r} = \int_0^\pi [a^2 \cos^2(\theta) + a^2 \sin^2(\theta)][-a\sin(\theta)\,d\theta]
$$

$$
- 2a^2 \cos(\theta) \sin(\theta)[a\cos(\theta)\,d\theta] + a\cos(\theta) \cdot 0 \quad (13.3.12)
$$

$$
= -a^3 \int_0^\pi \sin(\theta)\,d\theta - 2a^3 \int_0^\pi \cos^2(\theta) \sin(\theta)\,d\theta \quad (13.3.13)
$$

$$
= a^3 \cos(\theta)\big|_0^\pi + \tfrac{2}{3}a^3 \cos^3(\theta)\big|_0^\pi \quad (13.3.14)
$$

$$
= -2a^3 - \tfrac{4}{3}a^3 = -\tfrac{10}{3}a^3. \quad (13.3.15)
$$

• **Example 13.3.5: Circulation**

Let $\mathbf{v}(x, y, z)$ denote the velocity at the point (x, y, z) in a moving fluid. If it varies with time, this is the velocity at a particular instant of time. The integral $\oint_C \mathbf{v} \cdot d\mathbf{r}$ around a closed path C is called the *circulation* around that path. The average component of velocity along the path is

$$
\bar{v}_s = \frac{\oint_C v_s\,ds}{s} = \frac{\oint_C \mathbf{v} \cdot d\mathbf{r}}{s}, \quad (13.3.16)
$$

where s is the total length of the path. The circulation is thus $\oint_C \mathbf{v} \cdot d\mathbf{r} = \bar{v}_s s$, the product of the length of the path and the average velocity along the path. When the circulation is positive, the flow is more in the direction of integration than opposite to it. Circulation is thus an indication and to some extent a measure of motion around the path.

Problems

Evaluate $\int_C \mathbf{F} \cdot d\mathbf{r}$ for the following vector fields and curves:

1. $\mathbf{F} = y\sin(\pi z)\mathbf{i} + x^2 e^y \mathbf{j} + 3xz\mathbf{k}$ and C is the curve $x = t$, $y = t^2$, and $z = t^3$ from $(0,0,0)$ to $(1,1,1)$. Use MATLAB to illustrate the parametric curves.

2. $\mathbf{F} = y\mathbf{i} + z\mathbf{j} + x\mathbf{k}$ and C consists of the line segments from $(0,0,0)$ to $(2,3,0)$, and from $(2,3,0)$ to $(2,3,4)$. Use MATLAB to illustrate the parametric curves.

3. $\mathbf{F} = e^x\mathbf{i} + xe^{xy}\mathbf{j} + xye^{xyz}\mathbf{k}$ and C is the curve $x = t$, $y = t^2$, and $z = t^3$ with $0 \leq t \leq 2$. Use MATLAB to illustrate the parametric curves.

4. $\mathbf{F} = yz\mathbf{i} + xz\mathbf{j} + xy\mathbf{k}$ and C is the curve $x = t^3$, $y = t^2$, and $z = t$ with $1 \leq t \leq 2$. Use MATLAB to illustrate the parametric curves.

5. $\mathbf{F} = y\mathbf{i} - x\mathbf{j} + 3xy\mathbf{k}$ and C consists of the semicircle $x^2 + y^2 = 4$, $z = 0$, $y > 0$, and the line segment from $(-2,0,0)$ to $(2,0,0)$. Use MATLAB to illustrate the parametric curves.

6. $\mathbf{F} = (x+2y)\mathbf{i} + (6y - 2x)\mathbf{j}$ and C consists of the sides of the triangle with vertices at $(0,0,0)$, $(1,1,1)$, and $(1,1,0)$. Proceed from $(0,0,0)$ to $(1,1,1)$ to $(1,1,0)$ and back to $(0,0,0)$. Use MATLAB to illustrate the parametric curves.

7. $\mathbf{F} = 2xz\mathbf{i} + 4y^2\mathbf{j} + x^2\mathbf{k}$ and C is taken counterclockwise around the ellipse $x^2/4 + y^2/9 = 1$, $z = 1$. Use MATLAB to illustrate the parametric curves.

8. $\mathbf{F} = 2x\mathbf{i} + y\mathbf{j} + z\mathbf{k}$ and C is the contour $x = t$, $y = \sin(t)$, and $z = \cos(t) + \sin(t)$ with $0 \leq t \leq 2\pi$. Use MATLAB to illustrate the parametric curves.

9. $\mathbf{F} = (2y^2 + z)\mathbf{i} + 4xy\mathbf{j} + x\mathbf{k}$ and C is the spiral $x = \cos(t)$, $y = \sin(t)$, and $z = t$ with $0 \leq t \leq 2\pi$ between the points $(1,0,0)$ and $(1,0,2\pi)$. Use MATLAB to illustrate the parametric curves.

10. $\mathbf{F} = x^2\mathbf{i} + y^2\mathbf{j} + (z^2 + 2xy)\mathbf{k}$ and C consists of the edges of the triangle with vertices at $(0,0,0)$, $(1,1,0)$, and $(0,1,0)$. Proceed from $(0,0,0)$ to $(1,1,0)$

to $(0, 1, 0)$ and back to $(0, 0, 0)$. Use MATLAB to illustrate the parametric curves.

13.4 THE POTENTIAL FUNCTION

In §13.2 we showed that the curl operation applied to a gradient produces the zero vector: $\nabla \times \nabla \varphi = \mathbf{0}$. Consequently, if we have a vector field $\mathbf{F}$ such that $\nabla \times \mathbf{F} \equiv \mathbf{0}$ everywhere, then that vector field is called a *conservative* field and we can compute a potential φ such that $\mathbf{F} = \nabla \varphi$.

• **Example 13.4.1**

Let us show that the vector field $\mathbf{F} = ye^{xy}\cos(z)\mathbf{i} + xe^{xy}\cos(z)\mathbf{j} - e^{xy}\sin(z)\mathbf{k}$ is conservative and then find the corresponding potential function.

To show that the field is conservative, we compute the curl of $\mathbf{F}$ or

$$\nabla \times \mathbf{F} = \begin{vmatrix} \mathbf{i} & \mathbf{j} & \mathbf{k} \\ \frac{\partial}{\partial x} & \frac{\partial}{\partial y} & \frac{\partial}{\partial z} \\ ye^{xy}\cos(z) & xe^{xy}\cos(z) & -e^{xy}\sin(z) \end{vmatrix} = \mathbf{0}. \qquad (13.4.1)$$

To find the potential we must solve three partial differential equations:

$$\varphi_x = ye^{xy}\cos(z) = \mathbf{F} \cdot \mathbf{i}, \qquad (13.4.2)$$

$$\varphi_y = xe^{xy}\cos(z) = \mathbf{F} \cdot \mathbf{j}, \qquad (13.4.3)$$

and

$$\varphi_z = -e^{xy}\sin(z) = \mathbf{F} \cdot \mathbf{k}. \qquad (13.4.4)$$

We begin by integrating any one of these three equations. Choosing (13.4.2),

$$\varphi(x, y, z) = e^{xy}\cos(z) + f(y, z). \qquad (13.4.5)$$

To find $f(y, z)$ we differentiate (13.4.5) with respect to y and find that

$$\varphi_y = xe^{xy}\cos(z) + f_y(y, z) = xe^{xy}\cos(z) \qquad (13.4.6)$$

from (13.4.3). Thus, $f_y = 0$ and $f(y, z)$ can only be a function of z, say $g(z)$. Then,

$$\varphi(x, y, z) = e^{xy}\cos(z) + g(z). \qquad (13.4.7)$$

Finally,

$$\varphi_z = -e^{xy}\sin(z) + g'(z) = -e^{xy}\sin(z) \qquad (13.4.8)$$

from (13.4.4) and $g'(z) = 0$. Therefore, the potential is

$$\varphi(x, y, z) = e^{xy}\cos(z) + \text{constant}. \qquad (13.4.9)$$

Potentials can be very useful in computing line integrals because

$$\int_C \mathbf{F} \cdot d\mathbf{r} = \int_C \varphi_x \, dx + \varphi_y \, dy + \varphi_z \, dz = \int_C d\varphi = \varphi(B) - \varphi(A), \quad (\textbf{13.4.10})$$

where the point B is the terminal point of the integration while the point A is the starting point. Thus, any path integration between any two points is *path independent.*

Finally, if we close the path so that A and B coincide, then

$$\oint_C \mathbf{F} \cdot d\mathbf{r} = 0. \qquad (\textbf{13.4.11})$$

It should be noted that the converse is *not* true. Just because $\oint_C \mathbf{F} \cdot d\mathbf{r} = 0$, we do not necessarily have a conservative field $\mathbf{F}$.

In summary then, an irrotational vector in a given region has three fundamental properties: (1) its integral around every simply connected circuit is zero, (2) its curl equals zero, and (3) it is the gradient of a scalar function. For continuously differentiable vectors these properties are equivalent. For vectors which are only piece-wise differentiable, this is not true. Generally the first property is the most fundamental and taken as the definition of irrotationality.

- **Example 13.4.2**

Using the potential found in Example 13.4.1, let us find the value of the line integral $\int_C \mathbf{F} \cdot d\mathbf{r}$ from the point $(0,0,0)$ to $(-1,2,\pi)$.

From (13.4.9),

$$\int_C \mathbf{F} \cdot d\mathbf{r} = \left[e^{xy} \cos(z) + \text{constant} \right] \Big|_{(0,0,0)}^{(-1,2,\pi)} = -1 - e^{-2}. \qquad (\textbf{13.4.12})$$

Problems

Verify that the following vector fields are conservative and then find the corresponding potential:

1. $\mathbf{F} = 2xy\mathbf{i} + (x^2 + 2yz)\mathbf{j} + (y^2 + 4)\mathbf{k}$

2. $\mathbf{F} = (2x + 2ze^{2x})\mathbf{i} + (2y - 1)\mathbf{j} + e^{2x}\mathbf{k}$

3. $\mathbf{F} = yz\mathbf{i} + xz\mathbf{j} + xy\mathbf{k}$ 4. $\mathbf{F} = 2x\mathbf{i} + 3y^2\mathbf{j} + 4z^3\mathbf{k}$

5. $\mathbf{F} = [2x \sin(y) + e^{3z}]\mathbf{i} + x^2 \cos(y)\mathbf{j} + (3xe^{3z} + 4)\mathbf{k}$

6. $\mathbf{F} = (2x + 5)\mathbf{i} + 3y^2\mathbf{j} + (1/z)\mathbf{k}$ 7. $\mathbf{F} = e^{2z}\mathbf{i} + 3y^2\mathbf{j} + 2xe^{2z}\mathbf{k}$

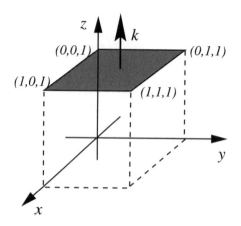

Figure 13.5.1: Diagram for the surface integration in Example 13.5.1.

8. $\mathbf{F} = y\mathbf{i} + (x+z)\mathbf{j} + y\mathbf{k}$ 9. $\mathbf{F} = (z+y)\mathbf{i} + x\mathbf{j} + x\mathbf{k}$.

13.5 SURFACE INTEGRALS

Surface integrals appear in such diverse fields as electromagnetism and fluid mechanics. For example, if we were oceanographers we might be interested in the rate of volume of seawater through an instrument which has the curved surface S. The volume rate equals $\iint_S \mathbf{v} \cdot \mathbf{n}\,d\sigma$, where $\mathbf{v}$ is the velocity and $\mathbf{n}\,d\sigma$ is an infinitesimally small element on the surface of the instrument. The surface element $\mathbf{n}\,d\sigma$ must have an orientation (given by $\mathbf{n}$) because it makes a considerable difference whether the flow is directly through the surface or at right angles. In the special case when the surface encloses a three-dimensional volume, then we have a *closed surface integral*.

To illustrate the concept of computing a surface integral, we will do three examples with simple geometries. Later we will show how to use surface coordinates to do more complicated geometries.

• **Example 13.5.1**

Let us find the flux out the top of a unit cube if the vector field is $\mathbf{F} = x\mathbf{i} + y\mathbf{j} + z\mathbf{k}$. See Figure 13.5.1.

The top of a unit cube consists of the surface $z = 1$ with $0 \le x \le 1$ and $0 \le y \le 1$. By inspection the unit normal to this surface is $\mathbf{n} = \mathbf{k}$, or $\mathbf{n} = -\mathbf{k}$. Because we are interested in the flux *out* of the unit cube, $\mathbf{n} = \mathbf{k}$, and

$$\iint_S \mathbf{F} \cdot \mathbf{n}\,d\sigma = \int_0^1 \int_0^1 (x\mathbf{i} + y\mathbf{j} + \mathbf{k}) \cdot \mathbf{k}\,dx\,dy = 1 \qquad (13.5.1)$$

because $z = 1$.

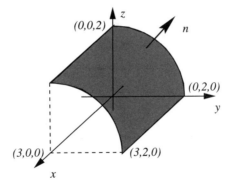

Figure 13.5.2: Diagram for the surface integration in Example 13.5.2.

• Example 13.5.2

Let us find the flux out of that portion of the cylinder $y^2 + z^2 = 4$ in the first octant bounded by $x = 0$, $x = 3$, $y = 0$, and $z = 0$. The vector field is $\mathbf{F} = x\mathbf{i} + 2z\mathbf{j} + y\mathbf{k}$. See Figure 13.5.2.

Because we are dealing with a cylinder, cylindrical coordinates are appropriate. Let $y = 2\cos(\theta)$, $z = 2\sin(\theta)$, and $x = x$ with $0 \le \theta \le \pi/2$. To find $\mathbf{n}$, we use the gradient in conjunction with the definition of the surface of the cylinder $f(x, y, z) = y^2 + z^2 = 4$. Then

$$\mathbf{n} = \frac{\nabla f}{|\nabla f|} = \frac{2y\mathbf{j} + 2z\mathbf{k}}{\sqrt{4y^2 + 4z^2}} = \frac{y}{2}\mathbf{j} + \frac{z}{2}\mathbf{k} \qquad (13.5.2)$$

because $y^2 + z^2 = 4$ along the surface. Since we want the flux *out* of the surface, then $\mathbf{n} = y\mathbf{j}/2 + z\mathbf{k}/2$ whereas the flux *into* the surface would require $\mathbf{n} = -y\mathbf{j}/2 - z\mathbf{k}/2$. Therefore,

$$\mathbf{F} \cdot \mathbf{n} = (x\mathbf{i} + 2z\mathbf{j} + y\mathbf{k}) \cdot \left(\frac{y}{2}\mathbf{j} + \frac{z}{2}\mathbf{k}\right) = \frac{3yz}{2} = 6\cos(\theta)\sin(\theta). \qquad (13.5.3)$$

What is $d\sigma$? Our infinitesimal surface area has a side in the x direction of length dx and a side in the θ direction of length $2\,d\theta$ because the radius equals 2. Therefore, $d\sigma = 2\,dx\,d\theta$.

Bringing all of these elements together,

$$\iint_S \mathbf{F} \cdot \mathbf{n}\,d\sigma = \int_0^3 \int_0^{\pi/2} 12\cos(\theta)\sin(\theta)\,d\theta\,dx \qquad (13.5.4)$$

$$= 6\int_0^3 \left[\sin^2(\theta)\big|_0^{\pi/2}\right] dx = 6\int_0^3 dx = 18. \qquad (13.5.5)$$

As counterpoint to this example, let us find the flux out of the pie-shaped surface at $x = 3$. In this case, $y = r\cos(\theta)$, $z = r\sin(\theta)$, and

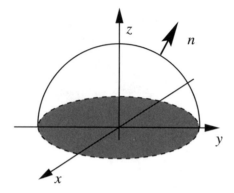

Figure 13.5.3: Diagram for the surface integration in Example 13.5.3.

$$\iint_S \mathbf{F} \cdot \mathbf{n} \, d\sigma = \int_0^{\pi/2} \int_0^2 [3\mathbf{i} + 2r\sin(\theta)\mathbf{j} + r\cos(\theta)\mathbf{k}] \cdot \mathbf{i}\, r \, dr \, d\theta \qquad (\mathbf{13.5.6})$$

$$= 3 \int_0^{\pi/2} \int_0^2 r \, dr \, d\theta = 3\pi. \qquad (\mathbf{13.5.7})$$

• **Example 13.5.3**

Let us find the flux of the vector field $\mathbf{F} = y^2\mathbf{i} + x^2\mathbf{j} + 5z\mathbf{k}$ out of the hemispheric surface $x^2 + y^2 + z^2 = a^2$, $z > 0$. See Figure 13.5.3.

We begin by finding the outwardly pointing normal. Because the surface is defined by $f(x, y, z) = x^2 + y^2 + z^2 = a^2$,

$$\mathbf{n} = \frac{\nabla f}{|\nabla f|} = \frac{2x\mathbf{i} + 2y\mathbf{j} + 2z\mathbf{k}}{\sqrt{4x^2 + 4y^2 + 4z^2}} = \frac{x}{a}\mathbf{i} + \frac{y}{a}\mathbf{j} + \frac{z}{a}\mathbf{k} \qquad (\mathbf{13.5.8})$$

because $x^2 + y^2 + z^2 = a^2$. This is also the outwardly pointing normal since $\mathbf{n} = \mathbf{r}/a$, where $\mathbf{r}$ is the radial vector.

Using spherical coordinates, $x = a\cos(\varphi)\sin(\theta)$, $y = a\sin(\varphi)\sin(\theta)$, and $z = a\cos(\theta)$, where φ is the angle made by the projection of the point onto the equatorial plane, measured from the x-axis, and θ is the colatitude or "cone angle" measured from the z-axis. To compute $d\sigma$, the infinitesimal length in the θ direction is $a\, d\theta$ while in the φ direction it is $a\sin(\theta)\, d\varphi$, where the $\sin(\theta)$ factor takes into account the convergence of the meridians. Therefore, $d\sigma = a^2 \sin(\theta)\, d\theta \, d\varphi$, and

$$\iint_S \mathbf{F} \cdot \mathbf{n}\, d\sigma = \int_0^{2\pi} \int_0^{\pi/2} (y^2\mathbf{i} + x^2\mathbf{j} + 5z\mathbf{k}) \left(\frac{x}{a}\mathbf{i} + \frac{y}{a}\mathbf{j} + \frac{z}{a}\mathbf{k}\right) a^2 \sin(\theta)\, d\theta \, d\varphi$$

$$(\mathbf{13.5.9})$$

$$\iint_S \mathbf{F} \cdot \mathbf{n}\, d\sigma = \int_0^{2\pi} \int_0^{\pi/2} \left(\frac{xy^2}{a} + \frac{x^2 y}{a} + \frac{5z^2}{a} \right) a^2 \sin(\theta)\, d\theta\, d\varphi \qquad (\mathbf{13.5.10})$$

$$= \int_0^{\pi/2} \int_0^{2\pi} \left[a^4 \cos(\varphi) \sin^2(\varphi) \sin^4(\theta) \right.$$
$$\left. + a^4 \cos^2(\varphi) \sin(\varphi) \sin^4(\theta) + 5a^3 \cos^2(\theta) \sin(\theta) \right] d\varphi\, d\theta \qquad (\mathbf{13.5.11})$$

$$= \int_0^{\pi/2} \left[\frac{a^4}{3} \sin^3(\varphi) \Big|_0^{2\pi} \sin^4(\theta) - \frac{a^4}{3} \cos^3(\varphi) \Big|_0^{2\pi} \sin^4(\theta) \right.$$
$$\left. + 5a^3 \cos^2(\theta) \sin(\theta)\varphi \Big|_0^{2\pi} \right] d\theta \qquad (\mathbf{13.5.12})$$

$$= 10\pi a^3 \int_0^{\pi/2} \cos^2(\theta) \sin(\theta)\, d\theta \qquad (\mathbf{13.5.13})$$

$$= -\frac{10\pi a^3}{3} \cos^3(\theta) \Big|_0^{\pi/2} = \frac{10\pi a^3}{3}. \qquad (\mathbf{13.5.14})$$

Although these techniques apply for simple geometries such as a cylinder or sphere, we would like a *general* method for treating any arbitrary surface. We begin by noting that a surface is an aggregate of points whose coordinates are functions of two variables. For example, in the previous example, the surface was described by the coordinates φ and θ. Let us denote these surface coordinates in general by u and v. Consequently, on any surface we can reexpress x, y, and z in terms of u and v: $x = x(u,v)$, $y = y(u,v)$, and $z = z(u,v)$.

Next, we must find an infinitesimal element of area. The position vector to the surface is $\mathbf{r} = x(u,v)\mathbf{i} + y(u,v)\mathbf{j} + z(u,v)\mathbf{k}$. Therefore, the tangent vectors along $v = $ constant, $\mathbf{r}_u$, and along $u = $ constant, $\mathbf{r}_v$, equal

$$\mathbf{r}_u = x_u \mathbf{i} + y_u \mathbf{j} + z_u \mathbf{k}, \qquad (\mathbf{13.5.15})$$

and

$$\mathbf{r}_v = x_v \mathbf{i} + y_v \mathbf{j} + z_v \mathbf{k}. \qquad (\mathbf{13.5.16})$$

Consequently, the sides of the infinitesimal area are $\mathbf{r}_u\, du$ and $\mathbf{r}_v\, dv$. Therefore, the vectorial area of the parallelogram that these vectors form is

$$\mathbf{n}\, d\sigma = \mathbf{r}_u \times \mathbf{r}_v\, du\, dv \qquad (\mathbf{13.5.17})$$

and is called the *vector element of area* on the surface. Thus, we may convert $\mathbf{F} \cdot \mathbf{n}\, d\sigma$ into an expression involving only u and v and then evaluate the surface integral by integrating over the appropriate domain in the uv-plane. Of course, we are in trouble if $\mathbf{r}_u \times \mathbf{r}_v = 0$. Therefore, we only treat regular

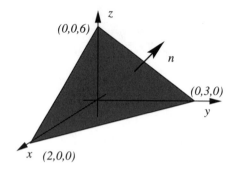

Figure 13.5.4: Diagram for the surface integration in Example 13.5.4.

points where $\mathbf{r}_u \times \mathbf{r}_v \neq \mathbf{0}$. In the next few examples, we show how to use these surface coordinates to evaluate surface integrals.

• Example 13.5.4

Let us find the flux of the vector field $\mathbf{F} = x\mathbf{i} + y\mathbf{j} + z\mathbf{k}$ through the top of the plane $3x + 2y + z = 6$ which lies in the first octant. See Figure 13.5.4.

Our parametric equations are $x = u$, $y = v$, and $z = 6 - 3u - 2v$. Therefore,

$$\mathbf{r} = u\mathbf{i} + v\mathbf{j} + (6 - 3u - 2v)\mathbf{k}, \tag{13.5.18}$$

so that

$$\mathbf{r}_u = \mathbf{i} - 3\mathbf{k}, \quad \mathbf{r}_v = \mathbf{j} - 2\mathbf{k}, \tag{13.5.19}$$

and

$$\mathbf{r}_u \times \mathbf{r}_v = 3\mathbf{i} + 2\mathbf{j} + \mathbf{k}. \tag{13.5.20}$$

Bring all of these elements together,

$$\iint_S \mathbf{F} \cdot \mathbf{n}\, d\sigma = \int_0^2 \int_0^{3-3u/2} (3u + 2v + 6 - 3u - 2v)\, dv\, du \tag{13.5.21}$$

$$= 6 \int_0^2 \int_0^{3-3u/2} dv\, du = 6 \int_0^2 (3 - 3u/2)\, du \tag{13.5.22}$$

$$= 6 \left(3u - \tfrac{3}{4}u^2\right)\big|_0^2 = 18. \tag{13.5.23}$$

To set up the limits of integration, we note that the area in u, v space corresponds to the xy-plane. On the xy-plane, $z = 0$ and $3u + 2v = 6$, along with boundaries $u = v = 0$.

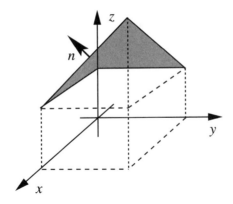

Figure 13.5.5: Diagram for the surface integration in Example 13.5.5.

- **Example 13.5.5**

Let us find the flux of the vector field $\mathbf{F} = x\mathbf{i} + y\mathbf{j} + z\mathbf{k}$ through the top of the surface $z = xy + 1$ which covers the square $0 \leq x \leq 1$, $0 \leq y \leq 1$ in the xy-plane. See Figure 13.5.5.

Our parametric equations are $x = u$, $y = v$, and $z = uv + 1$ with $0 \leq u \leq 1$ and $0 \leq v \leq 1$. Therefore,

$$\mathbf{r} = u\mathbf{i} + v\mathbf{j} + (uv + 1)\mathbf{k}, \qquad (13.5.24)$$

so that

$$\mathbf{r}_u = \mathbf{i} + v\mathbf{k}, \qquad \mathbf{r}_v = \mathbf{j} + u\mathbf{k}, \qquad (13.5.25)$$

and

$$\mathbf{r}_u \times \mathbf{r}_v = -v\mathbf{i} - u\mathbf{j} + \mathbf{k}. \qquad (13.5.26)$$

Bring all of these elements together,

$$\iint_S \mathbf{F} \cdot \mathbf{n}\, d\sigma = \int_0^1 \int_0^1 [u\mathbf{i} + v\mathbf{j} + (uv + 1)\mathbf{k}] \cdot (-v\mathbf{i} - u\mathbf{j} + \mathbf{k})\, du\, dv \quad (13.5.27)$$

$$= \int_0^1 \int_0^1 (1 - uv)\, du\, dv = \int_0^1 \left(u - \tfrac{1}{2}u^2 v\right)\big|_0^1 dv \qquad (13.5.28)$$

$$= \int_0^1 \left(1 - \tfrac{1}{2}v\right)\, dv = \left(v - \tfrac{1}{4}v^2\right)\big|_0^1 = \tfrac{3}{4}. \qquad (13.5.29)$$

- **Example 13.5.6**

Let us find the flux of the vector field $\mathbf{F} = 4xz\mathbf{i} + xyz^2\mathbf{j} + 3z\mathbf{k}$ through the exterior surface of the cone $z^2 = x^2 + y^2$ above the xy-plane and below $z = 4$. See Figure 13.5.6.

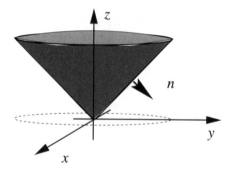

Figure 13.5.6: Diagram for the surface integration in Example 13.5.6.

A natural choice for the surface coordinates is polar coordinates r and θ. Because $x = r\cos(\theta)$ and $y = r\sin(\theta)$, $z = r$. Then

$$\mathbf{r} = r\cos(\theta)\mathbf{i} + r\sin(\theta)\mathbf{j} + r\mathbf{k} \qquad (13.5.30)$$

with $0 \le r \le 4$ and $0 \le \theta \le 2\pi$ so that

$$\mathbf{r}_r = \cos(\theta)\mathbf{i} + \sin(\theta)\mathbf{j} + \mathbf{k} \quad \mathbf{r}_\theta = -r\sin(\theta)\mathbf{i} + r\cos(\theta)\mathbf{j}, \qquad (13.5.31)$$

and

$$\mathbf{r}_r \times \mathbf{r}_\theta = -r\cos(\theta)\mathbf{i} - r\sin(\theta)\mathbf{j} + r\mathbf{k}. \qquad (13.5.32)$$

This is the unit area *inside* the cone. Because we want the exterior surface, we must take the negative of (13.5.32). Bring all of these elements together,

$$\iint_S \mathbf{F} \cdot \mathbf{n}\, d\sigma = \int_0^4 \int_0^{2\pi} \big\{ [4r\cos(\theta)]r[r\cos(\theta)]$$

$$+ [r^2 \sin(\theta)\cos(\theta)]r^2[r\sin(\theta)] - 3r^2 \big\}\, d\theta\, dr \quad (13.5.33)$$

$$= \int_0^4 \Big\{ 2r^3\big[\theta + \tfrac{1}{2}\sin(2\theta)\big]\big|_0^{2\pi} + r^5\tfrac{1}{3}\sin^3(\theta)\big|_0^{2\pi} - 3r^2\theta\big|_0^{2\pi} \Big\}\, dr$$

$$(13.5.34)$$

$$= \int_0^4 \big(4\pi r^3 - 6\pi r^2\big)\, dr = \big(\pi r^4 - 2\pi r^3\big)\big|_0^4 = 128\pi. \quad (13.5.35)$$

Problems

Compute the surface integral $\iint_S \mathbf{F} \cdot \mathbf{n}\, d\sigma$ for the following vector fields and surfaces:

1. $\mathbf{F} = x\mathbf{i} - z\mathbf{j} + y\mathbf{k}$ and the surface is the top side of the $z = 1$ plane where $0 \le x \le 1$ and $0 \le y \le 1$.

2. $\mathbf{F} = x\mathbf{i} + y\mathbf{j} + xz\mathbf{k}$ and the surface is the top side of the cylinder $x^2 + y^2 = 9$, $z = 0$, and $z = 1$.

3. $\mathbf{F} = xy\mathbf{i} + z\mathbf{j} + xz\mathbf{k}$ and the surface consists of both exterior *ends* of the cylinder defined by $x^2 + y^2 = 4$, $z = 0$, and $z = 2$.

4. $\mathbf{F} = x\mathbf{i} + z\mathbf{j} + y\mathbf{k}$ and the surface is the lateral and exterior sides of the cylinder defined by $x^2 + y^2 = 4$, $z = -3$, and $z = 3$.

5. $\mathbf{F} = xy\mathbf{i} + z^2\mathbf{j} + y\mathbf{k}$ and the surface is the curved exterior side of the cylinder $y^2 + z^2 = 9$ in the first octant bounded by $x = 0$, $x = 1$, $y = 0$, and $z = 0$.

6. $\mathbf{F} = y\mathbf{j} + z^2\mathbf{k}$ and the surface is the exterior of the semicircular cylinder $y^2 + z^2 = 4$, $z \geq 0$, cut by the planes $x = 0$ and $x = 1$.

7. $\mathbf{F} = z\mathbf{i} + x\mathbf{j} + y\mathbf{k}$ and the surface is the curved exterior side of the cylinder $x^2 + y^2 = 4$ in the first octant cut by the planes $z = 1$ and $z = 2$.

8. $\mathbf{F} = x^2\mathbf{i} - z^2\mathbf{j} + yz\mathbf{k}$ and the surface is the exterior of the hemispheric surface of $x^2 + y^2 + z^2 = 16$ above the plane $z = 2$.

9. $\mathbf{F} = y\mathbf{i} + x\mathbf{j} + y\mathbf{k}$ and the surface is the top of the surface $z = x + 1$, where $-1 \leq x \leq 1$ and $-1 \leq y \leq 1$.

10. $\mathbf{F} = z\mathbf{i} + x\mathbf{j} - 3z\mathbf{k}$ and the surface is the top side of the plane $x + y + z = 2a$ that lies above the square $0 \leq x \leq a$, $0 \leq y \leq a$ in the xy-plane.

11. $\mathbf{F} = (y^2 + z^2)\mathbf{i} + (x^2 + z^2)\mathbf{j} + (x^2 + y^2)\mathbf{k}$ and the surface is the top side of the surface $z = 1 - x^2$ with $-1 \leq x \leq 1$ and $-2 \leq y \leq 2$.

12. $\mathbf{F} = y^2\mathbf{i} + xz\mathbf{j} - \mathbf{k}$ and the surface is the cone $z = \sqrt{x^2 + y^2}$, $0 \leq z \leq 1$, with the normal pointing away from the z-axis.

13. $\mathbf{F} = y^2\mathbf{i} + x^2\mathbf{j} + 5z\mathbf{k}$ and the surface is the top side of the plane $z = y + 1$, where $-1 \leq x \leq 1$ and $-1 \leq y \leq 1$.

14. $\mathbf{F} = -y\mathbf{i} + x\mathbf{j} + z\mathbf{k}$ and the surface is the exterior or bottom side of the paraboloid $z = x^2 + y^2$, where $0 \leq z \leq 1$.

15. $\mathbf{F} = -y\mathbf{i} + x\mathbf{j} + 6z^2\mathbf{k}$ and the surface is the exterior of the paraboloids $z = 4 - x^2 - y^2$ and $z = x^2 + y^2$.

13.6 GREEN'S LEMMA

Consider a rectangle in the xy-plane which is bounded by the lines $x = a$, $x = b$, $y = c$, and $y = d$. We assume that the boundary of the rectangle is a piece-wise smooth curve which we denote by C. If we have a continuously differentiable vector function $\mathbf{F} = P(x, y)\mathbf{i} + Q(x, y)\mathbf{j}$ at each point of enclosed region R, then

$$\iint_R \frac{\partial Q}{\partial x}\, dA = \int_c^d \left[\int_a^b \frac{\partial Q}{\partial x}\, dx \right] dy \tag{13.6.1}$$

$$= \int_c^d Q(b, y)\, dy - \int_c^d Q(a, y)\, dy \tag{13.6.2}$$

$$= \oint_C Q(x, y)\, dy, \tag{13.6.3}$$

where the last integral is a closed line integral counterclockwise around the rectangle because the horizontal sides vanish since $dy = 0$. By similar arguments,

$$\iint_R \frac{\partial P}{\partial y}\, dA = - \oint_C P(x, y)\, dx \tag{13.6.4}$$

so that

$$\iint_R \left(\frac{\partial Q}{\partial x} - \frac{\partial P}{\partial y} \right) dA = \oint_C P(x, y)\, dx + Q(x, y)\, dy. \tag{13.6.5}$$

This result, often known as *Green's lemma*, may be expressed in vector form as

$$\oint_C \mathbf{F} \cdot d\mathbf{r} = \iint_R \nabla \times \mathbf{F} \cdot \mathbf{k}\, dA. \tag{13.6.6}$$

Although this proof was for a rectangular area, it can be generalized to *any* simply closed region on the xy-plane as follows. Consider an area which is surrounded by simply closed curves. Within the closed contour we can divide the area into an infinite number of infinitesimally small rectangles and apply (13.6.6) to each rectangle. When we sum up all of these rectangles, we find $\iint_R \nabla \times \mathbf{F} \cdot \mathbf{k}\, dA$, where the integration is over the entire surface area. On the other hand, away from the boundary, the line integral along any one edge of a rectangle cancels the line integral along the same edge in a contiguous

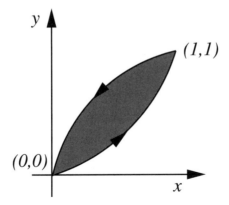

Figure 13.6.1: Diagram for the verification of Green's lemma in Example 13.6.1.

rectangle. Thus, the only nonvanishing contribution from the line integrals arises from the outside boundary of the domain $\oint_C \mathbf{F} \cdot d\mathbf{r}$.

● **Example 13.6.1**

Let us *verify* Green's lemma using the vector field $\mathbf{F} = (3x^2 - 8y^2)\mathbf{i} + (4y - 6xy)\mathbf{j}$ and the enclosed area lies between the curves $y = \sqrt{x}$ and $y = x^2$. The two curves intersect at $x = 0$ and $x = 1$. See Figure 13.6.1.

We begin with the line integral:

$$\oint_C \mathbf{F} \cdot d\mathbf{r} = \int_0^1 (3x^2 - 8x^4)\, dx + (4x^2 - 6x^3)(2x\, dx)$$

$$+ \int_1^0 (3x^2 - 8x)\, dx + (4x^{1/2} - 6x^{3/2})(\tfrac{1}{2}x^{-1/2}\, dx) \qquad (\mathbf{13.6.7})$$

$$= \int_0^1 (-20x^4 + 8x^3 + 11x - 2)\, dx = \tfrac{3}{2}. \qquad (\mathbf{13.6.8})$$

In (13.6.7) we used $y = x^2$ in the first integral and $y = \sqrt{x}$ in our return integration. For the areal integration,

$$\iint_R \nabla \times \mathbf{F} \cdot \mathbf{k}\, dA = \int_0^1 \int_{x^2}^{\sqrt{x}} 10y\, dy\, dx = \int_0^1 5y^2 \big|_{x^2}^{\sqrt{x}}\, dx \qquad (\mathbf{13.6.9})$$

$$= 5\int_0^1 (x - x^4)\, dx = \tfrac{3}{2} \qquad (\mathbf{13.6.10})$$

and Green's lemma is verified in this particular case.

• **Example 13.6.2**

Let us redo Example 13.6.1 except that the closed contour is the triangular region defined by the lines $x = 0$, $y = 0$, and $x + y = 1$.

The line integral is

$$\oint_C \mathbf{F} \cdot d\mathbf{r} = \int_0^1 (3x^2 - 8 \cdot 0^2)dx + (4 \cdot 0 - 6x \cdot 0) \cdot 0$$

$$+ \int_0^1 [3(1 - y)^2 - 8y^2](-dy) + [4y - 6(1 - y)y] \, dy$$

$$+ \int_1^0 (3 \cdot 0^2 - 8y^2) \cdot 0 + (4y - 6 \cdot 0 \cdot y) \, dy \tag{13.6.11}$$

$$= \int_0^1 3x^2 \, dx - \int_0^1 4y \, dy + \int_0^1 (-3 + 4y + 11y^2) \, dy \tag{13.6.12}$$

$$= x^3\big|_0^1 - 2y^2\big|_0^1 + \left(-3y + 2y^2 + \tfrac{11}{3}y^3\right)\big|_0^1 = \tfrac{5}{3}. \tag{13.6.13}$$

On the other hand, the areal integration is

$$\iint_R \nabla \times \mathbf{F} \cdot \mathbf{k} \, dA = \int_0^1 \int_0^{1-x} 10y \, dy \, dx = \int_0^1 5y^2\big|_0^{1-x} \, dx \tag{13.6.14}$$

$$= 5 \int_0^1 (1 - x)^2 \, dx = -\tfrac{5}{3}(1 - x)^3\big|_0^1 = \tfrac{5}{3} \tag{13.6.15}$$

and Green's lemma is verified in this particular case.

• **Example 13.6.3**

Let us verify Green's lemma using the vector field $\mathbf{F} = (3x + 4y)\mathbf{i} + (2x - 3y)\mathbf{j}$ and the closed contour is a circle of radius two centered at the origin of the xy-plane. See Figure 13.6.2.

Beginning with the line integration,

$$\oint_C \mathbf{F} \cdot d\mathbf{r} = \int_0^{2\pi} [6\cos(\theta) + 8\sin(\theta)][-2\sin(\theta) \, d\theta]$$

$$+ [4\cos(\theta) - 6\sin(\theta)][2\cos(\theta) \, d\theta] \tag{13.6.16}$$

$$= \int_0^{2\pi} [-24\cos(\theta)\sin(\theta) - 16\sin^2(\theta) + 8\cos^2(\theta)] \, d\theta \tag{13.6.17}$$

$$= 12\cos^2(\theta)\big|_0^{2\pi} - 8\left[\theta - \tfrac{1}{2}\sin(2\theta)\right]\big|_0^{2\pi} + 4\left[\theta + \tfrac{1}{2}\sin(2\theta)\right]\big|_0^{2\pi} \tag{13.6.18}$$

$$= -8\pi. \tag{13.6.19}$$

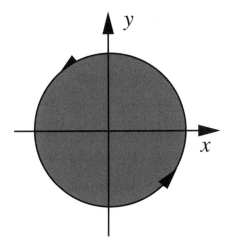

Figure 13.6.2: Diagram for the verification of Green's lemma in Example 13.6.3.

For the areal integration,

$$\iint_R \nabla \times \mathbf{F} \cdot \mathbf{k}\, dA = \int_0^2 \int_0^{2\pi} -2\, r\, d\theta\, dr = -8\pi \qquad (13.6.20)$$

and Green's lemma is verified in the special case.

Problems

Verify Green's lemma for the following two-dimensional vector fields and contours:

1. $\mathbf{F} = (x^2 + 4y)\mathbf{i} + (y - x)\mathbf{j}$ and the contour is the square bounded by the lines $x = 0$, $y = 0$, $x = 1$, and $y = 1$.

2. $\mathbf{F} = (x - y)\mathbf{i} + xy\mathbf{j}$ and the contour is the square bounded by the lines $x = 0$, $y = 0$, $x = 1$, and $y = 1$.

3. $\mathbf{F} = -y^2\mathbf{i} + x^2\mathbf{j}$ and the contour is the triangle bounded by the lines $x = 1$, $y = 0$, and $y = x$.

4. $\mathbf{F} = (xy - x^2)\mathbf{i} + x^2 y\mathbf{j}$ and the contour is the triangle bounded by the lines $y = 0$, $x = 1$, and $y = x$.

5. $\mathbf{F} = \sin(y)\mathbf{i} + x\cos(y)\mathbf{j}$ and the contour is the triangle bounded by the lines $x + y = 1$, $y - x = 1$, and $y = 0$.

6. $\mathbf{F} = y^2\mathbf{i} + x^2\mathbf{j}$ and the contour is the same contour used in Problem 4.

7. $\mathbf{F} = -y^2\mathbf{i} + x^2\mathbf{j}$ and the contour is the circle $x^2 + y^2 = 4$.

8. $\mathbf{F} = -x^2\mathbf{i} + xy^2\mathbf{j}$ and the contour is the closed circle of radius a.

9. $\mathbf{F} = (6y+x)\mathbf{i} + (y+2x)\mathbf{j}$ and the contour is the circle $(x-1)^2 + (y-2)^2 = 4$.

10. $\mathbf{F} = (x+y)\mathbf{i} + (2x^2 - y^2)\mathbf{j}$ and the contour is the boundary of the region determined by the curves $y = x^2$ and $y = 4$.

11. $\mathbf{F} = 3y\mathbf{i} + 2x\mathbf{j}$ and the contour is the boundary of the region determined by the curves $y = 0$ and $y = \sin(x)$ with $0 \le x \le \pi$.

12. $\mathbf{F} = -16y\mathbf{i} + (4e^y + 3x^2)\mathbf{j}$ and the contour is the pie wedge defined by the lines $y = x$, $y = -x$, $x^2 + y^2 = 4$, and $y > 0$.

13.7 STOKES' THEOREM

In §13.2 we introduced the vector quantity $\nabla \times \mathbf{v}$ which gives a measure of the rotation of a parcel of fluid lying within the velocity field $\mathbf{v}$. In this section we show how the curl can be used to simplify the calculation of certain closed line integrals.

This relationship between a closed line integral and a surface integral involving the curl is

Stokes' Theorem: *The circulation of $\mathbf{F} = P\mathbf{i} + Q\mathbf{j} + R\mathbf{k}$ around the closed boundary C of an oriented surface S in the direction counterclockwise with respect to the surface's unit normal vector $\mathbf{n}$ equals the integral of $\nabla \times \mathbf{F} \cdot \mathbf{n}$ over S or*

$$\oint_C \mathbf{F} \cdot d\mathbf{r} = \iint_S \nabla \times \mathbf{F} \cdot \mathbf{n} \, d\sigma. \tag{13.7.1}$$

Stokes' theorem requires that all of the functions and derivatives be continuous.

The proof of Stokes' theorem is as follows: Consider a finite surface S whose boundary is the loop C. We divide this surface into a number of small elements $\mathbf{n} \, d\sigma$ and compute the *circulation* $d\Gamma = \oint_L \mathbf{F} \cdot d\mathbf{r}$ around each element. When we add all of the circulations together, the contribution from an integration along a boundary line between two adjoining elements cancels out because the boundary is transversed once in each direction. For this reason, the only contributions that survive are those parts where the element

Figure 13.7.1: Sir George Gabriel Stokes (1819–1903) was Lucasian Professor of Mathematics at Cambridge University from 1849 until his death. Having learned of an integral theorem from his friend Lord Kelvin, Stokes included it a few years later among his questions on an examination that he wrote for the Smith Prize. It is this integral theorem that we now call Stokes' theorem. (Portrait courtesy of the Royal Society of London.)

boundaries form part of C. Thus, the sum of all circulations equals $\oint_C \mathbf{F} \cdot d\mathbf{r}$, the circulation around the edge of the whole surface.

Next, let us compute the circulation another way. We begin by finding the Taylor expansion for $P(x, y, z)$ about the arbitrary point (x_0, y_0, z_0):

$$
\begin{aligned}
P(x, y, z) = P(x_0, y_0, z_0) + (x - x_0)\frac{\partial P(x_0, y_0, z_0)}{\partial x} \\
+ (y - y_0)\frac{\partial P(x_0, y_0, z_0)}{\partial y} + (z - z_0)\frac{\partial P(x_0, y_0, z_0)}{\partial z} + \cdots
\end{aligned} \quad (\mathbf{13.7.2})
$$

with similar expansions for $Q(x, y, z)$ and $R(x, y, z)$. Then

$$
\begin{aligned}
d\Gamma = \oint_L \mathbf{F} \cdot d\mathbf{r} = P(x_0, y_0, z_0) \oint_L dx + \frac{\partial P(x_0, y_0, z_0)}{\partial x} \oint_L (x - x_0)\, dx \\
+ \frac{\partial P(x_0, y_0, z_0)}{\partial y} \oint_L (y - y_0)\, dy + \cdots \\
+ \frac{\partial Q(x_0, y_0, z_0)}{\partial x} \oint_L (x - x_0)\, dy + \cdots,
\end{aligned} \quad (\mathbf{13.7.3})
$$

where L denotes some small loop located in the surface S. Note that integrals such as $\oint_L dx$ and $\oint_L (x - x_0)\, dx$ vanish.

If we now require that the loop integrals be in the *clockwise* or *positive* sense so that we preserve the right-hand screw convention, then

$$\mathbf{n} \cdot \mathbf{k} \, \delta\sigma = \oint_L (x - x_0) \, dy = -\oint_L (y - y_0) \, dx, \qquad (13.7.4)$$

$$\mathbf{n} \cdot \mathbf{j} \, \delta\sigma = \oint_L (z - z_0) \, dx = -\oint_L (x - x_0) \, dz, \qquad (13.7.5)$$

$$\mathbf{n} \cdot \mathbf{i} \, \delta\sigma = \oint_L (y - y_0) \, dz = -\oint_L (z - z_0) \, dy, \qquad (13.7.6)$$

and

$$d\Gamma = \left(\frac{\partial R}{\partial y} - \frac{\partial Q}{\partial z}\right) \mathbf{i} \cdot \mathbf{n} \, \delta\sigma + \left(\frac{\partial P}{\partial z} - \frac{\partial R}{\partial x}\right) \mathbf{j} \cdot \mathbf{n} \, \delta\sigma$$
$$+ \left(\frac{\partial Q}{\partial x} - \frac{\partial P}{\partial y}\right) \mathbf{k} \cdot \mathbf{n} \, \delta\sigma = \nabla \times \mathbf{F} \cdot \mathbf{n} \, \delta\sigma. \qquad (13.7.7)$$

Therefore, the sum of all circulations in the limit when all elements are made infinitesimally small becomes the surface integral $\iint_S \nabla \times \mathbf{F} \cdot \mathbf{n} \, d\sigma$ and Stokes' theorem is proven. ☐

In the following examples we first apply Stokes' theorem to a few simple geometries. We then show how to apply this theorem to more complicated surfaces.[4]

• **Example 13.7.1**

Let us verify Stokes' theorem using the vector field $\mathbf{F} = x^2\mathbf{i} + 2x\mathbf{j} + z^2\mathbf{k}$ and the closed curve is a square with vertices at $(0, 0, 3)$, $(1, 0, 3)$, $(1, 1, 3)$, and $(0, 1, 3)$. See Figure 13.7.2.

We begin with the line integral:

$$\oint_C \mathbf{F} \cdot d\mathbf{r} = \int_{C_1} \mathbf{F} \cdot d\mathbf{r} + \int_{C_2} \mathbf{F} \cdot d\mathbf{r} + \int_{C_3} \mathbf{F} \cdot d\mathbf{r} + \int_{C_4} \mathbf{F} \cdot d\mathbf{r}, \qquad (13.7.8)$$

where C_1, C_2, C_3, and C_4 represent the four sides of the square. Along C_1, x varies while $y = 0$ and $z = 3$. Therefore,

$$\int_{C_1} \mathbf{F} \cdot d\mathbf{r} = \int_0^1 x^2 \, dx + 2x \cdot 0 + 9 \cdot 0 = \tfrac{1}{3}, \qquad (13.7.9)$$

---

[4] Thus, different Stokes for different folks.

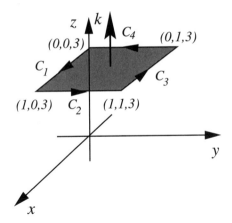

Figure 13.7.2: Diagram for the verification of Stokes' theorem in Example 13.7.1.

because $dy = dz = 0$, and $z = 3$. Along C_2, y varies with $x = 1$ and $z = 3$. Therefore,

$$\int_{C_2} \mathbf{F} \cdot d\mathbf{r} = \int_0^1 1^2 \cdot 0 + 2 \cdot 1 \cdot dy + 9 \cdot 0 = 2. \qquad (13.7.10)$$

Along C_3, x again varies with $y = 1$ and $z = 3$, and so,

$$\int_{C_3} \mathbf{F} \cdot d\mathbf{r} = \int_1^0 x^2 \, dx + 2x \cdot 0 + 9 \cdot 0 = -\tfrac{1}{3}. \qquad (13.7.11)$$

Note how the limits run from 1 to 0 because x is decreasing. Finally, for C_4, y again varies with $x = 0$ and $z = 3$. Hence,

$$\int_{C_4} \mathbf{F} \cdot d\mathbf{r} = \int_1^0 0^2 \cdot 0 + 2 \cdot 0 \cdot dy + 9 \cdot 0 = 0. \qquad (13.7.12)$$

Hence,

$$\oint_C \mathbf{F} \cdot d\mathbf{r} = 2. \qquad (13.7.13)$$

Turning to the other side of the equation,

$$\nabla \times \mathbf{F} = \begin{vmatrix} \mathbf{i} & \mathbf{j} & \mathbf{k} \\ \frac{\partial}{\partial x} & \frac{\partial}{\partial y} & \frac{\partial}{\partial z} \\ x^2 & 2x & z^2 \end{vmatrix} = 2\mathbf{k}. \qquad (13.7.14)$$

Our line integral has been such that the normal vector must be $\mathbf{n} = \mathbf{k}$. Therefore,

$$\iint_S \nabla \times \mathbf{F} \cdot \mathbf{n} \, d\sigma = \int_0^1 \int_0^1 2\mathbf{k} \cdot \mathbf{k} \, dx \, dy = 2 \qquad (13.7.15)$$

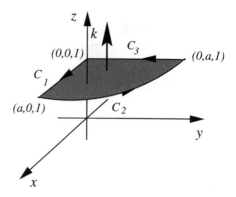

Figure 13.7.3: Diagram for the verification of Stokes' theorem in Example 13.7.2.

and Stokes' theorem is verified for this special case.

● **Example 13.7.2**

Let us verify Stokes' theorem using the vector field $\mathbf{F} = (x^2 - y)\mathbf{i} + 4z\mathbf{j} + x^2\mathbf{k}$, where the closed contour consists of the x and y coordinate axes and that portion of the circle $x^2 + y^2 = a^2$ that lies in the first quadrant with $z = 1$. See Figure 13.7.3.

The line integral consists of three parts:

$$\oint_C \mathbf{F} \cdot d\mathbf{r} = \int_{C_1} \mathbf{F} \cdot d\mathbf{r} + \int_{C_2} \mathbf{F} \cdot d\mathbf{r} + \int_{C_3} \mathbf{F} \cdot d\mathbf{r}. \tag{13.7.16}$$

Along C_1, x varies while $y = 0$ and $z = 1$. Therefore,

$$\int_{C_1} \mathbf{F} \cdot d\mathbf{r} = \int_0^a (x^2 - 0)\, dx + 4 \cdot 1 \cdot 0 + x^2 \cdot 0 = \frac{a^3}{3}. \tag{13.7.17}$$

Along the circle C_2, we use polar coordinates with $x = a\cos(t)$, $y = a\sin(t)$, and $z = 1$. Therefore,

$$\int_{C_2} \mathbf{F} \cdot d\mathbf{r} = \int_0^{\pi/2} [a^2 \cos^2(t) - a\sin(t)][-a\sin(t)\, dt]$$

$$+ 4 \cdot 1 \cdot a\cos(t)\, dt + a^2 \cos^2(t) \cdot 0, \tag{13.7.18}$$

$$= \int_0^{\pi/2} -a^3 \cos^2(t)\sin(t)\, dt + a^2 \sin^2(t)\, dt + 4a\cos(t)\, dt \tag{13.7.19}$$

$$= \frac{a^3}{3}\cos^3(t)\Big|_0^{\pi/2} + \frac{a^2}{2}\left[t - \frac{1}{2}\sin(2t)\right]\Big|_0^{\pi/2} + 4a\sin(t)\Big|_0^{\pi/2} \tag{13.7.20}$$

$$= -\frac{a^3}{3} + \frac{a^2\pi}{4} + 4a, \tag{13.7.21}$$

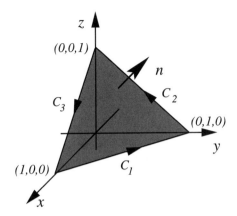

Figure 13.7.4: Diagram for the verification of Stokes' theorem in Example 13.7.3.

because $dx = -a\sin(t)\,dt$, and $dy = a\cos(t)\,dt$. Finally, along C_3, y varies with $x = 0$ and $z = 1$. Therefore,

$$\int_{C_3} \mathbf{F} \cdot d\mathbf{r} = \int_a^0 (0^2 - y) \cdot 0 + 4 \cdot 1 \cdot dy + 0^2 \cdot 0 = -4a, \qquad (13.7.22)$$

so that

$$\oint_C \mathbf{F} \cdot d\mathbf{r} = \frac{a^2 \pi}{4}. \qquad (13.7.23)$$

Turning to the other side of the equation,

$$\nabla \times \mathbf{F} = \begin{vmatrix} \mathbf{i} & \mathbf{j} & \mathbf{k} \\ \frac{\partial}{\partial x} & \frac{\partial}{\partial y} & \frac{\partial}{\partial z} \\ x^2 - y & 4z & x^2 \end{vmatrix} = -4\mathbf{i} - 2x\mathbf{j} + \mathbf{k}. \qquad (13.7.24)$$

From the path of our line integral, our unit normal vector must be $\mathbf{n} = \mathbf{k}$. Then,

$$\iint_S \nabla \times \mathbf{F} \cdot \mathbf{n}\, d\sigma = \int_0^a \int_0^{\pi/2} [-4\mathbf{i} - 2r\cos(\theta)\mathbf{j} + \mathbf{k}] \cdot \mathbf{k}\, r\, d\theta\, dr = \frac{\pi a^2}{4} \quad (13.7.25)$$

and Stokes' theorem is verified for this case.

● **Example 13.7.3**

Let us verify Stokes' theorem using the vector field $\mathbf{F} = 2yz\mathbf{i} - (x + 3y - 2)\mathbf{j} + (x^2 + z)\mathbf{k}$, where the closed triangular region is that portion of the plane $x + y + z = 1$ that lies in the first octant.

As shown in Figure 13.7.4, the closed line integration consists of three line integrals:

$$\oint_C \mathbf{F} \cdot d\mathbf{r} = \int_{C_1} \mathbf{F} \cdot d\mathbf{r} + \int_{C_2} \mathbf{F} \cdot d\mathbf{r} + \int_{C_3} \mathbf{F} \cdot d\mathbf{r}. \qquad (13.7.26)$$

Along C_1, $z = 0$ and $y = 1 - x$. Therefore, using x as the independent variable,

$$\int_{C_1} \mathbf{F} \cdot d\mathbf{r} = \int_1^0 2(1 - x) \cdot 0 \cdot dx - (x + 3 - 3x - 2)(-dx) + (x^2 + 0) \cdot 0$$

$$= -x^2 \big|_1^0 + x \big|_1^0 = 0. \qquad (13.7.27)$$

Along C_2, $x = 0$ and $y = 1 - z$. Thus,

$$\int_{C_2} \mathbf{F} \cdot d\mathbf{r} = \int_0^1 2(1 - z)z \cdot 0 - (0 + 3 - 3z - 2)(-dz) + (0^2 + z)\, dz$$

$$= -\tfrac{3}{2}z^2 + z + \tfrac{1}{2}z^2 \big|_0^1 = 0. \qquad (13.7.28)$$

Finally, along C_3, $y = 0$ and $z = 1 - x$. Hence,

$$\int_{C_3} \mathbf{F} \cdot d\mathbf{r} = \int_0^1 2 \cdot 0 \cdot (1 - x)\, dx - (x + 0 - 2) \cdot 0 + (x^2 + 1 - x)(-dx)$$

$$= -\tfrac{1}{3}x^3 - x + \tfrac{1}{2}x^2 \big|_0^1 = -\tfrac{5}{6}. \qquad (13.7.29)$$

Thus,

$$\oint_C \mathbf{F} \cdot d\mathbf{r} = -\tfrac{5}{6}. \qquad (13.7.30)$$

On the other hand,

$$\nabla \times \mathbf{F} = \begin{vmatrix} \mathbf{i} & \mathbf{j} & \mathbf{k} \\ \frac{\partial}{\partial x} & \frac{\partial}{\partial y} & \frac{\partial}{\partial z} \\ 2yz & -x - 3y + 2 & x^2 + z \end{vmatrix} = (-2x + 2y)\mathbf{j} + (-1 - 2z)\mathbf{k}. \quad (13.7.31)$$

To find $\mathbf{n}\, d\sigma$, we use the general coordinate system $x = u$, $y = v$, and $z = 1 - u - v$. Therefore, $\mathbf{r} = u\mathbf{i} + v\mathbf{j} + (1 - u - v)\mathbf{k}$ and

$$\mathbf{r}_u \times \mathbf{r}_v = \begin{vmatrix} \mathbf{i} & \mathbf{j} & \mathbf{k} \\ 1 & 0 & -1 \\ 0 & 1 & -1 \end{vmatrix} = \mathbf{i} + \mathbf{j} + \mathbf{k}. \qquad (13.7.32)$$

Thus,

$$\iint_S \nabla \times \mathbf{F} \cdot \mathbf{n}\, d\sigma = \int_0^1 \int_0^{1-u} [(-2u + 2v)\mathbf{j} + (-1 - 2 + 2u + 2v)\mathbf{k}]$$

$$\cdot [\mathbf{i} + \mathbf{j} + \mathbf{k}]\, dv\, du \qquad (13.7.33)$$

$$= \int_0^1 \int_0^{1-u} (4v - 3)\, dv\, du \qquad (13.7.34)$$

$$= \int_0^1 [2(1 - u)^2 - 3(1 - u)]\, du \qquad (13.7.35)$$

$$= \int_0^1 (-1 - u + 2u^2)\, du = -\tfrac{5}{6} \qquad (13.7.36)$$

and Stokes' theorem is verified for this case.

Problems

Verify Stokes' theorem using the following vector fields and surfaces:

1. $\mathbf{F} = 5y\mathbf{i} - 5x\mathbf{j} + 3z\mathbf{k}$ and the surface S is that portion of the plane $z = 1$ with the square at the vertices $(0, 0, 1)$, $(1, 0, 1)$, $(1, 1, 1)$, and $(0, 1, 1)$.

2. $\mathbf{F} = x^2\mathbf{i} + y^2\mathbf{j} + z^2\mathbf{k}$ and the surface S is the rectangular portion of the plane $z = 2$ defined by the corners $(0, 0, 2)$, $(2, 0, 2)$, $(2, 1, 2)$, and $(0, 1, 2)$.

3. $\mathbf{F} = z\mathbf{i} + x\mathbf{j} + y\mathbf{k}$ and the surface S is the triangular portion of the plane $z = 1$ defined by the vertices $(0, 0, 1)$, $(2, 0, 1)$, and $(0, 2, 1)$.

4. $\mathbf{F} = 2z\mathbf{i} - 3x\mathbf{j} + 4y\mathbf{k}$ and the surface S is that portion of the plane $z = 5$ within the cylinder $x^2 + y^2 = 4$.

5. $\mathbf{F} = z\mathbf{i} + x\mathbf{j} + y\mathbf{k}$ and the surface S is that portion of the plane $z = 3$ bounded by the lines $y = 0$, $x = 0$, and $x^2 + y^2 = 4$.

6. $\mathbf{F} = (2z + x)\mathbf{i} + (y - z)\mathbf{j} + (x + y)\mathbf{k}$ and the surface S is the interior of the triangularly shaped plane with vertices at $(1, 0, 0)$, $(0, 1, 0)$, and $(0, 0, 1)$.

7. $\mathbf{F} = z\mathbf{i} + x\mathbf{j} + y\mathbf{k}$ and the surface S is that portion of the plane $2x + y + 2z = 6$ in the first octant.

8. $\mathbf{F} = x\mathbf{i} + xz\mathbf{j} + y\mathbf{k}$ and the surface S is that portion of the paraboloid $z = 9 - x^2 - y^2$ within the cylinder $x^2 + y^2 = 4$.

13.8 DIVERGENCE THEOREM

Although Stokes' theorem is useful in computing closed line integrals, it is usually very difficult to go the other way and convert a surface integral into a closed line integral because the integrand must have a very special form, namely $\nabla \times \mathbf{F} \cdot \mathbf{n}$. In this section we introduce a theorem that allows with equal facility the conversion of a closed surface integral into a volume integral and *vice versa*. Furthermore, if we can convert a given surface integral into a closed one by the introduction of a simple surface (for example, closing a hemispheric surface by adding an equatorial plate), it may be easier to use the divergence theorem and subtract off the contribution from the new surface integral rather than do the original problem.

This relationship between a closed surface integral and a volume integral involving the divergence operator is

Figure 13.8.1: Carl Friedrich Gauss (1777–1855), the prince of mathematicians, must be on the list of the greatest mathematicians who ever lived. Gauss, a child prodigy, is almost as well known for what he did not publish during his lifetime as for what he did. This is true of Gauss's divergence theorem which he proved while working on the theory of gravitation. It was only when his notebooks were published in 1898 that his precedence over the published work of Ostrogradsky (1801–1862) was established. (Portrait courtesy of Photo AKG, London.)

The Divergence or Gauss's Theorem: *Let V be a closed and bounded region in three-dimensional space with a piece-wise smooth boundary S that is oriented outward. Let $\mathbf{F} = P(x, y, z)\mathbf{i} + Q(x, y, z)\mathbf{j} + R(x, y, z)\mathbf{k}$ be a vector field for which P, Q, and R are continuous and have continuous first partial derivatives in a region of three dimensional space containing V. Then*

$$\oiint_S \mathbf{F} \cdot \mathbf{n} \, d\sigma = \iiint_V \nabla \cdot \mathbf{F} \, dV. \tag{13.8.1}$$

Here, the circle on the double integral signs denotes a closed surface integral.

A nonrigorous proof of Gauss's theorem is as follows. Imagine that our volume V is broken down into small elements $d\tau$ of volume of any shape so long as they include all of the original volume. In general, the surfaces of these elements are composed of common interfaces between adjoining elements. However, for the elements at the periphery of V, part of their surface will be part of the surface S that encloses V. Now $d\Phi = \nabla \cdot \mathbf{F} \, d\tau$ is the net flux of the vector $\mathbf{F}$ out from the element $d\tau$. At the common interface between elements, the flux *out* of one element equals the flux *into* its neighbor.

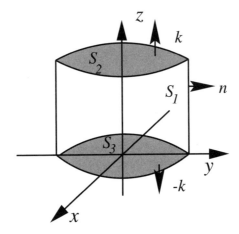

Figure 13.8.2: Diagram for the verification of the divergence theorem in Example 13.8.1.

Therefore, the sum of all such terms yields

$$\Phi = \iiint_V \nabla \cdot \mathbf{F} \, d\tau \qquad (13.8.2)$$

and all of the contributions from these common interfaces cancel; only the contribution from the parts on the outer surface S is left. These contributions, when added together, give $\oiint_S \mathbf{F} \cdot \mathbf{n} \, d\sigma$ over S and the proof is completed. $\square$

• **Example 13.8.1**

Let us verify the divergence theorem using the vector field $\mathbf{F} = 4x\mathbf{i} - 2y^2\mathbf{j} + z^2\mathbf{k}$ and the enclosed surface is the cylinder $x^2 + y^2 = 4$, $z = 0$, and $z = 3$. See Figure 13.8.2.

We begin by computing the volume integration. Because

$$\nabla \cdot \mathbf{F} = \frac{\partial(4x)}{\partial x} + \frac{\partial(-2y^2)}{\partial y} + \frac{\partial(z^2)}{\partial z} = 4 - 4y + 2z, \qquad (13.8.3)$$

$$\iiint_V \nabla \cdot \mathbf{F} \, dV = \iiint_V (4 - 4y + 2z) \, dV \qquad (13.8.4)$$

$$= \int_0^3 \int_0^2 \int_0^{2\pi} [4 - 4r\sin(\theta) + 2z] \, d\theta \, r \, dr \, dz \qquad (13.8.5)$$

$$= \int_0^3 \int_0^2 \left[4\theta \big|_0^{2\pi} + 4r\cos(\theta) \big|_0^{2\pi} + 2z\theta \big|_0^{2\pi} \right] r \, dr \, dz \qquad (13.8.6)$$

$$= \int_0^3 \int_0^2 (8\pi + 4\pi z) \, r \, dr \, dz \qquad (13.8.7)$$

$$\iiint_V \nabla \cdot \mathbf{F}\, dV = \int_0^3 4\pi(2+z)\tfrac{1}{2}r^2\big|_0^2\, dz \tag{13.8.8}$$

$$= 4\pi \int_0^3 2(2+z)\, dz = 8\pi(2z + \tfrac{1}{2}z^2)\big|_0^3 = 84\pi. \tag{13.8.9}$$

Turning to the surface integration, we have three surfaces:

$$\oiint_S \mathbf{F} \cdot \mathbf{n}\, d\sigma = \iint_{S_1} \mathbf{F} \cdot \mathbf{n}\, d\sigma + \iint_{S_2} \mathbf{F} \cdot \mathbf{n}\, d\sigma + \iint_{S_3} \mathbf{F} \cdot \mathbf{n}\, d\sigma. \tag{13.8.10}$$

The first integral is over the exterior to the cylinder. Because the surface is defined by $f(x, y, z) = x^2 + y^2 = 4$,

$$\mathbf{n} = \frac{\nabla f}{|\nabla f|} = \frac{2x\mathbf{i} + 2y\mathbf{j}}{\sqrt{4x^2 + 4y^2}} = \frac{x}{2}\mathbf{i} + \frac{y}{2}\mathbf{j}. \tag{13.8.11}$$

Therefore,

$$\iint_{S_1} \mathbf{F} \cdot \mathbf{n}\, d\sigma = \iint_{S_1} (2x^2 - y^3)\, d\sigma \tag{13.8.12}$$

$$= \int_0^3 \int_0^{2\pi} \left\{ 2[2\cos(\theta)]^2 - [2\sin(\theta)]^3 \right\} 2\, d\theta\, dz \tag{13.8.13}$$

$$= 8 \int_0^3 \int_0^{2\pi} \left\{ \tfrac{1}{2}[1 + \cos(2\theta)] - \sin(\theta) + \cos^2(\theta)\sin(\theta) \right\} 2\, d\theta\, dz \tag{13.8.14}$$

$$= 16 \int_0^3 \left[\tfrac{1}{2}\theta + \tfrac{1}{4}\sin(2\theta) + \cos(\theta) - \tfrac{1}{3}\cos^3(\theta) \right]\Big|_0^{2\pi} dz \tag{13.8.15}$$

$$= 16\pi \int_0^3 dz = 48\pi, \tag{13.8.16}$$

because $x = 2\cos(\theta)$, $y = 2\sin(\theta)$, and $d\sigma = 2\, d\theta\, dz$ in cylindrical coordinates.

Along the top of the cylinder, $z = 3$, the outward pointing normal is $\mathbf{n} = \mathbf{k}$, and $d\sigma = r\, dr\, d\theta$. Then,

$$\iint_{S_2} \mathbf{F} \cdot \mathbf{n}\, d\sigma = \iint_{S_2} z^2\, d\sigma = \int_0^{2\pi} \int_0^2 9r\, dr\, d\theta = 2\pi \times 9 \times 2 = 36\pi. \tag{13.8.17}$$

However, along the bottom of the cylinder, $z = 0$, the outward pointing normal is $\mathbf{n} = -\mathbf{k}$ and $d\sigma = r\, dr\, d\theta$. Then,

$$\iint_{S_3} \mathbf{F} \cdot \mathbf{n}\, d\sigma = \iint_{S_3} z^2\, d\sigma = \int_0^{2\pi} \int_0^2 0\, r\, dr\, d\theta = 0. \tag{13.8.18}$$

Consequently, the flux out the entire cylinder is

$$\oiint_S \mathbf{F} \cdot \mathbf{n} \, d\sigma = 48\pi + 36\pi + 0 = 84\pi, \qquad (13.8.19)$$

and the divergence theorem is verified for this special case.

• Example 13.8.2

Let us verify the divergence theorem given the vector field $\mathbf{F} = 3x^2y^2\mathbf{i} + y\mathbf{j} - 6xy^2z\mathbf{k}$ and the volume is the region bounded by the paraboloid $z = x^2 + y^2$ and the plane $z = 2y$. See Figure 13.8.3.

Computing the divergence,

$$\nabla \cdot \mathbf{F} = \frac{\partial(3x^2y^2)}{\partial x} + \frac{\partial(y)}{\partial y} + \frac{\partial(-6xy^2z)}{\partial z} = 6xy^2 + 1 - 6xy^2 = 1. \quad (13.8.20)$$

Then,

$$\iiint_V \nabla \cdot \mathbf{F} \, dV = \iiint_V dV \qquad (13.8.21)$$

$$= \int_0^\pi \int_0^{2\sin(\theta)} \int_{r^2}^{2r\sin(\theta)} dz \, r \, dr \, d\theta \qquad (13.8.22)$$

$$= \int_0^\pi \int_0^{2\sin(\theta)} [2r\sin(\theta) - r^2] \, r \, dr \, d\theta \qquad (13.8.23)$$

$$= \int_0^\pi \left[\frac{2}{3}r^3 \Big|_0^{2\sin(\theta)} \sin(\theta) - \frac{1}{4}r^4 \Big|_0^{2\sin(\theta)} \right] d\theta \qquad (13.8.24)$$

$$= \int_0^\pi \left[\frac{16}{3}\sin^4(\theta) - 4\sin^4(\theta) \right] d\theta \qquad (13.8.25)$$

$$= \int_0^\pi \frac{4}{3}\sin^4(\theta) \, d\theta \qquad (13.8.26)$$

$$= \frac{1}{3} \int_0^\pi [1 - 2\cos(2\theta) + \cos^2(2\theta)] \, d\theta \qquad (13.8.27)$$

$$= \frac{1}{3} \left[\theta \Big|_0^\pi - \sin(2\theta) \Big|_0^\pi + \frac{1}{2}\theta \Big|_0^\pi + \frac{1}{8}\sin(4\theta) \Big|_0^\pi \right] = \frac{\pi}{2}. \quad (13.8.28)$$

The limits in the radial direction are given by the intersection of the paraboloid and plane: $r^2 = 2r\sin(\theta)$, or $r = 2\sin(\theta)$, and y is greater than zero.

Turning to the surface integration, we have two surfaces:

$$\oiint_S \mathbf{F} \cdot \mathbf{n} \, d\sigma = \iint_{S_1} \mathbf{F} \cdot \mathbf{n} \, d\sigma + \iint_{S_2} \mathbf{F} \cdot \mathbf{n} \, d\sigma, \qquad (13.8.29)$$

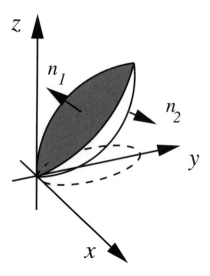

Figure 13.8.3: Diagram for the verification of the divergence theorem in Example 13.8.2. The dashed line denotes the curve $r = 2\sin(\theta)$.

where S_1 is the plane $z = 2y$, and S_2 is the paraboloid. For either surface, polar coordinates are best so that $x = r\cos(\theta)$, and $y = r\sin(\theta)$. For the integration over the plane, $z = 2r\sin(\theta)$. Therefore,

$$\mathbf{r} = r\cos(\theta)\mathbf{i} + r\sin(\theta)\mathbf{j} + 2r\sin(\theta)\mathbf{k}, \tag{13.8.30}$$

so that

$$\mathbf{r}_r = \cos(\theta)\mathbf{i} + \sin(\theta)\mathbf{j} + 2\sin(\theta)\mathbf{k}, \tag{13.8.31}$$

and

$$\mathbf{r}_\theta = -r\sin(\theta)\mathbf{i} + r\cos(\theta)\mathbf{j} + 2r\cos(\theta)\mathbf{k}. \tag{13.8.32}$$

Then,

$$\mathbf{r}_r \times \mathbf{r}_\theta = \begin{vmatrix} \mathbf{i} & \mathbf{j} & \mathbf{k} \\ \cos(\theta) & \sin(\theta) & 2\sin(\theta) \\ -r\sin(\theta) & r\cos(\theta) & 2r\cos(\theta) \end{vmatrix} = -2r\mathbf{j} + r\mathbf{k}. \tag{13.8.33}$$

This is an outwardly pointing normal so that we can immediately set up the surface integral:

$$\iint_{S_1} \mathbf{F} \cdot \mathbf{n}\, d\sigma = \int_0^\pi \int_0^{2\sin(\theta)} \left\{ 3r^4 \cos^2(\theta)\sin^2(\theta)\mathbf{i} + r\sin(\theta)\mathbf{j} \right.$$
$$\left. - 6[2r\sin(\theta)][r\cos(\theta)][r^2\sin^2(\theta)]\mathbf{k} \right\} \cdot \left(-2r\mathbf{j} + r\mathbf{k}\right)\, dr\, d\theta$$
$$\tag{13.8.34}$$

$$\iint_{S_1} \mathbf{F} \cdot \mathbf{n} \, d\sigma = \int_0^\pi \int_0^{2\sin(\theta)} \left[-2r^2 \sin(\theta) - 12r^5 \sin^3(\theta) \cos(\theta) \right] dr \, d\theta$$

$$(13.8.35)$$

$$= \int_0^\pi \left[-\tfrac{2}{3} r^3 \Big|_0^{2\sin(\theta)} \sin(\theta) - 2r^6 \Big|_0^{2\sin(\theta)} \sin^3(\theta) \cos(\theta) \right] d\theta$$

$$(13.8.36)$$

$$= \int_0^\pi \left[-\tfrac{16}{3} \sin^4(\theta) - 128 \sin^9(\theta) \cos(\theta) \right] d\theta \qquad (13.8.37)$$

$$= -\tfrac{4}{3} \left[\theta \Big|_0^\pi - \sin(2\theta) \Big|_0^\pi + \tfrac{1}{2}\theta \Big|_0^\pi + \tfrac{1}{8} \sin(4\theta) \Big|_0^\pi \right] - \tfrac{64}{5} \sin^{10}(\theta) \Big|_0^\pi$$

$$(13.8.38)$$

$$= -2\pi. \qquad (13.8.39)$$

For the surface of the paraboloid,

$$\mathbf{r} = r\cos(\theta)\mathbf{i} + r\sin(\theta)\mathbf{j} + r^2\mathbf{k}, \qquad (13.8.40)$$

so that

$$\mathbf{r}_r = \cos(\theta)\mathbf{i} + \sin(\theta)\mathbf{j} + 2r\mathbf{k}, \qquad (13.8.41)$$

and

$$\mathbf{r}_\theta = -r\sin(\theta)\mathbf{i} + r\cos(\theta)\mathbf{j}. \qquad (13.8.42)$$

Then,

$$\mathbf{r}_r \times \mathbf{r}_\theta = \begin{vmatrix} \mathbf{i} & \mathbf{j} & \mathbf{k} \\ \cos(\theta) & \sin(\theta) & 2r \\ -r\sin(\theta) & r\cos(\theta) & 0 \end{vmatrix} \qquad (13.8.43)$$

$$= -2r^2 \cos(\theta)\mathbf{i} - 2r^2 \sin(\theta)\mathbf{j} + r\mathbf{k}. \qquad (13.8.44)$$

This is an inwardly pointing normal so that we must take the negative of it before we do the surface integral. Then,

$$\iint_{S_2} \mathbf{F} \cdot \mathbf{n} \, d\sigma = \int_0^\pi \int_0^{2\sin(\theta)} \{ 3r^4 \cos^2(\theta) \sin^2(\theta)\mathbf{i} + r\sin(\theta)\mathbf{j}$$

$$- 6r^2 [r\cos(\theta)][r^2 \sin^2(\theta)]\mathbf{k} \}$$

$$\cdot \left[2r^2 \cos(\theta)\mathbf{i} + 2r^2 \sin(\theta)\mathbf{j} - r\mathbf{k} \right] dr \, d\theta \quad (13.8.45)$$

$$= \int_0^\pi \int_0^{2\sin(\theta)} \left[6r^6 \cos^3(\theta) \sin^2(\theta) + 2r^3 \sin^2(\theta) \right.$$

$$\left. + 6r^6 \cos(\theta) \sin^2(\theta) \right] dr \, d\theta \qquad (13.8.46)$$

$$\iint_{S_2} \mathbf{F} \cdot \mathbf{n}\, d\sigma = \int_0^\pi \left[\tfrac{6}{7} r^7 \Big|_0^{2\sin(\theta)} \cos^3(\theta)\sin^2(\theta) + \tfrac{1}{2} r^4 \Big|_0^{2\sin(\theta)} \sin^2(\theta) \right.$$

$$\left. + \tfrac{6}{7} r^7 \Big|_0^{2\sin(\theta)} \cos(\theta)\sin^2(\theta) \right] d\theta \tag{13.8.47}$$

$$= \int_0^\pi \left\{ \tfrac{768}{7} \sin^9(\theta)[1 - \sin^2(\theta)]\cos(\theta) + 8\sin^6(\theta) \right.$$

$$\left. + \tfrac{768}{7} \sin^9(\theta)\cos(\theta) \right\} d\theta \tag{13.8.48}$$

$$= \tfrac{1536}{70} \sin^{10}(\theta)\Big|_0^\pi - \tfrac{64}{7} \sin^{12}(\theta)\Big|_0^\pi + \int_0^\pi [1 - \cos(2\theta)]^3\, d\theta \tag{13.8.49}$$

$$= \int_0^\pi \left\{ 1 - 3\cos(2\theta) + 3\cos^2(2\theta) - \cos(2\theta)[1 - \sin^2(2\theta)] \right\} d\theta \tag{13.8.50}$$

$$= \theta\Big|_0^\pi - \tfrac{3}{2}\sin(2\theta)\Big|_0^\pi + \tfrac{3}{2}[\theta + \tfrac{1}{4}\sin(4\theta)]\Big|_0^\pi$$

$$- \tfrac{1}{2}\sin(2\theta)\Big|_0^\pi + \tfrac{1}{3}\sin^3(2\theta)\Big|_0^\pi \tag{13.8.51}$$

$$= \pi + \tfrac{3}{2}\pi = \tfrac{5}{2}\pi. \tag{13.8.52}$$

Consequently,

$$\oiint_S \mathbf{F} \cdot \mathbf{n}\, d\sigma = -2\pi + \tfrac{5}{2}\pi = \tfrac{1}{2}\pi, \tag{13.8.53}$$

and the divergence theorem is verified for this special case.

• Example 13.8.3: Archimedes' principle

Consider a solid[5] of volume V and surface S that is immersed in a vessel filled with a fluid of density ρ. The pressure field p in the fluid is a function of the distance from the liquid/air interface and equals

$$p = p_0 - \rho g z, \tag{13.8.54}$$

where g is the gravitational acceleration, z is the vertical distance measured from the interface (increasing in the $\mathbf{k}$ direction), and p_0 is the constant pressure along the liquid/air interface.

If we define $\mathbf{F} = -p\mathbf{k}$, then $\mathbf{F} \cdot \mathbf{n}\, d\sigma$ is the vertical component of the force on the surface due to the pressure and $\oiint_S \mathbf{F} \cdot \mathbf{n}\, d\sigma$ is the total lift. Using the divergence theorem and noting that $\nabla \cdot \mathbf{F} = \rho g$, the total lift also equals

$$\iiint_V \nabla \cdot \mathbf{F}\, dV = \rho g \iiint_V dV = \rho g V, \tag{13.8.55}$$

---

[5] Adapted from Altintas, A., 1990: Archimedes' principle as an application of the divergence theorem. *IEEE Trans. Educ.*, **33**, 222. ©IEEE.

which is the weight of the displaced liquid. This is *Archimedes' principle*: the buoyant force on a solid immersed in a fluid of constant density equals the weight of the fluid displaced.

• Example 13.8.4: Conservation of charge

Let a charge of density ρ flow with an average velocity $\mathbf{v}$. Then the charge crossing the element $d\mathbf{S}$ per unit time is $\rho\mathbf{v}\cdot d\mathbf{S} = \mathbf{J}\cdot d\mathbf{S}$, where $\mathbf{J}$ is defined as the conduction current vector or current density vector. The current across any surface drawn in the medium is $\iint_S \mathbf{J}\cdot d\mathbf{S}$.

The total charge inside the closed surface is $\iiint_V \rho\,dV$. If there are no sources or sinks inside the surface, the rate at which the charge decreases is $-\iiint_V \rho_t\,dV$. Because this change is due to the outward flow of charge,

$$-\iiint_V \frac{\partial\rho}{\partial t}\,dV = \oiint_S \mathbf{J}\cdot d\mathbf{S}. \tag{13.8.56}$$

Applying the divergence theorem,

$$\iiint_V \left(\frac{\partial\rho}{\partial t} + \nabla\cdot\mathbf{J}\right)dV = 0. \tag{13.8.57}$$

Because the result holds true for any arbitrary volume, the integrand must vanish identically and we have the equation of continuity or the *equation of conservation of charge*:

$$\frac{\partial\rho}{\partial t} + \nabla\cdot\mathbf{J} = 0. \tag{13.8.58}$$

Problems

Verify the divergence theorem using the following vector fields and volumes:

1. $\mathbf{F} = x^2\mathbf{i} + y^2\mathbf{j} + z^2\mathbf{k}$ and the volume V is the cube cut from the first octant by the planes $x = 1$, $y = 1$, and $z = 1$.

2. $\mathbf{F} = xy\mathbf{i} + yz\mathbf{j} + xz\mathbf{k}$ and the volume V is the cube bounded by $0 \le x \le 1$, $0 \le y \le 1$, and $0 \le z \le 1$.

3. $\mathbf{F} = (y - x)\mathbf{i} + (z - y)\mathbf{j} + (y - x)\mathbf{k}$ and the volume V is the cube bounded by $-1 \le x \le 1$, $-1 \le y \le 1$, and $-1 \le z \le 1$.

4. $\mathbf{F} = x^2\mathbf{i} + y\mathbf{j} + z\mathbf{k}$ and the volume V is the cylinder defined by the surfaces $x^2 + y^2 = 1$, $z = 0$, and $z = 1$.

5. $\mathbf{F} = x^2\mathbf{i} + y^2\mathbf{j} + z^2\mathbf{k}$ and the volume V is the cylinder defined by the surfaces $x^2 + y^2 = 4$, $z = 0$, and $z = 1$.

6. $\mathbf{F} = y^2\mathbf{i} + xz^3\mathbf{j} + (z-1)^2\mathbf{k}$ and the volume V is the cylinder bounded by the surface $x^2 + y^2 = 4$, and the planes $z = 1$ and $z = 5$.

7. $\mathbf{F} = 6xy\mathbf{i} + 4yz\mathbf{j} + xe^{-y}\mathbf{k}$ and the volume V is that region created by the plane $x + y + z = 1$, and the three coordinate planes.

8. $\mathbf{F} = y\mathbf{i} + xy\mathbf{j} - z\mathbf{k}$ and the volume V is that solid created by the paraboloid $z = x^2 + y^2$ and plane $z = 1$.

Chapter 14
Linear Algebra

Linear algebra involves the systematic solving of linear algebraic or differential equations that arise during the mathematical modeling of an electrical, mechanical, or even human system where two or more components are interacting with each other. In this chapter we present efficient techniques for expressing these systems and their solution.

14.1 FUNDAMENTALS OF LINEAR ALGEBRA

Consider the following system of m simultaneous linear equations in n unknowns $x_1, x_2, x_3, \ldots, x_n$:

$$a_{11}x_1 + a_{12}x_2 + \cdots + a_{1n}x_n = b_1,$$

$$a_{21}x_1 + a_{22}x_2 + \cdots + a_{2n}x_n = b_2,$$

$$\vdots \qquad\qquad (14.1.1)$$

$$a_{m1}x_1 + a_{m2}x_2 + \cdots + a_{mn}x_n = b_m,$$

where the coefficients a_{ij} and constants b_j denote known real or complex numbers. The purpose of this chapter is to show how *matrix algebra* can be used to solve these systems by first introducing succinct notation so that we can replace (12.1.1) with rather simple expressions and then employing a set of rules to manipulate these expressions. In this section we focus on developing these simple expressions.

The fundamental quantity in linear algebra is the *matrix*.[1] A matrix is an ordered rectangular array of numbers or mathematical expressions. We shall use upper case letters to denote them. The $m \times n$ matrix

$$
A = \begin{pmatrix}
a_{11} & a_{12} & a_{13} & \cdot & \cdot & \cdot & a_{1n} \\
a_{21} & a_{22} & a_{23} & \cdot & \cdot & \cdot & a_{2n} \\
\cdot & \cdot & \cdot & \cdot & \cdot & \cdot & \cdot \\
\cdot & \cdot & \cdot & \cdot & a_{ij} & \cdot & \cdot \\
\cdot & \cdot & \cdot & \cdot & \cdot & \cdot & \cdot \\
a_{m1} & a_{m2} & a_{m3} & \cdot & \cdot & \cdot & a_{mn}
\end{pmatrix}
\tag{14.1.2}
$$

has m *rows* and n *columns*. The *order* (or size) of a matrix is determined by the number of rows and columns; (12.1.2) is of order m by n. If $m = n$, the matrix is a *square* matrix; otherwise, A is *rectangular*. The numbers or expressions in the array a_{ij} are the *elements* of A and can be either real or complex. When all of the elements are real, A is a *real matrix*. If some or all of the elements are complex, then A is a *complex matrix*. For a square matrix, the diagonal from the top left corner to the bottom right corner is the *principal diagonal*.

From the limitless number of possible matrices, certain ones appear with sufficient regularity that they are given special names. A *zero* matrix (sometimes called a *null* matrix) has all of its elements equal to zero. It fulfills the role in matrix algebra that is analogous to that of zero in scalar algebra. The *unit* or *identity* matrix is a $n \times n$ matrix having 1's along its principal diagonal and zero everywhere else. The unit matrix serves essentially the same purpose in matrix algebra as does the number one in scalar algebra. A *symmetric* matrix is one where $a_{ij} = a_{ji}$ for all i and j.

• **Example 14.1.1**

Examples of zero, identity, and symmetric matrices are

$$
O = \begin{pmatrix} 0 & 0 & 0 \\ 0 & 0 & 0 \\ 0 & 0 & 0 \end{pmatrix}, \quad I = \begin{pmatrix} 1 & 0 \\ 0 & 1 \end{pmatrix}, \quad \text{and } A = \begin{pmatrix} 3 & 2 & 4 \\ 2 & 1 & 0 \\ 4 & 0 & 5 \end{pmatrix},
\tag{14.1.3}
$$

respectively.

A special class of matrices are *column vectors* and *row vectors*:

$$
\mathbf{x} = \begin{pmatrix} x_1 \\ x_2 \\ \vdots \\ x_m \end{pmatrix}, \quad \mathbf{y} = (y_1 \quad y_2 \quad \cdots \quad y_n).
\tag{14.1.4}
$$

---

[1] This term was first used by Sylvester, J. J., 1850: Additions to the articles, "On a new class of theorems," and "On Pascal's theorem." *Philos. Mag., Ser. 4*, **37**, 363–370.

We denote row and column vectors by lower case, boldfaced letters. The length or *norm* of the vector $\mathbf{x}$ of n elements is

$$||\mathbf{x}|| = \left(\sum_{k=1}^{n} x_k^2 \right)^{1/2}. \tag{14.1.5}$$

Two matrices A and B are equal if and only if $a_{ij} = b_{ij}$ for all possible i and j and they have the same dimensions.

Having defined a matrix, let us explore some of its arithmetic properties. For two matrices A and B with the same dimensions (conformable for addition), the matrix $C = A + B$ contains the elements $c_{ij} = a_{ij} + b_{ij}$. Similarly, $C = A - B$ contains the elements $c_{ij} = a_{ij} - b_{ij}$. Because the order of addition does not matter, addition is *commutative*: $A + B = B + A$.

Consider now a scalar constant k. The product kA is formed by multiplying every element of A by k. Thus the matrix kA has elements ka_{ij}.

So far the rules for matrix arithmetic conform to their scalar counterparts. However, there are several possible ways of multiplying two matrices together. For example, we might simply multiply together the corresponding elements from each matrix. As we will see, the multiplication rule is designed to facilitate the solution of linear equations.

We begin by requiring that the dimensions of A be $m \times n$ while for B they are $n \times p$. That is, the number of columns in A must equal the number of rows in B. The matrices A and B are then said to be *conformable* for multiplication. If this is true, then $C = AB$ is a matrix $m \times p$, where its elements equal

$$c_{ij} = \sum_{k=1}^{n} a_{ik} b_{kj}. \tag{14.1.6}$$

The right side of (14.1.6) is referred to as an *inner product* of the ith row of A and the jth column of B. Although (14.1.6) is the method used with a computer, an easier method for human computation is as a running sum of the products given by successive elements of the ith row of A and the corresponding elements of the jth column of B.

The product AA is usually written A^2; the product AAA, A^3, and so forth.

• Example 14.1.2

If

$$A = \begin{pmatrix} -1 & 4 \\ 2 & -3 \end{pmatrix}, \quad \text{and} \quad B = \begin{pmatrix} 1 & 2 \\ 3 & 4 \end{pmatrix}, \tag{14.1.7}$$

then

$$AB = \begin{pmatrix} [(-1)(1) + (4)(3)] & [(-1)(2) + (4)(4)] \\ [(2)(1) + (-3)(3)] & [(2)(2) + (-3)(4)] \end{pmatrix} \tag{14.1.8}$$

$$= \begin{pmatrix} 11 & 14 \\ -7 & -8 \end{pmatrix}. \tag{14.1.9}$$

Checking our results using MATLAB, we have that
```
>> A = [-1 4; 2 -3];
>> B = [1 2; 3 4];
>> C = A*B
C =

    11   14
    -7   -8
```

Note that there is a tremendous difference between the MATLAB command for matrix multiplication $*$ and element-by-element multiplication $.*$.

Matrix multiplication is associative and distributive with respect to addition:

$$(kA)B = k(AB) = A(kB), \qquad (14.1.10)$$

$$A(BC) = (AB)C, \qquad (14.1.11)$$

$$(A + B)C = AC + BC, \qquad (14.1.12)$$

and

$$C(A + B) = CA + CB. \qquad (14.1.13)$$

On the other hand, matrix multiplication is *not commutative*. In general, $AB \neq BA$.

• **Example 14.1.3**

Does $AB = BA$ if

$$A = \begin{pmatrix} 1 & 0 \\ 0 & 0 \end{pmatrix}, \quad \text{and} \quad B = \begin{pmatrix} 1 & 1 \\ 1 & 0 \end{pmatrix}? \qquad (14.1.14)$$

Because

$$AB = \begin{pmatrix} 1 & 0 \\ 0 & 0 \end{pmatrix} \begin{pmatrix} 1 & 1 \\ 1 & 0 \end{pmatrix} = \begin{pmatrix} 1 & 1 \\ 0 & 0 \end{pmatrix}, \qquad (14.1.15)$$

and

$$BA = \begin{pmatrix} 1 & 1 \\ 1 & 0 \end{pmatrix} \begin{pmatrix} 1 & 0 \\ 0 & 0 \end{pmatrix} = \begin{pmatrix} 1 & 0 \\ 1 & 0 \end{pmatrix}, \qquad (14.1.16)$$

$$AB \neq BA. \qquad (14.1.17)$$

• **Example 14.1.4**

Given

$$A = \begin{pmatrix} 1 & 1 \\ 3 & 3 \end{pmatrix}, \quad \text{and} \quad B = \begin{pmatrix} -1 & 1 \\ 1 & -1 \end{pmatrix}, \qquad (14.1.18)$$

find the product AB.

Performing the calculation, we find that

$$AB = \begin{pmatrix} 1 & 1 \\ 3 & 3 \end{pmatrix} \begin{pmatrix} -1 & 1 \\ 1 & -1 \end{pmatrix} = \begin{pmatrix} 0 & 0 \\ 0 & 0 \end{pmatrix}. \tag{14.1.19}$$

The point here is that just because $AB = 0$, this does *not* imply that either A or B equals the zero matrix.

We cannot properly speak of division when we are dealing with matrices. Nevertheless, a matrix A is said to be *nonsingular* or *invertible* if there exists a matrix B such that $AB = BA = I$. This matrix B is the multiplicative inverse of A or simply the *inverse* of A, written A^{-1}. A $n \times n$ matrix is *singular* if it does not have a multiplicative inverse.

• **Example 14.1.5**

If

$$A = \begin{pmatrix} 1 & 0 & 1 \\ 3 & 3 & 4 \\ 2 & 2 & 3 \end{pmatrix}, \tag{14.1.20}$$

let us verify that its inverse is

$$A^{-1} = \begin{pmatrix} 1 & 2 & -3 \\ -1 & 1 & -1 \\ 0 & -2 & 3 \end{pmatrix}. \tag{14.1.21}$$

We perform the check by finding AA^{-1} or $A^{-1}A$,

$$AA^{-1} = \begin{pmatrix} 1 & 0 & 1 \\ 3 & 3 & 4 \\ 2 & 2 & 3 \end{pmatrix} \begin{pmatrix} 1 & 2 & -3 \\ -1 & 1 & -1 \\ 0 & -2 & 3 \end{pmatrix} = \begin{pmatrix} 1 & 0 & 0 \\ 0 & 1 & 0 \\ 0 & 0 & 1 \end{pmatrix}. \tag{14.1.22}$$

In a later section we will show how to compute the inverse, given A.

Another matrix operation is transposition. The *transpose* of a matrix A with dimensions $m \times n$ is another matrix, written A^T, where we interchanged the rows and columns from A. In MATLAB, A^T is computed by typing `A'`. Clearly, $(A^T)^T = A$ as well as $(A + B)^T = A^T + B^T$, and $(kA)^T = kA^T$. If A and B are conformable for multiplication, then $(AB)^T = B^T A^T$. Note the reversal of order between the two sides. To prove this last result, we first show that the results are true for two 3×3 matrices A and B and then generalize to larger matrices.

Having introduced some of the basic concepts of linear algebra, we are ready to rewrite (14.1.1) in a canonical form so that we can present techniques for its solution. We begin by writing (14.1.1) as a single column vector:

$$
\begin{pmatrix}
a_{11}x_1 & + & a_{12}x_2 & + & \cdots & + & a_{1n}x_n \\
a_{21}x_1 & + & a_{22}x_2 & + & \cdots & + & a_{2n}x_n \\
\vdots & & \vdots & & \vdots & & \vdots \\
\vdots & & \vdots & & \vdots & & \vdots \\
a_{m1}x_1 & + & a_{m2}x_2 & + & \cdots & + & a_{mn}x_n
\end{pmatrix}
=
\begin{pmatrix}
b_1 \\
b_2 \\
\vdots \\
\vdots \\
b_m
\end{pmatrix}.
\tag{14.1.23}
$$

We now use the multiplication rule to rewrite (14.1.23) as

$$
\begin{pmatrix}
a_{11} & a_{12} & \cdots & a_{1n} \\
a_{21} & a_{22} & \cdots & a_{2n} \\
\vdots & \vdots & \vdots & \vdots \\
\vdots & \vdots & \vdots & \vdots \\
a_{m1} & a_{m2} & \cdots & a_{mn}
\end{pmatrix}
\begin{pmatrix}
x_1 \\
x_2 \\
\vdots \\
\vdots \\
x_n
\end{pmatrix}
=
\begin{pmatrix}
b_1 \\
b_2 \\
\vdots \\
\vdots \\
b_m
\end{pmatrix},
\tag{14.1.24}
$$

or

$$
A\mathbf{x} = \mathbf{b},
\tag{14.1.25}
$$

where $\mathbf{x}$ is the solution vector. If $\mathbf{b} = \mathbf{0}$, we have a *homogeneous* set of equations; otherwise, we have a *nonhomogeneous* set. In the next few sections, we will give a number of methods for finding $\mathbf{x}$.

• Example 14.1.6: Solution of a tridiagonal system

A common problem in linear algebra involves solving systems such as

$$
b_1 y_1 + c_1 y_2 = d_1,
\tag{14.1.26}
$$

$$
a_2 y_1 + b_2 y_2 + c_2 y_3 = d_2,
\tag{14.1.27}
$$

$$
\vdots
$$

$$
a_{N-1} y_{N-2} + b_{N-1} y_{N-1} + c_{N-1} y_N = d_{N-1},
\tag{14.1.28}
$$

$$
b_N y_{N-1} + c_N y_N = d_N.
\tag{14.1.29}
$$

Such systems arise in the numerical solution of ordinary and partial differential equations.

We begin by rewriting (14.1.26)–(14.1.29) in the matrix notation:

$$
\begin{pmatrix}
b_1 & c_1 & 0 & \cdots & 0 & 0 & 0 \\
a_2 & b_2 & c_2 & \cdots & 0 & 0 & 0 \\
0 & a_3 & b_3 & \cdots & 0 & 0 & 0 \\
\vdots & \vdots & \vdots & \vdots & \vdots & \vdots & \vdots \\
0 & 0 & 0 & \cdots & a_{N-1} & b_{N-1} & c_{N-1} \\
0 & 0 & 0 & \cdots & 0 & a_N & b_N
\end{pmatrix}
\begin{pmatrix}
y_1 \\
y_2 \\
y_3 \\
\vdots \\
y_{N-1} \\
y_N
\end{pmatrix}
=
\begin{pmatrix}
d_1 \\
d_2 \\
d_3 \\
\vdots \\
d_{N-1} \\
d_N
\end{pmatrix}.
$$

$$
\tag{14.1.30}
$$

The matrix in (14.1.30) is an example of a *banded matrix*: a matrix where all of the elements in each row are zero except for the diagonal element and a limited number on either side of it. In the present case, we have a *tridiagonal* matrix in which only the diagonal element and the elements immediately to its left and right in each row are nonzero.

Consider the nth equation. We can eliminate a_n by multiplying the $(n-1)$th equation by a_n/b_{n-1} and subtracting this new equation from the nth equation. The values of b_n and d_n become

$$b'_n = b_n - a_n c_{n-1}/b_{n-1}, \tag{14.1.31}$$

and

$$d'_n = d_n - a_n d_{n-1}/b_{n-1} \tag{14.1.32}$$

for $n = 2, 3, \ldots, N$. The coefficient c_n is unaffected. Because elements a_1 and c_N are never involved, their values can be anything or they can be left undefined. The new system of equations may be written

$$
\begin{pmatrix}
b'_1 & c_1 & 0 & \cdots & 0 & 0 & 0 \\
0 & b'_2 & c_2 & \cdots & 0 & 0 & 0 \\
0 & 0 & b'_3 & \cdots & 0 & 0 & 0 \\
\vdots & \vdots & \vdots & \vdots & \vdots & \vdots & \vdots \\
0 & 0 & 0 & \cdots & 0 & b'_{N-1} & c_{N-1} \\
0 & 0 & 0 & \cdots & 0 & 0 & b'_N
\end{pmatrix}
\begin{pmatrix}
y_1 \\ y_2 \\ y_3 \\ \vdots \\ y_{N-1} \\ y_N
\end{pmatrix}
=
\begin{pmatrix}
d'_1 \\ d'_2 \\ d'_3 \\ \vdots \\ d'_{N-1} \\ d'_N
\end{pmatrix}. \tag{14.1.33}
$$

The matrix in (14.1.33) is in *upper triangular* form because all of the elements below the principal diagonal are zero. This is particularly useful because y_n can be computed by *back substitution*. That is, we first compute y_N. Next, we calculate y_{N-1} in terms of y_N. The solution y_{N-2} can then be computed in terms of y_N and y_{N-1}. We continue this process until we find y_1 in terms of $y_N, y_{N-1}, \ldots, y_2$. In the present case, we have the rather simple:

$$y_N = d'_N/b'_N, \tag{14.1.34}$$

and

$$y_n = (d'_n - c_n d'_{n+1})/b'_n \tag{14.1.35}$$

for $n = N-1, N-2, \ldots, 2, 1$.

As we shall show shortly, this is an example of solving a system of linear equations by Gaussian elimination. For a tridiagonal case, we have the advantage that the solution can be expressed in terms of a recurrence relationship, a very convenient feature from a computational point of view. This algorithm is very robust, being stable[2] as long as $|a_i + c_i| < |b_i|$. By stability, we mean

---

[2] Torii, T., 1966: Inversion of tridiagonal matrices and the stability of tridiagonal systems of linear systems. *Tech. Rep. Osaka Univ.*, **16**, 403–414.

that if we change $\mathbf{b}$ by $\Delta\mathbf{b}$ so that $\mathbf{x}$ changes by $\Delta\mathbf{x}$, then $||\Delta\mathbf{x}|| < M\epsilon$, where $||\Delta\mathbf{b}|| \leq \epsilon$, $0 < M < \infty$, for *any* N.

Problems

Given $A = \begin{pmatrix} 3 & 4 \\ 1 & 2 \end{pmatrix}$, and $B = \begin{pmatrix} 1 & 1 \\ 2 & 2 \end{pmatrix}$, find

1. $A + B$, $B + A$ 2. $A - B$, $B - A$ 3. $3A - 2B$, $3(2A - B)$

4. $A^T, B^T, (B^T)^T$ 5. $(A + B)^T$, $A^T + B^T$ 6. $B + B^T$, $B - B^T$

7. $AB, A^T B, BA, B^T A$ 8. A^2, B^2 9. BB^T, $B^T B$

10. $A^2 - 3A + I$ 11. $A^3 + 2A$ 12. $A^4 - 4A^2 + 2I$

by hand and using MATLAB.

Can multiplication occur between the following matrices? If so, compute it.

13. $\begin{pmatrix} 3 & 5 & 1 \\ -2 & 1 & 2 \end{pmatrix} \begin{pmatrix} 2 & 1 \\ 4 & 1 \\ 1 & 3 \end{pmatrix}$ 14. $\begin{pmatrix} -2 & 4 \\ -4 & 6 \\ -6 & 1 \end{pmatrix} \begin{pmatrix} 1 & 2 & 3 \end{pmatrix}$

15. $\begin{pmatrix} 1 & 4 & 2 \\ 0 & 0 & 4 \\ 0 & 1 & 2 \end{pmatrix} \begin{pmatrix} 3 & 2 \\ 1 & 1 \\ 2 & 1 \end{pmatrix}$ 16. $\begin{pmatrix} 4 & 6 \\ 1 & 2 \end{pmatrix} \begin{pmatrix} 1 & 3 & 6 \\ 1 & 2 & 5 \end{pmatrix}$

17. $\begin{pmatrix} 6 & 4 & 2 \\ 1 & 2 & 3 \end{pmatrix} \begin{pmatrix} 3 & 1 & 4 \\ 2 & 0 & 6 \end{pmatrix}$

If $A = \begin{pmatrix} 1 & 1 \\ 1 & 2 \\ 3 & 1 \end{pmatrix}$, verify that

18. $7A = 4A + 3A$,   19. $10A = 5(2A)$,   20. $(A^T)^T = A$

by hand and using MATLAB.

If $A = \begin{pmatrix} 2 & 1 \\ 3 & 1 \end{pmatrix}$, $B = \begin{pmatrix} 1 & -2 \\ 4 & 0 \end{pmatrix}$, and $C = \begin{pmatrix} 1 & 1 \\ 1 & 1 \end{pmatrix}$, verify that

21. $(A + B) + C = A + (B + C)$,   22. $(AB)C = A(BC)$,

23. $A(B + C) = AB + AC$,   24. $(A + B)C = AC + BC$

by hand and using MATLAB.

Verify that the following A^{-1} are indeed the inverse of A:

25. $A = \begin{pmatrix} 3 & -1 \\ -5 & 2 \end{pmatrix}$ $A^{-1} = \begin{pmatrix} 2 & 1 \\ 5 & 3 \end{pmatrix}$

26. $A = \begin{pmatrix} 0 & 1 & 0 \\ 1 & 0 & 0 \\ 0 & 0 & 1 \end{pmatrix}$ $A^{-1} = \begin{pmatrix} 0 & 1 & 0 \\ 1 & 0 & 0 \\ 0 & 0 & 1 \end{pmatrix}$

by hand and using MATLAB.

Write the following linear systems of equations in matrix form: $A\mathbf{x} = \mathbf{b}$.

27.
$$x_1 - 2x_2 = 5 \qquad\qquad 3x_1 + x_2 = 1$$

28.
$$2x_1 + x_2 + 4x_3 = 2 \qquad 4x_1 + 2x_2 + 5x_3 = 6 \qquad 6x_1 - 3x_2 + 5x_3 = 2$$

29.
$$x_2 + 2x_3 + 3x_4 = 2 \qquad\qquad 3x_1 - 4x_3 - 4x_4 = 5$$

$$x_1 + x_2 + x_3 + x_4 = -3 \qquad\qquad 2x_1 - 3x_2 + x_3 - 3x_4 = 7.$$

14.2 DETERMINANTS

Determinants appear naturally during the solution of simultaneous equations. Consider, for example, two simultaneous equations with two unknowns x_1 and x_2,

$$a_{11}x_1 + a_{12}x_2 = b_1, \qquad\qquad (14.2.1)$$

and

$$a_{21}x_1 + a_{22}x_2 = b_2. \qquad\qquad (14.2.2)$$

The solution to these equations for the value of x_1 and x_2 is

$$x_1 = \frac{b_1 a_{22} - a_{12}b_2}{a_{11}a_{22} - a_{12}a_{21}}, \qquad\qquad (14.2.3)$$

and

$$x_2 = \frac{b_2 a_{11} - a_{21}b_1}{a_{11}a_{22} - a_{12}a_{21}}. \qquad\qquad (14.2.4)$$

Note that the denominator of (14.2.3) and (14.2.4) is the same. This term, which always appears in the solution of 2×2 systems, is formally given the name of *determinant* and written

$$\det(A) = \begin{vmatrix} a_{11} & a_{12} \\ a_{21} & a_{22} \end{vmatrix} = a_{11}a_{22} - a_{12}a_{21}. \qquad\qquad (14.2.5)$$

MATLAB provides a simple command det(A) which computes the determinant of A. For example, in the present case,

```
>> A = [2 -1 2; 1 3 2; 5 1 6];
>> det(A)

ans =

    0
```

Although determinants have their origin in the solution of systems of equations, any square array of numbers or expressions possesses a unique determinant, independent of whether it is involved in a system of equations or not. This determinant is evaluated (or expanded) according to a formal rule known as *Laplace's expansion of cofactors*.[3] The process revolves around expanding the determinant using any arbitrary column *or* row of A. If the ith row or jth column is chosen, the determinant is given by

$$\det(A) = a_{i1}A_{i1} + a_{i2}A_{i2} + \cdots + a_{in}A_{in} \tag{14.2.6}$$

$$= a_{1j}A_{1j} + a_{2j}A_{2j} + \cdots + a_{nj}A_{nj}, \tag{14.2.7}$$

where A_{ij}, the *cofactor* of a_{ij}, equals $(-1)^{i+j}M_{ij}$. The minor M_{ij} is the determinant of the $(n-1) \times (n-1)$ submatrix obtained by deleting row i, column j of A. This rule, of course, was chosen so that determinants are still useful in solving systems of equations.

• **Example 14.2.1**

Let us evaluate

$$\begin{vmatrix} 2 & -1 & 2 \\ 1 & 3 & 2 \\ 5 & 1 & 6 \end{vmatrix}$$

by an expansion in cofactors.

Using the first column,

$$\begin{vmatrix} 2 & -1 & 2 \\ 1 & 3 & 2 \\ 5 & 1 & 6 \end{vmatrix} = 2(-1)^2 \begin{vmatrix} 3 & 2 \\ 1 & 6 \end{vmatrix} + 1(-1)^3 \begin{vmatrix} -1 & 2 \\ 1 & 6 \end{vmatrix} + 5(-1)^4 \begin{vmatrix} -1 & 2 \\ 3 & 2 \end{vmatrix}$$

$$\tag{14.2.8}$$

$$= 2(16) - 1(-8) + 5(-8) = 0. \tag{14.2.9}$$

The greatest source of error is forgetting to take the factor $(-1)^{i+j}$ into account during the expansion.

---

[3] Laplace, P. S., 1772: Recherches sur le calcul intégral et sur le système du monde. *Hist. Acad. R. Sci., IIe Partie*, 267–376. *Œuvres*, **8**, pp. 369–501. See Muir, T., 1960: *The Theory of Determinants in the Historical Order of Development, Vol. I, Part 1, General Determinants Up to 1841*. Dover Publishers, pp. 24–33.

Although Laplace's expansion does provide a method for calculating $\det(A)$, the number of calculations equals $(n!)$. Consequently, for hand calculations, an obvious strategy is to select the column or row that has the greatest number of zeros. An even better strategy would be to manipulate a determinant with the goal of introducing zeros into a particular column or row. In the remaining portion of this section, we show some operations that may be performed on a determinant to introduce the desired zeros. Most of the properties follow from the expansion of determinants by cofactors.

- $\boxed{Rule\ 1}$: For every square matrix A, $\det(A^T) = \det(A)$.

The proof is left as an exercise.

- $\boxed{Rule\ 2}$: If any two rows or columns of A are identical, $\det(A) = 0$.

To see that this is true, consider the following 3×3 matrix:

$$\begin{vmatrix} b_1 & b_1 & c_1 \\ b_2 & b_2 & c_2 \\ b_3 & b_3 & c_3 \end{vmatrix} = c_1(b_2 b_3 - b_3 b_2) - c_2(b_1 b_3 - b_3 b_1)$$

$$+ c_3(b_1 b_2 - b_2 b_1) = 0. \tag{14.2.10}$$

- $\boxed{Rule\ 3}$: The determinant of a triangular matrix is equal to the product of its diagonal elements.

If A is lower triangular, successive expansions by elements in the first column give

$$\det(A) = \begin{vmatrix} a_{11} & 0 & \cdots & 0 \\ a_{21} & a_{22} & \cdots & 0 \\ \vdots & \vdots & \vdots & \vdots \\ a_{n1} & a_{n2} & \cdots & a_{nn} \end{vmatrix} = a_{11} \begin{vmatrix} a_{22} & \cdots & 0 \\ \vdots & \vdots & \vdots \\ a_{n2} & \cdots & a_{nn} \end{vmatrix} \tag{14.2.11}$$

$$= \cdots = a_{11} a_{22} \cdots a_{nn}. \tag{14.2.12}$$

If A is upper triangular, successive expansions by elements of the first row prove the property.

- $\boxed{Rule\ 4}$: If a square matrix A has either a row or a column of all zeros, then $\det(A) = 0$.

The proof is left as an exercise.

- $\boxed{Rule\ 5}$: If each element in one row (column) of a determinant is multiplied by a number c, the value of the determinant is multiplied by c.

Suppose that $|B|$ has been obtained from $|A|$ by multiplying row i (column j) of $|A|$ by c. Upon expanding $|B|$ in terms of row i (column j) each term in the expansion contains c as a factor. Factor out the common c, the result is just c times the expansion $|A|$ by the same row (column).

• $\boxed{Rule\ 6}$: If each element of a row (or a column) of a determinant can be expressed as a binomial, the determinant can be written as the sum of two determinants.

To understand this property, consider the following 3×3 determinant:

$$\begin{vmatrix} a_1 + d_1 & b_1 & c_1 \\ a_2 + d_2 & b_2 & c_2 \\ a_3 + d_3 & b_3 & c_3 \end{vmatrix} = \begin{vmatrix} a_1 & b_1 & c_1 \\ a_2 & b_2 & c_2 \\ a_3 & b_3 & c_3 \end{vmatrix} + \begin{vmatrix} d_1 & b_1 & c_1 \\ d_2 & b_2 & c_2 \\ d_3 & b_3 & c_3 \end{vmatrix}. \qquad (14.2.13)$$

The proof follows by expanding the determinant by the row (or column) that contains the binomials.

• $\boxed{Rule\ 7}$: If B is a matrix obtained by interchanging any two rows (columns) of a square matrix A, then $\det(B) = -\det(A)$.

The proof is by induction. It is easily shown for any 2×2 matrix. Assume that this rule holds of any $(n-1) \times (n-1)$ matrix. If A is $n \times n$, then let B be a matrix formed by interchanging rows i and j. Expanding $|B|$ and $|A|$ by a different row, say k, we have that

$$|B| = \sum_{s=1}^{n} (-1)^{k+s} b_{ks} M_{ks}, \quad \text{and} \quad |A| = \sum_{s=1}^{n} (-1)^{k+s} a_{ks} N_{ks}, \qquad (14.2.14)$$

where M_{ks} and N_{ks} are the minors formed by deleting row k, column s from $|B|$ and $|A|$, respectively. For $s = 1, 2, \ldots, n$, we obtain N_{ks} and M_{ks} by interchanging rows i and j. By the induction hypothesis and recalling that N_{ks} and M_{ks} are $(n-1) \times (n-1)$ determinants, $N_{ks} = -M_{ks}$ for $s = 1, 2, \ldots, n$. Hence, $|B| = -|A|$. Similar arguments hold if two columns are interchanged.

• $\boxed{Rule\ 8}$: If one row (column) of a square matrix A equals to a number c times some other row (column), then $\det(A) = 0$.

Suppose one row of a square matrix A is equal to c times some other row. If $c = 0$, then $|A| = 0$. If $c \neq 0$, then $|A| = c|B|$, where $|B| = 0$ because $|B|$ has two identical rows. A similar argument holds for two columns.

• $\boxed{Rule\ 9}$: The value of $\det(A)$ is unchanged if any arbitrary multiple of any line (row or column) is added to any other line.

To see that this is true, consider the simple example:

$$\begin{vmatrix} a_1 & b_1 & c_1 \\ a_2 & b_2 & c_2 \\ a_3 & b_3 & c_3 \end{vmatrix} + \begin{vmatrix} cb_1 & b_1 & c_1 \\ cb_2 & b_2 & c_2 \\ cb_3 & b_3 & c_3 \end{vmatrix} = \begin{vmatrix} a_1 + cb_1 & b_1 & c_1 \\ a_2 + cb_2 & b_2 & c_2 \\ a_3 + cb_3 & b_3 & c_3 \end{vmatrix}, \qquad (\mathbf{14.2.15})$$

where $c \neq 0$. The first determinant on the left side is our original determinant. In the second determinant, we again expand the first column and find that

$$\begin{vmatrix} cb_1 & b_1 & c_1 \\ cb_2 & b_2 & c_2 \\ cb_3 & b_3 & c_3 \end{vmatrix} = c \begin{vmatrix} b_1 & b_1 & c_1 \\ b_2 & b_2 & c_2 \\ b_3 & b_3 & c_3 \end{vmatrix} = 0. \qquad (\mathbf{14.2.16})$$

• **Example 14.2.2**

Let us evaluate

$$\begin{vmatrix} 1 & 2 & 3 & 4 \\ -1 & 1 & 2 & 3 \\ 1 & -1 & 1 & 2 \\ -1 & 1 & -1 & 5 \end{vmatrix}$$

using a combination of the properties stated above and expansion by cofactors. By adding or subtracting the first row to the other rows, we have that

$$\begin{vmatrix} 1 & 2 & 3 & 4 \\ -1 & 1 & 2 & 3 \\ 1 & -1 & 1 & 2 \\ -1 & 1 & -1 & 5 \end{vmatrix} = \begin{vmatrix} 1 & 2 & 3 & 4 \\ 0 & 3 & 5 & 7 \\ 0 & -3 & -2 & -2 \\ 0 & 3 & 2 & 9 \end{vmatrix} \qquad (\mathbf{14.2.17})$$

$$= \begin{vmatrix} 3 & 5 & 7 \\ -3 & -2 & -2 \\ 3 & 2 & 9 \end{vmatrix} = \begin{vmatrix} 3 & 5 & 7 \\ 0 & 3 & 5 \\ 0 & -3 & 2 \end{vmatrix} \qquad (\mathbf{14.2.18})$$

$$= 3 \begin{vmatrix} 3 & 5 \\ -3 & 2 \end{vmatrix} = 3 \begin{vmatrix} 3 & 5 \\ 0 & 7 \end{vmatrix} = 63. \qquad (\mathbf{14.2.19})$$

Problems

Evaluate the following determinants. Check your answer using MATLAB.

1. $\begin{vmatrix} 3 & 5 \\ -2 & -1 \end{vmatrix}$
2. $\begin{vmatrix} 5 & -1 \\ -8 & 4 \end{vmatrix}$

3. $\begin{vmatrix} 3 & 1 & 2 \\ 2 & 4 & 5 \\ 1 & 4 & 5 \end{vmatrix}$
4. $\begin{vmatrix} 4 & 3 & 0 \\ 3 & 2 & 2 \\ 5 & -2 & -4 \end{vmatrix}$

5.
$$\begin{vmatrix} 1 & 3 & 2 \\ 4 & 1 & 1 \\ 2 & 1 & 3 \end{vmatrix}$$

6.
$$\begin{vmatrix} 2 & -1 & 2 \\ 1 & 3 & 3 \\ 5 & 1 & 6 \end{vmatrix}$$

7.
$$\begin{vmatrix} 2 & 0 & 0 & 1 \\ 0 & 1 & 0 & 0 \\ 1 & 6 & 1 & 0 \\ 1 & 1 & -2 & 3 \end{vmatrix}$$

8.
$$\begin{vmatrix} 2 & 1 & 2 & 1 \\ 3 & 0 & 2 & 2 \\ -1 & 2 & -1 & 1 \\ -3 & 2 & 3 & 1 \end{vmatrix}$$

9. Using the properties of determinants, show that

$$\begin{vmatrix} 1 & 1 & 1 & 1 \\ a & b & c & d \\ a^2 & b^2 & c^2 & d^2 \\ a^3 & b^3 & c^3 & d^3 \end{vmatrix} = (b-a)(c-a)(d-a)(c-b)(d-b)(d-c).$$

This determinant is called *Vandermonde's determinant.*

10. Show that

$$\begin{vmatrix} a & b+c & 1 \\ b & a+c & 1 \\ c & a+b & 1 \end{vmatrix} = 0.$$

11. Show that if all of the elements of a row or column are zero, then $\det(A) = 0$.

12. Prove that $\det(A^T) = \det(A)$.

14.3 CRAMER'S RULE

One of the most popular methods for solving simple systems of linear equations is Cramer's rule.[4] It is very useful for 2×2 systems, acceptable for 3×3 systems, and of doubtful use for 4×4 or larger systems.

Let us have n equations with n unknowns, $A\mathbf{x} = \mathbf{b}$. Cramer's rule states that

$$x_1 = \frac{\det(A_1)}{\det(A)}, \quad x_2 = \frac{\det(A_2)}{\det(A)}, \quad \cdots, \quad x_n = \frac{\det(A_n)}{\det(A)}, \qquad (14.3.1)$$

where A_i is a matrix obtained from A by replacing the ith column with $\mathbf{b}$ and n is the number of unknowns and equations. Obviously, $\det(A) \neq 0$ if Cramer's rule is to work.

---

[4] Cramer, G., 1750: *Introduction à l'analyse des lignes courbes algébriques.* Geneva, p. 657.

To prove[5] Cramer's rule, consider

$$x_1 \det(A) = \begin{vmatrix} a_{11}x_1 & a_{12} & a_{13} & \cdots & a_{1n} \\ a_{21}x_1 & a_{22} & a_{23} & \cdots & a_{2n} \\ a_{31}x_1 & a_{32} & a_{33} & \cdots & a_{3n} \\ \vdots & \vdots & \vdots & \vdots & \vdots \\ a_{n1}x_1 & a_{n2} & a_{n3} & \cdots & a_{nn} \end{vmatrix} \qquad (14.3.2)$$

by Rule 5 from the previous section. By adding x_2 times the second column to the first column,

$$x_1 \det(A) = \begin{vmatrix} a_{11}x_1 + a_{12}x_2 & a_{12} & a_{13} & \cdots & a_{1n} \\ a_{21}x_1 + a_{22}x_2 & a_{22} & a_{23} & \cdots & a_{2n} \\ a_{31}x_1 + a_{32}x_2 & a_{32} & a_{33} & \cdots & a_{3n} \\ \vdots & \vdots & \vdots & \vdots & \vdots \\ a_{n1}x_1 + a_{n2}x_2 & a_{n2} & a_{n3} & \cdots & a_{nn} \end{vmatrix}. \qquad (14.3.3)$$

Multiplying each of the columns by the corresponding x_i and adding it to the first column yields

$$x_1 \det(A) = \begin{vmatrix} a_{11}x_1 + a_{12}x_2 + \cdots + a_{1n}x_n & a_{12} & a_{13} & \cdots & a_{1n} \\ a_{21}x_1 + a_{22}x_2 + \cdots + a_{2n}x_n & a_{22} & a_{23} & \cdots & a_{2n} \\ a_{31}x_1 + a_{32}x_2 + \cdots + a_{3n}x_n & a_{32} & a_{33} & \cdots & a_{3n} \\ \vdots & & \vdots & \vdots & \vdots & \vdots \\ a_{n1}x_1 + a_{n2}x_2 + \cdots + a_{nn}x_n & a_{n2} & a_{n3} & \cdots & a_{nn} \end{vmatrix}. $$
$$(14.3.4)$$

The first column of (14.3.4) equals $A\mathbf{x}$ and we replace it with $\mathbf{b}$. Thus,

$$x_1 \det(A) = \begin{vmatrix} b_1 & a_{12} & a_{13} & \cdots & a_{1n} \\ b_2 & a_{22} & a_{23} & \cdots & a_{2n} \\ b_3 & a_{32} & a_{33} & \cdots & a_{3n} \\ \vdots & \vdots & \vdots & \vdots & \vdots \\ b_n & a_{n2} & a_{n3} & \cdots & a_{nn} \end{vmatrix} = \det(A_1), \qquad (14.3.5)$$

or

$$x_1 = \frac{\det(A_1)}{\det(A)} \qquad (14.3.6)$$

provided $\det(A) \neq 0$. To complete the proof we do exactly the same procedure to the jth column. $\square$

---

[5] First proved by Cauchy, L. A., 1815: Mémoire sur les fonctions quine peuvent obtemir que deux valeurs égales et de signes contraires par suite des transportations opérées entre les variables qúelles renferment. *J. l'Ecole Polytech.*, **10**, 29–112.

• **Example 14.3.1**

Let us solve the following system of equations by Cramer's rule:

$$2x_1 + x_2 + 2x_3 = -1, \qquad\qquad (14.3.7)$$

$$x_1 + x_3 = -1, \qquad\qquad (14.3.8)$$

and

$$-x_1 + 3x_2 - 2x_3 = 7. \qquad\qquad (14.3.9)$$

From the matrix form of the equations,

$$\begin{pmatrix} 2 & 1 & 2 \\ 1 & 0 & 1 \\ -1 & 3 & -2 \end{pmatrix} \begin{pmatrix} x_1 \\ x_2 \\ x_3 \end{pmatrix} = \begin{pmatrix} -1 \\ -1 \\ 7 \end{pmatrix}, \qquad\qquad (14.3.10)$$

we have that

$$\det(A) = \begin{vmatrix} 2 & 1 & 2 \\ 1 & 0 & 1 \\ -1 & 3 & -2 \end{vmatrix} = 1, \qquad\qquad (14.3.11)$$

$$\det(A_1) = \begin{vmatrix} -1 & 1 & 2 \\ -1 & 0 & 1 \\ 7 & 3 & -2 \end{vmatrix} = 2, \qquad\qquad (14.3.12)$$

$$\det(A_2) = \begin{vmatrix} 2 & -1 & 2 \\ 1 & -1 & 1 \\ -1 & 7 & -2 \end{vmatrix} = 1, \qquad\qquad (14.3.13)$$

and

$$\det(A_3) = \begin{vmatrix} 2 & 1 & -1 \\ 1 & 0 & -1 \\ -1 & 3 & 7 \end{vmatrix} = -3. \qquad\qquad (14.3.14)$$

Finally,

$$x_1 = \frac{2}{1} = 2, \quad x_2 = \frac{1}{1} = 1, \quad \text{and} \quad x_3 = \frac{-3}{1} = -3. \qquad (14.3.15)$$

You can also use MATLAB to perform Cramer's rule. In the present example, the script is as follows:

```
clear; % clear all previous computations
A = [2 1 2; 1 0 1; -1 3 -2]; % input coefficient matrix
b = [-1 ; -1; 7]; % input right side
A1 = A; A1(:,1) = b; % compute A_1
A2 = A; A2(:,2) = b; % compute A_2
A3 = A; A3(:,3) = b; % compute A_3
% compute solution vector
```

```
x = [det(A1), det(A2), det(A3)] / det(A)
```

Problems

Solve the following systems of equations by Cramer's rule:

1. $x_1 + 2x_2 = 3, \quad 3x_1 + x_2 = 6$

2. $2x_1 + x_2 = -3, \quad x_1 - x_2 = 1$

3. $x_1 + 2x_2 - 2x_3 = 4, \quad 2x_1 + x_2 + x_3 = -2, \quad -x_1 + x_2 - x_3 = 2$

4. $2x_1 + 3x_2 - x_3 = -1, \quad -x_1 - 2x_2 + x_3 = 5, \quad 3x_1 - x_2 = -2.$

Check your answer using MATLAB.

14.4 ROW ECHELON FORM AND GAUSSIAN ELIMINATION

So far, we assumed that every system of equations has a unique solution. This is not necessary true as the following examples show.

• Example 14.4.1

Consider the system

$$x_1 + x_2 = 2, \tag{14.4.1}$$

and

$$2x_1 + 2x_2 = -1. \tag{14.4.2}$$

This system is inconsistent because the second equation does not follow after multiplying the first by 2. Geometrically (14.4.1) and (14.4.2) are parallel lines; they never intersect to give a unique x_1 and x_2.

• Example 14.4.2

Even if a system is consistent, it still may not have a unique solution. For example, the system

$$x_1 + x_2 = 2, \tag{14.4.3}$$

and

$$2x_1 + 2x_2 = 4 \tag{14.4.4}$$

is consistent, the second equation formed by multiplying the first by 2. However, there are an infinite number of solutions.

Our examples suggest the following:

Theorem: *A system of m linear equations in n unknowns may: (1) have no solution, in which case it is called an* inconsistent *system, or (2) have* *exactly one solution (called a unique solution), or (3) have an infinite number of solutions. In the latter two cases, the system is said to be* consistent.

Before we can prove this theorem at the end of this section, we need to introduce some new concepts.

The first one is equivalent systems. Two systems of equations involving the same variables are *equivalent* if they have the same solution set. Of course, the only reason for introducing equivalent systems is the possibility of transforming one system of linear systems into another which is easier to solve. But what operations are permissible? Also what is the ultimate goal of our transformation?

From a complete study of possible operations, there are only three operations for transforming one system of linear equations into another. These three *elementary row operations* are

---

(1) interchanging any two rows in the matrix,

(2) multiplying any row by a nonzero scalar, and

(3) adding any arbitrary multiple of any row to any other row.

Armed with our elementary row operations, let us now solve the following set of linear equations:

$$x_1 - 3x_2 + 7x_3 = 2, \tag{14.4.5}$$

$$2x_1 + 4x_2 - 3x_3 = -1, \tag{14.4.6}$$

and

$$-x_1 + 13x_2 - 21x_3 = 2. \tag{14.4.7}$$

We begin by writing (14.4.5)–(14.4.7) in matrix notation:

$$\begin{pmatrix} 1 & -3 & 7 \\ 2 & 4 & -3 \\ -1 & 13 & -21 \end{pmatrix} \begin{pmatrix} x_1 \\ x_2 \\ x_3 \end{pmatrix} = \begin{pmatrix} 2 \\ -1 \\ 2 \end{pmatrix}. \tag{14.4.8}$$

The matrix in (14.4.8) is called the *coefficient matrix* of the system.

We now introduce the concept of the *augmented matrix*: a matrix B composed of A plus the column vector $\mathbf{b}$ or

$$B = \begin{pmatrix} 1 & -3 & 7 & \bigm| & 2 \\ 2 & 4 & -3 & \bigm| & -1 \\ -1 & 13 & -21 & \bigm| & 2 \end{pmatrix}. \tag{14.4.9}$$

We can solve our original system by performing elementary row operations on the augmented matrix. Because x_i functions essentially as a placeholder, we can omit them until the end of the computation.

Returning to the problem, the first row can be used to eliminate the elements in the first column of the remaining rows. For this reason the first row is called the *pivotal* row and the element a_{11} is the *pivot*. By using the third elementary row operation twice (to eliminate the 2 and -1 in the first column), we have the equivalent system

$$B = \begin{pmatrix} 1 & -3 & 7 & 2 \\ 0 & 10 & -17 & -5 \\ 0 & 10 & -14 & 4 \end{pmatrix}. \qquad (14.4.10)$$

At this point we choose the second row as our new pivotal row and again apply the third row operation to eliminate the last element in the second column. This yields

$$B = \begin{pmatrix} 1 & -3 & 7 & 2 \\ 0 & 10 & -17 & -5 \\ 0 & 0 & 3 & 9 \end{pmatrix}. \qquad (14.4.11)$$

Thus, elementary row operations transformed (14.4.5)–(14.4.7) into the triangular system:

$$x_1 - 3x_2 + 7x_3 = 2, \qquad (14.4.12)$$
$$10x_2 - 17x_3 = -5, \qquad (14.4.13)$$
$$3x_3 = 9, \qquad (14.4.14)$$

which is *equivalent* to the original system. The final solution is obtained by *back substitution*, solving from (14.4.14) back to (14.4.12). In the present case, $x_3 = 3$. Then, $10x_2 = 17(3) - 5$, or $x_2 = 4.6$. Finally, $x_1 = 3x_2 - 7x_3 + 2 = -5.2$.

In general, if an $n \times n$ linear system can be reduced to triangular form, then it has a unique solution that we can obtain by performing back substitution. This reduction involves $n-1$ steps. In the first step, a pivot element, and thus the pivotal row, is chosen from the nonzero entries in the first column of the matrix. We interchange rows (if necessary) so that the pivotal row is the first row. Multiples of the pivotal row are then subtracted from each of the remaining $n - 1$ rows so that there are 0's in the $(2, 1), ..., (n, 1)$ positions. In the second step, a pivot element is chosen from the nonzero entries in column 2, rows 2 through n, of the matrix. The row containing the pivot is then interchanged with the second row (if necessary) of the matrix and is used as the pivotal row. Multiples of the pivotal row are then subtracted from the remaining $n - 2$ rows, eliminating all entries below the diagonal in the second column. The same procedure is repeated for columns 3 through $n - 1$. Note that in the second step, row 1 and column 1 remain unchanged, in the third step the first two rows and first two columns remain unchanged, and so on.

If elimination is carried out as described, we arrive at an equivalent upper triangular system after $n - 1$ steps. However, the procedure fails if, at any step, all possible choices for a pivot element equal zero. Let us now examine such cases.

Consider now the system

$$x_1 + 2x_2 + x_3 = -1, \qquad (14.4.15)$$

$$2x_1 + 4x_2 + 2x_3 = -2, \qquad (14.4.16)$$

$$x_1 + 4x_2 + 2x_3 = 2. \qquad (14.4.17)$$

Its augmented matrix is

$$B = \begin{pmatrix} 1 & 2 & 1 & -1 \\ 2 & 4 & 2 & -2 \\ 1 & 4 & 2 & 2 \end{pmatrix}. \qquad (14.4.18)$$

Choosing the first row as our pivotal row, we find that

$$B = \begin{pmatrix} 1 & 2 & 1 & -1 \\ 0 & 0 & 0 & 0 \\ 0 & 2 & 1 & 3 \end{pmatrix}, \qquad (14.4.19)$$

or

$$B = \begin{pmatrix} 1 & 2 & 1 & -1 \\ 0 & 2 & 1 & 3 \\ 0 & 0 & 0 & 0 \end{pmatrix}. \qquad (14.4.20)$$

The difficulty here is the presence of the zeros in the third row. Clearly any finite numbers satisfy the equation $0x_1 + 0x_2 + 0x_3 = 0$ and we have an infinite number of solutions. Closer examination of the original system shows a underdetermined system; (14.4.15) and (14.4.16) differ by a multiplicative factor of 2. An important aspect of this problem is the fact that the final augmented matrix is of the form of a staircase or *echelon form* rather than of triangular form.

Let us modify (14.4.15)–(14.4.17) to read

$$x_1 + 2x_2 + x_3 = -1, \qquad (14.4.21)$$

$$2x_1 + 4x_2 + 2x_3 = 3, \qquad (14.4.22)$$

$$x_1 + 4x_2 + 2x_3 = 2, \qquad (14.4.23)$$

then the final augmented matrix is

$$B = \begin{pmatrix} 1 & 2 & 1 & -1 \\ 0 & 2 & 1 & 3 \\ 0 & 0 & 0 & 5 \end{pmatrix}. \qquad (14.4.24)$$

We again have a problem with the third row because $0x_1 + 0x_2 + 0x_3 = 5$, which is impossible. There is no solution in this case and we have an *inconsistent system*. Note, once again, that our augmented matrix has a row echelon form rather than a triangular form.

In summary, to include all possible situations in our procedure, we must rewrite the augmented matrix in row echelon form. We have *row echelon form* when:

(1) The first nonzero entry in each row is 1.

(2) If row k does not consist entirely of zeros, the number of leading zero entries in row $k+1$ is greater than the number of leading zero entries in row k.

(3) If there are rows whose entries are all zero, they are below the rows having nonzero entries.

The number of nonzero rows in the row echelon form of a matrix is known as its *rank*. In MATLAB, the rank is easily found using the command `rank( )`. *Gaussian elimination* is the process of using elementary row operations to transform a linear system into one whose augmented matrix is in row echelon form.

• Example 14.4.3

Each of the following matrices is *not* of row echelon form because they violate one of the conditions for row echelon form:

$$\begin{pmatrix} 2 & 2 & 3 \\ 0 & 2 & 1 \\ 0 & 0 & 4 \end{pmatrix}, \begin{pmatrix} 0 & 0 & 0 \\ 0 & 2 & 0 \end{pmatrix}, \begin{pmatrix} 0 & 1 \\ 1 & 0 \end{pmatrix}. \tag{14.4.25}$$

• Example 14.4.4

The following matrices are in row echelon form:

$$\begin{pmatrix} 1 & 2 & 3 \\ 0 & 1 & 1 \\ 0 & 0 & 1 \end{pmatrix}, \begin{pmatrix} 1 & 4 & 6 \\ 0 & 0 & 1 \\ 0 & 0 & 0 \end{pmatrix}, \begin{pmatrix} 1 & 3 & 4 & 0 \\ 0 & 0 & 1 & 3 \\ 0 & 0 & 0 & 0 \end{pmatrix}. \tag{14.4.26}$$

• **Example 14.4.5**

Gaussian elimination can also be used to solve the general problem $AX = B$. One of the most common applications is in finding the inverse. For example, let us find the inverse of the matrix

$$A = \begin{pmatrix} 4 & -2 & 2 \\ -2 & -4 & 4 \\ -4 & 2 & 8 \end{pmatrix} \qquad (14.4.27)$$

by Gaussian elimination.

Because the inverse is defined by $AA^{-1} = I$, our augmented matrix is

$$\left(\begin{array}{ccc|ccc} 4 & -2 & 2 & 1 & 0 & 0 \\ -2 & -4 & 4 & 0 & 1 & 0 \\ -4 & 2 & 8 & 0 & 0 & 1 \end{array} \right). \qquad (14.4.28)$$

Then, by elementary row operations,

$$\left(\begin{array}{ccc|ccc} 4 & -2 & 2 & 1 & 0 & 0 \\ -2 & -4 & 4 & 0 & 1 & 0 \\ -4 & 2 & 8 & 0 & 0 & 1 \end{array} \right) = \left(\begin{array}{ccc|ccc} -2 & -4 & 4 & 0 & 1 & 0 \\ 4 & -2 & 2 & 1 & 0 & 0 \\ -4 & 2 & 8 & 0 & 0 & 1 \end{array} \right) \qquad (14.4.29)$$

$$= \left(\begin{array}{ccc|ccc} -2 & -4 & 4 & 0 & 1 & 0 \\ 4 & -2 & 2 & 1 & 0 & 0 \\ 0 & 0 & 10 & 1 & 0 & 1 \end{array} \right) \qquad (14.4.30)$$

$$= \left(\begin{array}{ccc|ccc} -2 & -4 & 4 & 0 & 1 & 0 \\ 0 & -10 & 10 & 1 & 2 & 0 \\ 0 & 0 & 10 & 1 & 0 & 1 \end{array} \right) \qquad (14.4.31)$$

$$= \left(\begin{array}{ccc|ccc} -2 & -4 & 4 & 0 & 1 & 0 \\ 0 & -10 & 0 & 0 & 2 & -1 \\ 0 & 0 & 10 & 1 & 0 & 1 \end{array} \right) \qquad (14.4.32)$$

$$= \left(\begin{array}{ccc|ccc} -2 & -4 & 0 & -2/5 & 1 & -2/5 \\ 0 & -10 & 0 & 0 & 2 & -1 \\ 0 & 0 & 10 & 1 & 0 & 1 \end{array} \right) \qquad (14.4.33)$$

$$= \left(\begin{array}{ccc|ccc} -2 & 0 & 0 & -2/5 & 1/5 & 0 \\ 0 & -10 & 0 & 0 & 2 & -1 \\ 0 & 0 & 10 & 1 & 0 & 1 \end{array} \right) \qquad (14.4.34)$$

$$= \left(\begin{array}{ccc|ccc} 1 & 0 & 0 & 1/5 & -1/10 & 0 \\ 0 & 1 & 0 & 0 & -1/5 & 1/10 \\ 0 & 0 & 1 & 1/10 & 0 & 1/10 \end{array} \right). \qquad (14.4.35)$$

Thus, the right half of the augmented matrix yields the inverse and it equals

$$A^{-1} = \begin{pmatrix} 1/5 & -1/10 & 0 \\ 0 & -1/5 & 1/10 \\ 1/10 & 0 & 1/10 \end{pmatrix}. \qquad (14.4.36)$$

MATLAB has the ability of doing Gaussian elimination step-by-step. We begin by typing

```
>>% input augmented matrix
>>aug = [4 -2 2 1 0 0 ; -2 -4 4 0 1 0;-4 2 8 0 0 1];
>>rrefmovie(aug);
```

The MATLAB command `rrefmovie(A)` produces the reduced row echelon form of A. Repeated pressing of any key gives the next step in the calculation along with a statement of how it computed the modified augmented matrix. Eventually you obtain

```
A =
    1      0      0     1/5   -1/10    0
    0      1      0      0    -1/5    1/10
    0      0      1    1/10     0     1/10
```

You can read the inverse matrix just as we did earlier.

Gaussian elimination may be used with overdetermined systems. *Overdetermined systems* are linear systems where there are more equations than unknowns $(m > n)$. These systems are usually (but not always) inconsistent.

• Example 14.4.6

Consider the linear system

$$x_1 + x_2 = 1, \tag{14.4.37}$$
$$-x_1 + 2x_2 = -2, \tag{14.4.38}$$
$$x_1 - x_2 = 4. \tag{14.4.39}$$

After several row operations, the augmented matrix

$$\begin{pmatrix} 1 & 1 & | & 1 \\ -1 & 2 & | & -2 \\ 1 & -1 & | & 4 \end{pmatrix} \tag{14.4.40}$$

becomes

$$\begin{pmatrix} 1 & 1 & | & 1 \\ 0 & 1 & | & 2 \\ 0 & 0 & | & -7 \end{pmatrix}. \tag{14.4.41}$$

From the last row of the augmented matrix (14.4.41) we see that the system is inconsistent.

If we test this system using MATLAB by typing

```
>>% input augmented matrix
>>aug = [1 1 1 ; -1 2 -2; 1 -1 4];
```

```
>>rrefmovie(aug);
```
eventually you obtain

```
A =
      1      0      0
      0      1      0
      0      0      1
```

Although the numbers have changed from our hand calculation, we still have an inconsistent system because $x_1 = x_2 = 0$ does not satisfy $x_1 + x_2 = 1$.

Considering now a slight modification of this system to

$$x_1 + x_2 = 1, \tag{14.4.42}$$
$$-x_1 + 2x_2 = 5, \tag{14.4.43}$$
$$x_1 = -1, \tag{14.4.44}$$

the final form of the augmented matrix is

$$\begin{pmatrix} 1 & 1 & 1 \\ 0 & 1 & 2 \\ 0 & 0 & 0 \end{pmatrix}, \tag{14.4.45}$$

which has the unique solution $x_1 = -1$ and $x_2 = 2$.

How does MATLAB handle this problem? Typing
```
>>% input augmented matrix
>>aug = [1 1 1 ; -1 2 5; 1 0 -1];
>>rrefmovie(aug);
```
we eventually obtain

```
A =
      1      0     -1
      0      1      2
      0      0      0
```

This yields $x_1 = -1$ and $x_2 = 2$, as we found by hand.

Finally, by introducing the set:

$$x_1 + x_2 = 1, \tag{14.4.46}$$
$$2x_1 + 2x_2 = 2, \tag{14.4.47}$$
$$3x_1 + 3x_3 = 3, \tag{14.4.48}$$

the final form of the augmented matrix is

$$\begin{pmatrix} 1 & 1 & 1 \\ 0 & 0 & 0 \\ 0 & 0 & 0 \end{pmatrix}. \tag{14.4.49}$$

There are an infinite number of solutions: $x_1 = 1 - \alpha$, and $x_2 = \alpha$.

Turning to MATLAB, we first type

```
>>% input augmented matrix
>>aug = [1 1 1 ; 2 2 2; 3 3 3];
>>rrefmovie(aug);
```

and we eventually obtain

```
A =
    1       1       1
    0       0       0
    0       0       0
```

This is the same as (14.4.49) and the final answer is the same.

Gaussian elimination can also be employed with underdetermined systems. An *underdetermined linear system* is one where there are fewer equations than unknowns $(m < n)$. These systems usually have an infinite number of solutions although they can be inconsistent.

• Example 14.4.7

Consider the underdetermined system:

$$2x_1 + 2x_2 + x_3 = -1, \qquad (14.4.50)$$
$$4x_1 + 4x_2 + 2x_3 = 3. \qquad (14.4.51)$$

Its augmented matrix can be transformed into the form:

$$\begin{pmatrix} 2 & 2 & 1 & -1 \\ 0 & 0 & 0 & 5 \end{pmatrix}. \qquad (14.4.52)$$

Clearly this case corresponds to an inconsistent set of equations. On the other hand, if (14.4.51) is changed to

$$4x_1 + 4x_2 + 2x_3 = -2, \qquad (14.4.53)$$

then the final form of the augmented matrix is

$$\begin{pmatrix} 2 & 2 & 1 & -1 \\ 0 & 0 & 0 & 0 \end{pmatrix} \qquad (14.4.54)$$

and we have an infinite number of solutions, namely $x_3 = \alpha$, $x_2 = \beta$, and $2x_1 = -1 - \alpha - 2\beta$.

Consider now one of the most important classes of linear equations: the homogeneous equations $A\mathbf{x} = \mathbf{0}$. If $\det(A) \neq 0$, then by Cramer's rule $x_1 =$

$x_2 = x_3 = \cdots = x_n = 0$. Thus, the only possibility for a nontrivial solution is $\det(A) = 0$. In this case, A is singular, no inverse exists, and nontrivial solutions exist but they are not unique.

• Example 14.4.8

Consider the two homogeneous equations:

$$x_1 + x_2 = 0, \tag{14.4.55}$$
$$x_1 - x_2 = 0. \tag{14.4.56}$$

Note that $\det(A) = -2$. Solving this system yields $x_1 = x_2 = 0$.

However, if we change the system to

$$x_1 + x_2 = 0, \tag{14.4.57}$$
$$x_1 + x_2 = 0, \tag{14.4.58}$$

which has the $\det(A) = 0$ so that A is singular. Both equations yield $x_1 = -x_2 = \alpha$, any constant. Thus, there is an infinite number of solutions for this set of homogeneous equations.

We close this section by outlining the proof of the theorem which we introduced at the beginning.

Consider the system $A\mathbf{x} = \mathbf{b}$. By elementary row operations, the first equation in this system can be reduced to

$$x_1 + \alpha_{12}x_2 + \cdots + \alpha_{1n}x_n = \beta_1. \tag{14.4.59}$$

The second equation has the form

$$x_p + \alpha_{2p+1}x_{p+1} + \cdots + \alpha_{2n}x_n = \beta_2, \tag{14.4.60}$$

where $p > 1$. The third equation has the form

$$x_q + \alpha_{3q+1}x_{q+1} + \cdots + \alpha_{3n}x_n = \beta_3, \tag{14.4.61}$$

where $q > p$, and so on. To simplify the notation, we introduce z_i where we choose the first k values so that $z_1 = x_1$, $z_2 = x_p$, $z_3 = x_q$, ... Thus, the question of the existence of solutions depends upon the three integers: m, n, and k. The resulting set of equations have the form:

$$\begin{pmatrix} 1 & \gamma_{12} & \cdots & \gamma_{1k} & \gamma_{1k+1} & \cdots & \gamma_{1n} \\ 0 & 1 & \cdots & \gamma_{2k} & \gamma_{2k+1} & \cdots & \gamma_{2n} \\ & & & \vdots & & & \\ 0 & 0 & \cdots & 1 & \gamma_{kk+1} & \cdots & \gamma_{kn} \\ 0 & 0 & \cdots & 0 & 0 & \cdots & 0 \\ & & & \vdots & & & \\ 0 & 0 & \cdots & 0 & 0 & \cdots & 0 \end{pmatrix} \begin{pmatrix} z_1 \\ z_2 \\ \vdots \\ z_k \\ z_{k+1} \\ \vdots \\ z_n \end{pmatrix} = \begin{pmatrix} \beta_1 \\ \beta_2 \\ \vdots \\ \beta_k \\ \beta_{k+1} \\ \vdots \\ \beta_m \end{pmatrix}. \tag{14.4.62}$$

Note that $\beta_{k+1}, \ldots, \beta_m$ need not be all zero.

There are three possibilities:

(a) $k < m$ and at least one of the elements $\beta_{k+1}, \ldots, \beta_m$ is nonzero. Suppose that an element β_p is nonzero $(p > k)$. Then the pth equation is

$$0z_1 + 0z_2 + \cdots + 0z_n = \beta_p \neq 0. \qquad (14.4.63)$$

However, this is a contradiction and the equations are inconsistent.

(b) $k = n$ and either (i) $k < m$ and all of the elements $\beta_{k+1}, \ldots, \beta_m$ are zero, or (ii) $k = m$. Then the equations have a unique solution which can be obtained by back-substitution.

(c) $k < n$ and either (i) $k < m$ and all of the elements $\beta_{k+1}, \ldots, \beta_m$ are zero, or (ii) $k = m$. Then, arbitrary values can be assigned to the $n - k$ variables $z_{k+1}, \ldots, z_n$. The equations can be solved for $z_1, z_2, \ldots, z_k$ and there is an infinity of solutions.

For homogeneous equations $\mathbf{b} = \mathbf{0}$, all of the β_i are zero. In this case, we have only two cases:

(b') $k = n$, then (14.4.62) has the solution $\mathbf{z} = \mathbf{0}$ which leads to the trivial solution for the original system $A\mathbf{x} = \mathbf{0}$.

(c') $k < n$, the equations possess an infinity of solutions given by assigning arbitrary values to $z_{k+1}, \ldots, z_n$. $\qquad \square$

Problems

Solve the following systems of linear equations by Gaussian elimination. Check your answer using MATLAB.

1. $2x_1 + x_2 = 4, \qquad\qquad 5x_1 - 2x_2 = 1$

2. $x_1 + x_2 = 0, \qquad\qquad 3x_1 - 4x_2 = 1$

3. $-x_1 + x_2 + 2x_3 = 0, \quad 3x_1 + 4x_2 + x_3 = 0, \quad -x_1 + x_2 + 2x_3 = 0$

4. $4x_1 + 6x_2 + x_3 = 2, \quad 2x_1 + x_2 - 4x_3 = 3, \quad 3x_1 - 2x_2 + 5x_3 = 8$

5. $3x_1 + x_2 - 2x_3 = -3, \quad x_1 - x_2 + 2x_3 = -1, \quad -4x_1 + 3x_2 - 6x_3 = 4$

6. $x_1 - 3x_2 + 7x_3 = 2, \qquad\qquad 2x_1 + 4x_2 - 3x_3 = -1,$
   $-3x_1 + 7x_2 + 2x_3 = 3$

7. $x_1 - x_2 + 3x_3 = 5, \qquad\qquad 2x_1 - 4x_2 + 7x_3 = 7,$
   $4x_1 - 9x_2 + 2x_3 = -15$

8. $x_1 + x_2 + x_3 + x_4 = -1, \qquad 2x_1 - x_2 + 3x_3 = 1,$
   $2x_2 + 3x_4 = 15, \qquad\qquad -x_1 + 2x_2 + x_4 = -2$

Find the inverse of each of the following matrices by Gaussian elimination. Check your answers using MATLAB.

9. $\begin{pmatrix} -3 & 5 \\ 2 & 1 \end{pmatrix}$

10. $\begin{pmatrix} 3 & -1 \\ -5 & 2 \end{pmatrix}$

11. $\begin{pmatrix} 19 & 2 & -9 \\ -4 & -1 & 2 \\ -2 & 0 & 1 \end{pmatrix}$

12. $\begin{pmatrix} 1 & 2 & 5 \\ 0 & -1 & 2 \\ 2 & 4 & 11 \end{pmatrix}$

13. Does $(A^2)^{-1} = (A^{-1})^2$? Justify your answer.

14.5 EIGENVALUES AND EIGENVECTORS

One of the classic problems of linear algebra[6] is finding all of the λ's which satisfy the $n \times n$ system

$$A\mathbf{x} = \lambda\mathbf{x}. \tag{14.5.1}$$

The nonzero quantity λ is the *eigenvalue* or *characteristic value* of A. The vector $\mathbf{x}$ is the *eigenvector* or *characteristic vector* belonging to λ. The set of the eigenvalues of A is called the *spectrum* of A. The largest of the absolute values of the eigenvalues of A is called the *spectral radius* of A.

To find λ and $\mathbf{x}$, we first rewrite (14.5.1) as a set of homogeneous equations:

$$(A - \lambda I)\mathbf{x} = \mathbf{0}. \tag{14.5.2}$$

From the theory of linear equations, (14.5.2) has trivial solutions unless its determinant equals zero. On the other hand, if

$$\det(A - \lambda I) = 0, \tag{14.5.3}$$

there are an infinity of solutions.

The expansion of the determinant (14.5.3) yields a nth-degree polynomial in λ, the *characteristic polynomial*. The roots of the characteristic polynomial are the eigenvalues of A. Because the characteristic polynomial has exactly n roots, A has n eigenvalues, some of which can be repeated (with multiplicity $k \leq n$) and some of which can be complex numbers. For each eigenvalue λ_i, there is a corresponding eigenvector $\mathbf{x}_i$. This eigenvector is the solution of the homogeneous equations $(A - \lambda_i I)\mathbf{x}_i = \mathbf{0}$.

An important property of eigenvectors is their *linear independence* if there are n distinct eigenvalues. Vectors are linearly independent if the equation

$$\alpha_1\mathbf{x}_1 + \alpha_2\mathbf{x}_2 + \cdots + \alpha_n\mathbf{x}_n = \mathbf{0} \tag{14.5.4}$$

---

[6] The standard reference is Wilkinson, J. H., 1965: *The Algebraic Eigenvalue Problem.* Oxford University Press, 662 pp.

can be satisfied only by taking *all* of the coefficients α_n equal to zero.

To show that this is true in the case of n distinct eigenvalues $\lambda_1, \lambda_2, \ldots,$ λ_n, each eigenvalue λ_i having a corresponding eigenvector $\mathbf{x}_i$, we first write down the linear dependence condition

$$\alpha_1 \mathbf{x}_1 + \alpha_2 \mathbf{x}_2 + \cdots + \alpha_n \mathbf{x}_n = \mathbf{0}. \tag{14.5.5}$$

Premultiplying (14.5.5) by A,

$$\alpha_1 A\mathbf{x}_1 + \alpha_2 A\mathbf{x}_2 + \cdots + \alpha_n A\mathbf{x}_n = \alpha_1 \lambda_1 \mathbf{x}_1 + \alpha_2 \lambda_2 \mathbf{x}_2 + \cdots + \alpha_n \lambda_n \mathbf{x}_n = \mathbf{0}. \tag{14.5.6}$$

Premultiplying (14.5.5) by A^2,

$$\alpha_1 A^2\mathbf{x}_1 + \alpha_2 A^2\mathbf{x}_2 + \cdots + \alpha_n A^2\mathbf{x}_n = \alpha_1 \lambda_1^2 \mathbf{x}_1 + \alpha_2 \lambda_2^2 \mathbf{x}_2 + \cdots + \alpha_n \lambda_n^2 \mathbf{x}_n = \mathbf{0}. \tag{14.5.7}$$

In a similar manner, we obtain the system of equations:

$$\begin{pmatrix} 1 & 1 & \cdots & 1 \\ \lambda_1 & \lambda_2 & \cdots & \lambda_n \\ \lambda_1^2 & \lambda_2^2 & \cdots & \lambda_n^2 \\ \vdots & \vdots & \vdots & \vdots \\ \lambda_1^{n-1} & \lambda_2^{n-1} & \cdots & \lambda_n^{n-1} \end{pmatrix} \begin{pmatrix} \alpha_1 \mathbf{x}_1 \\ \alpha_2 \mathbf{x}_2 \\ \alpha_3 \mathbf{x}_3 \\ \vdots \\ \alpha_n \mathbf{x}_n \end{pmatrix} = \begin{pmatrix} 0 \\ 0 \\ 0 \\ \vdots \\ 0 \end{pmatrix}. \tag{14.5.8}$$

Because

$$\begin{vmatrix} 1 & 1 & \cdots & 1 \\ \lambda_1 & \lambda_2 & \cdots & \lambda_n \\ \lambda_1^2 & \lambda_2^2 & \cdots & \lambda_n^2 \\ \vdots & \vdots & \vdots & \vdots \\ \lambda_1^{n-1} & \lambda_2^{n-1} & \cdots & \lambda_n^{n-1} \end{vmatrix} = \begin{matrix} (\lambda_2 - \lambda_1)(\lambda_3 - \lambda_2)(\lambda_3 - \lambda_1)(\lambda_4 - \lambda_3) \\ (\lambda_4 - \lambda_2) \cdots (\lambda_n - \lambda_1) \neq 0, \end{matrix} \tag{14.5.9}$$

since it is a Vandermonde determinant, $\alpha_1 \mathbf{x}_1 = \alpha_2 \mathbf{x}_2 = \alpha_3 \mathbf{x}_3 = \cdots = \alpha_n \mathbf{x}_n = 0$. Because the eigenvectors are nonzero, $\alpha_1 = \alpha_2 = \alpha_3 = \cdots = \alpha_n = 0$, and the eigenvectors are linearly independent. $\square$

This property of eigenvectors allows us to express any arbitrary vector $\mathbf{x}$ as a linear sum of the eigenvectors $\mathbf{x}_i$, or

$$\mathbf{x} = c_1 \mathbf{x}_1 + c_2 \mathbf{x}_2 + \cdots + c_n \mathbf{x}_n. \tag{14.5.10}$$

We will make good use of this property in Example 14.5.3.

• Example 14.5.1

Let us find the eigenvalues and corresponding eigenvectors of the matrix

$$A = \begin{pmatrix} -4 & 2 \\ -1 & -1 \end{pmatrix}. \tag{14.5.11}$$

We begin by setting up the characteristic equation:

$$\det(A - \lambda I) = \begin{vmatrix} -4 - \lambda & 2 \\ -1 & -1 - \lambda \end{vmatrix} = 0. \tag{14.5.12}$$

Expanding the determinant,

$$(-4 - \lambda)(-1 - \lambda) + 2 = \lambda^2 + 5\lambda + 6 = (\lambda + 3)(\lambda + 2) = 0. \tag{14.5.13}$$

Thus, the eigenvalues of the matrix A are $\lambda_1 = -3$, and $\lambda_2 = -2$.

To find the corresponding eigenvectors, we must solve the linear system:

$$\begin{pmatrix} -4 - \lambda & 2 \\ -1 & -1 - \lambda \end{pmatrix} \begin{pmatrix} x_1 \\ x_2 \end{pmatrix} = \begin{pmatrix} 0 \\ 0 \end{pmatrix}. \tag{14.5.14}$$

For example, for $\lambda_1 = -3$,

$$\begin{pmatrix} -1 & 2 \\ -1 & 2 \end{pmatrix} \begin{pmatrix} x_1 \\ x_2 \end{pmatrix} = \begin{pmatrix} 0 \\ 0 \end{pmatrix}, \tag{14.5.15}$$

or

$$x_1 = 2x_2. \tag{14.5.16}$$

Thus, any nonzero multiple of the vector $\begin{pmatrix} 2 \\ 1 \end{pmatrix}$ is an eigenvector belonging to $\lambda_1 = -3$. Similarly, for $\lambda_2 = -2$, the eigenvector is any nonzero multiple of the vector $\begin{pmatrix} 1 \\ 1 \end{pmatrix}$.

Of course, MATLAB will do all of the computations for you via the command eig which computes the eigenvalues and corresponding eigenvalues. In the present case, you would type

```
>> A = [-4 2; -1 -1]; % load in array A
>> % find eigenvalues and eigenvectors
>> [eigenvector,eigenvalue] = eig(A)
```

This yields

```
eigenvector =
     -0.8944      -0.7071
     -0.4472      -0.7071
```

and

```
eigenvalue =
     -3       0
      0      -2.
```

The eigenvalues are given as the elements along the principal diagonal of eigenvalue. The corresponding vectors are given by the corresponding column of eigenvector. As this example shows, these eigenvectors have been normalized so that their norm (14.1.5) equals one. Also their sign may be different than you would choose. We can recover our hand-computed results by dividing the first eigenvector by -0.4472 while in the second case we would divide by -0.7071.

● **Example 14.5.2**

Let us now find the eigenvalues and corresponding eigenvectors of the matrix

$$A = \begin{pmatrix} -4 & 5 & 5 \\ -5 & 6 & 5 \\ -5 & 5 & 6 \end{pmatrix}. \tag{14.5.17}$$

Setting up the characteristic equation:

$$\det(A - \lambda I)$$

$$= \begin{vmatrix} -4-\lambda & 5 & 5 \\ -5 & 6-\lambda & 5 \\ -5 & 5 & 6-\lambda \end{vmatrix} = \begin{vmatrix} -4-\lambda & 5 & 5 \\ -5 & 6-\lambda & 5 \\ 0 & \lambda-1 & 1-\lambda \end{vmatrix} \tag{14.5.18}$$

$$= (\lambda - 1) \begin{vmatrix} -4-\lambda & 5 & 5 \\ -5 & 6-\lambda & 5 \\ 0 & 1 & -1 \end{vmatrix} = (\lambda - 1)^2 \begin{vmatrix} -1 & 1 & 0 \\ -5 & 6-\lambda & 5 \\ 0 & 1 & -1 \end{vmatrix} \tag{14.5.19}$$

$$= (\lambda - 1)^2 \begin{vmatrix} -1 & 0 & 0 \\ -5 & 6-\lambda & 0 \\ 0 & 1 & -1 \end{vmatrix} = (\lambda - 1)^2 (6 - \lambda) = 0. \tag{14.5.20}$$

Thus, the eigenvalues of the matrix A are $\lambda_{1,2} = 1$ (twice), and $\lambda_3 = 6$.

To find the corresponding eigenvectors, we must solve the linear system:

$$(-4 - \lambda)x_1 + 5x_2 + 5x_3 = 0, \tag{14.5.21}$$

$$-5x_1 + (6 - \lambda)x_2 + 5x_3 = 0, \tag{14.5.22}$$

and

$$-5x_1 + 5x_2 + (6 - \lambda)x_3 = 0. \tag{14.5.23}$$

For $\lambda_3 = 6$, (14.5.21)–(14.5.23) become

$$-10x_1 + 5x_2 + 5x_3 = 0, \tag{14.5.24}$$

$$-5x_1 + 5x_3 = 0, \tag{14.5.25}$$

and

$$-5x_1 + 5x_2 = 0. \tag{14.5.26}$$

Thus, $x_1 = x_2 = x_3$ and the eigenvector is any nonzero multiple of the vector $\begin{pmatrix} 1 \\ 1 \\ 1 \end{pmatrix}$.

The interesting aspect of this example centers on finding the eigenvector for the eigenvalue $\lambda_{1,2} = 1$. If $\lambda_{1,2} = 1$, then (14.5.21)–(14.5.23) collapses into one equation

$$-x_1 + x_2 + x_3 = 0 \tag{14.5.27}$$

and we have *two* free parameters at our disposal. Let us take $x_2 = \alpha$, and $x_3 = \beta$. Then the eigenvector equals $\alpha \begin{pmatrix} 1 \\ 1 \\ 0 \end{pmatrix} + \beta \begin{pmatrix} 1 \\ 0 \\ 1 \end{pmatrix}$ for $\lambda_{1,2} = 1$.

In this example, we may associate the eigenvector $\begin{pmatrix} 1 \\ 1 \\ 0 \end{pmatrix}$ with $\lambda_1 = 1$, and $\begin{pmatrix} 1 \\ 0 \\ 1 \end{pmatrix}$ with $\lambda_2 = 1$ so that, along with the eigenvector $\begin{pmatrix} 1 \\ 1 \\ 1 \end{pmatrix}$ with $\lambda_3 = 6$, we still have n *linearly independent* eigenvectors for our 3×3 matrix. However, with repeated eigenvalues this is not always true. For example,

$$A = \begin{pmatrix} 1 & -1 \\ 0 & 1 \end{pmatrix} \tag{14.5.28}$$

has the repeated eigenvalues $\lambda_{1,2} = 1$. However, there is only a single eigenvector $\begin{pmatrix} 1 \\ 0 \end{pmatrix}$ for *both* λ_1 and λ_2.

What happens in MATLAB in the present case? Typing in

```
>> A = [-4 5 5; -5 6 5; -5 5 6]; % load in array A
>> % find eigenvalues and eigenvectors
>> [eigenvector,eigenvalue] = eig(A)
```

we obtain

```
eigenvector =
    -0.8165      0.5774       0.6345
    -0.4082      0.5774      -0.1278
    -0.4082      0.5774       0.7623
```

and

```
eigenvalue =
    1    0    0
    0    6    0
    0    0    1
```

The second eigenvector is clearly the same as the hand-computed one if you normalized it with 0.5774. The equivalence of the first and third eigenvectors is not as clear. However, if you choose $\alpha = \beta = -0.4082$, then the first eigenvector agrees with the hand-computed value. Similarly, taking $\alpha = -0.1278$ and $\beta = 0.7623$ result in agreement with the third MATLAB eigenvector.

• **Example 14.5.3**

When we discussed the stability of numerical schemes for the wave equation in §10.6, we examined the behavior of a prototypical Fourier harmonic to variations in the parameter $c\Delta t/\Delta x$. In this example we shall show another approach to determining the stability of a numerical scheme via matrices.

Consider the explicit scheme for the numerical integration of the wave equation (10.6.11). We can rewrite that single equation as the coupled difference equations:

$$u_m^{n+1} = 2(1 - r^2)u_m^n + r^2(u_{m+1}^n + u_{m-1}^n) - v_m^n, \tag{14.5.29}$$

and

$$v_m^{n+1} = u_m^n, \tag{14.5.30}$$

where $r = c\Delta t/\Delta x$. Let $u_{m+1}^n = e^{i\beta\Delta x}u_m^n$, and $u_{m-1}^n = e^{-i\beta\Delta x}u_m^n$, where β is real. Then (14.5.29)–(14.5.30) become

$$u_m^{n+1} = 2\left[1 - 2r^2\sin^2\left(\frac{\beta\Delta x}{2}\right)\right]u_m^n - v_m^n, \tag{14.5.31}$$

and

$$v_m^{n+1} = u_m^n, \tag{14.5.32}$$

or in the matrix form

$$\mathbf{u}_m^{n+1} = \begin{pmatrix} 2\left[1 - 2r^2\sin^2\left(\frac{\beta\Delta x}{2}\right)\right] & -1 \\ 1 & 0 \end{pmatrix}\mathbf{u}_m^n, \tag{14.5.33}$$

where $\mathbf{u}_m^n = \begin{pmatrix} u_m^n \\ v_m^n \end{pmatrix}$. The eigenvalues λ of this *amplification matrix* are given by

$$\lambda^2 - 2\left[1 - 2r^2\sin^2\left(\frac{\beta\Delta x}{2}\right)\right]\lambda + 1 = 0, \tag{14.5.34}$$

or

$$\lambda_{1,2} = 1 - 2r^2\sin^2\left(\frac{\beta\Delta x}{2}\right) \pm 2r\sin\left(\frac{\beta\Delta x}{2}\right)\sqrt{r^2\sin^2\left(\frac{\beta\Delta x}{2}\right) - 1}. \tag{14.5.35}$$

Because each successive time step consists of multiplying the solution from the previous time step by the amplification matrix, the solution is stable only if $\mathbf{u}_m^n$ remains bounded. This occurs only if all of the eigenvalues have a magnitude less or equal to one because

$$\mathbf{u}_m^n = \sum_k c_k A^n \mathbf{x}_k = \sum_k c_k \lambda_k^n \mathbf{x}_k, \qquad (14.5.36)$$

where A denotes the amplification matrix and $\mathbf{x}_k$ denotes the eigenvectors corresponding to the eigenvalues λ_k. Equation (14.5.36) follows from our ability to express any initial condition in terms of an eigenvector expansion

$$\mathbf{u}_m^0 = \sum_k c_k \mathbf{x}_k. \qquad (14.5.37)$$

In our particular example, two cases arise. If $r^2 \sin^2(\beta \Delta x/2) \leq 1$,

$$\lambda_{1,2} = 1 - 2r^2 \sin^2\left(\frac{\beta \Delta x}{2}\right) \pm 2ri \sin\left(\frac{\beta \Delta x}{2}\right) \sqrt{1 - r^2 \sin^2\left(\frac{\beta \Delta x}{2}\right)} \quad (14.5.38)$$

and $|\lambda_{1,2}| = 1$. On the other hand, if $r^2 \sin^2(\beta \Delta x/2) > 1$, $|\lambda_{1,2}| > 1$. Thus, we have stability only if $c\Delta t/\Delta x \leq 1$.

Problems

Find the eigenvalues and corresponding eigenvectors for the following matrices. Check your answers using MATLAB.

1. $A = \begin{pmatrix} 3 & 2 \\ 3 & -2 \end{pmatrix}$

   2. $A = \begin{pmatrix} 3 & -1 \\ 1 & 1 \end{pmatrix}$

3. $A = \begin{pmatrix} 2 & -3 & 1 \\ 1 & -2 & 1 \\ 1 & -3 & 2 \end{pmatrix}$

   4. $A = \begin{pmatrix} 0 & 1 & 0 \\ 0 & 0 & 1 \\ 0 & 0 & 0 \end{pmatrix}$

5. $A = \begin{pmatrix} 1 & 1 & 1 \\ 0 & 2 & 1 \\ 0 & 0 & 1 \end{pmatrix}$

   6. $A = \begin{pmatrix} 1 & 2 & 1 \\ 0 & 3 & 1 \\ 0 & 5 & -1 \end{pmatrix}$

7. $A = \begin{pmatrix} 4 & -5 & 1 \\ 1 & 0 & -1 \\ 0 & 1 & -1 \end{pmatrix}$

   8. $A = \begin{pmatrix} -2 & 0 & 1 \\ 3 & 0 & -1 \\ 0 & 1 & 1 \end{pmatrix}$

Project: Numerical Solution of the Sturm-Liouville Problem

You may have been struck by the similarity of the algebraic eigenvalue problem to the Sturm-Liouville problem. In both cases nontrivial solutions exist only for characteristic values of λ. The purpose of this project is to further deepen your insight into these similarities.

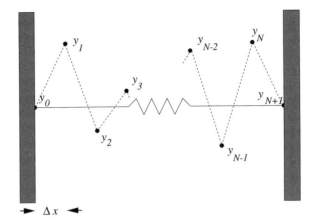

Figure 14.5.1: Schematic for finite-differencing a Sturm-Liouville problem into a set of difference equations.

Consider the Sturm-Liouville problem

$$y'' + \lambda y = 0, \qquad y(0) = y(\pi) = 0. \tag{14.5.39}$$

We know that it has the nontrivial solutions $\lambda_m = m^2$, $y_m(x) = \sin(mx)$, where $m = 1, 2, 3, \dots$.

Step 1: Let us solve this problem numerically. Introducing centered finite differencing and the grid shown in Figure 14.5.1, show that

$$y'' \approx \frac{y_{n+1} - 2y_n + y_{n-1}}{(\Delta x)^2}, \qquad n = 1, 2, \dots, N, \tag{14.5.40}$$

where $\Delta x = \pi/(N + 1)$. Show that the finite-differenced form of (14.5.39) is

$$-h^2 y_{n+1} + 2h^2 y_n - h^2 y_{n-1} = \lambda y_n \tag{14.5.41}$$

with $y_0 = y_{N+1} = 0$, and $h = 1/(\Delta x)$.

Step 2: Solve (14.5.41) as an algebraic eigenvalue problem using $N = 1, 2, \dots$. Show that (14.5.41) can be written in the matrix form of

$$\begin{pmatrix} 2h^2 & -h^2 & 0 & \cdots & 0 & 0 & 0 \\ -h^2 & 2h^2 & -h^2 & \cdots & 0 & 0 & 0 \\ 0 & -h^2 & 2h^2 & \cdots & 0 & 0 & 0 \\ \vdots & \vdots & \vdots & \vdots & \vdots & \vdots & \vdots \\ 0 & 0 & 0 & \cdots & -h^2 & 2h^2 & -h^2 \\ 0 & 0 & 0 & \cdots & 0 & -h^2 & 2h^2 \end{pmatrix} \begin{pmatrix} y_1 \\ y_2 \\ y_3 \\ \vdots \\ y_{N-1} \\ y_N \end{pmatrix} = \lambda \begin{pmatrix} y_1 \\ y_2 \\ y_3 \\ \vdots \\ y_{N-1} \\ y_N \end{pmatrix}. \tag{14.5.42}$$

Table 14.5.1: Eigenvalues Computed from (14.5.42) as a Numerical Approximation of the Sturm-Liouville Problem (14.5.39)

N	λ_1	λ_2	λ_3	λ_4	λ_5	λ_6	λ_7
1	0.81057						
2	0.91189	2.73567					
3	0.94964	3.24228	5.53491				
4	0.96753	3.50056	6.63156	9.16459			
5	0.97736	3.64756	7.29513	10.94269	13.61289		
6	0.98333	3.73855	7.71996	12.13899	16.12040	18.87563	
7	0.98721	3.79857	8.00605	12.96911	17.93217	22.13966	24.95100
8	0.98989	3.84016	8.20702	13.56377	19.26430	24.62105	28.98791
20	0.99813	3.97023	8.84993	15.52822	23.85591	33.64694	44.68265
50	0.99972	3.99498	8.97438	15.91922	24.80297	35.59203	48.24538

Note that the coefficient matrix is symmetric.

Step 3: You are now ready to compute the eigenvalues. For small N this could be done by hand. However, it is easier just to write a MATLAB program that will handle any $N \geq 2$. Table 14.5.1 has been provided so that you can check your program.

With your program, answer the following questions: How do your computed eigenvalues compare to the eigenvalues given by the Sturm-Liouville problem? What happens as you increase N? Which computed eigenvalues agree best with those given by the Sturm-Liouville problem? Which ones compare the worst?

Step 4: Let us examine the eigenfunctions now. Starting with the smallest eigenvalue, use MATLAB to plot Cy_j as a function of x_i where y_j is the jth eigenvector, $j = 1, 2, \ldots, N$, $x_i = i\Delta x$, $i = 1, 2, \ldots, N$, and C is chosen so that $C^2 \Delta x \sum_i y_j^2(x_i) = 1$. On the same plot, graph $y_j(x) = \sqrt{2/\pi}\,\sin(jx)$. Why did we choose C as we did? Which eigenvectors and eigenfunctions agree the best? Which eigenvectors and eigenfunctions agree the worst? Why? Why are there N eigenvectors and an infinite number of eigenfunctions?

Step 5: The most important property of eigenfunctions is orthogonality. But what do we mean by orthogonality in the case of eigenvectors? Recall from three-dimensional vectors we had the scalar dot product

$$\mathbf{a} \cdot \mathbf{b} = a_1 b_1 + a_2 b_2 + a_3 b_3. \tag{14.5.43}$$

For n-dimensional vectors, this dot product is generalized to the inner product

$$\mathbf{x} \cdot \mathbf{y} = \sum_{k=1}^{n} x_k y_k. \tag{14.5.44}$$

Orthogonality implies that $\mathbf{x} \cdot \mathbf{y} = 0$ if $\mathbf{x} \neq \mathbf{y}$. Are your eigenvectors orthogonal? How might you use this property with eigenvectors?

14.6 SYSTEMS OF LINEAR DIFFERENTIAL EQUATIONS

In this section we show how we may apply the classic algebraic eigenvalue problem to solve a system of ordinary differential equations.

Let us solve the following system:

$$x_1' = x_1 + 3x_2, \qquad (14.6.1)$$

and

$$x_2' = 3x_1 + x_2, \qquad (14.6.2)$$

where the primes denote the time derivative.

We begin by rewriting (14.6.1)–(14.6.2) in matrix notation:

$$\mathbf{x}' = A\mathbf{x}, \qquad (14.6.3)$$

where

$$\mathbf{x} = \begin{pmatrix} x_1 \\ x_2 \end{pmatrix}, \qquad \text{and} \qquad A = \begin{pmatrix} 1 & 3 \\ 3 & 1 \end{pmatrix}. \qquad (14.6.4)$$

Note that

$$\begin{pmatrix} x_1' \\ x_2' \end{pmatrix} = \frac{d}{dt}\begin{pmatrix} x_1 \\ x_2 \end{pmatrix} = \mathbf{x}'. \qquad (14.6.5)$$

Assuming a solution of the form

$$\mathbf{x} = \mathbf{x}_0 e^{\lambda t}, \qquad \text{where} \qquad \mathbf{x}_0 = \begin{pmatrix} a \\ b \end{pmatrix} \qquad (14.6.6)$$

is a constant vector, we substitute (14.6.6) into (14.6.3) and find that

$$\lambda e^{\lambda t}\mathbf{x}_0 = A e^{\lambda t}\mathbf{x}_0. \qquad (14.6.7)$$

Because $e^{\lambda t}$ does not generally equal zero, we have that

$$(A - \lambda I)\mathbf{x}_0 = \mathbf{0}, \qquad (14.6.8)$$

which we solved in the previous section. This set of homogeneous equations is the *classic eigenvalue problem*. In order for this set not to have trivial solutions,

$$\det(A - \lambda I) = \begin{vmatrix} 1 - \lambda & 3 \\ 3 & 1 - \lambda \end{vmatrix} = 0. \qquad (14.6.9)$$

Expanding the determinant,

$$(1 - \lambda)^2 - 9 = 0 \quad \text{or} \quad \lambda = -2, 4. \qquad (14.6.10)$$

Thus, we have two real and distinct eigenvalues: $\lambda = -2$ and 4.

We must now find the corresponding $\mathbf{x}_0$ or *eigenvector* for each eigenvalue. From (14.6.8),

$$(1 - \lambda)a + 3b = 0, \tag{14.6.11}$$

and

$$3a + (1 - \lambda)b = 0. \tag{14.6.12}$$

If $\lambda = 4$, these equations are consistent and yield $a = b = c_1$. If $\lambda = -2$, we have that $a = -b = c_2$. Therefore, the general solution in matrix notation is

$$\mathbf{x} = c_1 \begin{pmatrix} 1 \\ 1 \end{pmatrix} e^{4t} + c_2 \begin{pmatrix} 1 \\ -1 \end{pmatrix} e^{-2t}. \tag{14.6.13}$$

To evaluate c_1 and c_2, we must have initial conditions. For example, if $x_1(0) = x_2(0) = 1$, then

$$\begin{pmatrix} 1 \\ 1 \end{pmatrix} = c_1 \begin{pmatrix} 1 \\ 1 \end{pmatrix} + c_2 \begin{pmatrix} 1 \\ -1 \end{pmatrix}. \tag{14.6.14}$$

Solving for c_1 and c_2, $c_1 = 1$, $c_2 = 0$, and the solution with this particular set of initial conditions is

$$\mathbf{x} = \begin{pmatrix} 1 \\ 1 \end{pmatrix} e^{4t}. \tag{14.6.15}$$

• **Example 14.6.1**

Let us solve the following set of linear ordinary differential equations

$$x_1' = -x_2 + x_3, \tag{14.6.16}$$

$$x_2' = 4x_1 - x_2 - 4x_3, \tag{14.6.17}$$

and

$$x_3' = -3x_1 - x_2 + 4x_3; \tag{14.6.18}$$

or in matrix form,

$$\mathbf{x}' = \begin{pmatrix} 0 & -1 & 1 \\ 4 & -1 & -4 \\ -3 & -1 & 4 \end{pmatrix} \mathbf{x}, \qquad \mathbf{x} = \begin{pmatrix} x_1 \\ x_2 \\ x_3 \end{pmatrix}. \tag{14.6.19}$$

Assuming the solution $\mathbf{x} = \mathbf{x}_0 e^{\lambda t}$,

$$\begin{pmatrix} 0 & -1 & 1 \\ 4 & -1 & -4 \\ -3 & -1 & 4 \end{pmatrix} \mathbf{x}_0 = \lambda \mathbf{x}_0, \tag{14.6.20}$$

or

$$\begin{pmatrix} -\lambda & -1 & 1 \\ 4 & -1-\lambda & -4 \\ -3 & -1 & 4-\lambda \end{pmatrix} \mathbf{x}_0 = \mathbf{0}. \qquad (14.6.21)$$

For nontrivial solutions,

$$\begin{vmatrix} -\lambda & -1 & 1 \\ 4 & -1-\lambda & -4 \\ -3 & -1 & 4-\lambda \end{vmatrix} = \begin{vmatrix} 0 & 0 & 1 \\ 4-4\lambda & -5-\lambda & -4 \\ -3+4\lambda-\lambda^2 & 3-\lambda & 4-\lambda \end{vmatrix} = 0, \quad (14.6.22)$$

and

$$(\lambda-1)(\lambda-3)(\lambda+1) = 0, \quad \text{or} \quad \lambda = -1, 1, 3. \qquad (14.6.23)$$

To determine the eigenvectors, we rewrite (14.6.21) as

$$-\lambda a - b + c = 0, \qquad (14.6.24)$$

$$4a - (1+\lambda)b - 4c = 0, \qquad (14.6.25)$$

and

$$-3a - b + (4-\lambda)c = 0. \qquad (14.6.26)$$

For example, if $\lambda = 1$,

$$-a - b + c = 0, \qquad (14.6.27)$$

$$4a - 2b - 4c = 0, \qquad (14.6.28)$$

and

$$-3a - b + 3c = 0; \qquad (14.6.29)$$

or $a = c$, and $b = 0$. Thus, the eigenvector for $\lambda = 1$ is $\mathbf{x}_0 = \begin{pmatrix} 1 \\ 0 \\ 1 \end{pmatrix}$. Similarly,

for $\lambda = -1$, $\mathbf{x}_0 = \begin{pmatrix} 1 \\ 2 \\ 1 \end{pmatrix}$; and for $\lambda = 3$, $\mathbf{x}_0 = \begin{pmatrix} 1 \\ -1 \\ 2 \end{pmatrix}$. Thus, the most general

solution is

$$\mathbf{x} = c_1 \begin{pmatrix} 1 \\ 0 \\ 1 \end{pmatrix} e^t + c_2 \begin{pmatrix} 1 \\ 2 \\ 1 \end{pmatrix} e^{-t} + c_3 \begin{pmatrix} 1 \\ -1 \\ 2 \end{pmatrix} e^{3t}. \qquad (14.6.30)$$

• **Example 14.6.2**

Let us solve the following set of linear ordinary differential equations:

$$x_1' = x_1 - 2x_2, \qquad (14.6.31)$$

and

$$x_2' = 2x_1 - 3x_2; \tag{14.6.32}$$

or in matrix form,

$$\mathbf{x}' = \begin{pmatrix} 1 & -2 \\ 2 & -3 \end{pmatrix} \mathbf{x}, \qquad \mathbf{x} = \begin{pmatrix} x_1 \\ x_2 \end{pmatrix}. \tag{14.6.33}$$

Assuming the solution $\mathbf{x} = \mathbf{x}_0 e^{\lambda t}$,

$$\begin{pmatrix} 1-\lambda & -2 \\ 2 & -3-\lambda \end{pmatrix} \mathbf{x}_0 = \mathbf{0}. \tag{14.6.34}$$

For nontrivial solutions,

$$\begin{vmatrix} 1-\lambda & -2 \\ 2 & -3-\lambda \end{vmatrix} = (\lambda+1)^2 = 0. \tag{14.6.35}$$

Thus, we have the solution

$$\mathbf{x} = c_1 \begin{pmatrix} 1 \\ 1 \end{pmatrix} e^{-t}. \tag{14.6.36}$$

The interesting aspect of this example is the single solution that the traditional approach yields because we have repeated roots. To find the second solution, we try the solution

$$\mathbf{x} = \begin{pmatrix} a+ct \\ b+dt \end{pmatrix} e^{-t}. \tag{14.6.37}$$

We guessed (14.6.37) using our knowledge of solutions to differential equations when the characteristic polynomial has repeated roots. Substituting (14.6.37) into (14.6.33), we find that $c = d = 2c_2$, and $a - b = c_2$. Thus, we have one free parameter, which we choose to be b, and set it equal to zero. This is permissible because (14.6.37) can be broken into two terms: $b \begin{pmatrix} 1 \\ 1 \end{pmatrix} e^{-t}$ and $c_2 \begin{pmatrix} 1+2t \\ 2t \end{pmatrix} e^{-t}$. The first term can be incorporated into the $c_1 \begin{pmatrix} 1 \\ 1 \end{pmatrix} e^{-t}$ term. Thus, the general solution is

$$\mathbf{x} = c_1 \begin{pmatrix} 1 \\ 1 \end{pmatrix} e^{-t} + c_2 \begin{pmatrix} 1 \\ 0 \end{pmatrix} e^{-t} + 2c_2 \begin{pmatrix} 1 \\ 1 \end{pmatrix} te^{-t}. \tag{14.6.38}$$

• **Example 14.6.3**

Let us solve the system of linear differential equations:

$$x_1' = 2x_1 - 3x_2, \qquad (14.6.39)$$

and

$$x_2' = 3x_1 + 2x_2; \qquad (14.6.40)$$

or in matrix form,

$$\mathbf{x}' = \begin{pmatrix} 2 & -3 \\ 3 & 2 \end{pmatrix} \mathbf{x}, \qquad \mathbf{x} = \begin{pmatrix} x_1 \\ x_2 \end{pmatrix}. \qquad (14.6.41)$$

Assuming the solution $\mathbf{x} = \mathbf{x}_0 e^{\lambda t}$,

$$\begin{pmatrix} 2 - \lambda & -3 \\ 3 & 2 - \lambda \end{pmatrix} \mathbf{x}_0 = \mathbf{0}. \qquad (14.6.42)$$

For nontrivial solutions,

$$\begin{vmatrix} 2 - \lambda & -3 \\ 3 & 2 - \lambda \end{vmatrix} = (2 - \lambda)^2 + 9 = 0, \qquad (14.6.43)$$

and $\lambda = 2 \pm 3i$. If $\mathbf{x}_0 = \begin{pmatrix} a \\ b \end{pmatrix}$, then $b = -ai$ if $\lambda = 2 + 3i$, and $b = ai$ if $\lambda = 2 - 3i$. Thus, the general solution is

$$\mathbf{x} = c_1 \begin{pmatrix} 1 \\ -i \end{pmatrix} e^{2t + 3it} + c_2 \begin{pmatrix} 1 \\ i \end{pmatrix} e^{2t - 3it}, \qquad (14.6.44)$$

where c_1 and c_2 are arbitrary complex constants. Using Euler relationships, we can rewrite (14.6.44) as

$$\mathbf{x} = c_3 \begin{bmatrix} \cos(3t) \\ \sin(3t) \end{bmatrix} e^{2t} + c_4 \begin{bmatrix} \sin(3t) \\ -\cos(3t) \end{bmatrix} e^{2t}, \qquad (14.6.45)$$

where $c_3 = c_1 + c_2$ and $c_4 = i(c_1 - c_2)$.

Problems

Find the general solution of the following sets of ordinary differential equations using matrix technique. You may find the eigenvalues and eigenvectors either by hand or use MATLAB.

1. $x_1' = x_1 + 2x_2$ $x_2' = 2x_1 + x_2.$

2. $x_1' = x_1 - 4x_2$ $x_2' = 3x_1 - 6x_2.$

3. $x_1' = x_1 + x_2$ $x_2' = 4x_1 + x_2.$

4. $x_1' = x_1 + 5x_2$ $x_2' = -2x_1 - 6x_2.$

5. $x_1' = -\frac{3}{2}x_1 - 2x_2$ $x_2' = 2x_1 + \frac{5}{2}x_2.$

6. $x_1' = -3x_1 - 2x_2$ $x_2' = 2x_1 + x_2.$

7. $x_1' = x_1 - x_2$ $x_2' = x_1 + 3x_2.$

8. $x_1' = 3x_1 + 2x_2$ $x_2' = -2x_1 - x_2.$

9. $x_1' = -2x_1 - 13x_2$ $x_2' = x_1 + 4x_2.$

10. $x_1' = 3x_1 - 2x_2$ $x_2' = 5x_1 - 3x_2.$

11. $x_1' = 4x_1 - 2x_2$ $x_2' = 25x_1 - 10x_2.$

12. $x_1' = -3x_1 - 4x_2$ $x_2' = 2x_1 + x_2.$

13. $x_1' = 3x_1 + 4x_2$ $x_2' = -2x_1 - x_2.$

14. $x_1' + 5x_1 + x_2' + 3x_2 = 0$ $2x_1' + x_1 + x_2' + x_2 = 0.$

15. $x_1' - x_1 + x_2' - 2x_2 = 0$ $x_1' - 5x_1 + 2x_2' - 7x_2 = 0.$

16. $x_1' = x_1 - 2x_2$ $x_2' = 0$ $x_3' = -5x_1 + 7x_3.$

17. $x_1' = 2x_1$ $x_2' = x_1 + 2x_3$ $x_3' = x_3.$

18. $x_1' = 3x_1 - 2x_3$ $x_2' = -x_1 + 2x_2 + x_3$ $x_3' = 4x_1 - 3x_3.$

19. $x_1' = 3x_1 - x_3$ $x_2' = -2x_1 + 2x_2 + x_3$ $x_3' = 8x_1 - 3x_3.$

Answers
To the Odd-Numbered Problems

Section 1.1

1. $1 + 2i$

3. $-2/5$

5. $2 + 2i\sqrt{3}$

7. $4e^{\pi i}$

9. $5\sqrt{2}e^{3\pi i/4}$

11. $2e^{2\pi i/3}$

Section 1.2

1. $\pm\sqrt{2}, \quad \pm\sqrt{2}\left(\dfrac{1}{2} + \dfrac{\sqrt{3}i}{2}\right), \quad \pm\sqrt{2}\left(-\dfrac{1}{2} + \dfrac{\sqrt{3}i}{2}\right)$

3. $i, \quad -\dfrac{\sqrt{3}}{2} - \dfrac{i}{2}, \quad z_2 = \dfrac{\sqrt{3}}{2} - \dfrac{i}{2}$

5. $w_1 = \dfrac{1}{\sqrt{2}}\left(-\sqrt{\sqrt{a^2 + b^2} + a} + i\sqrt{\sqrt{a^2 + b^2} + a}\right), \qquad w_2 = -w_1$

7. $z_{1,2} = \pm(1 + i); \quad z_{3,4} = \pm 2(1 - i)$

Section 1.3

1. $u = 2 - y, \ v = x$

3. $u = x^3 - 3xy^2, \ v = 3x^2y - y^3$

5. $f'(z) = 3z(1 + z^2)^{1/2}$

7. $f'(z) = 2(1 + 4i)z - 3$

9. $f'(z) = -3i(iz - 1)^{-4}$ 11. $1/6$

13. $v(x, y) = 2xy + \text{constant}$

15. $v(x, y) = x\sin(x)e^{-y} + ye^{-y}\cos(x) + \text{constant}.$

Section 1.4

1. 0 3. $2i$ 5. $14/15 - i/3$

Section 1.5

1. $(e^{-2} - e^{-4})/2$ 3. $\pi/2$

Section 1.6

1. $\pi i/32$ 3. $\pi i/2$ 5. $-2\pi i$ 7. $2\pi i$ 9. -6π

Section 1.7

1. $\displaystyle\sum_{n=0}^{\infty}(n + 1)z^n$

3. $f(z) = z^{10} - z^9 + \dfrac{z^8}{2} - \dfrac{z^7}{6} + \cdots - \dfrac{1}{11!z} + \cdots$
We have an essential singularity and the residue equals $-1/11!$

5. $f(z) = \dfrac{1}{2!} + \dfrac{z^2}{4!} + \dfrac{z^4}{6!} + \cdots$
We have a removable singularity where the value of the residue equals zero.

7. $f(z) = -\dfrac{2}{z} - 2 - \dfrac{7z}{6} - \dfrac{z^2}{2} - \cdots$
We have a simple pole and the residue equals -2.

9. $f(z) = \dfrac{1}{2}\dfrac{1}{z - 2} - \dfrac{1}{4} + \dfrac{z - 2}{8} - \cdots$
We have a simple pole and the residue equals $1/2$.

Section 1.8

1. $-3\pi i/4$ 3. $-2\pi i.$ 5. $2\pi i$ 7. $2\pi i$

Section 2.1

1. first-order, linear   3. first-order, nonlinear

5. second-order, linear   7. third-order, nonlinear

9. second-order, nonlinear

11. first-order, nonlinear

13. first-order, nonlinear

15. second-order, nonlinear

Section 2.2

1. $y = -\ln(C - x^2/2)$

3. $y^2(x) - \ln^2(x) = 2C$

5. $2 + y^2(x) = C(1 + x^2)$

7. $y(x) = -\ln(C - e^x)$

9. $\dfrac{ay^3 - b}{ay_0^3 - b} = e^{-3at}$

13. $V(t) = \dfrac{V_0 S e^{-t/(RC)}}{S + RV_0\left[1 - e^{-t/(RC)}\right]}$

15. $N(t) = N(0)\exp\{\ln[K/N(0)]\left(1 - e^{-bt}\right)\}$

17.
$$\frac{1}{([A]_0 - [B]_0)([A]_0 - [C]_0)}\ln\left(\frac{[A]_0}{[A]_0 - [X]}\right)$$
$$+ \frac{1}{([B]_0 - [A]_0)([B]_0 - [C]_0)}\ln\left(\frac{[B]_0}{[B]_0 - [X]}\right)$$
$$+ \frac{1}{([C]_0 - [A]_0)([C]_0 - [B]_0)}\ln\left(\frac{[C]_0}{[C]_0 - [X]}\right) = kt$$

Section 2.3

1. $\ln|y| - x/y = C$

3. $|x|(x^2 + 3y^2) = C$

5. $y = x\left(\ln|x| + C\right)^2$

7. $\sin(y/x) - \ln|x| = C$

Section 2.4

1. $xy^2 - \frac{1}{3}x^3 = C$

3. $xy^2 - x + \cos(y) = C$

5. $y/x + \ln(y) = C$

7. $\cos(xy) = C$

9. $x^2y^3 + x^5y + y = C$

11. $xy\ln(y) + e^x - e^{-y} = C$

13. $y - x + \frac{1}{2}\sin(2x + 2y) = C$

Section 2.5

1. $y = \frac{1}{2}e^x + Ce^{-x}$, $\quad x \in (-\infty, \infty)$ 3. $y = \ln(x)/x + Cx^{-1}$, $\quad x \neq 0$

5. $y = 2x^3\ln(x) + Cx^3$, $\quad x \in (-\infty, \infty)$

7. $e^{\sin(2x)}y = C$, $\quad n\pi + \varphi < 2x < (n+1)\pi + \varphi$, where φ is any real and n is any integer.

9. $y(x) = \frac{4}{3} + \frac{11}{3}e^{-3x}$, $\quad x \in (-\infty, \infty)$

11. $y(x) = (x + C)\csc(x)$

13. $y(x) = \dfrac{\cos^a(x)\,y(0)}{[\sec(x) + \tan(x)]^b} + \dfrac{c\,\cos^a(x)}{[\sec(x) + \tan(x)]^b}\displaystyle\int_0^x \dfrac{[\sec(\xi) + \tan(\xi)]^b}{\cos^{a+1}(\xi)}\,d\xi$

15. $y(x) = \dfrac{2ax - 1}{8a^2} + \dfrac{\omega^2 e^{-2ax}}{8a^2(a^2 + \omega^2)} - \dfrac{a\sin(2\omega x) - \omega\cos(2\omega x)}{8\omega(a^2 + \omega^2)}$

17. $y^2(x) = 2(x - x^{2/k})/(2 - k)$ if $k \neq 2$; $y^2(x) = x\ln(1/x)$ if $k = 2$

19. $[A] = [A]_0 e^{-k_1 t}$, $[B] = \dfrac{k_1[A]_0}{k_2 - k_1}\left[e^{-k_1 t} - e^{-k_2 t}\right]$,

$[C] = [A]_0\left(1 + \dfrac{k_1 e^{-k_2 t} - k_2 e^{-k_1 t}}{k_2 - k_1}\right)$

21. $y(x) = [Cx + x\ln(x)]^{-1}$ 23. $y(x) = \left[Cx^2 + \frac{1}{2}x^2\ln(x)\right]^2$

25. $y(x) = [Cx - x\ln(x)]^{1/2}$

Section 2.6

5. The equilibrium points are $x = 0, \frac{1}{2}$, and 1. The equilibrium at $x = \frac{1}{2}$ is unstable while the equilibriums at $x = 0$ and 1 are stable.

7. The equilibrium point for this differential equation is $x = 0$, which is stable.

Section 2.7

1. $x(t) = e^t + t + 1$ 2. $x(t) = e^{t^2/2}$

3. $x(t) = [1 - \ln(t + 1)]^{-1}$ 4. $x(t) = \frac{1}{2}\left(e^{t-2} - e^{-t}\right)$

Section 3.0

1. $y_2(x) = A/x$ 3. $y_2(x) = Ax^{-4}$

5. $y_2(x) = A(x^2 - x + 1)$ 7. $y_2(x) = A\sin(x)/\sqrt{x}$

9. $y(x) = C_2 e^{C_1 x}$ 11. $y = \left(1 + C_2 e^{C_1 x}\right)/C_1$

13. $y = -\ln|1 - x|$ 15. $y = C_1 - 2\ln(x^2 + C_2)$

Section 3.1

1. $y(x) = C_1 e^{-x} + C_2 e^{-5x}$ 3. $y(x) = C_1 e^x + C_2 x e^x$

5. $y(x) = C_1 e^{2x}\cos(2x) + C_2 e^{2x}\sin(2x)$ 7. $y(x) = C_1 e^{-10x} + C_2 e^{4x}$

9. $y(x) = e^{-4x}[C_1\cos(3x) + C_2\sin(3x)]$ 11. $y(x) = C_1 e^{-4x} + C_2 x e^{-4x}$

13. $y(x) = C_1 + C_2 x + C_3\cos(2x) + C_4\sin(2x)$

15. $y(x) = C_1 e^{2x} + C_2 e^{-x}\cos(\sqrt{3}\,x) + C_3 e^{-x}\sin(\sqrt{3}\,x)$

17. $y(t) = e^{-t/(2\tau)} \left\{ A \exp\left[t\sqrt{1 - 2A\tau}/(2\tau) \right] + B \exp\left[-t\sqrt{1 - 2A\tau}/(2\tau) \right] \right\}$

Section 3.2

1. $x(t) = 2\sqrt{26} \, \sin(5t + 1.7682)$ 3. $x(t) = 2\cos(\pi t - \pi/3)$

5. $x(t) = s_0 \cos(\omega t) + \dfrac{v_0}{\omega} \sin(\omega t)$ and $v(t) = v_0 \cos(\omega t) - \omega s_0 \sin(\omega t)$, where $\omega^2 = Mg/mL$.

Section 3.3

1. $x(t) = 4e^{-2t} - 2e^{-4t}$ 3. $x(t) = e^{-5t/2} \left[4\cos(6t) + \frac{13}{3} \sin(6t) \right]$

5. The roots are equal when $c = 4$ when $m = -2$.

Section 3.4

1. $y(x) = Ae^{-3x} + Be^{-x} + \frac{1}{3}x - \frac{1}{9}$

3. $y(x) = e^{-x}[A\cos(x) + B\sin(x)] + x^2 - x + 2$

5. $y(x) = A + Be^{-2x} + \frac{1}{2}x^2 + 2x + \frac{1}{2}e^{-2x}$

7. $y(x) = (A + Bx)e^{-2x} + \left(\frac{1}{9}x - \frac{2}{27} \right)e^x$

9. $y(x) = A\cos(3x) + B\sin(3x) + \frac{1}{12}x^2 \sin(3x) + \frac{1}{36}x\cos(3x)$

11. $y(x) = \dfrac{2ax - 1}{8a^2} + \dfrac{\omega^2 e^{-2ax}}{8a^2(a^2 + \omega^2)} - \dfrac{a\sin(2\omega x) - \omega\cos(2\omega x)}{8\omega(a^2 + \omega^2)}$

Section 3.5

1. $\gamma = 3$

5. $x(t) = e^{-ct/(2m)} \left(A\cos(\omega_0 t) + B\sin(\omega_0 t) \right) + \dfrac{F_0 \, \sin(\omega t - \varphi)}{\sqrt{c^2\omega^2 + (k - m\omega^2)^2}}$

Section 3.6

1. $y(x) = Ae^x + Be^{3x} + \frac{1}{8}e^{-x}$

3. $y(x) = Ae^{2x} + Be^{-2x} - (3x + 2)e^x/9$

5. $y(x) = (A + Bx)e^{-2x} + x^3 e^{-2x}/6$

7. $y(x) = Ae^{2x} + Bxe^{2x} + \left(\frac{1}{2}x^2 + \frac{1}{6}x^3 \right)e^{2x}$

9. $y(x) = Ae^x + Bxe^x + x\ln(x)e^x$

Section 3.7

1. $y(x) = C_1 x + C_2 x^{-1}$ 3. $y(x) = C_1 x^2 + C_2/x$

5. $y(x) = C_1/x + C_2 \ln(x)/x$

7. $y(x) = C_1 x \cos[2\ln(x)] + C_2 x \sin[\ln(x)]$

9. $y(x) = C_1 \cos[\ln(x)] + C_2 \sin[\ln(x)]$

11. $y(x) = C_1 x^2 + C_2 x^4 + C_3/x$

Section 3.8

1. The trajectories spirals outward from $(0,0)$.

3. The equilibrium points are $(x,0)$; they are unstable.

5. The equilibrium points are $v = 0$ and $|x| < 2$; they are unstable.

Section 4.1

1. $f(t) = \dfrac{1}{2} - \dfrac{2}{\pi} \displaystyle\sum_{m=1}^{\infty} \dfrac{\sin[(2m-1)t]}{2m-1}$

3. $f(t) = -\dfrac{\pi}{4} + \displaystyle\sum_{n=1}^{\infty} \dfrac{(-1)^n - 1}{n^2 \pi} \cos(nt) + \dfrac{1 - 2(-1)^n}{n} \sin(nt)$

5. $f(t) = \dfrac{\pi}{8} + \dfrac{2}{\pi} \displaystyle\sum_{n=1}^{\infty} \dfrac{2\cos(n\pi/2)\sin^2(n\pi/4)}{n^2} \cos(nt) + \dfrac{\sin(n\pi/2)}{n^2} \sin(nt)$

7. $f(t) = \dfrac{\sinh(aL)}{aL} + 2aL\sinh(aL) \displaystyle\sum_{n=1}^{\infty} \dfrac{(-1)^n}{a^2 L^2 + n^2 \pi^2} \cos\left(\dfrac{n\pi t}{L}\right)$
$$-2\pi\sinh(aL) \sum_{n=1}^{\infty} \dfrac{n(-1)^n}{a^2 L^2 + n^2 \pi^2} \sin\left(\dfrac{n\pi t}{L}\right)$$

9. $f(t) = \dfrac{1}{\pi} + \dfrac{1}{2}\sin(t) - \dfrac{2}{\pi} \displaystyle\sum_{m=1}^{\infty} \dfrac{\cos(2mt)}{4m^2 - 1}$

11. $f(t) = \dfrac{a}{2} - \dfrac{4a}{\pi^2} \displaystyle\sum_{m=1}^{\infty} \dfrac{1}{(2m-1)^2} \cos\left[\dfrac{(2m-1)\pi t}{a}\right] - \dfrac{2a}{\pi} \displaystyle\sum_{n=1}^{\infty} \dfrac{(-1)^n}{n} \sin\left(\dfrac{n\pi t}{a}\right)$

13. $f(t) = \dfrac{\pi - 1}{2} + \dfrac{1}{\pi} \displaystyle\sum_{n=1}^{\infty} \dfrac{\sin(n\pi t)}{n}$

15. $f(t) = \dfrac{4a\cosh(a\pi/2)}{\pi} \displaystyle\sum_{m=1}^{\infty} \dfrac{\cos[(2m-1)t]}{a^2 + (2m-1)^2}$

Section 4.3

1. $f(x) = \dfrac{\pi}{2} - \dfrac{4}{\pi} \displaystyle\sum_{m=1}^{\infty} \dfrac{\cos[(2m-1)x]}{(2m-1)^2}, \quad f(x) = \dfrac{2}{\pi} \displaystyle\sum_{n=1}^{\infty} \dfrac{(-1)^{n+1}\sin(nx)}{n}$

3. $f(x) = \dfrac{a^3}{6} - \dfrac{a^2}{\pi^2} \displaystyle\sum_{m=1}^{\infty} \dfrac{1}{m^2} \cos\left(\dfrac{2m\pi x}{a}\right),$

$f(x) = \dfrac{8a^2}{\pi^3} \displaystyle\sum_{m=1}^{\infty} \dfrac{1}{(2m-1)^3} \sin\left[\dfrac{(2m-1)\pi x}{a}\right]$

5. $f(x) = \dfrac{1}{4} - \dfrac{2}{\pi^2} \displaystyle\sum_{m=1}^{\infty} \dfrac{\cos[2(2m-1)\pi x]}{(2m-1)^2},$

$f(x) = \dfrac{4}{\pi^2} \displaystyle\sum_{m=1}^{\infty} \dfrac{(-1)^{m+1}\sin[(2m-1)\pi x]}{(2m-1)^2}$

7. $f(x) = \dfrac{2\pi^2}{3} - 4 \displaystyle\sum_{n=1}^{\infty} \dfrac{(-1)^n}{n^2} \cos(nx),$

$f(x) = 2\pi \displaystyle\sum_{n=1}^{\infty} \dfrac{\sin(nx)}{n} + \dfrac{8}{\pi} \displaystyle\sum_{m=1}^{\infty} \dfrac{\sin[(2m-1)x]}{(2m-1)^3}$

9. $f(x) = \dfrac{a}{6} + \dfrac{4a}{\pi^2} \displaystyle\sum_{m=1}^{\infty} \dfrac{(-1)^m \sin[(2m-1)\pi/6]}{(2m-1)^2} \cos\left[\dfrac{(2m-1)\pi x}{a}\right]$

$f(x) = \dfrac{a}{\pi^2} \displaystyle\sum_{m=1}^{\infty} \dfrac{(-1)^m \sin(m\pi/3)}{m^2} \sin\left(\dfrac{2m\pi x}{a}\right) - \dfrac{2a}{3\pi} \displaystyle\sum_{n=1}^{\infty} \dfrac{(-1)^n}{n} \sin\left(\dfrac{n\pi x}{a}\right)$

11. $f(x) = \dfrac{3}{4} + \dfrac{1}{\pi} \displaystyle\sum_{m=1}^{\infty} \dfrac{(-1)^m}{2m-1} \cos\left[\dfrac{(2m-1)\pi x}{a}\right],$

$f(x) = \dfrac{1}{\pi} \displaystyle\sum_{n=1}^{\infty} \dfrac{1 + \cos(n\pi/2) - 2(-1)^n}{n} \sin\left(\dfrac{n\pi x}{a}\right)$

13. $f(x) = \dfrac{3a}{8} + \dfrac{2a}{\pi^2} \displaystyle\sum_{n=1}^{\infty} \dfrac{\cos(n\pi/2) - 1}{n^2} \cos\left(\dfrac{n\pi x}{a}\right),$

$f(x) = \dfrac{a}{\pi} \displaystyle\sum_{n=1}^{\infty} \left[\dfrac{2}{n^2\pi} \sin\left(\dfrac{n\pi}{2}\right) - \dfrac{(-1)^n}{n}\right] \sin\left(\dfrac{n\pi x}{a}\right)$

Section 4.4

1. $f(t) = \dfrac{1}{2} + \dfrac{2}{\pi} \displaystyle\sum_{n=1}^{\infty} \dfrac{\sin[(2n-1)t]}{2n-1}, \quad f(t) = \dfrac{1}{2} + \dfrac{2}{\pi} \displaystyle\sum_{n=1}^{\infty} \dfrac{\cos[(2n-1)t - \pi/2]}{2n-1}$

3. $f(t) = 2 \displaystyle\sum_{n=1}^{\infty} \dfrac{1}{n} \cos\left[nt + (-1)^n\dfrac{\pi}{2}\right], \quad f(t) = 2 \displaystyle\sum_{n=1}^{\infty} \dfrac{1}{n} \sin\left\{nt + [1 + (-1)^n]\dfrac{\pi}{2}\right\}$

Section 4.5

1. $f(t) = \dfrac{\pi}{2} - \dfrac{2}{\pi} \displaystyle\sum_{m=-\infty}^{\infty} \dfrac{e^{i(2m-1)t}}{(2m-1)^2}$ 3. $f(t) = 1 + \dfrac{i}{\pi} \displaystyle\sum_{\substack{n=-\infty \\ n \neq 0}}^{\infty} \dfrac{e^{n\pi it}}{n}$

5. $f(t) = \dfrac{1}{2} - \dfrac{i}{\pi} \displaystyle\sum_{m=-\infty}^{\infty} \dfrac{e^{2(2m-1)it}}{2m-1}$

Section 4.6

1. $y(t) = A\cosh(t) + B\sinh(t) - \dfrac{1}{2} - \dfrac{2}{\pi} \displaystyle\sum_{n=1}^{\infty} \dfrac{\sin[(2n-1)t]}{(2n-1) + (2n-1)^3}$

3. $y(t) = Ae^{2t} + Be^t + \dfrac{1}{4} + \dfrac{6}{\pi} \displaystyle\sum_{n=1}^{\infty} \dfrac{\cos[(2n-1)t]}{[2-(2n-1)^2]^2 + 9(2n-1)^2}$

$\qquad + \dfrac{2}{\pi} \displaystyle\sum_{n=1}^{\infty} \dfrac{[2-(2n-1)^2]\sin[(2n-1)t]}{(2n-1)\{[2-(2n-1)^2]^2 + 9(2n-1)^2\}}$

5. $y_p(t) = \dfrac{\pi}{8} - \dfrac{2}{\pi} \displaystyle\sum_{n=-\infty}^{\infty} \dfrac{e^{i(2n-1)t}}{(2n-1)^2[4-(2n-1)^2]}$

7. $q(t) = \displaystyle\sum_{n=-\infty}^{\infty} \dfrac{\omega^2 \varphi_n}{(in\omega_0)^2 + 2i\alpha n\omega_0 + \omega^2} e^{in\omega_0 t}$

Section 4.7

1. $f(t) = \frac{3}{2} - \cos(\pi x/2) - \sin(\pi x/2) - \frac{1}{2}\cos(\pi x)$

Section 5.3

1. $\pi e^{-|\omega/a|}/|a|$

Section 5.4

1. $-t/(1+t^2)^2$ 3. $f(t) = \frac{1}{2}e^{-t}H(t) + \frac{1}{2}e^t H(-t)$

5. $f(t) = e^{-t}H(t) - e^{-t/2}H(t) + \frac{1}{2}te^{-t/2}H(t)$

7. $f(t) = ie^{-at}H(t)/2 - ie^{at}H(-t)/2$

9. $f(t) = (1 - a|t|)e^{-a|t|}/(4a)$

11. $f(t) = (-1)^{n+1}t^{2n+1}e^{-at}H(t)/(2n+1)!$

13. $f(t) = e^{2t}H(-t) + e^{-t}H(t)$

15. $f_+(t) = \dfrac{i\,e^{-at}}{R^2 - e^{2a}} + \dfrac{1}{2R^{t+2}} \displaystyle\sum_{n=-\infty}^{\infty} \dfrac{e^{in\pi t}}{n\pi + [\ln(R) - a]\,i}$

$f_-(t) = \dfrac{i\,e^{-at}}{R^2 e^{-2a} - 1} H(t-2) + \dfrac{H(t-2)}{2R^t} \displaystyle\sum_{n=-\infty}^{\infty} \dfrac{e^{in\pi t}}{n\pi + [\ln(R) - a]\,i}$

17. $f(t) = -\dfrac{2\beta}{L} H(t) \displaystyle\sum_{n=1}^{\infty} (-1)^n e^{-(2n-1)\beta\gamma\pi t/2L}$

$\times \{\gamma \cos[(2n-1)\beta\pi t/2L] + \sin[(2n-1)\beta\pi t/2L]\}$

Section 5.6

1. $y(t) = [(t-1)e^{-t} + e^{-2t}]H(t)$ 3. $y(t) = \frac{1}{9}e^{-t}H(t) + \left[\frac{1}{9}e^{2t} - \frac{1}{3}te^{2t}\right]H(-t)$

Section 6.1

1. $F(s) = s/(s^2 - a^2)$ 3. $F(s) = 1/s + 2/s^2 + 2/s^3$

5. $F(s) = \left[1 - e^{-2(s-1)}\right]/(s-1)$

7. $F(s) = 2/(s^2 + 1) - s/(s^2 + 4) + \cos(3)/s - 1/s^2$

9. $f(t) = e^{-3t}$ 11. $f(t) = \frac{1}{3}\sin(3t)$

13. $f(t) = 2\sin(t) - \frac{15}{2}t^2 + 2e^{-t} - 6\cos(2t)$

17. $F(s) = 1/(2s) - sT^2/[2(s^2T^2 + \pi^2)]$

Section 6.2

1. $f(t) = (t-2)H(t-2) - (t-2)H(t-3)$

3. $y'' + 3y' + 2y = H(t-1)$ 5. $y'' + 4y' + 4y = tH(t-2)$

7. $y'' - 3y' + 2y = e^{-t}H(t-2)$ 9. $y'' + y = \sin(t)[1 - H(t-\pi)]$

Section 6.3

1. $F(s) = 2/(s^2 + 2s + 5)$

3. $F(s) = 1/(s-1)^2 + 3/(s^2 - 2s + 10) + (s-2)/(s^2 - 4s + 29)$

5. $F(s) = 2/(s+1)^3 + 2/(s^2 - 2s + 5) + (s+3)/(s^2 + 6s + 18)$

7. $F(s) = e^6 e^{-3s}/(s-2)$

9. $F(s) = 2e^{-s}/s^3 + 2e^{-s}/s^2 + 3e^{-s}/s + e^{-2s}/s$

11. $F(s) = (1 + e^{-s\pi})/(s^2 + 1)$ 13. $F(s) = 4(s+3)/(s^2 + 6s + 13)^2$

15. $f(t) = \frac{1}{2}t^2 e^{-2t} - \frac{1}{3}t^3 e^{-2t}$ 17. $f(t) = e^{-t}\cos(t) + 2e^{-t}\sin(t)$

19. $f(t) = e^{-2t} - 2te^{-2t} + \cos(t)e^{-t} + \sin(t)e^{-t}$

21. $f(t) = e^{t-3}H(t-3)$

23. $f(t) = e^{-(t-1)}[\cos(t-1) - \sin(t-1)]H(t-1)$

25. $f(t) = \cos[2(t-1)]H(t-1) + \frac{1}{6}(t-3)^3 e^{2(t-3)}H(t-3)$

27. $f(t) = \{\cos[2(t-1)] + \frac{1}{2}\sin[2(t-1)]\}H(t-1) + \frac{1}{6}(t-3)^3 H(t-3)$

29. $f(t) = t[H(t) - H(t-a)]; F(s) = 1/s^2 - e^{-as}/s^2 - ae^{-as}/s$

31. $F(s) = 1/s^2 - e^{-s}/s^2 - e^{-2s}/s$

33. $F(s) = e^{-s}/s^2 - e^{-2s}/s^2 - e^{-3s}/s$

35. $Y(s) = s/(s^2 + 4) + 3e^{-4s}/[s(s^2+4)]$

37. $Y(s) = e^{-(s-1)}/[(s-1)(s+1)(s+2)]$

39. $Y(s) = 5/[(s-1)(s-2)] + e^{-s}/[s^3(s-1)(s-2)]$
$\qquad + 2e^{-s}/[s^2(s-1)(s-2)] + e^{-s}/[s(s-1)(s-2)]$

41. $Y(s) = 1/[s^2(s+2)(s+1)] + ae^{-as}/[(s+1)^2(s+2)]$
$\qquad - e^{-as}/[s^2(s+1)(s+2)] - e^{-as}/[s(s+1)(s+2)]$

43. $f(0) = 1$ 45. $f(0) = 0$ 47. Yes   49. No   51. No

Section 6.4

1. $F(s) = \coth\left(\dfrac{s\pi}{2}\right)/(s^2 + 1)$ 3. $F(s) = \dfrac{1 - (1+as)e^{-as}}{s^2(1 - e^{-2as})}$

Section 6.5

1. $f(t) = e^{-t} - e^{-2t}$ 3. $f(t) = \frac{5}{4}e^{-t} - \frac{6}{5}e^{-2t} - \frac{1}{20}e^{3t}$

5. $f(t) = e^{-2t}\cos\left(t + \frac{3\pi}{2}\right)$ 7. $f(t) = 2.3584\cos(4t + 0.5586)$

9. $f(t) = \frac{1}{2} + \frac{\sqrt{2}}{2}\cos\left(2t + \frac{5\pi}{4}\right)$

Section 6.6

11. $f(t) = e^t - t - 1$

Section 6.7

1. $f(t) = 1 + 2t$ 3. $f(t) = t + t^2/2$

5. $f(t) = t^3 + t^5/20$ 7. $f(t) = t^2 - t^4/3$

9. $f(t) = 5e^{2t} - 4e^t - 2te^t$ 11. $f(t) = (1-t)^2 e^{-t}$

13. $f(t) = a\left[1 - e^{\pi t}\,\mathrm{erfc}(\sqrt{\pi t})\right]/2$ 15. $f(t) = \frac{1}{2}t^2$

17. $x(t) = \left\{ e^{c^2t} \left[1 + \text{erf}\left(c\sqrt{t}\right)\right] - c^2 t - 1 - 2c\sqrt{t/\pi} \right\}/c^2$

19. $f(t) = a + a^2 t + \frac{1}{2}a\,\text{erf}(\sqrt{at}) + a^3 t\,\text{erf}(\sqrt{at}) + a^2\sqrt{t}e^{-a^2t}/\sqrt{\pi}$

Section 6.8

1. $y(t) = \frac{5}{4}e^{2t} - \frac{1}{4} + \frac{1}{2}t$

3. $y(t) = e^{3t} - e^{2t}$

5. $y(t) = -\frac{3}{4}e^{-3t} + \frac{7}{4}e^{-t} + \frac{1}{2}te^{-t}$

7. $y(t) = \frac{3}{4}e^{-t} + \frac{1}{8}e^t - \frac{7}{8}e^{-3t}$

9. $y(t) = (t-1)H(t-1)$

11. $y(t) = e^{2t} - e^t + \left[\frac{1}{2} + \frac{1}{2}e^{2(t-1)} - e^{t-1}\right]H(t-1)$

13. $y(t) = \left[1 - e^{-2(t-2)} - 2(t-2)e^{-2(t-2)}\right]H(t-2)$

15. $y(t) = \left[\frac{1}{3}e^{2(t-2)} - \frac{1}{2}e^{t-2} + \frac{1}{6}e^{-(t-2)}\right]H(t-2)$

17. $y(t) = 1 - \cos(t) - \left[1 - \cos(t-T)\right]H(t-T)$

19. $y(t) = e^{-t} - \frac{1}{4}e^{-2t} - \frac{3}{4} + \frac{1}{2}t - \left[e^{-(t-a)} - \frac{1}{4}e^{-2(t-a)} - \frac{3}{4} + \frac{1}{2}(t-a)\right]H(t-a)$
$\quad\quad + a\left[\frac{1}{2}e^{-2(t-a)} + (t-a)e^{-(t-a)} - \frac{1}{2}\right]H(t-a)$

21. $y(t) = te^t + 3(t-2)e^{t-2}H(t-2)$

23. $y(t) = 3\left[e^{-2(t-2)} - e^{-3(t-2)}\right]H(t-2)$
$\quad\quad + 4\left[e^{-3(t-5)} - e^{-2(t-5)}\right]H(t-5)$

25. $x(t) = \cos(\sqrt{2}t)e^{3t} - \frac{1}{\sqrt{2}}\sin(\sqrt{2}t)e^{3t}; \quad y(t) = \frac{3}{\sqrt{2}}\sin(\sqrt{2}t)e^{3t}$

27. $x(t) = t - 1 + e^{-t}\cos(t), \quad y(t) = t^2 - t + e^{-t}\sin(t)$

29. $x(t) = 3F_1 - 2F_2 - F_1\cosh(t) + F_2 e^t - 2F_1\cos(t) + F_2\cos(t) - F_2\sin(t)$
$\quad\quad y(t) = F_2 - 2F_1 + F_1 e^{-t} - F_2\cos(t) + F_1\cos(t) + F_1\sin(t)$

Section 6.9

1. $G(s) = 1/(s+k)$ $\quad\quad g(t) = e^{-kt}$ $\quad\quad a(t) = \left(1 - e^{-kt}\right)/k$

3. $G(s) = 1/(s^2 + 4s + 3)$ $\quad g(t) = \frac{1}{2}\left(e^{-t} - e^{-3t}\right)$ $\quad a(t) = \frac{1}{6}e^{-3t} - \frac{1}{2}e^{-t} + \frac{1}{3}$

5. $G(s) = 1/[(s-2)(s-1)]$ $\quad g(t) = e^{2t} - e^t$ $\quad\quad a(t) = \frac{1}{2} + \frac{1}{2}e^{2t} - e^t$

7. $G(s) = 1/(s^2 - 9)$ $\quad\quad g(t) = \frac{1}{6}\left(e^{3t} - e^{-3t}\right)$ $\quad a(t) = \frac{1}{18}\left(e^{3t} + e^{-3t} - 2\right)$

9. $G(s) = 1/[s(s-1)]$ $\quad\quad g(t) = e^t - 1$ $\quad\quad a(t) = e^t - t - 1$

Section 6.10

1. $f(t) = (2-t)e^{-2t} - 2e^{-3t}$

3. $f(t) = \left(\frac{1}{4}t^2 - \frac{1}{4}t + \frac{1}{8}\right)e^{2t} - \frac{1}{8}$

5. $f(t) = \left[\frac{1}{2}(t-1) - \frac{1}{4} + \frac{1}{4}e^{-2(t-1)}\right]H(t-1)$

7. $f(t) = \dfrac{e^{-bt}}{\cosh(ab)} - 8ab \displaystyle\sum_{n=1}^{\infty} (-1)^n \dfrac{\sin[(2n-1)\pi t/(2a)]}{4a^2b^2 + (2n-1)^2\pi^2}$

$\qquad\qquad + 4 \displaystyle\sum_{n=1}^{\infty} (-1)^n \dfrac{(2n-1)\pi \cos[(2n-1)\pi t/(2a)]}{4a^2b^2 + (2n-1)^2\pi^2}$

Section 7.1

1. $F(z) = 2z/(2z-1)$ if $|z| > 1/2$ 3. $F(z) = (z^6 - 1)/(z^6 - z^5)$ if $|z| > 0$

5. $F(z) = (a^2 + a - z)/[z(z-a)]$ if $|z| > a$.

Section 7.2

1. $F(z) = zTe^{aT}/(ze^{aT} - 1)^2$ 3. $F(z) = z(z+a)/(z-a)^3$

5. $F(z) = [z - \cos(1)]/\{z[z^2 - 2z\cos(1) + 1]\}$

7. $F(z) = z[z\sin(\theta) + \sin(\omega_0 T - \theta)]/[z^2 - 2z\cos(\omega_0 T) + 1]$

9. $F(z) = z/(z+1)$ 11. $f_n * g_n = n+1$ 13. $f_n * g_n = 2^n/n!$

Section 7.3

1. $f_0 = 0.007143$, $f_1 = 0.08503$, $f_2 = 0.1626$, $f_3 = 0.2328$

3. $f_0 = 0.09836$, $f_1 = 0.3345$, $f_2 = 0.6099$, $f_3 = 0.7935$

5. $f_n = 8 - 8\left(\frac{1}{2}\right)^n - 6n\left(\frac{1}{2}\right)^n$ 7. $f_n = (1 - \alpha^{n+1})/(1 - \alpha)$

9. $f_n = \left(\frac{1}{2}\right)^{n-10} H_{n-10} + \left(\frac{1}{2}\right)^{n-11} H_{n-11}$

11. $f_n = \frac{1}{9}(6n - 4)(-1)^n + \frac{4}{9}\left(\frac{1}{2}\right)^n$ 13. $f_n = a^n/n!$

Section 7.4

1. $y_n = 1 + \frac{1}{6}n(n-1)(2n-1)$ 3. $y_n = \frac{1}{2}n(n-1)$

5. $y_n = \frac{1}{6}[5^n - (-1)^n]$ 7. $y_n = (2n-1)\left(\frac{1}{2}\right)^n + \left(-\frac{1}{2}\right)^n$

9. $y_n = 2^n - n - 1$ 11. $x_n = 2 + (-1)^n; y_n = 1 + (-1)^n$

13. $x_n = 1 - 2(-6)^n; y_n = -7(-6)^n$

Section 7.5

1. marginally stable   3. unstable

Section 8.1

7. $\hat{x}(t) = \dfrac{1}{\pi} \ln \left| \dfrac{t+a}{t-a} \right|$

Section 8.2

5. $w(t) = u(t) * v(t) = \pi e^{-1} \sin(t)$

Section 8.3

1. $z(t) = e^{i\omega t}$

Section 8.4

3. $x(t) = \dfrac{1 - t^2}{(1 + t^2)^2}; \quad \hat{x}(t) = \dfrac{2t}{(1 + t^2)^2}$

Section 9.1

1. $\lambda_n = (2n - 1)^2 \pi^2 / (4L^2)$, $y_n(x) = \cos[(2n - 1)\pi x / (2L)]$

3. $\lambda_0 = -1$, $y_0(x) = e^{-x}$ and $\lambda_n = n^2$, $y_n(x) = \sin(nx) - n\cos(nx)$

5. $\lambda_n = -n^4 \pi^4 / L^4$, $y_n(x) = \sin(n\pi x / L)$

7. $\lambda_n = k_n^2$, $y_n(x) = \sin(k_n x)$ with $k_n = -\tan(k_n)$

9. $\lambda_0 = -m_0^2$, $y_0(x) = \sinh(m_0 x) - m_0 \cosh(m_0 x)$ with $\coth(m_0 \pi) = m_0$; $\lambda_n = k_n^2$, $y_n(x) = \sin(k_n x) - k_n \cos(k_n x)$ with $k_n = -\cot(k_n \pi)$

11.

    (a) $\lambda_n = n^2 \pi^2$, $y_n(x) = \sin[n\pi \ln(x)]$

    (b) $\lambda_n = (2n - 1)^2 \pi^2 / 4$, $y_n(x) = \sin[(2n - 1)\pi \ln(x) / 2]$

    (c) $\lambda_0 = 0$, $y_0(x) = 1$; $\lambda_n = n^2 \pi^2$, $y_n(x) = \cos[n\pi \ln(x)]$

13. $\lambda_n = n^2 + 1$, $y_n(x) = \sin[n \ln(x)] / x$

15. $\lambda = 0$, $y_0(x) = 1$; $y_n(x) = \cosh(\lambda_n x) + \cos(\lambda_n x) - \tanh(\lambda_n)[\sinh(\lambda_n x) + \sin(\lambda_n x)]$, where $n = 1, 2, 3, \ldots$, and λ_n is the nth root of $\tanh(\lambda) = -\tan(\lambda)$.

Section 9.3

1. $f(x) = \dfrac{2}{\pi} \displaystyle\sum_{n=1}^{\infty} \dfrac{(-1)^{n+1}}{n} \sin\left(\dfrac{n\pi x}{L}\right)$

3. $f(x) = \dfrac{8L}{\pi^2} \displaystyle\sum_{n=1}^{\infty} \dfrac{(-1)^{n+1}}{(2n - 1)^2} \sin\left[\dfrac{(2n - 1)\pi x}{2L}\right]$

Section 9.4

1. $f(x) = \frac{1}{4} P_0(x) + \frac{1}{2} P_1(x) + \frac{5}{16} P_2(x) + \cdots$

3. $f(x) = \frac{1}{2}P_0(x) + \frac{5}{8}P_2(x) - \frac{3}{16}P_4(x) + \cdots$

5. $f(x) = \frac{3}{2}P_1(x) - \frac{7}{8}P_3(x) + \frac{11}{16}P_5(x) + \cdots$

Section 10.3

1. $u(x,t) = \dfrac{4L}{c\pi^2} \displaystyle\sum_{m=1}^{\infty} \dfrac{1}{(2m-1)^2} \sin\left[\dfrac{(2m-1)\pi x}{L}\right] \sin\left[\dfrac{(2m-1)\pi ct}{L}\right]$

3. $u(x,t) = \dfrac{9h}{\pi^2} \displaystyle\sum_{n=1}^{\infty} \dfrac{1}{n^2} \sin\left(\dfrac{2n\pi}{3}\right) \sin\left(\dfrac{n\pi x}{L}\right) \cos\left(\dfrac{n\pi ct}{L}\right)$

5. $u(x,t) = \sin\left(\dfrac{\pi x}{L}\right) \sin\left(\dfrac{\pi ct}{L}\right)$

$\qquad + \dfrac{4aL}{\pi^2 c} \displaystyle\sum_{n=1}^{\infty} \dfrac{(-1)^{n+1}}{(2n-1)^2} \sin\left[\dfrac{(2n-1)\pi}{4}\right] \sin\left[\dfrac{(2n-1)\pi x}{L}\right] \sin\left[\dfrac{(2n-1)\pi ct}{L}\right]$

7. $u(x,t) = \dfrac{4L}{\pi^2} \displaystyle\sum_{n=1}^{\infty} \dfrac{(-1)^{n+1}}{(2n-1)^2} \sin\left[\dfrac{(2n-1)\pi x}{L}\right] \cos\left[\dfrac{(2n-1)\pi ct}{L}\right]$

9. $u(x,t) = \sqrt{a} \displaystyle\sum_{n=1}^{\infty} \dfrac{J_1(2k_n\sqrt{a}) J_0(2k_n\sqrt{x})}{k_n^2 J_1^2(2k_n)} \sin(k_n t)$,

where k_n is the nth solution of $J_0(2k) = 0$.

11. $u(x,t) = \dfrac{8Lcs_0}{\pi^2} \displaystyle\sum_{n=1}^{\infty} \dfrac{(-1)^n}{(2n-1)^2} \cos\left[\dfrac{(2n-1)\pi x}{2L}\right] \sin\left[\dfrac{(2n-1)\pi ct}{2L}\right]$

Section 10.4

1. $u(x,t) = \sin(2x)\cos(2ct) + \cos(x)\sin(ct)/c$

3. $u(x,t) = \dfrac{1+x^2+c^2t^2}{(1+x^2+c^2t^2)^2 + 4x^2c^2t^2} + \dfrac{e^x \sinh(ct)}{c}$

5. $u(x,t) = \cos\left(\dfrac{\pi x}{2}\right) \cos\left(\dfrac{\pi ct}{2}\right) + \dfrac{\sinh(ax)\sinh(act)}{ac}$

Section 10.5

1. $u(x,t) = \dfrac{4}{\pi^2} \displaystyle\sum_{m=1}^{\infty} \dfrac{\sin[(2m-1)\pi x]\sin[(2m-1)\pi t]}{(2m-1)^2}$

3. $u(x,t) = \sin(\pi x)\cos(\pi t) - \sin(\pi x)\sin(\pi t)/\pi$

5. $u(x,t) = c \sum_{n=0}^{\infty} f(t - x/c - 2nL/c)H(t - x/c - 2nL/c)$

$\quad + c \sum_{m=1}^{\infty} f(t + x/c - 2mL/c)H(t + x/c - 2mL/c)$

7. $u(x,t) = xt - te^{-x} + \sinh(t)e^{-x} + \left[1 - e^{-(t-x)} + t - x - \sinh(t-x)\right]H(t-x)$

9. $u(x,t) = \dfrac{gx}{\omega^2} - \dfrac{2g\omega^2}{L} \sum_{n=1}^{\infty} \dfrac{\sin(\lambda_n x)\cos(\lambda_n t)}{\lambda_n^2(\omega^4 + \omega^2/L + \lambda_n^2)\sin(\lambda_n L)}$

11. $u(x,t) = E - \dfrac{4E}{\pi} \sum_{n=1}^{\infty} \dfrac{1}{2n-1} \sin\left[\dfrac{(2n-1)\pi x}{2\ell}\right] \cos\left[\dfrac{(2n-1)c\pi t}{2\ell}\right]$ or

$\quad u(x,t) = E \sum_{n=0}^{\infty} (-1)^n H\left(t - \dfrac{x+2n\ell}{c}\right) + E \sum_{n=0}^{\infty} (-1)^n H\left\{t - \dfrac{[(2n+2)\ell - x]}{c}\right\}$

13. $p(x,t) = p_0 - \dfrac{4\rho u_0 c}{\pi} \sum_{n=1}^{\infty} \dfrac{(-1)^n}{2n-1} \sin\left[\dfrac{(2n-1)\pi x}{2L}\right] \sin\left[\dfrac{(2n-1)c\pi t}{2L}\right]$

15. $u(r,t) = \dfrac{ap_0}{\rho r[(\beta/\sqrt{2} - \alpha)^2 + \beta^2]} \left\{ e^{-\beta\tau/\sqrt{2}}\left[\left(\dfrac{1}{\sqrt{2}} - \dfrac{\alpha}{\beta}\right)\sin(\beta\tau) + \cos(\beta\tau)\right] \right.$

$\quad \left. - e^{-\alpha\tau}\right\} H(\tau)$, where $\tau = t - (r-a)/c$.

17. $u(x,t) = \dfrac{2t}{3} + \dfrac{x^2}{2} - \dfrac{1}{6} - 2 \sum_{n=1}^{\infty} (-1)^n \cos(n\pi x)\left[\dfrac{\cos(n\pi t)}{n^2\pi^2} + \dfrac{2\sin(n\pi t)}{n^3\pi^3}\right]$

Section 11.3

1. $u(x,t) = \dfrac{4A}{\pi} \sum_{m=1}^{\infty} \dfrac{\sin[(2m-1)x]}{2m-1} e^{-a^2(2m-1)^2 t}$

3. $u(x,t) = -2 \sum_{n=1}^{\infty} \dfrac{(-1)^n}{n} \sin(nx)e^{-a^2 n^2 t}$

5. $u(x,t) = \dfrac{4}{\pi} \sum_{m=1}^{\infty} \dfrac{(-1)^{m+1}}{(2m-1)^2} \sin[(2m-1)x]e^{-a^2(2m-1)^2 t}$

7. $u(x,t) = \dfrac{\pi}{2} - \dfrac{4}{\pi} \sum_{m=1}^{\infty} \dfrac{\cos[(2m-1)x]}{(2m-1)^2} e^{-a^2(2m-1)^2 t}$

9. $u(x,t) = \dfrac{\pi}{2} - \dfrac{4}{\pi} \sum_{n=1}^{\infty} \dfrac{\cos[(2n-1)x]}{(2n-1)^2} e^{-a^2(2n-1)^2 t}$

11. $u(x,t) = \dfrac{32}{\pi} \sum_{n=1}^{\infty} \dfrac{(-1)^n}{(2n-1)^3} \cos\left[\dfrac{(2n-1)x}{2}\right] e^{-a^2(2n-1)^2 t/4}$

13. $u(x,t) = \dfrac{4}{\pi} \displaystyle\sum_{n=1}^{\infty} \dfrac{\sin[(2n-1)x/2]}{2n-1} e^{-a^2(2n-1)^2 t/4}$

15. $u(x,t) = \displaystyle\sum_{n=1}^{\infty} \left[\dfrac{4}{2n-1} - \dfrac{8(-1)^{n+1}}{(2n-1)^2 \pi} \right] \sin\left[\dfrac{(2n-1)x}{2} \right] e^{-a^2(2n-1)^2 t/4}$

17. $u(x,t) = \dfrac{T_0 x}{\pi} + \dfrac{2T_0}{\pi} \displaystyle\sum_{n=1}^{\infty} \dfrac{1}{n} \sin(nx) e^{-a^2 n^2 t}$

19. $u(x,t) = h_1 + \dfrac{(h_2 - h_1)x}{L}$
$\qquad + \dfrac{2(h_2 - h_1)}{\pi} \displaystyle\sum_{n=1}^{\infty} \dfrac{(-1)^n}{n} \sin\left(\dfrac{n\pi x}{L} \right) \exp\left(-\dfrac{a^2 n^2 \pi^2}{L^2} t \right)$

21. $u(x,t) = h_0 - \dfrac{4h_0}{\pi} \displaystyle\sum_{n=1}^{\infty} \dfrac{1}{2n-1} \sin\left[\dfrac{(2n-1)\pi x}{L} \right] \exp\left[-\dfrac{(2n-1)^2 \pi^2 a^2 t}{L^2} \right]$

23. $u(x,t) = \dfrac{1}{3} - t - \dfrac{2}{\pi^2} \displaystyle\sum_{n=1}^{\infty} \dfrac{(-1)^n}{n^2} \cos(n\pi x) e^{-a^2 n^2 \pi^2 t}$

25. $u(x,t) = \dfrac{4}{\pi} \displaystyle\sum_{n=1}^{\infty} \dfrac{(-1)^{n+1}}{(2n-1)^4} \sin[(2n-1)x] \left[1 - e^{-(2n-1)^2 t} \right]$

27. $u(x,t) = \dfrac{A_0(L^2 - x^2)}{2\kappa} + \dfrac{A_0 L}{h}$
$\qquad - \dfrac{2L^2 A_0}{\kappa} \displaystyle\sum_{n=1}^{\infty} \dfrac{\sin(\beta_n)}{\beta_n^4 [1 + \kappa \sin^2(\kappa)/hL]} \cos\left(\dfrac{\beta_n x}{L} \right) \exp\left(-\dfrac{a^2 \beta_n^2 t}{L^2} \right),$
where β_n is the nth root of $\beta \tan(\beta) = hL/\kappa$.

29. $u(x,t) = 4u_0 \displaystyle\sum_{n=1}^{\infty} \dfrac{\sin(k_n L) \cos(k_n x)}{2k_n L + \sin(2k_n L)} \cos(k_n x) \exp\left[-(k_1 + a^2 k_n^2)t \right],$
where k_n denotes the nth root of $k \tan(kL) = k_2/a^2$

31. $u(r,t) = \dfrac{2}{\pi r} \displaystyle\sum_{n=1}^{\infty} \dfrac{(-1)^{n+1}}{n} \sin(n\pi r) e^{-a^2 n^2 \pi^2 t}$

33. $u(r,t) = \dfrac{4bu_0}{r} \displaystyle\sum_{n=1}^{\infty} \dfrac{\sin(k_n) - k_n \cos(k_n)}{k_n[2k_n - \sin(2k_n)]} \sin\left(\dfrac{k_n r}{b} \right) e^{-a^2 k_n^2 t/b^2},$
where k_n is the nth root of $k \cot(k) = 1 - A$, $n\pi < k_n < (n+1)\pi$.

35. $u(r,t) = \theta + 2(1 - \theta) \displaystyle\sum_{n=1}^{\infty} \dfrac{J_0(k_n r/b)}{k_n J_1(k_n)} e^{-a^2 k_n^2 t/b^2},$
where k_n is the nth root of $J_0(k) = 0$.

37. $u(r,t) = \dfrac{G}{4\rho\nu}(b^2 - r^2) - \dfrac{2Gb^2}{\rho\nu} \displaystyle\sum_{n=1}^{\infty} \dfrac{J_0(k_n r/b)}{k_n^3 J_1(k_n)} e^{-\nu k_n^2 t/b^2},$

where k_n is the nth root of $J_0(k) = 0$.

39. $u(r,t) = \dfrac{2T_0}{L^2} e^{-\kappa t} \displaystyle\sum_{n=1}^{\infty} \dfrac{[LJ_1(k_n L) - bJ_1(k_n b)]J_0(k_n r)}{k_n[J_0^2(k_n L) + J_1^2(k_n L)]} e^{-a^2 k_n^2 t}$,

where k_n is the nth root of $kJ_1(kL) = hJ_0(kL)$.

Section 11.4

1. $u(x,t) = T_0 \left(1 - e^{-a^2 t}\right)$

3. $u(x,t) = x + \dfrac{2}{\pi} \displaystyle\sum_{n=1}^{\infty} \dfrac{(-1)^n}{n} \sin(n\pi x) e^{-n^2\pi^2 t}$

5. $u(x,t) = \dfrac{x(1-x)}{2} - \dfrac{4}{\pi^3} \displaystyle\sum_{m=1}^{\infty} \dfrac{\sin[(2m-1)\pi x]}{(2m-1)^3} e^{-(2m-1)^2\pi^2 t}$

7. $u(x,t) = x\,\mathrm{erfc}\left(\dfrac{x}{2\sqrt{t}}\right) - 2\sqrt{\dfrac{t}{\pi}} \exp\left(-\dfrac{x^2}{4t}\right)$

9. $u(x,t) = \dfrac{u_0}{2} e^{-\delta x} \mathrm{erfc}\left(\dfrac{x}{2a\sqrt{t}} + \dfrac{a(1-\delta)\sqrt{t}}{2}\right)$

$\qquad + \dfrac{u_0}{2} e^{-x} \mathrm{erfc}\left(\dfrac{x}{2a\sqrt{t}} - \dfrac{a(1-\delta)\sqrt{t}}{2}\right)$

11. $u(x,t) = \tfrac{1}{2} e^{\sigma^2 t - \beta t} \left[\dfrac{e^{-\sigma x}}{\rho + \sigma} \mathrm{erfc}\left(\dfrac{x}{2\sqrt{t}} - \sigma\sqrt{t}\right) + \dfrac{e^{\sigma x}}{\rho - \sigma} \mathrm{erfc}\left(\dfrac{x}{2\sqrt{t}} + \sigma\sqrt{t}\right)\right]$

$\qquad - \dfrac{\rho}{\rho^2 - \sigma^2} e^{\rho x + \rho^2 t - \beta t} \mathrm{erfc}\left(\dfrac{x}{2\sqrt{t}} + \rho\sqrt{t}\right)$

13. $u(x,t) = \dfrac{t(L-x)}{L} + \dfrac{Px(x-L)}{2a^2} - \dfrac{x(x-L)(x-2L)}{6a^2 L}$

$\qquad - \dfrac{2PL^2}{a^2\pi^3} \displaystyle\sum_{n=1}^{\infty} \dfrac{(-1)^n}{n^3} \sin\left(\dfrac{n\pi x}{L}\right) \exp\left(-\dfrac{a^2 n^2 \pi^2 t}{L^2}\right)$

$\qquad + \dfrac{2(P+1)L^2}{a^2\pi^3} \displaystyle\sum_{n=1}^{\infty} \dfrac{1}{n^3} \sin\left(\dfrac{n\pi x}{L}\right) \exp\left(-\dfrac{a^2 n^2 \pi^2 t}{L^2}\right)$

15. $u(x,t) = \dfrac{4q}{\pi} \displaystyle\sum_{n=1}^{\infty} \dfrac{(-1)^n \cos[(2n-1)\pi x/2L]}{(2n-1)[\alpha q - (2n-1)^2\pi^2 a^2/4L^2]}$

$\qquad \times \left\{1 - \exp[\alpha q t - (2n-1)^2\pi^2 a^2 t/4L^2]\right\}$

17. $u(r,t) = u_0 \left[1 - \dfrac{b-\beta}{r} f(r,t)\right] + \displaystyle\int_0^t \left[1 - \dfrac{b-\beta}{r} f(r, t-\tau)\right] q(\tau)\, d\tau$,

where $f(r,t) = \mathrm{erfc}\left(\dfrac{r-b}{2a\sqrt{t}}\right) - \exp\left(\dfrac{r-b}{\beta} + \dfrac{a^2 t}{\beta^2}\right) \mathrm{erfc}\left(\dfrac{a\sqrt{t}}{\beta} + \dfrac{r-b}{2a\sqrt{t}}\right)$.

19. $y(t) = \dfrac{4\mu A\omega^2}{mL} \displaystyle\sum_{n=1}^{\infty} \dfrac{\lambda_n e^{\lambda_n t}}{\lambda_n^4 - (\frac{2\mu}{mL})(1 + \frac{2\mu L}{m\nu})\lambda_n^3 + 2\omega^2\lambda_n^2 + \frac{6\omega^2\mu}{mL}\lambda_n + \omega^4}$,

where λ_n is the nth root of $\lambda^2 + 2\mu\lambda^{3/2}\coth\left(L\sqrt{\lambda/\nu}\right)/(m\sqrt{\nu}) + \omega^2 = 0$.

21. $u(x,t) = \dfrac{x}{a+1} + 2\displaystyle\sum_{n=1}^{\infty} \dfrac{\sin(\lambda_n x)\exp(-\lambda_n^2 t)}{[3a + 3 + \lambda_n^2/(3a)]\sin(\lambda_n)}$,

where λ_n is the nth root of $\lambda\cot(\lambda) = (3a + \lambda^2)/3a$.

23. $u(r,t) = -2\displaystyle\sum_{n=1}^{\infty} \dfrac{J_0(k_n r/a)}{k_n J_1(k_n)} e^{-k_n^2 t/a^2}$,

where k_n is the nth root of $J_0(k) = 0$.

25. $u(r,t) = e^{-t/\tau_0} + 2a^2\displaystyle\sum_{n=1}^{\infty} \dfrac{J_0(k_n r/a)}{k_n(a^2 - k_n^2\tau_0)J_1(k_n)} \left(e^{-k_n^2 t/a^2} - e^{-t/\tau_0}\right)$,

where k_n is the nth root of $J_0(k) = 0$.

Section 11.5

1. $u(x,t) = \frac{1}{2}\operatorname{erf}\left(\dfrac{b-x}{\sqrt{4a^2 t}}\right) + \frac{1}{2}\operatorname{erf}\left(\dfrac{b+x}{\sqrt{4a^2 t}}\right)$

3. $u(x,t) = \frac{1}{2}T_0\operatorname{erf}\left(\dfrac{b-x}{\sqrt{4a^2 t}}\right) + \frac{1}{2}T_0\operatorname{erf}\left(\dfrac{x}{\sqrt{4a^2 t}}\right)$

Section 12.3

1. $u(x,y) = \dfrac{4}{\pi}\displaystyle\sum_{m=1}^{\infty} \dfrac{\sinh[(2m-1)\pi(a-x)/b]\sin[(2m-1)\pi y/b]}{(2m-1)\sinh[(2m-1)\pi a/b]}$

3. $u(x,y) = -\dfrac{2a}{\pi}\displaystyle\sum_{n=1}^{\infty} \dfrac{\sinh(n\pi y/a)\sin(n\pi x/a)}{n\,\sinh(n\pi b/a)}$

5. $u(x,y) = \dfrac{4}{\pi}\displaystyle\sum_{n=1}^{\infty}(-1)^{n+1}\dfrac{\sinh[(2n-1)\pi y/2a]\cos[(2n-1)\pi x/2a]}{(2n-1)\sinh[(2n-1)\pi b/2a]}$

7. $u(x,y) = 1$

9. $u(x,y) = 1 - \dfrac{4}{\pi}\displaystyle\sum_{m=1}^{\infty} \dfrac{\cosh[(2m-1)\pi y/a]\sin[(2m-1)\pi x/a]}{(2m-1)\cosh[(2m-1)\pi b/a]}$

11. $u(x,y) = 1$

13. $u(x,y) = T_0 + \Delta T\cos(2\pi x/\lambda)e^{-2\pi y/\lambda}$

15. $u(r,z) = 2a\displaystyle\sum_{n=1}^{\infty} \dfrac{\sinh(k_n z/a)\,J_0(k_n r/a)}{k_n^2\cosh(k_n L/a)J_1(k_n)}$,

where k_n is the nth root of $J_0(k) = 0$.

17. $u(r, z) = A - \dfrac{2aB}{b^2} \displaystyle\sum_{n=1}^{\infty} \dfrac{J_1(k_n a) J_0(k_n r)}{k_n^2 J_0^2(k_n b)} \dfrac{\sinh[k_n(L-z)]}{\cosh(k_n L)}$,

where k_n is the nth root of $J_1(kb) = 0$.

19. $u(r, z) = \dfrac{2}{b^2 - a^2} \displaystyle\sum_{n=1}^{\infty} \dfrac{[bJ_1(k_n b) - aJ_1(k_n a)]J_0(k_n r)\cosh(k_n z)}{k_n \cosh(k_n d)J_0^2(k_n)}$,

where k_n is the nth root of $J_1(k) = 0$.

21. $u(r, z) = -\dfrac{2}{\pi} \displaystyle\sum_{n=1}^{\infty} (-1)^n \dfrac{\sin(n\pi z)I_1(n\pi r)}{n\, I_1(n\pi a)}$

23. $u(r, z) = 2 \displaystyle\sum_{n=1}^{\infty} \dfrac{\sinh(k_n z) J_1(k_n r)}{k_n^3 \cosh(k_n a)J_1(k_n)}$,

where k_n is the nth root of $kJ_0(k) = J_1(k)$.

25. $u(r, z) = 2B \displaystyle\sum_{n=1}^{\infty} \dfrac{\exp[z(1 - \sqrt{1 + 4k_n^2})/2]J_0(k_n r)}{(k_n^2 + B^2)J_0(k_n)}$,

where k_n is the nth root of $kJ_1(k) = BJ_0(k)$.

29. $u(r, \theta) = 50 \displaystyle\sum_{m=1}^{\infty} [P_{2m-2}(0) - P_{2m}(0)] \left(\dfrac{r}{a}\right)^{2m-1} P_{2m-1}[\cos(\theta)]$

31. $u(r, \varphi) = T_0$

Section 12.4

1. $u(x, y) = \dfrac{1}{\pi}\left[\tan^{-1}\left(\dfrac{1-x}{y}\right) + \tan^{-1}\left(\dfrac{x}{y}\right)\right]$

3. $u(x, y) = \dfrac{T_0}{\pi}\left[\dfrac{\pi}{2} - \tan^{-1}\left(\dfrac{x}{y}\right)\right]$

5. $u(x, y) = \dfrac{T_0}{\pi}\left[\tan^{-1}\left(\dfrac{1-x}{y}\right) + \tan^{-1}\left(\dfrac{1+x}{y}\right)\right]$
$\quad + \dfrac{T_1 - T_0}{2\pi} y \ln\left[\dfrac{(x-1)^2 + y^2}{x^2 + y^2}\right]$
$\quad + \dfrac{T_1 - T_0}{\pi} x \left[\tan^{-1}\left(\dfrac{1-x}{y}\right) + \tan^{-1}\left(\dfrac{x}{y}\right)\right]$

Section 12.5

1. $u(x, y) = \dfrac{64R}{\pi^4 T} \displaystyle\sum_{n=1}^{\infty}\sum_{m=1}^{\infty} \dfrac{(-1)^{n+1}(-1)^{m+1}}{(2n-1)(2m-1)}$
$\quad \times \dfrac{\cos[(2n-1)\pi x/2a]\cos[(2m-1)\pi y/b]}{(2n-1)(2m-1)[(2n-1)^2/a^2 + (2m-1)^2/b^2]}$

Section 12.6

1. $u(x,y) = \dfrac{4}{\pi} \displaystyle\sum_{m=1}^{\infty} \dfrac{1}{2m-1} \exp\left[-\dfrac{(2m-1)\pi x}{a}\right] \sin\left[\dfrac{(2m-1)\pi y}{a}\right]$

Section 13.1

1. $\mathbf{a} \times \mathbf{b} = -3\mathbf{i} + 19\mathbf{j} + 10\mathbf{k}$ 3. $\mathbf{a} \times \mathbf{b} = \mathbf{i} - 8\mathbf{j} + 7\mathbf{k}$

5. $\mathbf{a} \times \mathbf{b} = -3\mathbf{i} - 2\mathbf{j} - 5\mathbf{k}$

9. $\nabla f = y\cos(yz)\mathbf{i} + [x\cos(yz) - xyz\sin(yz)]\mathbf{j} - xy^2\sin(yz)\mathbf{k}$

11. $\nabla f = 2xy^2(2z+1)^2\mathbf{i} + 2x^2y(2z+1)^2\mathbf{j} + 4x^2y^2(2z+1)\mathbf{k}$

13. Plane parallel to the xy plane at height of $z = 3$, $\mathbf{n} = \mathbf{k}$

15. Paraboloid,
$$\mathbf{n} = -\frac{2x}{\sqrt{1+4x^2+4y^2}}\mathbf{i} - \frac{2y}{\sqrt{1+4x^2+4y^2}}\mathbf{j} + \frac{1}{\sqrt{1+4x^2+4y^2}}\mathbf{k}$$

17. A plane, $\mathbf{n} = \mathbf{j}/\sqrt{2} - \mathbf{k}/\sqrt{2}$

19. A parabola of infinite extent along the y-axis,
$$\mathbf{n} = -2x\mathbf{i}/\sqrt{1+4x^2} + \mathbf{k}/\sqrt{1+4x^2}$$

21. $y = 2/(x+1)$; $z = \exp[(y-1)/y]$

23. $y = x$; $z^2 = y/(3y-2)$

Section 13.2

1. $\nabla \cdot \mathbf{F} = 2xz + z^2$, $\nabla \times \mathbf{F} = (2xy - 2yz)\mathbf{i} + (x^2 - y^2)\mathbf{j}$,
$\nabla(\nabla \cdot \mathbf{F}) = 2z\mathbf{i} + (2x + 2z)\mathbf{k}$

3. $\nabla \cdot \mathbf{F} = 2(x-y) - xe^{-xy} + xe^{2y}$,
$\nabla \times \mathbf{F} = 2xze^{2y}\mathbf{i} - ze^{2y}\mathbf{j} + [2(x-y) - ye^{-xy}]\mathbf{k}$,
$\nabla(\nabla \cdot \mathbf{F}) = \left(2 - e^{-xy} + xye^{-xy} + e^{2y}\right)\mathbf{i} + \left(x^2e^{-xy} + 2xe^{2y} - 2\right)\mathbf{j}$

5. $\nabla \cdot \mathbf{F} = 0$, $\nabla \times \mathbf{F} = -x^2\mathbf{i} + (5y - 9x^2)\mathbf{j} + (2xz - 5z)\mathbf{k}$,
$\nabla(\nabla \cdot \mathbf{F}) = \mathbf{0}$

7. $\nabla \cdot \mathbf{F} = e^{-y} + z^2 - 3e^{-z}$, $\nabla \times \mathbf{F} = -2yz\mathbf{i} + xe^{-y}\mathbf{k}$,
$\nabla(\nabla \cdot \mathbf{F}) = -e^{-y}\mathbf{j} + (2z + 3e^{-z})\mathbf{k}$

9. $\nabla \cdot \mathbf{F} = yz + x^3ze^z + xye^z$,
$\nabla \times \mathbf{F} = (xe^z - x^3ye^z - x^3yze^z)\mathbf{i} + (xy - ye^z)\mathbf{j} + (3x^2yze^z - xz)\mathbf{k}$,
$\nabla(\nabla \cdot \mathbf{F}) = \left(3x^2ze^z + ye^z\right)\mathbf{i} + \left(z + xe^z\right)\mathbf{j} + \left(y + x^3e^z + x^3ze^z + xye^z\right)\mathbf{k}$,

11. $\nabla \cdot \mathbf{F} = y^2 + xz^2 - xy\sin(z)$,
$\nabla \times \mathbf{F} = [x\cos(z) - 2xyz]\mathbf{i} - y\cos(z)\mathbf{j} + (yz^2 - 2xy)\mathbf{k}$,
$\nabla(\nabla \cdot \mathbf{F}) = [z^2 - y\sin(z)]\mathbf{i} + [2y - x\sin(z)]\mathbf{j} + [2xz - xy\cos(z)]\mathbf{k}$

13. $\nabla \cdot \mathbf{F} = y^2 + xz - xy \sin(z),$
$\quad \nabla \times \mathbf{F} = [x \cos(z) - xy]\mathbf{i} - y \cos(z)\mathbf{j} + (yz - 2xy)\mathbf{k},$
$\quad \nabla(\nabla \cdot \mathbf{F}) = [z - y\sin(z)]\mathbf{i} + [2y - x\sin(z)]\mathbf{j} + [x - xy\cos(z)]\mathbf{k}$

Section 13.3

1. $16/7 + 2/(3\pi)$ 3. $e^2 + 2e^8/3 + e^{64}/2 - 13/6$ 5. -4π

7. 0   9. 2π

Section 13.4

1. $\varphi(x, y, z) = x^2 y + y^2 z + 4z + \text{constant}$

3. $\varphi(x, y, z) = xyz + \text{constant}$

5. $\varphi(x, y, z) = x^2 \sin(y) + xe^{3z} + 4z + \text{constant}$

7. $\varphi(x, y, z) = xe^{2z} + y^3 + \text{constant}$

9. $\varphi(x, y, z) = xy + xz + \text{constant}$

Section 13.5

1. $1/2$ 3. 0   5. $27/2$

7. 5   9. 0   11. $40/3$

13. $86/3$ 15. 96π

Section 13.6

1. -5 3. 1   5. 0

7. 0   9. -16π 11. -2

Section 13.7

1. -10 3. 2   5. π 7. $45/2$

Section 13.8

1. 3   3. -16 5. 4π 7. $5/12$

Section 14.1

1. $A + B = \begin{pmatrix} 4 & 5 \\ 3 & 4 \end{pmatrix} = B + A$

3. $3A - 2B = \begin{pmatrix} 7 & 10 \\ -1 & 2 \end{pmatrix}, \qquad 3(2A - B) = \begin{pmatrix} 15 & 21 \\ 0 & 6 \end{pmatrix}$

5. $(A+B)^T = \begin{pmatrix} 4 & 3 \\ 5 & 4 \end{pmatrix}$, $A^T + B^T = \begin{pmatrix} 4 & 3 \\ 5 & 4 \end{pmatrix}$

7. $AB = \begin{pmatrix} 11 & 11 \\ 5 & 5 \end{pmatrix}$, $A^T B = \begin{pmatrix} 5 & 5 \\ 8 & 8 \end{pmatrix}$, $BA = \begin{pmatrix} 4 & 6 \\ 8 & 12 \end{pmatrix}$, $B^T A = \begin{pmatrix} 5 & 8 \\ 5 & 8 \end{pmatrix}$

9. $BB^T = \begin{pmatrix} 2 & 4 \\ 4 & 8 \end{pmatrix}$, $B^T B = \begin{pmatrix} 5 & 5 \\ 5 & 5 \end{pmatrix}$

11. $A^3 + 2A = \begin{pmatrix} 65 & 100 \\ 25 & 40 \end{pmatrix}$

13. yes $\begin{pmatrix} 27 & 11 \\ 2 & 5 \end{pmatrix}$,   15. yes $\begin{pmatrix} 11 & 8 \\ 8 & 4 \\ 5 & 3 \end{pmatrix}$,   17. no

19. $5(2A) = \begin{pmatrix} 10 & 10 \\ 10 & 20 \\ 30 & 10 \end{pmatrix} = 10A$

21. $(A+B) + C = \begin{pmatrix} 4 & 0 \\ 8 & 2 \end{pmatrix} = A + (B+C)$

23. $A(B+C) = \begin{pmatrix} 9 & -1 \\ 11 & -2 \end{pmatrix} = AB + AC$

25. $\begin{pmatrix} 3 & -1 \\ -5 & 2 \end{pmatrix} \begin{pmatrix} 2 & 1 \\ 5 & 3 \end{pmatrix} = \begin{pmatrix} 1 & 0 \\ 0 & 1 \end{pmatrix}$

27. $\begin{pmatrix} 1 & -2 \\ 3 & 1 \end{pmatrix} \begin{pmatrix} x_1 \\ x_2 \end{pmatrix} = \begin{pmatrix} 5 \\ 1 \end{pmatrix}$

29. $\begin{pmatrix} 0 & 1 & 2 & 3 \\ 3 & 0 & -4 & -4 \\ 1 & 1 & 1 & 1 \\ 2 & -3 & 1 & -3 \end{pmatrix} \begin{pmatrix} x_1 \\ x_2 \\ x_3 \\ x_4 \end{pmatrix} = \begin{pmatrix} 2 \\ 5 \\ -3 \\ 7 \end{pmatrix}$

Section 14.2

1. 7   3. 1   5. −24   7. 3

Section 14.3

1. $x_1 = \frac{9}{5}$, $x_2 = \frac{3}{5}$ 3. $x_1 = 0$, $x_2 = 0$, $x_3 = -2$

Section 14.4

1. $x_2 = 2$, $x_1 = 1$ 3. $x_3 = \alpha$, $x_2 = -\alpha$, $x_1 = \alpha$

5. $x_3 = \alpha$, $x_2 = 2\alpha$, $x_1 = -1$ 7. $x_3 = 2.2$, $x_2 = 2.6$, $x_1 = 1$

9. $A^{-1} = \begin{pmatrix} -1/13 & 5/13 \\ 2/13 & 3/13 \end{pmatrix}$

11. $A^{-1} = \begin{pmatrix} 1 & 2 & 5 \\ 0 & -1 & 2 \\ 2 & 4 & 11 \end{pmatrix}$

Section 14.5

1. $\lambda = 4, \quad \mathbf{x}_0 = \alpha \begin{pmatrix} 2 \\ 1 \end{pmatrix}; \qquad \lambda = -3 \quad \mathbf{x}_0 = \beta \begin{pmatrix} 1 \\ -3 \end{pmatrix}$

3. $\lambda = 1 \quad \mathbf{x}_0 = \alpha \begin{pmatrix} -1 \\ 0 \\ 1 \end{pmatrix} + \beta \begin{pmatrix} 3 \\ 1 \\ 0 \end{pmatrix}; \qquad \lambda = 0, \quad \mathbf{x}_0 = \gamma \begin{pmatrix} 1 \\ 1 \\ 1 \end{pmatrix}$

5. $\lambda = 1, \quad \mathbf{x}_0 = \alpha \begin{pmatrix} 1 \\ 0 \\ 0 \end{pmatrix} + \beta \begin{pmatrix} 0 \\ 1 \\ -1 \end{pmatrix}; \qquad \lambda = 2, \quad \mathbf{x}_0 = \gamma \begin{pmatrix} 1 \\ 1 \\ 0 \end{pmatrix}$

7. $\lambda = 0, \quad \mathbf{x}_0 = \alpha \begin{pmatrix} 1 \\ 1 \\ 1 \end{pmatrix}; \lambda = 1, \quad \mathbf{x}_0 = \beta \begin{pmatrix} 3 \\ 2 \\ 1 \end{pmatrix}; \lambda = 2, \quad \mathbf{x}_0 = \gamma \begin{pmatrix} 7 \\ 3 \\ 1 \end{pmatrix}$

Section 14.6

1. $\mathbf{x} = c_1 \begin{pmatrix} 1 \\ -1 \end{pmatrix} e^{-t} + c_2 \begin{pmatrix} 1 \\ 1 \end{pmatrix} e^{3t}$

3. $\mathbf{x} = c_1 \begin{pmatrix} 1 \\ 2 \end{pmatrix} e^{3t} + c_2 \begin{pmatrix} 1 \\ -2 \end{pmatrix} e^{-t}$

5. $\mathbf{x} = c_1 \begin{pmatrix} 1 \\ -1 \end{pmatrix} e^{t/2} + c_2 \begin{pmatrix} t \\ -1/2 - t \end{pmatrix} e^{t/2}$

7. $\mathbf{x} = c_1 \begin{pmatrix} 1 \\ -1 \end{pmatrix} e^{2t} + c_2 \begin{pmatrix} -1+t \\ -t \end{pmatrix} e^{2t}$

9. $\mathbf{x} = c_3 \begin{pmatrix} -3\cos(2t) - 2\sin(2t) \\ \cos(2t) \end{pmatrix} e^t + c_4 \begin{pmatrix} 2\cos(2t) - 3\sin(2t) \\ \sin(2t) \end{pmatrix} e^t$

11. $\mathbf{x} = c_3 \begin{pmatrix} 2\cos(t) \\ 7\cos(t) + \sin(t) \end{pmatrix} e^{-3t} + c_4 \begin{pmatrix} 2\sin(t) \\ 7\sin(t) - \cos(t) \end{pmatrix} e^{-3t}$

13. $\mathbf{x} = c_3 \begin{pmatrix} -\cos(2t) + \sin(2t) \\ \cos(2t) \end{pmatrix} e^t + c_4 \begin{pmatrix} -\cos(2t) - \sin(2t) \\ \sin(2t) \end{pmatrix} e^t$

15. $\mathbf{x} = c_1 \begin{pmatrix} -1 \\ 2 \end{pmatrix} e^{3t} + c_2 \begin{pmatrix} -3 \\ 2 \end{pmatrix} e^{-t}$

17. $\mathbf{x} = c_1 \begin{pmatrix} 0 \\ 1 \\ 0 \end{pmatrix} + c_2 \begin{pmatrix} 0 \\ 2 \\ 1 \end{pmatrix} e^t + c_3 \begin{pmatrix} 2 \\ 1 \\ 0 \end{pmatrix} e^{2t}$

19. $\mathbf{x} = c_1 \begin{pmatrix} 3 \\ -2 \\ 12 \end{pmatrix} e^{-t} + c_2 \begin{pmatrix} 1 \\ 0 \\ 2 \end{pmatrix} e^{t} + c_3 \begin{pmatrix} 0 \\ 1 \\ 0 \end{pmatrix} e^{2t}$

Index